Contents

Elements of Mechanics of Materials

(Courtesy of Baldwin-Lima-Hamilton Corp.)

Elements of Mechanics of Materials

THIRD EDITION

GERNER A. OLSEN

Professor of Civil Engineering
The City College of the City University of New York

PRENTICE-HALL, INC., *Englewood Cliffs, New Jersey*

Library of Congress Cataloging in Publication Data

OLSEN, GERNER A.
Elements of mechanics of materials.

Bibliography: p.
1. Strength of materials. I. Title.
TA405.037–1974 620.1'12 73–5694
ISBN 0-13-266999-4

Printed in the United States of America

10 9 8 7 6 5 4 3 2 1

PRENTICE-HALL, INC., *London*
PRENTICE-HALL OF AUSTRALIA, PTY. LTD., *Sydney*
PRENTICE-HALL OF CANADA LTD., *Toronto*
PRENTICE-HALL OF INDIA PRIVATE LIMITED, *New Delhi*
PRENTICE-HALL OF JAPAN, INC., *Tokyo*

To my wonderful
wife, family, and friends

The Simultaneous Application of a Tension (or Compression) and Shearing Load

Matrix and Tensor Analysis Applied to Combined Stress and Strain

Experimental Determination of Stress

Preface

The study of mechanics of materials involves the development of basic design relationships and the application of these relationships to the more simple stress-resisting bodies. The theory is generally based on material having a straight line proportionality of stress to strain and is limited by properly confining assumptions and limitations. Whenever the linear relationship of stress to strain is not or cannot be maintained and the assumptions and limitations are modified, changes must be made in the basic theory. The degree to which these modifications must be included in design reflects the extent to which theory must be modified—and included in a basic mechanics of materials text.

Furthermore, the introduction of new concepts is vital to the advancement and interesting presentation of any subject. Knowledge is never static; new pedagogical methods used in teaching "old" subjects can make those subjects more understandable, more exciting, and simpler to use.

It is with these previously mentioned views in mind that I have written this new edition. My overwhelming desire has been to maintain and, if possible, improve upon the simple, yet thorough, approach of the previous editions. Applications have been noted which make the theory alive and purposeful; new theoretical approaches have been included, e.g., the step-function method of finding beam deflections, the grid method in experimental stress analysis, etc. These concepts, added to those introduced in the second edition—e.g., compatibility, a general torsion development, membrane analogy, the electrical analogy of simple beams, finite difference methods applied to beams and columns, etc.—provide a wealth of meaningful material for the solution of complex problems. Many new problems have been added involving applications of the latest AISC and ACI codes.

No text can be said to be the exclusive product of its author. The revision of this text is no exception. Many interchanges of thoughts and ideas of professors and students alike are included in its pages. I wish to acknowledge with thanks the assistance of all those who have had any direct or indirect share in placing it in its final form. Special mention must be made, however, of the invaluable secretarial and editorial assistance of my wife, and of the publishers, in particular Mrs. Virginia Huebner of their editorial staff, for their valued encouragement, confidence, and many courtesies.

GERNER A. OLSEN

Elements of Mechanics of Materials

1

Introduction

1-1 The Purpose of the Study of Mechanics of Materials

The study of mechanics of materials concerns the theory involved in either the *selection or investigation of members* subjected to various loading conditions. To illustrate the use of such theory in the *selection of members*, consider a man who is about to choose a plank on which to walk across a ravine. It must be both safe and economical. Two methods of selection present themselves. First, he could try planks of different thicknesses until, by trial and error, he obtained one capable of sustaining his weight with a minimum of material. This method would undoubtedly prove time consuming and expensive, both of material and limb, should one of the planks fail during the experiment. On the other hand, even before touching a plank, he could compute, from the basic theoretical laws of design, the exact dimensions required. Such an analysis could also include a determination of the maximum deflection, the effect of using different materials, their cost, and their resistance to impact loading and temperature change. The latter approach provides a safe, economical design, eliminating costly experimentation and delay.

A closely allied function of mechanics of materials, and a very important one, is the *investigation of* internal conditions of existing members. To illustrate, suppose a member is to be subjected to a revised system of loads. To insure its continued safety, the basic relationships of mechanics of materials are applied to prevent the occurrence of any dangerous internal conditions. Sometimes investigations are conducted on members to explain their persistent failure when subjected to operating loads. Whatever the reasons for an investigation, the *fundamentals* involved in the study of mechanics of materials will assume a major role in providing a proper solution to the problem.

1-2 Classification of Loads and Reactions

Concentrated loads, distributed loads, uniform loads. Loads applied to engineering members may be of various types. They may be *concentrated loads*, which act at one point, such as loads produced by a cable tied to a beam. Or they may be *distributed loads*, such as those produced by grain spread over the floor of a granary or by wind applied to the side of a building. If a distributed load is applied evenly, it is called a *uniform load.*

In the design of load-resisting members, such as tension and compression bars, beams, and so forth, uniform loads are often considered advantageously as concentrated loads acting at their centers of gravity. However, a note of warning is in order. When replacing a uniform load with a concentrated load, one must remember that such a concentrated load is hypothetical and can be used only when the effect of the concentration is similar to that of the uniform load which it replaces. This can be illustrated in the following manner: A man carrying a beam on his shoulder would feel its entire weight concentrated at the point of contact. No difference would be felt in the amount of load carried if the board were cut into several pieces and piled upon the man's shoulder. Therefore, when the magnitude of the thrust on the shoulder is desired, it is permissible to concentrate the load. However, if the deflection of the ends of the beam were to be determined, the weight of the beam could not be concentrated at its center of gravity because the maximum deflection of a uniformly loaded cantilevered beam varies as the cube of its length.

Classification of reactions. A loaded member may be supported by three types of reactions: *roller*, *pin or hinge*, and *fixed.*

The *roller reaction* has a fixed point of application and a fixed line of action—perpendicular to the roller plane (Fig. 1-1a and b). It may act in either direction (Fig. 1-1b). If components are drawn for a roller reaction, they bear a definite relationship to each other.

The *hinge or pin reaction* has a fixed point of application but no fixed direction (Fig. 1-1a and b). It provides a horizontal and vertical component of resistance. These components will be unrelated to each other until other data are obtained to fix the direction of the total reaction.

A *fixed reaction* resists movement in any direction, and rotationally

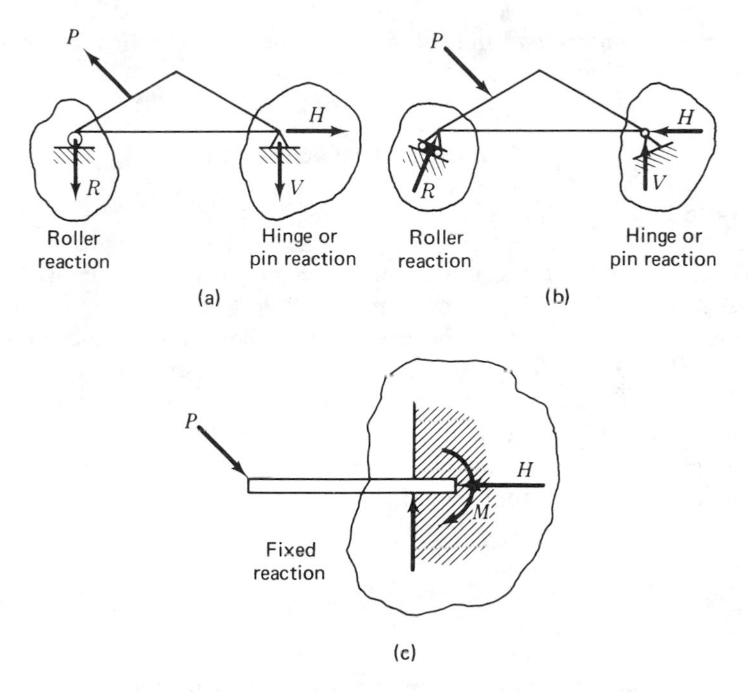

Figure 1-1

as well (Fig. 1-1c). It consists of three parts or components, a horizontal and vertical component and a resisting couple.

1-3 Dead Load; Live Load; Impact

In the study of statics, the probability of a force being applied continuously or intermittently was not considered. In the study of mechanics of materials, however, we must distinguish between permanent and occasional loads. Those loads that are continuously applied and never change position or magnitude are called *dead loads*. The weight of a structure or any piece of permanent equipment is such a load. Dead loads always act vertically downward.

Temporarily applied loads are called *live loads*. Live loads can be subdivided into two types—movable and moving. *Movable loads*, such as the lumber placed in a warehouse, the furniture in an office, and the wheat placed in a bin, are considered as steady loads and offer no difficulty in computation. *Moving loads*, such as trains on bridges, falling weights, and so forth, may involve considerable difficulty, especially when the movement is fast. Loads of this type will generally produce a much greater effect than that obtained by the same loads considered fixed in position. An even more quickly applied load, such as that produced by hitting a ball with a baseball bat, is further qualified as an *impact load* and will always require an investigation based on energy relationships.

FREE-BODY DIAGRAMS

1-4 Introduction

In the analysis of engineering problems much use is made of the *free-body diagram*. Such a diagram represents a whole body, or a portion thereof, separated from its adjacent supporting bodies. The forces produced by these supporting bodies, as well as any other external forces, are shown on the free-body diagram at their exact points of application and along their given lines of action. Because the free-body diagrams used in this text are derived from engineering members considered at rest or traveling at constant speed, the laws of equilibrium are applicable to them.

1-5 General Principles of Constructing Free-body Diagrams

In the study of mechanics, free-body diagrams are generally drawn for an entire body. *In the study of strength of materials, however, free-body diagrams are more often drawn using only a portion of a body*, because of the desire to find the internal forces acting on the interior sections exposed by the cutting planes used in obtaining the free body. In other words, that portion of the complete body selected for the free-body diagram is obtained by passing one or more imaginary planes through the body at the section or sections where the forces are desired. The complexity of the free body increases with the number of planes used to disengage it from the parent body. Regardless of its complexity, however, a free body will always have acting upon it the external forces which acted upon it when it was a part of the entire body—as well as the internal forces acting on the sections exposed by the cutting planes.

When a free-body diagram is drawn for a portion of an entire body, the premise is made that it will be in equilibrium* if the body from which it is cut is in equilibrium. After cutting the free body and depriving it of the restraining forces furnished by adjacent parts of the body, the same forces that existed on the cutting planes in the member before cutting must be placed on the exposed surfaces in order to reproduce the former state of equilibrium. Only one particular force system is possible on any section of a

*If a body is in equilibrium, the summation of the forces in any direction and the summation of the moments in any plane equal zero. Expressed theoretically, this condition exists when $\sum F_x = 0$, $\sum F_y = 0$, $\sum F_z = 0$, $\sum M_x = 0$, $\sum M_y = 0$, and $\sum M_z = 0$. For a coplanar force system, equilibrium exists when $\sum F_x = 0$, $\sum F_y = 0$, and $\sum M = 0$.

loaded member for the establishment of equilibrium. Consequently, the forces acting on an exposed surface of a free body are identical to the forces acting on the corresponding section of the loaded member before the free body was formed. The determination of these forces by the use of the laws of equilibrium forms the basis upon which further strength computations are made.

To help clarify this discussion, let us solve the following problems involving the development of the more simple free-body diagrams.

Illustrative Problem 1. Two tug-of-war teams are pulling valiantly, but neither side is able to budge the other. If each team consists of six men, each pulling with a force of 200 lb, find the tension in the rope at the two sections, *A-A* and *B-B* (Fig. 1-2).

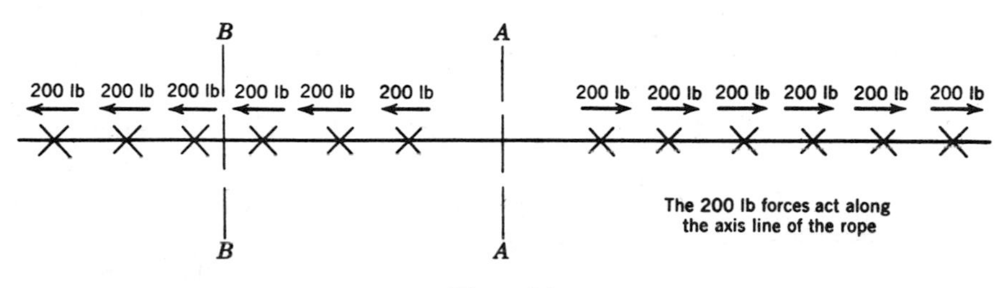

Figure 1-2

SOLUTIONS. *Section A-A.* To follow the principles laid down in our discussion, let us first obtain our free body. Inasmuch as the tensile force on section *A-A* is required first, let us pass a plane through the rope at this section, cutting the rope into two parts. Instantly, the two grunting teams fall backward. The rope recoils with tremendous force and the 200-lb forces are immediately reduced to zero. Certainly there is very definite evidence that some restraining force must have existed on section *A-A* before severance took place.

To find this force on section *A-A*, we must first reestablish equilibrium for one of the two portions of the rope adjoining this section. Let us choose the left-hand portion (Fig. 1-3a).

Since each 200-lb applied force acted along the axis of the member, the restraining force acting on section *A-A* must also have been axial. To restore

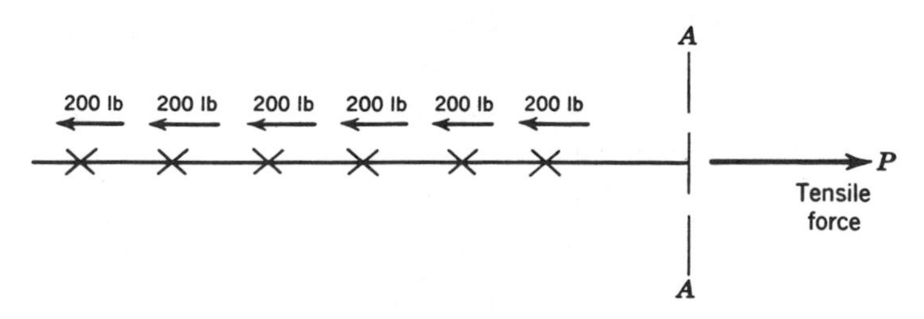

Figure 1-3a

the left-hand section of rope to its original loading condition, let us grasp the rope at section *A-A* and supply an axial load at that section directed toward the right. Assuming that the men never released their grasp on the rope, their restraining force will increase in proportion to our increasing pull. At the very moment that our pull requires each man to restrain with his capacity 200-lb force, we have restored the rope to its exact condition prior to severance. Let us now proceed to determine the force on section *A-A*. Since all the forces acting on the rope act along its axis, only one law of equilibrium ($\sum H = 0$) need be used to establish equilibrium. Hence, we can write

$$200 \times 6 - P = 0, \qquad P = 1200 \text{ lb}$$

This force of 1200 lb was the exact force acting on the fibers of the rope's cross section before cutting took place, because the restored condition of equilibrium in the cut rope matches exactly the condition of equilibrium of the uncut rope.

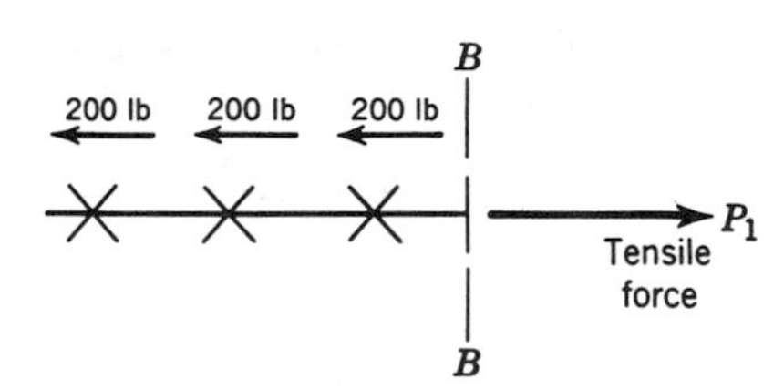

Figure 1-3b

Section B-B. To analyze the force on section *B-B*, let us choose the simplest free body adjoining this section (Fig. 1-3b):

Here again all the 200-lb forces are acting axially. The restraining force on section *B-B* must, therefore, also be axial. As the force on section *B-B* is restored, the three men increase their resisting power. At the exact moment they attain 200 lb of restraining force, the conditions previously maintained in the uncut rope will have been realized, and the force *P* must be such that

$$200 \times 3 - P_1 = 0, \qquad P_1 = 600 \text{ lb}$$

The forces *P* and P_1 found in this illustrative problem are further designated as *internal forces*, since they act *within* the body of the rope. They are further classified as *tensile forces* since they act out of the section and produce elongation.

Illustrative Problem 2. Determine the reactive forces at *B*, *D*, *E*, *F*, and *G* of the frame shown in Fig. 1-4a. All its joints are pinned.

SOLUTION. Consider the entire frame as a free body, replacing its roller reaction at *G* with the horizontal reaction force R_G and the pin reaction at *E* with its equivalent components E_V and E_H. Taking the algebraic sum of moments about *E* provides the solution to R_G:

$$\sum M_E = 4 \times 16 + 7 \times 8 - R_G \times 10 = 0$$

$$R_G = 12^k \longrightarrow$$

This reaction at *G* prevents the frame from rotating. To secure equi-

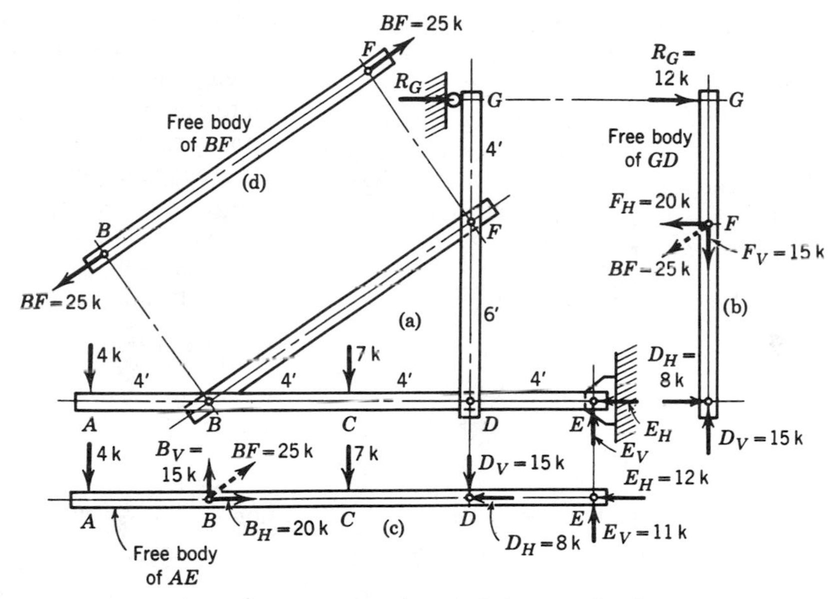

All reaction components shown in their correct direction

Figure 1-4

librium for the entire frame, it is also necessary that the external forces balance horizontally as well as vertically. Inasmuch as R_G and E_H are the only horizontal forces acting on the frame, E_H must be equal and opposite to R_G. Similarly, E_V is the only force that can balance the downward acting 4^k and 7^k applied forces; therefore, $E_V = 4 + 7 = 11^k$ acts upward.

The other required reactions occur between members. It is necessary therefore that we consider each of the three members of the frame as a separate free body.

Before separating the members of the frame, however, it is noted that member *BF* differs from the other two in that it has forces acting only at its ends. If this element is disengaged from the frame (Fig. 1-4d) and restored to equilibrium, the forces at the ends, whether single forces or the resultants of several forces, must be equal, opposite, and collinear. Furthermore, because *BF* is a straight member, these reactive forces will not only be collinear but will be coaxial as well, producing either tension or compression, with no bending. Such a member is called a *two-force member*.*

In contrast, the other two members, *AE* and *GD*, are acted upon by three or more forces, at least one of which acts across the longitudinal axis. Whenever such a member has a nonaxially directed force acting unopposed

*Theoretically, a homogeneous column acted upon by axial loads at either end could not bend. Practically, however, no column is perfectly homogeneous and some bending would, of course, occur.

somewhere along its length, bending is inevitable if equilibrium is to be maintained. Such a member is known as a *three-force member*. The direction of reactive force at a pin is initially indeterminate, and its effect is generally indicated by using horizontal and vertical components, e.g., D_H and D_V, of indefinite magnitude and assumed direction.

Proceeding now to the free body GD (Fig. 1-4b), a three-force member, and taking the sum of moments about D, we obtain

$$12 \times 10 - F_H \times 6 = 0$$

$$F_H = +20^k \longleftarrow \qquad \text{(the + checks the assumed direction)}$$

Because BF is a two-force member, and its reactions are directed axially,

$$F_V = +20 \times \tfrac{3}{4} = +15^k \downarrow$$

$$B_V = 15^k \uparrow, \qquad B_H = 20^k \rightarrow$$

and

$$BF = 20 \times \tfrac{5}{4} = +25^k$$

The sum of the horizontal and vertical forces on GD now yields

$$\sum H = 0, \qquad 12 - 20 + D_H = 0, \qquad D_H = +8^k \rightarrow$$

$$\sum V = 0, \qquad -15 + D_V = 0, \qquad D_V = +15^k \uparrow$$

Because the values of BF, D_H, and D_V as computed were positive, their assumed directions were confirmed. To have obtained negative values for either D_V or D_H would have indicated that the direction of the component was incorrectly assumed; thus, they should have acted in the opposite direction.

Proceeding next to the free body AE (Fig. 1-4c), the other three-force member,

$$\sum H = 0, \qquad B_H - D_H - E_H = 0$$

$$20 - 8 - E_H = 0, \qquad E_H = 12^k \leftarrow$$

$$\sum V = 0, \qquad -4 + 15 - 7 - 15 + E_V = 0, \qquad E_V = 11^k \uparrow$$

Static indeterminacy. If in Illustrative Problem 2 the reactions at G and E had been buried in concrete, each would have been able to restrain horizontally, vertically, and rotationally. There would have been developed E_V, E_H, and M_E at E and G_V, G_H, and M_G at G. These six unknown reaction components would have been in excess of the three available equilibrium equations, making the solution of the reactions impossible. Whenever such a situation arises, i.e., where the number of unknown reaction components exceeds the number of applicable equilibrium equations, the member is said to be *statically indeterminate* with respect to its reactions and will require additional relationships—generally derived from internal deformations—before a solution can be obtained.

A structure or frame may be statically determinate with respect to its reactions but indeterminate with respect to internal forces. This condition also becomes apparent when an insufficient number of equilibrium equations is available for the solution. *The degree of indeterminacy is equal to the number of unknowns in excess of the number of equilibrium equations applicable.*

Illustrative Problem 3. Determine the resisting forces acting on a vertical section 3 ft from the free end of a cantilever beam 10 ft long, loaded as shown in Fig. 1-5. The load is in the beam's vertical plane of symmetry.

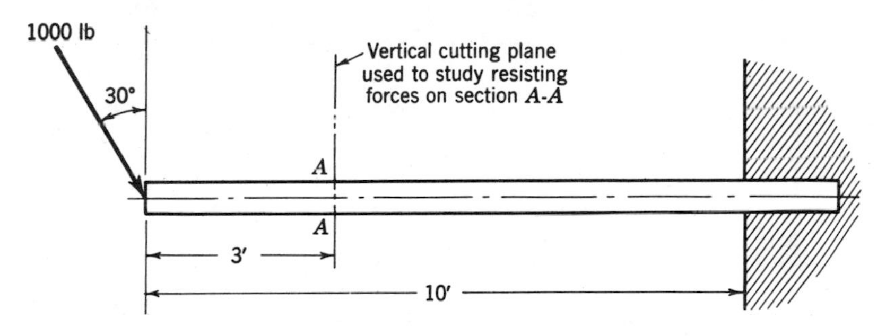

Figure 1-5

SOLUTION. As before, our first job must be to obtain the simplest free body that will enable us to obtain the required resisting forces. Such a free body must adjoin the section in which we are interested. Furthermore, it should have the least number of external forces. In keeping with these stipulations, our free-body diagram will be drawn for that portion of the beam to the left of the 3-ft section, thereby eliminating the necessity for analyzing and computing the reaction at the right end (Fig. 1-6).

It is quite obvious that this 3-ft section will collapse the instant the sectioning plane is passed through the beam. The three laws of equilibrium, $\sum V = 0$, $\sum H = 0$, and $\sum M = 0$, which are symbolic of perfect balance,

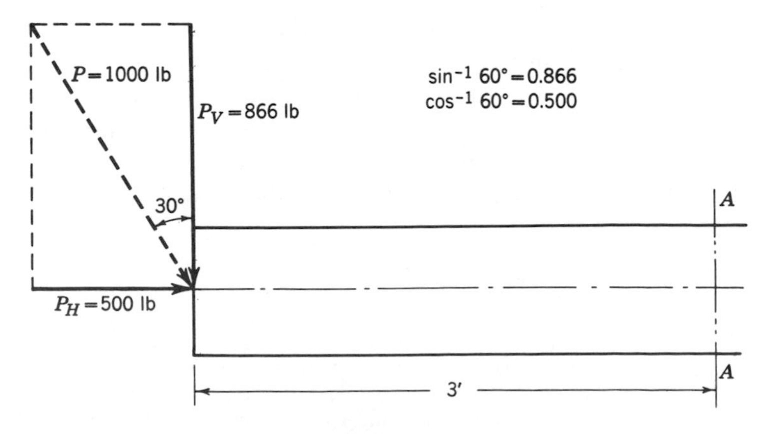

Figure 1-6

will then have been broken. To reestablish the condition of equilibrium, those forces which had been acting on the section and were released when the cut was made will have to be restored.

Let us apply these three laws of equilibrium to the free body in order that these forces may be found.

Consider first the horizontal forces.* If their summation must equal zero, there must have been a horizontal force acting on the cut section in opposition to the P_H force of 500 lb—i.e., a compression force, R_H, acting at the centroid of the section. See Fig. 1-7.

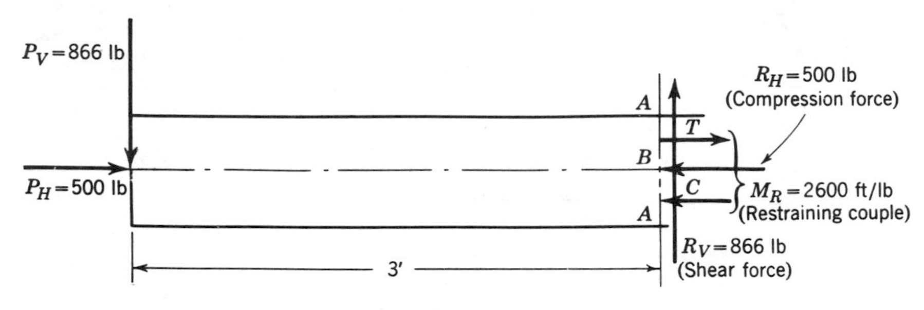

Figure 1-7

Our equation of horizontal forces will then read

$$\sum H = 500 - R_H = 0, \qquad R_H = 500 \text{ lb}$$

Let us now consider the vertical forces. As before, only one external force needs to be balanced—component P_V. Since the balancing force must act on the exposed section, there must exist an equal and opposite parallel force, R_V, acting in the plane, i.e., a *shear force*.

Writing the equation, we obtain

$$\sum V = 866 - R_V = 0, \qquad R_V = 866 \text{ lb}$$

Because the two vertical forces, P_V and R_V, are equal, parallel, and opposite in direction, they form a *couple* and would cause the body to rotate. It is necessary, therefore, to place a restraining couple M_R on the exposed surface to balance the P_V and R_V forces. This couple must act in a clockwise direction to oppose the counterclockwise $P_V - R_V$ couple.

By taking moments about any point B in the plane,

$$\sum M = 866 \times 3 - M_R = 0$$
$$M_R = +2600 \text{ ft-lb}$$

*When balancing either horizontal or vertical forces, it is well to imagine the free body acting in parallel guides placed in the direction of the equalizing operation under consideration. Thus, forces or components of forces acting perpendicular to the guides will have no effect on the motion of the free body acting in the guides. This aid is especially valuable in dealing with the vertical forces in this illustrative problem.

The plus sign before our numerical value indicates that the clockwise direction of the restraining couple was assumed correctly.

Observe that the two forces constituting the restraining couple were designated by the letters T and C. Owing to the applied moment, the beam bent downward, thereby placing the top fibers of the beam in tension and the bottom fibers in compression. Thus, the reason for making this particular designation becomes clear.

As far as the results are concerned, it makes no difference which of the above equations is used first. But whenever possible, it is more advantageous to begin with a force equation that has only one unknown.

Now that we have obtained the forces to complete our free body (Fig. (1-7), the perfectly balanced free body duplicates its original state when it was a portion of the whole beam.

Illustrative Problem 4. (a) Develop the free-body diagrams for the portions of the beam shown in Fig. 1-8a on either side of sections A and B. (b) Draw the free-body diagram for the portion of the beam located between sections A and B. (c) Draw the free-body diagram for the upper half of that portion of the beam described in (b).

SOLUTION. (a) Using the techniques previously discussed, the required free-body diagrams are shown in Fig. 1-8b and c. (b) The free-body diagram

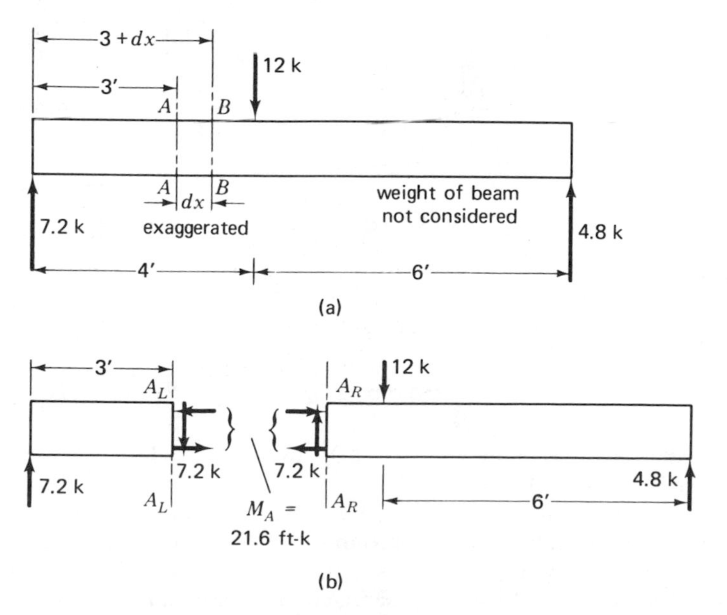

Figure 1-8a, b

of that portion of the beam located between section A and B (actually A_R and B_L) is shown in Fig. 1-8d. Note that under the action of the forces shown the free body is in perfect equilibrium, just as it was when it was a part of the whole beam. The forces of the M couples are located the same distance apart. (c) The free-body diagram shown in Fig. 1-8d may yet be subdivided. The free body of the upper half is obtained by using cutting plane C-C and is shown in Fig. 1-8e. Because the horizontal compressive forces are different by the value of 7.2 dx/a, equilibrium can only be restored by applying an equal and opposite force located in the plane exposed by plane C-C.

It is interesting to note that the free body formed by planes A, B, and C is a small part of the total beam. It could in fact be made infinitesimally small if desired through the use of other cutting planes, and still retain its equilibrium under the action of the external and internal forces acting upon it.

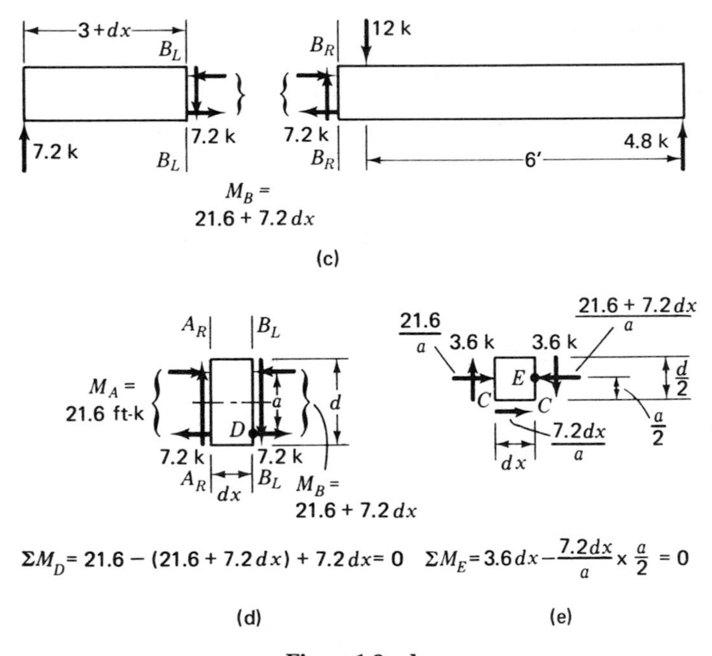

Figure 1-8c, d, e

PROBLEMS

1-1. For the truss shown in Fig. 1-9, solve for the items indicated in the order presented. All joints are pinned.

(a) Find the magnitude and direction of the vertical reaction at A.

(b) Find the magnitude and direction of the horizontal and vertical reaction components acting at E.

(c) Find the magnitude and character of force in members CD and CB starting at joint C.

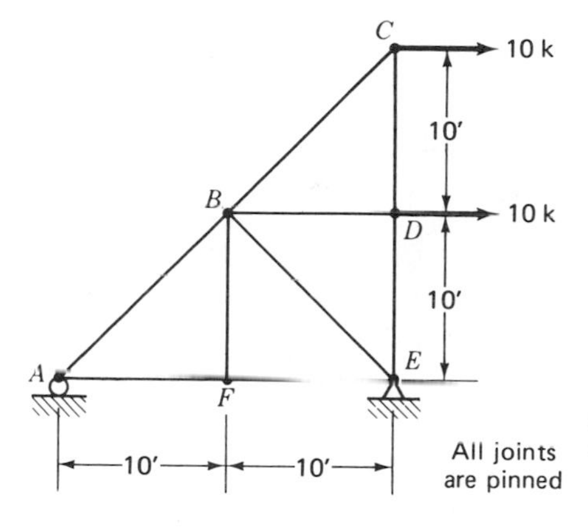

Figure 1-9

(d) Find the magnitude and character of force in members *AB*, *AF*, *FE*, and *BF*, in that order, starting at joint *A*.

(e) Find the magnitude and character of force in members *EB*, *ED*, and *DC*, in that order, starting at joint *E*.

(f) What is the magnitude and character of the force in member *BD*?

1-2. After finding the reactions of the pinned truss shown in Fig. 1-10,

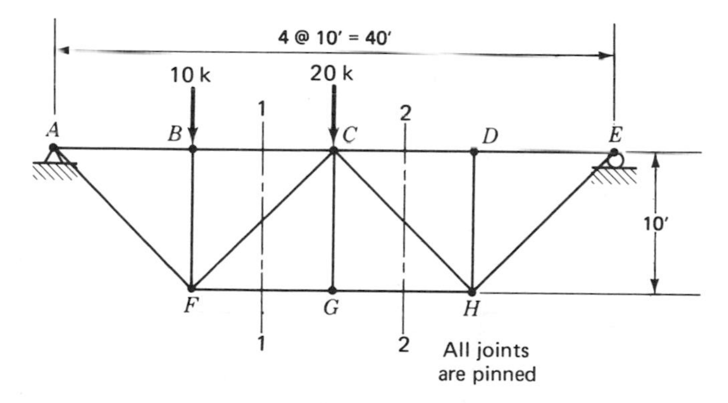

Figure 1-10

(a) Find the magnitude and character of force in members *FC* and *CH*, using free bodies separated by sections 1-1 and 2-2, respectively.

(b) Using a free body separated by using section 1-1, find the magnitude and character of force in members *BC* and *FG*.

(c) Using section 2-2, find the magnitude and character of force in members *CD* and *GH*.

1-3. Calculate the forces in the members of the truss shown in Fig. 1-11. All joints are pinned.

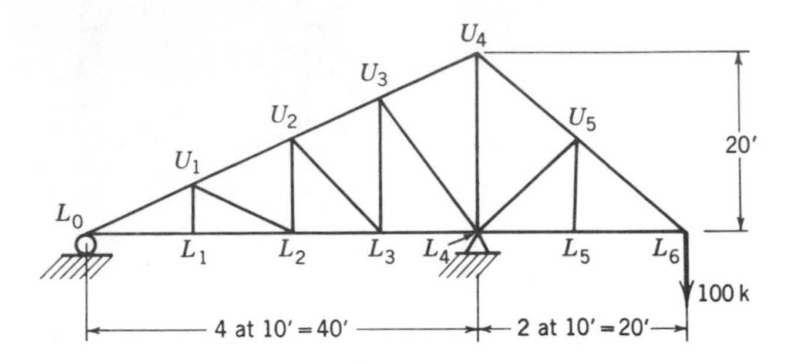

Figure 1-11

1-4. The free-body diagram shown in Fig. 1-12 is not complete. What must be added to the figure? *Answer:* 2000 lb acting vertically downward, 4 ft from R_1.

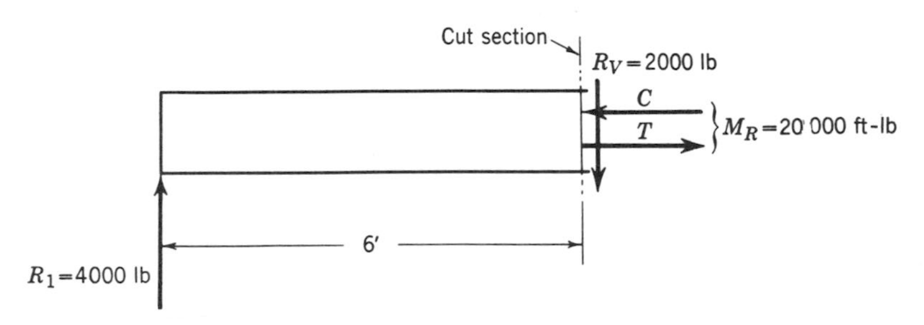

Figure 1-12

1-5. Draw the free-body diagrams for those portions of the beams shown in Fig. 1-13, both to the left and right of section A-A. Disregard the weight of the beam. 1 k = 1 kip = 1000 lb.

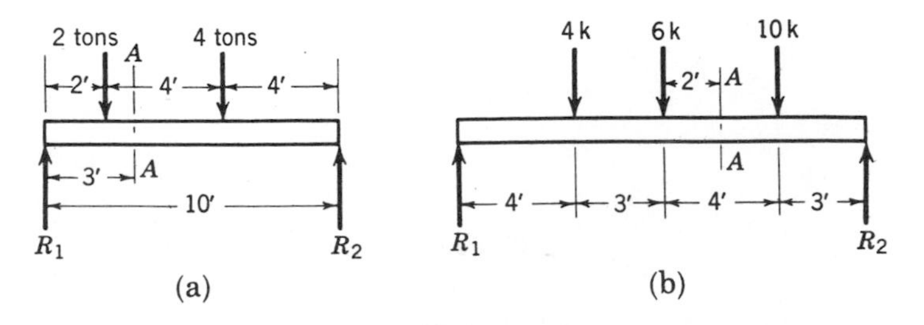

Figure 1-13

1-6. Draw the free-body diagram for those portions of the beams shown in Fig. 1-14, both to left and right of section A-A. Disregard the weights of the beams. One kip = 1000 lb. *Answers:* (a) $V_A = -1$ kip, $M_A = +33.5$ ft-kips; (b) $V_A = +6.4$ kips, $M_A = +7.7$ ft-kips.

1-7. In the beams shown in Fig. 1-14, consider additional sections B-B to be drawn 3 ft from the right end. Draw the free-body diagram of those portions of

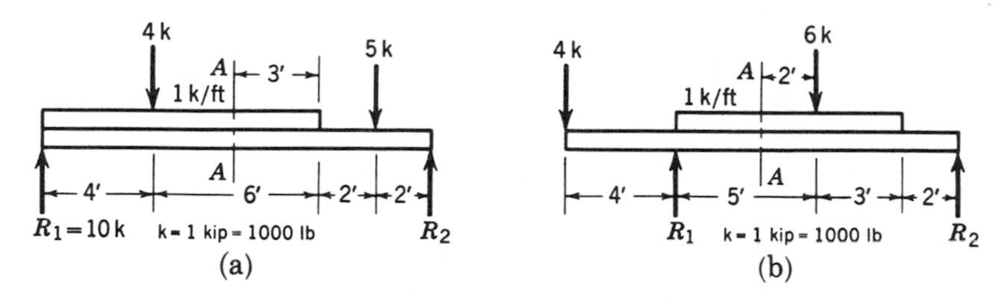

Figure 1-14

the beams between sections *A-A* and *B-B*. Disregard weights of beams. *Hint:* Use the forces acting on section *A-A* as obtained from Problem 1-6 to find those acting on section *B-B*. *Answers:* (a) $V_B = -4$ kips, $M_B = +22$ ft-kips; (b) $V_B = -3.6$ kips, $M_B = +13.3$ ft-kips.

1-8. In Fig. 1-15, beams *AB* and *CD* are in contact with one another at point *E*. Draw the free-body diagram of each beam. Consider the beams to be weightless. $P = 2^k$.

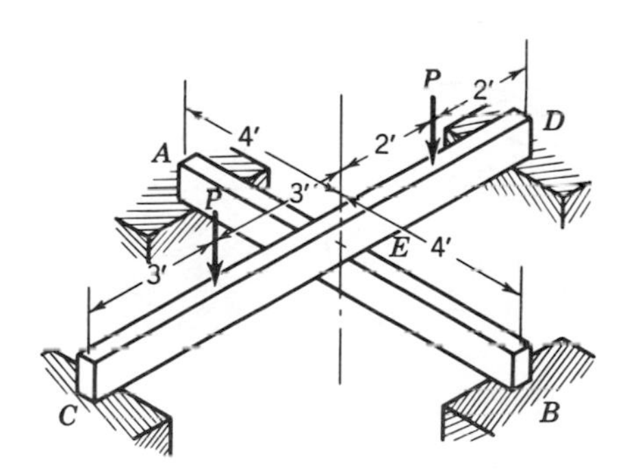

Figure 1-15

1-9a, b. Find the resistive forces in parts I, II, and III of the bars shown in Fig. 1-16.

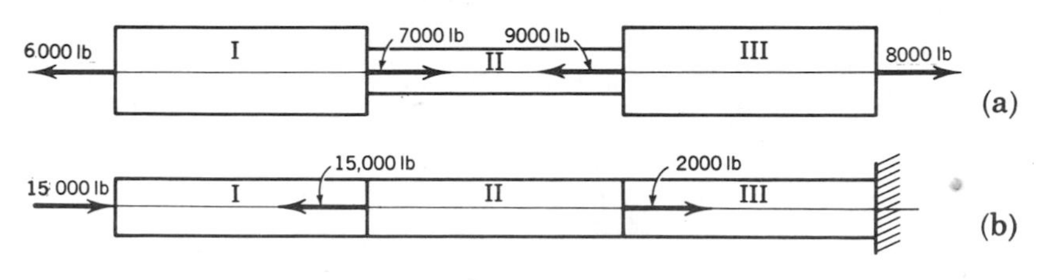

Figure 1-16

1-10. In the wavelike member of Fig. 1-17, draw the free-body diagrams of those portions to the left of vertical sections *A-A*, *B-B*, and *C-C*. The entire member lies in a vertical plane.

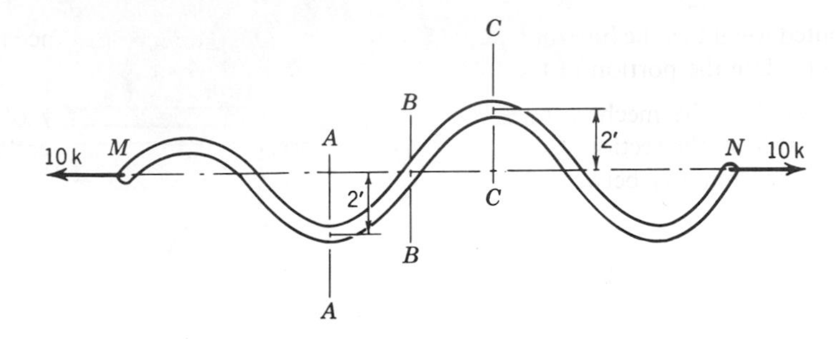

Figure 1-17

1-11. (a) Draw the free-body diagram of that portion of the axially loaded spring shown in Fig. 1-18 above the enlarged vertical section. The mean diameter of the spring is 3 in. *Answers:* $V_A = 100$ lb, $T_A = 150$ in.-lb. (b) Consider a section parallel to the one used in (a) but removed by a differential distance. Does the shear force and torsional couple change from section to section? (c) From the analysis of these sections, is there any bending effect in the spring?

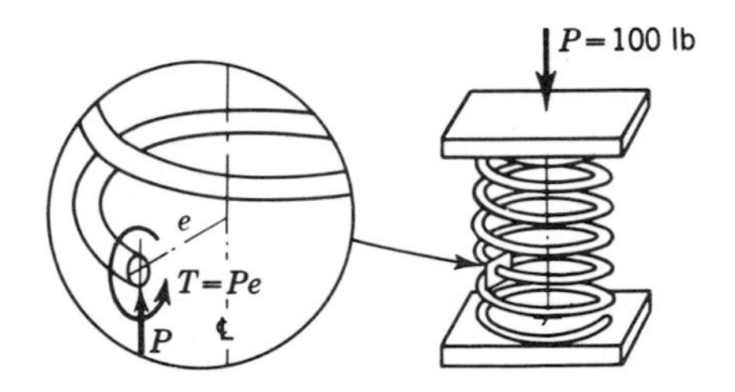

Figure 1-18

1-12. Two points are 6 ft apart horizontally and 3 ft apart vertically. A flexible cord 10 ft long is tied between the two points and a weight is hooked over the cord so it can slide until equilibrium is reached. Disregarding the weight and stretch of the cord, what is the location of the low point of the cord with respect to its ends? *Answer:* 2.50 ft below, and 1.88 ft horizontally from the lowest point.

1-13. A cable is supported as shown in Fig. 1-19. If its weight is assumed to be uniformly distributed horizontally, determine its sag h_2 in terms of the uniformly

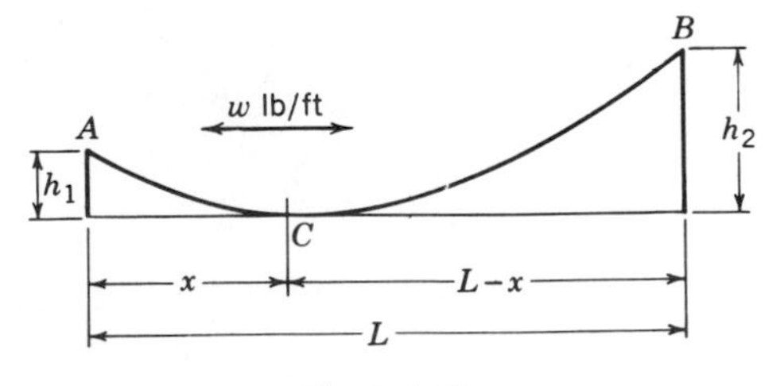

Figure 1-19

distributed weight w, the horizontal component of tension T, and the distance to the low point. Use the portion of the cable from C to B as a free body.

1-14. For the mechanism shown in Fig. 1-20 draw the free body of that portion (a) from the sectioning plane A-A to F, (b) from the horizontal sectioning plane located midway between B and C to point F.

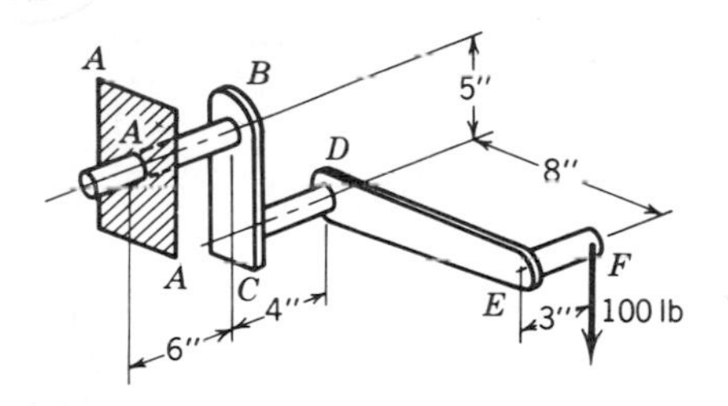

Figure 1-20

1-15. (a) If a couple of 2000 in.-lb is applied to the gear train (Fig. 1-21) at A, what couple is required at H to maintain equilibrium? Use a free body for each component to obtain your answer. Each shaft is held in alignment by properly placed bearings. (b) If gear A is rotated 90°, through what angle is gear H rotated? (c) What is the relationship between the couple and angle of rotation at gears A and H if the system rotates at uniform speed?

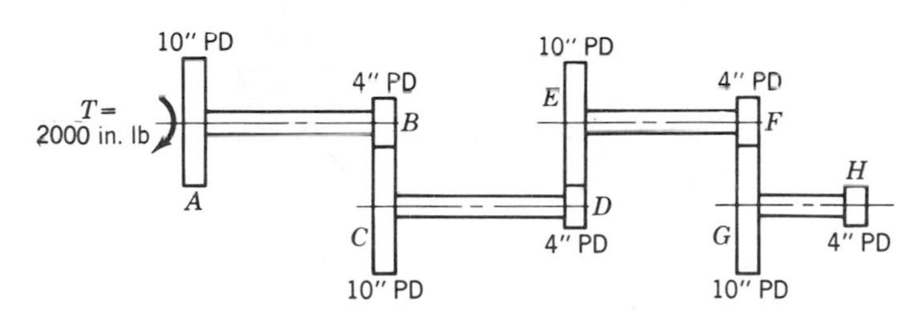

Figure 1-21

1-16. Draw the free bodies of the pin-connected members AB, CD, and BF of the frame shown in Fig. 1-22. Calculate the reactive components at B, C, E, F, and D. *Answers:* $C_v = 20.0^k$, $C_H = 8.0^k$, $E_v = 40.0^k$, $E_H = 26.7^k$.

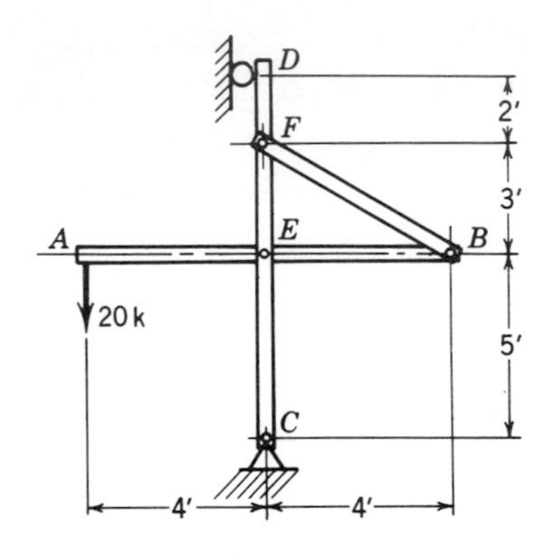

Figure 1-22

1-17. Draw the complete free body of that portion of the beam shown in Fig. 1-23 fıom *A* to *C*.

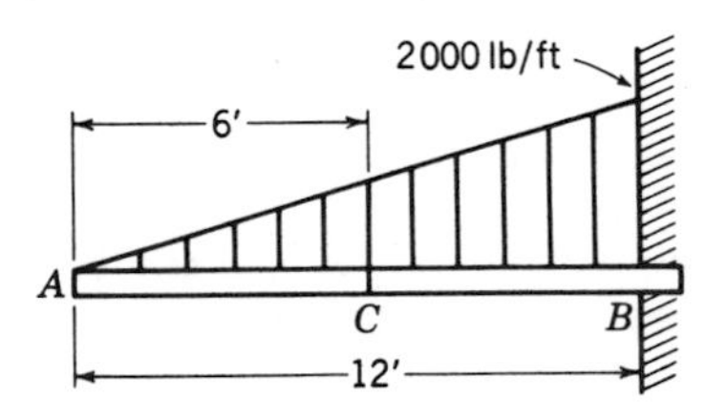

Figure 1-23

1-18. Draw free bodies of sections *A* to 1, *A* to 2, and *A* to 3 of the frame shown in Fig. 1-24.

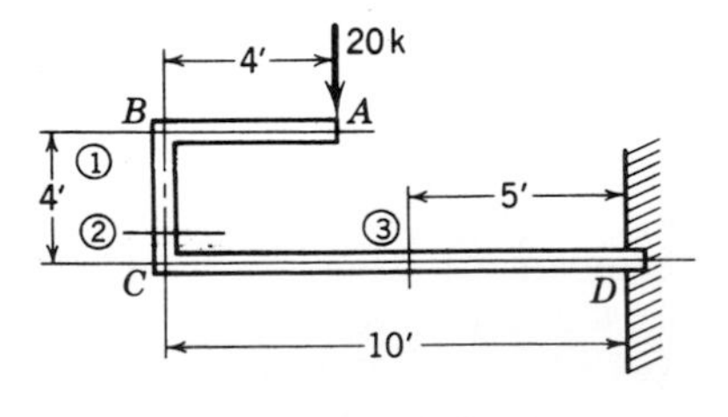

Figure 1-24

1-19. (a) In the beam assembly shown in Fig. 1-25, determine the roller reaction at *B* and the linkage reaction at *C*. (b) Determine the reaction at *D*. (c) Determine the forces acting on sections 1-1 and 2-2.

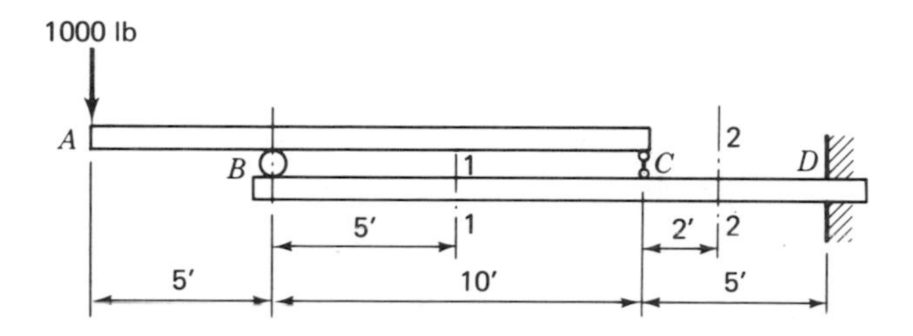

Figure 1-25

1-20. The two square frames shown in Fig. 1-26 are held together by a bolt tightened to a tension of 2000 lb. Subsequently, the two 1000-lb forces and two springs compressed to 500 lb each are added. (a) Does the tensile force in the bolt change due to the subsequently applied forces? (b) What effect does the addition of the spring have on the tensile force in the bolt? (c) What resistive forces are acting on the sections 1-1 equally distant from the center line of the assembly?

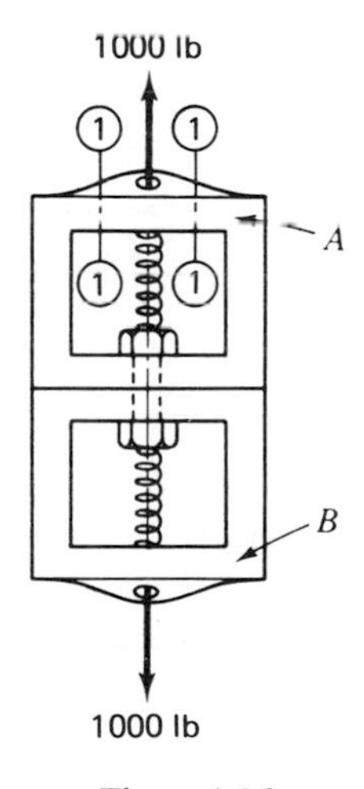

Figure 1-26

1-21a, b. In the beams shown in Fig. 1-27, draw the free bodies located between the parallel cutting planes located dx distance apart, as well as that portion of those free bodies located above the horizontal center line.

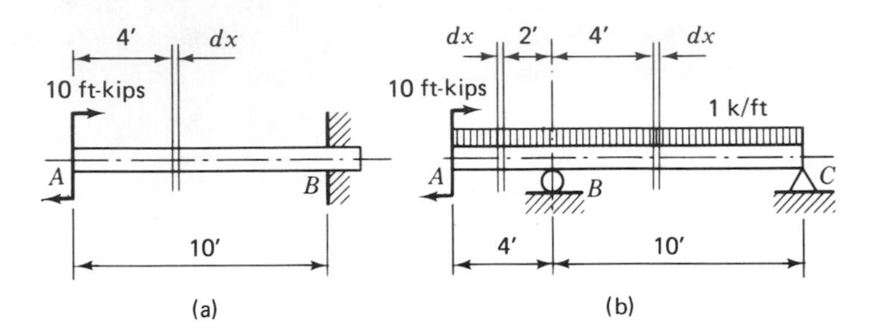

Figure 1-27

1-22. A bolt 12 in. long has one end held horizontally in a vise. The other end is acted upon by a monkey wrench 20 in. long located in a horizontal plane and perpendicular to the longitudinal axis of the bolt. If a force of 100 lb is applied perpendicular to the handle of the wrench (20 in. from the center line of the bolt), draw a free body of that half of the bolt including the monkey wrench. Draw in contrast the free body of the other half. *Answers:* $V_R = 100$ lb, $T_R = 2000$ in.-lb, $M_R = 600$ in.-lb.

1-23. Two equal cylinders, each weighing W lb, are placed in a container as shown in Fig. 1-28. Assuming that all contacts are frictionless, determine the reactions at A, B, and C. *Answers:* $R_A = W$, $R_B = 2W$, $R_C = W$.

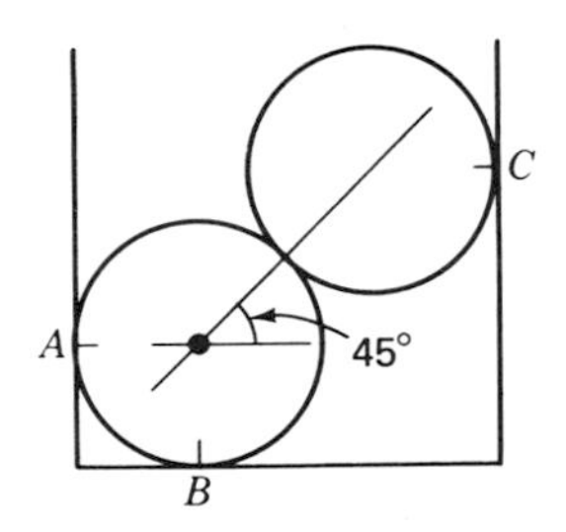

Figure 1-28

1-24. Determine the reactions at A, B, and C on the beam shown in Fig. 1-29. *Answers:* $R_A = 10$ kips, $R_C = 45.5$ kips.

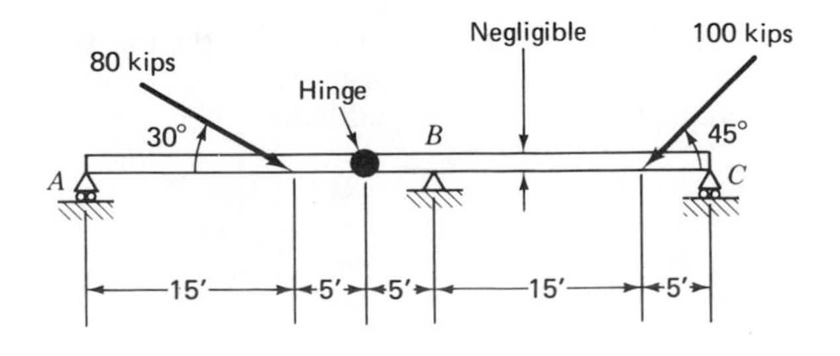

Figure 1-29

1-25. Draw a free-body diagram of the bar AD of the structure shown in Fig. 1-30. Determine the forces in the other members of this truss.

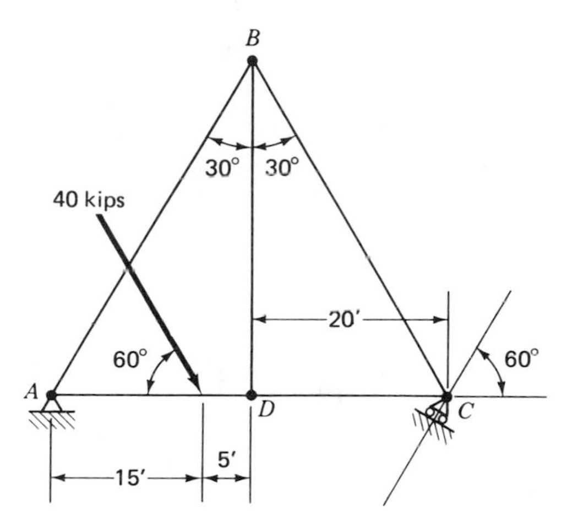

Figure 1-30

1-26a, b. Determine the reactions of the beams shown in Fig. 1-31. *Answers:* (a) $R_1 = 4580$ lb, $R_2 = 5220$ lb.

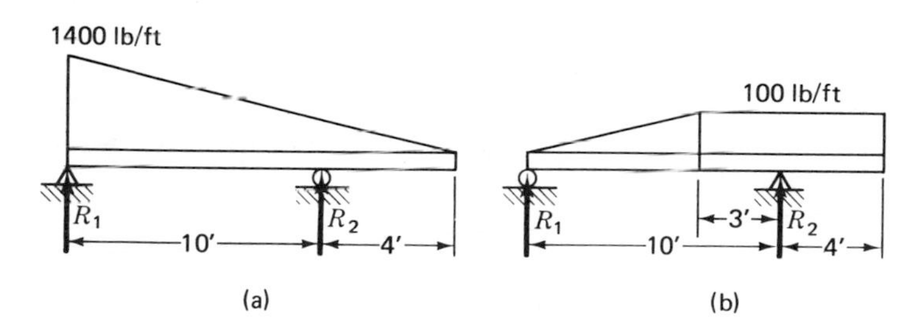

Figure 1-31

1-27a, b. Determine the reactions of the beams shown in Fig. 1-32. *Answers:* (a) $R_A = 5250$ lb, $R_E = 25{,}875$ lb.

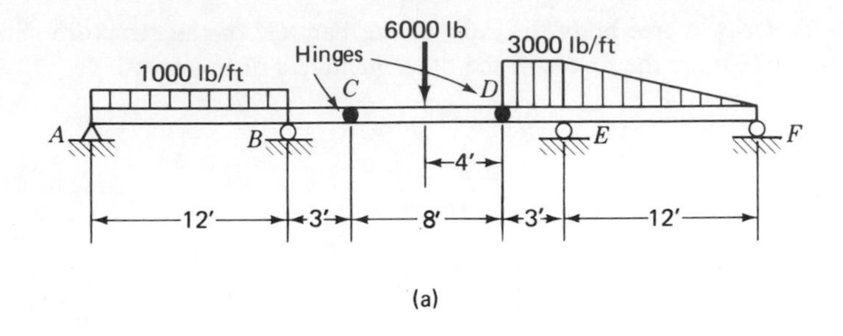

(a)

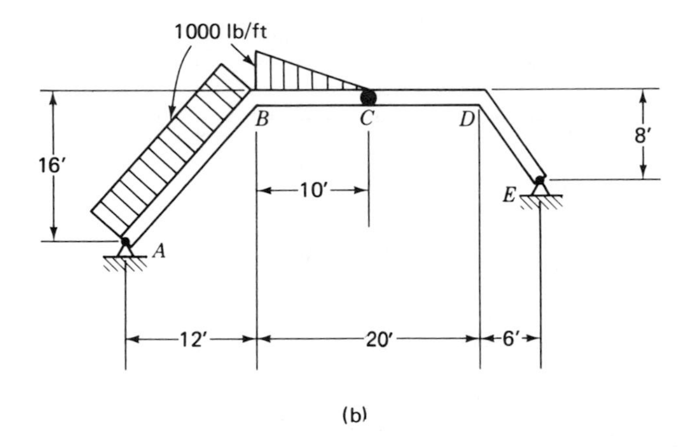

(b)

Figure 1-32

2

Members Subjected to Axial and Central Loads

2-1 Introduction

Much of the fundamental theory of engineering design concerns the action of axial or central loads on simple engineering members. The development of these fundamental theoretical principles, as well as their use and importance in design, is discussed in this chapter.

2-2 Central Loads; Axial Loads

If the action line of a concentrated load were to pass through the centroid of a cross section of any solid body (Fig. 2-1), it would be called a *central load* for that cross section. Its line of action would not need to be perpendicular to the section or coincident with the longitudinal axis of the body. Q is a central load for section AB.

However, if the line of action of the load acted through the center of gravity of *every* cross section in the member, it would

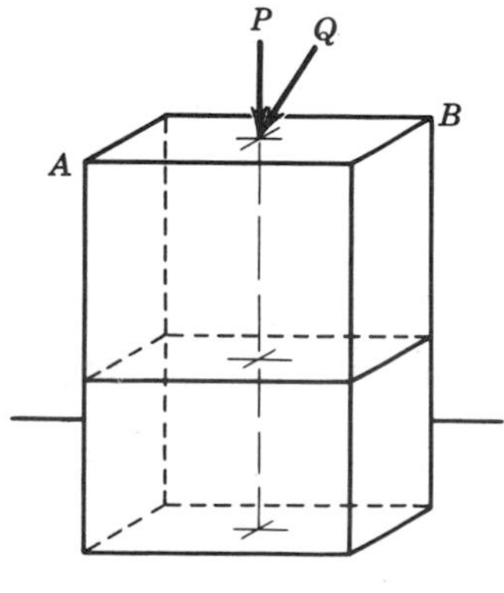

Figure 2-1

be called an *axial load* and would be said to act axially upon the member. Force P is an axial load acting on the body shown in Fig. 2-1.

NORMAL STRESS—SHEAR STRESS

2-3 Stress on Sections Perpendicular to the Longitudinal Axis of Axially Loaded Members

Consider a *short* compression member with a 2- by 2-in. cross section (Fig. 2-2a) to be loaded with a 40,000-lb axial force. It is desired to find the distribution and intensity of the resistance developed on section A-A perpendicular to the longitudinal axis.

Let us first pass a plane through section A-A, and consider the top portion as our free body (Fig. 2-2b). Since the load is axial, balance can be restored only by placing an equal, opposite, and collinear resisting force of 40,000 lb *at the center of gravity* of section A-A. This force, however, is not concentrated but is the *resultant* of the distributed internal resistance offered by that section.

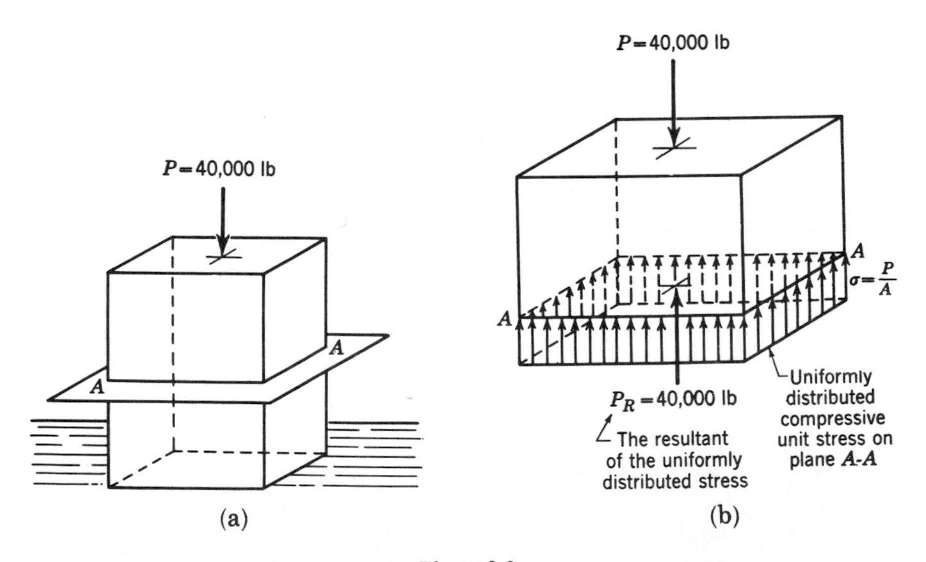

Figure 2-2

The question which then arises is whether this internal resistance is uniformly or nonuniformly distributed. Under the circumstances presented, where the body is homogeneous and of constant cross section, it is customary to assume a uniformly distributed resistance. Thus, the load distributed equally to each of the 4 in.^{2} in our cross section is 40,000/4 or 10,000 psi. The intensity of this resistive force per unit area (an average intensity) is commonly called a *unit stress** and is defined mathematically by

$$\sigma = \frac{P}{A} \tag{2.1}$$

where σ = unit stress in psi, P = axial load in pounds, and A = cross-sectional area in in.^{2} It is also a *normal unit stress* in that it acts perpendicularly to the section A-A.†

Had the resisting forces on section A-A been unevenly distributed, the value of $\sigma = P/A$ would not be exact stress for every point within area A but an average for all points. To increase the accuracy of the stress calculation at any point would require the use of a smaller and smaller area enclosing that point and the increment of load acting on it. The use of an infinitesimally small area would provide the exact stress. Expressed mathematically, this stress would equal:

$$\sigma = \lim_{\Delta A \to 0} \frac{\Delta P_R}{\Delta A} = \frac{dP_R}{dA} \tag{2.1a}$$

To return to our previous illustration, the resultant resisting force of 40,000 lb acting on the exposed section is called a *total stress*. In this case, the unit stress is further defined as being a *compressive unit stress*, owing to the crushing action of the applied force. If the applied force had acted in the opposite direction to tear the member apart, the stress induced on the section would have been a *tensile unit stress*.

Except for axially loaded members, unit stresses are not uniformly distributed over each cross section, but vary from point to point. It is the study of these variations of stress that occupies such a great portion of the study of mechanics of materials.

*The term *unit stress* is often used in its shortened form *stress*. The phrase *strength of a material* refers to the ultimate or *limiting unit* stress at rupture.

†The study of photoelasticity has revealed that the degree of uniformity of normal stress distribution due to an axial load depends somewhat on the shape and length of the member and the magnitude of the load applied. The stress will become more uniformly distributed as (1) the cross sections become more uniform, (2) the length of the member increases, and (3) the magnitude of the load decreases. Although the unit stress directly under a concentrated load is greatly intensified, this intensification is rapidly reduced as the distance from the load increases. According to St. Venant, one may consider the stress to be uniformly distributed over a cross section when the distance from the point of application of the axial load is equal to the largest dimension of that cross section. This statement assumes that the largest dimension of the cross section is small in comparison to the length of the member.

In elementary studies, such as ours, the normal unit stress on any cross section of an axially loaded, two-force member is considered to be uniformly distributed unless otherwise noted.

Notes on the use of the $\sigma = P/A$ formula. Figure 2-3 shows a curved compression member with a rectangular cross section. Suppose the stress on section *A-A* were desired and the suggestion was made to use the formula $\sigma = P/A$. Would the suggestion have been correct?

For our investigation, let us draw a free body of the upper portion of the member (Fig. 2-4). The only way to balance force P would be to add

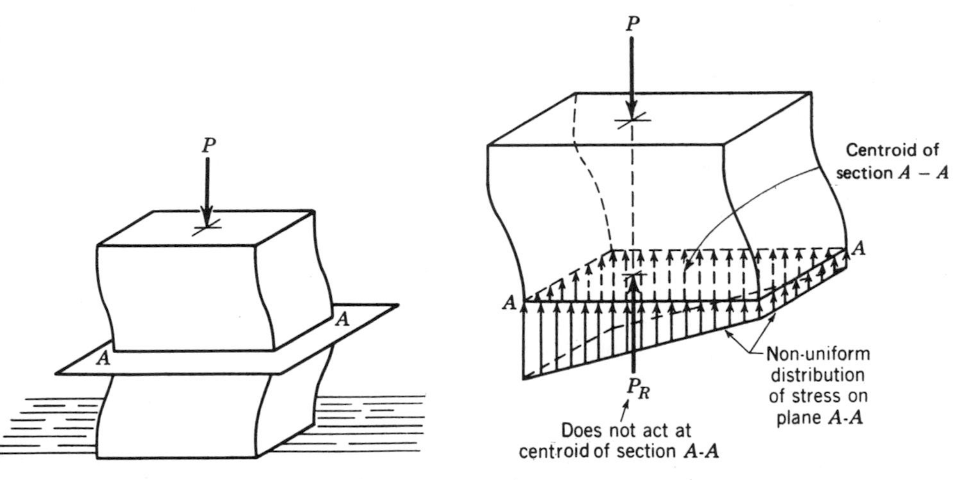

Figure 2-3

Figure 2-4

P_R, equal, opposite, and collinear to force P. Since P_R is the resultant internal force acting on section *A-A* and does not pass through the centroid of the section, the stress cannot be uniformly distributed. To use the $\sigma = P/A$ formula to determine the stress on section *A-A* would, therefore, yield an incorrect result, since dividing the load P by the number of units of area in section *A-A* would erroneously apportion the load to each of them equally, or, in other words, assume a constant unit stress.

To use the formula $\sigma = P/A$ correctly, it is well to bear the following points in mind:

1. The applied load P must pass through the centroid of the section under investigation. If it is inclined to the section, its component acting normal to the section produces a normal unit stress σ_n equal to P_n/A.

2. If the maximum tensile or compressive unit stress is desired in a prismatic axially loaded member, area A must be the smallest cross section obtainable. The area to be a minimum must, therefore, be perpendicular to the action line of the applied force.

3. Area A must be that area on which the stress is desired.

PROBLEMS

2-1. A square steel plate, 12 in. on a side and 1 in. thick, is placed symmetrically over the end of a hollow vertical cylinder with an 8-in. outside diameter and walls 1 in. thick. If a uniform load of 2000 psi is applied in the downward direction on the steel plate, what maximum compressive unit stress is acting in the walls of the hollow cylinder? Disregard the bending action in the plate.

2-2. Find the unit stress at sections 1, 2, 3, and 4 of the axially loaded assembly shown in Fig. 2-5. Find also the bearing stress between the wall and the washer. *Answers:* 2000 psi, 4000 psi, 2210 psi, 6000 psi.

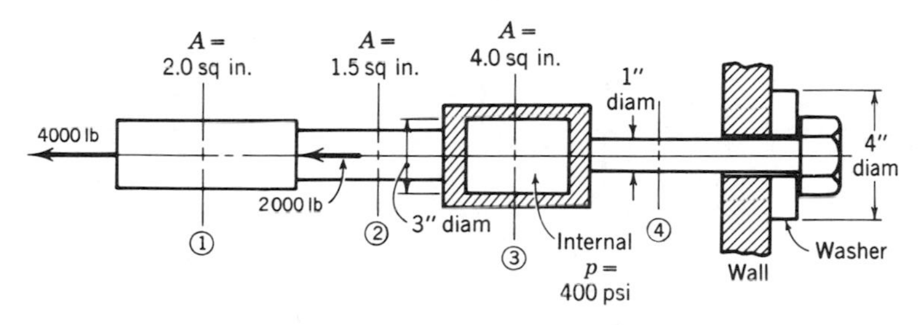

Figure 2-5

2-3. How high can a chimney be built before the compressive unit stress in the brick from which it is made exceeds 300 psi? Brick weighs 150 lb/ft.[3] *Answer:* 288 ft.

2-4. In Fig. 2-6 a bolt is used to hold members *A* and *B* together. The force in each spring is 3000 lb. What force *P* is required to produce a bearing stress of 50 psi between members *A* and *B* if their contact area is 20 in.[2] and the force in the bolt at that time is 5000 lb? *Answer:* 4000 lb.

2-5. What resisting areas are required for members *AB* and *AC* of Fig. 2-7 if their allowable stresses are 30,000 and 20,000 psi, respectively?

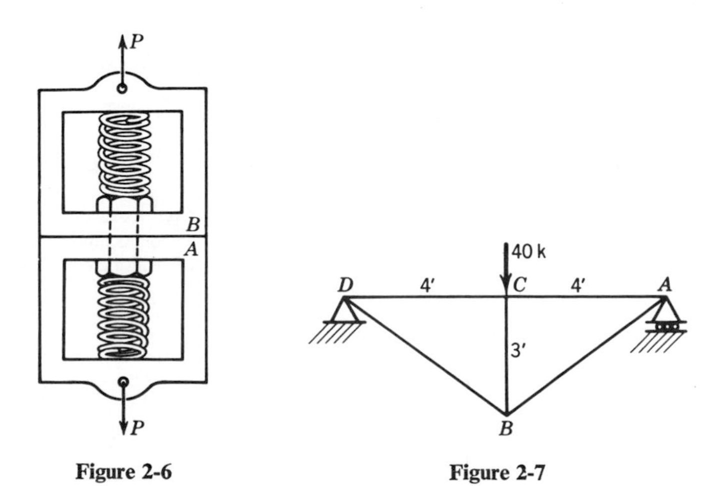

Figure 2-6

Figure 2-7

2-6. A 1-in.-diameter bolt runs vertically through an oak beam. If the unit stress in the bolt is 12,000 psi, what will the minimum diameter of the washer under the head of the bolt have to be if the bearing stress in the oak is not to exceed 500 psi? *Answer:* 5 in.

2-7. Determine the maximum tensile unit stress in bar AB of Fig. 2-8 when $P = 1000$ lb. The cross-sectional dimensions of AB are $1\frac{1}{4}$ in. wide by $\frac{1}{4}$ in. thick.

2-8. (a) Determine the theoretical relationship between the load P and the tension in rod AB of the scissor jack shown in Fig. 2-9. *Answer:* $P = T \tan \theta$. (b) What would the minimum cross-sectional area of AB have to be to hold a load of 10,000 lb where $\theta = 45°$? Use $\sigma = 10,000$ psi. *Answer:* 1.00 in.^{2}

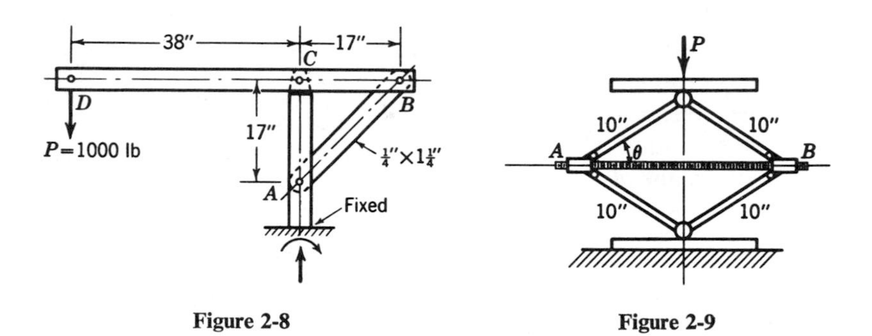

Figure 2-8

Figure 2-9

2-9. (a) What force P is required for equilibrium in the assembly shown in Fig. 2-10? Pulleys are placed between double-beam members AC and BD. (b) What is the unit stress in member AB? Area $= 0.5$ in.^{2}

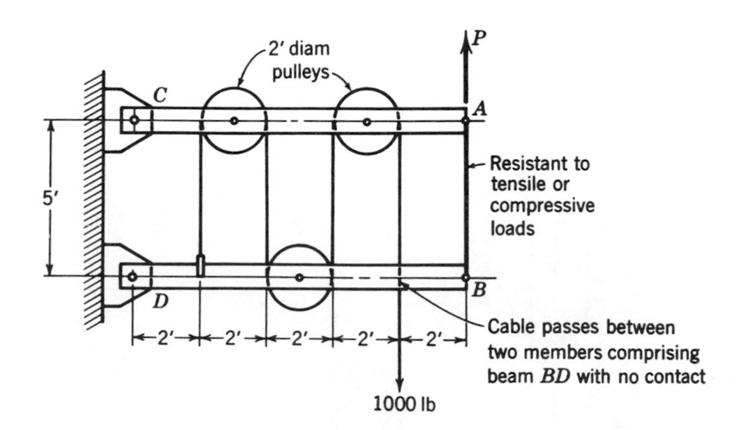

Figure 2-10

2-10. It is desired to place a 100-psi compressive unit stress on the top of a 4-in.-diameter cylinder, 6 in. high, through the use of spring pressure. Available for the job are a sufficient number of 2-in. I. D. tension springs of various lengths, all having a spring modulus of 500 lb/in. and an ample assortment of steel nuts, bolts, plates, and bars. The maximum elongation permitted for each spring is 15 per cent of its original length. Arrange and detail, using a simple sketch, a layout suitable for applying the required stress. Provide for an easy adjustment of pressure and a positive means of applying the pressure uniformly.

2-4 Shear Stress

Consider the block shown in Fig. 2-11a acted upon by the horizontal force P located in a vertical plane which is perpendicular to section *abcd* and passes through its centroid. Now if that portion of the block above section *abcd* were cut free, it would move under the action of force P. To restore its equilibrium condition necessitates, first, the application of a horizontal force P_R in the plane *abcd* (Fig. 2-11b) and, second, a couple T-C acting perpendicular to that plane to balance couple P-P_R (Fig. 2-12).

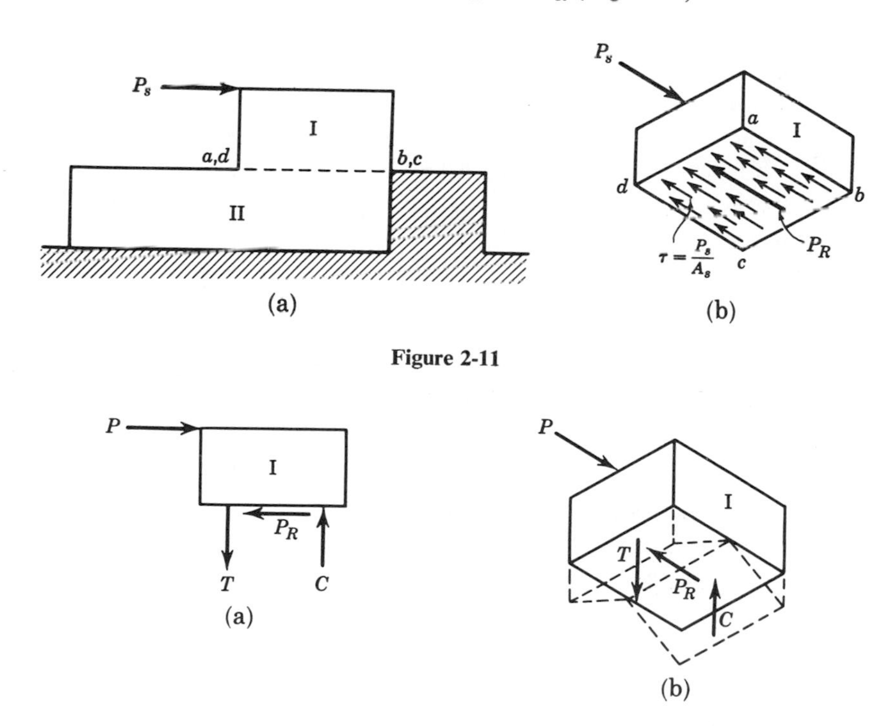

Figure 2-11

Figure 2-12

Because P_R acts *in the plane abcd* to prevent a sliding action between free bodies I and II, it is known as an *internal shearing force*. Of course it is not a concentrated force, but is the resultant of a distributed resistance acting on plane *abcd*. Because P_R passes through the centroid of area *abcd*, it is a central load and produces a shearing unit stress which is assumed to be *uniformly distributed*. Under this assumption, the resistance per unit area or *shearing unit stress* is equal to

$$\tau = \frac{P}{A} \tag{2.2}$$

Since the applied and resisting forces P and P_R are parallel, all horizontal planes located between them have the same tendency to slide with respect to one another and incur the same intensity of shearing unit stress.

Had the distribution been uneven, the exact determination of shear stress at any point, like the normal stress determination in the previous section, could have been obtained by considering the shear force acting on a differential area dA:

$$\tau = \lim_{\Delta A \to 0} \frac{\Delta P_R}{\Delta A} = \frac{dP_R}{dA} \tag{2.2a}$$

Bending stresses accompanying shear stresses. The couple T-C produced by couple P-P_R acts in the same plane as P and P_R and on the exposed section to produce tension and compression bending stresses (Fig. 2-12b). With a much taller block I (Fig. 2-13), the magnitude of the P-P_R couple, and hence its tilting tendency, would be increased. To balance this increase, couple T-C would have to be made larger by the same amount. With no increase in distance between the T-C forces possible, the only other alternative would

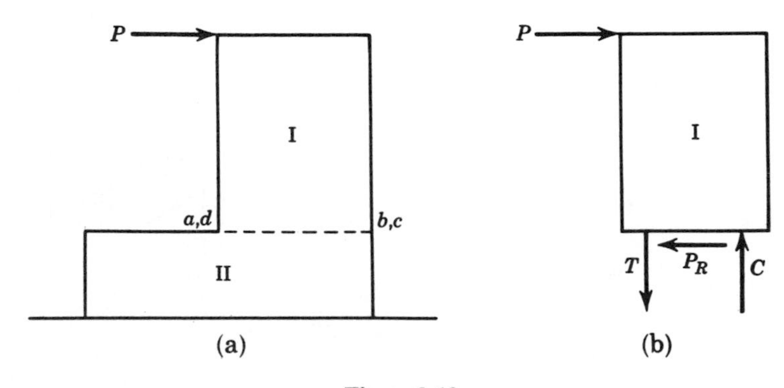

Figure 2-13

be that the T and C forces must themselves increase, with a corresponding intensification of the stresses they produce.

Should it be desirable to eliminate the tensile and compressive stresses from the desired shearing unit stress, the distance between forces P and P_R would have to be made infinitesimally small. That plane immediately below the top plane of block I in Fig. 2-11 would be such a plane. The distortion of the affected solid material during the stressing operation produces practical difficulties which make the production of absolutely *pure shear* by this method an impossibility. Whenever this method is used to produce so-called *pure shear*, care must be taken to make sure the distance between the applied force and the shearing resistance plane is so small as to be of insignificant importance in the results.

Shear stresses may be produced by still other means, such as by loading beams, twisting shafts, by combining twisting, bending, and direct forces, and so forth. Some of these more familiar methods will be handled in detail in subsequent chapters.

PROBLEMS

2-11. A small clip angle (Fig. 2-14) is used to support a load of 12,000 lb. If two $\frac{7}{8}$-in.-diameter rivets are used to resist this load, what shearing unit stress is developed in the rivets?

2-12. A heavy underslung trailer (Fig. 2-15) is used to carry a large excavator weighing 60 tons. If 2-in.-diameter axles support the 4 wheels, what shearing unit stress is developed in the axles? *Answer:* 4770 psi.

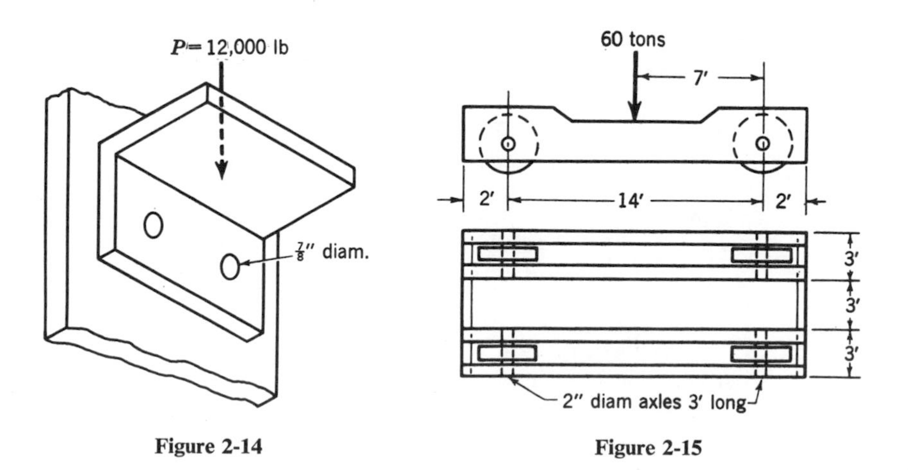

Figure 2-14

Figure 2-15

2-13. A 16-in. pulley is turned by means of a leather belt whose opposite tensions are 2000 and 1600 lb, respectively. If the pulley is keyed to a 2-in. shaft with a single rectangular key $\frac{1}{4}$ by $\frac{1}{2}$ by 4 in., what is the shearing unit stress developed on the cross section of the key (Fig. 2-16)? *Answer:* 3200 psi.

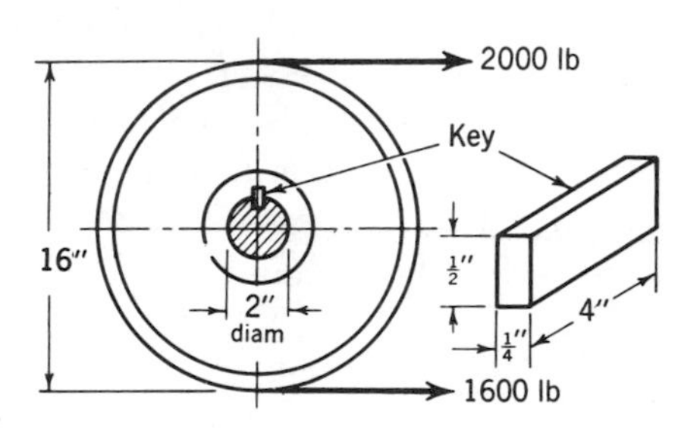

Figure 2-16

2-14. In Fig. 2-17 is shown the lower joint of a single timber truss. If the reaction is 10,000 lb and the members are 8 by 10 in., determine the shearing unit stress on the horizontal plane *abc*.

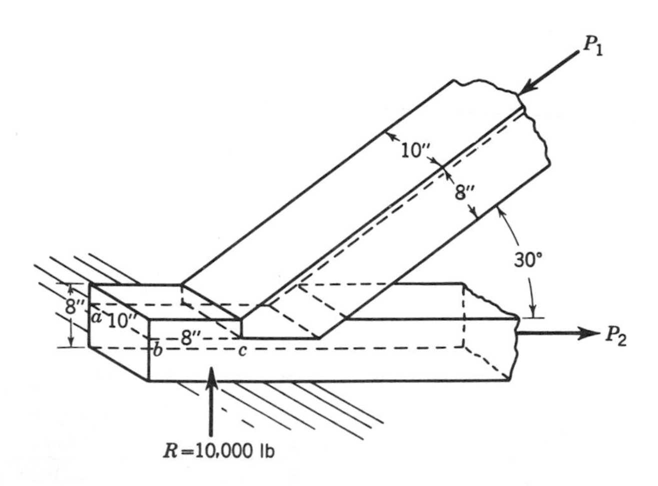

Figure 2-17

2-15. In the frame of Fig. 2-18, a 1-in.-diameter pin is used to provide the joint at *A*. What load *P* may be applied if shear in the pin is critical? Maximum allowable shear stress = 15,000 psi.

2-16. A plywood fabricator uses the arrangement shown in Fig. 2-19 to find the shear strength of a glued joint. If *P* at failure is 3000 lb, what is the shearing unit stress at this load? *Answer:* 1500 psi.

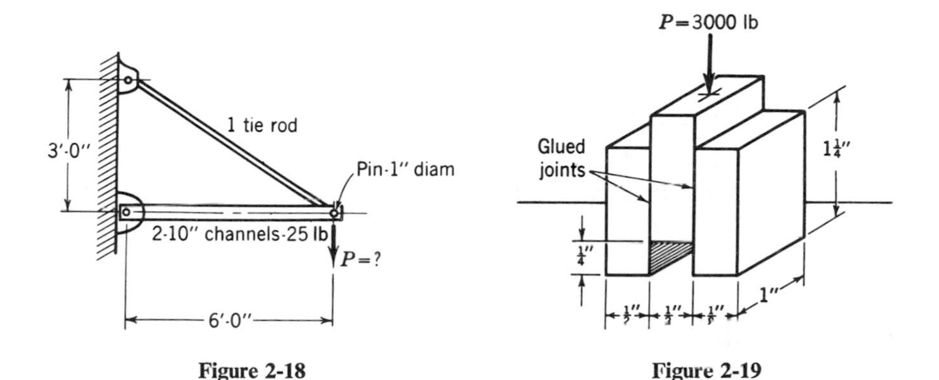

Figure 2-18 Figure 2-19

2-17. A specification on plywood construction offers the arrangement shown in Fig. 2-20 to test the shear strength of a glued joint. What is the shearing unit stress obtained at a failure load of 2000 lb? Compare the advantages or disadvantages of this joint with that of Problem 2-16. Do these two joints present conditions of pure shear?

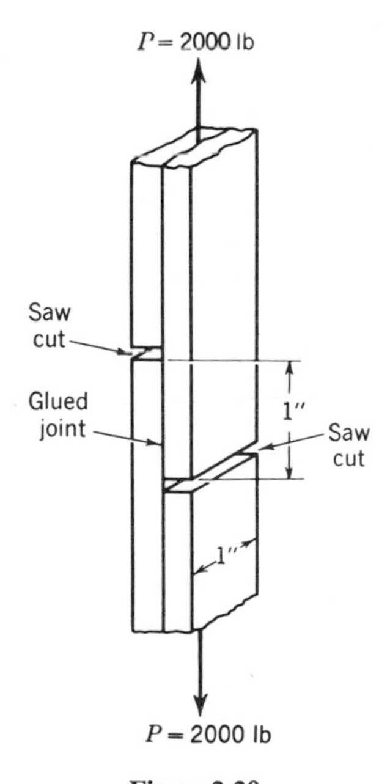

Figure 2-20

2-18. Two tubes and a solid circular bar are telescoped into the position shown in Fig. 2-21. What lengths of overlap are required at A and B to transfer the twisting action of the couple T equal to 2000 in.-lb? Maximum allowable shear stress at interfaces equals 400 psi. *Answers:* 3.18 in., 1.42 in.

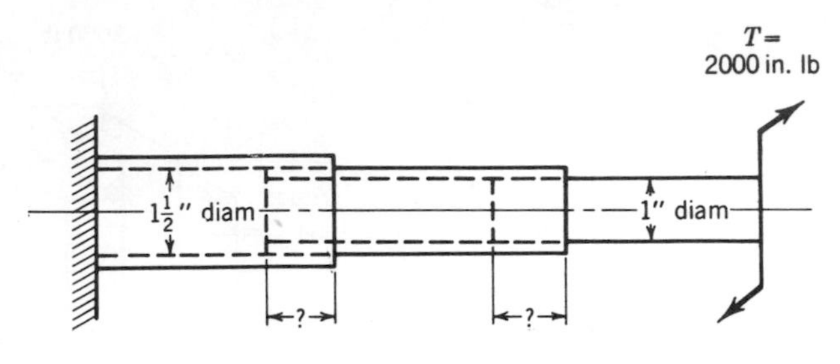

Figure 2-21

2-19. A projectile weighing 0.05 lb is fired at the center of a solid cylindrical mass of energy-absorbing material at a velocity 6000 ft/sec. If it emerges at a speed of 2780 ft/sec., what average shear stress did it produce on the internal cylindrical area—concentric with the axis of the projectile and 3 in. in diameter (Fig. 2-22)?

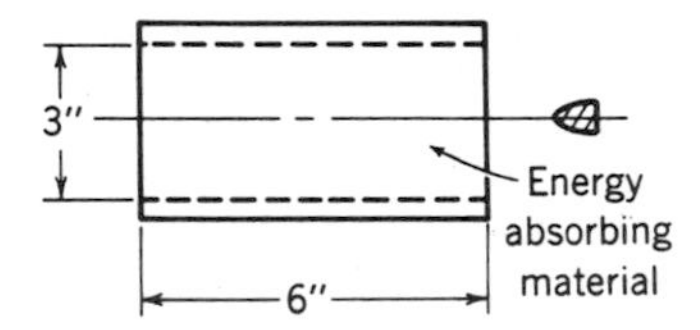

Figure 2-22

2-20. Simultaneous tests are to be performed on eight single shear test specimens similar to those shown in Fig. 2-20 (specimen length = 8 in.) at a temperature of 250°F in an oven whose interior dimensions are 2 ft wide by 3 ft high and 5 ft long. The unit stress on these specimens will be different, ranging from 100 to 800 psi in increments of 100 psi. Slotted weights are available, 6 in. in diameter by 1 in. thick, which weigh 10 lb each. Scale pans plus their holding rod weigh 10 lb each. Devise an apparatus that can be placed in the oven which will perform the necessary tests. Design all the members placed in shear and tension for their maximum stress condition. No attachments are to be made to the oven.

STRESSES ON DIAGONAL PLANES OF AXIALLY LOADED MEMBERS

2-5 Normal and Shearing Unit Stresses

An axial load applied to a compression block will be held in equilibrium only if the material on *all* planes passing through the block is capable of resisting the load. Suppose section *A-A* is passed diagonally through the compression block of Fig. 2-23a, reducing the resistance on this plane to zero. Obviously, the top portion would slide. The question then arises: What resisting stresses must have been present on plane *A-A* to have overcome the tendency to slide before the block was cut in two?

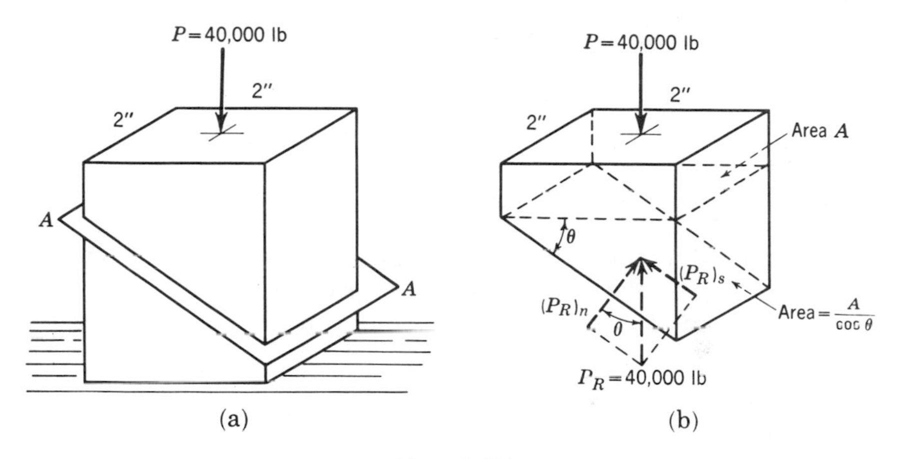

Figure 2-23

To answer this question, let us use the top portion of Fig. 2-23a as our free body. To balance the effect of the applied 40,000-lb load requires a 40,000-lb resistive force acting on section *A-A* placed opposite to, and collinear with, the applied load. This requirement also locates the force at the center of gravity of the exposed section, making it a central load producing uniform stresses.

Because load P_R is diagonal to the plane, it is expedient to resolve it into its components, $(P_R)_s$ and $(P_R)_n$, both of which are also central loads. Component $(P_R)_n$, which acts normal to the plane, will produce a uniformly distributed compressive unit stress, or, generally speaking, a normal unit stress. On the other hand, component $(P_R)_s$, which acts *in the plane* to prevent any sliding action, produces a uniformly distributed *shearing unit stress.* The component forces $(P_R)_s$ and $(P_R)_n$ may be expressed mathematically as

$$(P_R)_s = P_R \sin\theta = P\sin\theta$$

$$(P_R)_n = P_R \cos\theta = P\cos\theta$$

If the minimum cross-sectional area of the block is A, the area on which these component forces act is

$$A' = \frac{A}{\cos\theta}$$

Dividing the component forces by this area, we obtain the following unit stresses.

Shearing unit stress:

$$\tau = \frac{(P_R)_s}{A'} = \frac{P\sin\theta}{A/\cos\theta} = \frac{P}{2A}\sin 2\theta \tag{2.3}$$

Normal unit stress:

$$\sigma = \frac{(P_R)_n}{A'} = \frac{P\cos\theta}{A/\cos\theta} = \frac{P}{A}\cos^2\theta \tag{2.4}$$

The normal and shearing stresses on the plane for which $\theta = 30°$ would therefore be

$$\tau = \frac{40{,}000}{2\text{ in.} \times 2\text{ in.}} \times \frac{\sin 60°}{2} = 4330\text{ psi}$$

$$\sigma = \frac{40{,}000}{2\text{ in.} \times 2\text{ in.}} \times \cos^2 30° = 10{,}000 \times (0.866)^2 = 7500\text{ psi}$$

Interestingly enough, this normal stress computed is less than the normal stress previously obtained for the section perpendicular to the longitudinal axis. The variation of the shearing and normal unit stresses with the angle θ will provide a clearer insight into this and other problems concerning axially loaded members. To assist in this analysis, since $\cos^2 \theta$ is equal to $\frac{1}{2} + \frac{1}{2} \cos 2\theta$, we can write Eq. (2.4) as

$$\sigma = \frac{P}{2A} + \frac{P}{2A} \cos 2\theta \tag{2.5}$$

2-6 Analysis of Equations for Normal and Shear Stresses on Diagonal Planes of Axially Loaded Members

Normal stress. Analyzing the variation in normal stresses from Eq. (2.5), we note that when $\theta = 0°$, the normal stress equals

$$\sigma = \frac{P}{2A} + \frac{P \times 1}{2A} \quad \text{or} \quad \sigma = \frac{P}{A}$$

This analysis not only corroborates the statements made in the beginning of this chapter, but emphasizes the fact that the normal stress in an axially loaded member will be equal to P/A only if the plane under investigation is perpendicular to the longitudinal axis of the member. It also reveals that the *maximum normal stress* is obtained only on a plane perpendicular to the axis, since the value of $\cos \theta$ will then have its highest value.

As θ increases to 45°, the value of $\cos 2\theta$ becomes equal to zero, giving the normal stress a value of only half the maximum.

On a plane parallel to the direction of the axial load ($\theta = 90°$), there will theoretically be no normal stress, since the latter term of the equation becomes $-P/2A$ and cancels the first term.

Shear stress. The equation $\tau = (P/2A) \sin 2\theta$ indicates that the value of the shearing unit stress for any one loading condition varies in accordance with magnitude of θ. Since $\sin 2\theta$ attains its greatest value of 1 when $2\theta = 90°$ or $\theta = 45°$, the maximum value of the shearing unit stress will be

$$\tau_{\max} = \frac{P}{2A} \sin(2 \times 45°) = \frac{P}{2A} \tag{2.6}$$

Note that the maximum shearing unit stress equals only half the maximum normal unit stress:

$$\tau_{\max} = \tfrac{1}{2}\sigma_{\max}$$

If the value of θ were to be 0 or 90°, the shearing unit stress calculated would be zero. Recalling the fact that the normal stress is maximum when $\theta = 0°$, it is interesting to note in contrast that the shearing unit stress on this same plane equals zero. This is a recurring truth which will be established as a fundamental fact in a later chapter. Going to the other extreme, when θ is 90°, we note that both the shearing and normal unit stresses theoretically equal zero.

The graph shown in Fig. 2-24 reveals the variation of the shear and normal stresses with the change in angle θ.

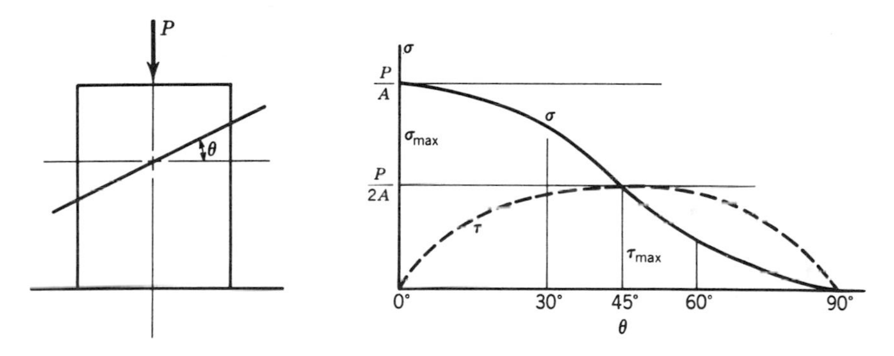

Figure 2-24

2-7 Mohr's Circle Diagram

To assist in remembering the foregoing analysis of normal and shearing unit stresses in axially loaded members, the use of Mohr's circle is invaluable.

Using a pair of rectangular axes, draw a circle on the x axis, tangent to the y axis, with a radius equal to $P/2A$ (Fig 2-25). Turn off angle 2θ clockwise from that portion of the horizontal diameter to the left of center O, the turning radius intersecting the circumference of the circle at D.

Note. Angle θ is always to be taken as the angle between the plane perpendicular to the longitudinal axis of the member and the plane under consideration.

Observation reveals that the abscissa and ordinate of point D as it travels about the circle are equal, respectively, to the normal and shearing unit stresses for the particular value of 2θ used.

Illustrative Problem 1. A compression block 2 by 4 by 8 in. long is axially loaded on one of its ends by a force of 16,000 lb. (a) What are the normal and shearing unit stresses acting on a plane making 60° to the vertical (Fig.

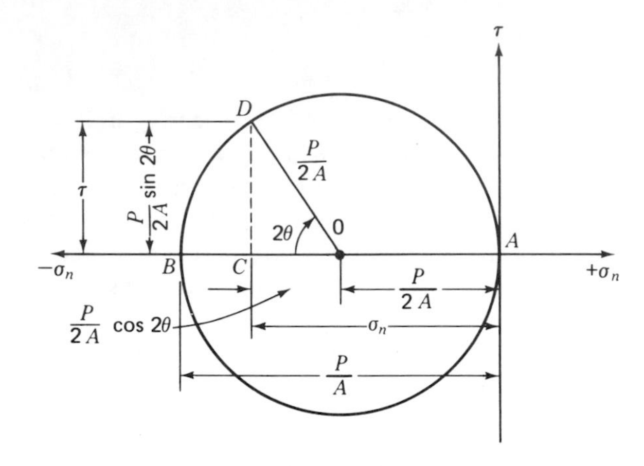

Figure 2-25

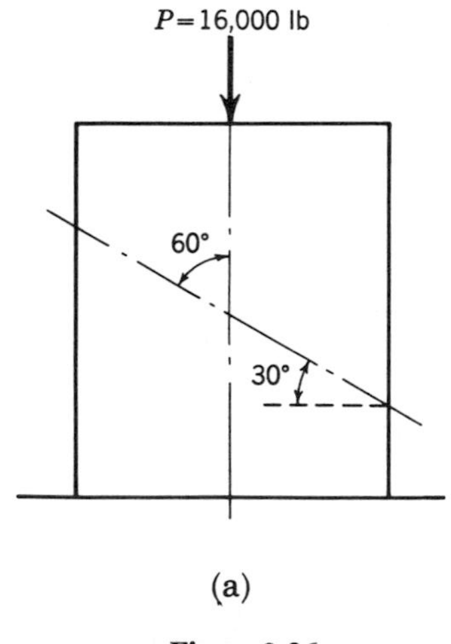

(a)

Figure 2-26a

2-26a)? (b) Compute the value of the maximum shearing unit stress and indicate on what plane it acts.

SOLUTION. (a) From previous comments, θ, the angle between the plane perpendicular to the axis and the plane upon which the desired stresses act, must be equal to 30°.

The values necessary for the construction of Mohr's circle are

$$\sigma_{\max} = \frac{P}{A} = \frac{16{,}000}{8} = 2000 \text{ psi}$$

$$\frac{1}{2}\frac{P}{A} = 1000 \text{ psi}$$

The Mohr's circle is shown plotted in Fig. 2-26b, with a radius of 1000 psi. Laying off an angle of $2\theta = 60°$, there are obtained an abscissa and ordinate to point D, equal respectively to the normal and shearing unit stress desired. Thus

$$AC = \sigma = 1000 + 1000 \times \cos 60° = 1000 + 500 = 1500 \text{ psi}$$

$$DC = \tau = 1000 \times \sin 60° = 866 \text{ psi}$$

(b) The maximum shearing stress always occurs on a plane making 45° with the longitudinal axis. If, therefore, an angle of 2θ of 90° is laid off on Mohr's circle, we obtain line OE, which, by inspection, is the longest ordinate possible of construction within the circle. The ordinate to point E is, of course, 1000 psi.

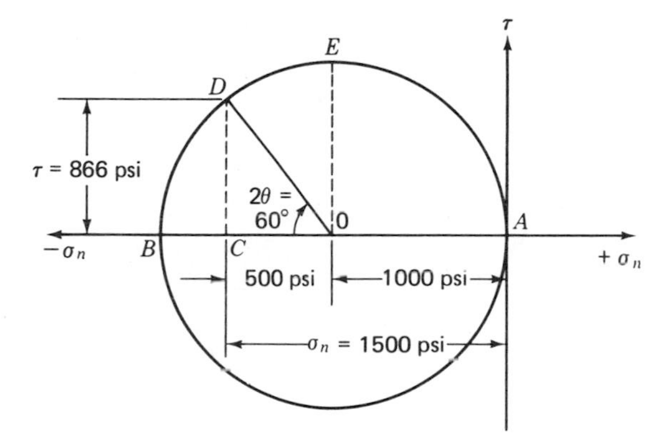

Figure 2-26b

2-8 Importance of Determining Diagonal Stresses

The many materials available for the design of load-resisting members have different limiting values of unit stress for tension, compression, and shear. Some materials are particularly strong in compression, but not so strong in shear. Others, again, are just the reverse: strong in shear and weak in compression.

Should the design of an axially loaded member involve a material whose limiting or rupture strength in shear is less than half the corresponding strength in tension or compression, it would be necessary to base this design on the maximum shear stress formula given in Eq. (2.6). The validity of this statement is understood when it is recalled that upon the gradual application of an axial load, the maximum tensile or compression unit stress in a prismatic member develops at a rate twice as fast as the maximum shearing unit stress. Certainly, then, the shearing unit stress would attain its limiting value first, since it would be less than one half the normal stress value.

Illustrative Problem 2. How would an axially loaded 2- by 2- by 8-in. compression block fail if it had the following rupture strengths: compression, 10,000 psi; shear, 4000 psi?

SOLUTION. The increase of the maximum normal and shearing unit stresses with increase in applied load is given in the following tabulation:

Applied Load (lb)	Max Normal Unit Stress P/A (psi)	Max Shearing Unit Stress $\frac{1}{2}(P/A)$ (psi)
0	0	0
10,000	2500	1250
20,000	5000	2500
30,000	7500	3750
31,992	7998	3999

None of the loads so far applied has created a normal or shearing unit stress above its respective rupture level. However, the 31,992-lb load has produced a shear stress that is dangerously near this rupture point. In fact, there are only 8 lb of load needed to raise the shear stress to the 4000-psi point where rupture theoretically will take place. On the other hand, the induced maximum and normal unit stress is well below its rupture strength.

The compression block used in this problem would, therefore, fail in shear by an applied load of 32,000 lb. Its design would of necessity be based upon shear.

The conditions requiring a design based upon shear stresses are not so frequent as may have been intimated. The majority of engineering materials used today have shear strengths above one half the tensile strength. Designs for these materials are based on their maximum normal stress.

However, when materials with low shear strengths, such as cast iron, wood, concrete, powder iron, tile, and so forth, are to be designed, it is well to investigate the rupture strengths carefully to determine what stress will determine the basis of design.

PROBLEMS

2-21. A porcelain test specimen 1 in. in diameter and 1.25 in. high fails on a plane making 45° to its longitudinal axis when subjected to a load of 80,000 lb. What was the magnitude of the shearing and normal unit stresses acting on the plane at the time of failure? *Answer:* $\tau = \sigma = 51{,}000$ psi.

2-22. During one phase of the tensile test of a piece of structural steel, there appears on its surface diagonal lines (Leuder lines) roughly 45° to its longitudinal axis. What is the magnitude of the shearing and normal stresses in the plane of these lines if the load at their occurrence is 24,000 lb and the cross section of the bar is 1.5 in. wide by 0.375 in. thick?

2-23. In Problem 2-22 what are the normal and shearing unit stresses acting on a plane making 60° with the axis? *Answers:* $\tau = 18{,}450$ psi, $\sigma = 31{,}950$ psi.

2-24. A short cylindrical compression member, $d = 6$ in., is to be made of a new material whose compressive and shearing rupture strengths are 14,000 and 6000 psi, respectively. What axial load can be placed on this member before failure will take place? *Answer:* $P = 339{,}000$ lb.

2-25. A short square column 6 in. on a side carries an axial load of 14,400 lb. Find the normal and shearing unit stresses on planes making angles of 30, 45, and 60° to the vertical (*side view*). Use Mohr's circle, indicating clearly thereon those lines representative of the stresses required.

2-26. A steel bar whose diameter equals $\frac{3}{4}$ in. is loaded with 10,000 lb in tension. Compute the normal and shearing unit stresses on planes making angles of 30, 45, and 75° with the cross section of the bar. Use Mohr's circle, indicating clearly thereon those lines representative of the stresses required. *Answers:* (30°) $\tau = 9830$ psi, $\sigma = 17{,}030$ psi; (45°) $\tau = \sigma = 11{,}350$ psi; (75°) $\tau = 5680$ psi, $\sigma =$ 1520 psi.

2-27. Two $\frac{1}{2}$-in.-diameter bolts are used in the connection shown in Fig. 2-27. If P is 10,000 lb, what are the axial and shearing unit stresses in the bolts?

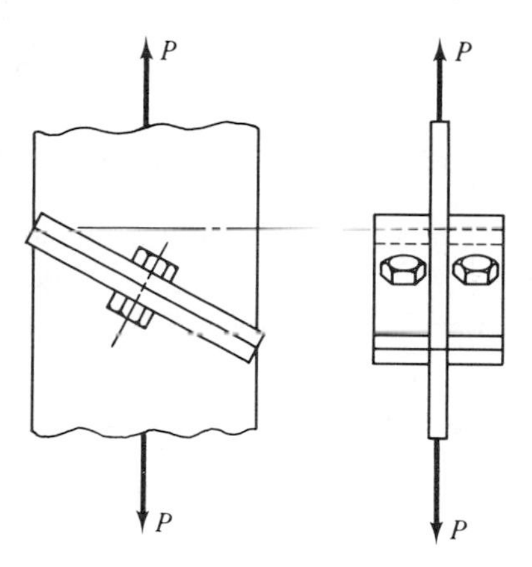

Figure 2-27

2-28. A rectangular block of wood, 4 by 4 by 10 in. high, has its grain running at an angle of 60° to the vertical axis. If the maximum allowable shearing unit stress parallel to the grain is 200 psi, what axial load may the block safely carry? *Answer:* 7380 lb.

2-29. Two wooden wedges are glued together to form a rectangular block 2 by 4 by 8 in. high (Fig. 2-28). What normal and shear stresses are developed on the glued plane by the application of an axial load of 10,000 lb?

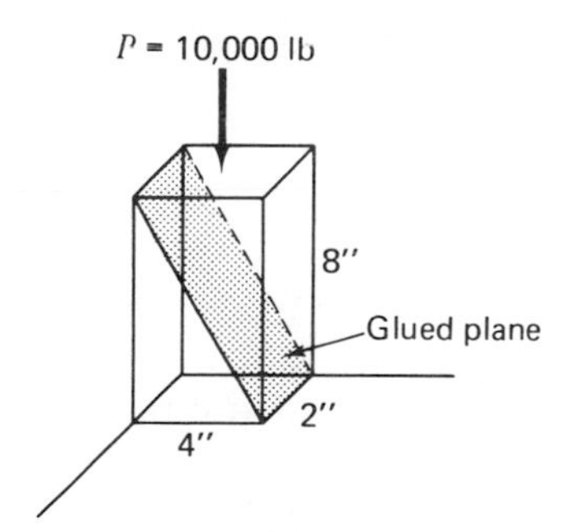

Figure 2-28

NORMAL STRAINS—SHEAR STRAINS IN AXIALLY LOADED MEMBERS

2-9 Unit Strain—Total Strain—True Strain

Deformation, more commonly referred to as *strain*, is the inevitable result of stress. All materials are deformable by load. Some materials are elastic within a certain range of stress (i.e., up to their *elastic limit*—see

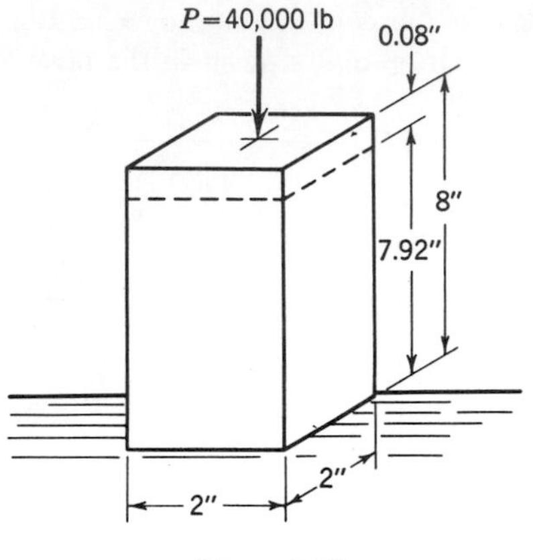

Figure 2-29

Chapter 3) and are able to recover their original size and shape; others show no elasticity whatsoever. But of one thing we are absolutely sure: a stress, no matter how small, incurs a strain.

Referring to Fig. 2-2a, suppose the load of 40,000 lb applied to the end of the 2- by 2- by 8-in. compression block shortened its length to 7.92 in. The *total strain* in the direction of the load would then be equal to 0.08 in. (Fig. 2-29). Assuming the deformation to have been spread evenly over its entire length, the *unit strain* would be equal to 0.08/8 in., or 0.01. In all problems of elongation or contraction the longitudinal or normal unit strain is defined as

$$\epsilon = \frac{e}{l} \tag{2.7}$$

where ϵ is the strain per unit of length, e is the total strain, and l is the original length of the member. This value would have been called a *unit elongation* had the load applied been tensile, creating a lengthening of the member.

Where deformation in an axially loaded member is not uniformly distributed, the unit strain is expressed using an infinitesimal length and deformation:

$$\epsilon = \frac{de}{dl}$$

The sidewise expansion or contraction taking place during the application of axial load will generally be about 25 to 35 per cent of the longitudinal deformation within the design range of any engineering material. This *ratio of lateral unit strain to longitudinal unit strain* is known as *Poisson's ratio* and is of great importance in the design of members subjected to combinations of various stresses. (See Chapter 3 for further discussion.)

Discussion of units. Note that the value of unit strain is obtained by dividing a length by a length, and if both are in feet or inches, the result is an abstract number or a ratio. Thus, if a bar is 100 in. long and has a unit strain of 0.01 imposed upon it, its total deformation would be equal to 1 in. If the bar had a length of 100 ft, with the same amount of unit strain imposed, the total strain would be equal to 1 ft.

Illustrative Problem 3. A bridge tie rod has dimensions as given in Fig. 2-30. Upon tightening, it is found that an elongation of 0.400 in. was created.

If the unit strain on the interior portion equaled 0.0026, what was the unit strain on the two end portions?

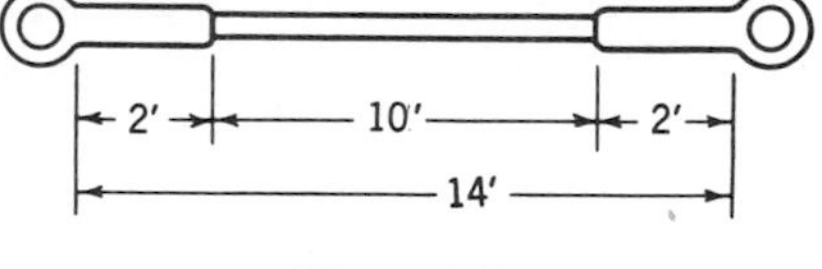

Figure 2-30

SOLUTION. The interior portion elongated a total of

$$e_i = \epsilon l = 0.0026 \times 120 = 0.312 \text{ in.}$$

Subtracting this value from the total elongation gives the total elongation of the end portions:

$$e_e = 0.400 \text{ in.} - 0.312 \text{ in.} = 0.088 \text{ in.}$$

Since the total length of the end sections is 4 ft, or 48 in., the unit strain in those sections is

$$\epsilon_e = \frac{e_e}{l} = \frac{0.088}{48} = 0.00183$$

PROBLEMS

2-30. A bar, 5 ft long and 2 in. in diameter, is elongated 0.24 in. (a) What is the unit strain in the bar? (b) If its lateral deformation equals 0.0024 in., what is Poisson's ratio? *Answers:* (a) 0.004 (b) 0.3.

2-31. The unit strain in the tensile member BC (Fig. 2-31) is to be set at 0.0005. Considering the bar AB to be so stiff as to produce an insignificant amount of bending, how far will point A have to be moved to give the desired strain? Give answer in inches.

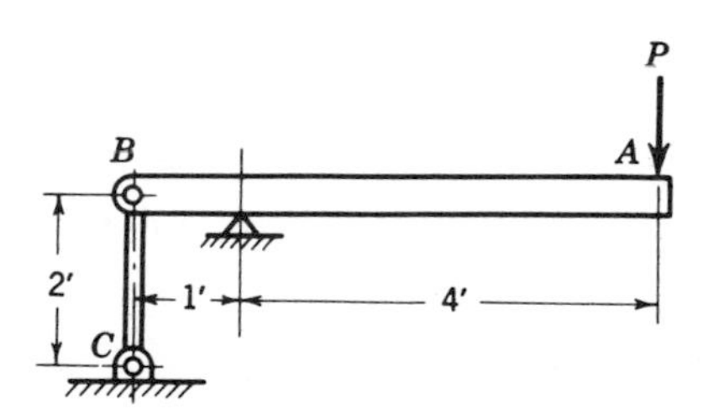

Figure 2-31

2-32. The unit strain in a 1-in.-diameter 100-ft rod, suspended from one end, varies in direct proportion to the distance from its free end. If the total elongation is 0.600 in., what is the maximum unit strain in the rod? *Answer:* 0.00100.

2-33. A suspended steel rod of constant cross section, 1000 ft long, has a unit strain of 0.0008 at 800 ft from its free end. What is the total elongation?

2-34. A suspended rod has a nonlinear deformation recorded as $\epsilon = 0.0005y^{1/2}$, where y is the distance from the free end. (a) What is the total deformation if the rod is 10,000 ft long? (b) What is the unit strain at the fixed end? *Answers:* (a) 333 ft. (b) 0.05.

2-35. The unit strain of a wire, 200 in. long, varies linearly with the distance from one end of the wire. What is its total elongation if at the center of the wire the unit strain is equal to 0.01 in./in.?

2-36. The unit strain of a wire varies with the square of the distance from one end. If the unstressed length of the wire was 100 in., compute the unit strain at the center and the total increase in the length of the wire if its maximum unit strain is 0.05 in./in. *Answers:* 0.0125, 1.67 in.

True strain. When a strain becomes relatively large, a somewhat different computation is often employed for its evaluation. This so-called *true strain* is considered to be the sum of the strains obtained for the successive increments of stress in the total where each increment of strain is obtained by using the actual or extended length prior to the application of a stress increment. As a somewhat oversimplified illustration, let us assume that under each of three successive stress increments of 100 psi an original length of 1 in. elongated $\frac{1}{4}$ in. The true strain for 300 psi would be the sum of the strains for each of the three psi increments, namely, 0.25/1.00, 0.25/1.25, and 0.25/1.50 or the total of 0.250 + 0.200 + 0.167 = 0.617.

If the load and corresponding deformation increments were respectively dP and dL, the true strain would be defined as

$$\epsilon = \int_{L_0}^{L_f} \frac{dL}{L} = \log_e \frac{L_f}{L_0} = \log_e \frac{1.75}{1} = 2.306 \times 0.233 = 0.538$$

In contrast, the conventional strain would be $3 \times 0.25/1 = 0.75$.

2-10 Shear Strains

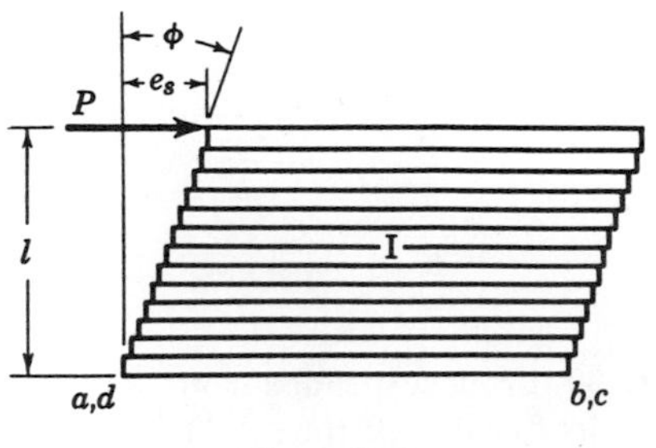

Figure 2-32

When block I was acted upon by force P in Fig. 2-11, it was deformed. Magnified, its distortion appears as shown in Fig. 2-32, and can be explained if block I is considered as made up of several thin layers, each layer sliding a small amount over the layer below.

It has been indicated that a unit strain is equal to the magnitude of the total strain divided by the length over which the deformation takes place. This is true regardless of whether the stress be tensile, compressive, or shear. Therefore, to evaluate the average unit strain in shear we must divide e_s by the depth of block I. Thus,

$$\gamma = \frac{e_s}{l}$$

Since the deformation and length are at right angles to one another, the above ratio is also equal to the tan ϕ, where ϕ is the angular change in the edge of block I. In fact, *the shear strain at any point on the surface of a loaded member is equal to the change of a right angle at that point.*

Inasmuch as ϕ is generally very small, tan ϕ may be expressed as the angle ϕ expressed in radians. Hence,

$$\gamma = \frac{e_s}{l} = \tan\phi = \phi \tag{2.8}$$

When the shear strain is not uniformly distributed over the length l,

$$\gamma = \frac{de_s}{dl}$$

where de_s is the shear deformation over an infinitesimal length dl.

Illustrative Problem 4. Suppose in Fig. 2-11 force P displaces the top of block I a horizontal distance of 0.0026 in. with respect to section *abcd*. What is the average shear strain, assuming the height of block I to be 1.3 in.?

SOLUTION
$$\epsilon_s = \frac{0.0026}{1.3} = 0.0020$$

STRESS-STRAIN RELATIONSHIPS AXIAL AND SHEAR LOADS

2-11 Tensile or Compressive Modulus of Elasticity

In 1678, Sir Robert Hooke observed that when rolled materials were subjected to equal increments of stress, they suffered equal increments of strain. The ratio formed by dividing a unit stress by its corresponding value of unit strain is called the *modulus of elasticity*, and is here represented by the capital letter E. It is written as

$$E = \frac{\sigma}{\epsilon} \tag{2.9}$$

where E = psi (lb/in.^{2}), σ = unit stress in psi, and ϵ = unit strain in in./in. This ratio of stress to strain remains constant for all steels and many other structural materials within their useful range. The maximum stress at which the proportionality of stress to strain remains a constant is known as the *proportional limit.*

The above expression may also be written in a very useful expanded form whenever the stress and deformation are caused by axial loads.

$$E = \frac{P/A}{e/l} = \frac{Pl}{Ae} \quad \text{or} \quad e = \frac{Pl}{AE} \tag{2.10}$$

One characteristic of steel that allows it to predominate as a structural material is its high modulus of elasticity. In contrast to aluminum, its modulus is about three times as great. This means that steel will allow about three times as much load as aluminum on a given area for a given deformation. However, the lightness of aluminum, in addition to its other advantages, has created many new markets—notably in the aircraft industry—in spite of its higher cost and lower modulus.

2-12 Shear Modulus of Elasticity

The shear modulus of elasticity is similar to the tension and compression modulus of elasticity insofar as it is a ratio of unit stress to unit strain. Expressed mathematically,

$$G = \frac{\tau}{\gamma} \tag{2.11}$$

where τ is the shearing unit stress in psi and γ is the shearing unit strain. It is different in that the total strain e_s is measured perpendicular to the height of the shear zone, making the value of γ equal to e_s/l or tan ϕ. Values of shear moduli of elasticity will be found in Table 2 (p. 75).

PROBLEMS

2-37. Determine the total elongation of a bronze bar 60 in. long when subjected to a unit stress of 10,000 psi. What would the elongation be if the bar were made of steel? Duralumin? Copper?

2-38. Two bars of exactly the same size, one of steel and the other of copper, are elongated the same amount. How do the stresses in the bars compare?

2-39. A steel ring, having an inside diameter of 19.995 in. and a 1- by $\frac{1}{4}$-in. cross section at room temperature, is to be heated and, in its expanded state, placed over a cast-iron cylinder 20 in. in diameter. What will be the unit stress developed in the steel ring upon cooling, assuming the cast-iron cylinder will not deform? *Answer:* $\sigma = 7500$ psi.

2-40. A brass ring 1 by $\frac{1}{4}$ in. in cross section is to be heated and fitted over a nondeformable cylinder 24 in. in diameter. What should the initial inside diameter of this ring be if the unit stress in the ring when cool is to be 10,000 psi?

2-41. Find the expression for the total elongation of a bar of constant cross-sectional area A and length L due to its own weight W when suspended from one end. *Answer:* $WL/2AE$.

2-42. Determine the expression for the total axial elongation of a vertically hung right circular cone of weight W, suspended from its base. The cone has a base area A, length L, density δ, and modulus of elasticity E. *Hint:* Consider the origin of the X and Y axes at the vertex of the cone. Assume uniformly distributed tensile stress on each cross section, and each disk element to be cylindrical. *Answer:* $WL/2AE$.

2-43. What is the elongation of a steel rod $\frac{3}{4}$ in. in diameter, 100 ft long, carrying a suspended load of 4000 lb? Include the elongation of the rod due to its own weight. *Hint:* The elongation of the rod due to its own weight and the 4000-lb load are found separately and added together. Steel weighs 490 lb/ft^3. *Answer:* 0.370 in.

2-44. Three steel tie rods, 3 in. wide and $\frac{1}{2}$ in. thick, are to be placed side by side and used as a composite tensile member in a Pratt truss. The rods were each supposed to be 15 ft long, but through an error in fabrication the middle bar was made 0.02 in. short. What will the stresses be in the rods if rigid pins are to be placed in the holes located at the ends of each rod? *Answers:* Outside rods, 1110 psi (C); inside rods, 2220 psi (T).

2-45. What will the stresses be in the tie rods of Problem 2-44 if the center bar is 0.03 in. too long?

2-46. Compute the total elongation of the bar shown in Fig. 2-33 if $E = 10 \times 10^6$ psi. *Answer:* 0.1800 in.

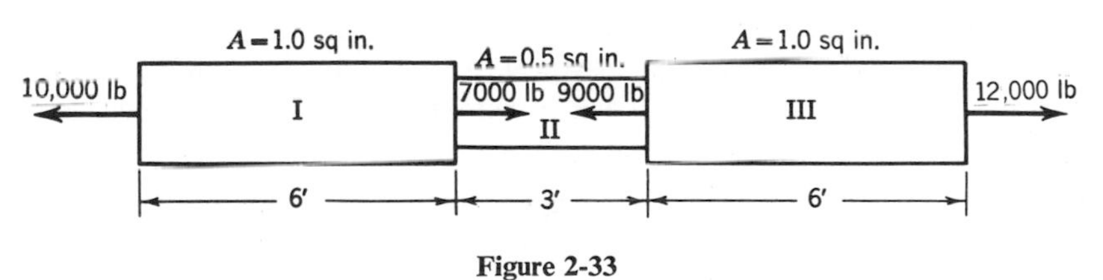

Figure 2-33

2-47. A horizontal beam 6 ft long is supended by two vertical rods, one placed at each end of the beam. One of the rods is steel, $\frac{1}{2}$ in. in diameter and 3 ft long; the other is bronze, $\frac{5}{8}$ in. in diameter and 8 ft long. Where must a load of 8000 lb be placed for the beam to retain its horizontal position after application of the load? Disregard the weight of the beam.

2-48. In analyzing a possible oil field, a geologist estimates the location of oil at 10,000 ft below the level of the earth. If a $1\frac{1}{2}$-in. steel rod is to be used by the well driller, how much will the rod elongate due to its own weight if the entire length of 10,000 ft is required? *Hint:* The unit stress in the rod varies directly with the distance from the free end. Steel weighs 490 lb/ft^3. *Answer:* 68.0 in.

2-49. Determine the magnitude of the total load P which will produce an axial compressive stress of 2000 psi in member C, if the clearance between the rigid member and C is 0.002 in. when P is zero (Fig. 2-34).

Member A: $1\frac{1}{2}$- by 2-in. brass rod, $E = 15 \times 10^6$ psi.
Member B: 2- by 1-in. Duralumin rod, $E = 10.5 \times 10^6$ psi.
Member C: 3- by 4-in. concrete block, $E = 3.6 \times 10^6$ psi. *Answer:* 29,350 lb.

2-50. Beam OA (Fig. 2-35) is supported on a pin joint at O, and by bronze rods AB and CD, each rod having a cross-sectional area of 0.75 in^2. When load W is applied, a strain gage placed on AB shows that the longitudinal strain in that

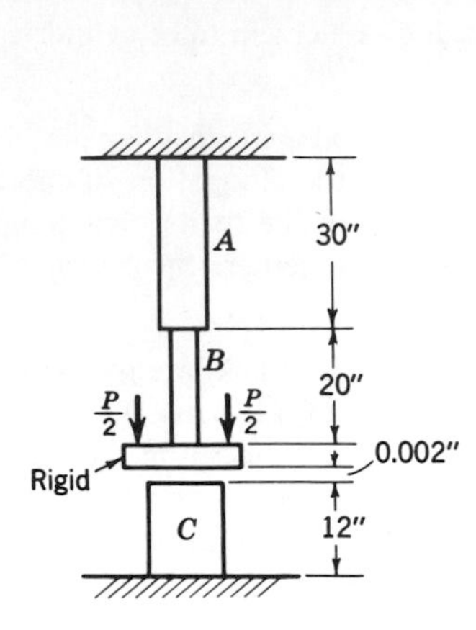

Figure 2-34

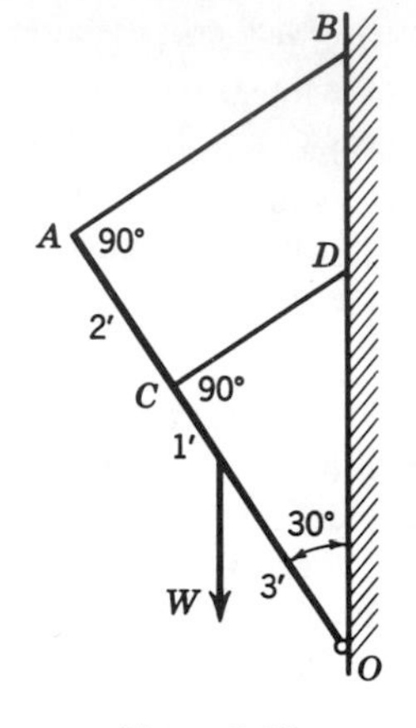

Figure 2-35

member is 0.000200 in./in. Determine the magnitude of load W, assuming that beam OA does not bend. $E = 12 \times 10^6$ psi. *Answer:* 12,000 lb.

2-51. A homogeneous bar (area = 1 in.^{2}) is rigidly fixed to D (Fig. 2-36). If $P_1 = 20{,}000$ lb and $P_2 = 10{,}000$ lb, what will the stress be in BC if the gap between A and the wall equals 0.086 in. before the loads are applied? *Answer:* 15,740 psi.

2-52. Three wires connect two rings as shown in Fig. 2-37. Their characteristics are as follows:

Wire	Area (in.^{2})	E (psi)	Length Before Load Applied (ft)
a	0.06	30×10^6	100.000
b	0.06	15×10^6	99.930
c	0.06	10×10^6	99.900

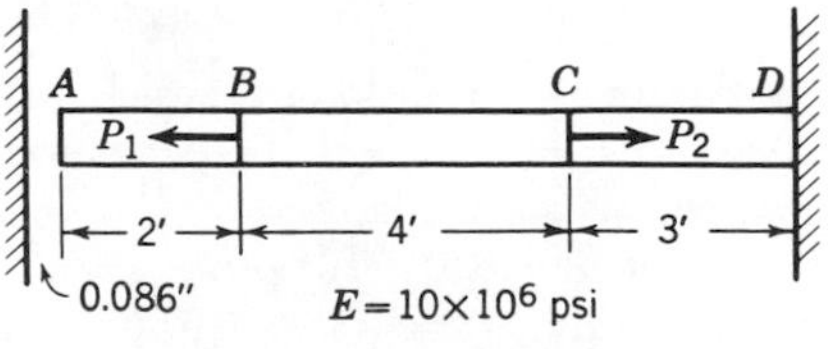

A B C D
P_1 P_2
2′ 4′ 3′
0.086″ $E = 10 \times 10^6$ psi

Figure 2-36

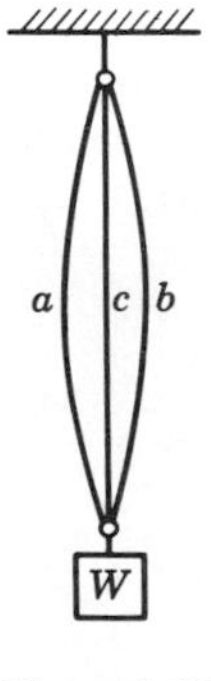

a c b
W

Figure 2-37

Determine the tensile unit stress in, and the final length of, wire a after a load of $W = 1500$ lb is applied at the lower ring.

2-53. In Fig. 2-38, the wood compression block is stressed to a value of 10,000 psi. (a) What is the stress in member *BC*? (b) What is the load *P*? *Answers:* (a) 90,000 psi. (b) 187,000 lb.

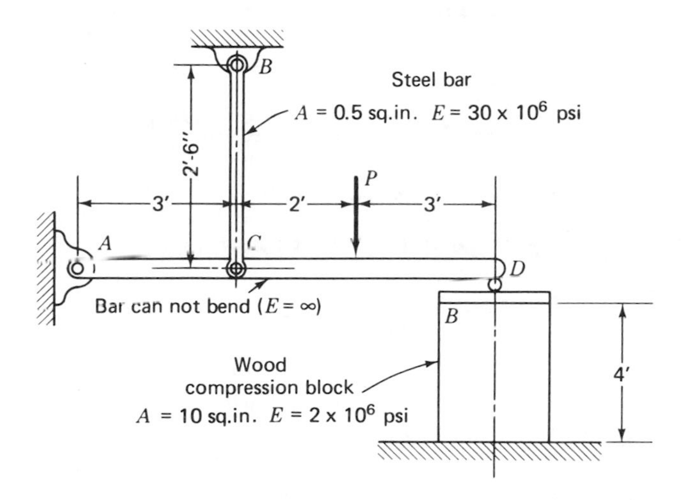

Figure 2-38

2-54. In Fig. 2-39, beam *AB* is horizontal while supported by two cables *CE* and *DF* and a hinge at *A*. If *CE* is made of aluminum and *DF* is made of steel, what load *P* is the beam carrying when the unit tensile strain in *DF* is equal to 0.0003? Assume bar *AB* to be of infinite stiffness. $E_{st.} = 30 \times 10^6$ psi. $E_{al.} = 10 \times 10^6$ psi. *Answer:* 7710 lb.

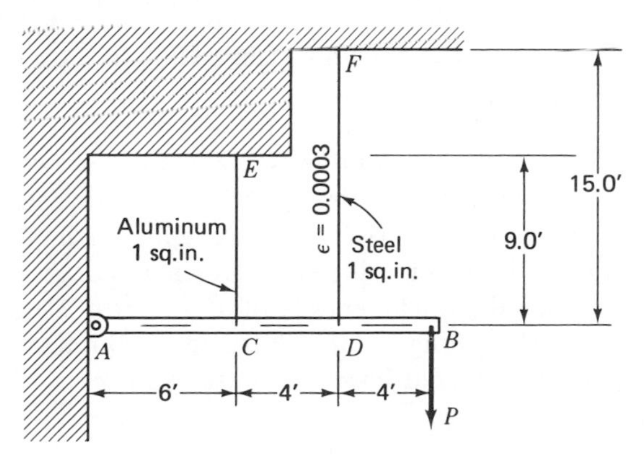

Figure 2-39

2-55. A force $P = 200$ lb is applied to the lever *AD* (Fig. 2-40). Wires *BE* and *CF* are each of area 0.01 in.^{2} and are unstressed before load *P* is applied. Neglecting any bending of the lever or sag effects in the wires, what will be the final load in each wire and how much will point *A* move to the right? $E = 30 \times 10^6$ psi.

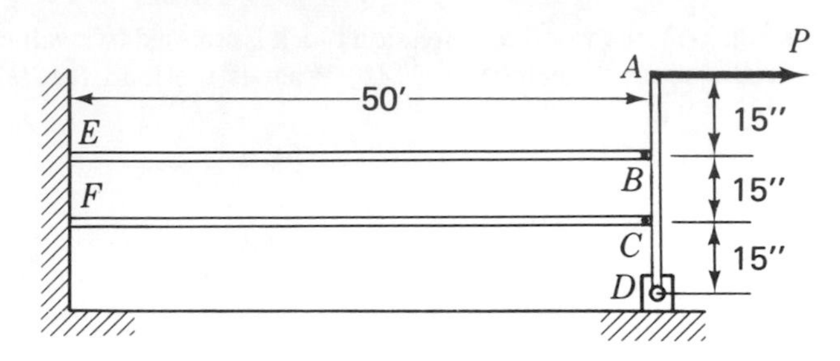

Figure 2-40

TEMPERATURE STRESSES

2-13 Effect of Temperature on Engineering Design

Almost all structural materials increase in volume when subjected to heat and decrease in volume when cooled. Whenever a design prevents the change in length of a member subjected to temperature variation, the stresses which develop may be sufficiently high to exceed the elastic limit and cause serious damage. For example, it is common practice to allow one end of a bridge to move over rollers. On very long bridges, where the expansion might measure several feet, the bridge is built in sections, each having a roller at one end. To prevent the action of free expansion on a hot day would undoubtedly cause severe buckling and probable failure. Cooling effects may be equally dangerous.

2-14 Coefficient of Linear Expansion—Dimensional Changes Due to Temperature

To facilitate the calculation of dimensional changes owing to temperature variations, the *coefficient of linear expansion* has been determined for all engineering materials (see Table 1). This coefficient is defined as the change in length, per unit of length per 1 degree Fahrenheit change in temperature. For example, if we raised the temperature of a 1-ft bar of steel through 1°F, we would find that the bar elongated 0.0000065 ft. Or, if it had been only 1 in. in length, the elongation would have been just one-twelfth of this amount, or 0.0000065 in.

TABLE 1

COEFFICIENTS OF LINEAR EXPANSION (per °F)

Material	Coefficient	Material	Coefficient
Steel	0.0000065	Concrete	0.0000062
Cast iron	0.0000062	Bronze	0.0000100
Brass	0.0000092	Invar	0.0000006
Duralumin	0.0000128	Copper	0.0000093

Had the bar previously used been 100 ft long, each 1-ft length in this bar would have expanded 0.0000065 ft, giving a total expansion of 0.00065 ft.

Since the coefficient of linear expansion remains constant over a considerable range of the solid state of the material, a rod subjected to a temperature change of 100°F will expand 100 times that obtained for a 1°F temperature change. Thus, the 100-ft bar previously mentioned would, if it were increased in temperature 100°F, expand 0.065 ft.

To summarize the calculations necessary to obtain a change in length of a bar, we may write the following expression:

$$e = \alpha l \, \Delta t \tag{2.12}$$

where e = the total change in length of bar l (values of e and l must have same units), α = the coefficient of linear expansion, Δt = the change in temperature in degrees Fahrenheit.

2-15 Stresses Caused by Temperature Variations

Of perhaps even greater importance in engineering design are the stresses developed by restraining the free expansion and contraction of members subjected to temperature variations. To calculate these temperature stresses, it is well to determine first the free expansion or contraction of the member involved and, second, the force and unit stress developed in forcing the member to attain its original length. The problem from this point on is exactly the same as those solved in the earlier portions of this chapter dealing with axial stresses and strains. The amount of stress developed by restoring a bar to its original length l is

$$\sigma = \epsilon E = \frac{e}{l} E = \frac{\alpha l \, \Delta t \, E}{l} = \alpha \, \Delta t \, E \tag{2.13}$$

Illustrative Problem 5. A steel I beam 10 ft long is placed between two immovable walls when the temperature is 60°F. What compressive unit stress will be developed in the I beam if the temperature rises to 120°F (α = 0.0000065)?

SOLUTION. Had no restraints been placed at the ends of the I beam, the beam would have elongated

$$e = \alpha l \, \Delta t = 0.0000065 \times 120 \text{ in.} \times 60°\text{F} = 0.0468 \text{ in.}$$

The stress developed by compressing the beam to its original length is, therefore,

$$\sigma = \epsilon E = \frac{e}{l} E = \frac{0.0000065 \times 120 \times 60 \times 30 \times 10^6}{120} = 11{,}700 \text{ psi}$$

Although the computation performed here in finding the stress in the beam may be somewhat lengthy, the procedure is safer and eliminates the necessity of memorizing Eq. (2.13).

PROBLEMS*

2-56. A stress of 10,000 psi is to be placed in a horizontal steel cable tied between two immovable points by heating the cable during its installation and then allowing it to cool to room temperature of 68°F. If the cable is 8 ft long, to what temperature must it be heated to attain the desired stress upon cooling?

2-57. A thin steel hoop of 21.96-in. inside diameter is to be shrunk over a cylinder 22 in. in diameter. To what temperature must the hoop be raised so that it will just slip over the cylinder, assuming the room temperature to be 70°F? What average stress will be induced in the ring when it cools to room temperature, assuming that the cylinder does not contract during cooling? *Answers:* 351°F, 54,600 psi.

2-58. A steel measuring tape measures exactly 100 ft at 60°F when subjected to a pull of 20 lb. If the tape is $\frac{1}{32}$ in. thick and $\frac{1}{2}$ in. wide, what correction should be made in the reading of the tape if it is subjected to a temperature of 120°F and a pull of 30 lb?

2-59. A wood stave penstock 10 ft in diameter is held together by $\frac{3}{4}$-in.-diameter steel hoops tightened to a tension of 12,000 psi when the temperature was 80°F. What will be the unit stress in the hoops at 0°F? at 120°F? Assume the penstock to be rigid during these temperature changes. *Answers:* 27,600 psi, 4200 psi.

2-60. Three vertical steel wires A, B, and C are loaded through a horizontal yoke. Wires A and C are 60 in. long and have a cross section of 0.1 in^2. Wire B is 30 in. long and has a cross section of 0.2 in^2. At the normal temperature of 68°F, the bottom ends of the wires were at the same elevation with no application of load. If load P of 6000 lb was then applied and the temperature increased by 100°F, what load is carried by each wire? (Fig. 2-41.)

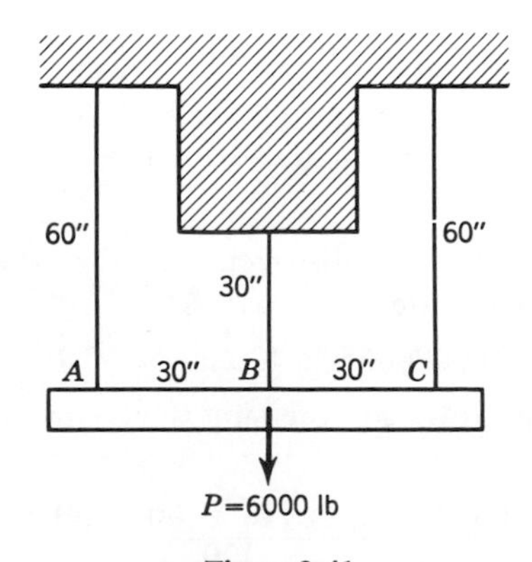

Figure 2-41

*The coefficients of linear expansion for various materials will be found in Table 1. The moduli of elasticity are found in Table 2.

2-61. If in the previous problem the same load is applied, to what temperature must the wires be heated before the load is entirely taken by wire *B*? *Answer:* 221.9°F.

2-62. A weight is supported by two taut steel wires, 3 ft and 4 ft long, which form a triangle with a level ceiling to which they are fastened at points 5 ft apart. If the wires are subjected to a temperature rise of 60°F, how much will the weight be lowered? Ceiling expansion is negligible. *Answer:* 0.0224 in.

2-63. Two rods are joined together and the ends attached to supports as shown in Fig. 2-42. If initially there is no stress in the rods, what is the unit stress developed in the steel bar if the temperature drops 80°F and the supports do not yield?

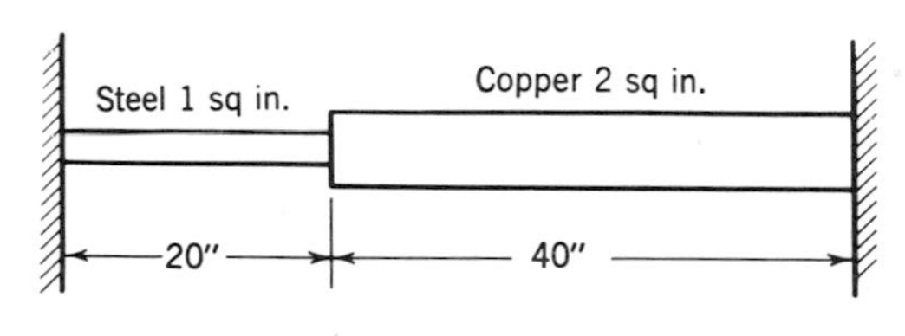

Figure 2-42

2-64. In Fig. 2-43, the left bar is copper, the right bar is steel, and the area of each bar is 4 in.². $\Delta = 0.010$ in. (a) Determine the total force in each bar when the temperature is raised 100°F. (b) Determine the deformation in each bar. The distance between walls remains constant. *Answers:* (a) 46,900 lb; (b) $e_{st.} = 0.00782$ in.; (c) $e_{cu.} = 0.01380$ in.

2-65. A bar of Invar steel and a bar of copper are placed end to end at an initial temperature of 32°F, and the temperature is raised to 232°F (Fig. 2-44). If the end supports are unyielding, determine the total force on the end of each bar. Cross-sectional area of each bar = 4.0 in². Invar $E = 21 \times 10^6$ psi. *Answer:* 93,000 lb.

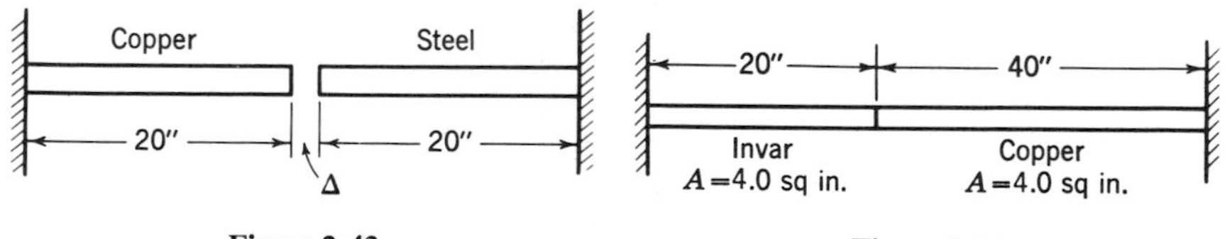

Figure 2-43 **Figure 2-44**

2-66. A Duralumin band (Fig. 2-45) is placed on an Invar steel core 12 in. in diameter, and a nut on the band is then tightened until the unit stress in the band is 10,000 psi. Determine the rise in temperature in the Duralumin band in order to just slip it off the core. The core is incompressible. *Answer:* 82.0°F.

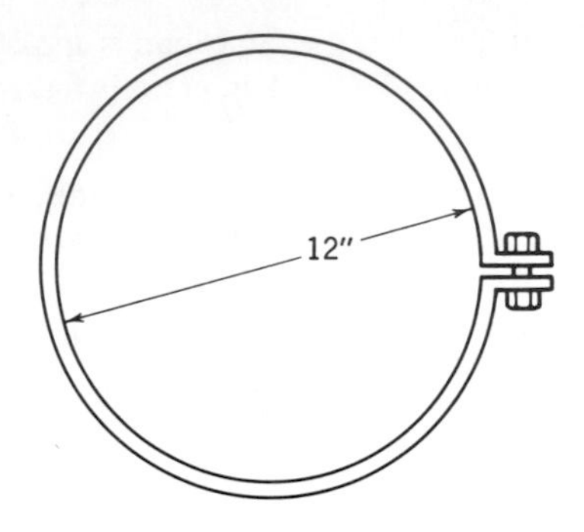

Figure 2-45

2-67. A copper hoop is 1 in. wide, 1 in. thick, and has a 20-in. inside diameter. This hoop fits over a steel hoop 1 in. wide, 1.5 in. thick with a 20-in. outside diameter, and zero clearance at room temperature (70°F). When the temperature is lowered to 0°F, determine the circumferential unit stress in each material. *Answers:* $\sigma_{cu.} = 2420$ psi (T), $\sigma_{st.} = 1610$ psi (C).

2-68. In Fig. 2-46 compression blocks A and B have been placed between two immovable surfaces and develop a compression force between them of 50 kips. It is desired to free block A. Two 20-kip forces are available for this purpose, to be applied at C and D as shown. What minimum temperature change is also required to permit A to be removed without any frictional resistance on its end surfaces? $E_A = 30 \times 10^6$ psi, $\alpha_A = 0.0000065/°F$, $E_B = 10 \times 10^6$ psi, $\alpha_B = 0.0000100/°F$. Disregard weights of members. *Answer:* 10.1°F.

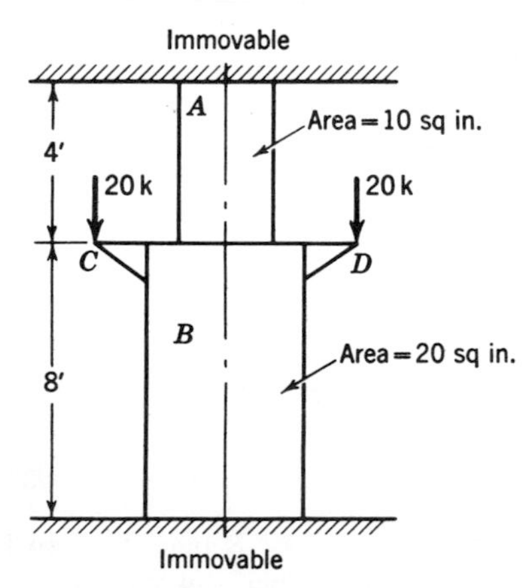

Figure 2-46

2-69. What will the unit stress be in block C (Fig. 2-34) if P equals zero and the temperature increases 50°F? $\alpha_A = 0.0000092$, $\alpha_B = 0.000012$, $\alpha_C = 0.000006$.

2-70. A steel bar, 40 in. in length and 2 in.^{2} in cross section (Fig. 2-47), is rigidly attached to a wall at its right end. Its other end is 0.020 in. from another rigid wall. If a 30,000-lb axial force is attached to the bar at its midpoint and the

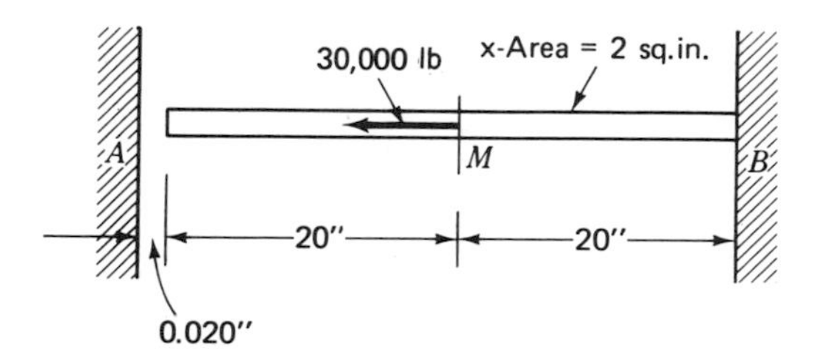

Figure 2-47

temperature is increased 100°F, what will the unit stress be in each portion of the bar? $E = 30 \times 10^6$ psi, $\alpha = 0.0000065$. *Answers:* $\sigma_{AM} = 12{,}000$ psi, $\sigma_{MB} = 3000$ psi.

STATICALLY INDETERMINATE AXIALLY LOADED MEMBERS

2-16 Stresses Due to Loads and Temperature Variations

When a load produces axial stress on two or more members simultaneously, it may not be possible to find the stresses in each of the component members through use of the simple $\sigma = P/A$ formula. In other words, the problem may be a statically indeterminate one, necessitating the incorporation of additional equations for a solution. These equations are usually based upon a relationship of the deformations in the component members.

Illustrative Problem 6. A short, hollow cast-iron cylinder is filled with concrete. If the outside diameter of the cylinder is 10 in. and the wall thickness is 1 in., find the unit stress transmitted to the cast iron and concrete by an axial load of 150,000 lb uniformly distributed over the cross section by means of a thick steel plate. $E_{\text{c.i.}} = 15 \times 10^6$ psi and $E_{\text{c.}} = 3 \times 10^6$ psi.

SOLUTION. That portion of the total 150,000-lb load acting on either the cast iron or concrete is impossible to find directly. Because the loads are unattainable, the stresses cannot be found by sustitution in $\sigma = P/A$. The problem is therefore *statically indeterminate*.

However, an investigation of the deformations produced may provide the necessary information to find the loads and stresses.

Because the cast iron is five times as rigid as the concrete, it must have a unit stress five times as great. This is shown by equating the total deformation of the cast iron to the total deformation of the concrete [see Eq. (2.10)].

$$e_{\text{c.}} = e_{\text{c.i.}}$$

$$\left(\frac{Pl}{AE}\right)_{\text{c.}} = \left(\frac{Pl}{AE}\right)_{\text{c.i.}}$$

In the foregoing equation, the values of $P_{\text{c.}}$ and $P_{\text{c.i.}}$ are the only unknowns. Since $\sigma = P/A$, and the problem asks for the unit stresses in the two materials, and the length of each material is the same, we may rewrite the equation as

$$\frac{\sigma_{c.}}{E_{c.}} = \frac{\sigma_{c.i.}}{E_{c.i.}}$$

or

$$\frac{\sigma_{c.i.}}{\sigma_{c.}} = \frac{E_{c.i.}}{E_{c.}} = \frac{15 \times 10^6}{3 \times 10^6} = 5$$

Thus, in an axially loaded member of two materials *of the same length* the unit stress in each is directly proportional to their moduli of elasticity.

Obviously, the stresses cannot be found without the assistance of another equation. This we may obtain from equating the partial loads $P_{c.}$ and $P_{c.i.}$.

$$P_{c.} + P_{c.i.} = 150{,}000$$

$$A_{c.} \times \sigma_{c.} + A_{c.i.} \times \sigma_{c.i.} = 150{,}000$$

But

$$\sigma_{c.i.} = 5\sigma_{c.}$$

Hence

$$A_{c.}\sigma_{c.} + A_{c.i.} \times 5\sigma_{c.} = 150{,}000$$

$$\frac{\pi 8^2}{4} \times \sigma_{c.} + \frac{\pi}{4}(\overline{10^2} - 8^2)5\sigma_{c.} = 150{,}000$$

$$50.2\sigma_{c.} + 141.5\sigma_{c.} = 150{,}000$$

$$191.7\sigma_{c.} = 150{,}000$$

$$\sigma_{c.} = 783 \text{ psi}$$

$$\sigma_{c.i.} = 5\sigma_{c.} = 3915 \text{ psi}$$

The loads acting on the concrete and cast iron are, therefore,

$$P_{c.} = \frac{\pi 8^2}{4} \times 783 = 39{,}200 \text{ lb}$$

$$P_{c.i.} = \frac{\pi}{4}(\overline{10^2} - 8^2) \times 3915 = 110{,}800$$

$$P = P_{c.} + P_{c.i.} = 150{,}000 \text{ lb}$$

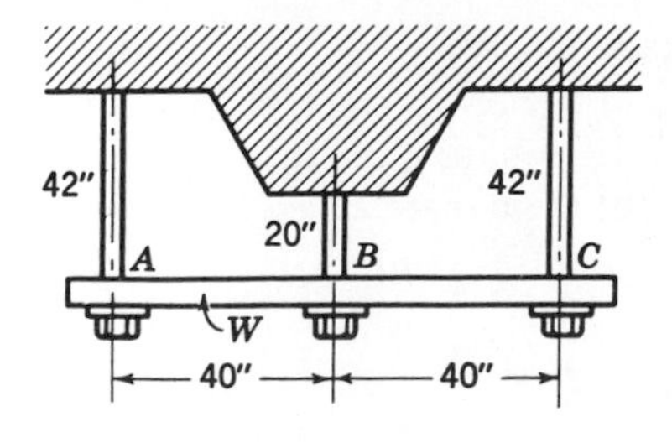

Figure 2-48

Illustrative Problem 7. In Fig. 2-48, the weight W of 40,000 lb is being held in a horizontal position by three rods, A, B, and C. Rods A and C are 42 in. long, 1 in. in diameter, and are made of steel. Rod B is 20 in. long, $1\frac{1}{2}$ in. in diameter, and is made of brass. If the nuts on the three bolts were in the same horizontal plane before the loading took place, determine the load placed on each rod. $E_{st.} = 30 \times 10^6$ psi, $E_{b.} = 14 \times 10^6$ psi.

SOLUTION. Upon the application of weight W, all three rods elongated the same amount, Δ. The relationship between the elongations of the rods is

$$e_{st.} = e_{b.}$$

or

$$\left(\frac{Pl}{AE}\right)_{st.} = \left(\frac{Pl}{AE}\right)_{b.}$$

Substituting, we obtain

$$\frac{P_{st.} \times 42}{0.785 \times 30 \times 10^6} = \frac{P_{b.} \times 20}{1.76 \times 14 \times 10^6}$$

$$1.785P_{st.} = 0.812P_{b.}$$

$$P_{st.} = 0.455P_{b.}$$

The relationship of the loads provides the second equation necessary for the solution. Since

$$2P_{st.} + P_{b.} = 40{,}000 \quad \text{and} \quad P_{st.} = 0.455\,P_{b.}$$

we may write

$$0.91P_{b.} + P_{b.} = 40{,}000$$

whence

$$P_{b.} = \frac{4{,}000}{1.91} = 21{,}000 \text{ lb}$$

Also,

$$P_{st.} = \frac{40{,}000 - 21{,}000}{2} = 9500 \text{ lb}$$

Illustrative Problem 8. If in Illustrative Problem 7 the temperature were to increase 50°F, what would the distribution of the loads be? $\alpha_{st.} = 0.0000065$, $\alpha_{b.} = 0.0000092$.

SOLUTION. In those statically indeterminate problems involving temperature changes, it is well to determine first the change in length of the rods with no load applied.

Since at the original temperature the nuts on the rods were in a horizontal plane, the difference in the elongations due to the temperature rise would equal the difference in elevation of the nuts.

$$e_{st.} = 0.0000065 \times 42 \times 50 = 0.0136 \text{ in.}$$

$$e_{b.} = 0.0000092 \times 20 \times 50 = 0.0092 \text{ in.}$$

$$e_{st.} - e_{b.} = 0.0044 \text{ in.}$$

Thus we note that the steel rods elongate more than the brass by 0.0044 in. This means that resistance to the entire load of 40,000 lb will be taken by the brass rod alone until it has stretched 0.0044 in.

The load that is necessary for a stretch of 0.0044 in. in the brass rod is

$$P'_{\text{b.}} = A\sigma = A\epsilon E = A\frac{e}{l}E = 1.76 \times \frac{0.0044}{20} \times 14 \times 10^6 = 5420 \text{ lb}$$

Upon the application of this load, contact is made with the nuts on the steel rods. Thereafter, the load distribution is obtained by the method given in Illustrative Problem 7. Since the elongation from this point will be the same for the three rods,

$$\left(\frac{Pl}{AE}\right)_{\text{st.}} = \left(\frac{P''l}{AE}\right)_{\text{b.}}$$

$$\frac{P_{\text{st.}} \times 42}{0.785 \times 30 \times 10^6} = \frac{P_{\text{b.}} \times 20}{1.76 \times 14 \times 10^6}$$

$$P_{\text{st.}} = 0.455P''_{\text{b.}}$$

Also,

$$2P_{\text{st.}} + P''_{\text{b.}} = 40{,}000 - 5420$$

$$0.91P''_{\text{b.}} + P''_{\text{b.}} = 34{,}580$$

$$P''_{\text{b.}} = \frac{34{,}580}{1.91} = 18{,}100$$

$$P_{\text{st.}} = 0.455 \times 18{,}100 = 8240 \text{ lb}$$

To obtain the total load on the brass rod $P_{\text{b.}}$ we must add the 5420-lb force required to stretch it prior to contacting the steel to the 18,100-lb load applied as its proportional part of the load when acting with the steel.

Thus, $P_{\text{b.}} = P'_{\text{b.}} + P''_{\text{b.}} = 5420 + 18{,}100 = 23{,}520$ lb.

Illustrative Problem 9. If, in Illustrative Problem 7, the point of attachment of the brass rod to the ceiling could be heightened or lowered, what length of brass rod would be required to create no difference in load between the rods as the temperature varied?

Solution. To cause no change in the loads applied to the three rods during temperature variations would require the temperature elongations of the rods to remain equal under all conditions. Hence,

$$e_{\text{st.}} = e_{\text{b.}}$$

or

$$0.0000065 \times 42 \times \Delta T = 0.0000092 \times l \times \Delta T$$

$$l = \frac{0.0000065}{0.0000092} \times 42 = 29.7 \text{ in.}$$

PROBLEMS

2-71. A steel bolt 1 in. in diameter is passed through a bronze tube 36 in. long, $1\frac{1}{2}$-in. outside diameter and 1-in. inside diameter. (a) If the pitch of the bolt threads is $\frac{1}{8}$ in., what stresses will be induced in the bronze tube and steel bolt by tightening the nut one quarter of a turn, assuming contact of nut to bolt just before tightening? (b) What change in temperature would be necessary to decrease the stress in the steel bolt to zero? Indicate whether the temperature must increase or decrease. *Answers:* (a) $\sigma_{br.} = 6940$ psi, $\sigma_{st.} = 8670$ psi; (b) 248°F decrease.

2-72. Two steel plates $\frac{1}{4}$ in. thick by 8 in. wide are placed on either side of a block of oak 4 in. thick by 8 in. wide, as shown in Fig. 2-49. If the allowable stress in the steel is 20,000 psi and in the oak is 1200 psi, what total load may be applied?

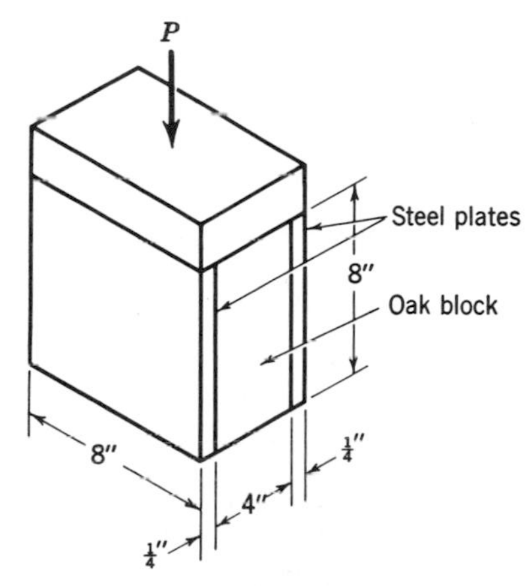

Figure 2-49

2-73. If, in the previous problem, both plates were placed on one side of the oak block, what total load creating equal deformation in the steel and oak could be applied if the unit stress in the steel was maintained at 20,000 psi? How far from the center line of the block must P be applied? *Answer:* 122,700 lb acting 3.47 in. from outside of oak block.

2-74. A horizontal beam carrying a uniformly distributed load of 50,000 lb is supported by three rods, A, B, and C (Fig. 2-50). Rods A and C are of steel, each 1 in.^{2} in cross section and 40 in. long. Rod B is of bronze, 1.2 in.^{2} in cross section and 60 in. long. Before the load was applied, the nuts on the rods were at the same elevation. Determine the load carried by each bar immediately after loading and also after the temperature had increased 100°F. Assume beam to slide freely over rods A, B, and C, and to be inflexible.

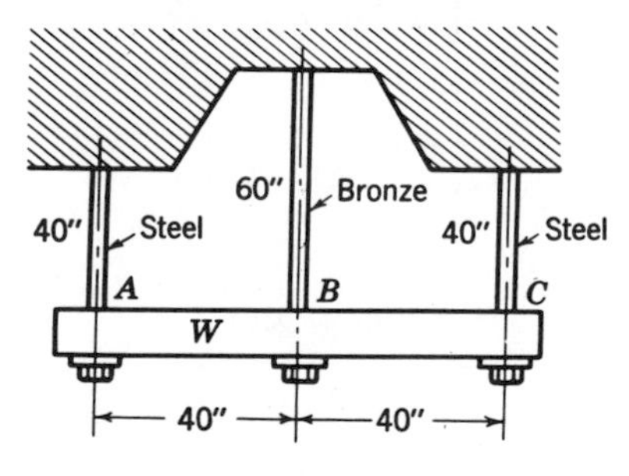

Figure 2-50

2-75. While a steel tensile bar with two flanges (Fig. 2-51) is subjected to a force of 40,000 lb and a temperature of 100°F above room temperature, a bronze sleeve at room temperature is fastened to the flanges. What are the stresses in the bar and sleeve after the assembly attains room temperature and the force is removed? Consider a 10-in. length of sleeve and bar in the computations.

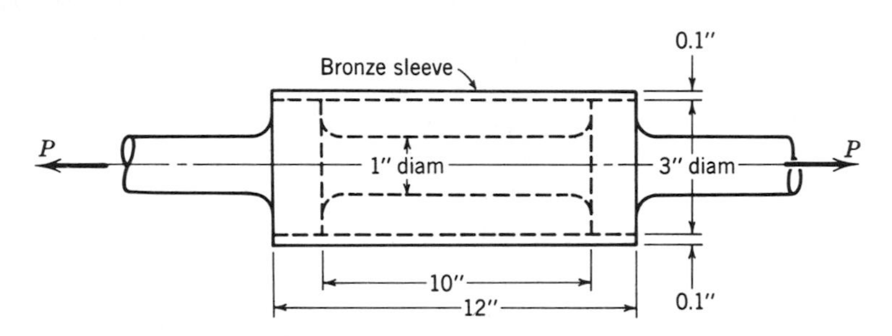

Figure 2-51

2-76. A close-fitting bronze sleeve $1\frac{1}{2}$ in. in diameter fits over a steel bolt 1 in. in diameter. If at 60°F the shoulder of the nut fits snugly against the sleeve, what will be the intensity of stress in the sleeve and bolt when the temperature rises to 160°F? *Answers:* $\sigma_{\text{br.}} = 4600$ psi, $\sigma_{\text{st.}} = 5700$ psi.

2-77. A horizontal beam 10 ft long is suspended by two vertical rods, one at each end of the beam (Fig. 2-52). One of the rods is steel $\frac{1}{2}$ in. in diameter and 4 ft long; the other is brass $\frac{7}{8}$ in. in diameter and 8 ft long. If a load of 6000 lb is applied to the beam, where must it be placed in order that the beam remain absolutely horizontal? $E_{\text{b.}} = 14 \times 10^6$ psi. $E_{\text{st.}} = 30 \times 10$ psi. Disregard the weight of the beam. *Answer:* 49.9 in. from steel rod.

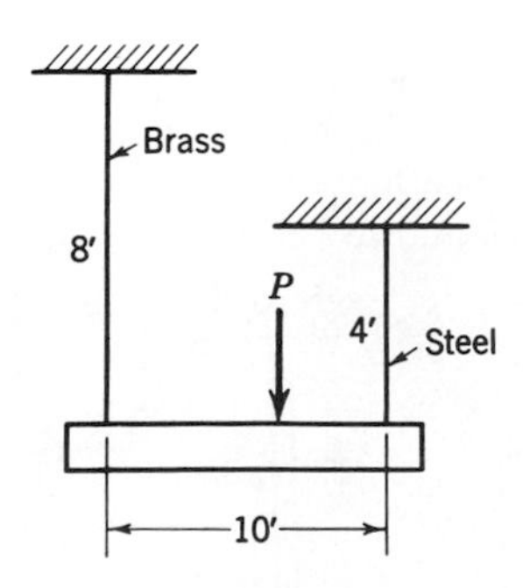

Figure 2-52

2-78. The cables of a power line are copper-coated steel wire. The overall diameter of the wire is $\frac{3}{4}$ in. The steel core has a diameter of $\frac{5}{8}$ in. If the maximum tension in a wire is 10,000 lb, what are the unit stresses in the steel and in the copper?

2-79. Assuming FG to be initially horizontal (Fig. 2-53), what variation of temperature could be tolerated so that point F would not sag more than 0.075 in. when subjected also to $P = 20$ kips? Load and temperature change applied simultaneously. Consider bar BC inflexible. *Answer:* 100°F.

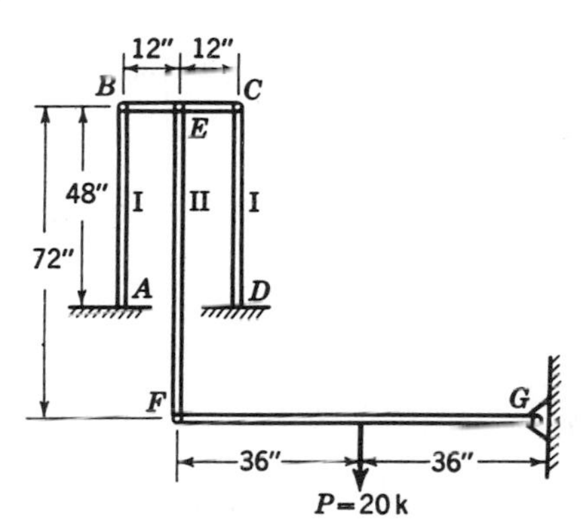

Figure 2-53

2-80. A rigid block weighing 3000 lb is supported in a horizontal position by two bronze rods as shown in Fig. 2-54. If the temperature rises 100°F, determine the load P required to restore the block to a horizontal position after the temperature rise. *Answer:* 1760 lb.

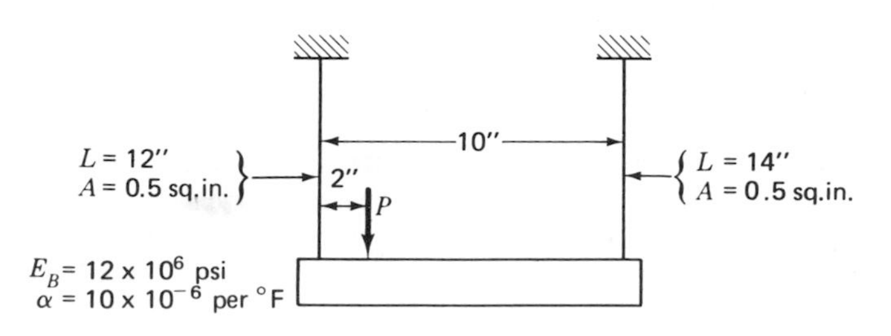

Figure 2-54

2-81. Two aluminum wires each 0.2 in.^{2} in cross section are 100.000 in. long, and one steel wire 0.2 in.^{2} in cross section is 100.010 in. long. The bottom end of the steel wire is 0.010 in. below the bottom of the aluminum wires at 70°F under no load. A rigid block weighing 5000 lb is then symmetrically attached to the ends, as shown in Fig. 2-55. Determine the unit stress in the steel wire when the temperature is 40°F. $E_{al.} = 10 \times 10^6$ psi, $\alpha_{al.} = 12 \times 10^{-6}$/°F, $E_{st.} = 30 \times 10^6$ psi, $\alpha_{st.} = 6.5 \times 10^{-6}$/°F. *Answer:* 11,850 psi.

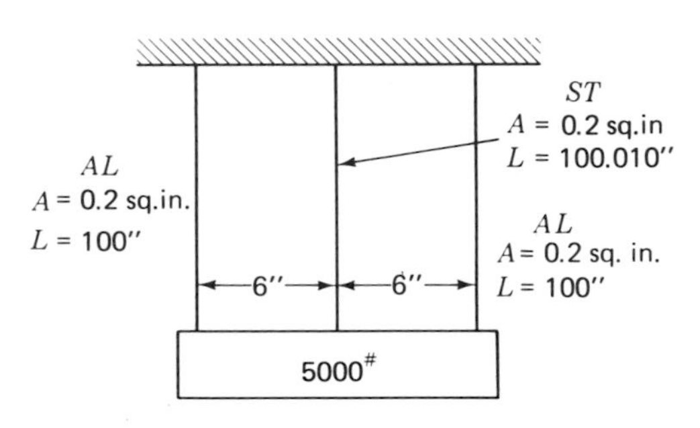

Figure 2-55

2-82. An unloaded rigid bar AC of negligible weight is supported in a horizontal position by a steel rod and an aluminum rod each 10 ft long, as shown in Fig. 2-56. A load of 4800 lb is then applied at point C. (a) Determine the vertical displacement of point C. (b) Determine the change in temperature of the rods which will cause the bar with the applied load to become horizontal.

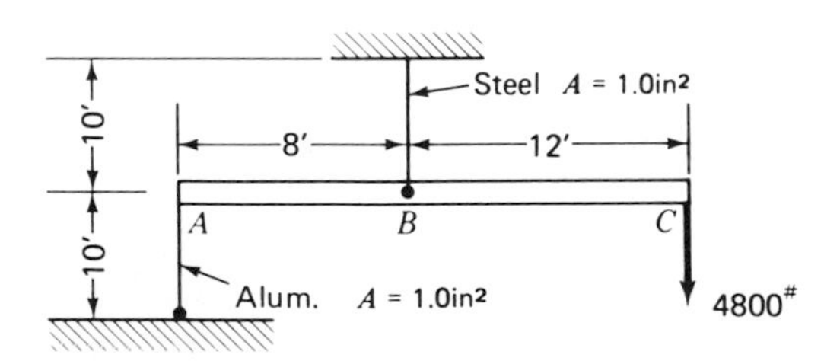

Figure 2-56

2-83. Two aluminum blocks, A and B, are placed on either side of an oak block C, as shown in Fig. 2-57. If a heavy steel plate is placed over their top to ensure uniform deformation under load, (a) what unit stresses are developed in these blocks by a load of 200,000 lb? (b) If the stresses are not uniform, what adjustment in the length of C will be required to make the stresses under the plate uniform? $E_A = E_B = 10 \times 10^6$ psi, $E_C = 2 \times 10^6$ psi. *Answers:* (a) $\sigma_A = \sigma_B = 3170$ psi, $\sigma_C = 317$ psi.

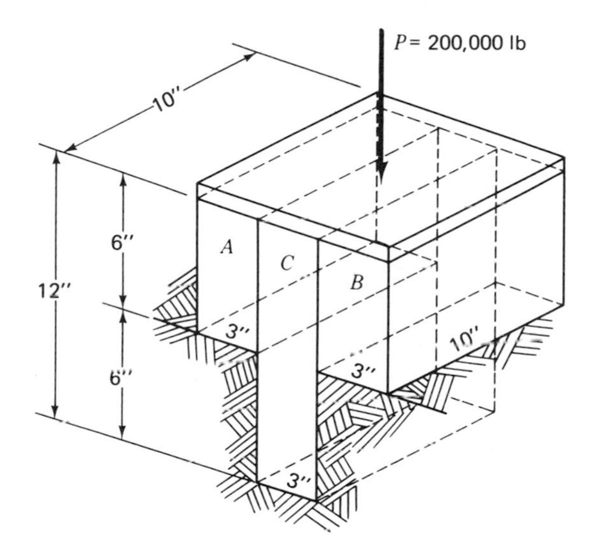

Figure 2-57

2-84. A copper strip is soldered between two steel strips as shown in Fig. 2-58. (a) Find the minimum value of a gradually increasing axial load P at which one of the materials will yield. The yield strengths of steel and copper are 30,000 and 20,000 psi, respectively. $E_{st.} = 30 \times 10^6$ psi and $E_{cu.} = 15 \times 10^6$ psi. (b) If the coefficients of thermal expansion are $6.5 \times 10^{-6}/°F$ for steel and $10 \times 10^{-6}/°F$ for copper, determine the longitudinal normal stresses caused in each material by a temperature decrease of 100°F. [Consider the load $P = 0$ in part (b).]

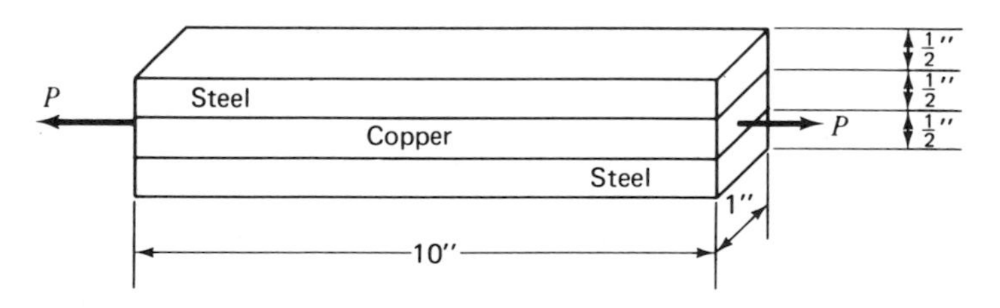

Figure 2-58

2-85. Two aluminum bars and one steel bar are fastened together by two pins 1 in. in diameter (area = 0.785 in.^{2}), as shown in Fig. 2-59, the pins fitting snugly in the holes at 100°F. Determine the shearing unit stress in the pins when the temperature drops to 0°F. $E_{st.} = 30 \times 10^6$ psi, $E_{al.} = 10 \times 10^6$ psi, $\alpha_{st.} = 6.5 \times 10^{-6}/°F$, $\alpha_{al.} = 12.5 \times 10^{-6}/°F$. *Answer:* 10,500 psi.

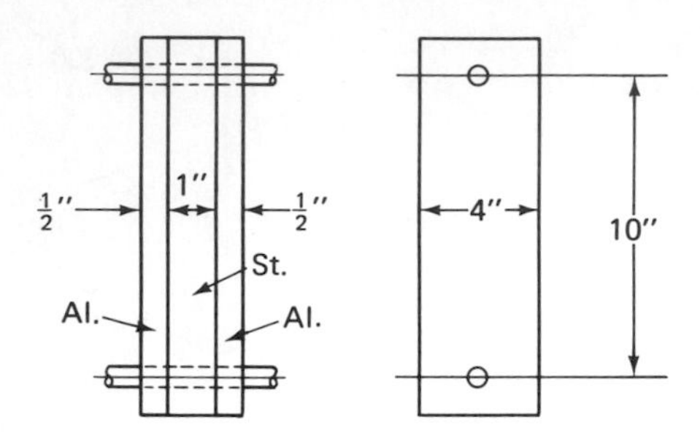

Figure 2-59

2-86. In the assembly shown in Fig. 2-60, find the unit stress in the steel sleeve due to a temperature drop of 50°F. $E_{st.} = 30 \times 10^6$ psi, $\alpha_{st.} = 6.5 \times 10^{-6}$/°F.

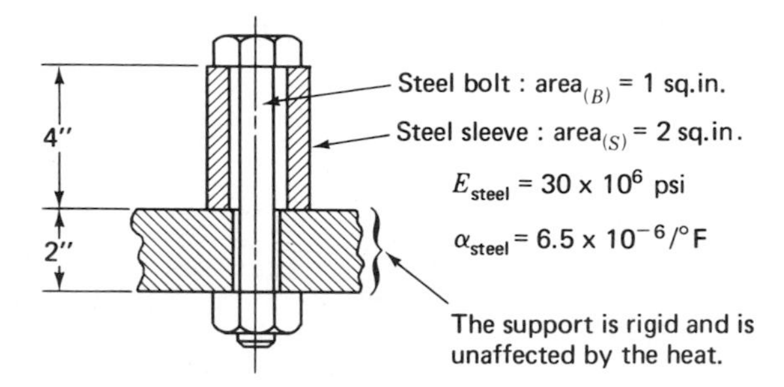

Figure 2-60

2-87. The bar shown in Fig. 2-61 is rigidly attached to its supports and free of stress before the application of an axial force P at its upper third point C. The material of the bar is elastoplastic (Fig. 2-61b). In terms of A, L, E, and σ_y, determine the displacement δ of point C as a function of the force P, as P varies from zero to its ultimate value. Present the results in a graph of P versus δ and give the coordinates of all angular points in the diagram.

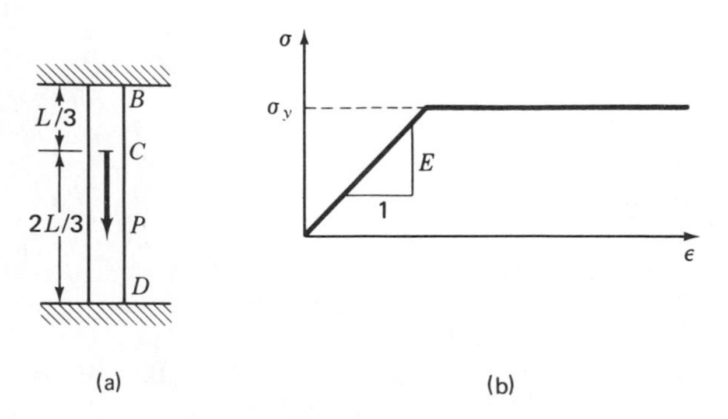

Figure 2-61

2-88. The beam of Fig. 2-62 is held in position by cables AB and CD and by the pin at E. Determine the force in AB and CD due to the 100-kip force applied as shown. Beam BE is assumed not to bend and to be of negligible weight. *Answer:* $P_B = P_D = 50^k$.

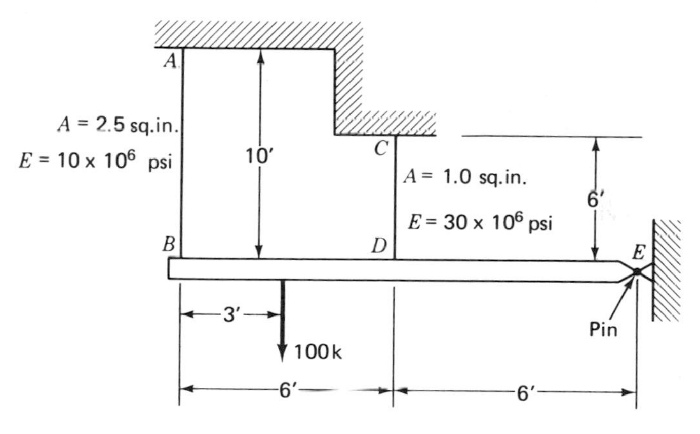

Figure 2-62

2-89. It is desired to maintain member B of Fig. 2-63 at a constant compressive stress of 10,000 psi. If member B is subjected to temperature changes what compensating temperature changes would have to be applied to member BC to maintain the desired stress. Show relationship. $\epsilon_B = 0.00000217$.

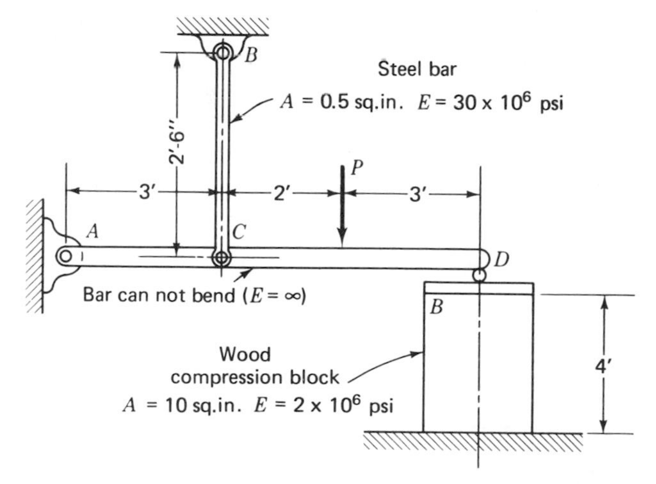

Figure 2-63

2-90. The inflexible bar AC of Fig. 2-64 is horizontal under the imposed load. It is desired to keep it horizontal under all temperature conditions by modifying the lengths of the alminum and steel strut and enabling the aluminum strut to take compression. What relationship of lengths of struts is required to attain the desired objective.

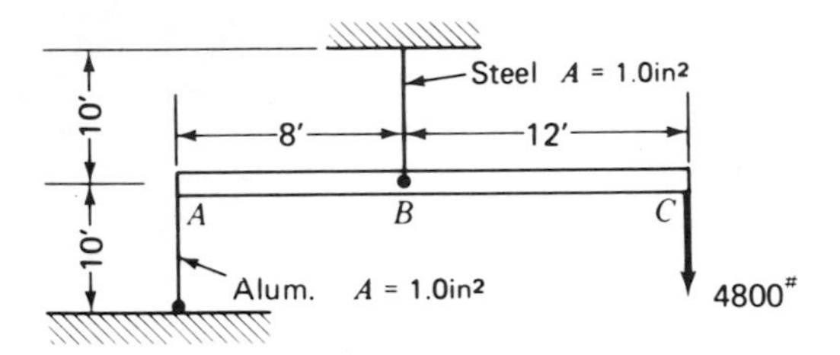

Figure 2-64

2-91. A steel exhaust stack AE, 140 ft high, is held in a vertical position by two steel guy wires BC and BD attached at B (Fig. 2-65). Determine the increase in stress developed in BC and BD if the stack is subjected to a temperature rise of 200°F. Consider the 45° angle not to vary during the temperature increase. $\alpha =$ 0.0000065. *Answer:* $P = 18{,}000$ lb.

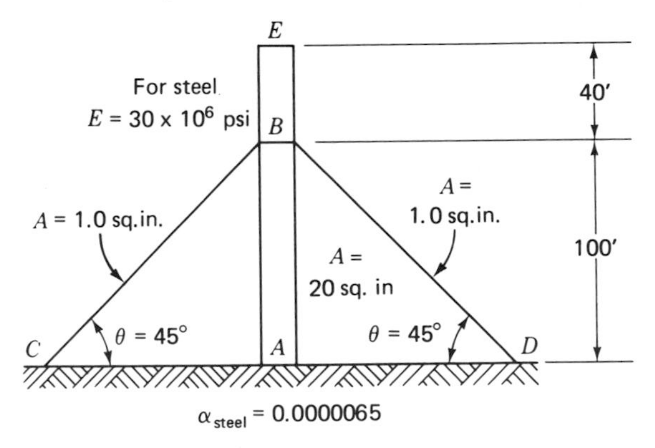

Figure 2-65

2-17 Concept of Compatibility

In every theoretical stress analysis, the application of three fundamental conditions must be included. They are

1. Equilibrium must be maintained.
2. All boundary conditions must be observed.
3. The strains must be compatible with the stresses imposed. It is the third condition of compatibility which we would emphasize at this point.

To illustrate the importance of this condition, let us find the stress in the members of the frame shown in Fig. 2-66. Each member has the same value of E.

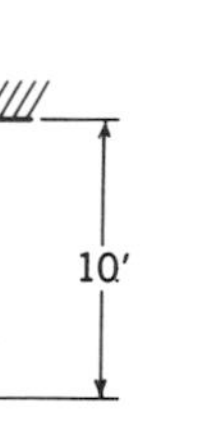
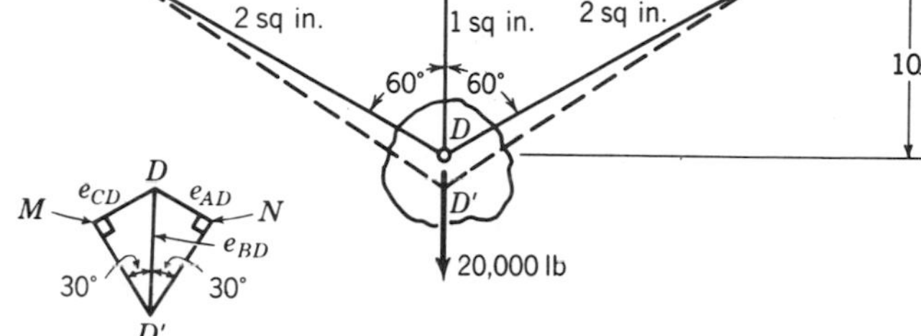

Figure 2-66

The first condition imposed requires that the three members must resist the downward pull of the 20,000-lb load.

$$(P_{AD})_V + P_{BD} + (P_{CD})_V = 20{,}000 \text{ lb}$$

However, the structure is statically indeterminate, and additional information, secured through the use of the second and third conditions, must be involved before the solution can be obtained.

The second condition refers to the final location of panel point D on the deformed structure. Because of symmetry it is obvious that it must move vertically downward, with all three members attached, to a point where equilibrium is reached. The imposition of this condition requires that

$$e_{CD} = e_{AD} = \frac{e_{BD}}{2} \tag{2.15}$$

The third condition—that of compatibility of strain—acknowledges the resisting capability of the three members and the boundary distortion; but it insists as well that the strains in each member be consistent with the forces required for equilibrium. Stated negatively, if the forces in the three members are calculated incorrectly, it would be impossible to maintain their attachment at D while they are suspended from points A, B, and C.

Continuing the solution, we have from the deformation triangle

$$2e_{AD} = e_{BD} = 2e_{CD} \qquad \text{(condition 2)}$$

from which

$$2\epsilon_{AD} l_{AD} = \epsilon_{BD} l_{BD} = 2\epsilon_{CD} l_{CD} \qquad \text{(condition 3)}$$

and

$$2\frac{\sigma_{AD}}{E} \times 20 = \frac{\sigma_{BD}}{E} \times 10 = \frac{2\sigma_{CD}}{E} \times 20$$

Simplifying,

$$4\sigma_{AD} = \sigma_{BD} = 4\sigma_{CD}$$

In terms of the forces (see areas of members)

$$2P_{AD} = P_{BD} = 2P_{CD} \tag{2.16}$$

This relationship of forces must be maintained for the strains to be compatible.

From the equilibrium relationship, Eq. (2.14),

$$(P_{AD})_V + P_{BD} + (P_{CD})_V = 20{,}000 \qquad \text{(condition 1)}$$

But

$$(P_{AD})_V = \frac{P_{AD}}{2}, \quad (P_{CD})_V = \frac{P_{CD}}{2}, \quad \text{and} \quad P_{AD} = P_{CD}$$

Thus,

$$P_{AD} + P_{BD} = 20{,}000$$

Inserting now the relationships of Eq. (2.16),

$$P_{AD} + 2P_{AD} = 20{,}000$$

$$P_{AD} = 6670 \text{ lb} \qquad \sigma_{AD} = 3330 \text{ psi}$$

and

$$P_{AD} = 13{,}330 \text{ lb} \qquad \sigma_{BD} = 13{,}330 \text{ psi}$$

The importance of the compatibility of strain cannot be too strongly emphasized. It is a requirement for every stress analysis, and in particular for those stresses which are statically indeterminate.

2-18 Effect of Changing the Cross Section of an Axially Loaded Member

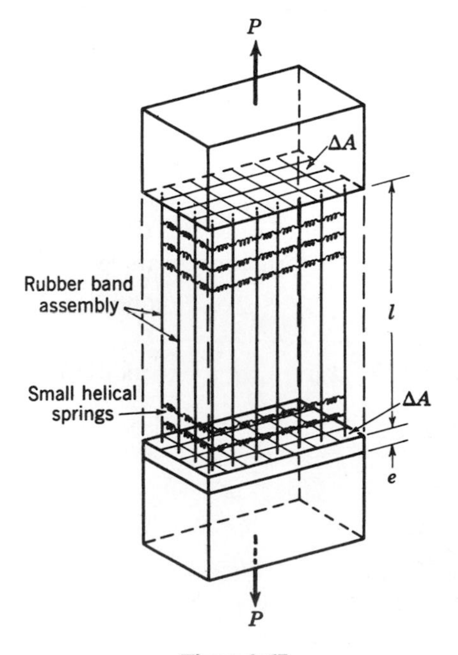

Figure 2-67

In order to ascertain qualitatively the effect of changing the cross section of an axially loaded member, let us consider its central portion, 1 by 5 in. in cross section, to be represented by a uniformly distributed, cross-linked family of rubber bands arranged parallel to the applied load (Fig. 2-67). The cross linkage may be considered to be provided by small helical springs having stress-strain characteristics similar to the rubber bands. Because of the uniform distribution of the rubber bands, the ΔA areas on which they act are equal. If these ΔA areas are projected vertically onto the top and bottom loading blocks, they will present a duplicate set of areas at the centers of which the rubber bands act as concentrated forces.

If a tensile force P is now applied to the member through the loading blocks, each rubber band will stretch the same amount and will apply the same load to each ΔA area. This condition is representative of uniformly distributed stress in the end pieces as well as in the rubber-band assembly, inasmuch as the equal forces are equally spaced.

It is to be noted that any *lateral* strain of the end pieces is reflected in a slightly contracted position of the rubber bands.

Now suppose we taper the central portion of our member to a cross section of 1 by $2\frac{1}{2}$ in. at its lower end (Fig. 2-68). The concentric areas on the lower end will be half their previous value, or $\Delta A/2$. Because the density of the rubber bands on the lower area is twice that on the top area, the stress produced on the lower area will be twice that on the top area. If now the rubber bands are adjusted to zero stress with no slack and load P is reapplied, the rubber bands will stretch through approximately the same vertical distance. However, their unit strain will vary because of (1) their different initial length and (2) the lessened effect of the vertical displacement on rubber bands with increased slope (Fig. 2-69). Thus, the strain, i.e., load, in the outer

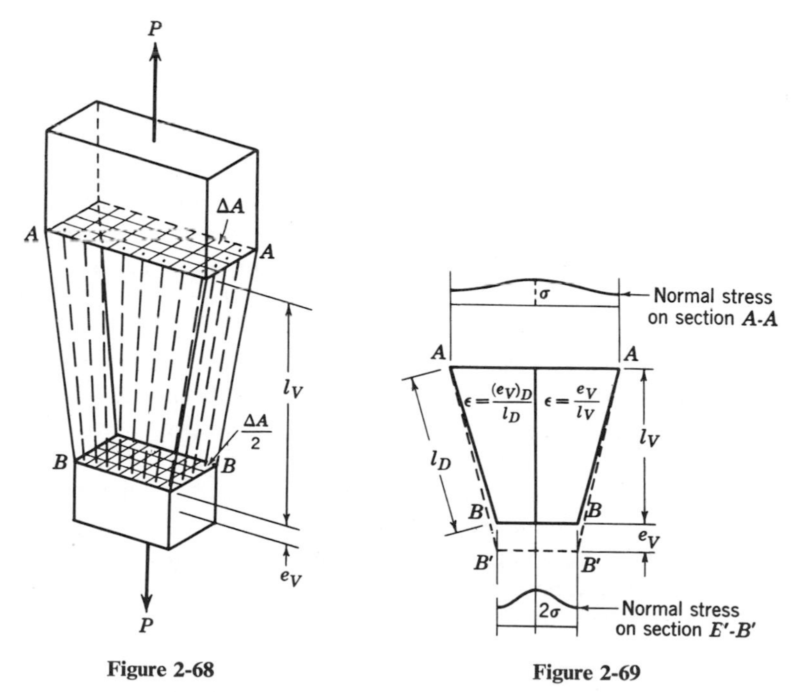

Figure 2-68 **Figure 2-69**

inclined rubber bands will be less than on the shorter vertical rubber bands. Furthermore, the normal stress placed on the horizontal concentric areas will reflect an even greater variation due to their inclination to the outer rubber bands. Thus, the variation of normal stress on the top and bottom cross sections will appear as shown in Fig. 2-69.

Let us now consider the previously discussed straight tensile member with a $2\frac{1}{2}$ in-diameter hole in it. Its rubber-band system is arranged around the hole as though a circular bar of the same diameter had been forced through it. The diameter of the hole is elastically maintained by a readjustment of the cross-linkage system. The entire imaginary system of rubber bands and springs is at zero stress.

With the application of force P, another previously unencountered phenomenon occurs. Those bands that have been curved around the hole attempt to straighten. In so doing, they apply lateral force against the cross-linkage springs, crowd themselves toward the side of the hole, and attempt to make it egg shaped. This crowding action does two things: (1) it decreases the size of the areas on which the rubber bands act, and (2) it prevents the free extension of the rubber bands. The latter effect causes a greater strain in the rubber bands around the hole, which, coupled with the effect of the decreased areas, increases the stress at the hole to a value much greater than the average stress. The normal stress distribution on sections A-A and B-B will therefore be somewhat as shown in Fig. 2-70. The ratio of the maximum stress at the hole to the average stress on cross section obtained as $\sigma = P/A_2$ is equal to the *stress concentration factor*.

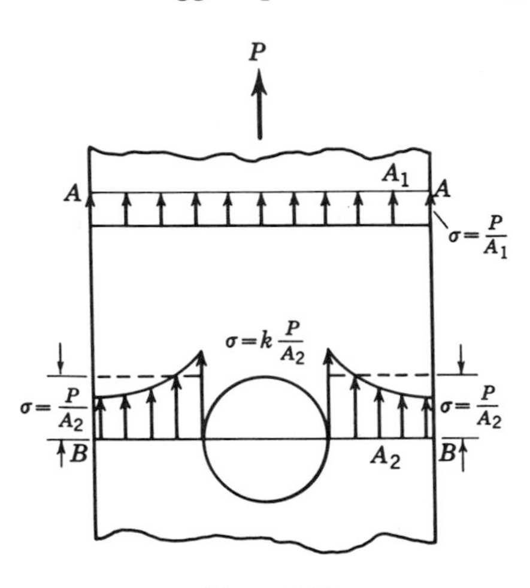

Figure 2-70

The effect of other variations of cross sections on axially loaded members can be analyzed in a manner similar to that already shown. As indicated from the beginning, this study is intended for qualitative reasoning alone. In order for the analogy between the actual member and the rubber-band assembly to be exact, their ratios of lateral to longitudinal unit strain would have to be the same. Actual stress concentrations for various types of discontinuities and the method of including them in design will be considered in Chapter 13. Up to that point, no consideration will be given to stress concentration unless otherwise directed.

3

Design Data Obtained by Experimentation

STRESS-STRAIN CURVES

3-1 General Discussion

The numerical data required for the design of engineering members are obtained from standard tests performed in a materials-testing laboratory. These tests include tension, compression, torsion, bending, etc., and are performed on the materials from which the members are to be made. During the course of each test, corresponding readings of load and deformation are plotted directly to give *load-deformation curves*, or, in the case of axial loading, are converted to unit stresses and unit strains and plotted to form *stress-strain curves*. In either case, the information obtained is invaluable to efficient and economical design.

The stress-strain curves obtained from tension or compression tests conducted on various materials reveal several characteristic patterns (Fig. 3-1). Ductile* rolled steels, such as ordinary structural steel, stretch considerably after following a straight-line variation of stress and strain. For steels alloyed with increasing amounts of carbon and other strengthening material, such as chromium, nickel, silicon, manganese, and so forth, the tendency to produce such an intermediate stretching point becomes increasingly remote. The stress-strain curves for heavily alloyed steels are generally straight to a point a short distance from the rupture point.

*Ductile materials are defined as materials that can undergo considerable deformation before actual rupture.

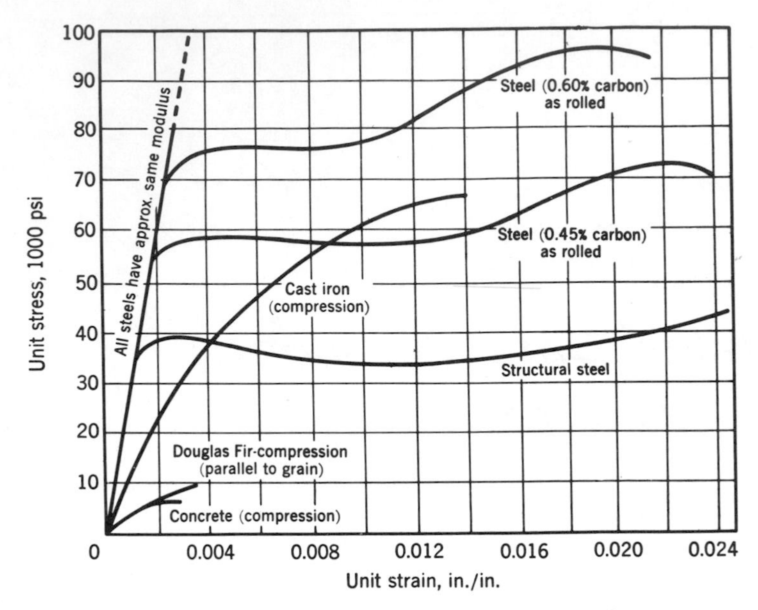

Figure 3-1 Stress-strain diagram for various materials.

In contrast to the straight-line stress-strain curves mentioned above are those obtained for materials such as cast iron, brass, concrete, wood, and so forth, which are often curved throughout most of their length.

If the slope of the initial portion of a stress-strain curve is constant, a constant *modulus of elasticity* is obtained. Indeed, the slope of the initial portion of the curve is equal to the modulus. This important fact allows us to obtain with a glance at the graph a knowledge of the comparative *rigidity* of the material tested, since the steeper the slope, the more stress required to produce a given deformation. A knowledge of the curvature of the initial portion of the curve also permits some estimate of the accuracy of deformation computations. If the initial portion of the curve is straight, the possibility of obtaining accurate deformations is excellent. If the curve is rounded, the possibility of obtaining accuracy is more remote, since the modulus of elasticity for such a material is usually an average for its working range.

3-2 Important Strength Values

The stress-strain curve for low-carbon steel (Fig. 3-2) will form the basis for the ensuing remarks concerning several familiar strength values. Low-carbon steel is the most commonly used steel for structural purposes.

Proportional limit. The *proportional limit* is a strength value often used in technical literature and discussion and, as such, has a specific meaning. The

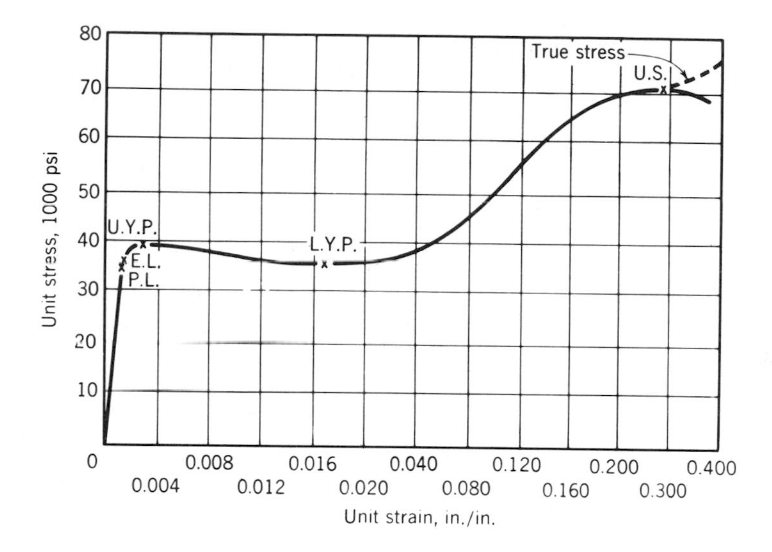

Figure 3-2 Detailed stress-strain diagram for structural steel.

proportional limit is *that unit stress beyond which the ratio of stress to strain no longer remains constant.* Its exact location on a stress-strain graph is not easily determined, because it often depends upon the judgment of the draftsman as to where the curve starts to bend. The location of the proportional limit in Fig. 3-2 is at point P.L.

Elastic limit. Located close to the proportional limit, yet entirely different in meaning, is the *elastic limit.* The elastic limit (point E.L. in Fig. 3-2) is *that maximum unit stress that can be developed in a material without causing a permanent set.* In other words, a specimen stressed to a point below its elastic limit will assume its original dimensions when the stress is released. If the stress should exceed its elastic limit, the specimen will deform *plastically* and will no longer attain its original dimensions when unloaded. It is then said to have incurred a *permanent set.*

The elastic limit is found by placing increasing loads on a specimen and noting the presence of any permanent deformation after the release of each load. That maximum unit stress, produced by the increasing loads, which causes no permanent set will be the *elastic limit.*

The determination of the elastic-limit stress is seldom made, since the time consumed for the test is prohibitive. This stress is usually limited to academic discussions where its concept plays an important part in comparing the relative merits of the various materials.

Yield point. Soon after the stress passes the elastic limit, low-carbon steel attains its yield-point stress. *The yield point of a material is defined as that* *unit stress that will cause an increase of deformation without an increase in load.* Upon the arrival of the yield point, a ductile material such as low-carbon steel stretches an almost unbelievable amount, frequently 10 per cent of the original length. During the yielding period, there is usually a slight relaxation of stress to a *lower yield point* (point L.Y.P. in Fig. 3-2). This is differentiated from the more familiar *upper yield point*, which has already been mentioned (point U.Y.P. in Fig. 3-2).

In contrast to the two previous strength values, the yield point is very easily determined by a tension test. For this reason, it is one of the critical strength values used in writing the specifications for steel and is, consequently, a much-discussed term.

Ultimate strength. An easily obtainable and very important strength value is the *ultimate unit stress*, frequently called the *ultimate strength.* It is defined as being the maximum unit stress which the material is capable of withstanding (point U.S. in Fig. 3-2). The load creating this stress is readily indicated by the testing machine (Fig. 3-3).

Figure 3-3 Universal testing machine. (Courtesy of Baldwin-Lima-Hamilton Corp.)

TABLE 2

AVERAGE STRENGTH VALUES FOR COMMON ENGINEERING MATERIALS

Materials	Yield Stress or Proportional Limit (psi)			Ultimate Strength (psi)			Modulus of Elasticity (psi)	
	Tension	*Compress.*	*Shear*	*Tension*	*Compress.*	*Shear*	*Tension or Compress.*	*Shear*
Steel								
A-36 Structural	36,000*	36,000*	22,000*	58,000	—	40,000	30,000,000	12,000,000
A-242 Structural	50,000*	50,000*	30,000*	70,000	—	45,000	30,000,000	12,000,000
High carbon, 0.7% carbon	65,000	65,000	42,000	110,000	90,000	80,000	30,000,000	12,000,000
Nickel steel, 3.5% carbon	50,000	50,000	38,000	100,000		75,000	30,000,000	12,000,000
Iron								
Gray cast	†	25,000	‡	20,000	80,000	20,000	15,000,000	6,000,000
Malleable	25,000	25,000	12,000	50,000		40,000	24,000,000	11,000,000
Wrought	30,000*	30,000*	13,000*	58,000		38,000	25,000,000	10,000,000
Aluminum alloy								
6061-T6—rolled or extruded shapes	35,000	35,000	20,000	38,000	—	24,000	10,000,000	3,800,000
6063-T6—extruded tubes	25,000	25,000	14,000	30,000	—	19,000	10,000,000	3,800,000
Duralumin, rolled, 2017-T3	32,000		21,000	56,000	—	32,000	10,000,000	4,000,000
Brass, rolled, 70% copper, 30% zinc	25,000		15,000	55,000		48,000	14,000,000	6,000,000
Bronze, cast heat treated	55,000		37,000	75,000		56,000	12,000,000	5,000,000
Copper, drawn	40,000		24,000	52,000		36,000	17,000,000	6,000,000
Timber, air dry								
Yellow pine		6,200			8,400	1,000	1,600,000	
White oak		5,000			7,400	1,300	2,000,000	
Douglas fir		5,400			6,800	800	1,600,000	
Spruce		4,000			5,000	750	1,200,000	
Concrete, 1:2:4 mix, 28 days old					2,500		2,900,000	

*Yield stress. Others in column are proportional limit stresses.

†Not well defined.

‡Greater than tensile strength.

Rupture strength. True stress. In a ductile material, rupture does not usually occur at the ultimate unit stress. After the ultimate unit stress has been obtained, the material will generally *neck down*, and its rapidly increasing elongation will be accompanied by a decrease in load. This decrease becomes more rapid as the rupture point is approached. The rupture strength, obtained by dividing the load at rupture by the original area, has little or no value in design.

A more correct evaluation of the variation of stress following the attainment of the ultimate unit stress is obtained by dividing the loads by the simultaneously occurring decreasing areas. As plotted on Fig. 3-2, these so-called *true stresses* cause the stress-strain graph to veer sharply upward.

STRESS-STRAIN CURVES—MATERIALS WITH NO YIELD POINT

3-3 General Discussion

Judicious engineering design never permits a member to have a general stress condition approaching its yield point. To the contrary, elastic analyses of members are generally based on stresses considerably lower than the yield-point stress, even though at times this stress may be exceeded at points of high stress concentration.

In the light of these statements, no difficulties in determining design stresses would be encountered if all stress-strain curves had a yield point. Unfortunately, not all structural materials show this phenomenon. Some are of the type whose characteristic stress-strain curve is shown in Fig. 3-4.

It has, therefore, been necessary to adopt an arbitrary method of locating on the stress-strain curve a point that would designate the beginning of the plastic range and serve as an *upper limit of usefulness*. Two such methods are commonly used whose upper limit values serve not only for comparative purposes, but also as ceiling values in determining allowable working stresses. They are the *Johnson's Apparent Elastic Limit* (*JAEL*) and the *yield strength* or *offset stress*.

Johnson's apparent elastic limit. This limiting stress value, supposedly indicative of a general plastic yielding, was proposed by Professor J. B. Johnson and is defined as *that unit stress at which the unit strain increases at a rate of 50 per cent greater than shown by the initial tangent to the stress-strain curve*. The construction necessary for the procurement of this value is shown in Fig. 3-5.

Yield strength or offset stress. When a test specimen is stressed beyond its elastic limit and then has its load released, it is revealed that during the

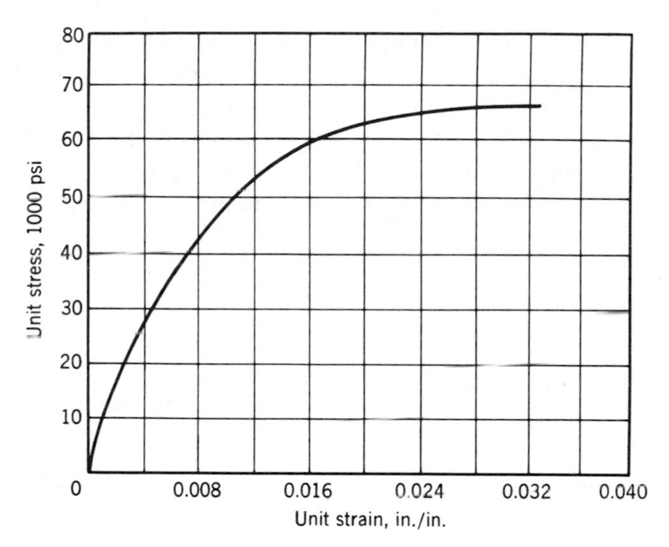

Figure 3-4 Stress-strain diagram for cast iron (compression).

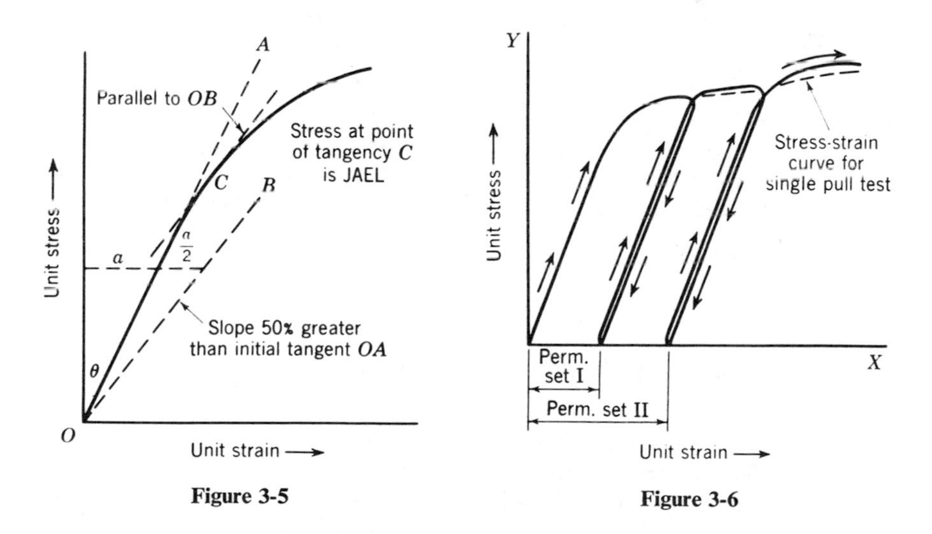

Figure 3-5

Figure 3-6

load-reducing stage the stress-strain curve parallels the initial portion of the curve, as shown in Fig. 3-6. The horizontal intercept on the x axis is the permanent set.

Such a load cycle does not necessarily damage a material even if the imposed stress exceeds the elastic limit. On the contrary, the material may be benefited. True, its ductility may be lowered, but the hardness and elastic limit stress of the material will generally increase. The large amount of ductility possessed by some materials is not always essential to good service.

To derive the maximum efficient use of a material exhibiting no yield-point stress, an upper useful limit of the material is frequently obtained by locating the unit stress that will produce an arbitrarily assigned permanent set. This unit stress is called the *yield strength* or *offset stress*. Yield strengths are now universally employed for materials without yield points. The permanent set values range from about 0.1 to 0.3 per cent of the original gage length. The construction necessary for locating the yield strength is shown in Fig. 3-7.

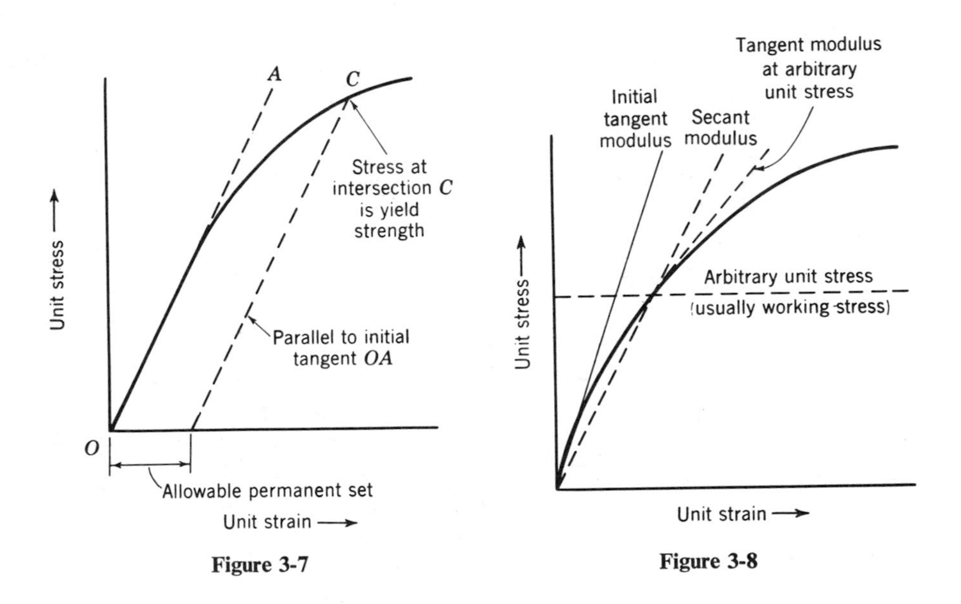

Figure 3-7

Figure 3-8

3-4 Tangent and Secant Modulus of Elasticity

As already discussed, some stress-strain graphs are continuous sweeping curves, whereas others contain an initial straight-line portion. In the latter type, the slope of the initial portion is equal to the modulus of elasticity. In the curved type, however, the stress developed in a material having a curved stress-strain diagram is often high enough on the curve to create at that point a much flatter slope than found at the origin. To offset this condition, a secant modulus of elasticity is frequently used in addition to the already discussed tangent modulus. The secant modulus (Fig. 3-8) is the slope of a line joining the point of zero stress to that point on the curve whose ordinate equals the allowable working stress, or some other arbitrarily selected stress. Its use is an attempt to average the moduli obtainable over the working portion of the stress-strain curve.

OTHER IMPORTANT PROPERTIES OF STRUCTURAL MATERIALS

3-5 Behavior of Materials Under Repetition of Stress: Fatigue

Data obtained from simple tension, compression, and torsion tests provide basic information used in the design of members subjected to steady load conditions. Should these conditions vary and include many repetitions of stress, the design might prove to be inacceptable, for it was not based on data secured under similar operating conditions.

Fortunately, most structural members do not encounter enough repetitions of stress to warrant any special attention to their design. Those structural and machine members that do undergo many thousands of repetitions of stress in their lifetime are subject to *fatigue* and require special care in their design, as will be described in Chapter 13. We shall concern ourselves at this point with a comparatively few, rather than millions, of stress cycles as applied to the ordinary ductile engineering materials.

When a specimen of ductile steel is stressed a small amount beyond its elastic limit and then has its load released, the return stress-strain graph, as already explained, travels toward the zero-stress axis on a line parallel to the initial stress-strain curve shown in Fig. 3-6. Subsequent repetitions of stress above the elastic limit will duplicate the previous action. After the stress has passed beyond the lower yield point, as obtained from a single pull test, the stress-strain graph is identical to that of a cold-rolled steel. The material in this state will usually be assigned a higher allowable design stress, creating lighter members. This process of continued repetition of stress above the elastic limit of a material is called *cold working*.

3-6 Behavior of Materials Under Sustained Stress: Creep

The deformation which most structural materials undergo when stressed up to their allowable limit at room temperature is a *completed* deformation and will not increase no matter how long the stress is applied. At some higher temperature, however, these same materials will reveal a continuing deformation or *creep* which, if permitted to continue, will ultimately lead to excessive displacements or rupture (Fig. 3-9). The length of time that elapses before failure or unsatisfactory perfor-

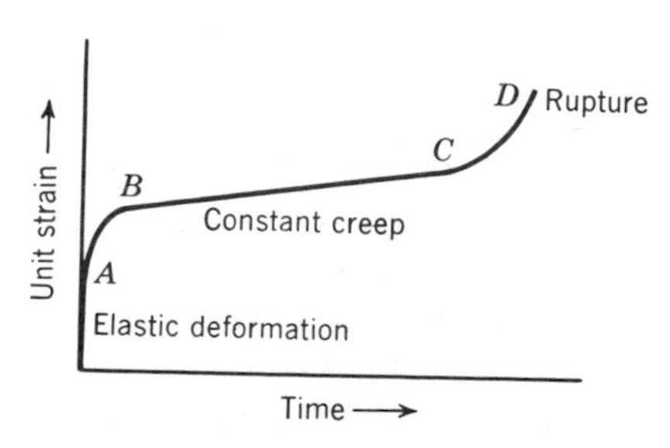

Figure 3-9

mance occurs may extend beyond the ordinary useful life of the member, causing no concern. On the other hand, if the temperature at this maximum stress condition is increased, failure due to creep is hastened and may occur before the termination of its expected normal life.

When such a possibility exists, the maximum allowable stress must be reduced in order to establish a rate of creep that will insure satisfactory performance at the desired temperature over a specified length of time. This reduced maximum stress is known as the *creep limit*. In this design of machine parts, the creep limit is generally set at that stress which will produce a total elongation of 1 per cent in 100,000 hr (11.4 years). Curves showing the reduction of the creep limit with temperature are given for three steels in Fig. 3-10.

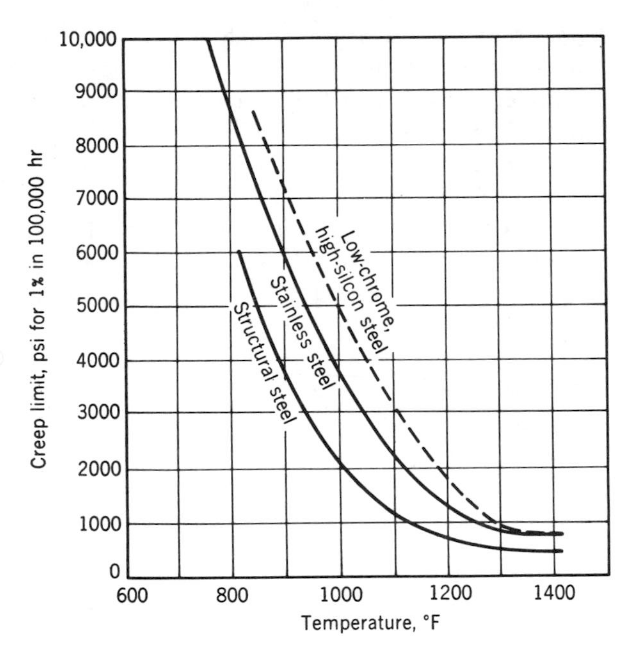

Figure 3-10

3-7 Ductility

Ductility is that property of a material that allows a load to develop large plastic deformations without rupture. There are two general methods used for the determination of ductility. The first is based upon the percentage of elongation in a specified length. If e equals the total deformation in length l, the percentage of elongation would be

$$Q = \frac{e}{l} \times 100 \tag{3.1}$$

The second method used in determining ductility employs the percentage of reduction in area (sometimes called *contraction*). If A is the original cross section and A' the cross section of the ruptured area, the percentage of reduction of area is

$$R = \frac{A - A'}{A} \times 100 \tag{3.2}$$

It is generally true that if the figures for percentage of elongation and reduction of area are large, the material has the ability to stand abuse in fabrication and service. It is also true that in many instances ductility is necessary to relieve localized and initial stresses and to act as insurance against destruction and complete collapse due to excessive loads.

Metallurgical explanation of ductility—Cryogenics. The theory of elasticity used in the study of strength of materials treats solids as continuous elastic media. However, solids are not continuous media but consist of atoms bound together by interatomic forces. Metals, more specifically, consist of irregularly shaped grains whose fundamental units of matter are symmetrical *atomic* structures known as crystal lattices. The three most common crystal lattices are the face-centered cubic, the body-centered cubic and the close-packed hexagonal. These are shown in Fig. 3-11. Because of their arrange-

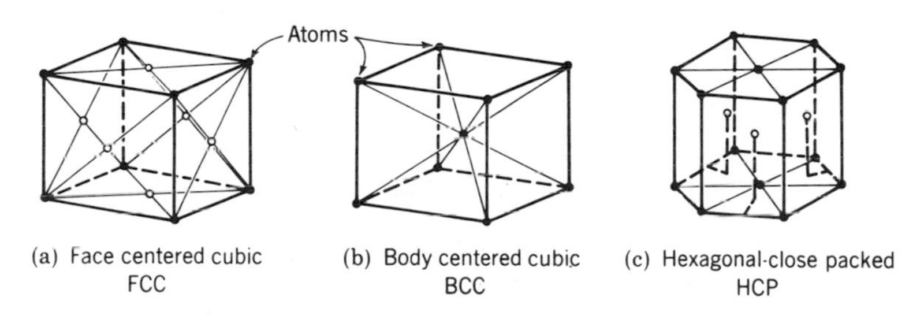

(a) Face centered cubic FCC (b) Body centered cubic BCC (c) Hexagonal-close packed HCP

Figure 3-11

ment these structures have planes of potential weakness, known as *slip planes*, over which the atoms can move more easily than over others. The perfect crystal or grain consists of many of these lattice structures stacked perfectly in regular array. In a real crystal, however, there are always some irregularities which destroy the orderly array of its lattices. These irregularities are classified as

1. Dislocations—revealed by atomic planes which have slipped over each other.

2. Interstitials—displaced atoms located in a crevice of the regular lattice structure.

3. Vacancies—atomic "holes" left by vacated atoms.

4. Impurities—foreign atoms occurring in the lattice.
5. Stacking faults—lack of perfect repetition of atomic layers.

It is the interaction of these defects in the direction of the slip planes that accounts for the *yield stress* and *ductility* of metals, whereas it is the resistance of the more orderly regular arrangement of the atoms in the lattice which contributes to the development of the *modulus of elasticity*.

Grain boundaries are also discontinuities which resist plastic yielding. Because cold working results in smaller grain sizes, ductility is decreased.

The effect of low temperatures on the ductility of materials, included in the science called *cryogenics*,* reveals changes in behavior largely dependent on the lattice structure with its previously indicated defects. For instance, a face-centered-cubic crystal structure is relatively insensitive to decreases in temperature because it has more potential slip planes than other crystal structures. The tendency of body-centered and hexagonal-close-packed structures to modify the orientation of their slip planes at low temperatures results in increased slipping difficulty and a tendency toward brittleness. Thus, copper and nickel and their alloys, aluminum and its alloys, and austenitic stainless steel containing more than 7 per cent nickel, all having a face-centered-cubic structure, remain ductile to near absolute zero, whereas iron and low alloy steels which are body centered become brittle at low temperatures.

3-8 Resilience

When a tension member is acted on by an axial load P and undergoes a corresponding axial deformation Δ, an amount of external work equal to $\frac{1}{2}P\Delta$ is performed *on the member*.† This external work, as dictated by the conservation of energy theorem, is transformed into an equivalent amount of internal energy created by the elastic or semielastic distortion of the interior elements and is equal to a corresponding amount of area under the stress-strain curve.

The ability of a member to release this captured energy quickly as the load is removed is a measure of its *resilience*. It is measured as the area under the released stress-strain curve. For purposes of comparison and design, the value of *modulus of resilience* is used. It is equal to the area under the stress-strain curve up to the elastic limit. For materials having a straight-line variation of stress to strain up to the elastic limit, and whose released stress-strain curve is almost identical to it (Fig. 3-6), the modulus of resilience is

*The science dealing with the properties of materials at low temperatures—generally below $-150°C$.

†Assuming a linear relationship of load to deflection, the work equals $\frac{1}{2}P\Delta$ rather than $P\Delta$ because the maximum deformation Δ was obtained only after P had increased from zero to its full value.

equal to $\frac{1}{2}\sigma_E \varepsilon_E$ or $\sigma_E^2/2E$, where σ_E is the elastic-limit stress of the material. It might also be thought of as the amount of energy stored in each cubic inch of material at the elastic limit.

3-9 Toughness-Impact

The total energy absorbed by a material before failure is known as its toughness, and, although mostly unrecoverable, is important in determining the maximum resistance of a material to impact. The unit measure of this maximum resistance (toughness) is called the *modulus of toughness* and is defined as *the amount of energy absorbed per unit volume up to the breaking point*. It is equal to the area under the stress-strain curve to rupture.

Moduli of toughness, however, are strength values obtained from static tests. The use of quickly applied loads can produce entirely different results, depending on the speed of loading, character of material, geometry of the specimen, etc. The great difficulty in obtaining true resistance values under impact loads has caused empirical tests to be developed, the exact interpretation of which are often debatable.

One of the most frequently used measures of impact resistance is the Charpy test, which employs the swing of a heavy pendulum to break a notched beam (Fig. 3-12). The difference in height between the initial and final position of the swing in feet, times the weight of the pendulum, is taken as the energy in foot-pounds required to break the specimen. This test finds its greatest usefulness in the metallurgical field where differences in heat treatment, alloy content, and susceptibility to brittle fracture can be detected. The absolute values of energy obtained from the test are, unfortunately, only qualitative—and even then close correlation of test results to service behavior of some metals is almost impossible.

The presence of a notch in the specimen causing extreme variation of stress over a confined area and the necessity of using a hammer make it impossible to obtain an accurate value of energy that can be assigned to any one unit volume. To overcome these difficulties, tests have been made at the City College of New York with spherical specimens where stress concentrations are largely eliminated and where an internal explosive charge is used as a prime mover. The results of these tests have revealed stresses higher than those obtained under static conditions. Computed values of impact resistance in foot-pounds per cubic inch give promise of providing a rational design for members subjected to impact loading.

3-10 Poisson's Ratio

Poisson, a French engineer of the early 19th century, found that the lateral unit strain of structural materials bore an almost constant ratio to the longitudinal unit strain when subjected to either tension or compression

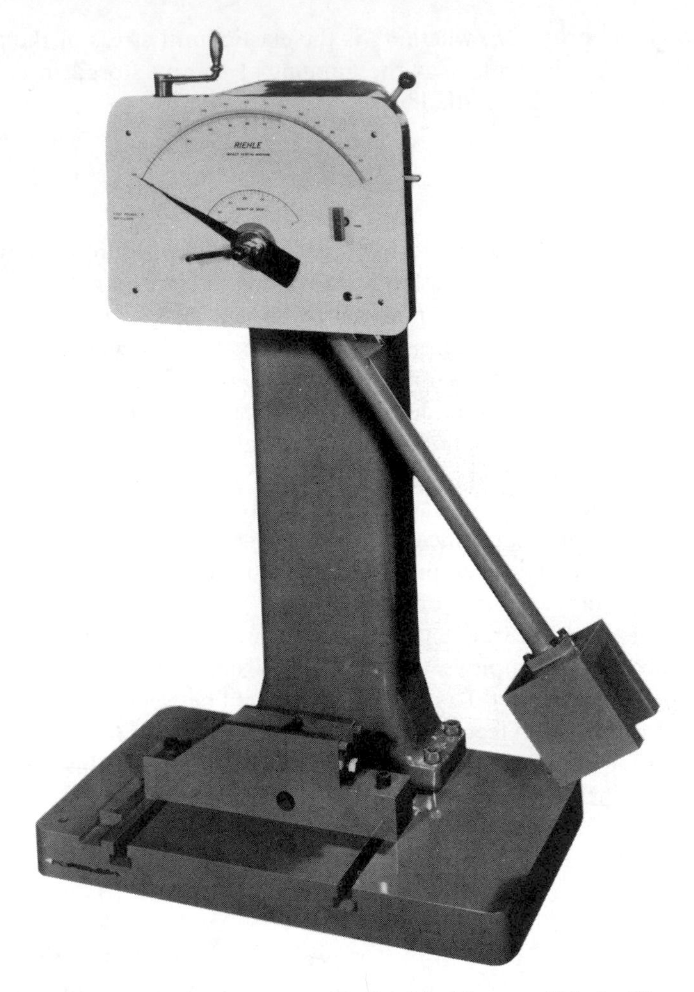

Figure 3-12 Combination Impact Testing Machine on which the Charpy impact test can be performed. (Photo courtesy of Riehle Testing Machines, a Division of Ametek, Inc., E. Moline, Illinois.)

stress within the elastic limit. The Poisson ratios for metals vary from about 0.25 to 0.36 and are denoted by the letter μ.

An important stress-strain relationship. Because of the presence of lateral strain, a cubical element (Fig. 3-13a) stressed in tension in the y direction produces strains along the three axes equal to

$$\varepsilon_y' = \frac{\sigma_y}{E}, \qquad \varepsilon_x' = -\mu\varepsilon_y', \qquad \varepsilon_z' = -\mu\varepsilon_y' \tag{3.3}$$

the negative sign indicating a contraction.

Similarly, were a tensile stress acting along the x direction (Fig. 3-13b), the strains would be

$$\varepsilon_x'' = \frac{\sigma_x}{E}, \qquad \varepsilon_y'' = -\mu\varepsilon_x'', \qquad \varepsilon_z'' = -\mu\varepsilon_x'' \tag{3.4}$$

However, should there be tensile stresses acting in *both* the x and y directions (Fig. 3-13c), their corresponding strains would have an additive

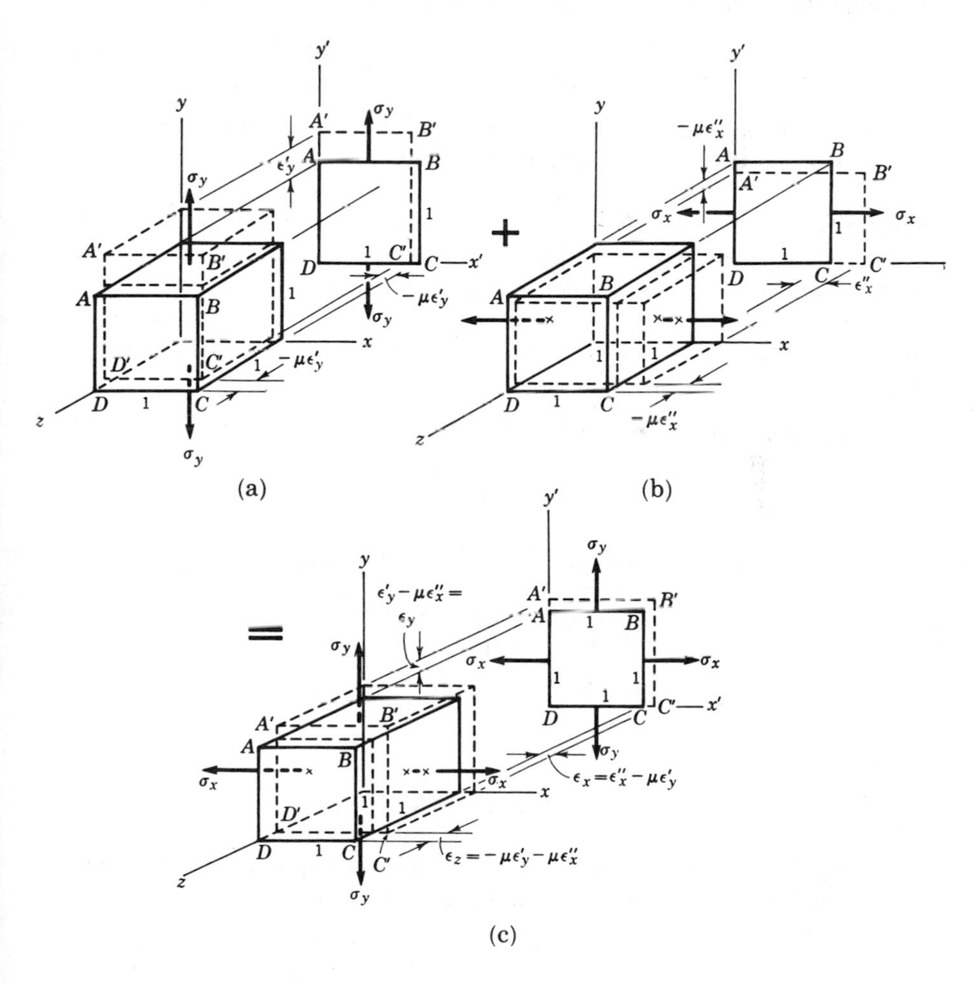

Figure 3-13

effect on one another, so that

$$\varepsilon_x = \varepsilon_x'' - \mu\varepsilon_y' \tag{3.5}$$

$$\varepsilon_y = \varepsilon_y' - \mu\varepsilon_x'' \tag{3.6}$$

$$\varepsilon_z = -\mu\varepsilon_y' - \mu\varepsilon_x'' \tag{3.7}$$

where the values of ε_x, ε_y, and ε_z are the actual experimental strains measured in the x, y, and z directions. If the biaxial strain equations above are placed in terms of the stresses in the x and y directions, they will read as follows:

$$\varepsilon_x = \frac{\sigma_x}{E} - \frac{\mu\sigma_y}{E}, \qquad \varepsilon_y = \frac{\sigma_y}{E} - \frac{\mu\sigma_x}{E}, \qquad \varepsilon_z = -\frac{\mu\sigma_y}{E} - \frac{\mu\sigma_x}{E} \tag{3.8}$$

Simultaneous solution for σ_x and σ_y obtains

$$\sigma_x = \frac{(\varepsilon_x + \mu\varepsilon_y)E}{1 - \mu^2}, \qquad \sigma_y = \frac{(\varepsilon_y + \mu\varepsilon_x)E}{1 - \mu^2} \tag{3.9}$$

These equations are used in the experimental determination of stresses and will be discussed in more detail in a later chapter.

Illustrative Problem 1. An engine housing is to be tested under load. Two strain gages placed at a certain point on its surface perpendicular to one another and in directions parallel to applied stresses σ_x and σ_y record unit strains of $\varepsilon_x = +0.00080$ and $\varepsilon_y = -0.00050$. If Poisson's ratio equals 0.3 and $E = 15{,}000{,}000$ psi, what are the stresses in the directions of the gages? No shear stresses accompany σ_x and σ_y.

SOLUTION. Substituting the given data in Eq. (3.9) gives

$$\sigma_x = \frac{(0.00080 - 0.3 \times 0.00050)}{1 - 0.09} 15 \times 10^6$$

$$= \frac{0.00065 \times 15 \times 10^6}{0.91} = 10{,}700 \text{ psi}$$

$$\sigma_y = \frac{(-0.00050 + 0.3 \times 0.00080)}{0.91} 15 \times 10^6$$

$$= \frac{-0.00026 \times 15 \times 10^6}{0.91} = -4280 \text{ psi}$$

PROBLEMS

3-1. A 2-in. Duralumin test specimen ($d = 0.505$ in.), when subjected to tensile load, elongates in accordance with the following data:

Unit Stress	Unit Strain	Unit Stress	Unit Strain	Unit Stress	Unit Strain
5,000	0.0005	30,000	0.003	50,000	0.125
10,000	0.0010	32,500	0.00325	54,000	0.300
15,000	0.0015	35,000	0.00375	45,000	0.480 (Ruptured)
20,000	0.0020	40,000	0.0065	Reduced diameter = 0.320 in.	
25,000	0.0025	45,000	0.0110		

Plot the stress-strain curve and determine the proportional limit, the modulus of elasticity, the yield strength (offset = 0.1 per cent strain), the percentage of elongation, and the percentage of reduction of area.

3-2. From the tension test on a piece of structural steel, the following data are obtained: original diameter, 0.505 in.; diameter of ruptured area, 0.346 in.; original test length, 2 in.; elongated test length, 2.88. Determine the ductility of the steel based on percentage of elongation and percentage of reduction in area. *Answers:* 44.0 per cent, 53.0 per cent.

3-3. In the tension test of a cellulose acetate butyrate plastic, the following stress-strain readings were obtained:

Unit Stress (psi)	Unit Strain	Unit Stress (psi)	Unit Strain	Unit Stress (psi)	Unit Strain
0	0.00	2800	0.035	3720	0.310
400	0.005	3200	0.040	3780	0.370
800	0.010	3600	0.048	3950	0.440
1200	0.015	3900	0.060	4150	0.500
1600	0.020	4380	0.120	4450	0.560
2000	0.025	3750	0.185	4800	0.620
2400	0.030	3700	0.250	5100	0.660

Determine the following: (a) proportional limit, (b) upper yield point stress, (c) lower yield point stress, (d) modulus of elasticity, (e) modulus of resilience, (f) modulus of toughness, (g) ultimate strength. *Answers:* (a) 3200 psi, (b) 4380 psi, (c) 3700 psi, (d) 80,000 psi, (e) 64 psi, (f) 2580 psi, (g) 5100 psi.

3-4. A gray cast-iron specimen, with a 2-in. gage length tested in tension, gave the following stress-strain readings:

Unit Stress	Unit Strain
5,000	0.0003
10,000	0.0007
15,000	0.0012
20,000	0.0020
25,000	0.0032
30,000	0.0055
32,500	0.008 (failure)

Determine the following: (a) tangent modulus of elasticity, (b) secant modulus of elasticity (use stress = 15,000 psi), (c) Johnson's apparent elastic limit, (d) yield strength (offset equal to 0.001), (e) modulus of toughness. *Answers:* (a) 16.67×10^6 psi, (b) 12.50×10^6 psi, (c) 7500 psi, (d) 21,000 psi, (e) 172 in.-lb/in.^{3}.

3-5. In a tensile test of an aluminum specimen 1 in. in diameter, the total elongation in 8 in. recorded at a load of 46,000 lb is 0.04680 in. At the same moment, the diameter measures 0.9985 in. Compute the modulus of elasticity and Poisson's ratio, assuming the applied stress to be below the elastic limit. *Answers:* 10×10^6 psi, 0.258.

3-6. If Poisson's ratio is equal to 0.30 for the specimen of Problem 3-1, determine the lateral contraction of the diameter at a unit stress of 25,000 psi. *Answer:* 0.000379 in.

3-7. If in Illustrative Problem 1 the strains recorded were $\epsilon_x = -0.00080$ and $\epsilon_y = +0.00050$, what would the corresponding stresses be?

3-8. Assume $\epsilon_x = +0.00080$ and $\epsilon_y = +0.00080$ in Illustrative Problem 1 and solve for the corresponding stresses. *Answers:* $\sigma_x = \sigma_y = 17{,}150$ psi.

ALLOWABLE WORKING STRESS—FACTOR OF SAFETY

3-11 Allowable Working Stress

An allowable working unit stress is that maximum unit stress that can be safely applied to a material subjected to a particular loading condition. It is a unit stress derived from the results of many tests and the accumulated experience obtained from many years of firsthand observation in the performance of members in actual service. The magnitude of the allowable working unit stress will control the cross-sectional area and, consequently, the weight and cost of the members designed.

3-12 Factor of Safety

Owing to the possible inconsistency of structural strengths, the variance in calculated and exact stress, and the difficulty of ascertaining the magnitude of the applied loads, a reduction of stress from the ultimate, or yield-point, stresses is necessary for design purposes. The ratio of the ultimate unit stress (or yield-point stress) to the allowable working unit stress is called the *factor of safety*.

The actual determination of the value of the factor of safety to be used, say, for structural steel for buildings, is usually delegated to a group of engineers appointed by some of the larger national engineering societies or by a large city for use in its building code. These engineers, in addition to determining the uncertainties of structural properties, stresses, and applied loads, consider such matters as the depreciation or useful life of the structure, the loss in life and property should the structure collapse, the effectiveness of regular painting, as well as the increase in cost as the allowable stresses are reduced.

The specifications for machine members are perhaps even more rigid than those for structural steel. Impact, fatigue, and creep are items generally

TABLE 3

ALLOWABLE UNIT STRESSES FOR VARIOUS ENGINEERING MATERIALS

	Tension (psi)	Compression (psi)	Shear (psi)
Structural steel	22,000	22,000	14,500
Wrought iron	12,000	12,000	8,000
Aluminum alloy 6063 T-6	19,000	19,000	12,000
Duralumin	15,000	15,000	10,000
Cast iron	3,000	15,000	3,000
Yellow pine, air dry	—	1,200*	150†
White oak, air dry	—	1,000*	200†
Douglas fir, air dry	—	1,200*	150†
Spruce, air dry	—	700*	100†

*Parallel to grain.
†Shear, with the grain.

to be encountered. In each case, a reduction of the allowable working unit stress is required.

Values of allowable working unit stress for various materials under static load and normal working conditions are given in Table 3. This table should be committed to memory.

Illustrative Problem 2. The clevis joint shown in Fig. 3-14 has a capacity load of 6 tons. (a) If the material has a yield point in tension of 40,000 psi and 0.6 of that amount in shear, what are the factors of safety, based upon the yield points of the tensile stress in the clevis on section *A-A* and the shear stress in the pin? (b) If the allowable working unit stresses were increased to 14,000 psi in tension and 9000 psi in shear, what percentage of material would be saved? (c) Determine the factors of safety of the allowable working unit stresses given in (b) with the ultimate tensile unit stress of 66,000 psi and the ultimate shearing unit stress of 40,000 psi.

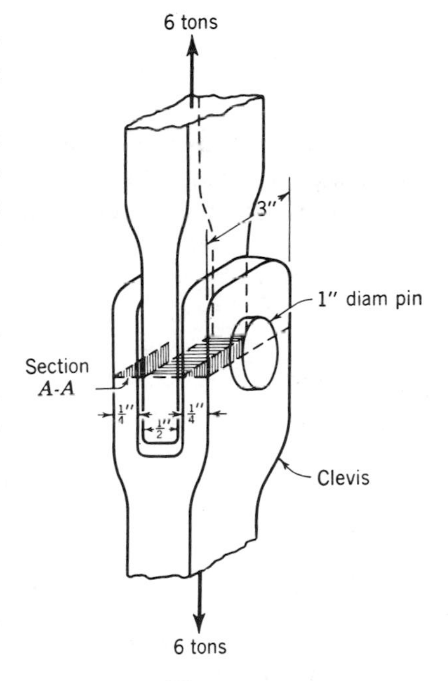

Figure 3-14

SOLUTION. (a) Factor of safety on section *A-A* of clevis:

$$\text{tensile unit stress used in design} = \frac{P}{A} = \frac{12{,}000}{2 \times (3 - 1) \times \frac{1}{4}} = 12{,}000 \text{ psi}$$

$$\text{factor of safety} = \frac{40{,}000}{12{,}000} = 3.33$$

Factor of safety on pin:

$$\text{shearing unit stress used in design} = \frac{P_s}{A_s} = \frac{12{,}000}{2 \times \pi 1^2/4} = 7640 \text{ psi}$$

$$\text{factor of safety} = \frac{24{,}000}{7640} = 3.14$$

(b) The percentage of material saved on section *A-A* would equal

$$\frac{\text{present cross section} - \text{new cross section}}{\text{present cross section}}$$

$$= \frac{2 \times \frac{1}{4} \times (3-1) - 12{,}000/14{,}000}{2 \times \frac{1}{4} \times (3-1)} = \frac{1 - 0.833}{1} \times 100$$

$$= 16.7 \text{ per cent}$$

The percentage of material saved on the pin would be equal to the reduction of the cross-sectional area:

$$\frac{2 \times 0.7854 - 12{,}000/9000}{2 \times 0.7854} = \frac{1.571 - 1.333}{1.571} \times 100 = 15.1 \text{ per cent}$$

Further computation would reveal that the amount of material used in the design of tension and shear members is inversely proportional to the allowable working unit stress. In critical periods of history, when shortages of steel have been keenly felt, the allowable working stress in steel has been increased up to 50 per cent, resulting in a substantial saving of this precious metal.

(c) The new factors of safety with respect to the given ultimate unit stresses are:

For tension,

$$\frac{66{,}000}{14{,}000} = 4.71$$

For shear,

$$\frac{40{,}000}{9000} = 4.44$$

These values are considerably higher than those used for steel under conditions of steady load and normal temperatures (factor of safety = 3 to 4). However, should impact, fatigue, high temperatures, or combinations of two or three of these items be involved in the design of a member, the factor of safety could easily reach the values obtained above.

PROBLEMS

3-9. The allowable soil-bearing pressure at the site of a contemplated foundation is 4 tons/ft². What size of square footing would be required to resist a load of 372,000 lb? The weight of the footing is 20,000 lb. *Answer:* 7 by 7 ft.

3-10. A certain building code requires that an allowable stress of 22,000 psi be used in the design of a certain member. If the material of which this member is made has an ultimate strength of 70,000 psi and a yield-point strength of 38,000 psi, what will be the factor of safety (a) based on the ultimate strength, (b) based on the yield-point strength?

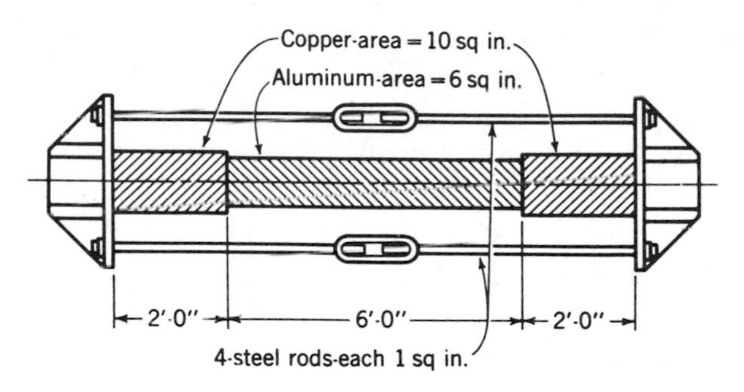

Figure 3-15

3-11. In order to stress the aluminum bar shown in Fig. 3-15 to its maximum allowable value 15,000 psi, how much will the four steel cables have to be shortened? $E_{cu.} = 17 \times 10^6$ psi, $E_{al.} = 10 \times 10^6$ psi, and $E_{st.} = 30 \times 10^6$ psi. *Answer:* 0.223 in.

3-12. A pulley 20 in. in diameter (Fig. 3-16) is fastened to a 2-in. shaft with a $\frac{1}{4}$- by $\frac{1}{2}$-in. key, 4 in. long. While operating at constant speed, the belt driving the pulley has opposite tensions of 300 and 200 lb. What factor of safety is obtained in the key if the ultimate shearing unit stress is 30,000 psi?

3-13. A roll of paper weighing 6000 lb has passed through its center a 2-in. steel shaft supported by two angle frames, one on each end (Fig. 3-17). (a) If the shearing ultimate strength of the steel is 30,000 psi, what is the factor of safety of the shaft based on shear? (b) The end supports rest on timber bearing blocks. If the contact surface between one metal frame and its bearing block is 20 in.^{2},

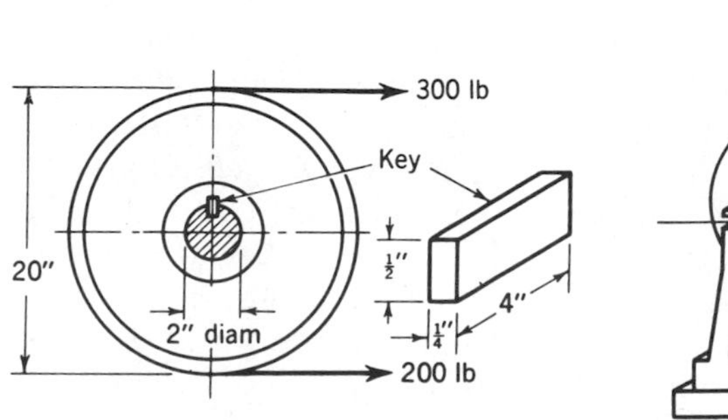

Figure 3-16

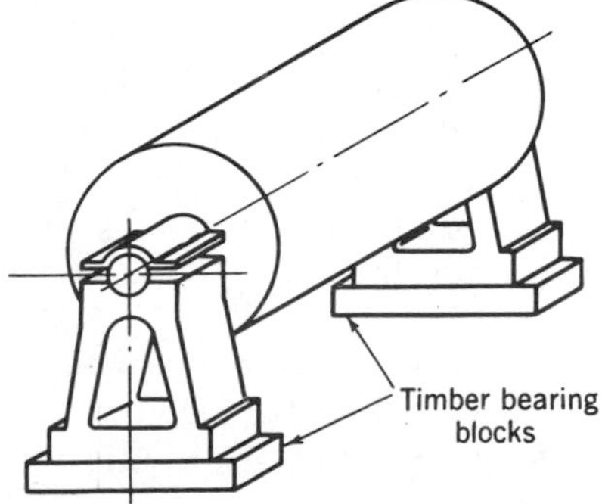

Figure 3-17

what factor of safety is obtained if the timber has a bearing ultimate strength of 750 psi? *Answers:* (a) 31.4; (b) 5.0.

3-14. A hotel canopy is 25 by 20 ft, the 25-ft dimension being parallel to the wall. The canopy is held by two round tie bars fastened to the canopy 12 ft from the wall and making an angle of 45° with the horizontal. The maximum total load is 100 lb/ft^2. What diameter tie bar is required at an allowable tensile stress of 10,000 psi, assuming 20 percent of the section to be lost because of rivet holes? Consider the canopy to be free to rotate at the wall. *Answer:* 2.16 in.

3-15. What cross-sectional area will be required for member U_5L_6 of the truss shown in Fig. 1-11 if its maximum allowable unit stress is 20,000 psi? *Answer:* 7.07 in.^{2}

3-16. The member *FC* of the truss shown in Fig. 1-10 is fabricated from A-36 steel and has a cross-sectional area of 2 in.^{2}. (a) What is its unit stress? (b) What is its factor of safety based upon the yield stress?

3-13 Plastic Yielding at Stress Concentrations

In the analysis of the axially loaded tensile member made of rubber bands (§2-18), it was shown that the insertion of a hole produced a non-uniform stress distribution on the transverse section passing through its center (Fig. 3-18a). More specifically, the stress at the edge of the hole was substantially higher than the average normal stress on the section. Forming the basis of this analysis was the assumption that stress and strain were proportional; in other words, that all deformations were perfectly elastic.

Let us now assume that the material from which the member is made has the stress-strain diagram shown in Fig. 3-18b and that the stress at the edge of the hole has just attained the yield point. If the load is now removed, the stresses would reduce to zero, elastically restoring the hole to its original shape. If, on the other hand, the load is increased beyond the previous elastic-limit point, the fibers near the edge of the hole will find it impossible to increase their resistance and will try to yield instead in accordance with line *YF* on Fig. 3-18b. Because of the restraint offered by other fibers on the section, such as *B*, *C*, and *D*, still stressed below the yield point and deformed elastically, yielding at the edge of the hole is restrained, attaining a point *R* instead of *F* (Fig. 3-18c). Were the load now removed, all fibers on the section would try to resume their original position. Fiber *A*, however, stretched through a portion of the yield range, has undergone a permanent set *OR′* and cannot return completely. Fiber *B*, still elastic, tries to assume its normal zero-strain position but, because adjacent fiber *A* has a permanent set, finds it impossible to do so. Fiber *B* is therefore still under tension, compressing fiber *A*. The approximate stress distribution curve after this last removal of load is shown in Fig. 3-18d.

Subsequent static tensile loadings of the same member would find its resistance improved rather than lessened. This apparent inconsistency is

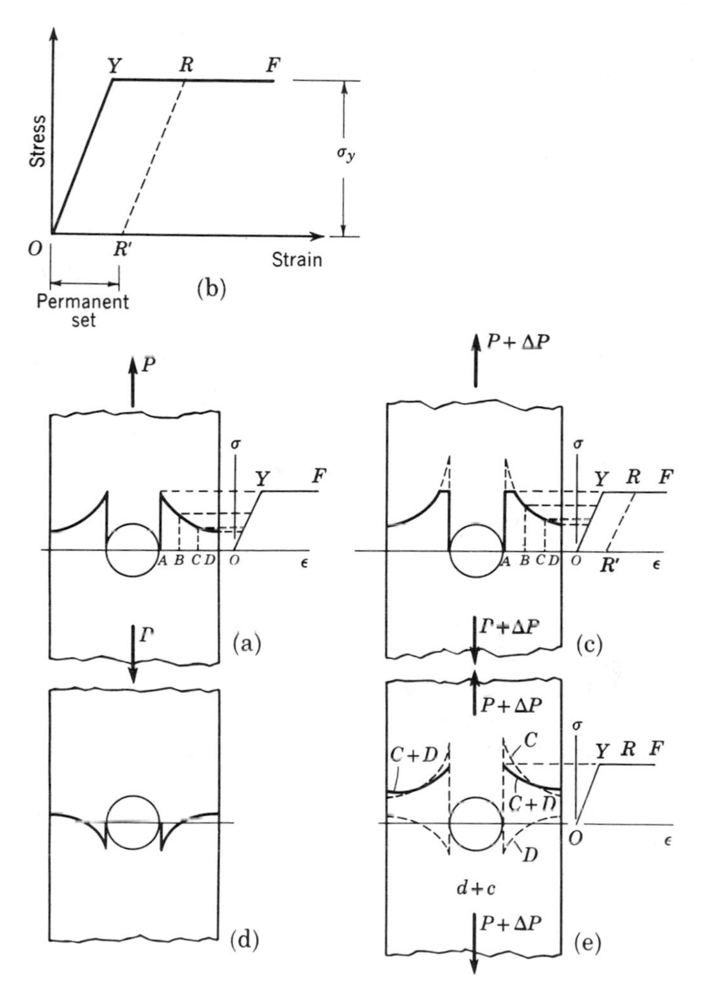

Figure 3-18

substantiated when stress diagrams 3-18c and 3-18d are superimposed on one another, revealing a decrease of stress concentration and perfect elastic action of all fibers (Fig. 3-18e). Compression loadings, however, are therefore more critical while repeated tension and compression loads reveal the inelastic stretch to have weakened the material considerably.

3-14 Plastic Yield in Statically Indeterminate Combinations of Two-Force Members

Suppose we are required to find the unit stress in each of the resisting members of the steel frame shown in Fig. 3-19a due to the load of 70,000 lb acting vertically downward at D. The stress-strain diagram of the steel in tension is that shown in Fig. 3-19b.

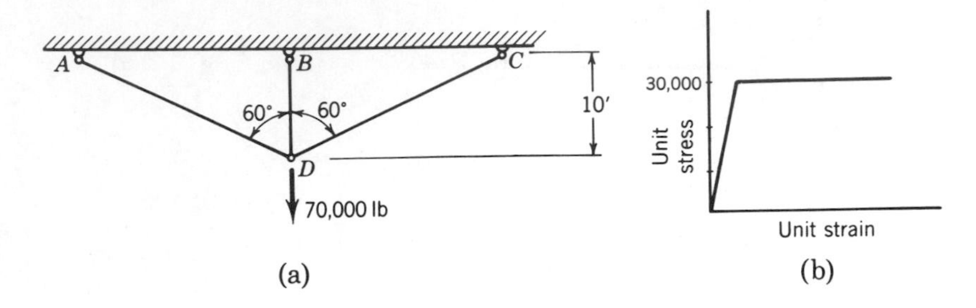

Figure 3-19

This is the same frame previously considered in §2-17. Under the action of 20,000 lb the stress which had been calculated for members AD and CD was 3330 psi, and that for member BD was 13,330 psi. Because the load is now increased to 70,000 lb, the stresses should be $3\frac{1}{2}$ times higher—or 11,670 and 46,700 psi, respectively—assuming that the members have retained their elasticity. The calculated stress in member BD, however, exceeds the yield-point stress of the material, viz., 30,000 psi. This destroys the correctness of the elastic analysis.

When member BD attains this unit stress of 30,000 psi and loses its elasticity, it deforms considerably while maintaining the same stress. This deformation is sufficiently great to continue the 30,000-lb force in BD, while members AD and CD contribute their elastic resistance to the total load of 70,000 lb.

The equation of the vertical components of force at point D is therefore

$$\sum V = (AD)_v + (CD)_v + 30{,}000 = 70{,}000$$

But

$$(AD)_v = (CD)_v$$

$$\therefore (AD)_v = (CD)_v = 20{,}000 \text{ lb}$$

Hence

$$AD = CD = 40{,}000 \text{ lb}$$

The unit stress in each of these members is therefore

$$\sigma_{AD} = \sigma_{CD} = \frac{40{,}000}{2} = 20{,}000 \text{ psi}$$

PROBLEMS

3-17. The bar ABC shown in Fig. 3-20a is attached to its supports and free of stress before the application of an axial force $F = 128^K$ at midheight. The cross section of the bar is a 2- by 2-in. square. If the material of the bar is linearly elastic in compression and follows the σ-ε curve in tension, as shown in Fig. 3-20b, determine the stress in AB and BC and also the displacement of point B. *Answers:* $\sigma_{AB} = 11{,}100$ psi, $\sigma_{BC} = 20{,}900$ psi.

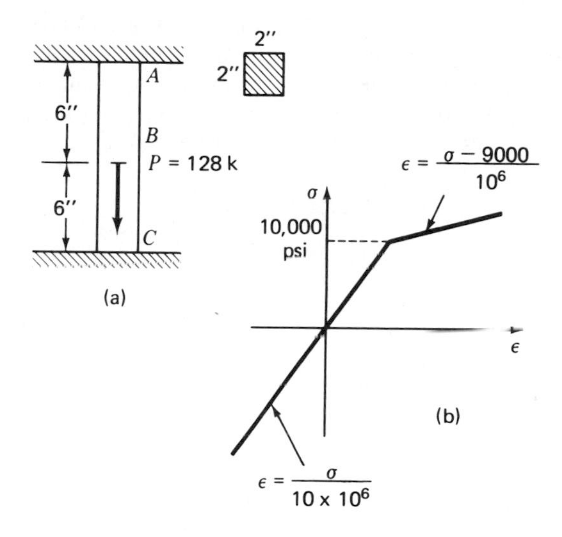

Figure 3-20

3-18. The rigid bar BCF shown in Fig. 3-21a is supported at B by a hinge and at C and F by two identical bars CD and FG having a cross-sectional area equal to A. The material of CD and FG is linearly elastic, ideally plastic with the stress-strain diagram shown in Fig. 3-21b. A force P acts at point F. In terms of L, A, E, and σ_0 determine (a) The value of the load $P = P_E$ at which yielding begins. (b) The displacement of point F when $P = P_E$. (c) The limit (ultimate) value of the load P.

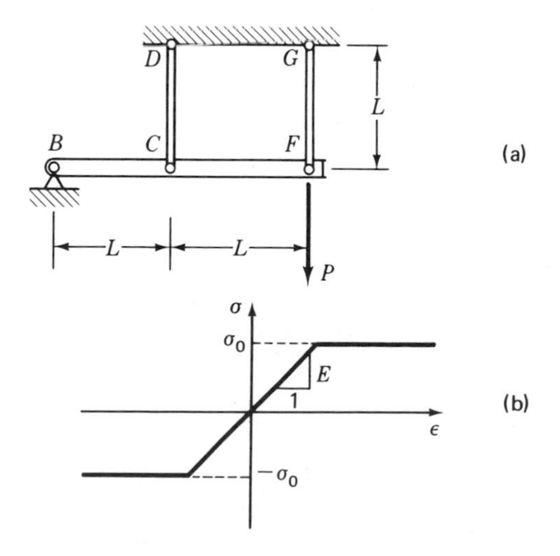

Figure 3-21

3-19. The truss shown in Fig. 3-22 is fabricated from structural steel members, each having a cross section of 2 in.^{2}. If the steel has an idealized stress-strain curve similar to that shown in Fig. 3-19b, (a) Calculate the stresses in the members due to the load P of 40,000 lb. (b) Calculate the stresses in the members due to the load P of 80,000 lb. (c) Calculate its limiting load before failure.

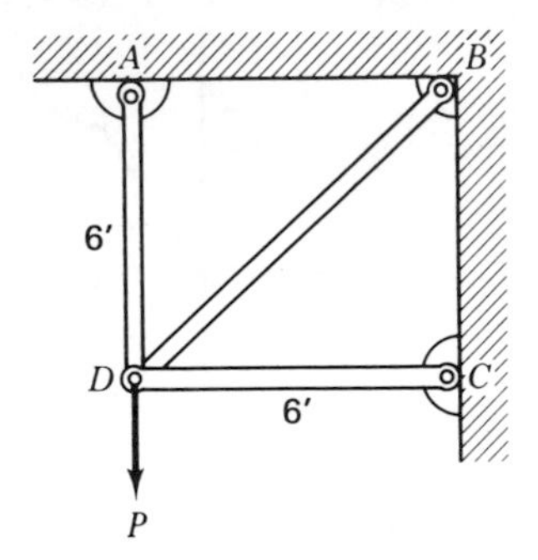

Figure 3-22

3-20. A steel rod (Fig. 3-23a) having an area $A_s = 0.1$ in.^{2} is inserted in an aluminum sleeve having an area $A_{al} = 0.2$ in.^{2}, and both members are welded to rigid plates at both ends. The (idealized) mechanical properties of both materials are given in Fig. 3-23b. Determine the temperature rise ΔT (the same in both materials) at which yielding will first occur. $\alpha_{st.} = 6.5 \times 10^{-6}$ in./in./°F, $\alpha_{al.} = 13.0 \times 10^{-6}$ in./in./°F.

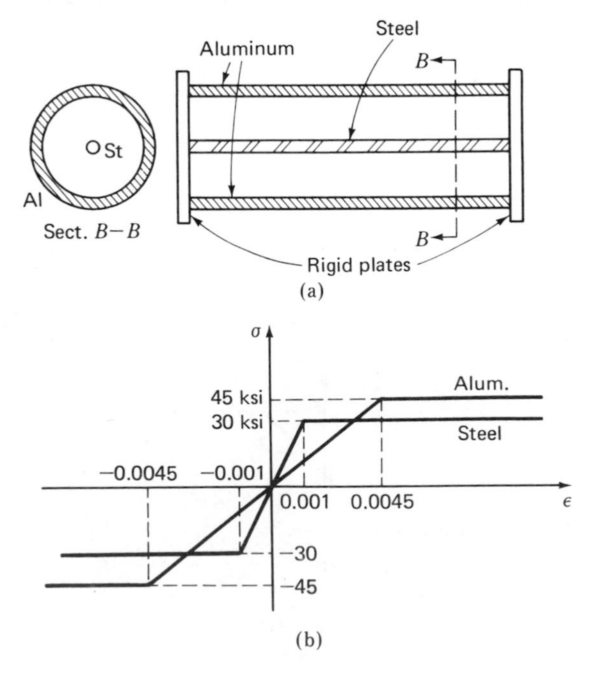

Figure 3-23

4

Stresses in Thin-Walled Cylinders and Spheres, Fabricated Joint Design

4-1 Stresses in Thin-Walled Cylinders and Spheres—General Discussion

The formulas of stresses in the walls of thin-walled cylinders and spheres subjected to uniform internal pressure are based upon the assumption of a uniform stress distribution over the thickness of the wall (Fig. 4-1). The close agreement of this assumption with the actual stress distribution in such vessels permits the application of these formulas to the design of all common types of

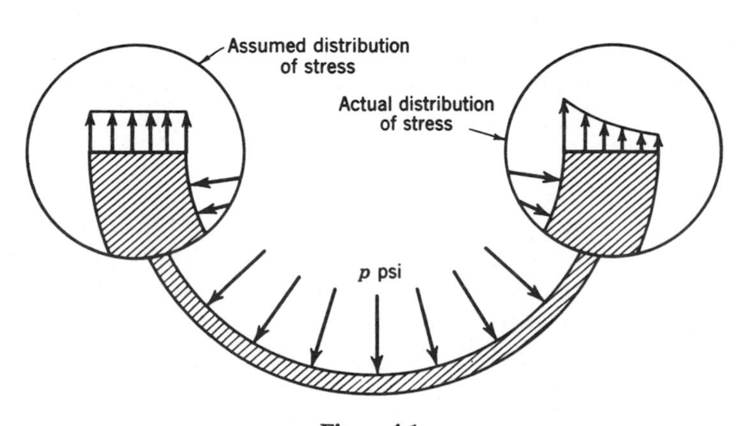

Figure 4-1

cylinders and spheres, such as storage tanks, water tanks, steam boilers, and so forth.

Because the modifier *thin-walled* is a relative term, care must be taken not to use the following stress formulas for all cylinders and spheres. The effect of wall thickness on stress variation will be discussed in §4-3.

4-2 Derivation of Formula for Stress on Longitudinal Section of Thin-Walled Cylinders Subjected to Uniform Interior Pressure

The longitudinal section on which the maximum stress will occur is found by passing a plane through the cylinder that will expose a minimum cross section of wall and provide a maximum thrust. Such a combination is obtained when the cutting plane passes through the longitudinal axis (see Fig. 4-2b).

In Fig. 4-2a is shown a section of a thin-walled cylinder subjected to a uniform interior pressure. An end view of one half of this section is shown in Fig. 4-2b and will serve as the free body for this discussion.

Since the internal pressure acting on the interior of our circular-sectioned free body is uniformly distributed and symmetrical about the horizontal center line, it produces a perfect balance in the vertical direction. The resultant must, consequently, be horizontal and act to the left of the figure along the center line.

To balance this resultant internal pressure, resisting forces must act on the two exposed vertical surfaces. Since these two forces are equidistant from the resultant, they must be equal and horizontal. The resisting forces found for similar free bodies of the same cylinder cut by axial sectioning planes at other angles will also be tangent to the circumference and equal to one another. In fact, Fig. 4-2d, a free body of a very small portion of the cylinder wall, shows that the effect of the internal pressure is to change the direction of the resisting force in the shell so that it is tangent at all points on the circumference. Assuming these resisting forces to act at the centroids, they will develop uniformly distributed tensile unit stresses over the wall sections. If the average unit stress thus developed is σ_t, each resisting force will be equal to

$$P = \sigma_t \times t \times l \tag{4.1}$$

To ascertain the magnitude of the force created by the internal pressure p psi, let us refer to Figs. 4-2a and b. Directing our attention more specifically to dA, one of the many very small areas of ds width and length l shown on Fig. 4-2a, we may write the following:

The force on dA equals

$$p \times dA = p \times l \times ds$$

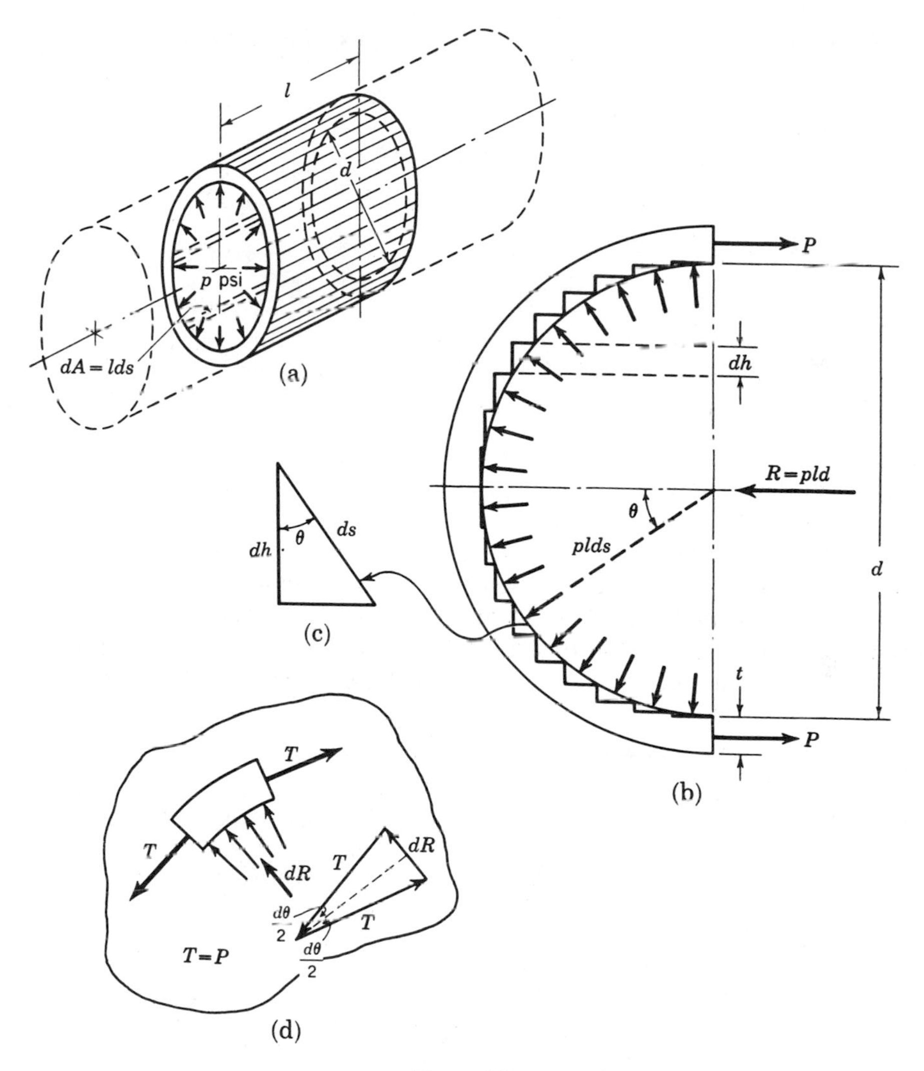

Figure 4-2

The horizontal component of force acting on dA equals

$$p \times l \times ds \times \cos\theta$$

To obtain the total horizontal effect R, we must add all the horizontal components acting on the various dA's.

$$R = \int pl \cos\theta \, ds$$

But since p and l are constant quantities appearing in every horizontal component, and the value of $\cos\theta \, ds$ is equal to the vertical projection of the arc distance ds or dh (Fig. 4-2c), we may write the expression for R as

$$R = pl \int dh = pld \tag{4.2}$$

Since the horizontal forces must balance,

$$P + P = R, \qquad 2P = pld$$

But $P = \sigma_t tl$; therefore,

$$t = \frac{pd}{2\sigma_t}, \quad \text{or} \quad \sigma_t = \frac{pd}{2t} \tag{4.3}$$

where σ_t = unit stress on longitudinal section of wall in psi, p = internal uniform fluid pressure in psi, d = inside diameter of cylinder in inches, and t = wall thickness in inches.

4-3 Effect of Cylinder Wall Thickness on Stress Variation

The graph shown in Fig. 4-3 reveals the variation of the maximum stress to the uniform stress assumed in the thin-cylinder formula as the thickness of the wall increases. To illustrate its use, let us determine the ratio of the maximum unit stress to the assumed uniform stress in a cylinder wall whose ratio of inside to outside radius is equal to 0.7. The value of the ordinate

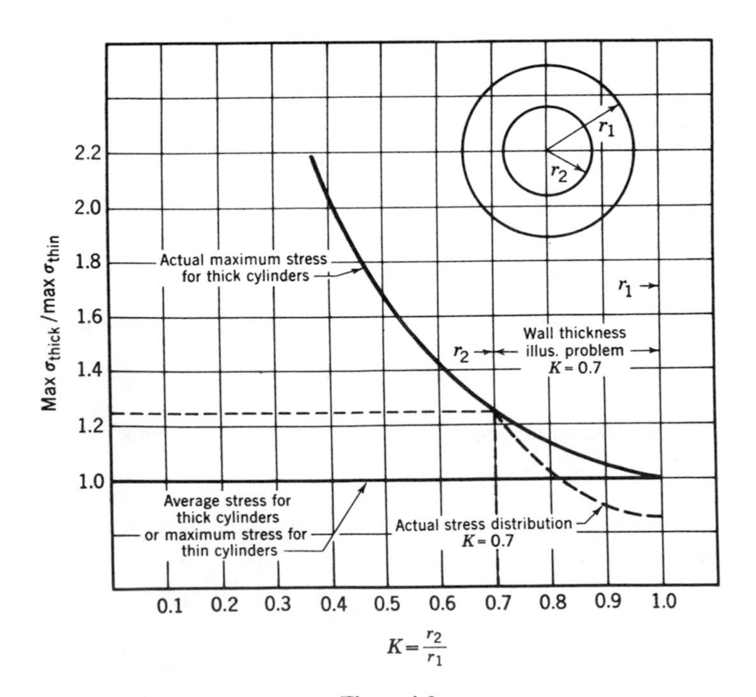

Figure 4-3

at the 0.7 abscissa is 1.25, indicating that this is the ratio of the maximum stress to the uniform stress assumed in the thin-cylinder formula. The dash line has been added to show the actual stress distribution over the cross section of the shell.

An analysis of this graph further reveals that when the wall thickness is 20 per cent of the internal radius, the maximum stress computed by exact theory is about 12 per cent greater than the uniform stress assumed in the thin-wall formula. Reducing the wall thickness to 10 per cent of the internal radius reduces the difference between the stresses to about 4 per cent.

The thin-cylinder formula, consequently, has a field of application as wide as that allowed by the permissible stress variation. If the ratio of the radii, r_1/r_2, becomes so large as to increase the stress variation above the allowable, the constant found in the graph of Fig. 4-3 when multiplied by the stress obtained from the formula will give the theoretical maximum stress.

4-4 Stress on Transverse Section of Cylinders and Spheres

Cylinders. In addition to the tendency of a cylinder to burst open along a longitudinal section when subjected to an internal pressure, there is also the tendency for the ends to be blown off. To combat this latter tendency, tensile stresses are developed along every transverse section of the cylinder. Significantly enough, the cylinder, owing to the pressure acting on each end plate, is in reality an *axially loaded tension member*. To determine these tensile stresses, therefore, we need only apply the now familiar $P = A\sigma$ formula.

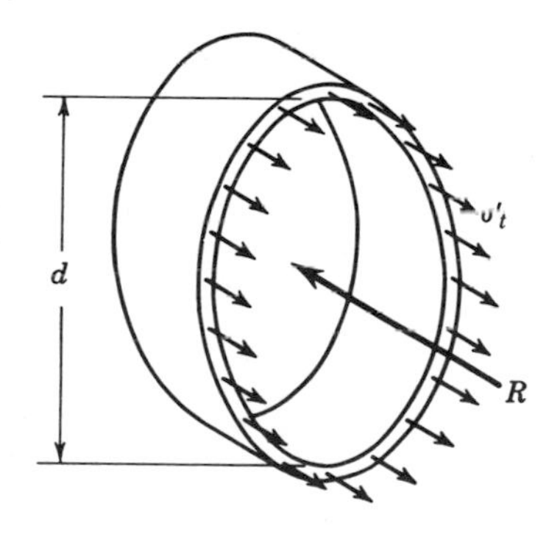

Figure 4-4

Derivation. In Fig. 4-4, a free body of the end portion of a cylinder is shown exposing a section perpendicular to the longitudinal axis. The resultant of the uniformly distributed internal pressure acting on the inside of the end bulkhead is an axial force equal to

$$R = \frac{\pi d^2}{4} \times p$$

Opposing this internal pressure, and of necessity equal to R, is the restraining force on the cross section of the cylinder wall. Equating both in the $P = A\sigma$ formula, there is obtained

$$\frac{\pi d^2}{4} \times p = (t \times \pi d)\sigma'_t$$

from which

$$\sigma'_t = \frac{pd}{4t} \tag{4.4}$$

A comparison of the stresses on the longitudinal and transverse sections reveals that the stress on the longitudinal section is just twice the value of the stress on the transverse section. This observation leads to an important conclusion in that *if joints are required in a cylinder, twice the amount of reinforcement in either rivets, or bolts, or weld material is required for a longitudinal joint as compared to that for a transverse joint.*

Spherical shells. The hemisphere shown in Fig. 4-5 is used as a free body in the determination of the maximum tensile stresses occurring in spherical

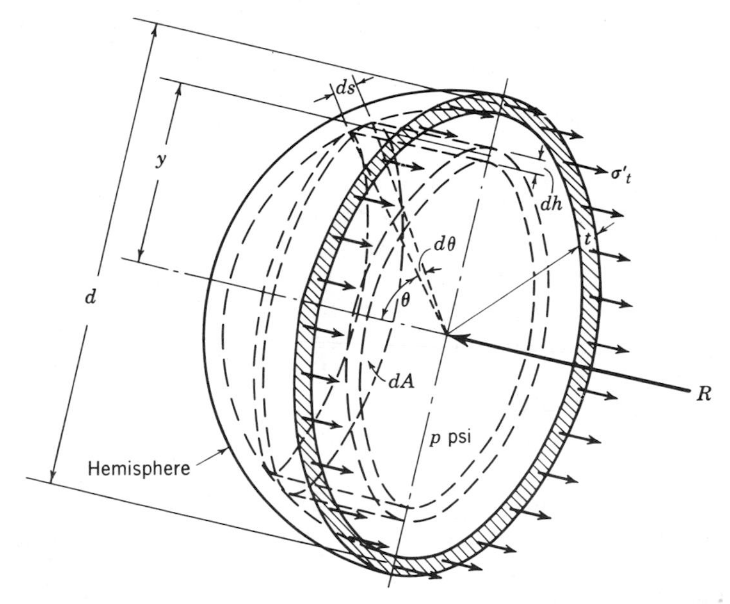

Figure 4-5

shells. The ratio of the internal force acting on this free body to the area of its metallic cross section, being greater than for any other possible free body, makes the stresses on the section a maximum. Because an infinite number of different pairs of hemispherical free bodies can be formed by bisecting a sphere, it follows that every point in the spherical shell is subjected to the same maximum tensile unit stress. Furthermore, since the number of great circles passing through any point on the sphere is unlimited, each point has impressed upon it equal maximum tensile stresses occurring in all tangential directions.

Derivation. From Pascal's law, we know that the internal pressure acts normal to the surface. The force acting on an elementary circular strip dA, shown in Fig. 4-5, is therefore

$$dF = p \times dA = p \times 2\pi y\, ds$$

From the symmetry of the hemisphere about the horizontal axis, we are assured that the components of forces perpendicular to the axis annul one another. It is only those components acting parallel to the axis that produce tensile stress on the exposed section.

The horizontal component of force acting on dA equals

$$dR = 2\pi y\, ds \times p \times \cos\theta$$

The sum of all such horizontal components is

$$R = \int 2\pi y\, ds \times p \times \cos\theta$$

But $ds \cos\theta$ or dh is equal to the vertical projection of the intercepted arc ds, making $2\pi y\, dh$ the vertical projection of area dA. The value of R may, therefore, be further simplified as

$$R = p \int 2\pi y\, dh$$

To project every possible dA area on the vertical cutting plane would completely cover the circular section, and the value of R may be written as

$$R = p \times \frac{\pi d^2}{4} \tag{4.5}$$

Note that this resultant force is equal to that already found for a flat-ended cylinder. Thus, as a further development of these studies, we may conclude that *the resultant internal pressure on any symmetrical enclosure is equal to the pressure contact area of that sectioning plane oriented perpendicular to the axis of symmetry and used to separate it as a free body, multiplied by the internal pressure.*

Referring again to the sphere (Fig. 4-5), it is shown that the resultant of the internal pressure must be balanced by the internal tensile forces created on the exposed circumferential section. Writing the equation for horizontal forces, there is obtained

$$\pi d \times t \times \sigma_t' = \frac{\pi d^2}{4} \times p$$

from which

$$\sigma_t' = \frac{pd}{4t} \tag{4.6}$$

Thus, the tensile unit stress developed on a diametral section of a sphere is exactly the same as that developed on the transverse section of a cylinder of equal diameter.

Illustrative Problem 1. A 50-ft-diameter sphere is to be used to contain helium gas under 100-psi pressure. If the joints in the sphere reduce the carrying capacity of the plates by 80 per cent, what thickness of plate will be required? Allowable tensile unit stress is 20,000 psi.

SOLUTION

$$20{,}000 = \frac{50 \times 12 \times 100}{4t \times \frac{4}{5}}, \quad t = \frac{5000 \times 12 \times 5}{4 \times 20{,}000 \times 4}, \quad t = \frac{15}{16}\text{ in.}$$

4-5 Weight of Fluid

In all our discussions on stresses in cylinders and spheres, no consideration was taken of the weight of the fluid producing the internal pressure. This factor may become very important, especially if the cylinder is lying horizontally and carries a heavy fluid. However, if the fluid creating the pressure is light, such as steam or air, its weight may be disregarded because its effect is of little or no consequence.

PROBLEMS

4-1. If the boiler in Fig. 4-2 is 60 in. in diameter and 8 ft long, and is subjected to an internal pressure of 100 psi, determine (a) the total stress acting on the two exposed longitudinal cross sections of the free body ($l = 4$ in.), and (b) the unit stress acting on these sections if their thickness is $\frac{1}{2}$ in.

4-2. A water supply pipe is 6 ft in diameter and built of longitudinal wood staves held by hoops of $\frac{7}{8}$-in. round steel rod. What spacing of hoops is required when the pressure is 50 psi? Allowable unit stress in steel rod is 20,000 psi. *Answer:* 6.67 in.

4-3. A wood stave pipe, 4 ft in diameter, is wrapped spirally with a flat steel band $\frac{1}{2}$ in. wide by $\frac{1}{8}$ in. thick. If the maximum allowable unit tensile stress in the steel band is 50,000 psi and the maximum allowable unit crushing pressure on the wood under the band is 200 psi, what actual unit tensile stress may safely be used in the steel band? *Answer:* 38,400 psi.

4-4. If over the end of a 6-in.-diameter steel pipe there is placed a rubber diaphragm which expands to a radius of 2 ft upon the application of 20-psi pressure, what is the tensile stress in the rubber if its expanded thickness is equal to 0.1 in.? What is the tensile unit stress in the rubber if it is deflated to a 1-ft radius with one half the above pressure, and if the thickness increases to 0.12 in.? What is the resultant pressure acting on the diaphragm in each case? *Answers:* 2400 psi, 500 psi, 566 lb, 283 lb.

4-5. A hydroelectric plant has a welded surge tank connected to its penstocks that is 10 ft in diameter and 50 ft high. What is the maximum tensile unit stress in the bottom vertical plate if the plate thickness is $\frac{1}{4}$ in. and the tank is completely filled? Water weighs 62.5 lb/ft^3.

4-6. A spherical shell 40 ft in diameter is subjected to an internal pressure of 3 atmospheres. Find the tensile stress in the shell if its thickness is equal to $\frac{1}{2}$ in. One atmosphere is 14.7 psi. *Answer:* 10,600 psi.

4-7. A steam boiler is to have a 12-in.-diameter hand hole placed on one end. If the maximum internal pressure is to be 200 psi, how many $\frac{5}{8}$-in. bolts will be required to keep the cover plate in position? Root diameter of a $\frac{5}{8}$-in. bolt is $\frac{1}{2}$ in. Allowable tensile unit stress is 10,000 psi. *Answer:* 12 bolts.

4-8. A bronze hoop $\frac{1}{2}$ in. thick and with an inside diameter of 25 in. fits snugly over a steel hoop $\frac{3}{4}$ in. thick when the two hoops are at an initial temperature of 370°F. Both hoops are of width 4 in. $\alpha_{st} = 0.0000065$, $\alpha_{br} = 0.0000100$. Compute the contact pressure between the hoops when the temperature drops to 70°F. *Answer:* 398 psi.

4-9. A 90° elbow b by h ft in cross section (Fig. 4-6) is connected to the end of a rectangular conduit of the same cross section. If a temporary wall is to be placed across the open end of the elbow, (a) what is the moment of the forces acting on the interior surfaces of the elbow about the vertical center line passing through the center of the connected end? (b) Determine the normal force acting on the vertical radial plane making 45° with the entering cross section. (c) Draw free-body diagrams of portions of the vertical corners at the temporary wall. (d)

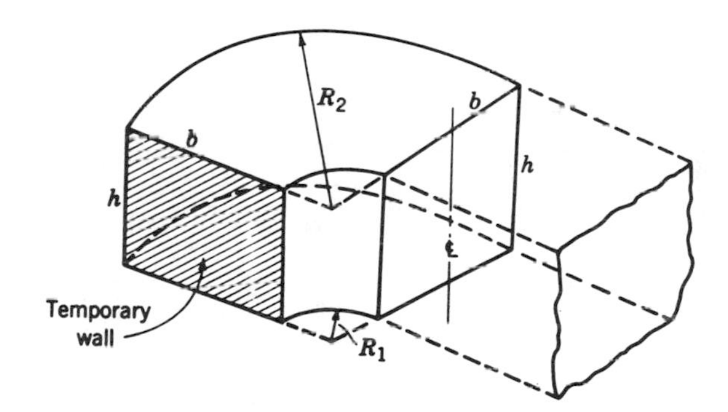

Figure 4-6

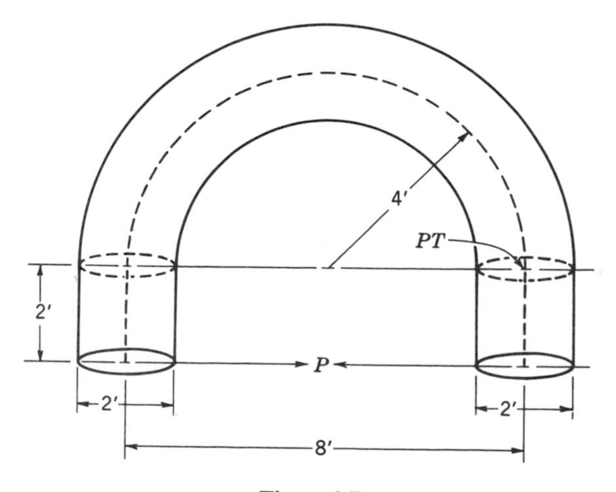

Figure 4-7

If the wall of the conduit is thin, what tendency would be observed at the corners of the temporary wall? *Answers:* (a) 0, (b) *pbh*.

4-10. A U-shaped gas container (Fig. 4-7) is to be subjected to an internal pressure of 200 psi. What force P is required to prevent the legs of the container from moving apart? Use only internal pressure forces.

4-6 Types of Joints—Definitions

Structural joints are joined together either by the use of fasteners, such as bolts or rivets*, or welds. Two main types of joint arrangement are used: the lap joint (Fig. 4-8a) and the butt joint (Fig. 4-8b).

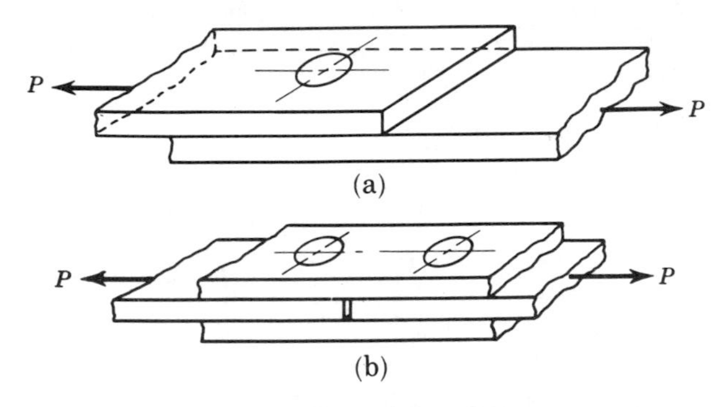

Figure 4-8

Lap and butt joints are identified by their complexity of structure. If, for instance, a lap joint has but one row of bolts, it is called a *single-bolted lap joint* (Fig. 4-8a). If a lap joint has two rows of bolts, it is called a *double-bolted lap joint* (Fig. 4-9); if it has four rows, a *quadruple-bolted lap joint;* and so forth. Butt joints are similarly identified by the number of rows located

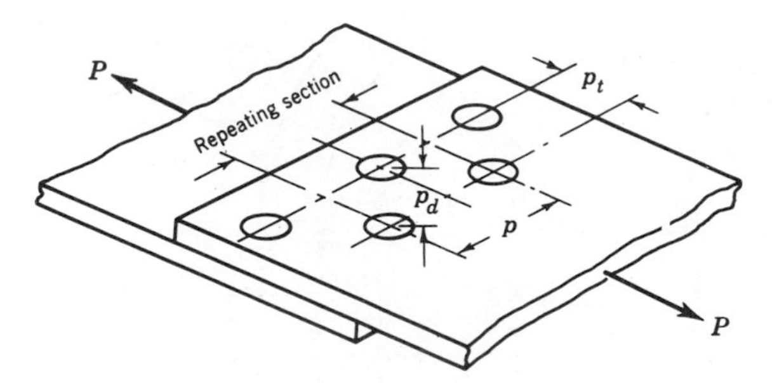

Figure 4-9

*Bolted joints have almost superseded the use of riveted joints. Their design is almost identical to that of riveted joints.

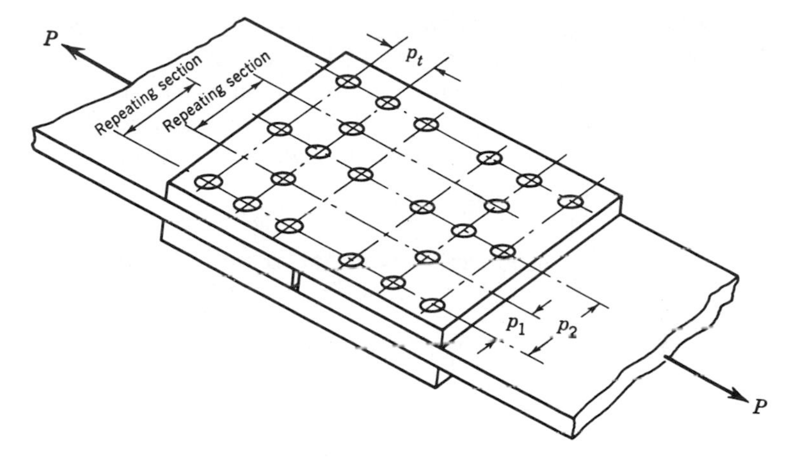

Figure 4-10

on one side of the joint. In Fig. 4-8b is shown a single-bolted butt joint. A triple-bolted butt joint is shown in Fig. 4-10. In some instances, such as shown in Fig. 4-9, adjoining rows of bolts may be staggered.

Pitch. The distance between two adjacent fasteners on any row is called the *pitch.* It is not necessary that the pitch be the same for every row, as is shown in Fig. 4-10.

The distance between adjacent rows is called the *transverse pitch* p_t. If the fasteners are staggered, the diagonal distance between adjacent fasteners on adjoining rows is called the *diagonal pitch* p_d.

Repeating section. Important in the analysis of every long bolted or riveted joint is the selection of the repeating section (Figs. 4-9 and 4-10). This section is the smallest symmetrical grouping of fasteners that repeats itself along the length of a joint. It is customary to use the repeating section rather than the entire length of a joint when determining its strength characteristics.

4-7 Assumptions in the Design of Bolted or Riveted Joints

Load distribution. One of the major assumptions made in the design of bolted or riveted structural joints is that the load transfer of the fasteners is proportioned directly on the basis of their resisting cross-sectional areas. In other words, each resisting cross section of equal size transfers the same amount of load. Based on this assumption, the resultant of the resisting forces passes through the centroid of the rivet cross sections. If the applied load also passes through the centroid of the rivets, no eccentricity is experienced.

Unfortunately, this assumption, which would provide a uniform distribution of stress, is an approximation. However, it should not be construed as meaning to eliminate all rigor from the usual stress calculations, because the approximation approaches the assumed condition after the rivets have been loaded and the stresses at the critical points have been relieved by exceeding their yield points. The present method of calculating rivet stresses is a simple and practical one which, in the light of past experience, might better be continued than modified.

Friction. Although a considerable amount of friction can be developed on the interfaces of a bolted or riveted joint, its magnitude generally decreases with repeated loadings.

4-8 Modes of Failure

To design a bolted or riveted joint properly, one must anticipate and control the maximum stresses developed at critical sections. Since it is to be expected that failure will occur at one of these critical sections, any knowledge as to where these sections may be located provides information for a successful design. Several common critical stress conditions developed in all joints, each of which is capable of producing failure, form the basis of our study of the double-bolted lap joint shown in Fig. 4-11.

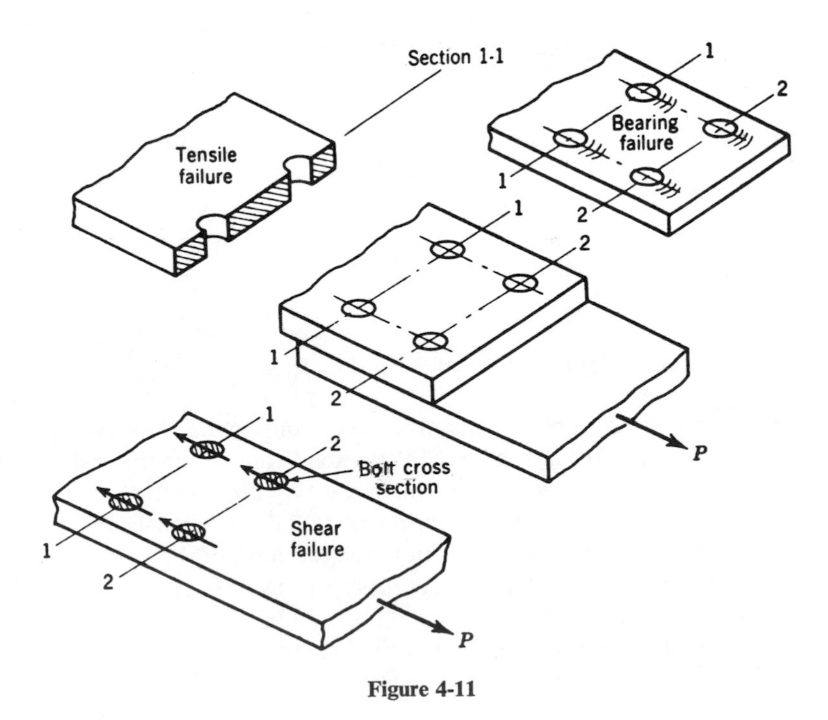

Figure 4-11

Three basic types of failure present themselves in our study of this joint:

1. The net section through a row of bolts may fail in *tension.*
2. The bolts may fail in *shear* along the plane of sliding.
3. The material immediately in back of the joint may wrinkle because of an intensive compressive unit stress and cause failure of the joint in *bearing*.

Other methods of joint failure may be possible but are not likely to occur, owing to imposed design specifications.

In larger bolted or riveted joints, failure may be produced by the simultaneous action of two or more basic modes of failure acting at adjoining critical sections. In Fig. 4-12, it is possible to obtain failure by a simultaneous

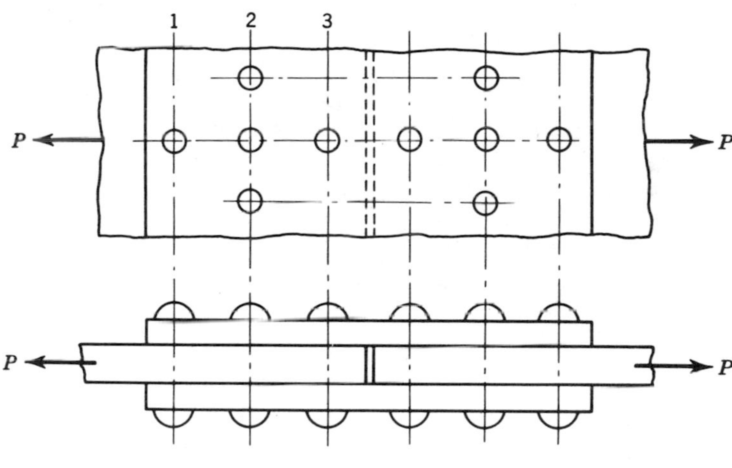

Figure 4-12

shear of the fastener on line 1-1 and tension of the main plate on line 2-2. To ensure a safe design, other combinations on this and other joints should be analyzed carefully for every possibility of failure.

4-9 Efficiency of a Bolted or Riveted Joint

The efficiency of a bolted or riveted joint is defined as its effectiveness in transferring load across the joint. It is measured by the percentage of the allowable load that may be applied to the solid plate. Thus

$$\text{per cent efficiency} = \frac{\text{maximum allowable load on joint}}{\text{allowable load on main plate with no joint}} \times 100 \tag{4.7}$$

In some instances the per cent efficiency is obtained on the basis of ultimate loads. Generally, little difference is realized in changing from one method to the other.

4-10 Design and Investigation of Bolted and Riveted Joints

Illustrative Problem 2. In the double-bolted butt joint shown in Fig. 4-13, determine the shearing unit stress in the bolts, the bearing unit stress between

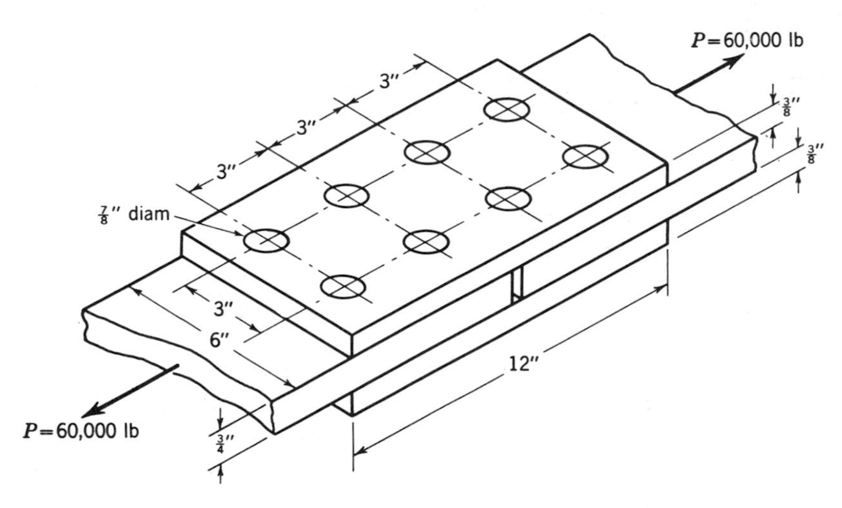

Figure 4-13

the bolts and the plate, and the maximum tensile unit stress on the net sections passing through the rows of bolts when the applied load is 60,000 lb. Calculate also the efficiency of the joint.

SOLUTION

Shearing unit stress. As discussed in §4-7, bolted joints are designed on the basis of equal distribution of stress to all resisting bolt cross sections. The 60,000-lb load placed on the joint now under consideration must, therefore, be distributed equally to the four bolts on each side of the joint.

Consider the free body of each half of the joint (Fig. 4-14). From this

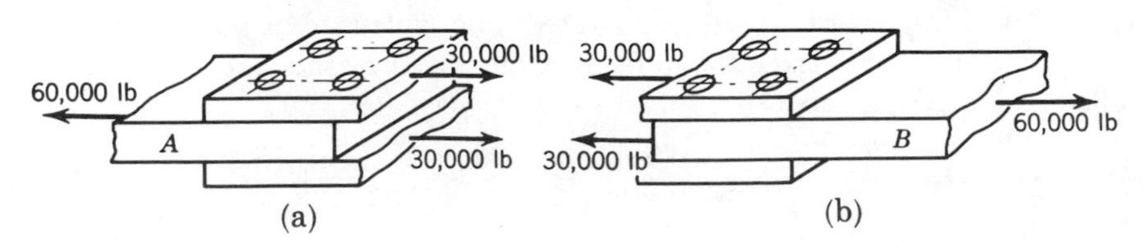

Figure 4-14

figure, it is evident that free body A can be balanced only by placing a load of 30,000 lb on each strap plate. Likewise, free body B must have the same forces on its two strap plates in order to be perfectly balanced. Consequently, only those bolts on one side of the joint need be involved in any investigation of stresses, since they restrain the entire load and produce the same stress condition as that found on the opposite side.

Upon application of the 60,000-lb load, the main plate in each free body described above would tend to slip out from between the two strap plates. Of course, the bolts restrain the plates from moving, but the tendency to slip is still there. Figure 4-15 shows the position of the various parts of free

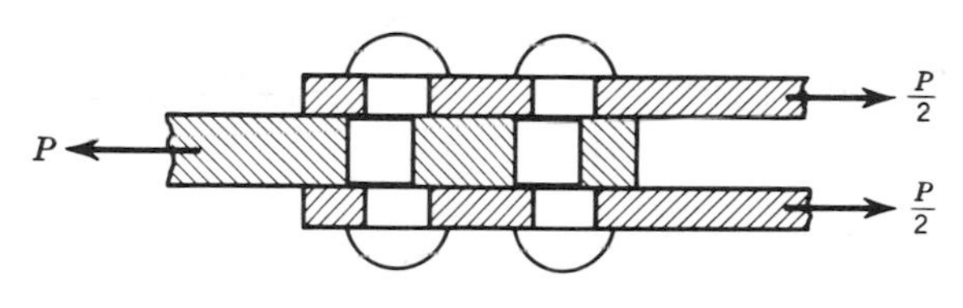

Figure 4-15

body A if the rivets should fail. Note that failure would occur on two sections of each bolt simultaneously. These bolts are, therefore, in *double shear.* Since the load passes through the center of gravity of the bolt group, the shearing unit stress developed on the cross sections of these bolts is

$$\tau = \frac{P}{A_s} = \frac{P}{n \times \pi d^2/4} = \frac{60{,}000}{2 \times 4 \times 0.60} = 12{,}000 \text{ psi} \tag{4.8}$$

Bearing unit stress. As previously described, a failure in bearing implies a compression or wrinkling failure in either the main plate or butt straps of a joint. A bolt producing a bearing unit stress is assumed to bear upon one half the bolt hole. It is further assumed that the bolt imposes on that half of the bolt hole with which it theoretically comes in contact a uniformly distributed compression stress σ_b. By a method analogous to that used in finding the resultant pressure on thin cylinders, we can obtain the resultant force of this bearing by projecting the induced compressive stresses onto the vertical diametrical plane $ABCD$ (Fig. 4-16). This force is equal to

$$P = A_{ABCD} \times \sigma_b = A_b\sigma_b = \sigma_b td \tag{4.9}$$

If several bolts are effective in resisting the bearing stress, as is the case in our present problem, their combined projected areas must be included in the above formula in the following manner:

$$P = \sum A_b\sigma_b = \sigma_b \sum td \quad \text{or} \quad \sigma_b = \frac{P}{\sum td} \tag{4.10}$$

Figure 4-17 indicates the areas that are subjected to bearing unit stress under the action of force P. Since the strap plates as well as the main plate are involved, the proper evaluation of t in the above formula is essential.

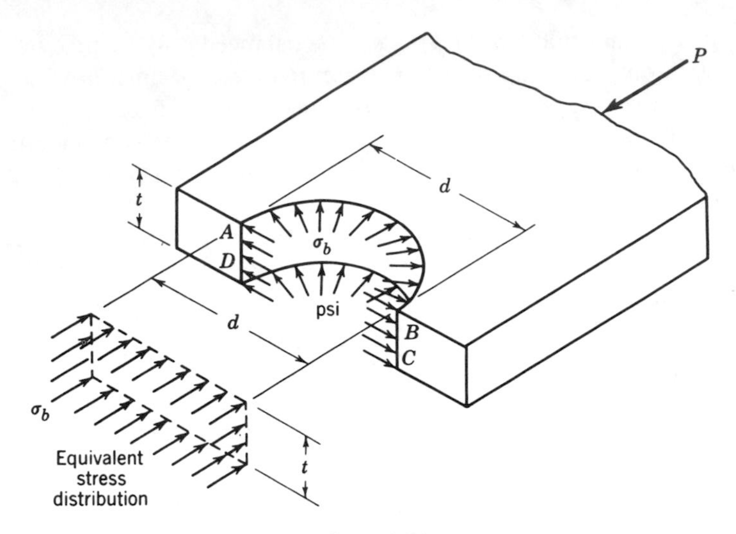

Figure 4-16

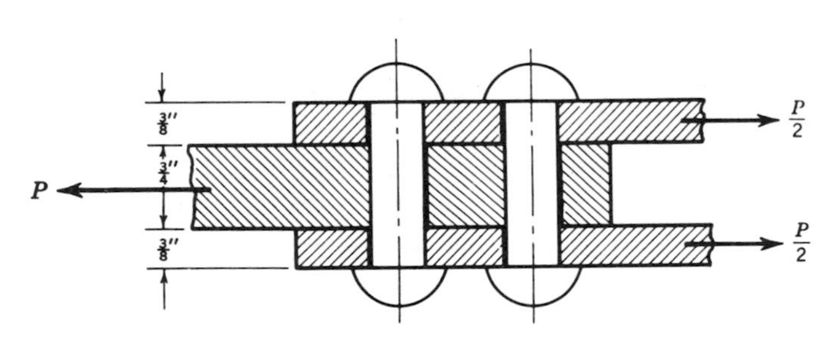

Figure 4-17

When the load is applied, the restraining forces in the main plate and butt straps are simultaneously induced. Since all bolts take an equal amount of load and are of the same diameter throughout the length of the grip, the question as to whether the maximum bearing stress is located in the butt straps or in the main plate depends upon a comparison of their thicknesses. If the thickness of the main plate should exceed the combined thickness of the two butt straps, the maximum bearing unit stress would be located in the butt straps. If the thickness of the main plate is the same as the combined thickness of the two butt straps, the bearing unit stress would be the same in each one. Such is the case in our present problem. It is well to investigate the thickness of the various plates before a bearing stress investigation is made, even though in practice the thickness of butt straps generally exceeds one half of the main-plate thickness.

The bearing unit stress between the bolts and the main plate of our present problem is, therefore,

$$\sigma_b = \frac{P}{\sum td} = \frac{60{,}000}{4 \times \frac{7}{8} \times \frac{3}{4}} = 22{,}800 \text{ psi}$$

Tensile unit stress on net section. To investigate the tensile unit stress on the net section of any row of bolts, a free-body analysis is important, since ultimate failure may occur in either the main or strap plates, as was indicated for the bearing unit stress in the previous section. Figures 4-18a

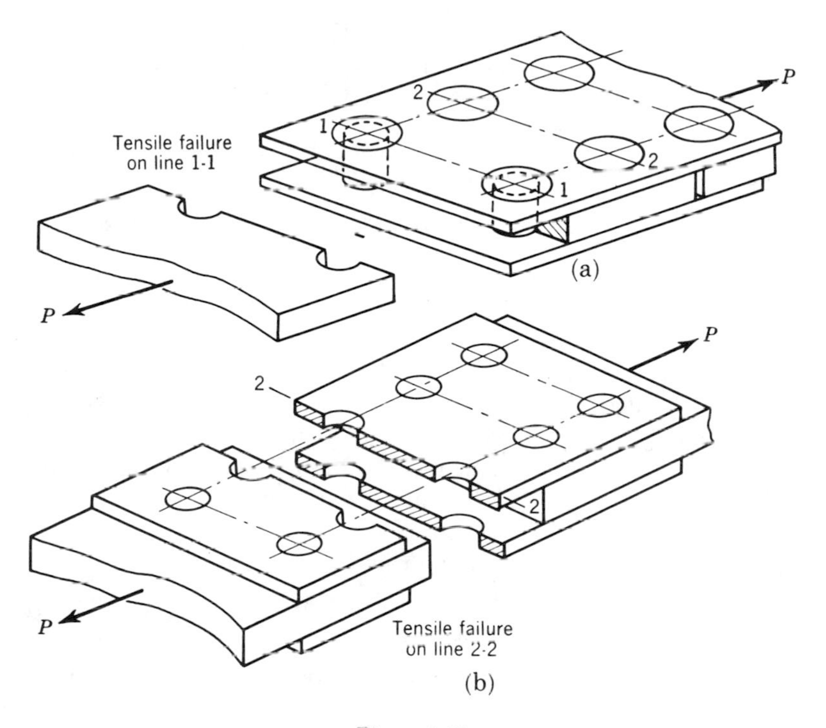

Figure 4-18

and b show how these failures may take place in the joint used for illustration. To verify the possibility of such failures, let us study the flow of load through the bolted joint.

The flow chart in Fig. 4-19 is based on two facts:

1. The strap plates *transfer* load from one main plate to another.
2. This transferral takes place on the assumption that each bolt cross section transfers its proportional share of the total load.

Because eight equal bolt cross sections resist the full load, each transfers one eighth of the load. Thus, at row 1, where each of the two bolts has two such cross sections, one on top and one on the bottom, loads of $P/4$ pass into the top and bottom butt strap. The total load transferred at row 1 is then equal to $P/2$, leaving the remaining load of $P/2$ in the main plate between rows 1 and 2. The same transference of load prevails at row 2, leaving no

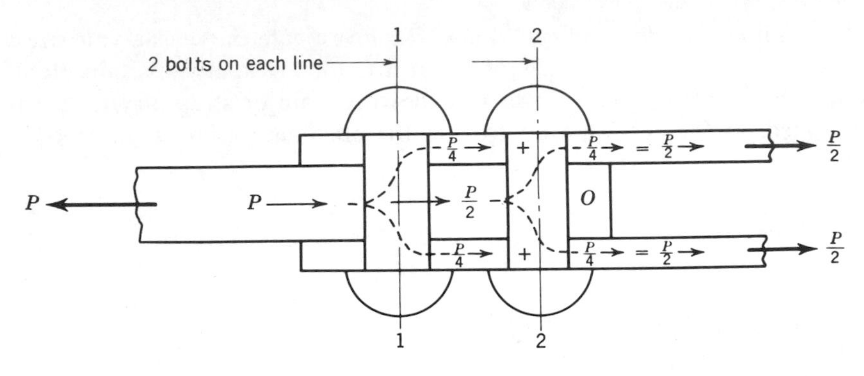

Figure 4-19

load in the main plate and placing the entire load on the butt straps. No tension exists in the margins of the main plate and butt straps.

Since P is an axial load, the tensile unit stress acting on any net section may be computed by the familiar $P = A_t\sigma_t$ formula. Those conditions that will produce the desired maximum unit stress, which in turn govern the design, depend upon a combination of high load and small area. Therefore, it is best to begin the stress investigation at those rows of bolts acted upon by the entire load. From Fig. 4-19, these are shown to be at (1) row 1 in the main plate and at (2) row 2 in the butt straps.

The tensile unit stress acting on row 1 of the main plate is

$$\sigma_{t_{1-1}} = \frac{P}{A_t} = \frac{P}{(w - n \times d')t} = \frac{60{,}000}{(6 - 2 \times 1)\frac{3}{4}} = 20{,}000 \text{ psi} \tag{4.11}$$

where w = width of plate or repeating section, n = number of bolts, d' = diameter of bolt plus $\frac{1}{8}$ in.*, t = thickness of plate on which the load is impressed.

Considering row 2 as mentioned above, note that while each butt strap carries only a load of $P/2$ lb and has a thickness of $\frac{3}{8}$ in., the intensity of stress is the same as that in the main plate at row 1, since both their load and thickness are one half the main-plate values. The computation for tensile unit stress acting in the butt straps on row 2 is

$$\sigma_{t_{2-2}} = \frac{P}{A_t} = \frac{P}{2(w - nd')t} = \frac{30{,}000}{(6 - 2 \times 1)\frac{3}{8}} = 20{,}000 \text{ psi}$$

Had the butt straps been thinner, the maximum tensile stress in the joint would have occurred in the butt straps on the row-2 section.

Efficiency of joint. The efficiency of a bolted or riveted joint is equal to the ratio of its allowable strength to the allowable tensile strength of the

*Each bolt hole is made $\frac{1}{16}$ in. larger than the nominal size of the volt for clearance purposes. However, when such holes are punched, they are somewhat ragged, and an additional $\frac{1}{16}$ in. is added to the diameter to compensate for possible variations in dimensions and stress concentrations.

solid plate of the same width and thickness. Therefore, it is necessary to obtain those allowable loads that may be applied to the joint based upon the allowable shearing, bearing, and tensile unit stresses. Dividing the least of these loads by the allowable tensile strength of the main plate with no bolt holes will give the efficiency of load transfer.

The allowable loads and efficiency for the joint under investigation are computed as follows, using $\tau = 15{,}000$ psi, $\sigma_t = 20{,}000$ psi, and $\sigma_b = 45{,}000$ psi.

Allowable shear load:

$$P_s = A_s\tau = 8 \times 0.60 \times 15{,}000 = 72{,}000 \text{ lb}$$

Allowable bearing load:

$$P_b = A_b\sigma_b = 4 \times \tfrac{7}{8} \times \tfrac{3}{4} \times 45{,}000 = 118{,}200 \text{ lb}$$

Allowable tensile load:

$$P_t = A_t\sigma_t = (6 - 2 \times 1)\tfrac{3}{4} \times 20{,}000 = 60{,}000 \text{ lb}$$

Allowable tensile load of solid main plate:

$$P = A_t\sigma_t = 6 \times \tfrac{3}{4} \times 20{,}000 = 90{,}000 \text{ lb}$$

Efficiency of joint:

$$\frac{60{,}000}{90{,}000} = 66.7 \text{ per cent}$$

Illustrative Problem 3. Determine the maximum load that can be applied to the riveted structural joint shown in Fig. 4-20. All holes are punched. Use $\tau = 15{,}000$ psi, $\sigma_b = 45{,}000$ psi, and $\sigma_t = 20{,}000$ psi.

SOLUTION. The loads that can be sustained by the shear and bearing resistance of the rivets are as follows:

$$\text{shear} = 12 \times 0.785 \times 15{,}000 = 141{,}300 \text{ lb}$$

$$\text{bearing} = 12 \times 1 \times \tfrac{3}{8} \times 45{,}000 = 202{,}500 \text{ lb}$$

There still remains the determination of the allowable tensile load that can be applied to the various net cross sections passing through the rivet holes. Once again, the flow of load diagram through the joint is drawn (Fig. 4-20b). Equating the maximum load applied to each cross section to the resistance offered, we obtain

$$\text{tension 1-1} = (18 - 1\tfrac{1}{8}) \times \tfrac{3}{8} \times 20{,}000 = P, \qquad P = 126{,}700 \text{ lb}$$

$$\text{tension 2-2} = (18 - 2\tfrac{1}{4}) \times \tfrac{3}{8} \times 20{,}000 = \tfrac{11}{12}P, \qquad P = 128{,}800 \text{ lb}$$

$$\text{tension 3-3} = (18 - 3\tfrac{3}{8}) \times \tfrac{3}{8} \times 20{,}000 = \tfrac{9}{12}P, \qquad P = 146{,}200 \text{ lb}$$

The maximum load that the joint can resist is therefore 126,700 lb.

$$\text{efficiency of joint} = \frac{126{,}700}{18 \times \tfrac{3}{8} \times 20{,}000} \times 100 = 93.8 \text{ per cent}$$

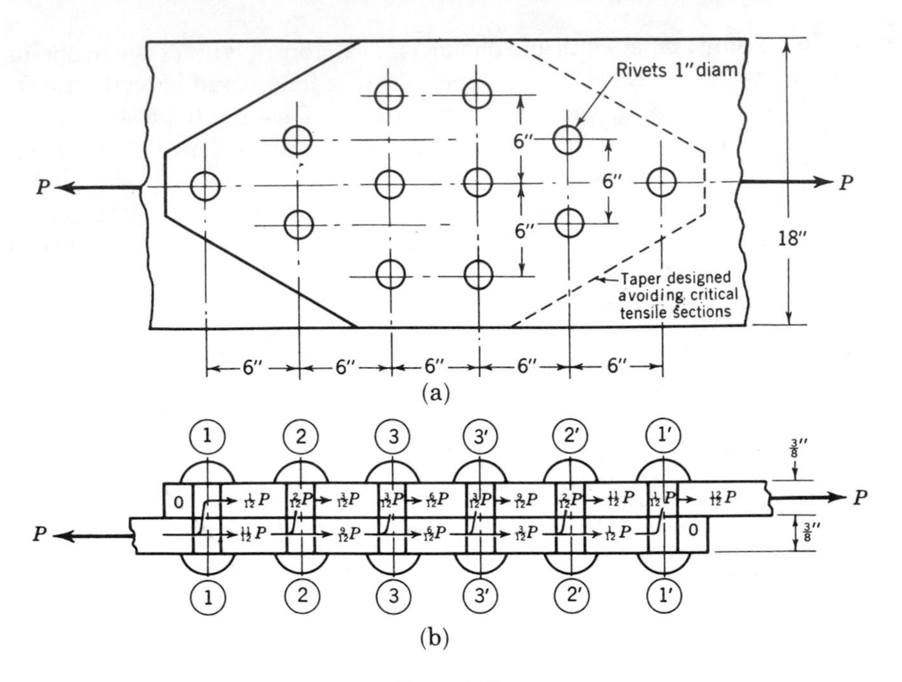

Figure 4-20

Note on compatibility of stress and strain. Between the rivet lines 1 and 2 of Fig. 4-20b there exists a load of $\frac{1}{12}P$ on the top plate and $\frac{11}{12}P$ on the bottom. If the average width of the top plate between lines 1 and 2 is taken as 9 in. its total strain $e_{TP} = (\frac{1}{12}P \times l)/(9 \times \frac{3}{8} \times E) = \frac{8}{324}(Pl/E)$. In contrast, the elongation in the lower plate is

$$e_{BP} = \frac{\frac{11}{12}P \times l}{18 \times \frac{3}{8} \times E} = \frac{44}{324}\frac{Pl}{E}$$

With this incompatibility of stress and strain it is obvious that some readjustment of the rivets is necessary to sustain the apportionment of rivet loads as indicated in Fig. 4-20. In fact, underlying all riveted joint design is the assumption that sufficient plastic action will validate the assumption of the proportional load-carrying capacity of each rivet.

PROBLEMS

Note: All bolt or rivet holes are assumed to be punched. Where maximum allowable stresses are required, use $\tau = 15{,}000$ psi, $\sigma_t = 20{,}000$ psi, and $\sigma_b = 45{,}000$ psi.

4-11. A single-bolted lap joint is made from two steel plates, $2\frac{1}{2}$ in. wide by $\frac{1}{2}$ in. thick, fastened together with a $\frac{3}{4}$-in. bolt. Compute the tensile unit stress on the net section through the bolt hole, the shearing unit stress in the bolt, and the bearing unit stress on the side of the bolt due to a tensile load of 8000 lb. *Answers:* $\tau = 18{,}200$ psi, $\sigma_b = 21{,}300$ psi, $\sigma_t = 9850$ psi.

4-12. If a single-bolted lap joint is made of $\frac{1}{2}$-in. steel plate and $\frac{3}{4}$-in.-diameter bolts, find that pitch that will make the shearing strength of the bolts equal to the tensile strength of the plate.

4-13. If the vertical joints of the surge tank of Problem 4-5 have an efficiency of 85 per cent, compute the required thickness of the bottom vertical plate. *Answer:* 0.077 in.

4-14. Determine the strength and efficiency of a double-bolted lap joint if the pitch on both rows is 5 in., bolt diameter is 1 in., and the plate thickness is $\frac{1}{2}$ in.

4-15. A 3- by $\frac{3}{8}$-in. flat, carrying a tensile load of 10,000 lb, is bolted to a $\frac{3}{8}$-in. gusset plate. If $\frac{5}{8}$-in. bolts are used, find the number of bolts required to transmit the load.

4-16a, b, c, d, e. Determine the number of rivets necessary for each member framing into joint E of the truss shown in Fig. 4-21. Rivets are $\frac{5}{8}$ in. *Answers:* (a) 6; (b) 3; (c) 3; (d) 5; (e) 4.

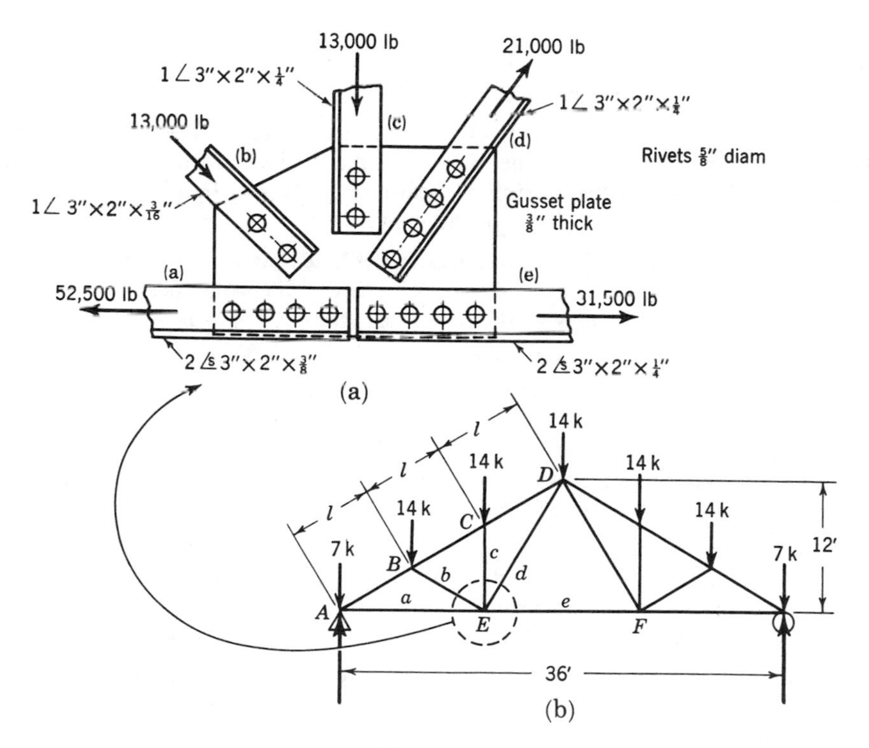

Figure 4-21

4-17. Two $\frac{3}{8}$-in. structural plates are joined by $\frac{3}{4}$-in. bolts to form a lap joint (Fig. 4-22). The bolts are placed in three rows, the pitch of the center row being $2\frac{1}{2}$ in. and the pitch of the outer rows 5 in. If the tension in the cross section of the plates is 5000 lb/in. of width, find the following unit stresses in the joint: (a) shear and bearing in the bolts and (b) tension in the net section of the plate at the

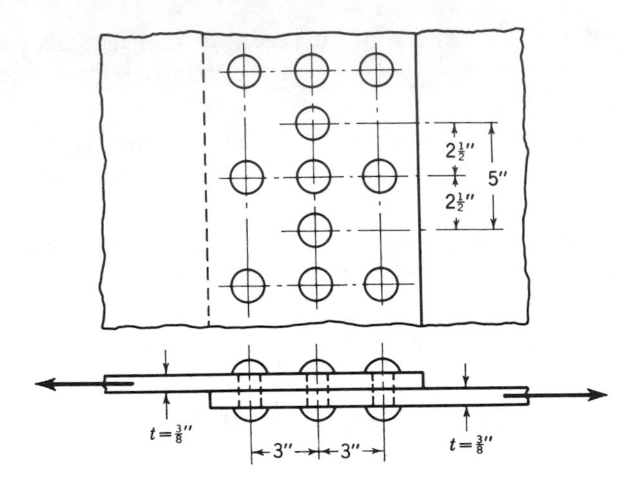

Figure 4-22

outer row of bolts and at the center row of bolts. *Answers:* (a) $\tau = 14{,}200$ psi, $\sigma_b = 22{,}200$ psi; (b) 16,150 psi, 15,340 psi.

4-18. Two ½-in. plates, each 14 in. wide, are united by five 1-in.-diameter bolts in a single row parallel to the joint to form a lap joint. If the joint transmits a pull of 52,000 lb, find the maximum shearing, tensile, and bearing unit stresses in the joint. What is the efficiency of the joint?

4-19. A spherical gas tank 50 ft in diameter holds gas at a maximum pressure of 50 psi. If it is to have double-riveted butt joints with a tensile efficiency of 85 per cent, what must the thickness of the plates be? *Answer:* 0.441 in.

4-20. A boiler 100 in. in diameter, having plates ¾ in. thick, is subjected to an internal pressure of 200 psi. The longitudinal joint is a double-riveted butt joint with two cover plates each $\frac{5}{16}$ in. thick. The diameter of the rivets is ⅞ in., the pitch is 3 in., and the distance between rows is 3 in. Find the maximum bearing, shearing, and tensile unit stresses in the joint. *Hint:* Use $2P = plD$ to obtain load on repeating section. *Answers:* 27,500 psi, 12,500 psi, 24,000 psi.

4-21. Two ⅜-in. boiler plates are united by a triple-riveted lap joint of ¾-in. rivets. The pitch is 4½ in. in the outer rows and 2¼ in. in the center rows. What is the strength in pounds per linear inch of joint?

4-22. A boiler 4 ft in diameter is made of ¼-in. steel plate. The longitudinal joints are fastened with a single row of ⅞-in. rivets placed with a pitch of 2½ in. What maximum pressure should be permitted? What is the efficiency of this joint? *Answers:* 125 psi, 60.0%

4-11 Maximum Design Efficiency

As has been indicated, every bolted or riveted joint owes its strength to its resistance in shear, bearing, and tension. The maximum load that may be applied to the joint is dependent on the least of these resistances, or any of their combinations. Should one of them have a low value, the additional strength offered by the other resistances would be wasted. It is therefore advisable to have each of the resistances in shear, bearing, or tension equal each other. In so doing, the joint will have been most efficiently designed.

WELDED JOINTS

4-12 General Discussion

The origin of welding dates back to the decade following 1880. Owing to several drawbacks, welding was not in common use until World War I, the decade of 1910-1920, when extensive research eliminated many of the difficulties hindering its growth. Since that time, the phenomenal rise of welding as a means of joining structural and machine members has caused a decrease in the use of rivets and bolts and has also precipitated the decline in the use of metal castings in preference to fabricated members (Fig. 4-23).

Figure 4-23 Field welding: connecting beam and column. (Courtesy of Lincoln Electric Co.)

Welding, as ordinarily considered for structural use, may be defined as a method of joining metals by fusion without the application of pressure. To accomplish the fusion process, heat is generally supplied by either an electric arc or an oxyacetylene torch. The metal at the joint, together with additional metal supplied in the form of filler metal, is melted, forming a small pool or crater. Upon cooling, the weld and base metal form a continuous and almost homogeneous joint.

The design of welded joints constitutes only one of the several very important problems arising in the application of welding to structural and machine parts. The procedure that must be followed, the proper amperage and voltage of the welding machine, the rate of cooling, and the development of internal stresses are some of the other important factors that must be considered.

These added considerations were, in the past, primarily the concern of the welder. With the inception of better methods and standardizations, much of this responsibility has now been shifted from the welder to the welding code. Seriously flawed work has been substantially eliminated by requiring each welder to pass rigid qualification tests and to submit his work to the careful scrutiny of a trained inspector. To further test the safety of welded joints, X rays and gamma rays are frequently used to locate internal flaws.

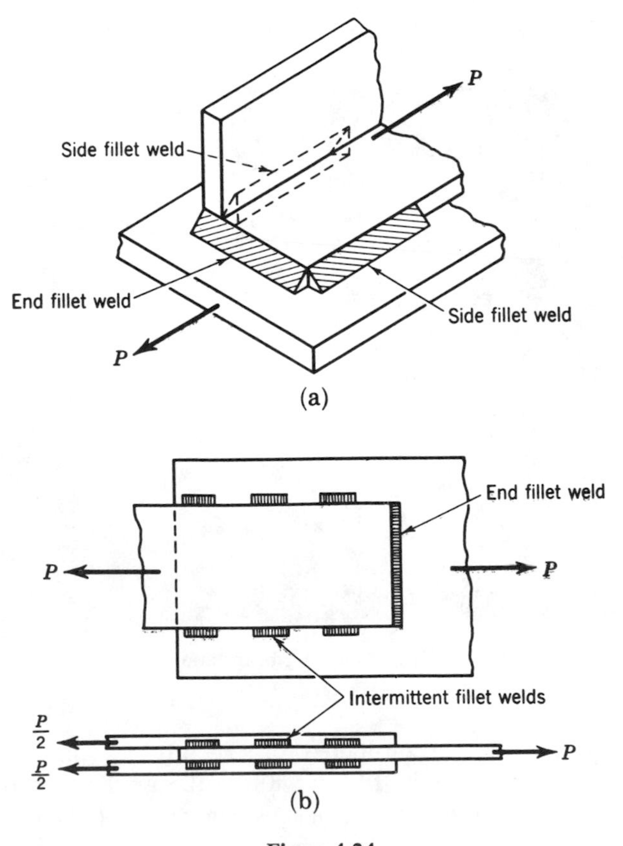

Figure 4-24

4-13 Types of Welded Joints—Definitions

There are various types of welded joints in common use. They are made by the use of (1) fillet welds, (2) butt welds, (3) plug or slot welds, (4) resistance welds, and (5) tack welds. A brief description of each follows.

Fillet welds. If a weld is placed in a corner made by two adjoining structural members, it is called a *fillet weld.* Several examples of fillet welds are shown in Fig. 4-24.

Size. Fillet welds are generally specified with equal legs. The length of these legs is, therefore, conveniently used to represent the *size of the fillet weld.*

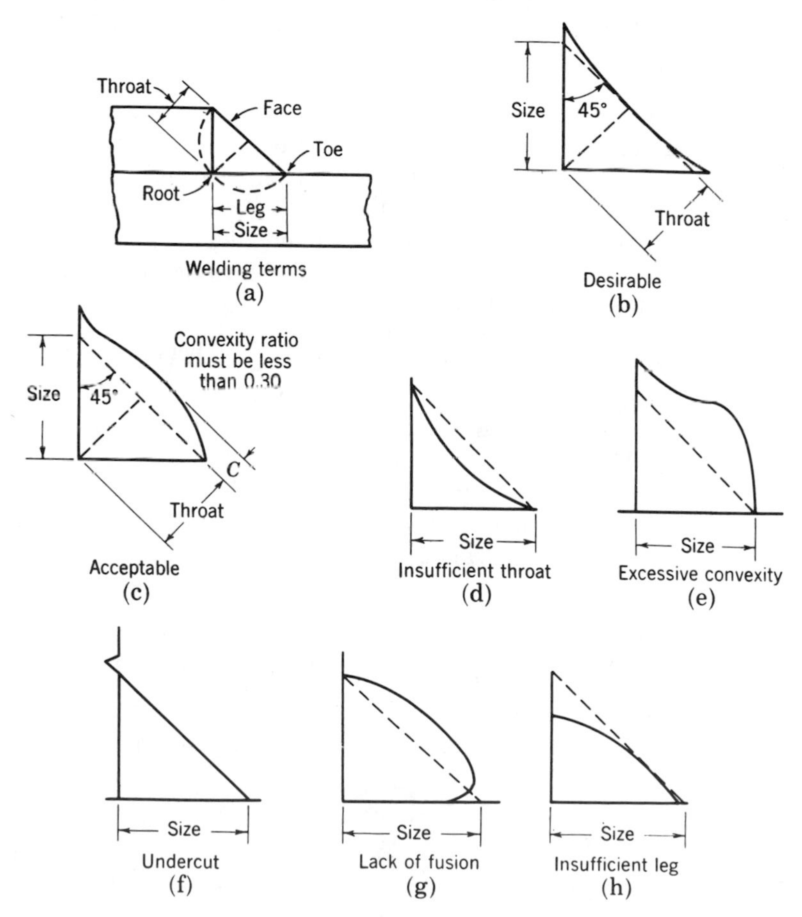

Figure 4-25 (a) Welding terms; (b–h) acceptable and defective fillet-weld profiles.

Throat. The minimum thickness of a weld measured in its cross section along a straight line passing through its *root* is called the *throat* of the weld (Fig. 4-25a). The effective throat thickness of an equal-leg 45° fillet weld shall be considered as 0.707 times the nominal size of the weld.

Butt welds. When the edges of two structural or machine members are butted together and welded, the weld formed is a *butt weld.* Butt welds are used more frequently for joining machine parts than for joining structural members. The frequent need for beveling the edges of a butt joint makes it more expensive. Figure 4-26 shows several complete-penetration groove butt-welded joints.

Throat. The effective throat thickness of a complete-penetration butt weld is considered to be the thickness of the thinner part joined. If a butt weld were to be made with an incomplete penetration, its strength would have to be reduced in accordance with the welding code used.

Plug and slot welds. Upon occasion, structural members are joined using plug welds (Fig. 4-27a). These joints are made by filling holes punched in one of the overlapping members of the joint with weld metal. Plug welds are used in joints where fillet welds are either insufficient to connect the members of the joint properly or the position of butt and fillet welds is impractical. Slot welds (Fig. 4-27b) are also used to overcome difficulties involved in providing sufficient strength in welded joints.

Resistance welds. Resistance welds, of which spot welds (Fig. 4-27c) are a type, are made by a simultaneous application of intense pressure and the passage of high current through two metal pieces at the point where they are to be joined. The intense heat created by the current causes the metal in both pieces to melt and merge with each other. Upon cooling, the two metal pieces are securely joined. They are frequently used in the assembly of small electrical and mechanical equipment.

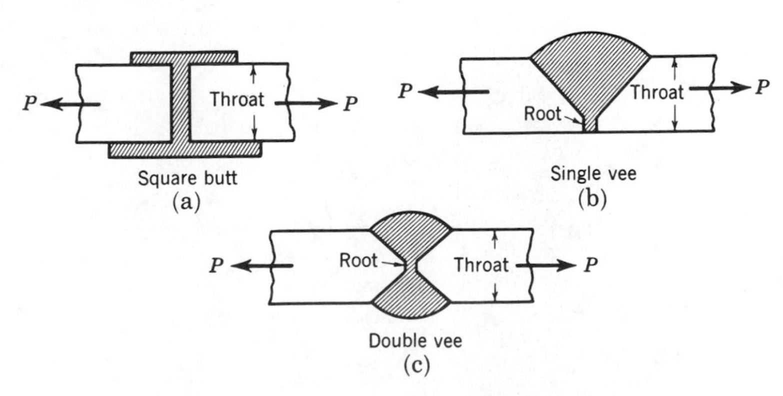

Figure 4-26 Types of complete-penetration groove welds.

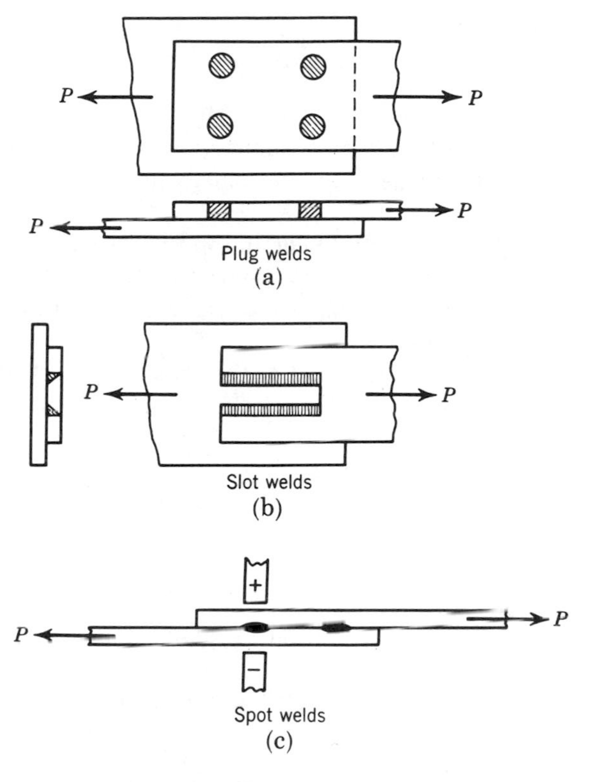

Figure 4-27

Tack welds. Tack welds are used, as their name implies, for holding metal parts in position while the major welding operation is in progress.

4-14 Welding Specifications and Allowable Unit Stresses

The more detailed requirements governing design and fabrication procedures involved in building construction are contained in the Specification for the Design, Fabrication, and Erection of Structural Steel for Buildings, prepared by the American Institute of Steel Construction. These specifications are under constant surveillance by engineers and designers and as a result are being constantly improved and revised to introduce advantages of new improvements and developments in techniques and materials.

The maximum allowable unit stresses presented below are for welds made manually by the shielded metal-arc process for the connection of low-carbon (A-36) and low-alloy (A-242) structural steel members. They are:

Shear on section through throat of fillet weld, or on faying surface area of filled plug or slot weld:

$$\tau = 21{,}000 \text{ psi}$$

For complete-penetration groove welds:

The allowable tensile and compressive stress parallel to the axis and the allowable tensile stress normal to its effective throat are the same as for the base metal.

For complete- or partial-penetration groove welds:

The allowable shear stress acting on and the allowable compressive stress acting normal to the effective throat are the same as for the base metal.

In other words, a butt weld of the same cross section as the connected members is assumed to be 100 per cent efficient in transferring stress.

To make sure of the proper maximum allowable stress for a particular welded steel project, it is advisable to consult the code for exact values and limitations.

4-15 Design of Welded Joints—Fillet-Welded Joints

The design of any fillet-welded joint requires the determination of three values: (1) the size of the weld, (2) the length of weld, and (3) the proper distribution of the weld.

The size of a fillet weld to be used depends upon the thickness of the parts to be joined, the strength desired, and the economy of making the joint. It is usual to limit the maximum size of a fillet weld applied to a square edge of a plate or section to $\frac{1}{16}$ in. less than the nominal thickness of the edge for thicknesses over $\frac{1}{4}$ in. A minimum size of weld generally bears a relationship with the maximum thickness of the member welded. The American Institute of Steel Construction specification, for instance, presently requires a minimum of a $\frac{3}{16}$-in. fillet weld to be used with thicknesses up to and including $\frac{1}{2}$ in. The minimum size increases to a value of $\frac{5}{8}$ in. required for thicknesses over 6 in.

Economy also dictates to a great extent the size of the fillet weld to be used. Owing to (1) the larger proportion of metal per inch of throat required for increasingly larger welds, (2) the increased labor cost in making welds of two or more passes, (3) the necessity of making connections in small space, and (4) the limitations of welding technique, the costs of welds might appear to vary directly with their size. In contradiction to this premise, however, the use of larger welds allows the use of larger welding rods, which in turn provide a greater rate of weld metal output in pounds per hour. The practical evaluation of these factors has established the $\frac{5}{16}$-in. weld as the most common size for structural welding. This size of weld is the largest weld that can be conveniently made with one pass of the welding rod. Larger fillet welds generally require two or more passes. Since the cost of fillet welds increases approximately in proportion to the square of their size, the use of fillet welds larger than $\frac{5}{16}$ in. should be based on structural reasons that warrant their higher cost.

The critical stress in a fillet weld shall be considered as the average shearing unit stress that occurs on its throat section. Although tests made through the elastic range show that the stress at the ends of a weld, pulled lengthwise, is higher than that near its midpoint, consequent yielding at points of high stress under higher loads produces fair uniformity. To determine the general formula for shear stress on the throat section of a fillet weld, therefore, let us use Fig. 4-28, assuming the stress to be uniformly distributed.

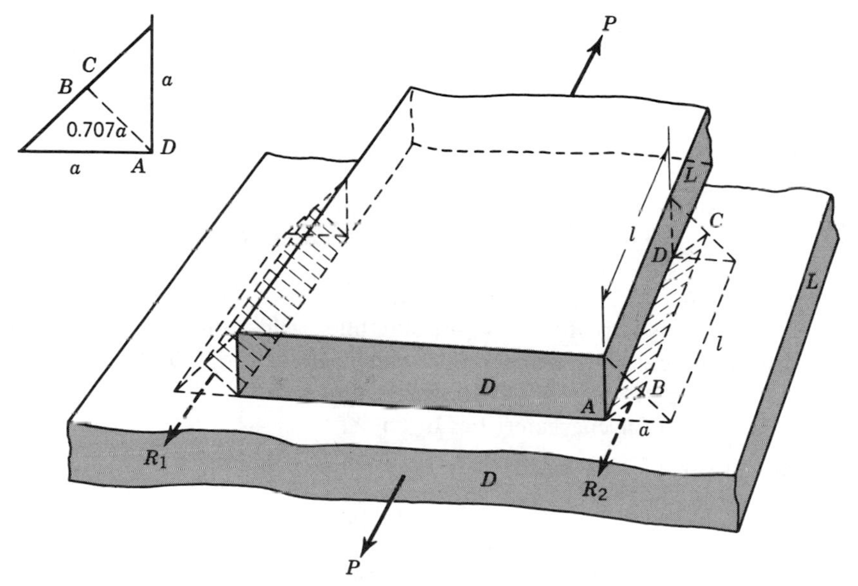

Figure 4-28

Considering the effective area of a fillet weld as the product of its effective length and effective throat thickness, the resisting force offered by one of the fillet welds is

$$R_1 = R_2 = 0.707a \times l \times \tau \tag{4.12}$$

Equating the horizontal forces on the joint, we obtain

$$P = 2 \times 0.707a \times l \times \tau$$

or if $2l = L$, the total length,

$$P = 0.707a \times L \times \tau \tag{4.13}$$

Should L be made equal to 1 in. and $\tau = 21{,}000$ lb/in.², the value of P in the above formula would be

$$P = 0.707a \times 1 \text{ in.} \times 21{,}000 \text{ psi} = 14{,}800a \qquad (4.14)$$

If, then, a were made equal to 1 in., the value of allowable load P would be equal to 14,800 lb. In other words, a 1-in. weld 1 in. long would resist 14,800 lb of force. A $\frac{1}{2}$-in. weld could carry only half as much, or 7400 lb. To obtain the strength of any weld 1 in. long merely requires the multiplication of the size of the weld by 14,800.

Illustrative Problem 4. If, in Fig. 4-28, load P were equal to 60,000 lb, what length of $\frac{5}{16}$-in. weld would be required on each side of the plate (A-36) to resist the load? Use the A.I.S.C. allowable welding stresses.

SOLUTION. Solving for L in Eq. (4.13) we get

$$L = \frac{P}{0.707a \times \tau}$$

$$= \frac{60{,}000}{0.707 \times \frac{5}{16} \times 21{,}000} = 12.9 \text{ in.}$$

or

$$l = 6.5 \text{ in.,} \qquad \text{effective length required on each side}$$

Illustrative Problem 5. The long leg of a single 5- by $3\frac{1}{2}$- by $\frac{3}{8}$-in. angle is to be welded to a gusset plate with $\frac{5}{16}$-in. side-fillet welds (Fig. 4-29). Determine the amount of weld needed at each side of the angle to develop the full capacity of the angle economically. Allowable unit stress in angle is 20,000 psi. Consult Table 16 for characteristics of angle.

SOLUTION. The load to be resisted is equal to the area of the angle times the allowable unit stress.

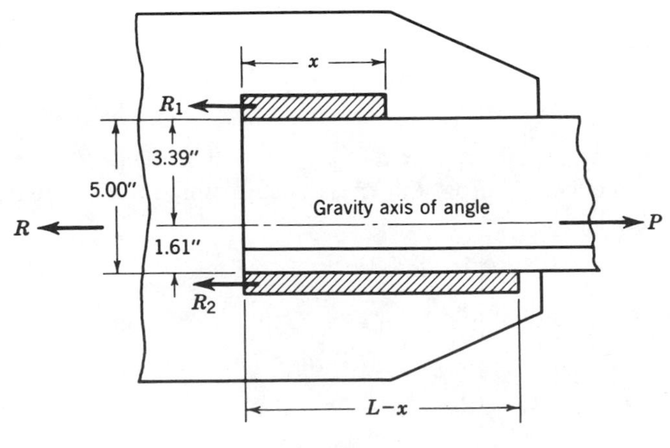

Figure 4-29

$$P = a\sigma = 3.05 \times 20{,}000 = 61{,}000 \text{ lb}$$

The total length of weld required to restrain this force is ($\frac{5}{16}$-in. fillet-weld capacity = 4640 lb/in. of weld)

$$L = \frac{61{,}000}{4640} = 13.2 \text{ in.}$$

Since the center of gravity of the angle is not on the center line of the attached leg, the amount of weld on each side must be proportioned in such a way as to make the *resultant* resistive force of both welds act collinear with the applied force of 61,000 lb. To perform this distribution, use is made of the theorem which states that the algebraic sum of the moments of component forces about a point is equal to the moment of their resultant. If, then, the moment center is selected on the action line of the resultant (the centroidal axis of the angle) projected into the plane including the resistive forces R_1 and R_2, its moment is zero, and the algebraic sum of the moments of the two weld forces must also be equal to zero. Stated differently, the moments of these two weld forces taken about the centroidal axis of the angle must balance and be equal to each other—R_1 and R_2 considered as acting on the sides of the angle.

We may therefore write

$$x \times 4640 \times 3.39 = (13.2 - x) \times 4640 \times 1.61$$

The allowable strength of the weld might have been omitted, as the cancellation indicates.

$$3.39x + 1.61x = 21.2$$

$$x = \frac{21.2}{5.0} = 4.3 \text{ in.,} \qquad \text{effective length at top of angle}$$

$$13.2 - x = 8.9 \text{ in.,} \qquad \text{effective length at bottom angle}$$

The effective length of a fillet weld is considered as the overall length of the full-sized fillet.

4-16 End Fillet Welds

From studies made by the American Bureau of Welding,* fillet welds placed at the ends of structural members transverse to the direction of loading average 35 per cent greater strength than side fillet welds of equal dimension. Test results on side fillet welds, however, show less variation in load resistance.

When end fillet welds are used together with side fillet welds, a statically

*H. M. Priest, "Strength of Structural Welds," *Engineering News Record*, Sept. 17, 1941.

indeterminate condition is imposed upon the joint. Not all designers are agreed that an equal load should be delegated to every inch of weld in such a joint.

In the opinion of the formulators of the code already discussed, however, sufficient similarity does exist, particularly after localized yielding, to warrant the assumption of equal distribution of stress to both side and end fillet welds. The effective throat section acted upon by the critical shear stress is again assumed to form the basis of the design.

Illustrative Problem 6. If, in Fig. 4-30, $\frac{5}{16}$-in. side and end fillet welds are to be used to fasten the 4-in.-wide tension member to the gusset plate, what distribution of welds will be necessary? $P = 60{,}000$ lb.

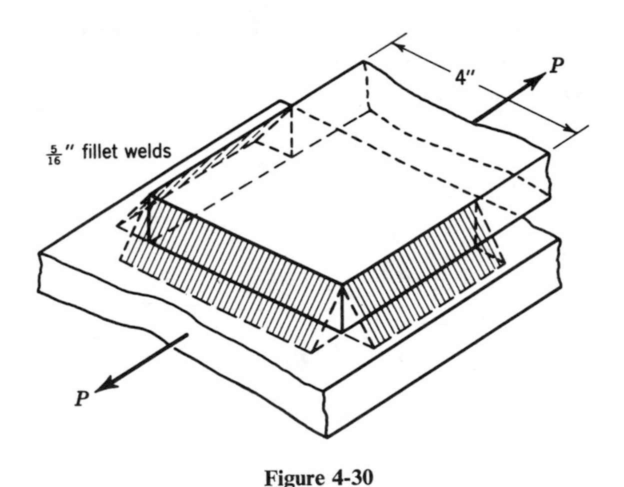

Figure 4-30

SOLUTION. The total length of weld necessary, inasmuch as a $\frac{5}{16}$-in. fillet weld 1 in. long can restrain a load of 4640 lb, is

$$L = \frac{60{,}000}{4640} = 12.90 \text{ in.}$$

The end fillet weld is then equal to 4 in., and the remainder is distributed equally between the two sides, 4.45 in. each.

Illustrative Problem 7. Referring to Illustrative Problem 5, what distribution of welds would be necessary if an end fillet weld were also used (Fig. 4-31)?

SOLUTION. Here the resistive forces are augmented by the resistance in the end fillet weld. If the end fillet weld is divided into its portions on either side of the centroidal axis and included in the moments of the resistive forces, this problem is not unlike that of its predecessor.

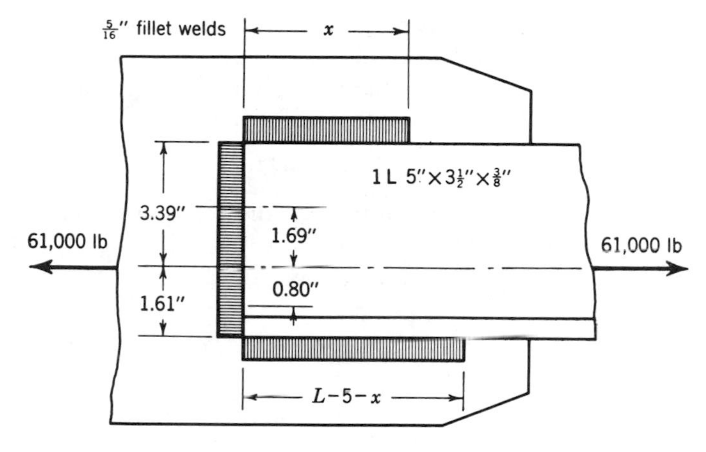

Figure 4-31

The total length of $\frac{5}{16}$-in. weld required is again

$$L - \frac{61{,}000}{4640} = 13.2 \text{ in.}$$

Of this amount, only 8.2 in. must be distributed to the two sides as 5 in. of weld is placed at the end of the angle. The method of distribution, as previously shown, is slightly different, however, since that portion of the end fillet weld above the gravity axis has a different resistive moment than that below. The proper distribution may then be made by taking the algebraic sum of the resisting moments about the centroidal axis, eliminating the allowable strengths which were previously shown to be unnecessary.

$$x \times 3.39 + 3.39 \times 1.69 = 1.61 \times 0.80 + (8.2 - x)1.61$$

$$3.39x + 1.61x = 13.2 + 1.29 - 5.73$$

$$x = \frac{8.76}{5} = 1.75 \text{ in.,} \quad \text{effective length, top weld}$$

$$8.2 - x = 6.45 \text{ in.,} \quad \text{effective length, bottom weld}$$

4-17 Design of Butt Joints

All butt joints placed perpendicular to the applied load are designed on the basis of the tensile or compressive stress induced on the effective throat section.

Illustrative Problem 8. Two 4- by $\frac{1}{2}$-in. structural-steel plates are to be butt-welded at their ends. How much tensile load is the joint capable of carrying?

SOLUTION. The effective weld area under load is equal to 2 in.^{2}. As previously noted, the allowable tensile unit stress applied to the throat of a butt weld is equal to the allowable tensile unit stress of the base metal, which for A-36 structural steel is 22,000 psi. Thus,

$$P = 2 \times 22{,}000 = 44{,}000 \text{ lb}$$

The 100 per cent transmission of load allowed by the welding code is entirely justifiable, since welded joints made with ordinary care and inspection are invariably stronger than the base metal.

PROBLEMS

(Use A.I.S.C. allowable stresses in the following problems where necessary.)

4-23. If the spherical gas tank in Problem 4-19 is composed of two hemispheres made of A-36 structural steel and butt-welded together, what internal pressure may be applied if the thickness of the plate is $\frac{3}{4}$ in.? *Answer:* 110 psi.

4-24. A 5- by $\frac{1}{2}$-in. steel plate is to be welded to a gusset plate with $\frac{5}{16}$-in. side fillet welds. What length welds are required to develop the maximum tensile unit stress of 18,000 psi in the plate?

4-25. An 8-in., 11.5-lb channel placed horizontally is to have its back welded to the flange of a vertical W beam (Fig. 4-32). If the flange of the W beam is 8 in.

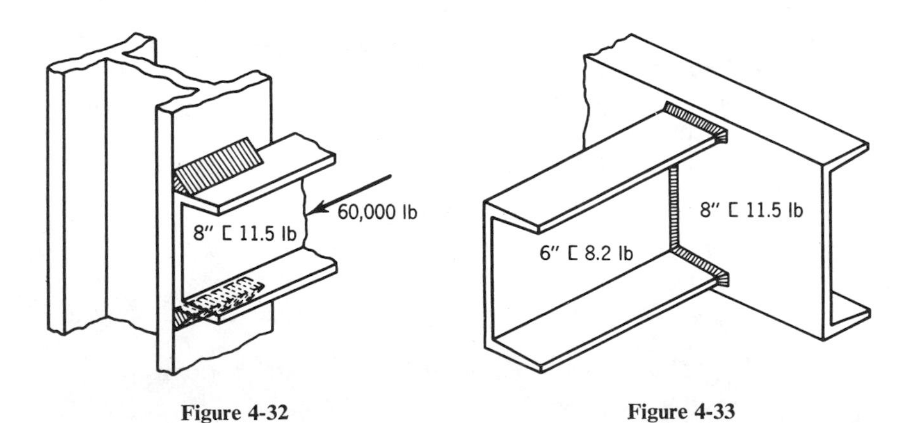

Figure 4-32

Figure 4-33

wide, will $\frac{5}{16}$-in. side fillet welds be sufficient to hold the compressive load of 92,800 lb placed on the channel? What most economical procedure should be followed if the side welds are insufficient? *Answer:* No. Use 8 in. on each side with 4 in. on end.

4-26. A 6-in., 8.2-lb channel is fastened to the back of an 8-in., 11.5-lb channel with a complete-penetration groove weld. What is the maximum permissible tension that may be applied to the 6-in. channel? What reaction load is the butt weld capable of taking in shear? Allowable shearing unit stress is 15,000 psi. Assume an average 0.2-in. throat thickness for the weld (Fig. 4-33).

4-27. A 4- by 3- by $\frac{3}{8}$-in. angle is to have its 4-in. leg welded to a gusset plate with $\frac{1}{4}$-in. welds. What length of weld is required and where must it be placed to restrain a tensile load of 50,000 lb? *Answers:* 4.30 in. at toe, 9.20 in. at heel.

4-28. If, in Problem 4-27, the weld were to be placed along the end of the 4-in. leg as well as on its sides, what most economical distribution could be made? *Answers:* 2.33 in. at toe, 4 in. on end, 7.17 in. at heel.

4-29. A hemispherical steel pot, 6 ft in diameter, is filled with mercury. If its circular edge is to be welded to a circular supporting ring with $\frac{3}{16}$-in. fillet tack welds, each 2 in. long and spaced uniformly around the circumference, determine the number and center-to-center spacing of the welds. The pot itself weighs 2000 lb. Specific gravity of mercury is 13.6.

5

Torsion

5-1 Introduction

If a member is subjected to the action of a pair of equal and oppositely directed couples lying in parallel planes, that portion of the member between the couples is said to be in torsion. Since couples always produce rotation, the portion of the member between the oppositely directed couples will be subjected to a twisting action. One of the objects of this chapter will be to determine the relationship between twisting couples and the resulting stresses and strains, particularly in circular shafts, since these are most frequently used to transmit torsion.

5-2 Development of Torsion

When two equal couples act in opposite directions on planes perpendicular to the axis of a shaft (Fig. 5-1a), the magnitude of the twisting moment is simply the moment, or *torque*, of one of the couples.

If only one of the couples (the applied torque) is visible and equilibrium is still maintained, it follows that the shaft is restrained by an equal and opposite, invisible couple developed by the body which holds it (Fig. 5-1b). The magnitude of the torque on the shafts of both Figs. 5-1a and b will, of course, be equal to the same Pp value.

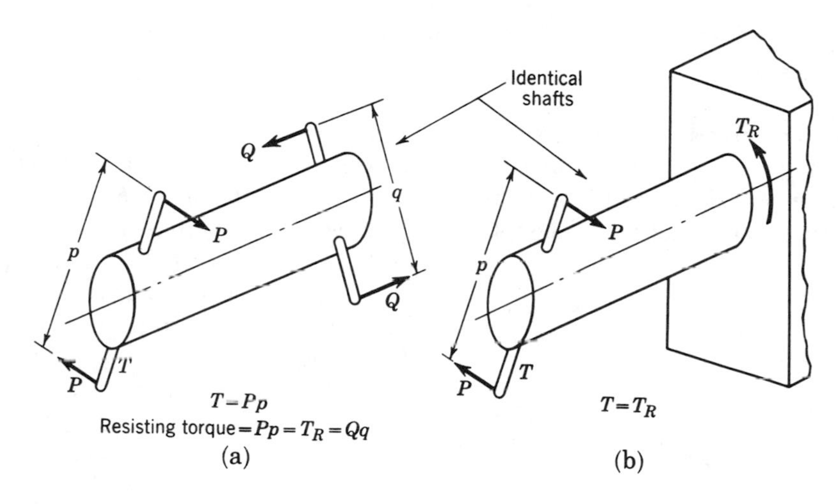

Figure 5-1

The effectiveness of a couple in producing torsion in a shaft is maximum when the couple acts in a plane perpendicular to the axis of the shaft. If the couple is rotated out of this plane, only that component of the couple that is in the plane perpendicular to the shaft produces torsion (Fig. 5-2). The other component produces only bending, because it lies in the horizontal plane containing the axis of the shaft. A couple lying in any plane containing the axis of the shaft can produce no twisting or rotation of the shaft. For example,

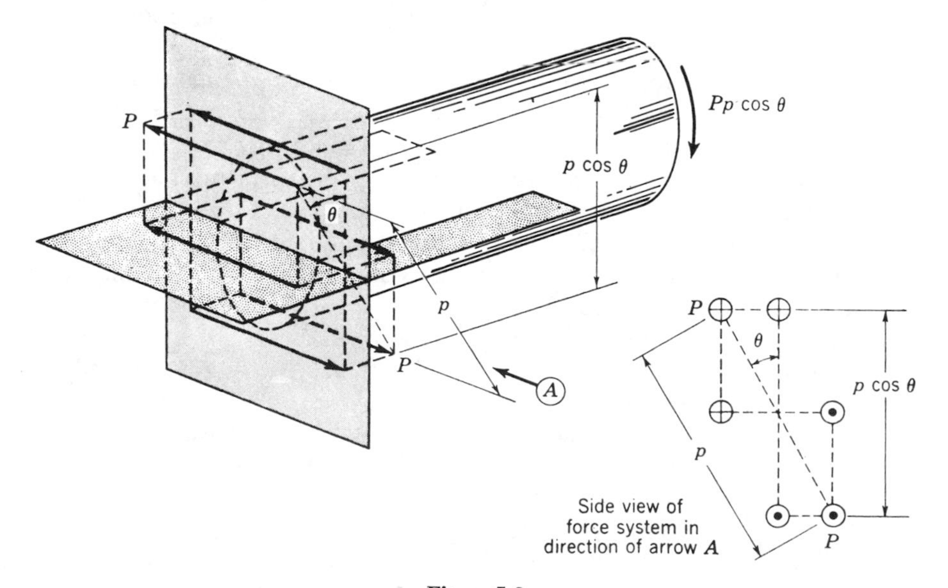

Figure 5-2

if a boy should hang from a doorknob, the door would not swing about its hinges.

Resolution of a force into a force and a couple. Often the presence of a couple acting on a shaft is obscured. This is especially true when a single force is applied to the end of a lever arm fastened to a shaft. Since no reaction is visible, the presence of a couple is not always detected. Take, for example, the case presented in Fig. 5-3. Only one force P is applied, and yet we know that the obvious purpose of the wrench is to twist the shaft.

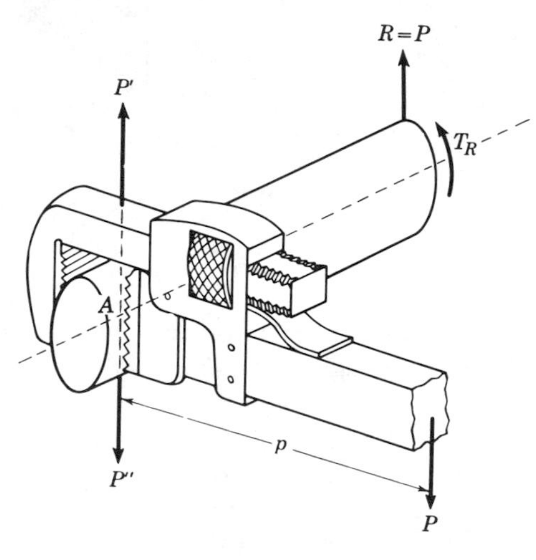

Figure 5-3

This problem can be clarified by recalling the method used in mechanics to resolve a force into a force and a couple. To illustrate its use, let us add at point A, located on the axis of the shaft in Fig. 5-3, two opposing forces equal and parallel to force P. Certainly their addition to the shaft causes no change in equilibrium, since they themselves are balanced. They alter neither the original conditions of the problem nor its final outcome. They do, however, simplify the problem by separating into convenient fractions the action of twisting and bending produced by force P.

Consider the forces P and P'. Since they are equal, parallel, and of opposite direction, they constitute a *couple*. Moreover, they are in a plane perpendicular to the axis of the shaft and *can produce only torsion*.

Force P'' acts as a *single force* located in the vertical plane passing through the longitudinal axis of the shaft and as such *can produce only bending*. The detailed study of its effects will be made in Chapter 6. This action of force P'' does not, however, influence the twisting action of the couple PP'. The complete study of problems of this type involves investigation of the combined effects of both the couple and the single force.

It is recommended that the above resolution of forces be made whenever a force acts off the center line of a shaft causing it to rotate about its longitudinal axis. *The resolution should be made on the axis line passing through the center of gravity of the resisting section* so that the two forces constituting the couple lie in a plane perpendicular to the shaft.

Shafts operating at constant speed. The study of mechanics reveals that any unbalanced system of forces or couples acting on a body produces acceleration. A shaft operating at constant speed is, therefore, theoretically acting under the same balanced conditions that prevail under static torque. Because of the maintenance of equilibrium, it is permissible to use the principles of torsion herein developed to determine the operational stresses of shafts rotating at constant speeds.

PROBLEMS

5-1. A cantilevered shaft, 3 ft long, is rotated by a lever having an effective moment arm of 25 in. (a) If the force applied at the handle of the lever is 40 lb, what will be the torque supplied to the shaft? (b) Is this the only effect? Explain. *Answers:* (a) 1000 in.-lb; (b) No.

5-2. A shaft is acted upon by forces A and B, as shown in Fig. 5-4. (a) Calculate the torque. What would be the torque on the shaft if only forces A and

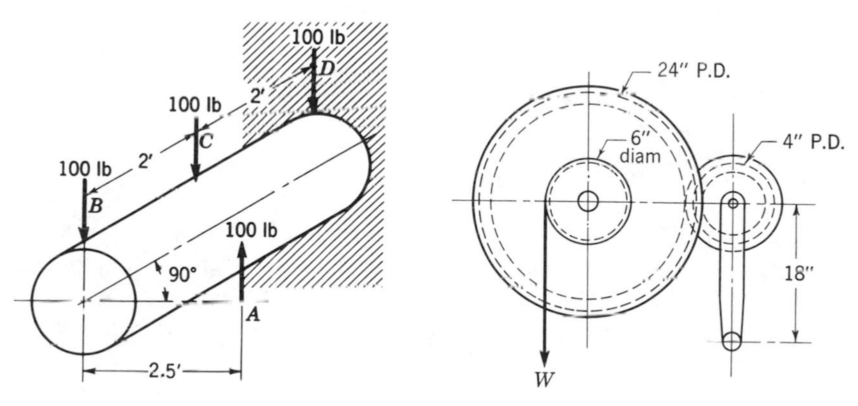

Figure 5-4 All forces shown are applied externally.

Figure 5-5

C were acting? Also, only forces A and D? (b) What, if any, bending effect would each of the couples AB, AC, and AD produce? *Answers:* (a) 250 ft-lb in each case; (b) Bending effect increases as force B moves toward D.

5-3. If force A in Fig. 5-4 were to act alone through a lever, what torque would be produced on the shaft? What dissimilarity in conditions would be produced by this single force as compared to the couples in the previous problem?

5-4. The windlass shown in Fig. 5-5 is operated at constant speed by a man applying a force of 100 lb to the crank. If the two gears are 4 and 24 in. in diameter, what maximum weight may be lifted by the cable passing over the 6-in.-diameter sheave? What is the torque acting on each of the shafts?

5-5. Two pulleys, 12 and 24 in. in diameter, respectively, are placed on opposite ends of a horizontal shaft of 2-in. diameter. (a) If the belt placed around the 24-in. pulley has 2500- and 500-lb tensile forces acting vertically downward on opposite sides of the pulley, what torque is acting on the shaft? (b) What must the torque be on the 12-in. pulley to maintain equilibrium? (c) What single force (other force of couple obscured) acting tangentially to the surface of the 12-in. pulley will keep the shaft in equilibrium?

5-6. In the gear train shown in Fig. 5-6, a clockwise torque of 2000 in.-lb is applied to wheel *A* without moving gear *E*. Calculate the torque in shafts *B*, *C*, and *D*. *Answers:* T_B = 1000 in.-lb, T_C = 500 in.-lb, T_D = 250 in.-lb.

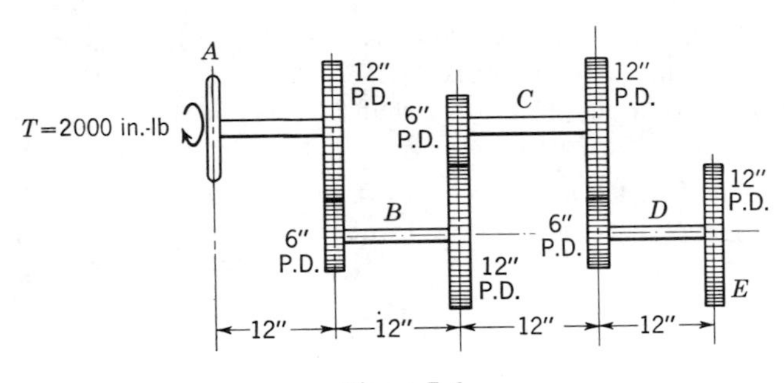

Figure 5-6

5-3 Development of Torsion Formula

Proper shaft design requires a knowledge of the relationship between the applied torque and the resulting stresses and strains. Therefore, it is important to discover first how these stresses and strains vary qualitatively on any one cross section. Let us consider the shaft shown in Fig. 5-7a acted

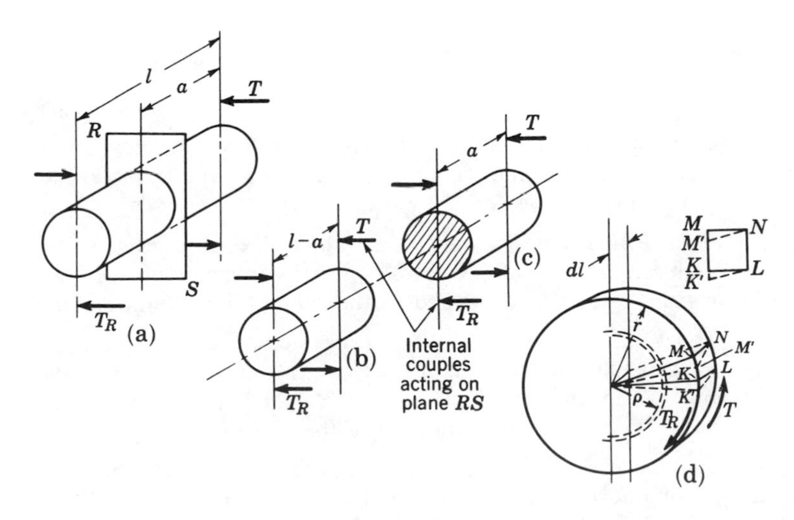

Figure 5-7

upon by an applied couple T and resisting couple T_R. Perfect balance exists over the full length of the shaft because the couples T and T_R at either end are equal to each other. Moreover, each intermediate section offers a resistance capable of maintaining this balance. To determine the couples involved in this balancing operation, let us pass plane RS (Fig. 5-7a) through the shaft perpendicular to its axis at any intermediate point. To rebalance the right-hand segment (Fig. 5-7c) a new T_R couple must act on its exposed section. The left-hand segment (Fig. 5-7b) is balanced by a couple T also acting on the exposed section. Thus each free-body segment is under the same torque condition that prevailed on the uncut shaft. Because these couples were required at a random intermediate section, they must also exist on every section.

It is a generally accepted fact that when a couple acts on any section it twists or rotates that section about its center of resistance. In the case of a circular section, the center of resistance and the rotation point are located at its center because of symmetry. Therefore, the axis of the shaft contains the rotation point of each section and remains straight, even though the shaft should twist.

Suppose, now, we consider the right-hand segment to be reduced in length to an infinitesimal dl, as shown in Fig. 5-7d. As the torques T and T_R act on either side of the segment and cause a relative motion of one side with respect to the other, the following deductions can be made:

1. If a point on the exposed face of the segment moves, it will travel in a circular path.

2. The farther the point is from the center of rotation, the greater will be the distance traveled.

3. The relative motion of initially opposite points increases proportionally with the distance from the center.

4. The direction of this relative motion or strain at any point is tangent to the circle passing through that point.

5. The strain, because of the parallel movement of the one end section relative to the other, is definitely of the shear type.

It was previously indicated that unit strain was the result of unit stress, and that the two were proportional to each other below the proportional limit. Based on these two facts, it must also be true that the shearing unit stress below the proportional limit varies directly with the distance from the center and acts in the direction of the strain, tangent to the circle passing through the point where the shearing unit stress is developed. Expressed mathematically,

$$\frac{\tau}{r} = \frac{\tau_\rho}{\rho} \tag{5.1}$$

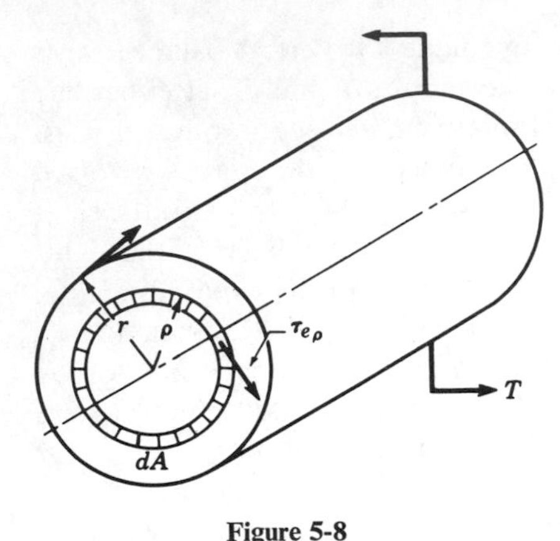

Figure 5-8

where r = radius of shaft and ρ = any radial distance shorter than the radius r.

Turning now to the free body shown in Fig. 5-8, let us consider the resistance offered by the shearing stresses developed on the exposed section.

At any distance ρ, let us draw a narrow circular band of very small width—so small that the unit shearing resistance acting thereon is a constant τ_ρ. If, then, the circular element whose area equals dA were divided into n parts, the shearing force on each part would be

$$\frac{dA}{n}\tau_\rho$$

The moment of this small resisting force will then be

$$\frac{dA}{n}\tau_\rho \rho$$

But there are n parts included in dA. The total resisting moment of the force developed on dA is, therefore,

$$dT_R = \frac{dA}{n}\tau_\rho \rho n = dA\,\tau_\rho \rho$$

Substituting for the value of τ_ρ its equal $(\tau/r)\rho$, we obtain

$$dT_R = \frac{\tau}{r}\rho^2 dA$$

The circular element chosen is only one of the many that can be included on the cross section. *If the resisting moment of every circular element could be added together, their sum would be equal to* T_R, *i.e., the resisting moment required to hold the free body in equilibrium*, and also to T, the applied torque. Thus, by integration,

$$T = T_R \int_0^r \frac{\tau}{r}\rho^2 dA \tag{5.2}$$

But since τ/r is included in the resisting torque of every circular element and $\int_0^r \rho^2\, dA$ is the general term for polar moment of inertia* expressed as J,

$$T = \frac{\tau}{r}J \tag{5.3}$$

*See Appendix for further discussion.

where τ = maximum shearing stress occurring on the outside fibers in psi, r = radius of shaft in inches, J = polar moment of inertia of cross section in in.^{4}, T = applied torque in in.-lb. This is known as the *torsion formula.* It is applicable to the design of all circular shafts.

5-4 Torsion Formula Adapted for Hollow Circular Shafts

In the development of the torsion formula above, the resisting moments of all the circular elements on the cross section were added to secure the relationship

$$T = T_R = \frac{\tau}{r}\int_0^r \rho^2 dA$$

If the cylindrical shaft were hollow, the number of circular elements that could be included in the cross section would be decreased by the number excluded by the hole. The polar moment of inertia of the net section would, therefore, include only those $\rho^2\, dA$ values for the actual circular elements that could be drawn within its boundaries. This summation can be performed by subtracting the $\rho^2\, dA$ values omitted by inserting the hole from the $\rho^2\, dA$ values included in the gross cross section. The above relationship may, therefore, be modified for hollow cylinders as follows:

$$T = T_R = \frac{\tau}{r}\left(\int_0^r \rho^2 dA - \int_0^{r_i} \rho^2 dA\right)$$

It follows that

$$T = T_R = \frac{\tau}{r}(J_o - J_i) \tag{5.4}$$

where J_o = the polar moment of inertia for the cross section included within the outside circle, $\pi d_o^4/32$, and J_i = the polar moment of inertia for the cross section included within the hole, or inside circle, $\pi d_i^4/32$.

5-5 Assumptions and Limitations of the Torsion Formula

The assumptions and limitations included in the development of the torsion formula are as follows:

1. The imposed shearing unit stress varies directly with the distance from the center of the shaft. Photoelastic experiments have substantiated this assumption.

2. A diameter before twisting remains a diameter after twisting. This assumption implies that the unit strain varies directly with the distance from the center of the shaft. In conjunction with the first assumption, it further

implies that the material must follow a straight stress-strain curve and that the shearing proportional limit of the material must not be exceeded.

3. A plane cross section before twisting remains plane after twisting. This is true only of circular cross sections.

4. The torsion formula is applicable only to circular or hollow circular sections if the hole therein is concentric.

Inasmuch as the stress on a rectangular or square section does not vary linearly from the center to the outside fibers, Eq. (5.1) does not apply to their design.

Illustrative Problem 1. What percentage of the strength of a solid-steel circular shaft 5 in. in diameter is lost by boring a 2.5-in. axial hole in it? Maximum allowable unit stress is 12,000 psi. Compare the weight of the shaft before and after boring.

SOLUTION. The maximum allowable torque that can be applied to the solid shaft is

$$T = \frac{\tau J}{r} = \frac{12{,}000 \times \pi \times 5^4}{2.5 \times 32} = 295{,}000 \text{ in.-lb}$$

The hollow shaft can be subjected to a torque equal to

$$T_H = \frac{\tau}{r}(J_o - J_i) = \frac{12{,}000}{2.5} \times \frac{\pi}{32}(5^4 - 2.5^4) = 276{,}000 \text{ in.-lb}$$

Percentage of reduction in strength equals

$$\frac{T - T_H}{T} \times 100 = \frac{295{,}000 - 276{,}000}{295{,}000} \times 100 = 6.45 \text{ per cent}$$

Since the length remained the same after boring, the percentage of reduction in weight is proportional to the reduction in area.

Percentage of reduction in weight equals

$$\frac{A_o - (A_o - A_i)}{A_o} \times 100 = \frac{\pi 5^2/4 - \pi 5^2/4 - \pi 2.5^2/4}{\pi 5^2/4} \times 100$$

$$= \frac{5^2 - 5^2 + 2.5^2}{5^2} \times 100$$

$$= \frac{6.25}{25} \times 100 = 25 \text{ per cent}$$

The reason for sometimes using hollow shafts is apparent when consideration is given to the amount of unit stress taken by the fibers of a solid shaft close to its axis. These fibers can take only a very small part of the maximum allowable stress taken by the outermost fiber, since the shearing stress is proportional to the distance from the center of the shaft, and the maximum stress is a fixed tabular value that cannot be exceeded. Capable of taking more stress than that which is placed upon them, these inside fibers often do not sustain a sufficient amount of load to justify their inclusion in the total weight to be carried.

Illustrative Problem 2. Compute the diameter of a hollow shaft with a drilled axial hole one half the outside diameter that will withstand the torque applied to the solid shaft in Illustrative Problem 1. Use same maximum allowable unit stress, 12,000 psi.

SOLUTION. Torque = 295,000 in.-lb. Substituting in the torsion formula, we obtain

$$T = \frac{\tau(J_o - J_i)}{r}$$

$$295{,}000 = \frac{12{,}000 \times \frac{\pi}{32}\left[d_o^4 - \left(\frac{d_o}{2}\right)^4\right]}{\frac{d_o}{2}}$$

$$295{,}000 = \frac{12{,}000\pi}{16d_o}\left(d_o^4 - \frac{d_o^4}{16}\right)$$

$$\frac{295{,}000 \times 256}{12{,}000 \times \pi \times 15} = d_o^3$$

$$d_o^3 = 133.50, \qquad d_o = 5.10 \text{ in.}, \qquad d_i = 2.55 \text{ in.}$$

Illustrative Problem 3. If the shear stress in a solid circular shaft varies in accordance with the equation $\tau_\rho = k\rho^2$, what will its total resisting torque be?

SOLUTION. From the derivation of the torsion equation

$$dT_R = \rho \cdot \tau_\rho \, dA$$

Inserting $dA = 2\pi\rho \cdot d\rho$

$$dT_R = \rho \cdot \tau_\rho \cdot 2\pi\rho \cdot d\rho = 2\pi\tau_\rho\rho^2 \cdot d\rho$$

from which $dT_R = 2\pi \cdot k\rho^2 \cdot \rho^2 d\rho$

and $T_R = 2\pi k \int_0^R \rho^4 d\rho = \frac{2\pi k R^5}{5} = \frac{4}{5}kJR$

PROBLEMS

5-7. What diameter of shaft is required to withstand a torque of 2360 in.-lb if the maximum allowable shearing unit stress is 12,000 psi? *Answer:* 1.00 in.

5-8. If a hollow shaft measuring 2 in. in outside diameter and $1\frac{1}{2}$ in. in inside diameter is to resist a torque of 10,000 in.-lb, what maximum shearing unit stress will be developed? What will the shearing stress be on the innermost fibers? *Answers:* 9300 psi, 6980 psi.

5-9. An extra-strong pipe having an outer diameter of 2.375 in. and an inside diameter of 1.939 in. is being screwed into a coupling by a plumber using a 24-in. pipe wrench. If the effective moment arm is 20 in. and the force applied is 150 lb, what is the maximum shearing unit stress developed in the pipe?

5-10. A solid steel shaft, 2 in. in diameter, is to be fitted and welded to the end of a hollow steel shaft having an inside diameter of 2 in. If the maximum allowable shear stress in steel is 15,000 psi, what will the minimum outside diameter have to be to transmit the maximum permissible torque? *Answer:* 2.38 in.

5-11. Two meshed gears of 8- and 12-in. pitch diameters rotate at a constant speed (Fig. 5-9). If the 12-in. gear is driven and develops a torque of 6000

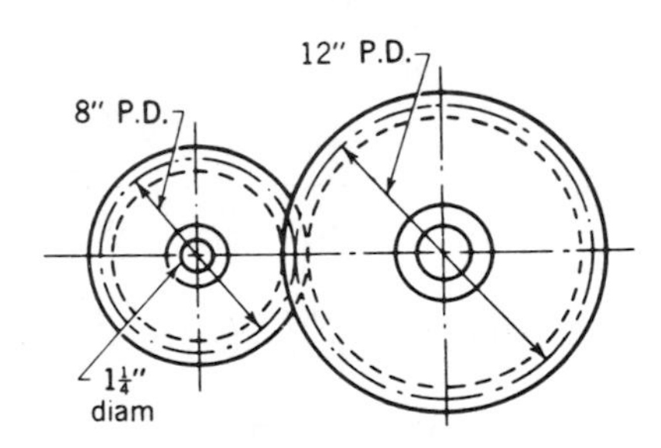

Figure 5-9

in.-lb, what maximum shearing unit stress will be found in the $1\frac{1}{4}$-in. shaft driven by the 8-in. gear? *Answer:* 10,400 psi.

5-12. In the solid steel stepped shaft of Fig. 5-10, the maximum allowable shearing unit stress is restricted to 15,000 psi. If the shaft is fixed at D, what will be the maximum torque T_A permitted on the shaft A? *Answer:* 1475 in.-lb.

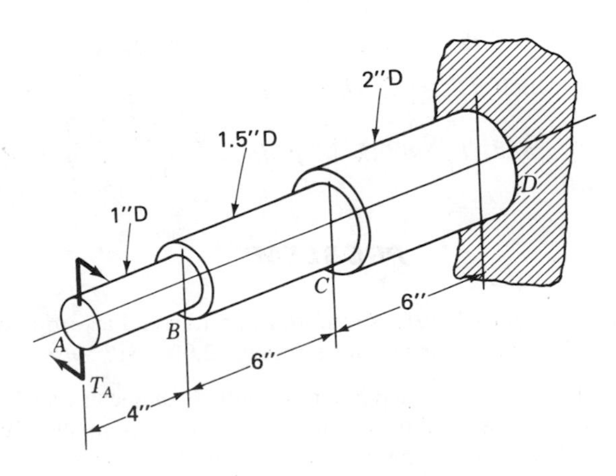

Figure 5-10

5-13. If, in Fig. 5-10, T_A was set at 2000 in.-lb and an opposing torque T_B set at 5000 in.-lb at B, what would the maximum shearing unit stresses be in the three segments of the shaft?

5-14. If, in Fig. 5-10, clockwise torques of 2000 and 5000 in.-lb were placed at A and B, respectively, with a counterclockwise torque of 20,000 in.-lb placed at C, what would the maximum shearing unit stresses be in the three segments of the shaft? *Answers:* 10,200 psi, 10,600 psi, 8280 psi.

5-15. What is the limit of torque that can be applied to a hollow circular shaft 2 in. O.D., $1\frac{1}{2}$ in. I.D. whose shearing yield stress is $\tau_y = 30{,}000$ psi?

5-16. What will the resisting torque be for a solid circular shaft whose shear stress varies with the distance from the center of rotation in accordance with $\tau_\rho = k\rho^{1/2}$, where k is a strength constant and ρ is the distance from the center? *Answer:* $T = \frac{4}{7}\pi k r^{7/2}$.

5-17. A solid composite shaft consists of two concentrically placed shafts A and B, one (A) solid, the other (B) hollow. If the inner shaft of diameter d_i has a shearing yield stress of 30,000 psi and the outer shaft has an outside diameter d_o and a shearing yield stress of 20,000 psi, what limiting torque may be obtained?

5-18. A hollow circular shaft having an outside and inside diameter of d_o and d_i, respectively, is made of a material which is linearly elastic, ideally plastic, having a τ-γ diagram similar in shape to the σ-ϵ diagram of Fig. 3-21b. Determine the ultimate torsional moment that the shaft can carry. *Answer:* $T_u = \pi\tau_0[d_o^3 - d_i^3]/12$

5-19. A hollow circular shaft whose outer radius is twice the inner radius is made up of an elastic, perfectly plastic material, having a τ-γ diagram similar in shape to the σ-ϵ diagram of Fig. 3-21b. Find the ratio of the ultimate torque to the yield torque (torque at which yielding will just begin).

5-6 Shaft Couplings

It often becomes necessary to join two shafts in order to transmit torque over a long distance. A common joint used for this purpose is a *bolted coupling*. In this connection, the torque from one shaft is passed through the bolts of the coupling by their resistance to shear, to the other shaft.

If the coupling shown in Fig. 5-11a is separated at its joint by means of a cutting plane and Fig. 5-11b taken as a free body, the exposed surfaces of the bolts must offer sufficient shearing resistance through their moment arm r to hold T in equilibrium. If τ is considered as the developed shearing unit stress spread uniformly over each rivet cross section A_s, and since $T = T_R$, we may equate the moments as follows:

$$T = T_R = n \times A_s\tau \times r \tag{5.5}$$

Illustrative Problem 4. Two shafts 5 in. in diameter are to be joined by a bolted coupling. If the maximum allowable shearing unit stress in the shafts is 10,000 psi, the diameter of the bolt circle is 10 in., and the allowable shear-

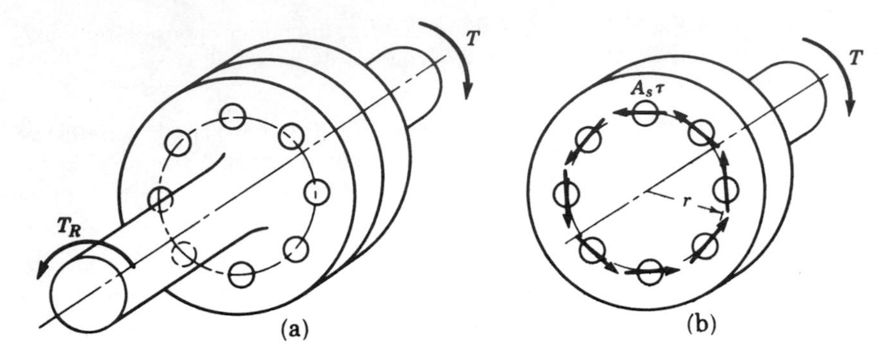

Figure 5-11

ing unit stress in the bolt is 9000 psi, find the number of $\frac{7}{8}$-in. bolts necessary for the connection.

SOLUTION. The torque to be transferred by the coupling is found from the torsion formula as applied to the 5-in. shafts.

$$T = \frac{\tau J}{r} = \frac{10{,}000 \times \pi 5^4}{2.5 \times 32} = 245{,}000 \text{ in.-lb}$$

To successfully transmit this torque, the resisting moment of the bolts must be equal to it.

Hence, the resisting force on one bolt equals

$$9000 \times \frac{\pi}{4}\left(\frac{7}{8}\right)^2 = 5400 \text{ lb}$$

and the resisting moment of the force on one bolt equals

$$5400 \times 5 = 27{,}000 \text{ in.-lb}$$

If, then, the resisting moment of one bolt is 27,000 in.-lb, and there are a total of 245,000 in.-lb to resist, the number of bolts required will be equal to

$$n = \frac{245{,}000}{27{,}000} = 9.08, \qquad \text{or 10 bolts}$$

This problem could, of course, be solved by direct substitution in Eq. (5.5) as follows:

$$n = \frac{T}{A_t \tau r} = \frac{245{,}000}{0.60 \times 9000 \times 5} = 9.08, \qquad \text{or 10 bolts}$$

It is suggested, however, that the first procedure, which depends on logic, be used rather than the second, which depends upon memorization of the formula.

PROBLEMS

5-20. Two 4-in.-diameter solid steel shafts are to be joined by means of flanged couplings and eight $\frac{1}{2}$-in.-diameter steel bolts set on a bolt circle d in. in diameter. Determine the diameter of the bolt circle in order that the bolts and the steel shafts be equally strong in resisting torsion. Maximum allowable $\tau = 8000$ psi. *Answer:* 16 in.

5-21. How many $\frac{3}{4}$-in. bolts will be required in a coupling used to join two 3-in. shafts if the bolt circle is 8 in. and the torque is 150,000 in.-lb? Maximum allowable shearing unit stress in bolts is 10,000 psi.

5-22. Two 4-in.-diameter shafts are to be connected by means of a coupling. What torque may safely be transmitted through this coupling if eight $\frac{7}{8}$-in. bolts having an allowable shearing unit stress of 12,000 psi are placed on a 10-in. bolt circle? *Answer:* 288,000 in.-lb.

5-23. To prevent damage to the engine and propeller of a tugboat, a short auxiliary shaft is placed between the main shaft and the propeller shaft to act as a *safety valve*. If the main and propeller shafts are each 6 in. in diameter and the auxiliary is 5 in., how many 1-in. bolts are required in each coupling if the bolt circle is 12 in. in diameter and the maximum torque to be restrained is computed on the basis of a proportional limit stress of 24,000 psi being attained in the auxiliary shaft? Maximum allowable shearing unit stress in bolts is 10,000 psi. *Answer:* Use 13 bolts.

5-24. A coupling flange (Fig. 5-12) is made by joining a hollow steel shaft, $3\frac{3}{4}$-in. O.D. by $2\frac{1}{2}$-in. I.D., to the center of a steel disk, 10 in. in diameter and

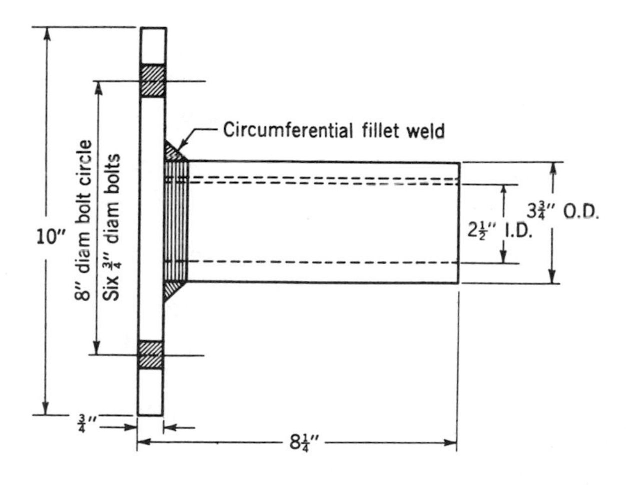

Figure 5-12

$\frac{3}{4}$ in. thick, with a circumferential fillet weld. What is the theoretical size of the 45° fillet weld that would be required to develop the maximum torque? Maximum allowable shear stress in shaft and bolts = 15,000 psi. Maximum allowable bearing stress of bolts against plate = 30,000 psi. Consider center of 45° shear plane in weld to be located 2 in. from center of shaft. *Answer:* 0.52 in.

5-7 Relationship Between Horsepower, Revolutions Per Minute, and Torque

One of the most useful relationships in mechanical and electrical design is the relationship between the horsepower, speed, and torque as applied to a shaft. With this relationship the horsepower delivered by a shaft can be computed, if we know the speed and torque supplied to the shaft. Or, the torque capable of being delivered by a shaft can be determined by using the speed and horsepower of the motor or engine employed. To obtain greater facility in the use of the relationship, let us first develop it step by step.

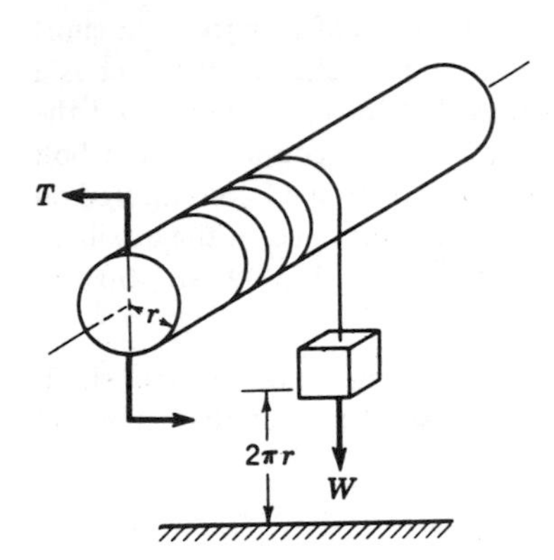

Figure 5-13

Figure 5-13 shows a shaft acted upon by a weight W placed at the end of a rope wrapped around the shaft several times.

Suppose the shaft is rotated so that weight W is lifted a distance of $2\pi r$; then the work done in lifting weight W a distance of $2\pi r$ (one revolution of the shaft, where r = inches) is

$$W \times 2\pi r \quad \text{in.-lb} \qquad \text{(force} \times \text{distance)}$$

The work done in lifting weight W through q revolutions of the shaft equals

$$W \times 2\pi r \times q \quad \text{in.-lb}$$

If weight W is to be lifted through q revolutions in t minutes, the rate of work performed equals

$$\frac{W \times 2\pi r \times q}{t} \quad \text{in.-lb/min}$$

But $q/t = n$, angular speed per minute, and $Wr = T$, the torque supplied to the shaft; therefore, the rate of work performed in the shaft $= 2\pi Tn$ in.-lb/min.

To convert inch-pounds per minute to horsepower (33,000 ft-lb/min = hp), we may write

$$\text{hp} = \frac{2\pi Tn}{12 \times 33{,}000} \tag{5.6}$$

It should be recalled that to use this formula properly, T must be in inch-pounds and n in revolutions per minute.

Illustrative Problem 5. A pulley 6 ft in diameter is mounted on a 2-in. shaft operating at 150 rpm. Placed around the pulley is a belt whose opposite tensions are 2500 and 1100 lb (Fig. 5-14). (a) Find the horsepower being transmitted. (b) What is the maximum shearing unit stress in the shaft?

SOLUTION. To find the horsepower being transmitted requires knowledge of both the speed and the torque. The torque, determined from the belt tensions and the radius of the pulley, is

$$T = (1500 - 1100) \times 36 = 14{,}400 \text{ in.-lb}$$

The horsepower being transmitted is, therefore,

$$\text{hp} = \frac{2\pi Tn}{12 \times 33{,}000} = \frac{2\pi \times 14{,}400 \times 150}{12 \times 33{,}000}$$

$$= 34.36$$

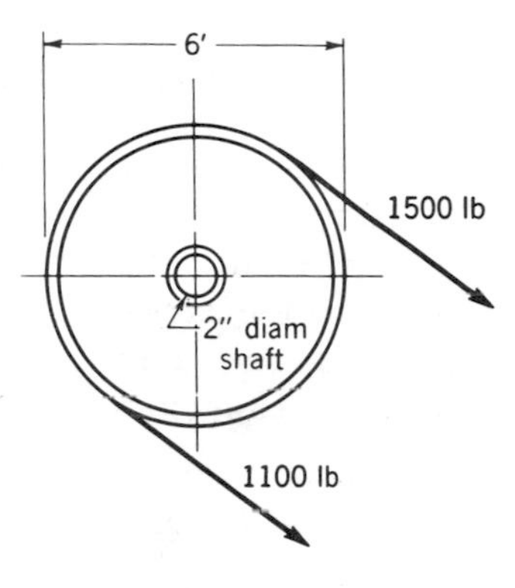

Figure 5-14

The maximum shearing unit stress in the shaft is

$$\tau = \frac{Tr}{J} = \frac{14{,}400 \times 1}{(\pi \times 2^4)/32} = 9170 \text{ psi}$$

PROBLEMS

5-25. A $1\frac{1}{2}$-in. shaft is operating at a speed of 350 rpm under a maximum shearing unit stress of 10,000 psi. Determine the horsepower being delivered by the shaft.

5-26. Determine the work produced by a 4-in.-diameter shaft per revolution while transmitting 400 hp at 180 rpm. Calculate also the torque and the maximum shear stress acting on the shaft. *Answers:* 11,680 ft-lb, 11,130 psi.

5-27. A $1\frac{1}{2}$-in.-diameter shaft operated by a motor whose shaft horsepower equals 20 rotates at a speed of 400 rpm. What is the maximum shear stress acting on the shaft?

5-28. What diameter of shaft will be necessary to transmit a shaft horsepower of 20 under a speed of 1200 rpm? The maximum allowable shearing unit stress is 10,000 psi. *Answer:* 0.81 in.

5-29. Find the outside diameter of a hollow shaft required to transmit a horsepower of 1000 while operating at a speed of 120 rpm, if its inner diameter is 4 in. and the maximum allowable shearing unit stress is 8000 psi.

5-30. Determine the speed of a motor, undergoing a dynamometer test, which is delivering 25 hp under a resistance of 6000 in.-lb. If the shaft is 2 in. in diameter, what is its maximum shear stress? *Answers:* 263 rpm, 3820 psi.

5-31. The wrapping and processing mechanisms of a textile machine are synchronized by a 40-ft steel shaft, $1\frac{1}{4}$ in. in diameter. If the torque necessary to operate the wrapping mechanism is 2000 in.-lb, what horsepower must be delivered to the shaft at the processing point to maintain a shaft speed of 800 rpm. What is the maximum shear stress developed in the shaft? *Answers:* 25.4 hp, 5240 psi.

5-8 Angle of Twist Produced by Torque

In some instances the controlling feature of the design of a shaft is based upon its resistance to twisting rather than its strength. Lathes, milling machines, and other precision tools usually have shafts which, if designed solely on the basis of strength, would twist too much for the high accuracy required. They are usually stiffened by increasing the diameter according to the formula about to be derived.

Derivation. Consider the strip element *AB* on the shaft of Fig. 5-15 and the rectangular element *KLMN*, a portion of element *AB*. Upon application of

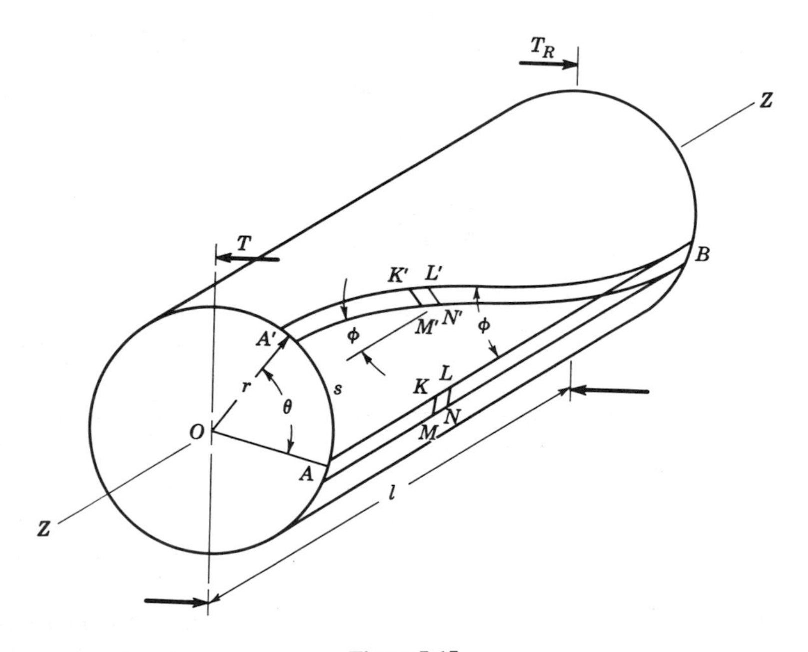

Figure 5-15

torque *T*, the shaft twists so that radius *OA* rotates to a new position *OA'*, and the rectangular element *KLMN* moves to *K'L'M'N'*. In producing this movement, element *KLMN* is distorted by the shearing unit stress. From §2-10 unit strain is

$$\gamma = \tan \phi$$

But
$$\tan \phi = \frac{s}{l}$$

Hence
$$\gamma = \frac{s}{l} \quad \text{or} \quad \frac{r\theta}{l}$$

where θ is expressed in radians.

It is common practice when determining relationships between a deformation and a force to apply Hooke's law for the particular type of stress under consideration.

$$G = \frac{\tau}{\gamma}$$

Since $$\tau = \frac{Tr}{J} \quad \text{and} \quad \gamma = \frac{r\theta}{l}$$

$$G = \frac{Tr/J}{r\theta/l} = \frac{Tl}{J\theta} \tag{5.7a}$$

or $$\theta = \frac{Tl}{GJ} \tag{5.7b}$$

where θ = angle of twist in radians, T = torque expressed in in.-lb, l = length of shaft in inches, G = shearing modulus of elasticity in psi, J = polar moment of inertia in in.^{4}.

If the above derivation were based on an infinitesimally small element required for more complex analyses of shafts, it would proceed with the inclusion of infinite rather than finite values, e.g., dz for l, $d\theta$ for θ, etc. Thus, since $\gamma = r(d\theta/dz)$, $\tau = Tr/J$, and $G = \tau/\gamma$,

$$G = \frac{Tr/J}{r(d\theta/dz)} = \frac{T\,dz}{J\,d\theta} \quad \text{or} \quad d\theta = \frac{T\,dz}{GJ}$$

Illustrative Problem 6. Compute the diameter of a solid steel shaft necessary to carry a torque of 400,000 in.-lb if the twist is not to exceed 1° in 5 ft, the maximum allowable shearing unit stress is to be held at 12,000 psi, and G = 12,000,000 psi.

SOLUTION. The diameter required to meet the twist specification is as follows:

$$\theta = \frac{Tl}{GJ}$$

$$\frac{1^\circ}{57.3} = \frac{400{,}000 \times 5 \times 12 \times 32}{12{,}000{,}000 \times \pi d^4}$$

$$d^4 = \frac{400{,}000 \times 5 \times 12 \times 32}{12{,}000{,}000 \times \pi} \times 57.3 = 1170$$

$$d = 5.85 \text{ in.}$$

The diameter required to hold the maximum shearing unit stress at 12,000 psi is

$$\tau = \frac{Tr}{J} = \frac{T(d/2)}{(\pi d^4)/32} = \frac{16T}{\pi d^3}$$

Hence, $$d^3 = \frac{16T}{\tau\pi} = \frac{16 \times 400{,}000}{12{,}000 \times \pi} = 170$$

$$d = 5.53 \text{ in.}$$

It is, therefore, necessary to use a diameter of at least 5.85 in. to meet the more stringent condition of twisting.

Illustrative Problem 7. A hollow steel cylindrical shaft 10 in. in diameter and 6 in. in outside diameter is stressed to 8000 psi. (a) What horsepower is being delivered when the shaft is turning at 1000 rpm? (b) What is the twist in degrees for 100 ft of this shaft? $G = 12 \times 10^6$ psi.

SOLUTION. (a) Since the torque is necessary to determine the horsepower, let us use the torsion formula for its determination.

$$T = \frac{\tau(J_o - J_i)}{r} = \frac{8000 \times (\pi/32)(10^4 - 6^4)}{5} = 1{,}368{,}000 \text{ in.-lb}$$

Inserting the torque in the horsepower formula, we obtain

$$\text{hp} = \frac{2\pi \times 1{,}368{,}000 \times 1000}{12 \times 33{,}000} = 21{,}700$$

(b) The twist of the shaft is

$$\theta = \frac{Tl}{GJ} = \frac{1{,}368{,}000 \times 1200}{12{,}000{,}000 \times (\pi/32)(10^4 - 6^4)}$$

$$= 0.16 \text{ rad} = 0.16 \times 57.3 = 9.2^\circ$$

PROBLEMS

5-32. If a 1-in.-diameter steel drill rod 8000 ft long is used in drilling an oil well, is it possible that the rod may twist a complete revolution without exceeding its maximum allowable shearing unit stress of 20,000 psi? Show figures. *Answer:* Yes. 51.0 revolutions.

5-33. Determine the limiting length of a $\frac{1}{2}$-in.-diameter Duralumin (2017-T3) shaft that will not exceed a twist of 10° when subjected to a torque of 500 in.-lb.

5-34. A 1-in.-diameter steel shaft is twisted 1° in a length of 4 ft. Determine the torque required. *Answer:* 428 in.-lb.

5-35. Determine the angle of twist of the shaft in Problem 5-31. *Answer:* 19.2°.

5-36. A solid circular steel shaft 4 in. in diameter is subjected to torsion as shown in Fig. 5-16. (a) Determine the angle of twist of a diameter in the right end with respect to the initial position for the unstrained shaft. (b) Determine the maximum shearing unit stress.

5-37. For the stepped steel shaft shown in Fig. 5-17, determine the total angle of twist of end C with respect to the fixed end A. Maximum allowable $\tau = 10{,}000$ psi. Modulus of rigidity $G = 12 \times 10^6$ psi. *Answer:* 0.0575 radians clockwise.

5-38. An aluminum circular shaft AD of radius $R = 2$ in. is subjected to external torques as shown in Fig. 5-18, where $T_o = 1000$ in.-lb. (a) Determine the

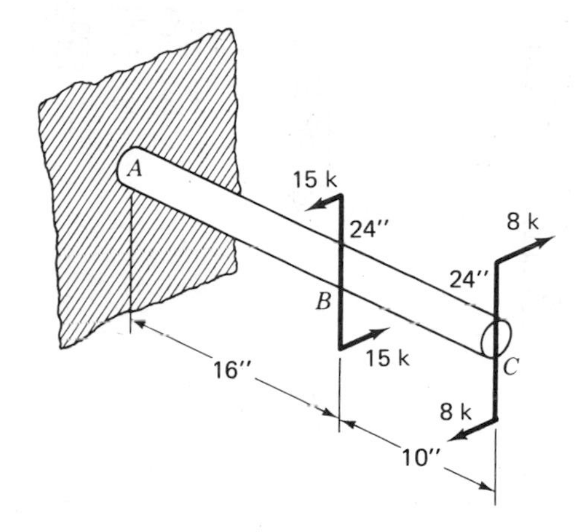

Figure 5-16

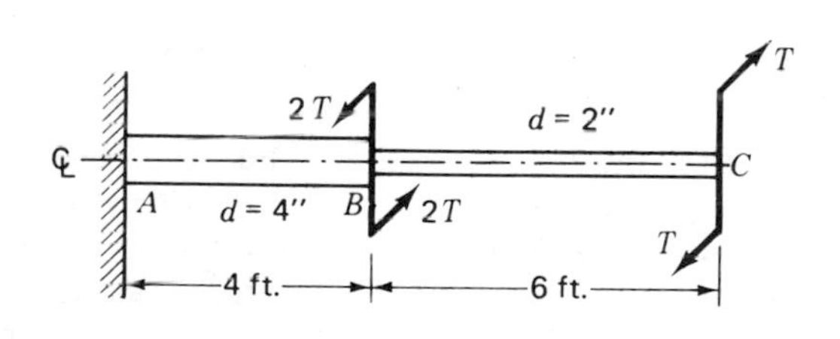

Figure 5-17

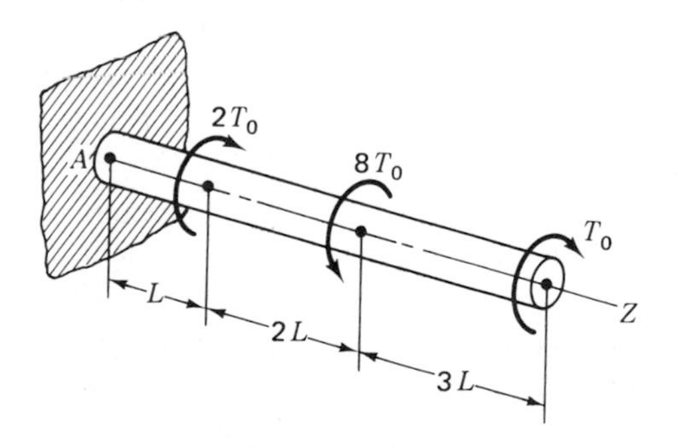

Figure 5-18

greatest shear stress existing in the shaft. (b) Determine the angle of rotation of section D if $L = 12$ in. Indicate direction of rotation. $G = 4 \times 10^6$ psi.

5-39. Four gears are arranged as shown in Fig. 5-19. If the torque supplied at the 24-in. gear A is 2000 in.-lb, through what angle could gear A be rotated if gear D were maintained stationary? Both shafts are made of steel. *Answer:* 9.18°.

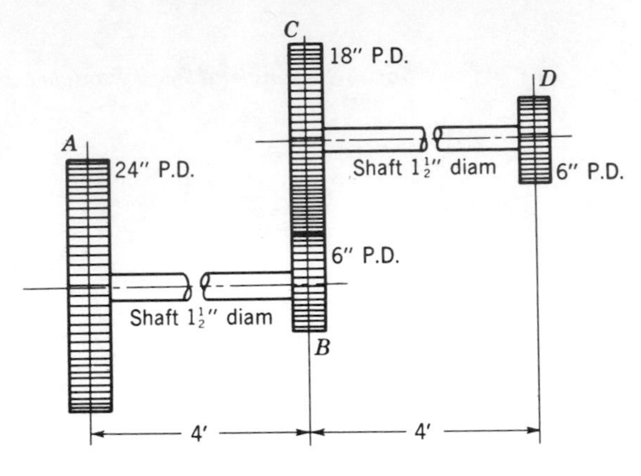

Figure 5-19

5-40. A solid circular shaft of steel is used to transmit 200 hp at 600 rpm. (a) What diameter of shaft is required if the allowable maximum shearing stress is 20,000 psi? (b) If a hollow steel shaft is used that has an inside diameter equal to the diameter of the solid shaft determined in part (a), what must be the outside diameter of this shaft if the angular twists of the two shafts are to be equal?

5-41. A hollow steel shaft 10 ft long, with 6-in. O.D. and 3-in. I.D., is driven by a diesel engine with 200 brake hp when the shaft is operating at 300 rpm. Calculate the shaft's angle of twist. *Answer:* 0.185°.

5-42. Pulleys A, B, and C are joined by solid steel shafting and are 30 and 18 in. apart, respectively. The torque applied at the driving pulley, A, is 3540 ft-lb; the torque delivered at pulley B is 2150 ft-lb; the torque delivered at pulley C is 1390 ft-lb. (a) If maximum allowable τ is 8000 psi for the steel shaft, determine the required shaft diameters for each section. (b) If the shaft has a 3-in. diameter over its entire length, find the angle of twist in degrees between A and C. *Note:* The shaft rests in bearings just beyond A and C. Disregard any beam action due to transverse loads. *Answers:* (a) $d_1 = 3.0$ in., $d_2 = 2.2$ in.; (b) 0.952°.

5-43. A shaft of length l, and fixed at one end, is acted upon by a torque T at a point $\frac{2}{3}l$ from the fixed end. If the opposite end is now fixed and T released, what are the end restraining moments in terms of T?

5-44. A shaft l ft long and fixed at each end has applied at a point a ft from one end a torque T. Determine the end restraining moments in terms of T, the segmental distances a and b, and length l. *Answer:* Ta/l, Tb/l.

5-45. The shaft shown in Fig. 5-20 consists of two cylindrical bars of torsional stiffness G_1J_1 and G_2J_2 respectively. The bars are welded to each other and to two fixed walls. The shaft is subjected to two torques T_o. Assuming linearly elastic behavior and $G_1J_1 = 2G_2J_2$, determine (a) the reactions at the wall, and (b) the angle of rotation of section C.

5-46. Gear A transmits a torque of 30,000 in.-lb to two steel shafts (Fig. 5-21). Gear B takes off 10,000 in.-lb from shaft AB, and gear C takes off 20,000 in.-lb from shaft AC. Diameter of shaft AC is 3 in. (a) Determine the diameter of shaft AB so that the angle of twist between gear A and the end gears is the same magnitude. (b) Determine the maximum shearing unit stress in shaft AC. $G = 12 \times 10^6$ psi. *Answers:* (a) 2.12 in., (b) 3780 psi.

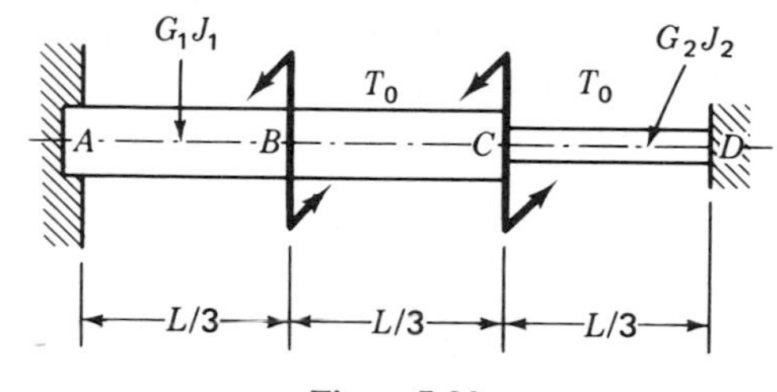

Figure 5-20

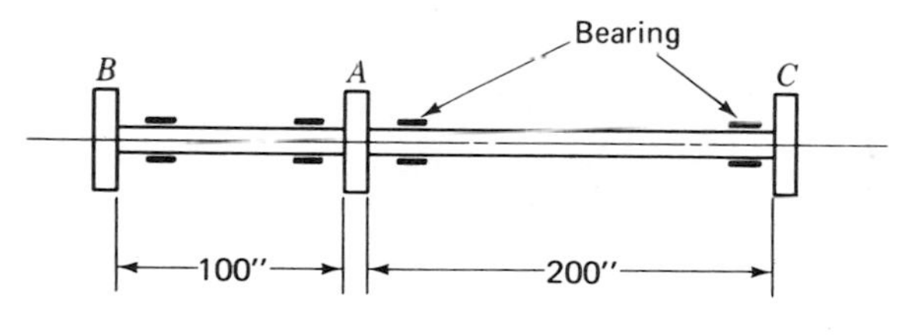

Figure 5-21

5-47. Three gears are attached to two 2-in.-diameter steel shafts as shown in Fig. 5-22. End D is fixed. If the gear at A is twisted through an angle of 11°, what is the maximum shearing stress in both shafts? $G = 12 \times 10^6$ psi.

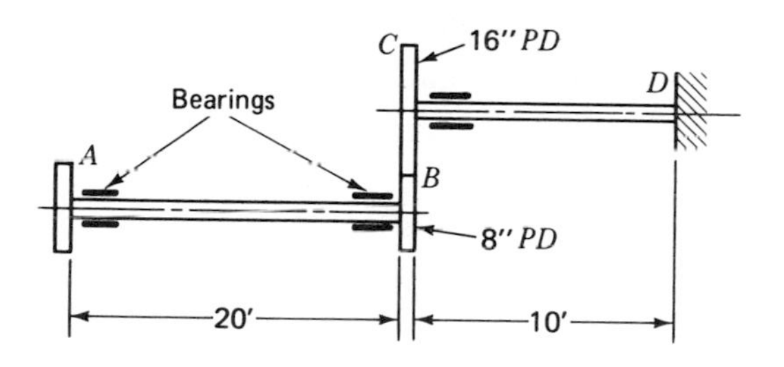

Figure 5-22

5-48. A 6-ft hollow aluminum shaft with 3-in. O.D., 2-in. I.D., has 2 in. of its length slipped over a 2-in.-diameter solid steel shaft 8 ft 2 in. long. Both are then fixed at their outer ends. If a torque of 20,000 in.-lb is applied to the aluminum shaft at its overlapped end, and then fastened rigidly in its distorted condition to the unstressed steel shaft, what will the maximum shear stress in the aluminum and steel shafts be when the applied torque is released? Consider the lengths of the aluminum and steel in the compound shaft to be 6 and 8 ft, respectively. $G_{al} = 4 \times 10^6$ psi. *Answers:* 1675 psi, 4540 psi.

5-49. A compound shaft AD fixed at each end consists of three segments. Segment AB, 2 in. in diameter and 2 ft long, is made of aluminum; BC, 3 in. in diameter and 3 ft long, is made of bronze; CD, 2 in. in diameter and 4 ft long, is made of steel. If similarly directed torques are applied at B and C equal to 5000 and 10,000 in.-lb, respectively, determine the restraining torques at the supports.

$G_{al.} = 4 \times 10^6$ psi, $G_{br.} = 6 \times 10^6$ psi, $G_{st.} = 12 \times 10^6$ psi. *Answers:* $T_L = 5890$ in.-lb, $T_R = 9110$ in.-lb.

5-50. A cantilevered circular shaft is subjected to a distributed torque which varies linearly from zero at its free end to a maximum value T at its fixed end. If the material is linearly elastic, find the twist of the shaft at its free end. *Answer:* $TL/2GJ$.

5-51. A cantilevered circular shaft is subjected to a distributed torque which varies in accordance with the equation $T_z = T\sqrt{z/L}$, where z is the distance from the free end. Find the twist of the shaft at its free end if the material is linearly elastic. *Answer:* $2TL/3GJ$.

5-52. In a cylindrical shaft of radius R subjected to end torques T_o, the stress-strain relation is $\tau = k\gamma^n$, where k and n are material constants. (a) Show that the relation between T_o and the relative angle of twist $d\theta$ is $T_o = (2\pi k/n + 3)R^{n+3}$ $(d\theta/dz)^n$. (b) Determine the maximum shearing stress.

5-9 Mechanical Properties in Torsion

Torque-twist curves (Fig. 5-23), similar in shape and character to the stress-strain curves discussed in Chapter 3, are drawn from data obtained from torsion tests. The data from such a test consist of corresponding values of torque and relative twist taken over a 6- to 10-in. test length. Those points on a torque-twist curve representing the *yield stress, proportional limit, and elastic limit* are located similarly and are identical in definition to those ob-

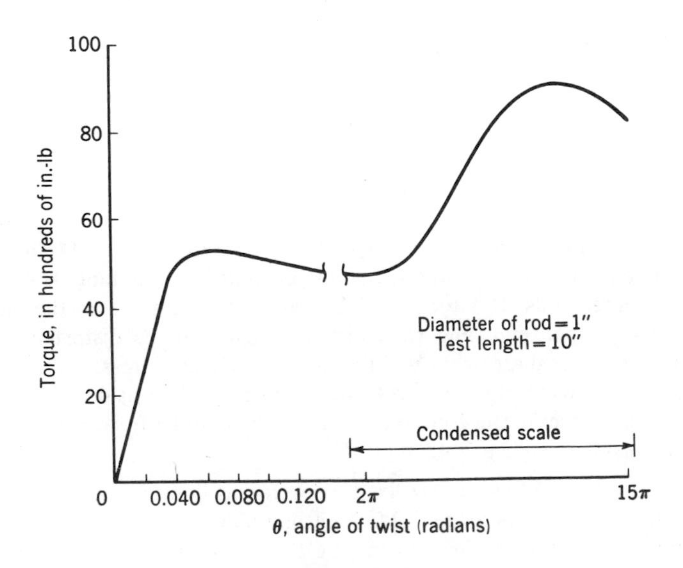

Figure 5-23

tained from a stress-strain curve obtained from an axial test. Their actual numerical values, however, must be computed from their corresponding torques using the torsion formula.

In order for the ultimate stress to be obtained with validity from the torsion formula, the variation of shear stress across the cross section must be linear throughout the entire torque-twist test. This is generally not the case. Only in the torsion tests of thin-walled hollow tubes and of isotropic, homogeneous materials with straight torque-twist curves is this a possibility. The stress obtained from the torsion formula using the maximum torque is a fictitious stress called the *modulus of rupture* (see §5-14).

The value of modulus of elasticity is computed from Eq. (5.7a),

$$G = \frac{Tl}{\theta J}$$

by substituting T/θ, the slope of the torque-twist curve below the proportional limit, l the test length, and J, its polar moment of inertia.

5-10 Longitudinal Shear Stresses Produced in Shafts by Torsion—Pure Shear

Figure 5-24a shows a very thin disk cut from a shaft subjected to torsion and placed in equilibrium as a free body (thickness exaggerated). If from the surface of this disk there is cut a free body consisting of a square element one unit deep (Fig. 5-24b) two equal areas $ABCD$ and $EFGH$ are exposed on which the shearing stress τ acts. Since these areas and the stresses acting upon them are equal, the forces so produced constitute a couple. But

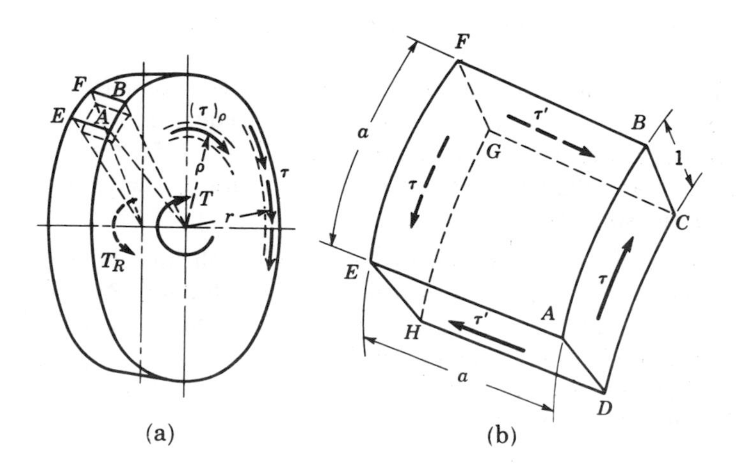

Figure 5-24

an element acted upon by only one couple cannot be in equilibrium: another couple is required to balance the free body.

This necessary balancing couple must be in the same plane as the applied couple. Its forces must act on the only remaining resisting planes, *BCGF* and *ADHE*.* the stresses on these planes must also be equal and consist of shear, since only the resultant shear forces provide the necessary couple. Let us temporarily designate the shearing unit stresses on planes *BCGF* and *ADHE* as τ'. From the geometry of the element, all four of the side areas are equal. We may, therefore, write the moment of equation of the shear forces acting on the element about line *AD* as

$$\tau \times \text{area } EFGH \times a = \tau' \times \text{area } BCGF \times a$$

But since area $EFGH =$ area $BCGF$ and $a = a$, then

$$\tau = \tau' \tag{5.8}$$

To generalize, whenever a circular shaft is twisted, two equal shearing stresses mutually perpendicular to each other are developed at every point on its surface, one on a plane passing through the axis, the other on a plane perpendicular to it. Because these shear stresses develop with no accompanying normal stresses, the planes on which they act are said to be subjected to *pure shear*.

It is of further interest to note that superimposed normal stresses do not interfere with the development or equality of these mutually perpendicular shear stresses. Thus, whatever the combination of normal and shear stresses which may occur at a point, the mutually perpendicular shear stresses will always be equal to each other.

5-11 Maximum Diagonal Stresses Produced by Torsion

In conjunction with the development of the two aforementioned equal-shearing unit stresses, maximum normal stresses equal in magnitude to the shear stresses are developed on the two planes making 45° to the shear planes. On one 45° plane, the normal stress is tensile in nature, whereas on the other 45° plane, the stress is compressive.

Proof. If a sectioning plane is passed through line *AD* of Fig. 5-24 at an angle θ to plane *ABCD*, and the portion including *JBA* (Fig. 5-25) is used as a free body, the stresses which act on the exposed plane *JADK* must recreate the condition of equilibrium. Because of the general nature of this analysis, both shear and normal stress, σ_n and τ_D, respectively, will be assumed to act on this plane.

*Because area *ABFE* is a free surface on which no shear forces act, no shear forces can theoretically develop on area *DCGH*.

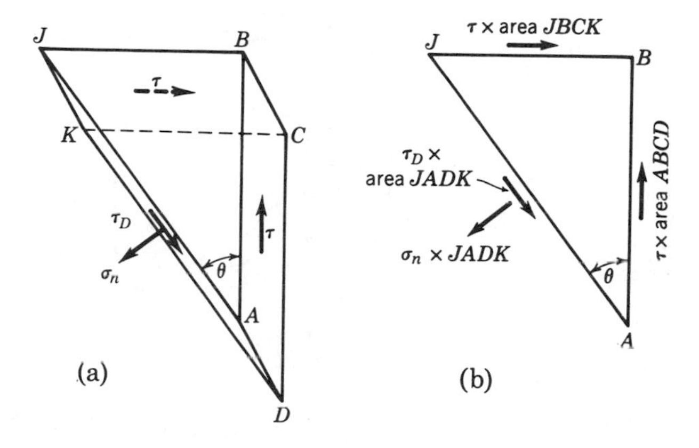

Figure 5-25

Summing the forces in the direction of τ_D (using line JA to represent area $JADK$ and so forth), it follows that

$$\tau_D \times JA = \tau \times BA \cos\theta - \tau \times JB \sin\theta$$

$$\tau_D = \tau \times \frac{BA}{JA}\cos\theta - \tau \times \frac{JB}{JA}\sin\theta$$

$$\tau_D = \tau(\cos^2\theta - \sin^2\theta) = \tau\cos 2\theta$$

Thus, when $\theta = 45°$, the value of $\tau_D = 0$. Furthermore, no shearing stress is developed which is greater than τ.

Performing a similar operation in the direction of the normal stress σ_n produces

$$\sigma_n \times JA = \tau \times JB \cos\theta + \tau \times BA \times \sin\theta$$

$$= \tau \times \frac{JB}{JA}\cos\theta + \tau\frac{BA}{JA}\sin\theta$$

$$= \tau \times 2\sin\theta\cos\theta = \tau \times \sin 2\theta$$

An analysis of this expression reveals that maximum normal stresses exist at values of $\theta = 45$ and $135°$. For $\theta = 45°$, when $\sin 2\theta = 1$, the maximum stress is tension—whereas, when $\theta = 135°$ ($\sin 2\theta = -1$), the maximum stress is compression.

It becomes evident, therefore, that the two equal and mutually perpendicular maximum shear stresses develop simultaneously with a complex of other shear and normal stresses, chief of which are the mutually perpendicular maximum tensile and compressive stresses which are equal to them (Fig. 5-26). From a quantitative viewpoint a condition of pure shear develops the following maximum stress relationship:

$$\tau = (\sigma_n)_{\max} = (\sigma_t)_{\max} = (\sigma_c)_{\max}$$

It should be noted that the maximum normal stresses are not accompanied

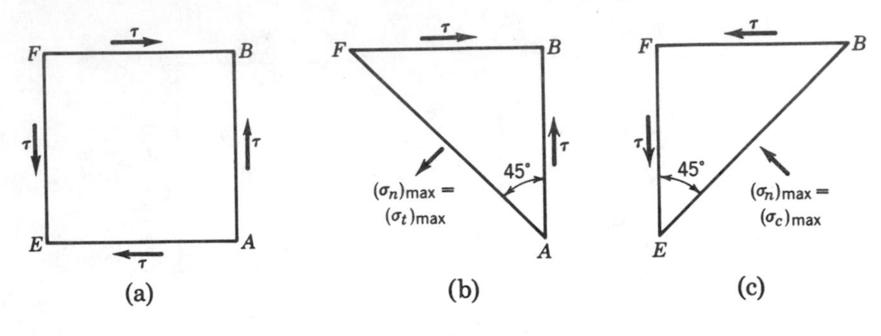

Figure 5-26

by shearing stresses. A further study of these conditions will be made in Chapter 12.

Thus, as a shaft is twisted, the critical stresses developed are as follows:

1. The maximum shearing unit stress τ equal to Tr/J acting at the surface of the shaft on planes perpendicular to the axis.

2. The maximum shearing unit stress τ acting at the surface of the shaft on planes passing through the longitudinal axis.

3. The maximum tensile unit stress $(\sigma_t)_{\max}$ acting at the surface of the shaft at an angle of 45° to the maximum shear stresses.

4. The maximum compressive unit stress $(\sigma_c)_{\max}$ acting at the surface of the shaft at an angle of 90° to the maximum tensile unit stress.

These four stresses are equal to one another and are produced simultaneously.

5-12 Importance of the Critical Stresses Developed in Shafts

It will be recalled from our previous discussion of the strengths of various materials that their resistances in tension, compression, and shear may vary considerably. Some materials may be weakest in shear, some in tension, and still others in compression. Failure will occur in torsional members on those planes for which the strength is least. Since all the maximum stresses develop equally and simultaneously, that stress first to arrive at its ceiling value will produce rupture.

The character of the failure of a shaft subjected to torque action may be forecast if the three strengths are known. For instance, a steel shaft will fail on a plane perpendicular to its longitudinal axis (Fig. 5-27a), because

1. Steel is weaker in shear than in tension or compression.
2. A maximum shearing stress is developed on this plane.

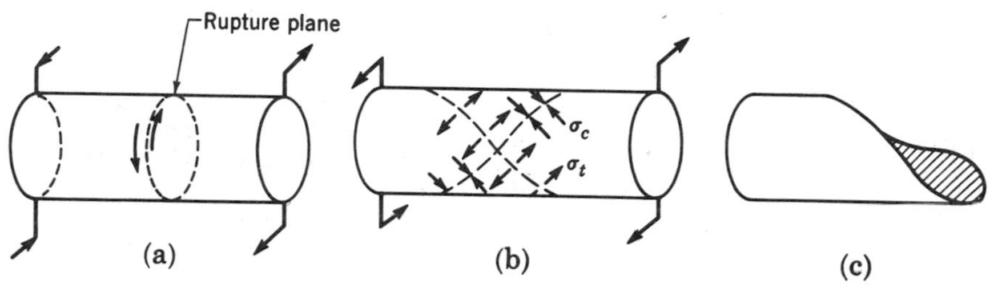

Figure 5-27

3. The gripping action at the shaft ends increases the resistance to a longitudinal shear failure.

In contrast, cast-iron shafts being weak in tension will fail along a helical plane inclined approximately 45° to the axis of the shaft (Fig. 5-27b and c). A piece of blackboard chalk will also rupture in the same characteristic manner.

In the light of the previous remarks, the design of a shaft must be governed by the weakest strength of the material to be used.

Illustrative Problem 8. A shaft is to be made from a new material having the following ultimate strengths: $\sigma_t = 20{,}000$ psi, $\sigma_c = 30{,}000$ psi, and $\tau = 30{,}000$ psi. If a safety factor of 4 is to be used, what diameter is required to resist a torque of 1000 ft-lb?

SOLUTION. A glance at the three ultimate strengths reveals the ultimate tensile unit stress to be the lowest. The design must, therefore, be made on the basis of a working unit stress of

$$(\sigma_t)_w = \frac{20{,}000}{4} = 5000 \text{ psi}$$

At the very moment that the torque has developed this maximum tensile unit stress of 5000 psi, the maximum shearing unit stress on the plane perpendicular to the axis is also 5000 psi. Since the relationship of diameter, torque, and maximum shearing unit stress is contained in the torsion equation, the use of the maximum shearing unit stress of 5000 psi as a limiting stress in the formula will provide the necessary diameter to also limit the tensile stress to 5000 psi. These working stresses are assumed to be below the proportional limit, so that the torsion formula will be applicable.

Hence, $$\tau = \frac{Tr}{J} \quad \text{and} \quad T = 1000 \text{ ft-lb} \quad \text{or} \quad 12{,}000 \text{ in.-lb}$$

$$5000 = \frac{1000 \times 12(d/2)}{\pi d^4/32} = \frac{12{,}000 \times 16}{\pi d^3}$$

$$d^3 = \frac{12{,}000 \times 16}{\pi \times 5000} = 12.20, \qquad d = 2.30 \text{ in.}$$

PROBLEMS

5-53. A high-carbon steel (0.7 per cent carbon) shaft is to be designed to take a torque of 40,000 in.-lb with a safety factor of 5. (a) Determine its diameter and the maximum shear, tensile, and compressive unit stresses. (b) Indicate by a sketch of the shaft the direction in which stresses will act. (c) If the torque were to be increased to cause rupture, how would the shaft fail? (d) Indicate the anticipated plane of failure on the sketch of (b). Refer to Table 2 for stresses. *Answer:* (a) 2.34 in.

5-54. A ceramic insulator will have imposed upon it a torque of 600 in.-lb. If the ultimate strengths in shear, tension, and compression are, respectively, 5000, 4000, and 50,000 psi, what must its diameter be? How would it fail? Use a safety factor of 5.

5-55. A cast-iron cylindrical supporting rod 1 in. in diameter is to be used occasionally to resist a twisting action. What amount of torque in inch-pounds can it safely withstand if a factor of safety of 5 is to be used? Ultimate unit stresses in shear, tension, and compression are, respectively, 30,000, 24,000, and 60,000 psi. *Answer:* 943 in.-lb.

5-56. A delicate mechanical apparatus requires a Duralumin (2017-T3) shaft $\frac{1}{4}$ in. in diameter. Determine the allowable torque in inch-pounds that may be applied to the shaft using factor of safety of 4. Consult Table 2.

5-13 Relationship Between G and E

A study of a square element of an elastic material subjected to pure shear provides a convenient relationship between the shear and normal moduli of that material.

Such an element $ABCD$ is shown in Fig. 5-28 and is distorted by pure shear forces into shape $AB'C'D$. The tensile strain along diagonal AC is

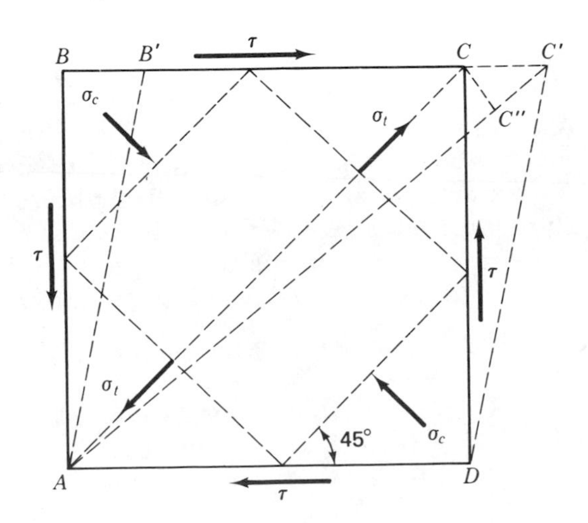

Figure 5-28

$$\epsilon_t = \frac{EC'}{AC} = \frac{CC' \cos 45°}{CD/\sin 45°} = \frac{CC'}{CD} \times \frac{1}{2}$$

But CC'/CD is equal to the shear strain γ, making

$$\epsilon_t = \frac{\gamma}{2} \quad \text{or} \quad \epsilon_t = \frac{\tau}{2G} \tag{5.9a}$$

From the study of normal stresses produced by pure shear (§5-11), the maximum stresses in shear, tension, and compression were found to be equal to each other. Thus, if another square element were to be imagined within $ABCD$, tilted 45° to the base AD, the diagonal strain ϵ_t expressed in terms of the normal stresses σ_t and σ_c (see §3-10) would be

$$\epsilon_t = \frac{\sigma_t - \mu\sigma_c}{E} \tag{5.9b}$$

But $\sigma_c = -\sigma_t$; therefore,

$$\epsilon_t = \frac{\sigma_t + \mu\sigma_t}{E} = \frac{\sigma_t(1 + \mu)}{E}$$

From Eq. (5.9a) it follows that

$$\epsilon_t = \frac{\sigma_t(1 + \mu)}{E} = \frac{\tau}{2G}$$

and since $\sigma_t = \tau$

$$G = \frac{E}{2(1 + \mu)} \tag{5.10}$$

With this relationship the value of μ can be determined from a knowledge of the experimental values of G and E. The equation also reveals that structural steel having a value of μ equal to 0.25 provides a theoretical ratio of $G/E = 2/5$, which agrees closely with its experimental determination.

5-14 Modulus of Rupture in Torsion

Should the value of the maximum torque required to rupture a shaft be placed in the torsion formula and a solution made for the corresponding shearing unit stress, the stress

$$\tau_r = \frac{T_{\max} r}{J} \tag{5.11}$$

thus obtained would be known as the shearing *modulus of rupture*. It is generally a fictitious stress, since it is ordinarily above the proportional limit. This fact, it will be recalled, invalidates the use of the torsion equation, because one of the assumptions made in its derivation requires a constant stress-strain ratio. In order that the shearing modulus of rupture be equal to the ultimate shearing unit stress, the stress-strain curve would have to be a straight line to the point of rupture. Few materials have this characteristic.

Although the shearing modulus of rupture may equal the ultimate shearing unit stress, it generally exceeds it. The truth of this statement may be verified by the following reasoning.

The variation in shear stress over the cross section will be constant, and the torsion equation will remain valid as long as the stress on the outermost fiber does not exceed the proportional limit. However, if the stress exceeds the proportional limit, the variation is no longer constant. Those outer elementary areas stressed beyond the proportional limit seek to deform at a correspondingly greater rate—that is, comparable to that indicated by the portion of shear stress-strain diagram above the proportional limit. This is impossible, since the outer areas are integrally connected to the inner areas that conform to the more slowly developing strain obtained below the proportional limit. Therefore, if the outer areas are prevented from deforming as much as they desire, the stress imposed upon them must remain at a lower value, while the inner areas must assume a proportionately greater part of the twisting restraint.

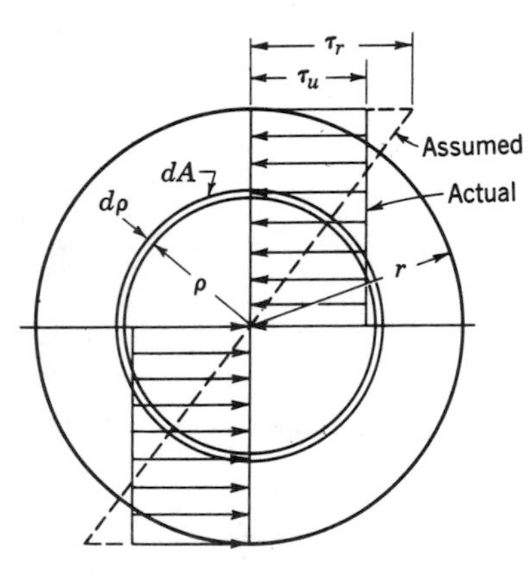

Figure 5-29

Thus, as the torque continues to increase beyond the proportional limit, the deviation of the stress variation from that of the straight line becomes more marked. Ultimately, nearly all the fibers on the cross section attain their ultimate strength τ_u* simultaneously (Fig. 5-29). The torque under this condition is approximately equal to

$$T_{\max} = \int_0^r \tau_u \times dA \times \rho = \tau_u \int_0^r 2\pi\rho \times d\rho \times \rho$$

$$= \tau_u \times \frac{2\pi r^3}{3} \times \frac{2r}{2r} = \frac{4}{3} \times \frac{\pi_u J}{r} \quad (5.12)$$

But, if from Eq. (5.11)

$$T_{\max} = \frac{\tau_r J}{r} \quad (5.13)$$

then

$$\tau_r = \frac{4}{3}\tau_u \quad (5.14)$$

When the torsion equation is used in finding the shear modulus of rupture, the resisting torque is the result of an assumed straight-line variation of stress from the center of the shaft. In order that the assumed stresses might provide the same resisting torque as the actual stresses, the maximum assumed stress τ_r must be greater than the constant actual stress τ_u. Thus, the maximum stress computed by the torsion equation (that is, the modulus

*Stress τ_u is usually found by testing a hollow cylinder to failure in torsion.

of rupture) is the result of an assumption known to be invalidated and is greater than the actual maximum stress.

Although moduli of rupture are fictitious values of stress, they can be and are used advantageously to determine maximum torques that may be applied to shafts.

Illustrative Problem 9. Between the crankshaft and propeller shaft on a small freighter there is to be placed an auxiliary shaft that will purposely fail when the torque attains a value of 200,000 in.-lb. If the modulus of rupture of the steel used in the shaft is 40,000 psi, how large must its diameter be?

SOLUTION

$$\tau_r = \frac{T_{\max}\, r}{J}$$

$$40{,}000 = \frac{200{,}000(d/2)}{(\pi d^4)32}$$

$$d^3 = 25.4, \qquad d = 2.94 \text{ in.}$$

PROBLEMS

5-57. A $\frac{1}{2}$-in. steel shaft attains a maximum torque of 800 in.-lb before failing. Determine its modulus of rupture. *Answer:* 32,600 psi.

5-58. Would a hollow torsion specimen give a modulus of rupture closer to the actual ultimate shearing unit stress than a solid torsion specimen? Give reason for your answer.

5-59. A solid circular shaft of radius R and length L is subjected to two end torques T (Fig. 5-30a). The material of the shaft reveals a stress-strain curve (Fig. 5-30b) with yield stress τ_o. Determine (a) the value of T for which yielding will just begin in the shaft. (b) The value of T for which yielding will extend through one half of the radius. (c) The angle of twist in both cases. *Answers:* (a) $\pi\tau_0 r^3/2$ (b) $\frac{31}{48}\pi\tau_0 r^3$

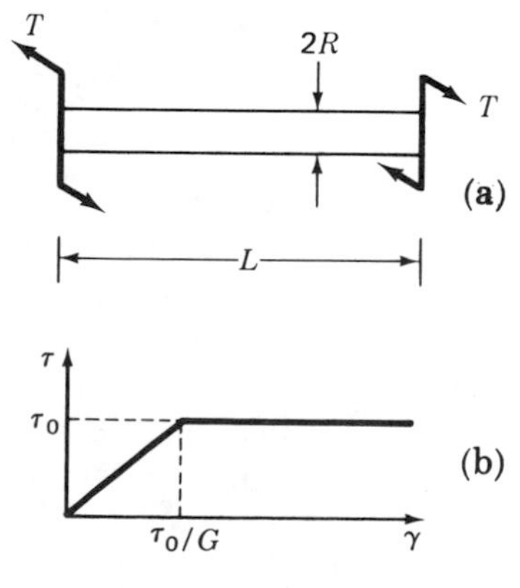

Figure 5-30

5-60. A pressure relief apparatus consists of a door affixed to a horizontal steel shaft of $\frac{1}{2}$-in. diameter located on its lower edge (Fig. 5-31). The ends of the shaft are also fixed in position. The other edges are sealed but do not prevent the outward movement of the door. If the modulus of rupture is 30,000 psi, what maximum uniform pressure in psi may be applied to the surface of the door before it opens and releases the contents? *Answer:* 0.513 psi.

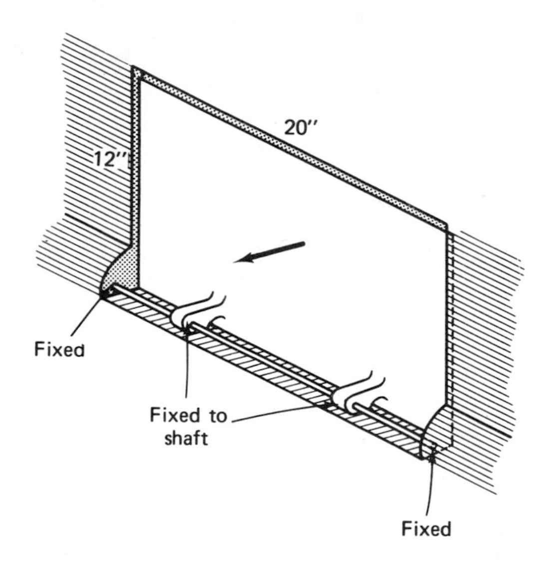

Figure 5-31

5-15 Noncircular Thin-Walled Tubes Subjected to Torsion—Shear Flow

In certain structures, notably those of the aircraft industry, stability is obtained through the torsional resistance of their membrane covering. These coverings will often take the shape of noncircular thin-walled tubes or shells. Some of these tubes may even have walls of varying thickness. The study of the shear stresses which are developed call for a special approach not heretofore considered.

Shearing stresses. A tube of uniform cross section but of unequal wall thickness, shown in Fig. 5-32, is subjected to torque T. Assuming the contour of the cross section to be retained as the torque is applied, resisting shearing stresses which are directed tangent to the periphery of the shell will be developed on the cross section. The moment of the resisting shear forces taken about any point O in the plane of the section will be equal to the applied couple T.

Consider now the variation of the shear stresses as they act on the ex-

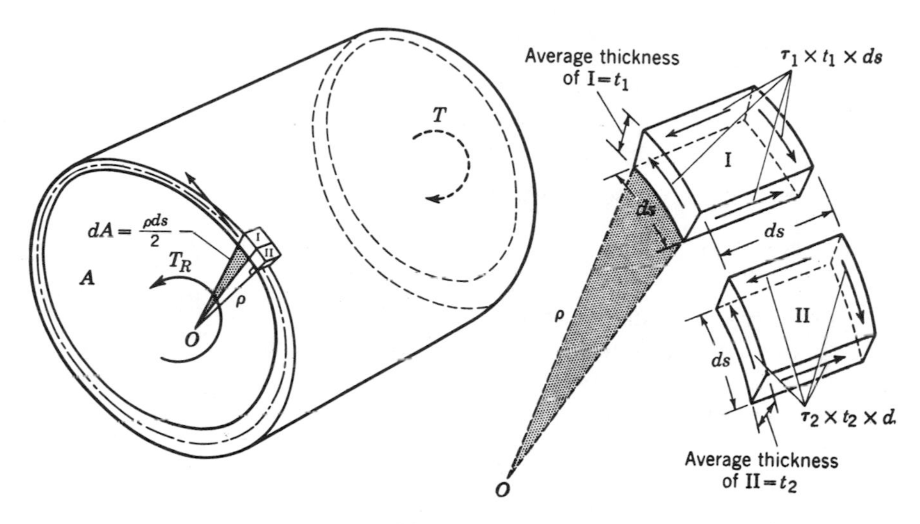

Figure 5-32

posed cross section of the shell. On an elementary free body I of equal width and length, shear stress τ_1 acts on its edge to produce a force equal to $\tau_1 t_1 ds$, where t_1 is the average thickness of I. As indicated in §5-10, such a force is restrained by an equal force on the opposite edge and an opposing couple of similar forces on the perpendicular edges. A similar arrangement of forces must act on the adjacent element II.

It now becomes apparent that the two shear forces on the common faces must be equal and that

$$\tau_1 \times t_1 = \tau_2 \times t_2 = \tau \times t = q \tag{5.15}$$

It follows also that the product of the stress and thickness at all points is equal. This continuity of equality about the entire shell is analogous to the uniformity of flow of water in a continuous pipe of varying diameter. Thus, the term *shear flow* (q) was originated to convey the equality of the product of stress and thickness at all points. Its use has been extended to include shear on thin sections regardless of how produced.

The moment of the shear force on any cross-sectional element taken about any point O is equal to $\tau \times t \times ds \times \rho$.* The total resistance torque is thus equal to

$$T_R = \int \tau \times t \times ds \times \rho = q \int \rho \, ds$$

But $\rho \, ds/2$ is equal to the area of the shaded elementary triangle in Fig. 5-32.

*Assuming the cross section to be in equilibrium, the forces acting thereon must balance in every direction. In this respect they comprise a system of couples, the resultant of which is T_R. The moment of this coplanar system of couples is the same about any point in the plane.

The total area of the cross section within the median line of the shell is therefore

$$\int \frac{\rho\, ds}{2}, \quad \text{making} \quad \int \rho\, ds = 2A$$

and
$$T_R = 2qA = T \tag{5.16}$$

from which
$$q = \frac{T}{2A} \quad \text{and} \quad \tau = \frac{T}{2At} \tag{5.17}$$

Angle of twist. The angle of twist of the shell may be determined by the use of the conservation of energy principle discussed in §3-8. We shall assume all the stresses to be below the proportional limit and that the torque T produces an angle of twist equal to θ. Thus, expressing the equality of external and internal energy, we have

$$\frac{1}{2}T\theta = \frac{\tau^2}{2G} \times \text{volume of shell}$$

$$= \int_0^s \frac{\tau^2}{2G} \times tL\, ds$$

where s is the length of the median line of the shell cross section.

Substituting the equivalent value of τ from Eq. (5.17), we obtain

$$\frac{1}{2}T\theta = \int_0^s \frac{T^2}{4A^2t^2} \times \frac{tL\, ds}{2G}$$

from which
$$\theta = \frac{TL}{4A^2G}\int_0^s \frac{ds}{t} \tag{5.18}$$

or
$$\theta = \frac{Lq}{2AG}\int_0^s \frac{ds}{t} \tag{5.19}$$

In using the equations developed above it is well to remember that while shearing stresses vary in a shell of unequal thickness, the shearing strains will also vary. That implies that the elementary volumes, such as I and II, located about the cross section will distort unequally, causing a warping of the section. This warping must be permitted without distorting the contour of the cross section to insure the validity of the derived equations. Sharp corners, abrupt changes in section, holes, and so forth will, of course, have a modifying effect on the stresses and twists computed.

Illustrative Problem 10. Determine the shearing stresses acting on the cross section of Fig. 5-33a due to a torque of 40,000 in.-lb. Calculate the twist of the tube if $L = 10$ ft and $G = 4 \times 10^6$ psi.

SOLUTION. The shear flow

$$q = \frac{T}{2A} = \frac{40{,}000}{2 \times 78.54} = 254.4 \text{ lb/in.}$$

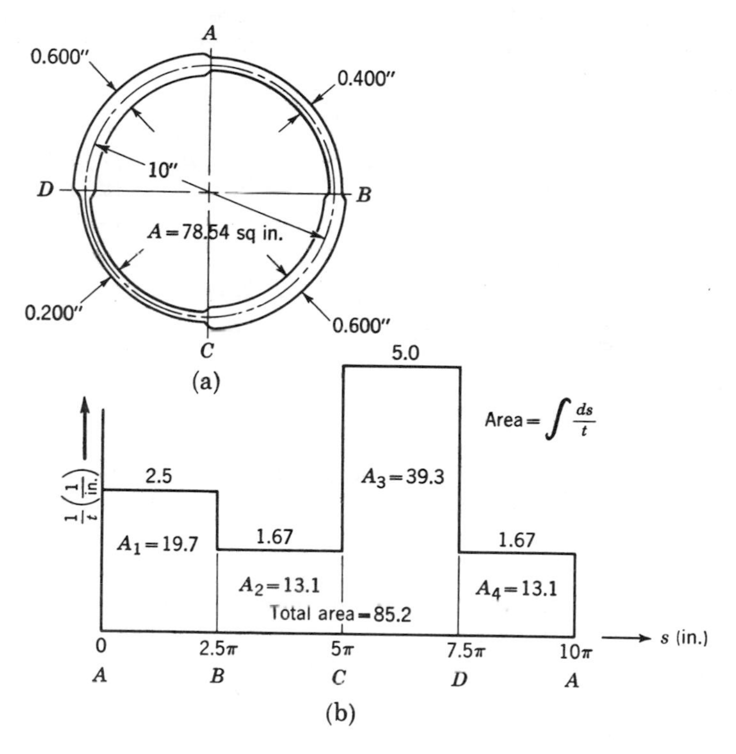

Figure 5-33

Therefore,

$$\tau_{AB} = \frac{254.4}{0.400} = 636 \text{ psi}$$

$$\tau_{BC} = \frac{254.4}{0.600} = 424 \text{ psi}$$

$$\tau_{CD} = \frac{254.4}{0.200} = 1272 \text{ psi}$$

$$\tau_{DA} = 424 \text{ psi}$$

The computation of the angle of twist is dependent upon our evaluation of $\int ds/t$. In view of the constant thicknesses we may use the finite summation $\Sigma\, \Delta s/t$ in its stead. This term is depicted graphically in Fig. 5-33b. The angle of twist is therefore

$$\theta = \frac{10 \times 12 \times 254.4}{2 \times 78.54 \times 4 \times 10^6} \times 85.2 = 0.00414 \text{ rad}$$

PROBLEMS

5-61. Show that the equation for shear flow [Eq. (5.17)] is identical to the torsion formula $\tau = Tr/J$ when applied to cylindrical thin-walled cylinders.

5-62. Compare the shear stresses in the hollow sections shown in Fig. 5-34 if the torque applied is equal to 6000 in.-lb. Wall thickness for each section equals $\frac{1}{4}$ in.

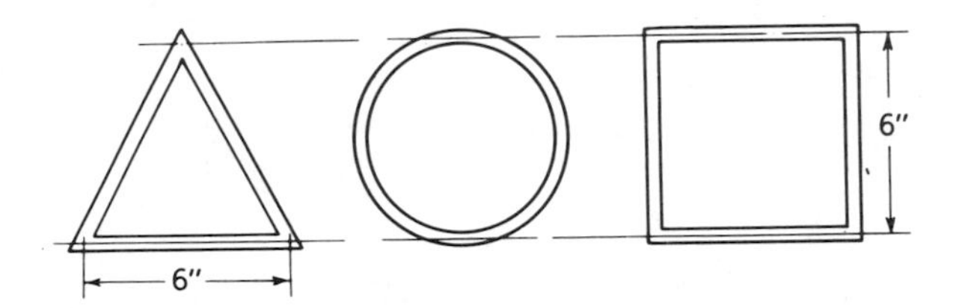

Figure 5-34

5-63. A tube with a hollow square cross section, measuring 10 in. between the center lines of walls, is subjected to a torque of 16,000 in.-lb. If the wall thicknesses for the four sides are successively $\frac{1}{4}$, $\frac{3}{8}$, $\frac{1}{2}$, and $\frac{3}{4}$ in., what are the shear stresses in the walls? What will the angle of twist be equal to if L is 10 ft and $G = 4 \times 10^6$ psi? *Answers:* $q = 80$ lb/in., 0.00120 rad.

5-64. In the wood spar section shown in Fig. 5-35 the maximum shear stress is 200 psi. Determine the maximum torque the section can carry. What will its angle of twist be if $L = 12$ ft and $G = 2 \times 10^6$ psi?

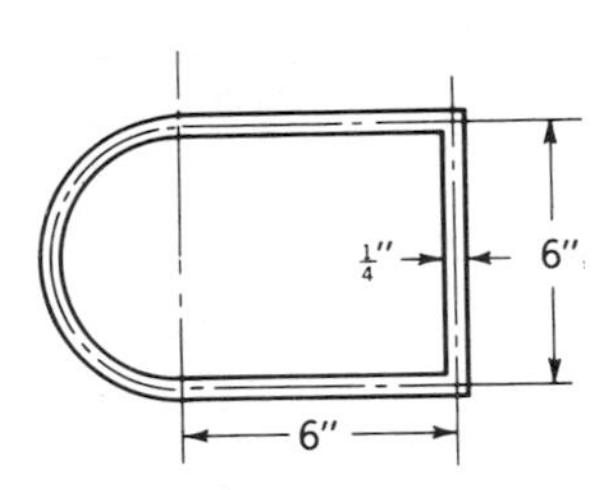

Figure 5-35

5-65. Find the shear distribution and the angle of twist for the section shown in Fig. 5-36 if $L = 12$ ft, $G = 4 \times 10^6$ psi, and $T = 10{,}000$ in.-lb? *Answer:* 0.001578 rad.

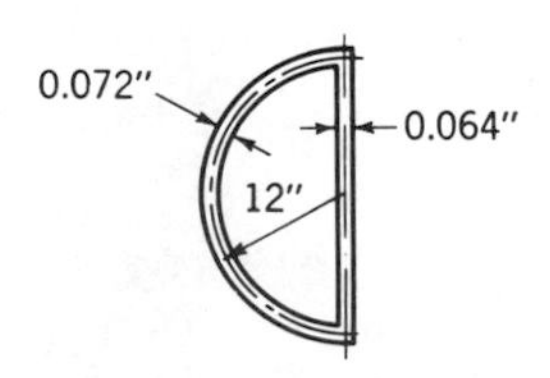

Figure 5-36

5-66. In the wood spar section shown in Fig. 5-37 the maximum permissible shear stress is 200 psi. (a) Determine the maximum permissible torque the section can carry. (b) What will its angle of twist be if $L = 16$ ft and $G = 2 \times 10^6$ psi. Give θ in degrees.

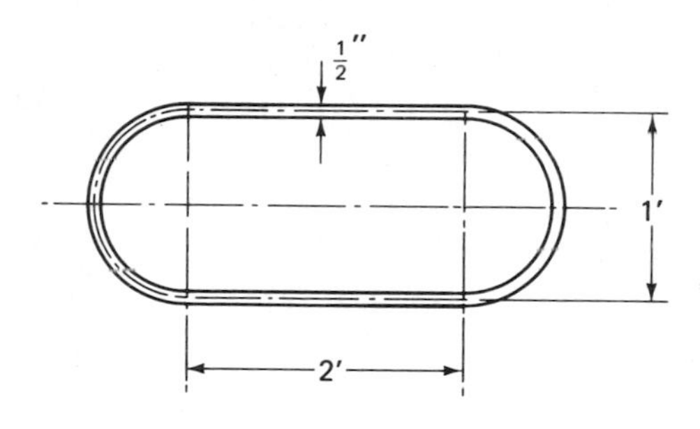

Figure 5-37

5-67. A hollow square tube is to replace a round rod of 1-in. diameter being used as a shaft. Determine the minimum outside dimension of the tube that would provide the same torque as the round rod if the same maximum allowable stress were used in both. Wall thickness of the tube is 0.1 of its lateral dimension. Maximum allowable shearing stress is 10,000 psi. *Answer:* 1.07 in.

5-68. Determine the torsional strength T and stiffness θ/l of a thin-walled tube of radius R and thickness t with and without a longitudinal crack.

5-16 Torsion of Noncircular Cross Sections—General Solution

The torsion formula [Eq. (5.3)] is used only to determine the torsional resistance of a straight shaft of circular cross section with or without a concentric circular core. This restricted usage was created by the assumption that the stresses acting on concentric circular elements varied directly with their distance from the center of rotation. The more unlike a cross section is to a circular section, the greater is the discrepancy between the actual and theoretical stress distribution assumed for the torsion formula. As one approaches a rectangular section, the similarity of stress distributions is considerably changed. This is also true of the strain.

Rubber models show clearly the nonuniform variation of surface strain in noncircular sections subjected to torsional forces. It will be observed in Fig. 5-38b and c that the elements on the surface of the elliptical and rectangular cross-sectioned rods are not equally distorted, and cause the section to warp. On the contrary, Fig. 5-38a, a rod of circular cross section, shows equal element distortion, and a perfectly plane cross-sectional periphery. Obviously, the unequal element distortions indicate nonuniformly varying shear strains and corresponding nonuniformly varying shear stresses.

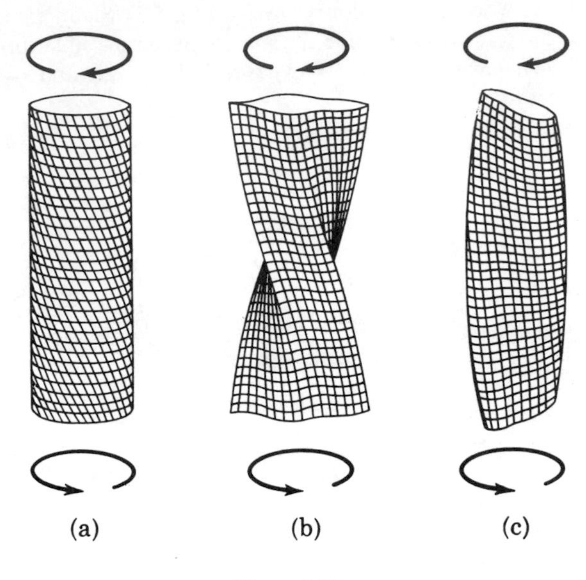

Figure 5-38

The torsion formula derived for shafts having circular sections cannot, therefore, be used for those having noncircular sections.

It is of interest to note that on both the elliptical and rectangular cross sections the maximum stress is not located at the fiber farthest removed from the center, but at the ends of the shortest axis of symmetry. Even more surprising is the fact that the corners of the rectangular cross section show no shear stress at all.

General solution—St. Venant's theory. The more general solution of the torsion problem as applied to shafts of *any* constant cross section was conceived by St. Venant in 1853. As in the more simple development applied to circular shafts, the material is assumed to be isotropic and homogeneous and stressed below its proportional limit. Couples are placed at either end of the shaft having a length L and located in planes perpendicular to the twisting axis, so that it might be reasonably assumed that no normal stresses act on the faces of the elementary prism oriented in the shaft as shown in Fig. 5-39a; i.e., $\sigma_x = \sigma_y = \sigma_z = 0$. It is further assumed that the resultant shearing action* acting on any cross section perpendicular to the twisting axis is equal to zero; i.e., $\int \tau_{xz}\, dA = 0$.

*A shear stress with double subscripts, such as τ_{xy}, is identified as the shear stress parallel to the x axis lying in a plane perpendicular to the y axis.

Let us then consider first the equilibrium of the forces acting on the surfaces of the typical element of the shaft shown in Fig. 5-39b, second, the distortions resulting from these forces, and finally their compatibility.

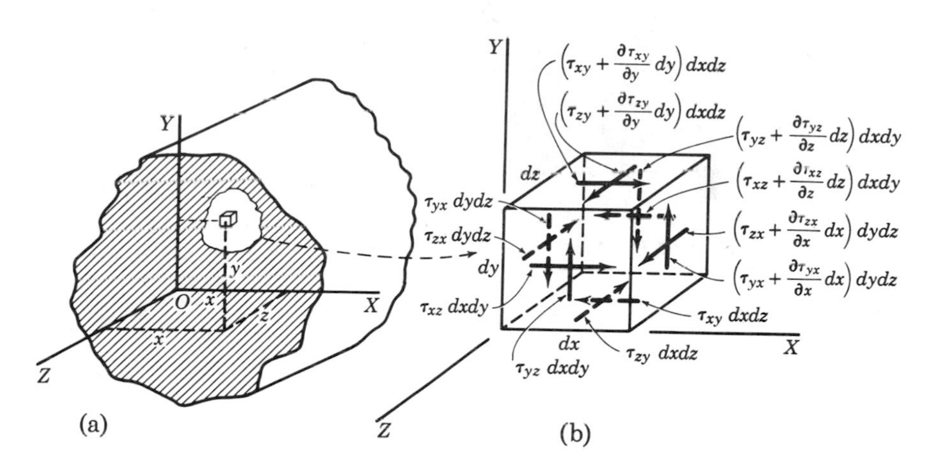

Figure 5-39

Equilibrium Considerations. The sum of the forces acting in the z direction is

$$\sum F_z = -\tau_{zy}\, dx\, dz + \left(\tau_{zy} + \frac{\partial \tau_{zy}}{\partial y} dy\right) dx\, dz$$
$$+ \left(\tau_{zx} + \frac{\partial \tau_{zx}}{\partial x} dx\right) dy\, dz - \tau_{zx}\, dy\, dz = 0$$

from which

$$\frac{\partial \tau_{zy}}{\partial y} + \frac{\partial \tau_{zx}}{\partial x} = 0 \tag{5.20}$$

Similarly, the summations in the x and y directions reveal that

$$\frac{\partial \tau_{xy}}{\partial y} + \frac{\partial \tau_{xz}}{\partial z} = 0 \tag{5.21}$$

and

$$\frac{\partial \tau_{yx}}{\partial x} + \frac{\partial \tau_{yz}}{\partial z} = 0 \tag{5.22}$$

Distortion Considerations. The second stage in the development of the general solution concerns the distortion of the element under the action of the impressed forces. Proceeding then to Fig. 5-40, let O be the instantaneous center of rotation of all the elements on the cross section as the shaft twists clockwise about the z axis. During the rotation, point A moves with respect

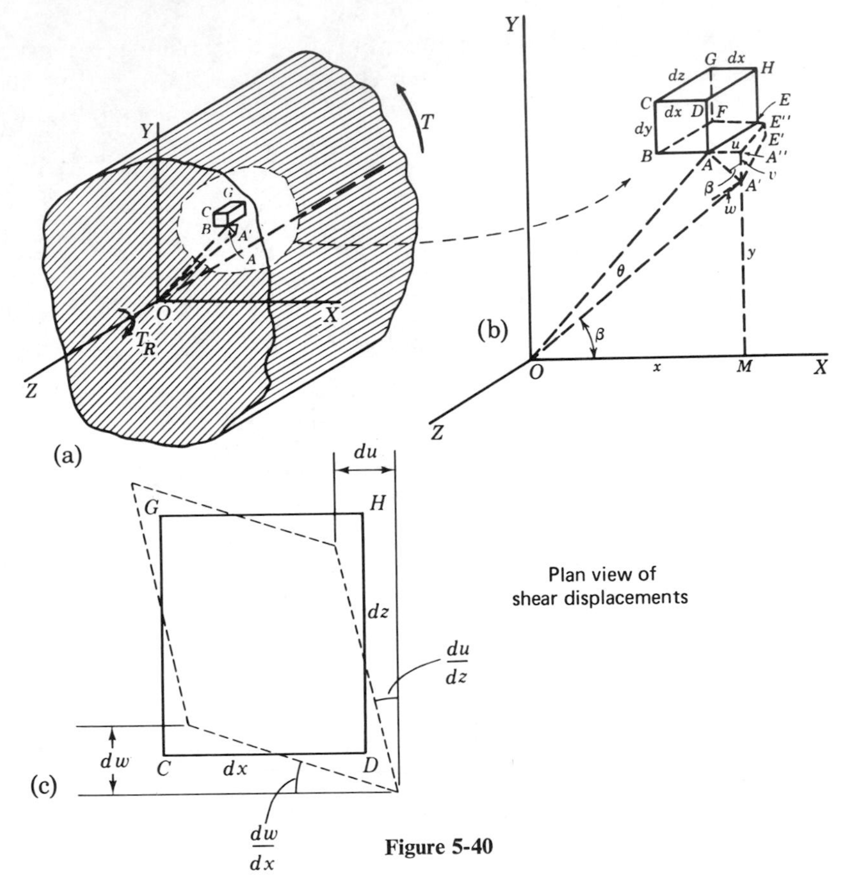

Figure 5-40

to corresponding points at the opposite end of the shaft the distances u, v, and w in the x, y, and z directions. Thus, in the xy plane, using similar triangles $AA'A''$ and $A'OM$,

$$\frac{AA'}{OA} = \frac{A'A''}{x} = \theta; \quad \text{also} \quad \frac{AA'}{OA} = \frac{AA''}{y} = \theta$$

from which we can write

$$v = -x\theta \quad \text{and} \quad u = +y\theta$$

the minus sign indicating that v decreases as point A moves to A'. These total movements have their similar infinitesimally small proportional counterparts as noted in Fig. 5-40c. Thus,

$$\frac{du}{dz} = \frac{u}{L} = \frac{+y\theta}{L} \quad \text{and} \quad \frac{du}{dz} = \frac{v}{L} = \frac{-x\theta}{L} \tag{5.23}$$

In the same figure, the shearing strains* in the xz and yz planes at point

*The shear strain γ_{xz} is the measure of angular deformation lying in the xz plane.

A (see §2-10) may be expressed as

$$\gamma_{xz} = \frac{du}{dz} + \frac{dw}{dx} = \frac{+y\theta}{L} + \frac{dw}{dx} \tag{5.24}$$

$$\gamma_{yz} = \frac{dv}{dz} + \frac{dw}{dy} = \frac{-x\theta}{L} + \frac{dw}{dy} \tag{5.25}$$

The presence of the dw/dx and dw/dy terms in the previous equations indicates warping of the cross section. It also indicates that the cross section must be noncircular. Let us explain: if for a *circular* cross section we were to assume the existence of an equal du/dx shear strain for each element located within one of its many concentric rings, the front face, e.g., $ABCD$ on Fig. 5-40b, of each successive matching element would pass farther and farther away from its initial unstressed plane as the strains accumulated, like coils of a spring, with no hope of matching the first and last elements as one completed the circuit of one ring. A discontinuity would have to exist under this assumption in order for torsional stress and strain to be compatible in a circular cross section. The value of dw/dx and dw/dy would have to equal zero for a plane section to remain plane after twisting.

Compatibility Considerations. The third stage in the development of the general torsion equation concerns the introduction of the properties of the material and the compatibility of stress and strain.

If the deformation equations just developed are now differentiated, the first with respect to y and the second with respect to x, and then subtracted one from the other, there results

$$\frac{\partial \gamma_{xz}}{\partial y} = \frac{\partial \tau_{xz}}{G\,\partial y} = +\frac{\theta}{L} + \frac{\partial^2 w}{\partial x\,\partial y}$$

$$\frac{\partial \gamma_{yz}}{\partial x} = \frac{\partial \tau_{yz}}{G\,\partial x} = -\frac{\theta}{L} + \frac{\partial^2 w}{\partial x\,\partial y}$$

$$\frac{\partial \tau_{xz}}{\partial y} - \frac{\partial \tau_{yz}}{\partial x} = +\frac{2G\theta}{L} \tag{5.26}$$

This equation provides the relationship which must exist between stress and strain to insure their compatibility in a twisted shaft.

Stress Function. The first two terms of Eq. (5.26) as well as those in the equilibrium equations (5.20), (5.21), and (5.23), reveal sets of equations whose slopes at points of intersection are mutually perpendicular to each other like lines of longitude and latitude on a globe. Collectively they suggest an inflated membrane blown over a hole exactly equal to the cross section in question, whose surface is defined by a *stress function* $\psi(x, y)$, the height at any point of which is equal to ψ, and which has mutually perpendicular slopes on its surface at any point equal to

$$\tau_{xz} = -\frac{\partial \psi}{\partial y} \quad \text{and} \quad \tau_{yz} = +\frac{\partial \psi}{\partial x} \tag{5.26a}$$

Inserted in Eq. (5.26), there results

$$\frac{\partial^2 \psi}{\partial x^2} + \frac{\partial^2 \psi}{\partial y^2} = -\frac{2G\theta}{L} \tag{5.27}$$

This is the basic differential equation for a member of any cross section subjected to pure torque. The determination of the proper stress function ψ is necessary before the solution can be obtained.

Development of general expression for torque. Referring to Fig. 5-39, the expression for the resisting torque applied to a cross section perpendicular to a twisting axis is equal to

$$T_R = \iint (\tau_{xz} y - \tau_{yz} x)\, dx\, dy \tag{5.28}$$

Inserting the values for the stresses as found in Eq. (5.26a),

$$T_R = -\iint \frac{\partial \psi}{\partial y} y\, dx\, dy - \iint \frac{\partial \psi}{\partial x} x\, dx\, dy \tag{5.28a}$$

Each of the integrals in this expression may be integrated by parts. The first integral, for instance, may be simplified as follows:

$$\iint \frac{\partial \psi}{\partial y} y\, dx\, dy = \int \left[\int \frac{\partial \psi}{\partial y} y\, dy\right] dx = \int \left[y\psi - \int \psi\, dy \right] dx$$

$$= \int \left[0 - \int \psi\, dy\right] dx = -\iint \psi\, dx\, dy$$

The integral $\int y\psi\, dx$ is a line integral, and since ψ is zero over the entire boundary, the integral is equal to zero. A similar simplification may be performed on the second integral, resulting in the same answer.

The torque is therefore

$$T = 2\iint \psi\, dx\, dy \quad \text{or} \quad 2\iint \psi\, da \tag{5.29}$$

Obtaining the General Solution. The solution of a torsion problem by the general approach is dependent upon the determination of the stress function ψ. Once this stress function is known, its differentiation with respect to x and y provides the two components of shear stress τ_{xz} and τ_{yz}, and its insertion into Eq. (5.29) provides the torque.

The determination of the stress function generally proceeds from a knowledge of the location of points on its envelope. As the envelope ordinarily passes through the boundary of the cross section under consideration, the equation of the boundary will satisfy points on the surface expressed by the stress function. Thus, it is frequently possible to develop the stress function from the equation of the boundary.

For example, the equation of the boundary of a circular cross section of radius r is

$$r^2 = x^2 + y^2 \quad \text{or} \quad r^2 - x^2 - y^2 = 0$$

The stress function, if now taken as $\psi = k_1 (r^2 - x^2 - y^2)$, will provide double differentials with respect to x and y equal to $-2k_1$. Substituting these differentials into Eq. (5.27) provides

$$-2k_1 - 2k_1 = -\frac{2G\theta}{L}$$

$$k_1 = +\frac{G\theta}{2L}$$

The stress function will then be equal to

$$\psi = +\frac{G\theta}{2L}(r^2 - x^2 - y^2)$$

Differentiating the stress function with respect to y gives

$$\frac{\partial \psi}{\partial y} = -\frac{G\theta y}{L} = -\tau_{xz}$$

which is a maximum when $y = r$, viz.,

$$(\tau_{xz})_{\max} = +\frac{G\theta r}{L} \quad \text{or} \quad (\tau_{xz})_{\max} = \frac{Tr}{J}$$

The torque is determined by substituting the stress function in Eq. (5.29).

$$T = 2\int \frac{G\theta}{2L}(r^2 - x^2 - y^2)\, da$$

$$= \frac{G\theta}{L}(ar^2 - I_y - I_x)$$

$$= \frac{G\theta}{L}\left(\pi r^4 - \frac{\pi r^4}{4} - \frac{\pi r^4}{4}\right)$$

$$= \frac{\pi G\theta r^4}{2L} = \frac{G\theta J}{L}$$

or

$$\theta = \frac{TL}{GJ}$$

The application of the exact theory to other than a few of the more simple cross sections becomes involved mathematically. Recourse is therefore made either to approximate mathematical solutions or to experimental analogies for solutions to these more complicated problems. One very useful experimental analogy will now be considered.

5-17 Membrane Analogy

Invaluable in the determination of the torsional characteristics of noncircular cross sections is the membrane analogy. This important experimental device, introduced by Prandtl in 1903, is based upon theory which reveals definite relationships between the torsional characteristics of a cross section and physical measurements of an inflated membrane or soap bubble placed over an opening of the same or geometrically similar section.

Let us assume that such a membrane, having no bending resistance, is inflated over an opening in a flat plate (Fig. 5-41). Let us further assume that its deflections are small and that its resistance is supplied by a surface tension F expressed in pounds per inch.

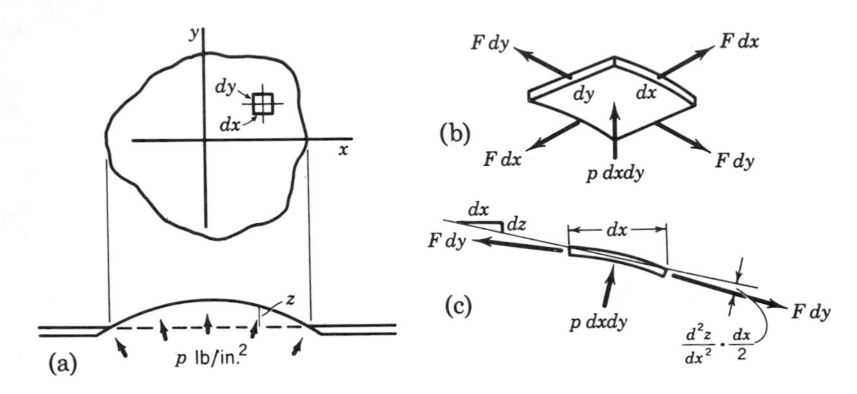

Figure 5-41

The general summation of forces normal to its surface (Fig. 5-40b and c) is, therefore,

$$p\,dx\,dy + F\frac{d^2z}{dx^2}dx\,dy + F\frac{d^2z}{dy^2}dx\,dy = 0$$

and

$$\frac{d^2z}{dx^2} + \frac{d^2z}{dy^2} = -\frac{p}{F} \tag{5.30}$$

Note the similarity between Eqs. (5.30) and (5.27), the basic differential equation for a member subjected to pure torque. In fact, if

$$\frac{p}{F} = \frac{2G\theta}{L} \tag{5.31}$$

and the boundary conditions of each equation are the same, the deflections of the membrane z will equal corresponding values of the stress function ψ.

To satisfy the boundary conditions set by theory, the boundary of the membrane must lie in a plane ($z = \psi = 0$) (see previous derivations) and must follow a similar slope to that of the cross section being considered. It is to be noted also that under these restrictions the slope of the tangent to the boundary line must be zero and, consequently, the shearing stress perpendicular to it. [See Eq. (5.26a).] Of even greater significance, the same equations indicate that if the equality of Eq. (5.31) is maintained, the shearing stress at any point on the cross section subjected torsion is equal to the slope at the corresponding point on the membrane measured perpendicular to the horizontal contour passing through that point.

From a practical point of view, the identity $p/F = 2G\theta/L$ is seldom realized. It is therefore common practice to write the equation with a proportionality factor k.

$$\frac{p}{F} = k\frac{2G\theta}{L} \tag{5.32}$$

Thus, the slopes of the ordinarily developed membranes taken perpendicular to their horizontal contours are proportional to the shearing stresses tangent to the contours at corresponding points.

If the same proportionality factor is introduced into Eq. (5.27) and (5.30), it can be shown that

$$z = k\psi$$

The volume under the membrane can then be expressed as

$$V = \int z\, dA = k \int \psi\, dA$$

But from Eq. (5-29)

$$T = 2 \int \psi\, dA$$

Thus,

$$V = k\frac{T}{2} \tag{5.33}$$

indicating that the volume under the membrane is proportional to the torque applied.

To summarize therefore:

1. The shear stress at any point A (Fig. 5-42) of the cross section is proportional to the slope at the corresponding point A of the membrane ($\tau = k\alpha$).

2. The torsional resistance of a member is proportional to twice the volume under the membrane ($T_R = k \times 2V$).

Note that the proportionality constant k is the same in relationships 1 and 2.

3. The direction of the shear stress at any point A of the given cross section is tangent to the contour passing through that point on the membrane.

4. The shear flows acting on those thin-walled tubes of varying cross section located between contours are equal.

The determination of the proportionality constant k (indicated in 1 and 2 above) for any one membrane is made by introducing the known theoretical relationship of shear stress to torque as obtained by the torsion formula for circular cross section. The ability to relate these stresses and torque to the corresponding slopes and volume of a membrane inflated over a circular hole enables the calculation of k required in the solution of shear stresses and

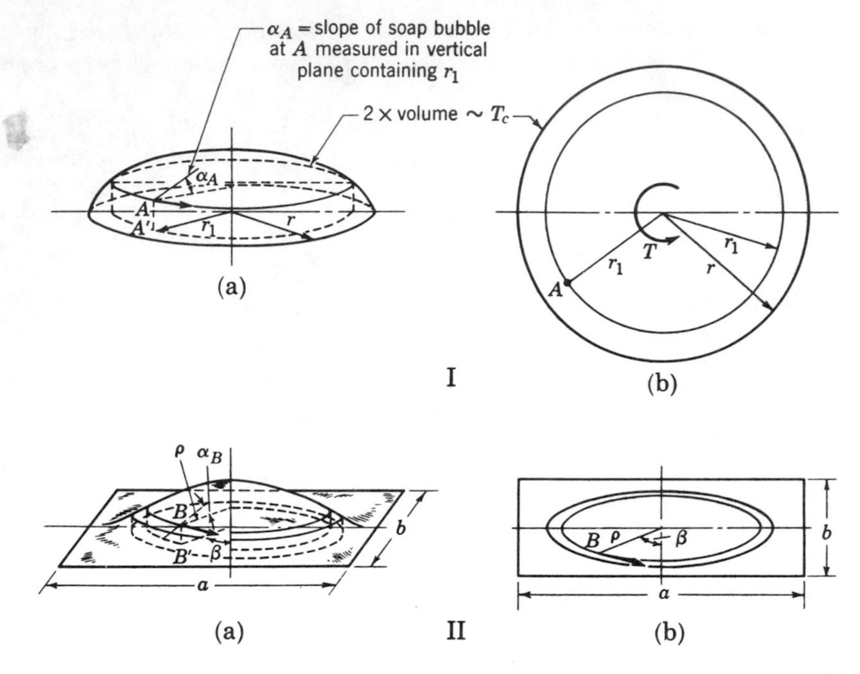

1. For given T_c and circular cross section, obtain $\tau_{\max}$ from torsion formula.
2. Obtain surface slopes and volume under circular membrane I(a).
3. Then $\tau_A = \tau_{\max} \times r_1/r = k\alpha_A$. Compute k.
4. Obtain volume under rectangular membrane.
5. Compute corresponding torque T_R on rectangular section II(a): $T_R = T_c(2V_R/2V_c)$.
6. Obtain slope at B (IIa). Stress at B is $\tau_B = k\alpha_B$ or $\tau_B = \tau_A \times \alpha_B/\alpha_A$.

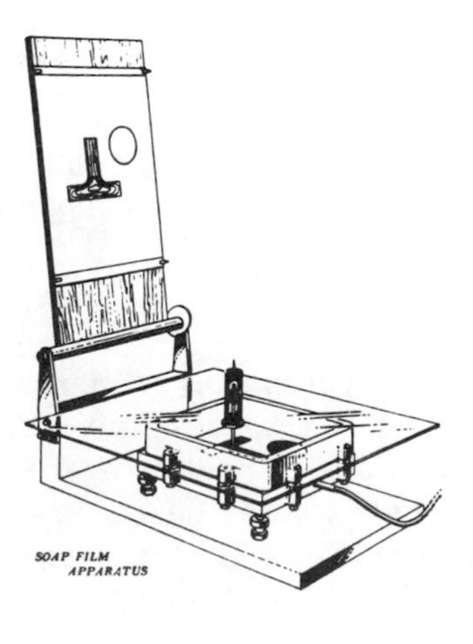

Figure 5-42 Membrane analogy procedure for finding shear stresses and torques of noncircular cross sections.

torques for noncircular sections. It is imperative for this analysis to have the inflation pressures under each membrane exactly the same.

It is more common, however, to avoid the determination of k by using a circular membrane adjoining the noncircular membrane and inflating each *simultaneously* with the same pressure. In so doing, the circular membrane acts as a calibrating agent and provides the simple relationships

$$\frac{T_c}{T_n} = \frac{V_c}{V_n} \quad \text{and} \quad \frac{\alpha_c}{\alpha_n} = \frac{\tau_c}{\tau_n}$$

where the subscripts c and n refer to the circular and noncircular cross sections. Suppose, then, that by careful micrometer measurements it is determined that the volume under a noncircular membrane is half that under a circular one. Assuming the circular shaft to be 2 in. in diameter and subjected to 10,000 in.-lb, its maximum stress is 6370 psi as obtained by the torsion formula. This stress would be proportional to the slope of the membrane at the edge of the circular hole. At a point on the noncircular membrane where the slope is exactly the same, the corresponding shear stress would be indenti-

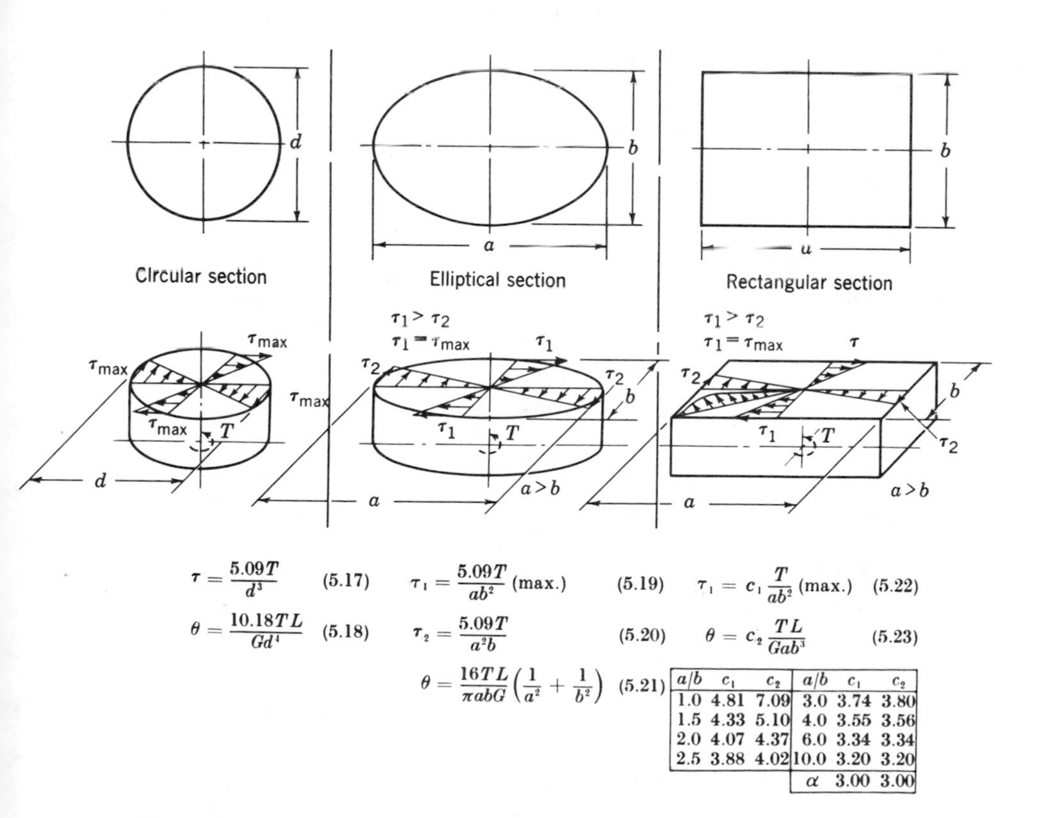

$$\tau = \frac{5.09T}{d^3} \qquad (5.17)$$

$$\theta = \frac{10.18TL}{Gd^4} \qquad (5.18)$$

$$\tau_1 = \frac{5.09T}{ab^2} \text{ (max.)} \qquad (5.19)$$

$$\tau_2 = \frac{5.09T}{a^2b} \qquad (5.20)$$

$$\theta = \frac{16TL}{\pi abG}\left(\frac{1}{a^2} + \frac{1}{b^2}\right) \qquad (5.21)$$

$$\tau_1 = c_1 \frac{T}{ab^2} \text{ (max.)} \qquad (5.22)$$

$$\theta = c_2 \frac{TL}{Gab^3} \qquad (5.23)$$

a/b	c_1	c_2	a/b	c_1	c_2
1.0	4.81	7.09	3.0	3.74	3.80
1.5	4.33	5.10	4.0	3.55	3.56
2.0	4.07	4.37	6.0	3.34	3.34
2.5	3.88	4.02	10.0	3.20	3.20
			α	3.00	3.00

Figure 5-43 Comparative shear stress distributions on circular, elliptical, and rectangular cross sections subjected to torsion.

cal if the torque on that section were 5000 in.-lb. Correspondingly, under the action of the same torque, points having twice the slope would have twice the shear stress. Thus by comparing the volume and slopes of the circular membrane to the volume and slopes of the membrane of a noncircular section, the shearing stresses can be obtained for any torque applied to the noncircular section. See Fig. 5-43.

Although such a quantitative comparison is invaluable, the membrane analogy may also be used qualitatively by making mental comparisons of circular and noncircular membranes. In so doing, values of maximum and minimum shear stress and comparative torques may be approximated.

5-18 Application of the Membrane Analogy

The membrane analogy has become an invaluable tool in obtaining approximate evaluations of stresses and angles of twist of torsional members developed from one or more thin rectangular sections. Consider the long rectangular cross section shown in Fig. 5-44 subjected to torsion. If a mem-

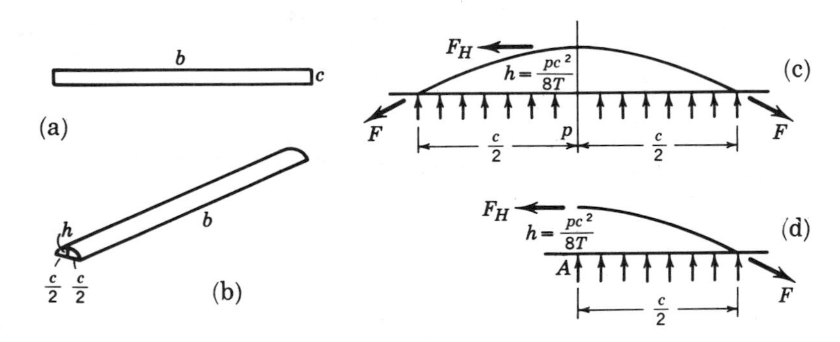

Figure 5-44

brane were inflated over an opening of similar shape, the cross section of the membrane would be constant except for a short distance at the ends. Assuming its cross section to be parabolic and sufficiently shallow so that $F = F_H$, the moment taken about point A of the half-section shown in Fig. 5-44d indicates that

$$F_H \times h \cong F \times h = p \times \frac{c}{2} \times \frac{c}{4}$$

$$h = \frac{pc^2}{8F}$$

Twice its volume would then equal

$$2V = 2 \times \frac{2}{3} \times \frac{pc^2}{8F} \times c \times b = \frac{p}{2F} \times \frac{bc^3}{3}$$

But if the height of the membrane z is equal to the stress function $p/2F = G\theta$ and $2V = T$,

$$T = 2V = G\theta\frac{bc^3}{3}$$

and

$$\theta = \frac{3T}{Gbc^3} \tag{5.34}$$

where θ is the twist per unit length.

It is also true that if $p/2F = G\theta$, then the stresses are equal to the slopes. Referring again to Fig. 5-44, we may write

$$2F\frac{dz}{dx}b = p \times c \times b$$

$$\frac{dz}{dx} = \frac{p}{2F} \times c$$

But $p/2F = G\theta$ when $2V = T$.

Therefore,

$$\frac{dz}{dx} = \tau = G\theta \times c = \frac{3T}{bc^2} \tag{5.35}$$

The shear stress τ is that taken at the point where dz/dx was determined —along the long side of the rectangle.

It is obvious that the volume and shape of the membrane placed over this narrow rectangle would not change to any great extent if it had a sharp angle in it, or if it was bent into a circular shape—or was rearranged into an H section. The maximum stresses and torques of these sections must therefore be approximately the same.

PROBLEMS

5-69. Compare the torsional resistance and twist of the following steel cross sections, using a maximum shear stress of 10,000 psi and $l = 10$ ft.

(a) A solid circular section 2 in. in diameter.

(b) A hollow circular section 2 in. O.D. and $1\frac{1}{2}$ in. I.D.

(c) A square section 2 in. on a side.

(d) A rectangular section 2 in. wide by 4 in. long.

(e) An elliptical section 2-in. minor diameter by 4-in. major diameter.

5-70. Compare the torsional rigidities of the sections shown in Fig. 5-45 by applying the membrane analogy qualitatively. The total length of $\frac{1}{4}$-in. thickness in each figure is equal to 10 in.

5-71. A T and circular membrane on a membrane analogy device showed volumes of 0.8 and 2.0 in.^{3}, and maximum slopes of 45° and 30°, respectively. If the circular section is 2 in. in diameter and its maximum allowed shear stress is 12,000 psi, what are the maximum shear stress and corresponding torque developed on the T section? *Answers:* 7550 in.-lb, 18,000 psi.

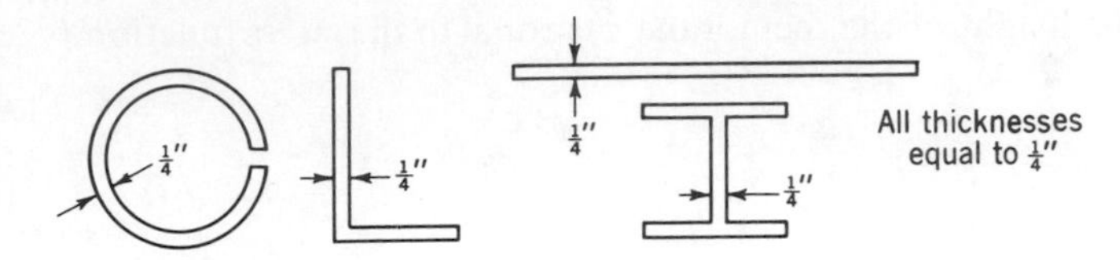

Figure 5-45

5-72. Two aluminum shafts have approximately the same total weight but different sectional shapes (Fig. 5-46). Calculate (a) the ratio of maximum shear stresses of the two shafts τ_1/τ_2. (b) The ratio of angles of twist of the two shafts θ_1/θ_2. $G = 4 \times 10^6$ psi; $E = 10.4 \times 10^6$ psi. Both shafts are 20 in. long and are acted upon by torques of 1000 in.-lb.

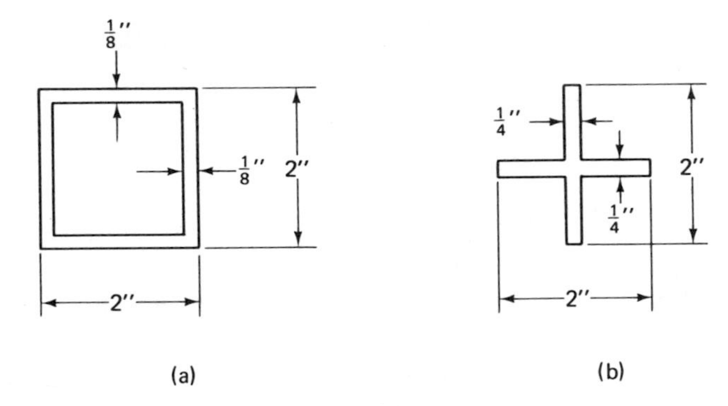

Figure 5-46

6

Shear- and Bending-Moment Diagrams

6-1 Different Types of Beams

A beam, generally speaking, is any bar or structural member that may be subjected to bending by transverse forces. This broad definition includes a great many irregularly shaped members acted upon by forces located in

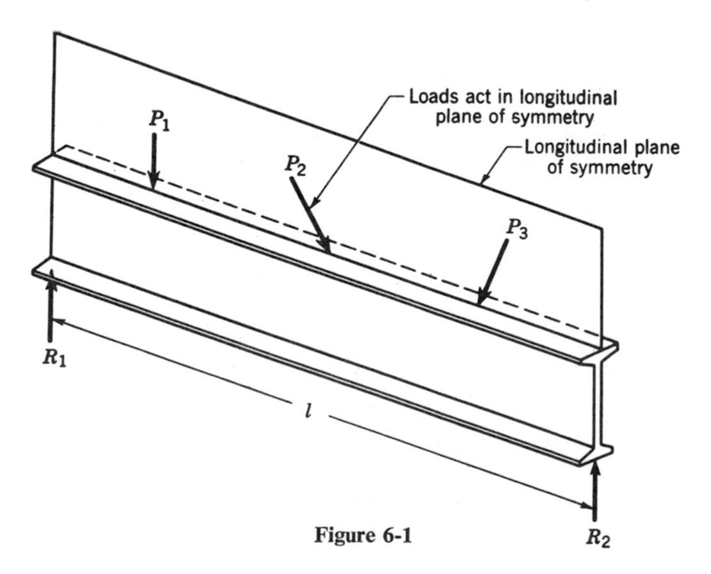

Figure 6-1

many planes. In order to simplify our presentation, therefore, it becomes necessary to confine this preliminary study to beams that are initially straight and of uniform symmetrical cross section and that have all applied loads acting in the plane of symmetry (Fig. 6-1).

Beams need not be horizontal; they may be vertical or on a slant. They may have either one, two, or several reactions. The type and number of their reactions provide the following classification:

Simple. Any beam whose ends rest on supports which do not restrain its deflection when it is subjected to load (Fig. 6-2a).

Continuous. Any beam resting on three or more supports (Fig. 6-2b).

Cantilever. Any beam supported at one end only (Fig. 6-2c).

Overhanging. Any beam with one or two supports not located at the ends of the beam (Fig. 6-2d).

Restrained. Any beam that has located at either one or more reaction points restraining forces which prevent its free deflection. These restraining forces may be supplied, for instance, by building one or both ends of a beam

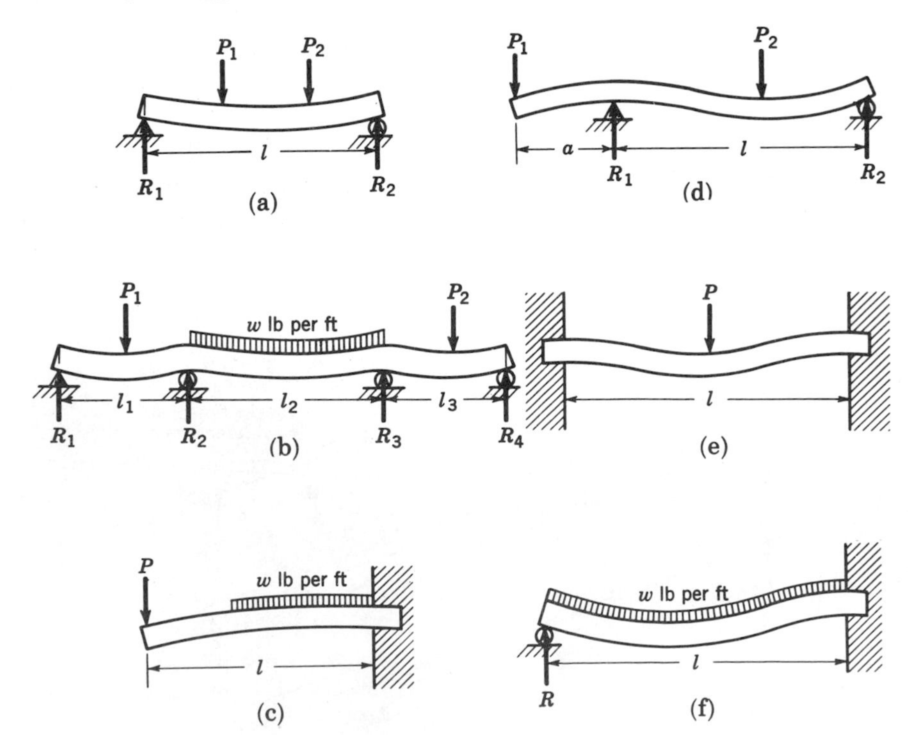

Figure 6-2

into a concrete wall (Figs. 6-2e and f). A continuous beam consists of a series of restrained beams.

6-2 Distributed and Concentrated Loads

Whenever a distributed load acts over a short distance of a beam relative to its entire length, it may be considered as a concentrated load. Such a load is shown on a loading diagram as a single force—for instance, P in Fig. 6-2c. Loads applied to the beam by other beams framing into it at right angles or by a loaded differential pulley mounted on a beam are representative of concentrated loads.

Frequently, loads are spread uniformly over the length of a beam, as in the case of a load of sand or wheat. These uniform loads are considered to bear on a beam with equal intensity on every unit length of contact. Only in the investigation of other external forces, such as reactions, can a uniform load ever be bunched together at its center of gravity and assumed to act as a concentrated load. (See Chapter 1.)

Distributed loads may also be *uniformly varying and nonuniform*. In a uniformly varying load, the load increases or decreases at a definite rate along the length of a beam, such as water pressure acting on a vertical beam. Where the load is nonuniform, no regularity exists in its distribution; for example, unequal piles of coal scattered about the floor of a storage bin.

6-3 Internal Resistance of a Beam

When a beam does not collapse under an applied load, it is because of the development of its internal resistance stresses. If for any reason these resisting stresses are subsequently prevented from acting, collapse would be inevitable. If the beam should break because of an excessive amount of applied load, it would necessarily be inferred that on the cross section where failure took place the maximum resisting stresses had been exceeded. It is the intent of the discussion that follows to establish where on a beam the most critically stressed section occurs and how, by the application of subsequent theory, the size of the beam can be made large enough to reduce the induced stresses to a safe value.

To commence our search for the most critically stressed section or sections, let us center our attention on the beam shown in Fig. 6-3. Its own weight will be considered negligible in comparison to the magnitude of the applied concentrated loads and, therefore, will not appear in the computations.

By passing a vertical section through the beam at a point 1 ft from R_1, the resisting stresses on that section are instantly reduced to zero and

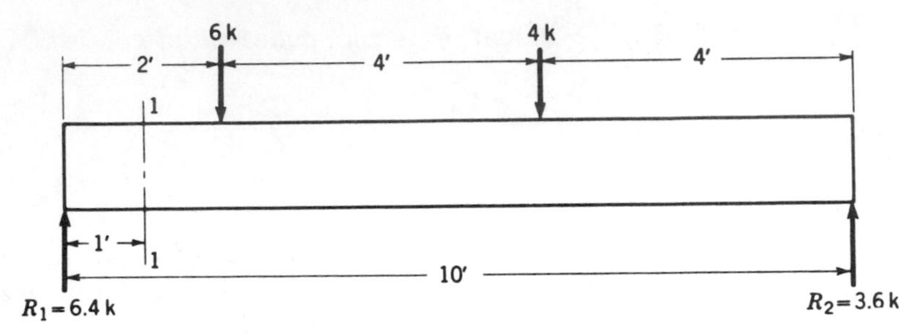

Figure 6-3

the beam collapses. To restore the original state of equilibrium to the two portions of the beam, the internal forces, which previously acted on the cut surfaces, must be replaced. These resisting forces are obtained by applying the three laws of equilibrium.

6-4 Vertical Shear, *C-T* Couple, and Bending Moment

Utilizing the smaller and more simple portion of the beam first, let us restore to the exposed section those forces that are required to balance the 6.4-kip reaction. Since the 6.4-kip reaction would tend to push the free body vertically upward, another 6.4-kip force (V_L) must act vertically downward on the exposed section to keep it in vertical balance (Fig. 6-4a). Immediately

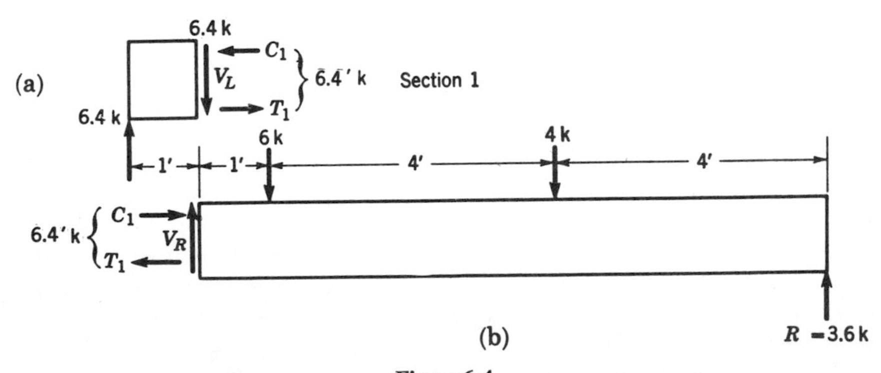

Figure 6-4

upon the application of this force, however, it is evident that the body is still not in equilibrium but is being acted upon by a couple. Another couple, the so-called *C-T* couple, is therefore required, and must, of course, be placed on the exposed section. Its action must be such as not to interfere with the 6.4-kip force already obtained, and should act opposite to the prevailing couple. These requirements necessitate the action of the forces of the couple perpendicular to the plane, with the *C* force acting in compression on the top of the beam and the T force acting in tension at the bottom. It is well to recall at this point that in all simple beams bent in a downward direction, the bottom fibers stretch and the upper fibers contract. These phenomena are effects produced by the internal C and T forces just mentioned and can be observed by bending any flexible material.

The magnitude of the C-T couple acting on the exposed section is, as previously stated, equal to the moment of the 6.4-kip forces, or $6.4 \times 1 = 6.4$ ft-kips. It is customary to take moments of these forces about a point on the action line of the restraining vertical force V_L.

Considering, now, the other longer portion of the beam (Fig. 6-4b), it is evident that a similar operation of balancing is required. Equating the vertical forces, including the assumed V_R force, we obtain

$$\Sigma V = 6 + 4 - 3.6 - V_R = 0$$

$$V_R = 6.4 \text{ kips acting upward}$$

This force is exactly equal to the vertical force on the exposed section of the shorter portion of the beam, but acts in the opposite direction. Considered together, these oppositely directed parallel forces, V_L and V_R, being of like magnitude and acting on opposite sides of the sectioning plane, produce an *internal-resisting shear action* on that plane. Although this shearing action has been revealed by the use of free bodies, it occurs in like magnitude on the corresponding section in the actual uncut beam, since the same equilibrium conditions must be maintained in both. It may further be concluded that shearing forces are developed on every vertical cross section of the actual beam where the forces on either side of the section are out of balance. The amount of this unbalance is equal *to the algebraic sum of the vertical forces on either the left or right side of the section, and is called the internal-resisting shear force or, more simply, the shear on the section.*

Although the above equation indicates that a 6.4-kip internal-resisting shear force was required to balance the vertical forces to the right of the section, equilibrium cannot be assured unless the moments also equal zero. Assuming a clockwise resisting moment M_R on the exposed section, the moment equation taken about a point on the action line of the internal shear force is

$$\Sigma M = 6 \times 1 + 4 \times 5 - 3.6 \times 9 + M_R = 0$$

$$M_R = 6.4 \text{ ft-kips}$$

It is interesting to note that this internal-resisting moment is exactly equal to the internal-resisting moment acting on the exposed section of the left-hand portion of the beam. Also, since this couple of 6.4 ft-kips must be directed clockwise, the upper force C of the couple must act into the section (compression), and the lower force T must act out of the section (tension). The stresses on the top and bottom of the beam on either side of the section are, therefore, of like character and are continuous across the section.

Although free bodies obtained by cutting the beam were used in analyzing the stress condition on the exposed surfaces, the uncut beam has the identical stress condition on the corresponding section, since both the free bodies and the beam are in equilibrium under the action of the same forces.

The magnitude of the internal-resisting moment (*or the C-T couple*) *on any section is*, as has been indicated in the previous moment equations, *equal to the algebraic sum of the moments of the forces to either the right or left side of the section.* This algebraic sum of moments produces the applied bending moment to the section and is differentiated from the *C-T* couple which, although equal to it in magnitude, acts oppositely to it. The *C-T* couple is, in other words, the equilibrant of the applied bending moment. The term *bending moment* is more frequently used than either *C-T* couple or internal-resisting moment.

6-5 Shear- and Bending-Moment Variations

Since the design of a beam depends on a knowledge of the variation of shear and bending moment along its length, it will be necessary to continue our study of internal shears and *C-T* couples at successive 1-ft intervals along the beam.

At the second-foot mark is located the 6-kip load. Because it is impossible to divide a concentrated load by the passage of a sectioning plane and because the magnitude of the resisting shear force varies in accordance with which side of the concentrated load we choose to locate our section, we must use two sections, each an infinitesimal distance to left and right of the

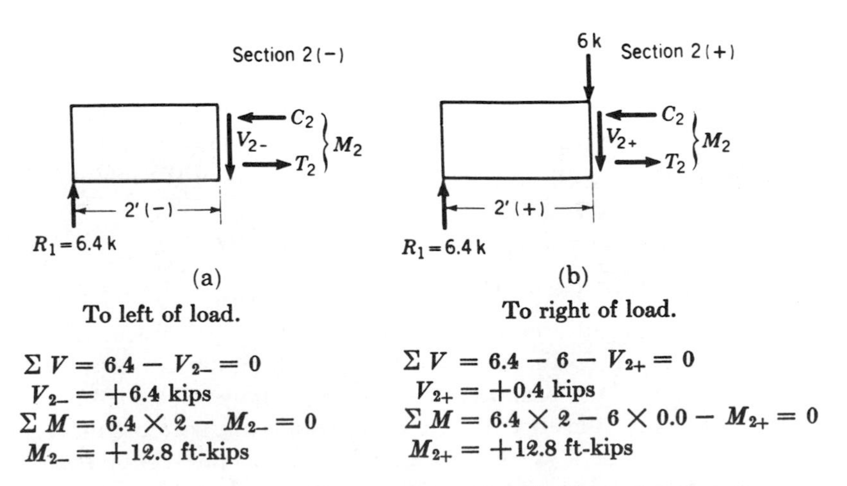

Figure 6-5 Shear- and bending-moment determinations of sections immediately to left and right of 6-kip concentrated load.

6-kip load. The two free bodies required are shown in Fig. 6-5. The right-hand portion of the beam is omitted in each case.

The magnitude of the bending moment on these two sections will, for all practical purposes, be the same, since the sections are separated by an infinitesimal distance, producing no appreciable change in moment arms. The equations and calculation of the bending moment, or C-T couple, at these two sections are given in Fig. 6-5.

The remainder of the shear- and bending-moment equations together with their corresponding free bodies are shown in Figs. 6-6 to 6-8.

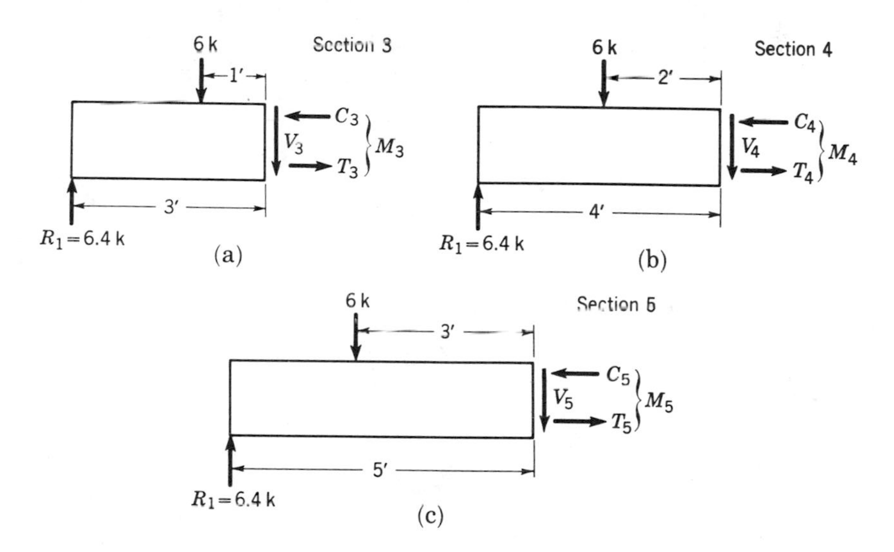

Figure 6-6

Section 3:

$$\Sigma V = 6.4 - 6 - V_3 = 0; \qquad V_3 = 0.4 \text{ kip}$$

$$\Sigma M = 6.4 \times 3 - 6 \times 1 - M_3 = 0; \qquad M_3 = 13.2 \text{ ft-kips}$$

Section 4:

$$\Sigma V = 6.4 - 6 - V_4 = 0; \qquad V_4 = 0.4 \text{ kip}$$

$$\Sigma M = 6.4 \times 4 - 6 \times 2 - M_4 = 0; \qquad M_4 = 13.6 \text{ ft-kips}$$

Section 5:

$$\Sigma V = 6.4 - 6 - V_5 = 0; \qquad V_5 = 0.4 \text{ kip}$$

$$\Sigma M = 6.4 \times 5 - 6 \times 3 - M_5 = 0; \qquad M_5 = 14.0 \text{ ft-kips}$$

Located 6 ft from R_1 is a concentrated load of 4 kips. Using the procedure explained for the 6-kip concentrated load, the necessary free bodies (Fig. 6-7) and equations are as follows:

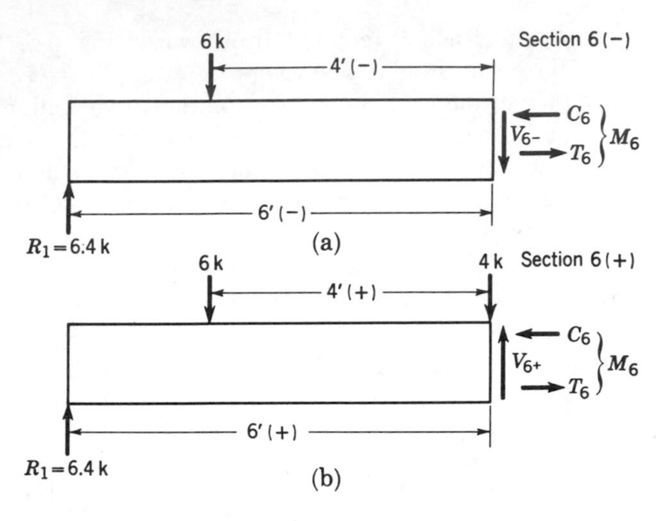

Figure 6-7

Section 6−:

$$\Sigma V = 6.4 - 6 - V_{6-} = 0; \qquad V_{6-} = 0.4 \text{ kip}$$

$$\Sigma M = 6.4 \times 6 - 6 \times 4 - M_{6-} = 0; \qquad M_{6-} = 14.4 \text{ ft-kips}$$

Section 6+:

$$\Sigma V = 6.4 - 6 - 4 + V_{6+} = 0; \qquad V_{6+} = -3.6 \text{ kips}$$

$$\Sigma M = 6.4 \times 6 - 6 \times 4 - 4 \times 0.0 - M_{6+} = 0; \qquad M_{6+} = 14.4 \text{ ft-kips}$$

In the above shear equation, attention should be called to the change in direction of the resisting shear force.

For convenience, the shear- and bending-moment evaluations for the remainder of the sections (Fig. 6-8) will be obtained using the right-hand portion of the beam as a free body. This procedure will eliminate the necessity of using a greater number of loads in the determination of shear and bending moment. An explanation of the conventional use of plus and minus signs will be given in the following section.

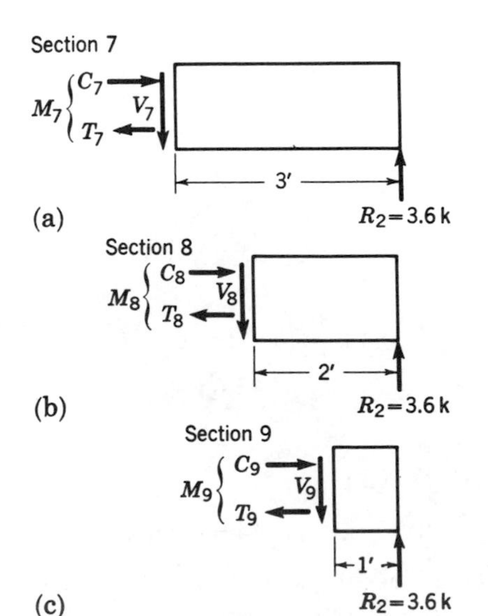

Figure 6-8

Section 7:

$$\Sigma V = V_7 - 3.6 = 0; \quad V_7 = 3.6 \text{ kips}$$

$$\Sigma M = 3.6 \times 3 - M_7 = 0;$$

$$M_7 = 10.8 \text{ ft-kips}$$

Section 8:

$$\Sigma V = V_8 - 3.6 = 0; \qquad V_8 = 3.6 \text{ kips}$$
$$\Sigma M = 3.6 \times 2 - M_8 = 0; \qquad M_8 = 7.2 \text{ ft-kips}$$

Section 9:

$$\Sigma V = V_9 - 3.6 = 0; \qquad V_9 = 3.6 \text{ kips}$$
$$\Sigma M = 3.6 \times 1 - M_9 = 0; \qquad M_9 = 3.6 \text{ ft-kips}$$

Since no intervening loads are present between forces V_9 and R_2, the shear on all sections between these two forces must also be equal to 3.6 kips. As one passes beyond R_2, the shear must return to zero, since the beam was previously placed in equilibrium by requiring the algebraic sum of all vertical forces to equal zero.

The bending moment between section 9 and R_2 decreases to zero as R_2 is approached, since the moment arm of the reaction becomes increasingly smaller.

6-6 Signs for Shear and Bending Moment

In order that a common understanding be developed concerning the direction of shear and bending moment, a conventional method of determining their proper signs must be consistently followed. That system most generally used is given below.

Shear. This criterion applies solely to the left-hand portion of the beam. Thus, as is indicated in Fig. 6-9a, if the algebraic sum of the vertical forces to the left of the section acts upward and the internal-resisting shear acts downward, the shear is positive. Conversely, the shear will be negative when the reverse is true, as shown in Fig. 6-9b.

Moment. The bending moment at any section is positive if the algebraic sum of the moments to the left of the section is directed clockwise. An algebraic sum of moments to the left of the section is negative if it is directed counterclockwise (Fig. 6-10a).

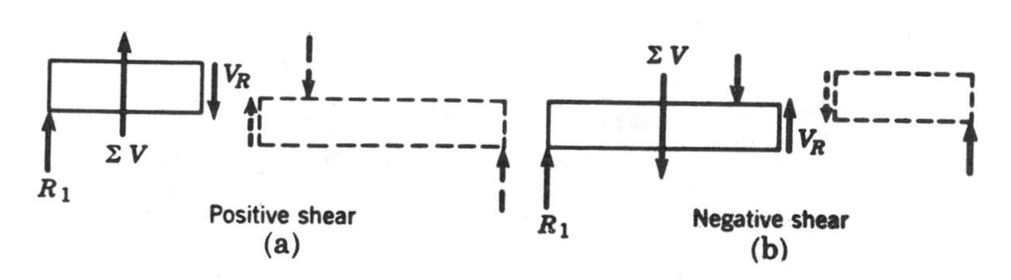

Figure 6-9

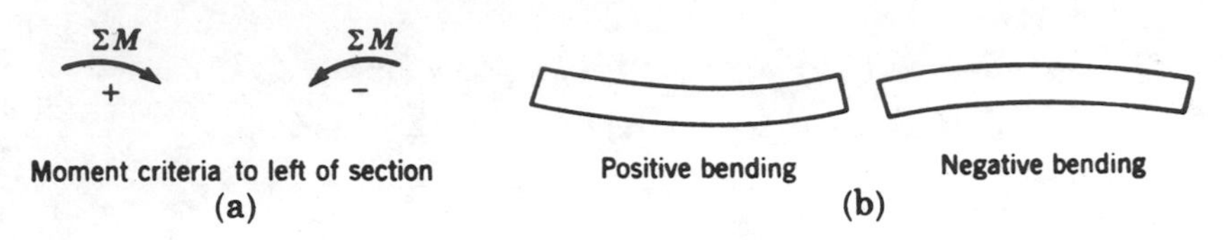

Figure 6-10

An even more general statement producing the same results is that a bending moment is positive if it produces compression on the upper fibers. Conversely, if tension occurs on the upper fibers, the bending moment is negative. This is illustrated in Fig. 6-10b.

PROBLEMS

6-1. Determine the magnitude of the internal-resisting forces and couple acting on section *A-A* of Fig. 6-11. *Hint:* Resolve the 10-kip force into components. *Answers:* Shear, 3.53 kips; moment, 28.24 ft-kips; compression, 20.24 kips.

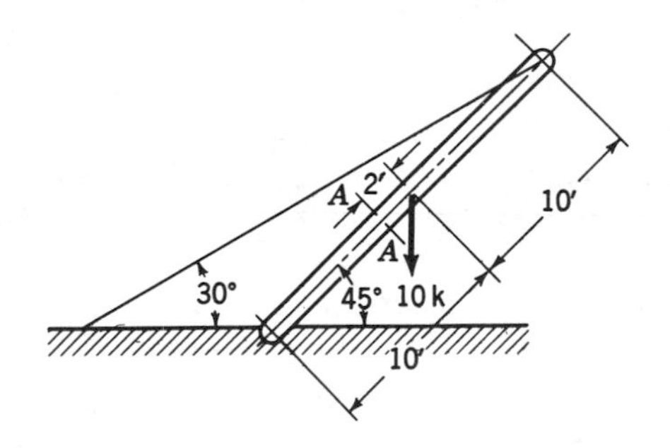

Figure 6-11

6-2. Determine the magnitude of the internal shear force and the bending moment acting on sections *A-A*, *B-B*, and *C-C* of the beam shown in Fig. 6-12. *Answers:* $V_A = +7.0$ kips, $M_A = +4.0$ ft-kips, $V_B = -3.0$ kips, $M_B = 9.0$ ft-kips, $V_C = -1.5$ kips, $M_C = +1.5$ ft-kips.

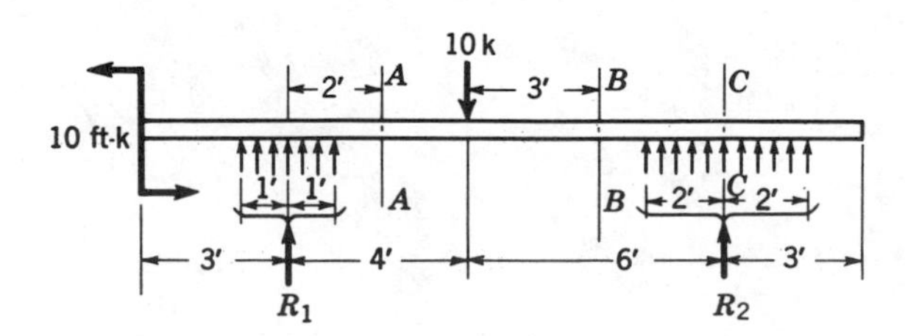

Figure 6-12

6-3a, b, c, d. Determine the magnitude of the internal shear force and the bending moment (*C-T* couple) acting on sections *A-A* and *B-B* of the beams shown in Fig. 6-13. Draw a free body for each section studied and show the resisting force and couple acting in their proper directions.

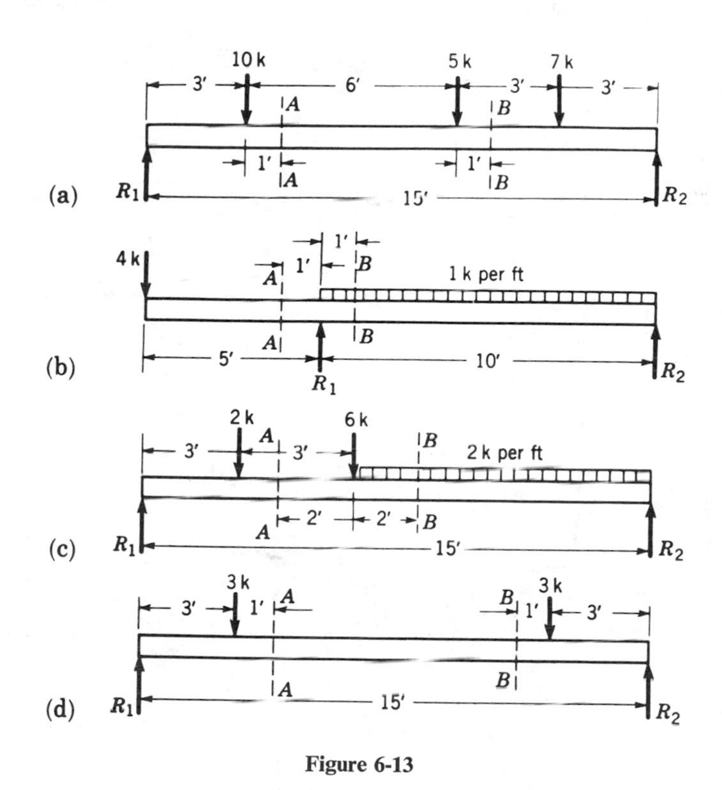

Figure 6-13

SHEAR- AND BENDING-MOMENT DIAGRAMS

6-7 Shear Diagram

A shear diagram is a graphical representation of the variation of the internal-resisting shear along the full length of the beam. Every properly conducted design of a beam should include the construction of its shear diagram. It reveals at a glance the complete story of the shear and, in particular, those sections that are critically stressed.

Referring to the problem used in the discussion of the internal-resisting forces in a beam, let us tabulate the values of internal shear force with their proper sign, as instructed in §6-6.

$V_0 = 0.0$ kip	$V_3 = +0.4$ kip	$V_7 = -3.6$ kips
$V_{0+} = +6.4$ kips	$V_4 = +0.4$ kip	$V_8 = -3.6$ kips
$V_1 = +6.4$ kips	$V_5 = +0.4$ kip	$V_9 = -3.6$ kips
$V_{2-} = +6.4$ kips	$V_6 = +0.4$ kip	$V_{10-} = -3.6$ kips
$V_{2+} = +0.4$ kip	$V_{6+} = -3.6$ kips	$V_{10} = 0.0$ kip

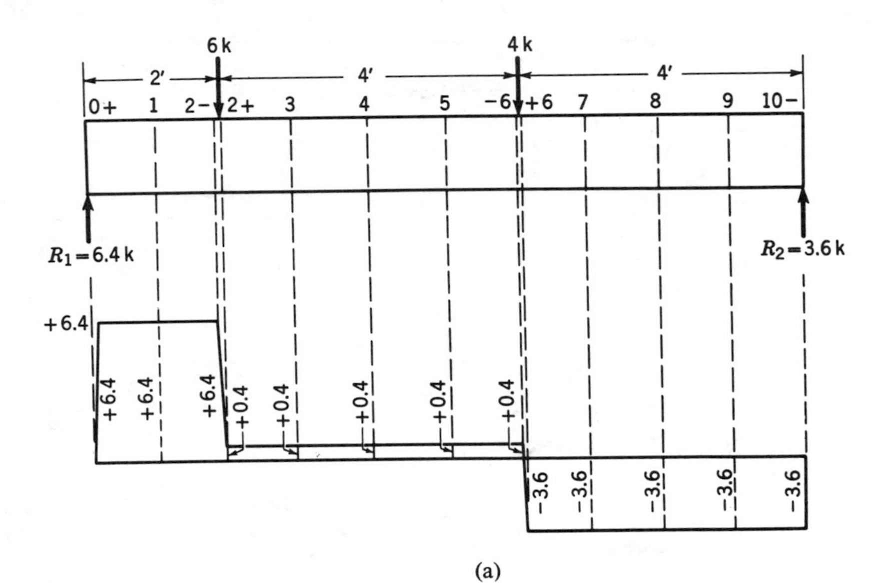

(a)

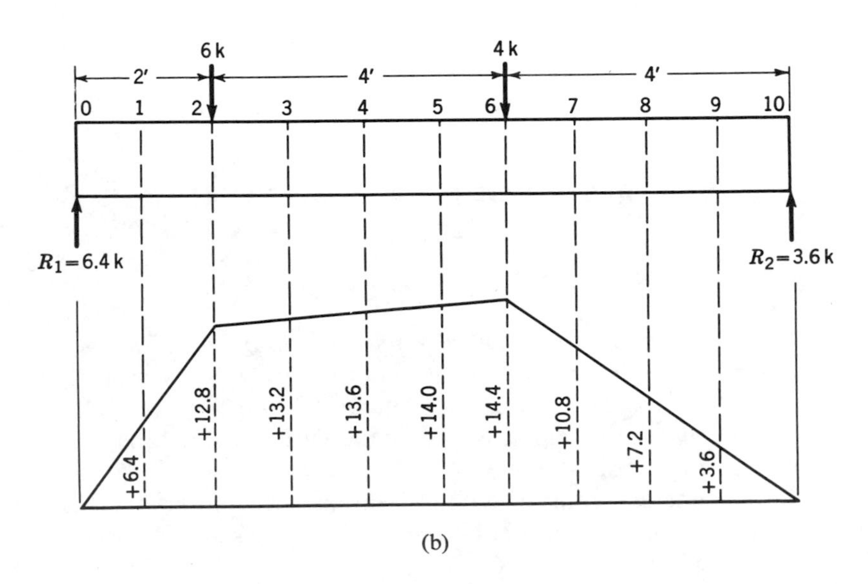

(b)

Figure 6-14

By plotting these values on an axis equal to the length of the beam, the shear diagram is obtained (Fig. 6-14a). The most critically stressed portion of the beam with reference to shear is shown to be included between R_1 and the 6-kip concentrated force.

Although the risers in Fig. 6-14a are shown slightly inclined for the sake of clarity, their horizontal projections represent minute distances between adjacent sections. In all subsequent shear diagrams, therefore, risers will be shown vertically, which indicates more accurately their proper position.

A simplification. Now that the groundwork of shear determination has been laid, an even more convenient method of analyzing the shear forces in a beam can be considered. This simple method conceives of the shear diagram as being an *accumulative algebraic sum* of the vertical forces to the left of a section as one proceeds from the left to the right end of the beam. Starting from zero shear force at the horizontal axis, the shear diagram passes through upward and downward jogs, eventually to end once more on the horizontal axis. Each upward jog is equivalent to an upward-acting force, and each downward jog is equivalent to a downward-acting force. The magnitude of the the load is represented by the length of the jog.

Thus, if we could consider the beam covered by a piece of paper and slowly exposed by moving the paper from left to right, the algebraic sum of the loads exposed would be equal to the ordinate to the shear diagram at the section located by the left edge of the paper.

6-8 Bending-Moment Diagram

A bending-moment diagram is a graphical representation of the variation of the bending moment or internal-resisting moment along the full length of the beam. The bending-moment diagram for the beam just discussed in §§6-3 to 6-7 is obtained by plotting the values computed in §6-5 on a horizontal axis equal to the length of the beam (Fig. 6-14b). These values with their proper sign are retabulated below for convenience.

$$M_0 = 0.0 \text{ ft-kip} \qquad M_6 = +14.4 \text{ ft-kips}$$
$$M_1 = +6.4 \text{ ft-kips} \qquad M_7 = +10.8 \text{ ft-kips}$$
$$M_2 = +12.8 \text{ ft-kips} \qquad M_8 = +7.2 \text{ ft-kips}$$
$$M_3 = +13.2 \text{ ft-kips} \qquad M_9 = +3.6 \text{ ft-kips}$$
$$M_4 = +13.6 \text{ ft-kips} \qquad M_{10} = 0.0 \text{ ft-kip}$$
$$M_5 = +14.0 \text{ ft-kips}$$

Since, by inspection, the beam under consideration must bend concave upward, the sign of all the calculated bending moments may be easily ascertained as being positive. The most critical bending moment to be resisted by the beam is indicated by the maximum ordinate as occurring under the 4-kip load.

6-9 Definitions—Shear and Bending Moment

Inasmuch as the determination of shear and bending moment in beams is imperative to the design of beams, it is important that their definitions, given below, be committed to memory.

The shearing force occurring on any section perpendicular to the longitudinal axis of a straight horizontal beam is equal to the algebraic sum of the vertical forces or components to either the left or to the right of that section.

The bending moment occurring on any section perpendicular to the longitudinal axis of a straight horizontal beam is equal to the algebraic sum of the moments of the forces to either the left or to the right of that section.

Illustrative Problem 1. Draw the shear and bending-moment diagrams of the beam shown in Fig. 6-15.

SOLUTION. The uniform load included on this beam will offer no difficulty in the determination of shear and bending moment if it is allowed to remain in its distributed position. Any attempt at concentrating a uniform load for the determination of shear and bending moment will generally result in confusion for the beginner. As indicated previously, only in the computation of *unknown external forces*—that is, reactions in this case—is it advisable to concentrate distributed loads.

Reactions. The reactions R_1 and R_2 are obtained by taking moments successively at points A and B. Too much emphasis cannot be placed upon their correct calculation, since an error is usually detected upon the completion of the moment diagram, and only after a considerable amount of work has been performed. The computations of reactions R_1 and R_2 follow.

$$\Sigma M_A$$

$$6 \times -3 = -18$$
$$12 \times 3 = +36$$
$$3 \times 8 = +24$$
$$+42 - 10R_2 = 0$$
$$R_2 = 4.2 \text{ kips}$$

$$\Sigma M_B$$

$$6 \times 13 = +78$$
$$12 \times 7 = +84$$
$$3 \times 2 = +6$$
$$+168 - 10R_1 = 0$$
$$R_1 = 16.8 \text{ kips}$$

Check:

$$\Sigma R = \Sigma P$$
$$4.2 + 16.8 = 6.0 + 12.0 + 3.0$$
$$21.0 = 21.0$$

Computation of shear forces. Although drawing free bodies to the left or right of every foot section (see §6-7) results in an absolutely sure method of gaining data for the shear diagram, it is a lengthy procedure. For the sake of simplicity let us think of the shear curve as being an accumulative algebraic

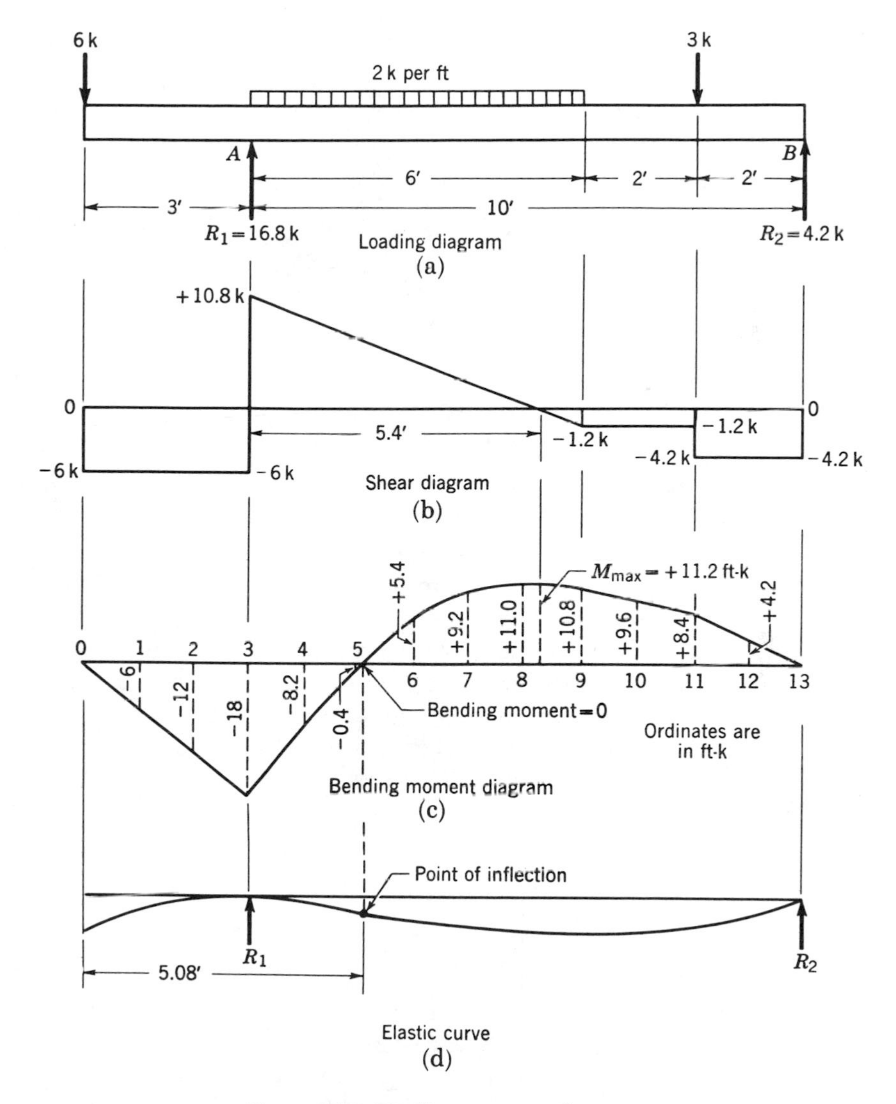

Figure 6-15 Bending-moment diagram.

sum of the vertical forces to the left of a section as we proceed from the left to the right of the beam. In tabular form, these algebraic sums are (in kips):

$$V_0 = 0$$
$$V_{0+} = -6$$
$$V_1 = -6$$
$$V_2 = -6$$
$$V_{3-} = -6$$

$$
\begin{aligned}
V_{3+} &= -6 + 16.8 = +10.8 \\
V_4 &= -6 + 16.8 - 2 \times 1 = +8.8 \\
V_5 &= -6 + 16.8 - 2 \times 2 = +6.8 \\
V_6 &= -6 + 16.8 - 2 \times 3 = +4.8 \\
V_7 &= -6 + 16.8 - 2 \times 4 = +2.8 \\
V_8 &= -6 + 16.8 - 2 \times 5 = +0.8 \\
V_9 &= -6 + 16.8 - 2 \times 6 = -1.2 \\
V_{10} &= -6 + 16.8 - 2 \times 6 = -1.2 \\
V_{11-} &= -6 + 16.8 - 2 \times 6 = -1.2 \\
V_{11+} &= -6 + 16.8 - 2 \times 6 - 3 = -4.2 \\
V_{12} &= -6 + 16.8 - 2 \times 6 - 3 = -4.2 \\
V_{13-} &= -6 + 16.8 - 2 \times 6 - 3 = -4.2 \\
V_{13+} &= -6 + 16.8 - 2 \times 6 - 3 + 4.2 = 0
\end{aligned}
$$

Note: $V = 0$: (1) at R_1 and R_2. (2) at 5.4 ft to right of R_1, or section 8.4.

As demonstrated above, the calculation of shear, where uniform load is encountered, proceeds with no difficulty if that load is left in its distributed form. The shear diagram is shown in Fig. 6-15b.

Computation of bending moments. The valuation of the bending moments, or the required *C-T* resistance couples, will be obtained by writing the algebraic sums of the moments of the forces to the left or right of the sections. A mental picture of the pertinent free body should be kept in mind while analyzing these equations. The proper sign of the bending moment should be checked in accordance with the criteria given in §6-6.

Taken to the left of section (in foot-kips):

$$
\begin{aligned}
M_0 &= 0 \\
M_1 &= -6 \times 1 = -6 \\
M_2 &= -6 \times 2 = -12 \\
M_3 &= -6 \times 3 = -18 \\
M_4 &= -6 \times 4 + 16.8 \times 1 - 2 \times 1 \times \tfrac{1}{2} = -8.2 \\
M_5 &= -6 \times 5 + 16.8 \times 2 - 2 \times 2 \times 1 = -0.4 \\
M_6 &= -6 \times 6 + 16.8 \times 3 - 2 \times 3 \times 1\tfrac{1}{2} = +5.4 \\
M_7 &= -6 \times 7 + 16.8 \times 4 - 2 \times 4 \times 2 = +9.2 \\
M_8 &= -6 \times 8 + 16.8 \times 5 - 2 \times 5 \times 2\tfrac{1}{2} = +11 \\
M_9 &= -6 \times 9 + 16.8 \times 6 - 2 \times 6 \times 3 = +10.8
\end{aligned}
$$

Taken to the right of section (in foot-kips):

$$
\begin{aligned}
M_{10} &= +4.2 \times 3 - 3 \times 1 = +9.6 \\
M_{11} &= +4.2 \times 2 = +8.4 \\
M_{12} &= +4.2 \times 1 = +4.2 \\
M_{13} &= 0
\end{aligned}
$$

As revealed by the moment diagram, the maximum bending moment or the maximum *C-T* couple of −18 ft-kips is located on that section immediately above R_1. There is, to be sure, a maximum positive moment acting on the beam whose approximate magnitude is equal to 11-ft-kips. But because most beams have uniform cross sections that are symmetrical to the neutral axis and are made of materials equally strong in tension and compression, the maximum intensity of stress is developed by the maximum numerical value of bending moment. It is for such beams that the −18 ft-kips bending moment would prove critical.

Beams of nonuniform or irregular cross section or of materials weak in either tension or compression do not necessarily have their designs controlled by the maximum bending moment, but by that moment that will produce the least favorable stress condition. A reversal of sign in the bending moment that produces an exchange in position of the compression and tension forces in the *C-T* couple may inadvertently place a force where the beam is least capable of carrying it. It is well, therefore, to know all the maximum moments (positive and negative) and the laws governing the variation of bending moments over the length of the beam. By writing these laws algebraically, the determination of moments anywhere on the beam may be simplified.

6-10 Algebraic Expressions of Shear and Moment

The algebraic expressions for the variations of shear and bending moment are portrayed graphically by the shear and moment curves already plotted in Fig. 6-15b and c. In this connection, it should be recalled that the straight lines and continuous curves comprising the shear- and bending-moment diagram obey definite yet altogether different algebraic laws. Referring again to the beam of Fig. 6-15 and using the same procedure as before for obtaining shear and bending moments, we shall separate and express algebraically each portion of its shear and moment diagram. The equations, *referred to the left-hand end of the beam*, are as follows:

For Shear

From section 0 to 3 $\quad V_x = -6.0$ kips

From section 3 to 9 $\quad V_x = 10.8 - 2(x - 3) = -2x + 16.8$

From section 9 to 11 $\quad V_x = -1.2$

From section 11 to 13 $\quad V_x = -4.2$

For Moment

From section 0 to 3, $M_x = -6x$

From section 3 to 9, $M_x = -6x + 16.8(x - 3) - \dfrac{2(x - 3)(x - 3)}{2} = -x^2 + 16.8x - 59.4$

From section 9 to 11, $M_x = -6x+16.8(x-3)-2\times 6\times (x-9+3) = -1.2x + 21.6$

From section 11 to 13, $M_x = -6x+16.8(x-3)-2\times 6\times (x-6)-3(x-11) = -4.2x + 54.6$

As may already have been surmised, these equations are generalizations of the expressions for shear and moment obtained previously for the shear- and bending-moment diagrams. To procure a shear or bending moment at a particular section, we need only insert the corresponding length into the proper equation, and solve.

It is important to note that for that portion of the beam between sections 3 and 9, the moment equation is of the second degree (that is, it contains an x^2 term) and, when plotted, produces a parabolic curve. All the other moment and shear equations are represented by straight lines.

Uniformly varying loads. The plotting of the algebraic expressions for shear and bending moment is particularly recommended for constructing shear- and bending-moment diagrams for beams subjected to uniformly varying loads. See Fig. 6-16.

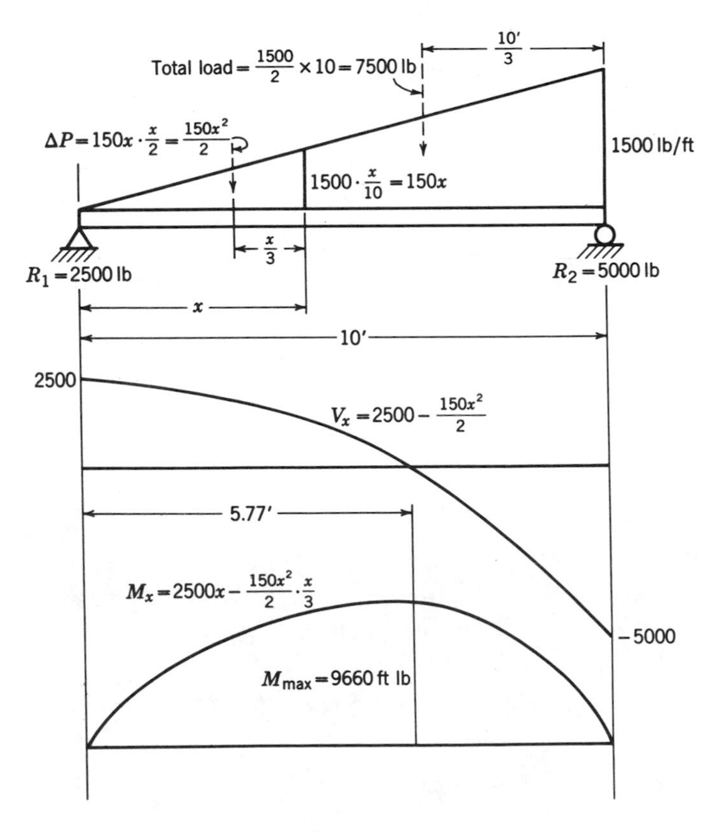

Figure 6-16

6-11 Locating a Maximum Bending Moment

When the maximum-moment ordinate occurs at a sharp point on the moment curve, the determination of its location is made by observation (see moment curve of Fig. 6-14b). However, if a maximum bending moment occurs on the top of a smooth parabolic curve, as in the case of the maximum positive moment of Fig. 6-15, the exact determination can be made by taking the derivative of the moment equation applicable to this section, equating the derivative to zero, and solving thus:

The equation

$$M_x = -x^2 + 16.8x - 59.4$$

when differentiated yields

$$\frac{dM_x}{dx} = -2x + 16.8$$

Placing this equation equal to zero and solving gives $x = 8.4$ ft.

An alternative method for obtaining the location of maximum moment can be made from the differentiated equation. Note that

$$\frac{dM}{dx} = -2x + 16.8 = V_x \qquad \text{(the shear equation at the point of maximum moment)}$$

To set the differentiated equation equal to zero and solve is tantamount to obtaining the value of x when V_x is equal to zero. Thus, the result of this illustration may be generalized as follows:

A maximum bending moment is obtained whenever the shear diagram cuts the horizontal axis.

Note that this statement does not allude only to *the* maximum bending moment. Since there may be several maximum bending moments for one beam depending on the magnitude and location of its loads and reactions, *the* maximum bending moment is the greatest of the several possible maximum moments.

The location of the maximum positive moment on the beam shown in Fig. 6-15 is 5.4 ft to the right of reaction R_1. The exact value of the bending moment is obtained by inserting the value of 8.4 ft (the value of x is determined from the left-hand end of the beam) into the bending moment equation applicable to that portion of the beam and solving for M_x.

$$M_x = -x^2 + 16.8x - 59.4 = -8.4^2 + 16.8 \times 8.4 - 59.4$$
$$= -70.6 + 141.2 - 59.4 = 11.2 \text{ ft-kips}$$

6-12 Point of Inflection

At the very beginning of our study in stresses and strains, it was stated that there could be no stress without an accompanying deformation. When, in the loading of a uniform cross-sectioned beam, internal-resisting *C-T* couples are produced, each section must deform in proportion to the magnitude of resisting moment acting upon it.* The combined effect of all the sectional deformations is found in the deflection curve of the beam. Those sections that resist the greatest amount of moment suffer the greatest deformation. Those sections that have no bending moment applied to them (that is, where the moment equals zero) are subject to no bending deformation and are consequently straight. When such a point lies between zones of positive and negative moment, it is known as a *point of inflection*. It might also be termed as that point on a beam where the curvature reverses itself (Fig. 6-17).

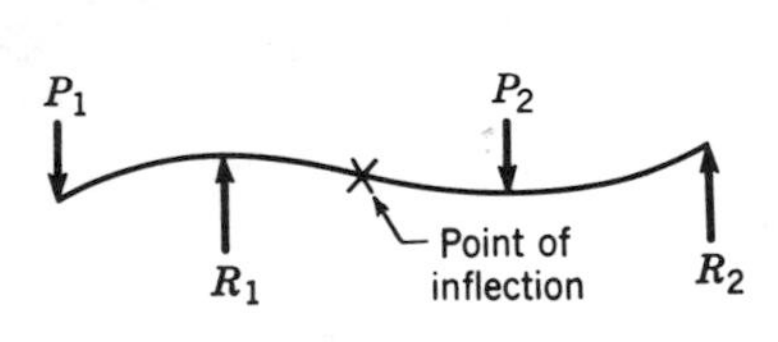

Figure 6-17

To locate the point of inflection in a beam requires the solution of the proper moment equation set equal to zero. For the beam of Fig. 6-15, this solution is as follows:

$$M_x = -x^2 + 16.8x - 59.4 = 0$$

from which

$$x = 5.08 \text{ ft} \quad \text{or} \quad 11.72 \text{ ft}$$

Obviously, the correct answer is 5.08 ft, since it falls within the limits of the curve on the moment diagram. The 11.72-ft answer refers to that intersection with the horizontal axis that would occur if the curve were extended. It has no meaning for the present problem.

The selection of the proper equation for the determination of a point of inflection is generally based upon the approximate drawing of a moment diagram. Usually, that curve which crosses the horizontal axis is not difficult to find. If, by chance, a selection is made that results in an imaginary answer, the most probable adjoining equation should be chosen. This equation will generally provide the desired result.

PROBLEMS

6-4a, b, c, d, e, f, g, h. Draw the shear- and bending-moment diagrams for the beams in Fig. 6-18a to h, indicating thereon all controlling ordinates. Determine the maximum shear and bending moment. Disregard the weight of the beam. *Answers:* (a) Maximum shear, +30 kips, maximum moment, +200 ft-kips; (c) +7.0 kips, +32.5 ft-kips; (e) +12.56 kips, +65.7 ft-kips; (g) +12.62 kips, +59.4 ft-kips.

6-5a, b, c, d, e, f, g, h. Draw the shear- and bending-moment diagrams for the beams in Fig. 6-19a to h, indicating thereon all controlling ordinates. Determine

*Deformations due to shearing stresses have been omitted from this discussion. For all beams, other than those that are short and stubby, the deflection due to the shearing stresses is so small as to be inconsequential.

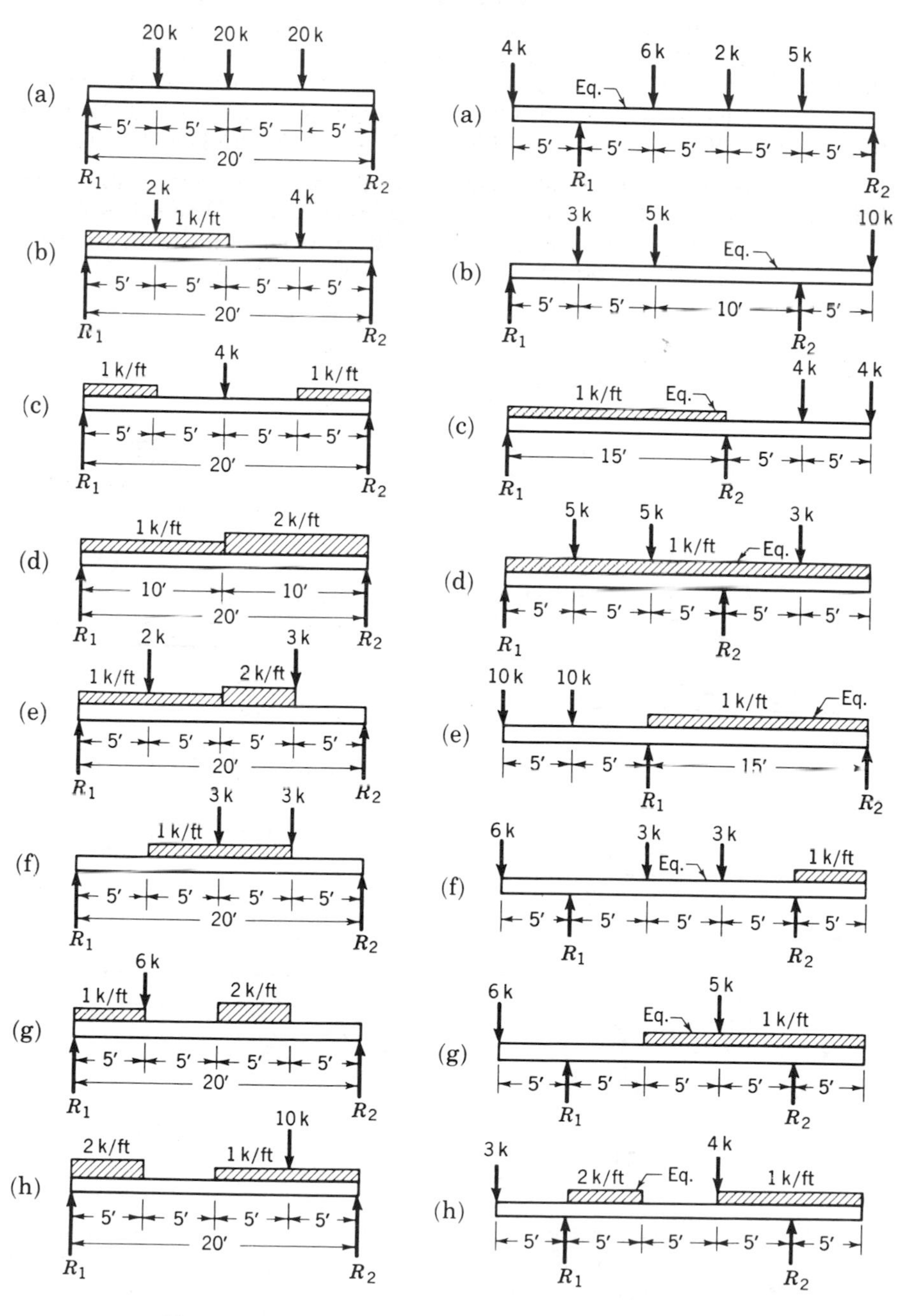

Figure 6-18

Figure 6-19

the shear-and bending-moment equations for portions of the beams marked "Eq." Calculate the position and the magnitude of maximum shear and bending moment. Locate the points of inflection. Disregard weight of beam. *Answers:* (a) Maximum shear, 7.75 kips, maximum moment, 27.50 ft-kips; (b) 10 kips, −50 ft-kips; (c) −11.5 kips, −60 ft-kips; (d) −16.8 kips, −65 ft-kips; (e) −20 kips, −150 ft-kips; (f) −6 kips, −30 ft-kips; (g) −8.83 kips, −30 ft-kips; (h) 10.7 kips, 17 ft-kips.

6-6a, b, c, d. Determine the magnitude of the maximum vertical shear and bending moment for the beams shown in Fig. 6-20. Locate these values on the shear- and bending-moment diagrams. Disregard the weight of the beam. *Answers:* (a) $P/2$, $Pl/4$; (b) $wl/2$, $wl^2/8$; (c) $-P$, $-Pl$; (d) $-wl$, $-(wl^2/2)$.

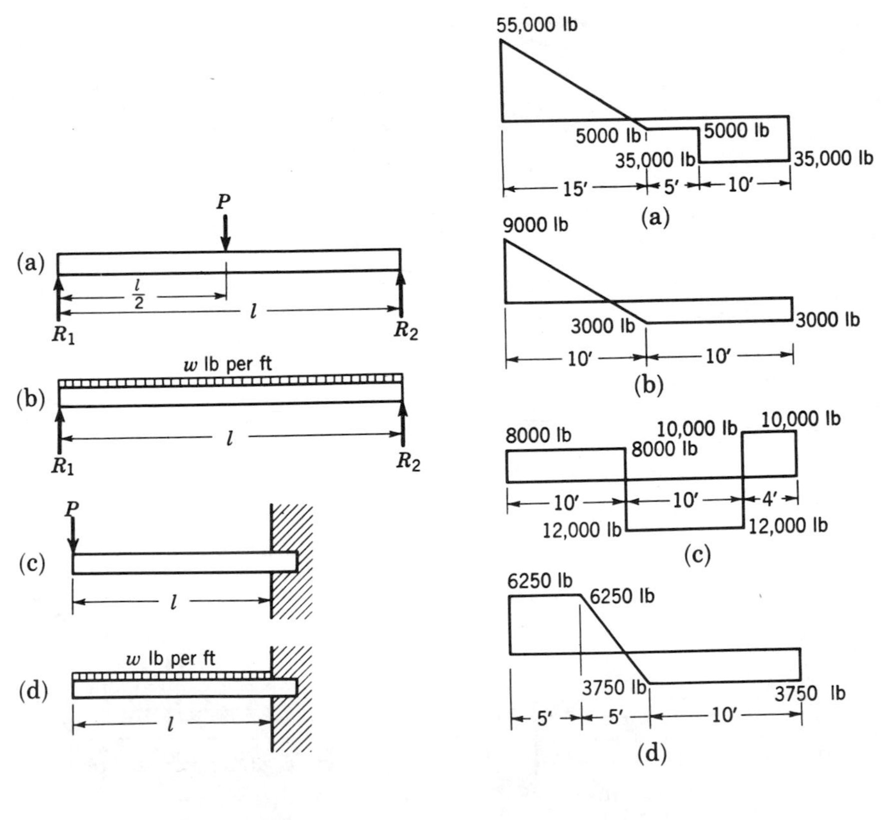

Figure 6-20

Figure 6-21

6-7a, b, c, d. The shear diagrams shown on Fig. 6-21 have been drawn for beams whose loading diagrams have been lost. Draw the picture of each beam with loads in their correct locations and the corresponding bending-moment diagram.

6-8. Draw the shear- and bending-moment diagrams for the beam shown in Fig. 6-22. *Hint:* The same rules for determining shear and bending moment at any section remain valid regardless of the type of loading.

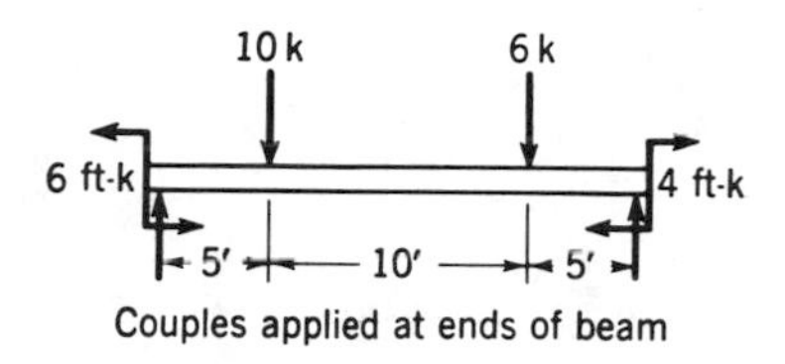

Figure 6-22

6-9a, b. Draw the shear- and bending-moment diagrams of the beams shown in Fig. 6-23, giving all critical values and determining all points of zero shear and bending moment.

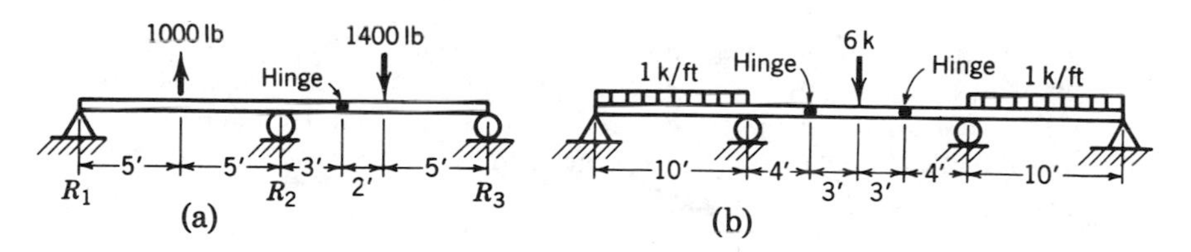

Figure 6-23

6-10. Three 12-ft lengths of 12-in. bell-and-spigot cast-iron pipe have been joined together with leaded joints. The assemblage is then lowered into a pipe trench by two slings so placed that there is no tendency to break the joints. Locate the positions of the slings. *Answer:* 8 ft from each end.

6-11. A long precast concrete pile, uniform in cross section, is to be stored in a horizontal position on two supports. Compute the position of the supports that will induce in the pile a minimum bending moment. What is the value of the minimum bending moment?

6-12. A barge 100 ft long and weighing 2 kips/ft of length is to carry a fabricated part for a bridge which will exert concentrated loads on the deck as shown in Fig. 6-24. Determine the distance *a* such that the barge will float with its bottom horizontal, i.e., the pressure on the bottom is uniform. Considering the barge as a longitudinal beam, draw its shear- and bending-moment diagrams. Give all critical values. *Answer:* 14 ft.

6-13. Draw the shear- and bending-moment diagrams for each straight portion of the beam shown in Fig. 6-25.

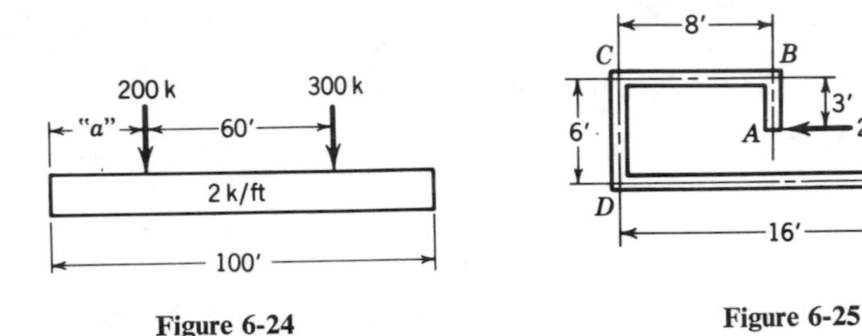

Figure 6-24

Figure 6-25

6-14. After drawing the shear- and bending-moment diagrams for the two cantilever beams shown in Fig. 6-26, discuss the differences imposed upon these diagrams by placing the couple on the beam in the two different ways.

6-15. A typical column supporting an overhead electrification system has loads applied to it as shown in Fig. 6-27. Draw the shear- and bending-moment diagrams giving all controlling ordinates. Assume the foundation to be capable of resisting forces P_1 and P_2, with a moment about the base of 10,000 ft-lb. *Answers:* Maximum shear, 2150 lb, maximum moment, 21,650 ft-lb.

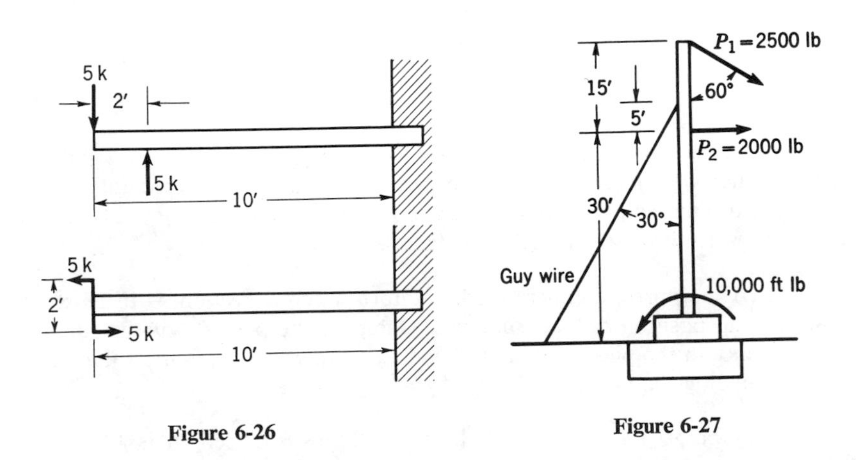

Figure 6-26

Figure 6-27

6-16a, b. Draw the shear- and bending-moment diagrams for the beams shown in Fig. 6-28. Determine the algebraic equations of the shear- and bending-moment curves. Determine the exact location and magnitude of the maximum bending moments. Disregard the weight of the beam.

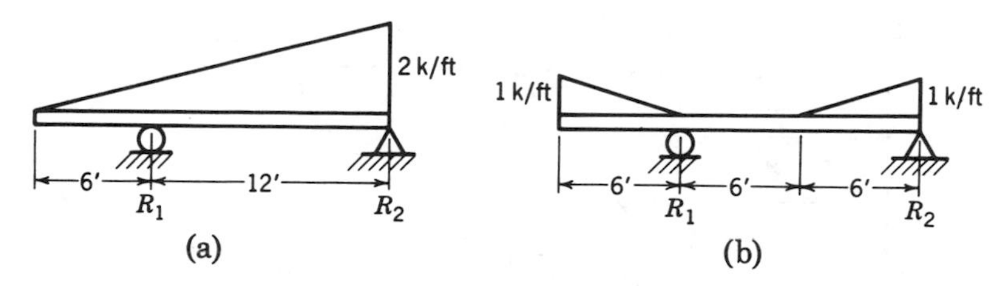

Figure 6-28

SHEAR-AREA METHOD OF CONSTRUCTING BENDING-MOMENT DIAGRAMS

6-13 Theory of the Shear-Area Method

In §6-11 it was shown that the derivative of the moment equation with respect to x was equal to the shear, or, mathematically stated,

$$\frac{dM_x}{dx} = V_x$$

This expression may be changed to read as follows:

$$dM_x = V_x\,dx \tag{6.1}$$

Stated simply, this equation says that the difference between the bending moments acting on two vertical sections of a beam separated by an infinitesimally small distance, dx, is equal to the product $V_x\,dx$. But $V_x\,dx$ is equal to the area of the shear diagram included between the extremes of this same distance dx (Fig. 6-29). Thus, we may conclude that the difference between these or any other two bending moments separated by a distance dx is equal to the area of the shear diagram between them (shown crosshatched).

This axiomatic statement may be expanded into a more useful form if we establish from it the difference in moments M_1 and M_2 between any two points on the beam. For this determination we must add all the $V_x\,dx$ areas between these two points. Thus, by integration we obtain

$$\int_{M_1}^{M_2} dM_x = \int_{x_1}^{x_2} V_x\,dx \tag{6.2}$$

which in turn equal $M_2 - M_1$, the sum of the $V_x\,dx$ areas between x_1 and x_2, or, simply stated, the area of the shear diagram between those two points. Because in this summation process some shear areas may be positive and some negative, the shear areas are added algebraically. It is also well to bear in mind that bending moments must carry their correct sign.

As shown in Fig. 6-29, the difference in bending moment between sections A and B of the previous beam is equal to the intervening shear diagram.

$$M_B - M_A = \frac{10.8 \times 5.4}{2} = 29.2 \text{ ft-kips}$$

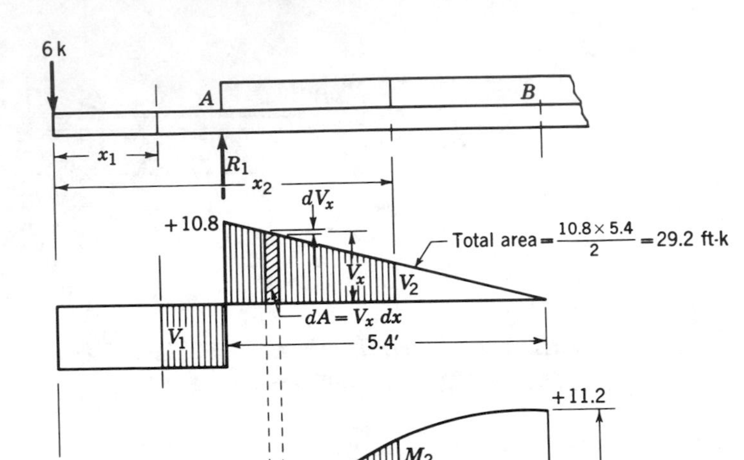

Figure 6-29

Added to the negative moment of −18 ft-kips at section A, the maximum moment at section B becomes 29.2 − 18 = 11.2 ft-kips.

Before going into further detail, let us place in definition form the shear-area method of constructing bending-moment diagrams. It is as follows: *The difference in bending moment between any two sections A and B on a straight beam loaded with concentrated* or uniform loads is equal to the algebraic sum of the shear area between those two sections.*

6-14 Application of the Shear-Area Method

Let us now proceed to unfold this method in more detail. For our illustration, let us return to the beam of Fig. 6-15, which we have already analyzed by the longer free-body method.

Starting from the left-hand end, we note that the shear area for each 1-ft interval of the overhang is equal to $-6 \times 1 = -6$ ft-kips (Fig. 6-30). Since the moment at the very end is zero, the addition of −6 ft-kips of shear area gives a moment at section 1 of −6 ft-kips. Adding the two subsequent −6-ft-kip shear areas successively produces the moments at the second- and third-foot sections of −12 and −18 ft-kips, respectively. Observe that the moment diagram is a straight line because of the addition of *equal* magnitudes of shear area.

*Does not include couples acting at a point (see §6-15).

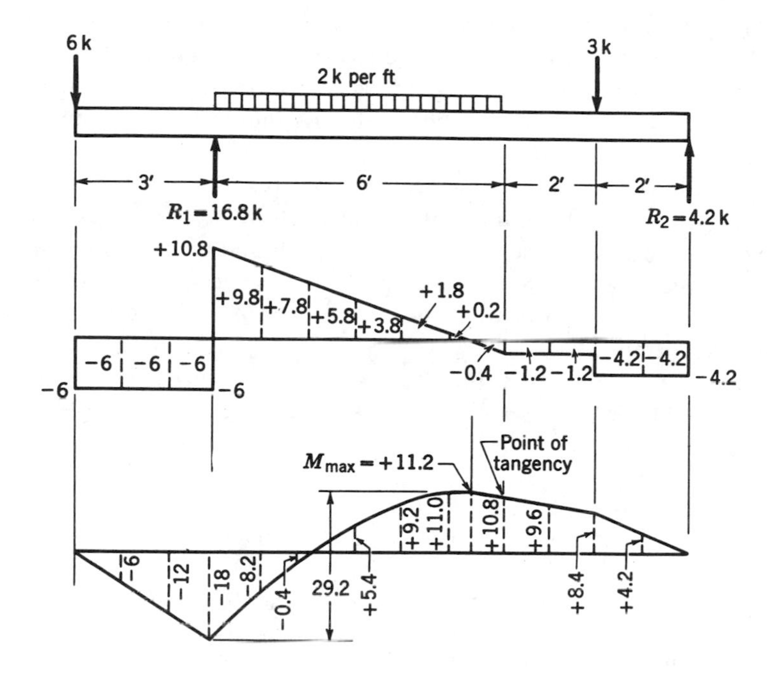

Figure 6-30 All values on shear curve are in kips; all values on moment curve are in ft-kips.

Continuing on toward the right end of the beam, positive-shear area is encountered. If this positive area is added algebraically to the −18 ft-kips acting at R_1, this moment will sharply decrease in negative magnitude, and, if enough positive-shear area is available, will itself ultimately become positive. Adding successively, therefore, the shear areas per foot of beam beyond R_1 to the moment of −18 ft-kips gives bending moments at the succeeding foot points on the beam, as illustrated below.

−18.0 ft-kips, moment at R_1
+9.8 ft-kips, shear area between 3 and 4
−8.2 ft-kips, moment at section 4
+7.8 ft-kips, shear area between 4 and 5
−0.4 ft-kip moment at section 5
+5.8 ft-kips, shear area between 5 and 6
+5.4 ft-kips, moment at section 6
+3.8 ft-kips, shear area between 6 and 7
+9.2 ft-kips, moment at section 7
+1.8 ft-kips, shear area between 7 and 8
+11.0 ft-kips, moment at section 8
+0.2 ft-kip shear area between 8 and 8.4
+11.2 ft-kips, moment at 8.4, maximum positive moment

In keeping with the previously proved parabolic equation for this portion of the moment curve, the positive-shear areas, because of their decreasing magnitude, cause the moment curve to rise sharply and then to *taper off*, becoming horizontal at the point of zero shear, or maximum bending moment.

From the zero-shear point located at 8.4 ft to the right of the left end to the right end of the beam, only negative-shear area is encountered. If, as the definition states, "the difference in bending moment between any two sections is the algebraic sum of the shear area between these sections," the ensuing moment diagram should become less positive and tend toward the zero-moment line.

Adding the subsequent increments of negative-shear area to the maximum positive moment of 11.2 ft-kips at section 8.4 gives the following bending moments.

+11.2	ft-kips, moment at section 8.4
−0.4	ft-kips, shear area between 8.4 and 9
+10.8	ft-kips, moment at section 9
−1.2	ft-kips, shear area between 9 and 10
+9.6	ft-kips, moment at section 10
−1.2	ft-kips, shear area between 10 and 11
+8.4	ft-kips, moment at section 11
−4.2	ft-kips, shear area between 11 and 12
+4.2	ft-kips, moment at section 12
−4.2	ft-kips, shear area between 12 and 13
0.0	ft-kips, moment at section 13

Note: These computations are based on the fact that the moment at the left end of the beam equals zero. Should the moment there not equal zero, the moments here computed would change by its magnitude, algebraically added.

Immediately after the maximum-moment point, the moment curve starts slowly on its downward path. Since the shear area starts from zero, we know that a point of tangency must exist at the point of maximum bending moment. The curve is still parabolic until section 9 is reached. Thereafter, because of the equal shear increments between sections 9 and 11, and 11 and 13, the curve moves downward along two successive *straight* lines of different slope.

As skill is acquired in the forming of these bending-moment diagrams, entire portions of positive- and negative-shear areas may be added instead of the smaller increments. Keeping in mind the relative magnitude of the shear-area increments, their sign and sequence, the proper curvature of the moment curves may be ascertained.

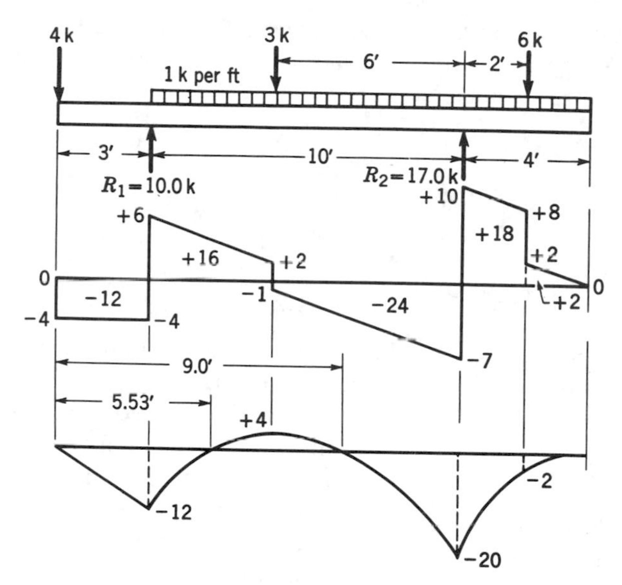

Figure 6-31

Illustrative Problem 2. Draw the shear- and bending-moment diagram for the beam shown in Fig. 6-31. Determine the maximum shear and the maximum positive and negative bending moments. Locate the inflection points.

SOLUTION

maximum shear $= +10^k$; maximum positive moment $= +4$ ft-kips
maximum negative moment $= -20$ ft-kips

Inflection points. In view of the fact that the bending-moment diagram passes through the zero axis twice during its traverse from one end of the beam to the other, two inflection points must occur. Their locations are indicated on Fig. 6-32, which represents the approximate deflected position of the beam (shown magnified).

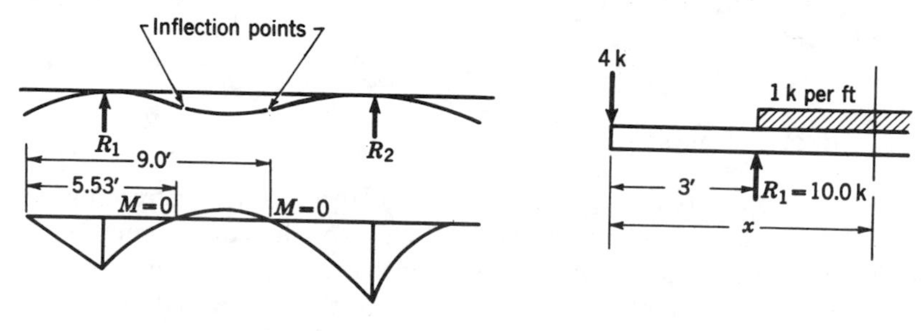

Figure 6-32

Figure 6-33

The inflection point, located between R_1 and the 3-kip force, is determined by setting the moment equation for that portion of the beam (Fig. 6-33) equal to zero and solving for x.

$$\Sigma M_x = -4x + 10(x-3) - 1(x-3)\frac{(x-3)}{2} = 0$$

$$x^2 - 18x + 69 = 0$$

$$x = 5.53 \text{ ft} \qquad \text{(correct)}$$

and

$$x = 12.4 \text{ ft} \qquad \text{(irrelevant)}$$

The application of the above moment equation, not extending beyond the point 7 ft from the left end, renders the 12.4-ft answer irrelevant. If the curve had extended beyond the 12.4-ft point, the answer would have been relevant and would have indicated the distance of the point of intersection of the moment curve with the axis line from the left-hand end of the beam.

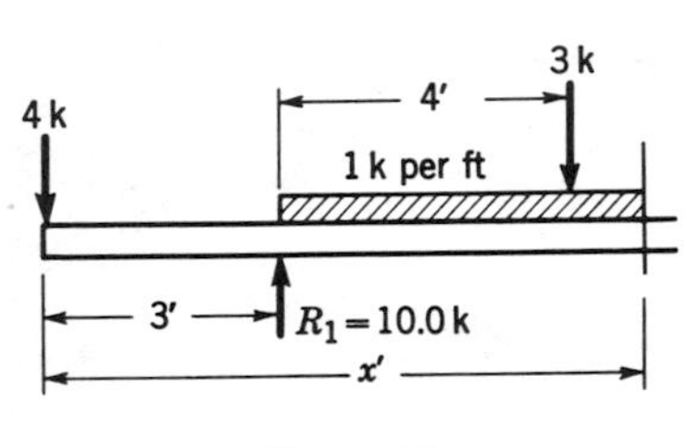

Figure 6-34

The second inflection point is obtained by solving the moment equation for that portion of the beam (Fig. 6-34) on which it occurs—that is, between the 3-kip load and R_2.

$$M_x = -4x + 10(x-3) - 1\frac{(x-3)^2}{2} - 3(x-7) = 0$$

$$x^2 - 12x + 27 = 0$$

$$(x-9)(x-3) = 0$$

$$x = 9 \text{ ft}$$

$$x = 3 \text{ ft} \qquad \text{(irrelevant)}$$

Obviously, the value of x equal to 3 ft is irrelevant to our problem, since it occurs outside the limits of the moment equation. The other answer, $x = 9$ ft, falls within the limits of the section under consideration and is, therefore, valid.

6-15 Effect of Couples

If a couple acts through a crossarm or some other restraining device fastened to a beam at one point, the full effect of the couple must be felt at that point. The total moment ordinate, located at the point of action of the couple, must always include the magnitude of the couple. Often a moment ordinate may straddle the zero axis. As already indicated, the sum of the positive and negative portions must equal the magnitude of the couple.

If the forces of the couple act as previously mentioned, at one point on the beam, their presence is not detected on the shear diagram. This gives rise to the one exception to be noted in following the shear-area method of conducting bending-moment diagrams: in determining the difference in bending moment between any two points A and B on a beam, the algebraic sum of the shear areas between these two points should include the concentrated moments produced by couples located between A and B.

Illustrative Problem 3. Determine the shear- and bending-moment diagrams for the cantilever and simple beams of Figs. 6-35 and 6-36 acted upon by the couples shown.

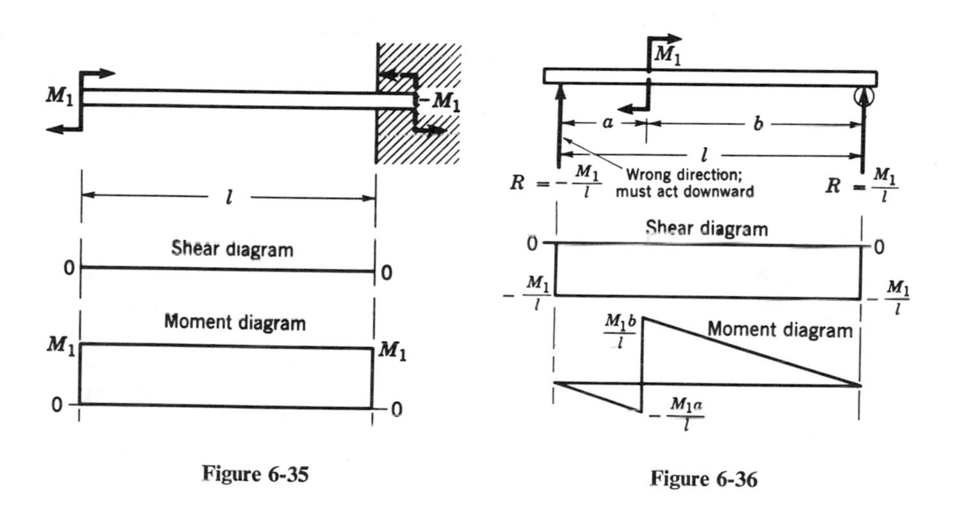

Figure 6-35

Figure 6-36

SOLUTION.

Cantilever beam. The only reaction required for the equilibrium of the cantilever beam shown in Fig. 6-35 is a couple of equal and opposite magnitude to the applied couple, located in the supporting wall. Since couples are acting at the ends of the beam, there is theoretically no shear developed on the beam. With no shear area present, there can be no difference in bending moment from point to point on the beam, with the result that the bending moment along the beam must be constant.

Simple beam. The reactions of the simple beam shown in Fig. 6-36 are the only means by which this beam may attain equilibrium. It is, therefore, necessary that these reactions form a couple, since only another couple can balance the applied couple M_1. Taking moments about R_1,

$$M_1 - R_2 \times l = 0, \qquad R_2 = \frac{M_1}{l}$$

Likewise, the moments taken about R_2 yield

$$M_1 - R_1 \times l = 0, \qquad R_1 = -\frac{M_1}{l}$$

Thus, R_1 must act downward and R_2 must act upward with equal intensity to balance couple M_1.

The shear diagram is entirely negative, since R_1 acts downward and forces any portion of the beam to the left of a section in a similar direction past the adjoining right-hand portion.

The bending-moment diagram illustrates the point previously made: that the concentrated moment must be included in the algebraic sum of the shear area for its proper construction.

PROBLEMS

Note: Use shear-area method in the solution of the following problems.

6-17. A simply supported beam 20 ft long overhangs the left support 4 ft. It carries the following: 400 lb/ft from left end to left support, 2 kips at 4 ft to the right of the left support, and 800 lb/ft from the 2-kip load to the right support. Construct shear and moment diagrams giving all controlling ordinates. *Answers:* $V_{max} = -6300$ lb, $M_{max} = +24{,}800$ ft-lb.

6-18. A beam 36 ft long is simply supported with one support at the right end and the other support 6 ft from the left end. A uniform load of 200 lb/ft begins at the right end and extends for 12 ft. There is a concentrated load of 1000 lb 18 ft from the right end and a second concentrated load of 1000 lb at the left end. A uniform load of 400 lb/ft begins 2 ft from the left end and extends to a point 22 ft from the right support. Sketch shear- and bending-moment diagrams giving values of controlling ordinates.

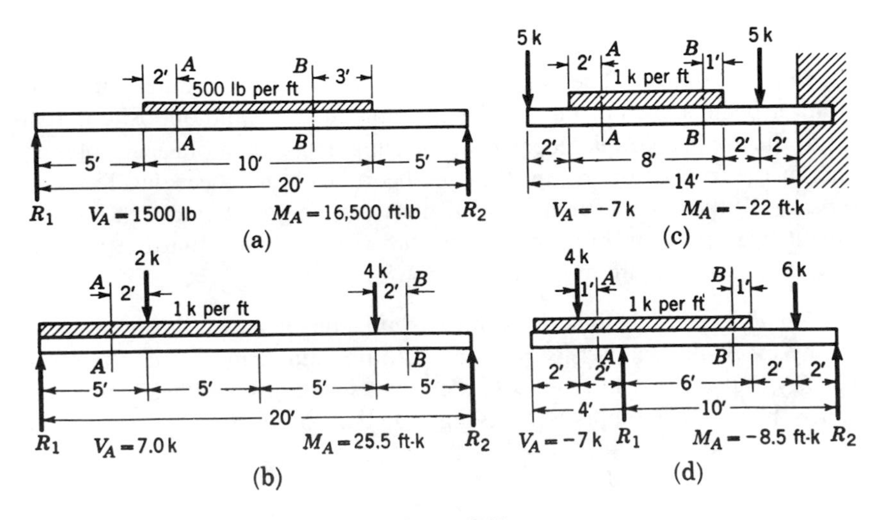

Figure 6-37

6-19a, b, c, d. Given the shear and bending moment at section *A-A* of the beams shown in Fig. 6-37, determine the bending moment at section *B-B*, using the shear-area method. *Answers:* (a) −1000 lb, 17,750 ft-lb; (b) −6.0 kips, 18.0 ft-kips; (c) −12 kips, −69.5 ft-kips; (d) 2 kips, 6.5 ft-kips.

6-20a, b, c, d. Given the shear diagrams of the various beams in Fig. 6-38, construct their bending-moment diagrams.

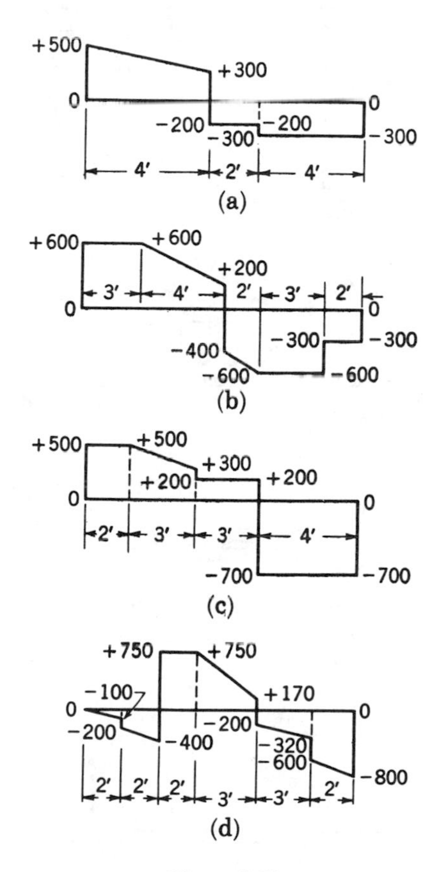

Figure 6-38

6-21a, b, c, d, e, f, g, h. Construct the bending-moment diagram of the beams shown in Fig. 6-18, using the shear-area method. Record the magnitude of the various elementary areas on the shear diagram.

6-22a, b, c, d, e, f, g, h. Construct the bending-moment diagram of the beams shown in Fig. 6-19, using the shear-area method. Record the magnitude of the various elementary areas on the shear diagram.

6-23. Draw the shear- and bending-moment diagrams for the beam shown in Fig. 6-39. Note the weight of the beam. *Answers:* $V_{max} = -1320$ lb, $M_{max} = -12{,}120$ ft-lb.

6-24. Draw the shear- and bending-moment diagrams for the beam shown in Fig. 6-40.

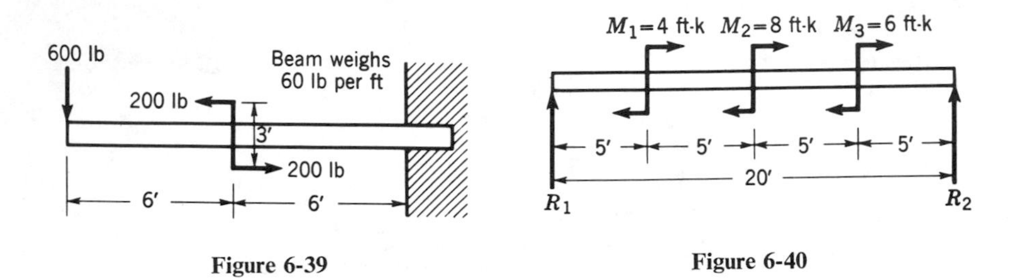

Figure 6-39 **Figure 6-40**

6-25. Draw the shear- and bending-moment diagrams of the beam shown in Fig. 6-41 giving all controlling ordinates. *Answers:* $V_{max} = \pm 12$ kips, $M_{max} = +36$ ft-kips.

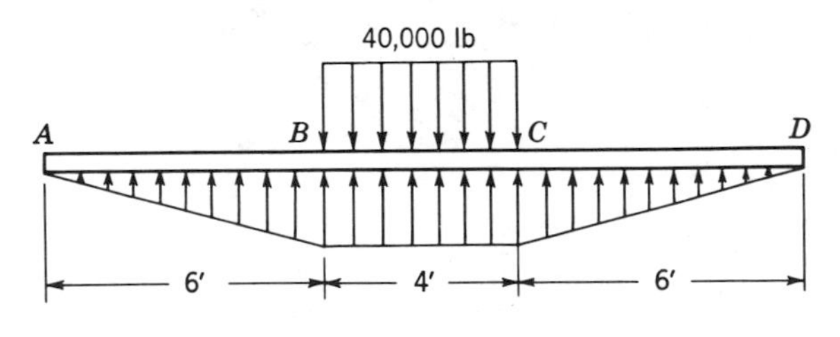

Figure 6-41

6-26. The transverse loads on a certain airplane wing are equivalent to those shown in Fig. 6-42. There is also a clockwise couple of 3315 in.-lb at *A*. Draw the shear- and bending-moment diagrams giving all controlling ordinates. *Answers:* $V_{max} = +1053$ lb, $M_{max} = -19{,}050$ in.-lb.

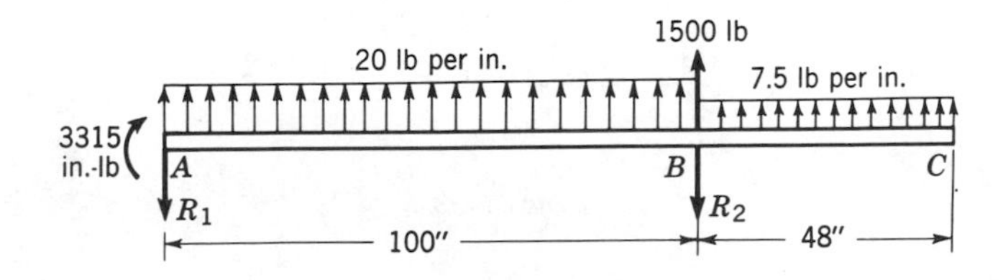

Figure 6-42

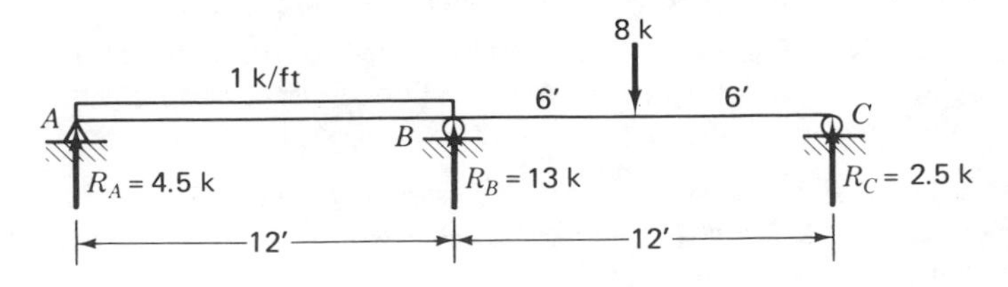

Figure 6-43

6-27. Draw the shear- and composite bending-moment diagrams for the continuous beam of two spans shown in Fig. 6-43. The bending moment at the center support B is -18.0 ft-kips and the reaction at B is 13 kips. Compute all critical or controlling ordinates. Locate the position of zero bending moment in span AB.

6-28. Draw complete shear- and bending-moment diagrams for the beam shown in Fig. 6-44. *Answers*: $V_{\max} = +9$ kips, $M_{\max} = -55$ ft-kips.

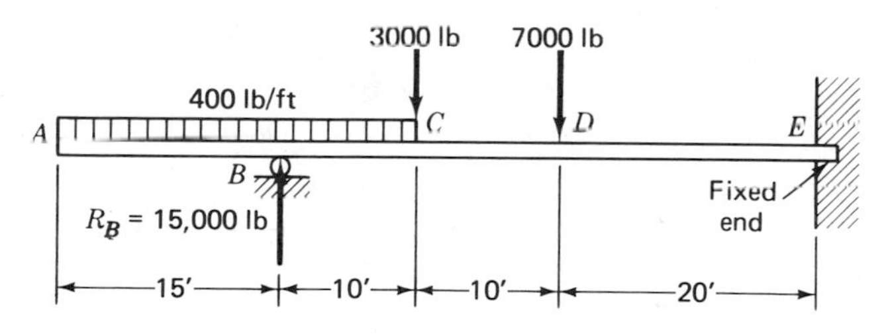

Figure 6-44

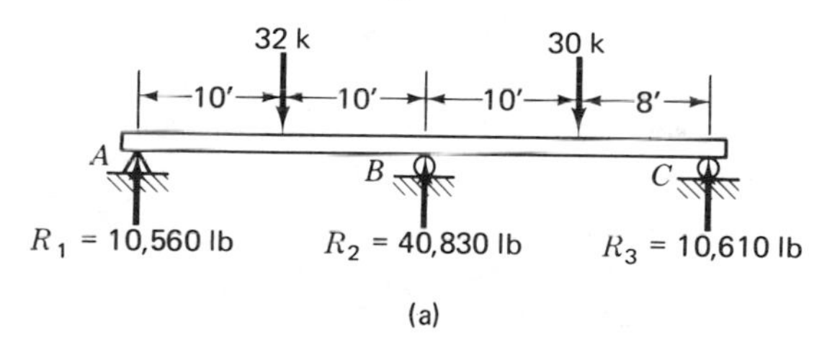

(a)

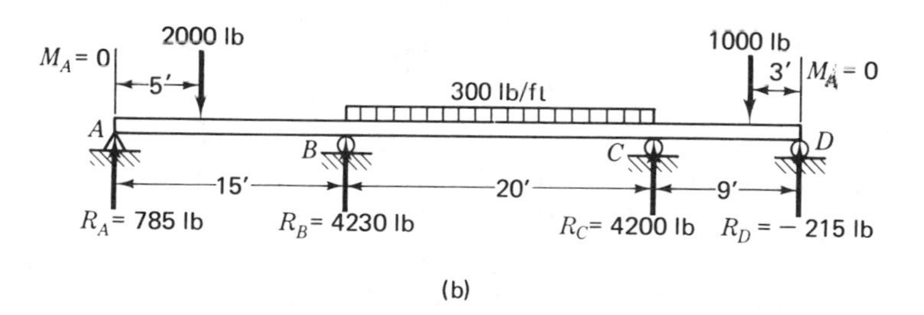

(b)

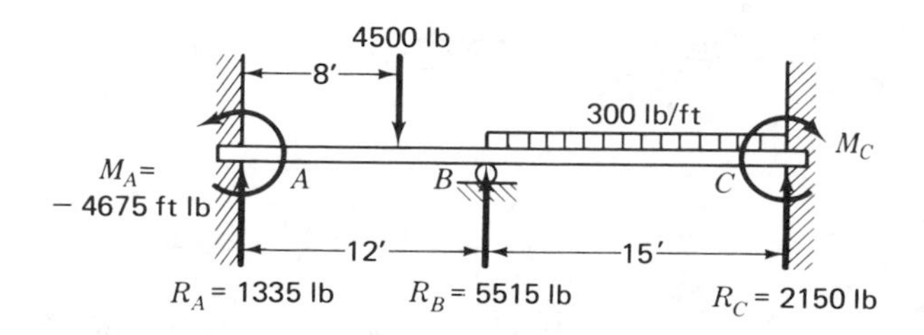

(c)

Figure 6-45

6-29a, b, c. Draw the shear- and bending-moment diagrams for the continuous beams shown in Fig. 6-45.

6-30. A bending-moment diagram for a beam 18 ft long is shown in Fig. 6-46 with ordinates in kip feet. Curved lines are parabolas, i.e., plots of equations of the second degree. Draw the shear diagram and the loading diagram for this beam, giving all important values on same. *Answers*: $V_{max} = 6.0$ kips, $R_E = 9.0$ kips.

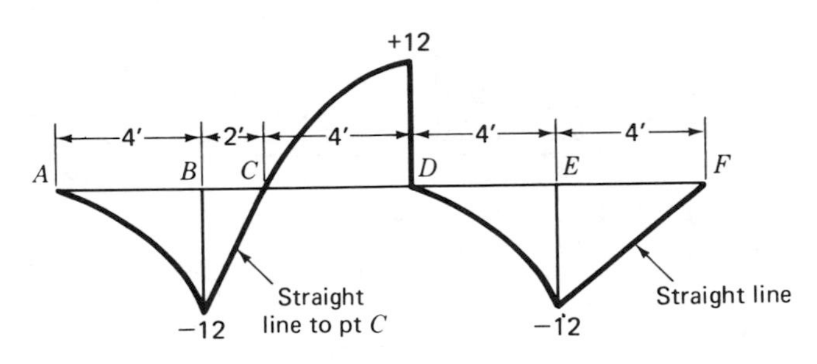

Figure 6-46

6-31. The beam shown in Fig. 6-47 is fixed at the right end, is simply supported at *B*, and is hinged 2 ft from the fixed end. Determine the reactions of the

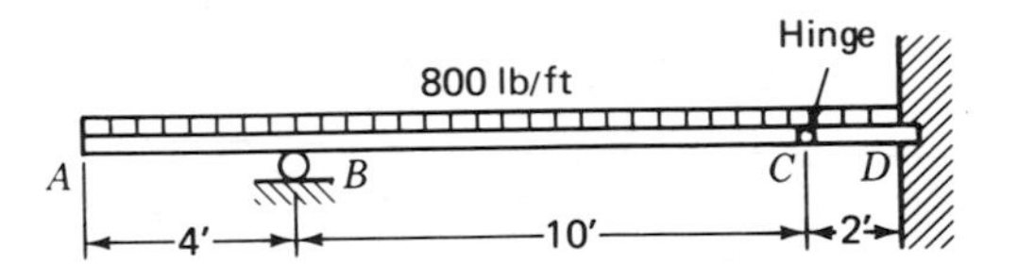

Figure 6-47

beam. Without writing shear or moment equations, draw the shear and moment diagrams, giving all important values on the diagrams. Locate all points of zero shear and zero bending moment. *Answers*: $V_{max} = -4960$ lb, $M_{max} = -8320$ ft-lb.

6-32. For the beam shown in Fig. 6-48, draw the shear- and bending-moment diagrams, giving all critical values on same. Locate all points of zero shear and zero bending moment. *Answers*: $V_{max} = +5.0$ kips, $M_{max} = -8.0$ ft-kips.

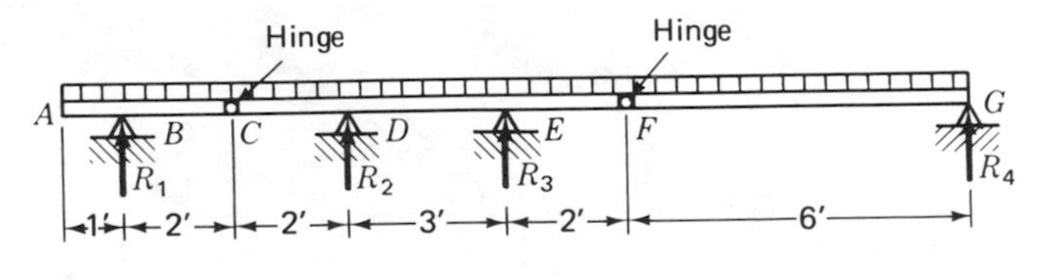

Figure 6-48

6-33. Draw the shear diagram and the bending-moment diagram (not by parts) for the continuous beam of two spans shown in Fig. 6-49. The bending

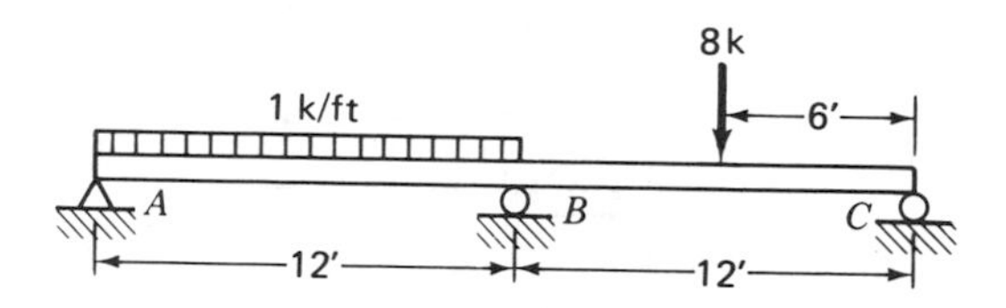

Figure 6-49

moment at the center support B is -36 ft-kips and the reaction at B is 26 kips. Give shear and moment values at all critical points.

6-34. A beam of length 33 ft is supported at points B and E (Fig. 6-50). (a) Determine the reactions. (b) Draw the shear diagram. (c) Using the shear diagram, determine values and plot the bending-moment diagram. All important

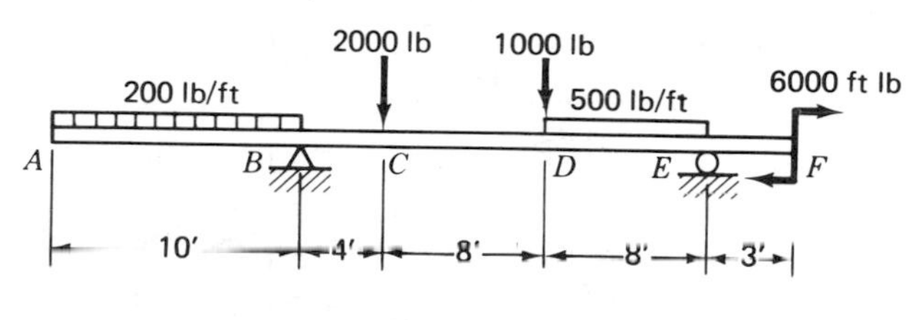

Figure 6-50

values on shear- and bending-moment diagrams are to be marked on the diagrams. Neglect the weight of the beam. *Answers*: $V_{max} = -4.0$ kips, $M_{max} = +10.0$ ft-kips.

ELECTRICAL ANALOGY OF BEAMS

6-16 General Considerations

When the basic mathematical expressions of two unrelated physical phenomena are of the same form, an analogy is said to exist between them. Such an analogy has been shown to exist between the currents and voltages of electric circuits and the shearing and bending resistances of beams.*

Consider the cantilever beam shown in Fig. 6-51 with its shear- and bending-moment diagrams. Its vertical shearing resistance at sections taken between the loads is

$$\begin{aligned} V_{12} &= P_1 \\ V_{2n} &= P_1 + P_2 \\ V_{nB} &= P_1 + P_2 + P_n \end{aligned} \tag{6.3a}$$

*S. Freiberg, "Electric Circuits for the Simulation of Bending in Beams," *Journal of the Society for Experimental Stress Analysis*, **2** (1962), 24.

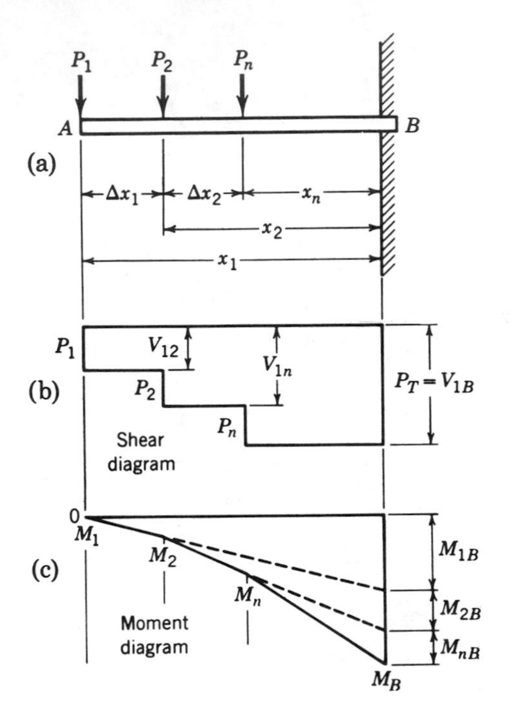

Figure 6-51

It also follows that

$$R_B = P_1 + P_2 + P_n = \sum_{k=1}^{k=n} P_k \tag{6.3b}$$

The bending moments at each of the loads and the reaction are

$$M_1 = 0$$

$$M_2 = -P_1(x_1 - x_2) = -P\,\Delta x_1$$

$$M_n = -P_1(x_1 - x_n) - P_2(x_2 - x_n) = -P_1(\Delta x_1 + \Delta x_2) - P_2\Delta x_2$$

$$M_B = -P_1x_1 - P_2x_2 - P_nx_n = -M_{1B} - M_{2B} - M_{nB} \tag{6.4}$$

Note that M_B is equal to the superimposed moments of P_1, P_2, and P_n taken about B.

Consider now the electrical circuit $ABCD$ shown in Fig. 6-52. The portion AB of the circuit includes a series of fixed resistances each of which is proportional to the corresponding length of beam between loads. Its total length is considered to be the *electrical beam* now to be proved to be analogous to the actual beam shown in Fig. 6-51. Note the comparisons.

First, by adjusting the variable resistors shown in the vertical legs of the circuit the currents I_1, I_2, and I_n are made proportional to the actual loads. According to Kirchhoff's law, they add their intensities as they enter the

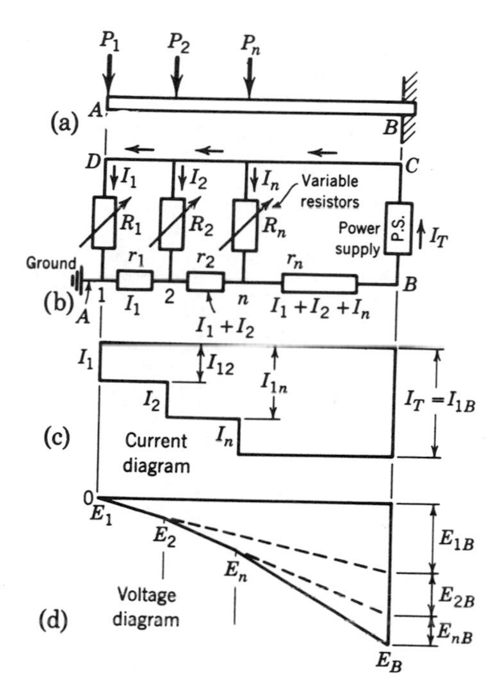

Figure 6-52

portion AB and proceed to B. Stated mathematically,

$$I_B = I_1 + I_2 + I_n = \sum_{k=1}^{k=n} I_k \tag{6.5}$$

The similarity to Eq. (6.3b) is noted immediately. It also follows that the currents flowing from A to B are successively

$$I_{r_1} = I_1$$
$$I_{r_2} = I_1 + I_2$$
$$I_{r_n} = I_1 + I_2 + I_n$$

These equations will be recognized as being analogous to the shear equations given in Eq. (6.3a).

Second, the voltage drops along the *electrical beam*,

$$\begin{aligned} E_{12} &= I_1 r_1 \\ E_{2n} &= I_1(r_1 + r_2) + I_2 r_2 \\ E_{nB} &= I_1(r_1 + r_2 + r_n) + I_2(r_2 + r_n) + I_n r_n = E_{1B} + E_{2B} + E_{nB} \end{aligned} \tag{6.6}$$

are analogs of Eq. (6.4).

Thus it is seen that the electrical circuit of Fig. 6-52 behaves like the cantilever beam of Fig. 6-51 if the forces are simulated by current increments, the bending moments by the voltage drops, and the lengths by the resistors.

Analogy for simple beam. A similar analogy may be made between a simple beam and its equivalent circuit (Figs. 6-53 and 6-54).

This analogy may be extended to include more involved statically determinate and indeterminate beams. Deflections may be studied in a similar way. The ease and simplicity of its application to more difficult problems suggest a continuing and widespread use.

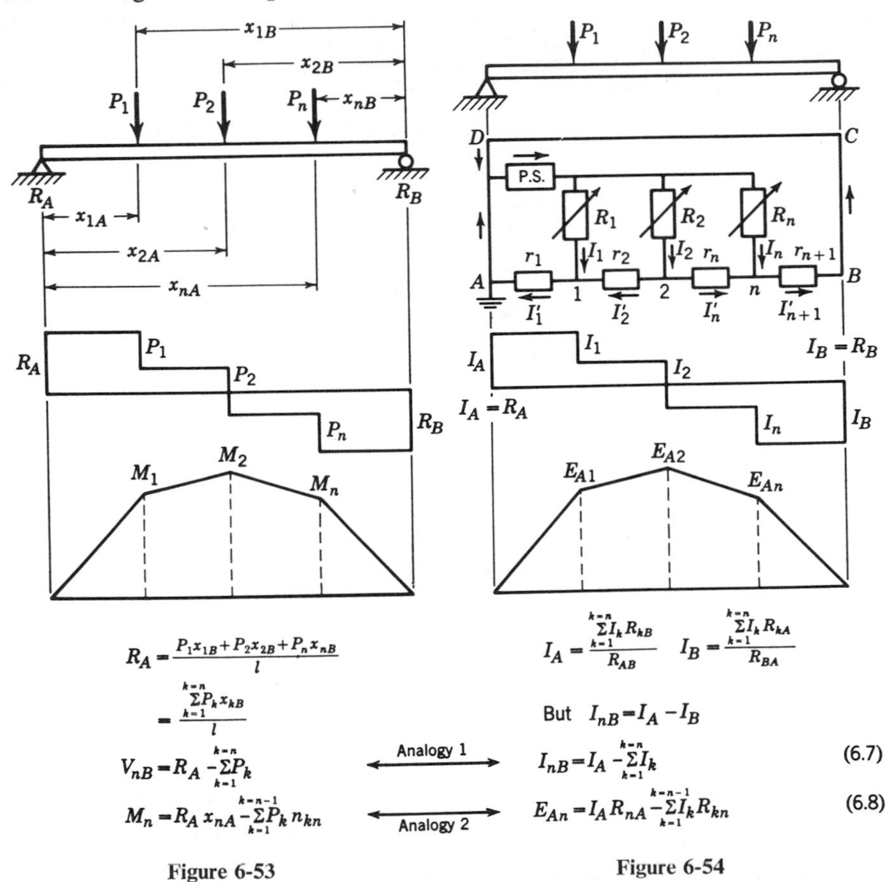

$$R_A = \frac{P_1 x_{1B} + P_2 x_{2B} + P_n x_{nB}}{l} = \frac{\sum_{k=1}^{k=n} P_k x_{kB}}{l}$$

$$V_{nB} = R_A - \sum_{k=1}^{k=n} P_k$$

$$M_n = R_A x_{nA} - \sum_{k=1}^{k=n-1} P_k n_{kn}$$

$$I_A = \frac{\sum_{k=1}^{k=n} I_k R_{kB}}{R_{AB}} \qquad I_B = \frac{\sum_{k=1}^{k=n} I_k R_{kA}}{R_{BA}}$$

But $I_{nB} = I_A - I_B$

$$I_{nB} = I_A - \sum_{k=1}^{k=n} I_k \tag{6.7}$$

$$E_{An} = I_A R_{nA} - \sum_{k=1}^{k=n-1} I_k R_{kn} \tag{6.8}$$

Figure 6-53

Figure 6-54

7

The Design of Beams

7-1 Fundamental Considerations

With the exception of those cases where the moment or shear is zero, there is present on every vertical section of a loaded horizontal beam an internal shear force, V, and an internal resisting couple, C-T, the magnitudes of which are recorded at corresponding positions on the shear- and bending-moment diagrams. As a result of these internal resistances, there are developed on the interior of the beam shear and bending unit stresses whose intensities are dependent upon the magnitude of the resisting forces, as well as on the size and shape of the cross section. Thus, if a cross-sectional area were to be made smaller while still retaining its characteristic shape, the internal stresses would be forced to increase to resist the total shear, V, and bending moment, C-T, acting on that section. In contrast, if the area were to be made larger, the stresses would decrease. It is the designer's responsibility to assign to a beam that cross-sectional area whose maximum stresses will not exceed the limit of the allowable working unit stresses. Naturally, the closer the maximum stresses approach the allowable values, the more economical the design will be.

Fortunately, the maximum shear and bending unit stresses which control the design of a beam act at different locations and develop with no interference from each other. Thus, an area designed on the basis of the maximum internal bending moment C-T may thereafter be checked for dangerous shear stresses. The opposite procedure of obtaining the area required for shear and of checking it for dangerous bending stresses might also be followed, although this practice is not quite so common. Regardless of what procedure is followed, it should be remembered that no beam design is ever complete without it being satisfactorily shown that the induced shear and bending unit stresses are within their respective allowable limits.

In order properly to adjust or design the cross-sectional area of a beam, separate relationships for shear and bending moment must be derived. The relationship between the cross-sectional area and the bending moment that will be derived first is known as the *flexure formula*, so called because the act of bending is also known as *flexing*.

7-2 Derivation of the Flexure Formula

Let us consider the beam of rectangular cross section shown in Fig. 7-1a, loaded by forces P_1 and P_2. A free body of a portion of this beam is shown in Fig. 7-1b on whose exposed section is acting an internal shear force

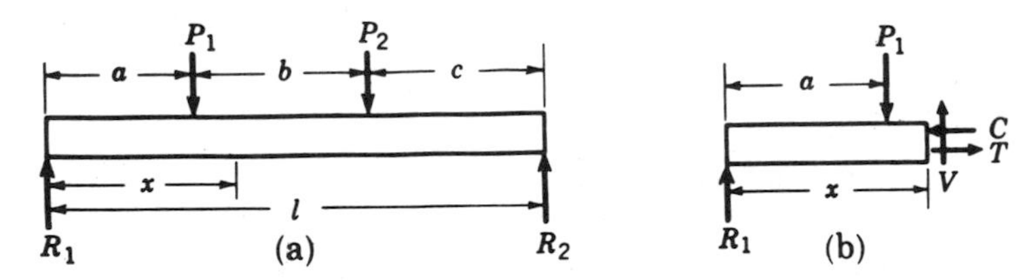

Figure 7-1

V and the internal bending moment C-T. Close observation of the bent beam shown in Fig. 7-2 reveals that the fibers in the upper half of the beam have contracted and those in the lower half have elongated. These deformations are the result of forces spread over their respective cross-sectional areas, the resultants of which are the C and T forces comprising the resisting C-T couple. It is imperative, therefore, that we investigate the variation and magnitude of the distributed forces and the location of their resultants.

Variation of the resisting flexural stresses. The variation of the resisting flexural stresses acting on those cross sections of a beam perpendicular to its longitudinal axis is based on *the fundamental assumption that these cross-sectional planes remain plane after the beam is bent.* Except for beams made from a certain few materials, such as gray cast iron, low-strength concrete, and so forth, this assumption has been essentially substantiated. If, therefore, the movement of two cross-sectional planes of the beam, shown in Fig. 7-2 as initially separated by a small distance dx, were studied during the applica-

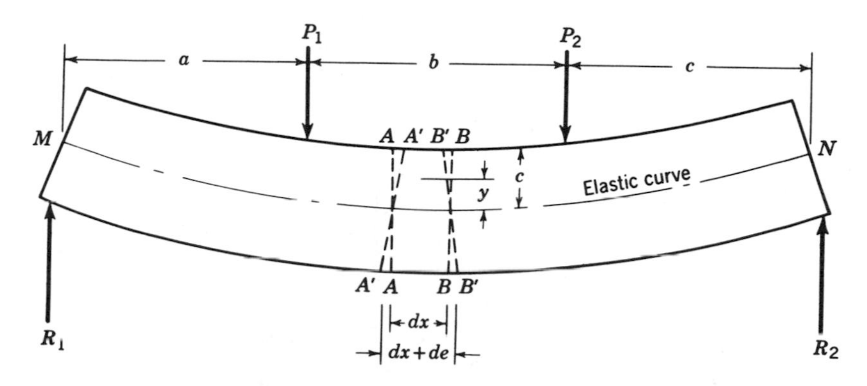

Figure 7-2

tion of loads P_1 and P_2, the vertical projection of these planes, on the basis of this assumption, would move from the unloaded positions *A-A* and *B-B* to positions *A'-A'* and *B'-B'*, respectively, when the beam was fully loaded.

The convergence of these two cross-sectional planes coupled with the presence of the *C* and *T* forces on the top and bottom of the beam, respectively, indicates that the upper fibers must have been shortened and the lower fibers lengthened in proportion to their distance from some neutral or unchanged plane of fibers. Because the length of each of the fibers between the two cross sections was *dx* before loading, *the unit strain, as well as the unit stress, varied with the distance from these neutral fibers.*

The location of the plane of *neutral fibers* or neutral plane (*MN*, Fig. 7-2) in a beam having a rectangular cross section passes through the center-of-gravity axis of every section, since the *C* and *T* forces must be equal and their contributing stresses and strains symmetrical with one another. When projected into a vertical plane, this neutral plane becomes a line and is known as the *elastic curve*.

The axis about which a cross section rotates, i.e., the line of zero flexural stress, is called the *neutral axis* (Fig. 7-4).

Summation of the resisting flexural stresses. If the proportional limit is not exceeded, the variation of flexural stress to the distance from the neutral axis (Fig. 7-3) may be written as

$$\frac{\sigma_y}{y} = \frac{\sigma}{c} \qquad (7.1)$$

Thus, considering again the exposed section of the free body of Fig. 7-1b (arranged isometrically in Fig. 7-4), the

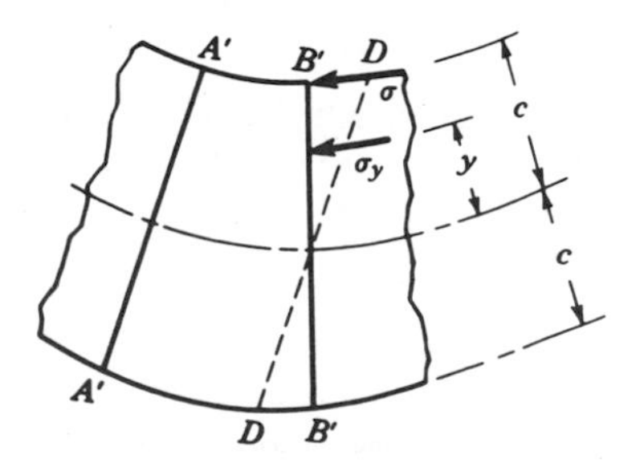

Figure 7-3

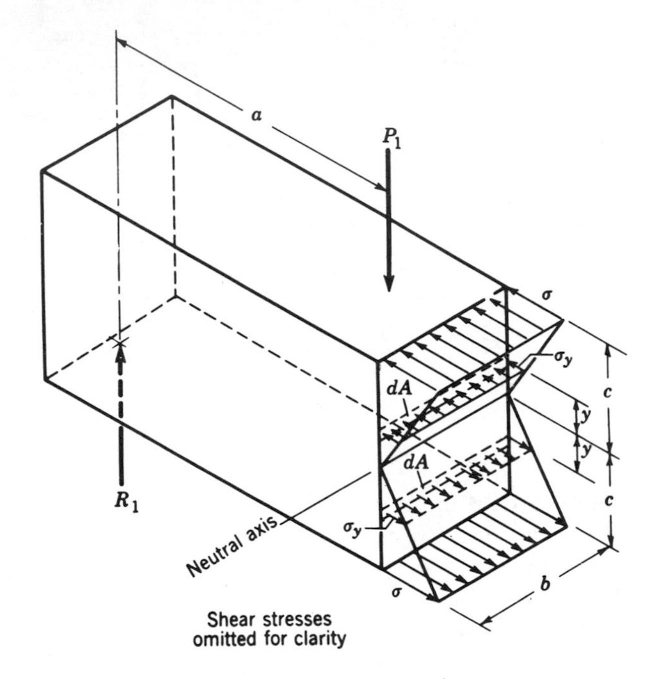

Figure 7-4

resisting force acting on the narrow horizontal strip dA located y distance from the neutral axis is equal to $\sigma_y\, dA$. Its resisting moment taken about the neutral axis must therefore be $\sigma_y\, dA \times y$.

Moreover, since there are a tremendously large number of dA strips included in the depth d, each of which is located a y distance from the neutral axis and acted upon by a unit stress σ_y, the resisting moments of the forces acting on the dA areas, collectively known as the *internal C-T resisting moment* M_R, are equal to $\int \sigma_y\, dA \cdot y$.

By substituting the value of σ_y as obtained from Eq. (7.1),

$$M_R = \int \frac{\sigma}{c} y\, dA \cdot y \quad \text{or} \quad \int \frac{\sigma}{c} y^2\, dA$$

Because in Eq. (7.1) the term σ/c equals σ_y/y (a constant) and because this ratio appears in the resisting moment of every dA area, it may be placed outside the integral sign. Thus,

$$M_R = \frac{\sigma}{c} \int y^2\, dA \tag{7.2}$$

The term $\int y^2\, dA$ will be recognized as the value of the *moment of inertia* I of the cross-sectional area taken about the centroidal axis. The value of M_R that is equal to the *C-T* couple must also equal, for equilibrium reasons, the value of the bending moment M at the same section. We may, therefore, write the completed equation, generally known as the *flexure formula*, as

$$M = \frac{\sigma I}{c} \tag{7.3}$$

in which σ is the maximum flexural unit stress, commonly expressed in psi at distance c in. from the neutral axis, and I is the moment of inertia of the cross section in in.^{4} taken about the neutral axis. Should these units be employed, the value of M would be in in.-lb.

Illustrative Problem 1. At a certain section of a 4- by 10-in. timber beam, the bending moment (C-T couple) is equal to 3500 ft-lb. Determine the maximum flexure stress on the section if the 10-in. side is vertical.

SOLUTION. Rearranging Eq. (7.3) to appear as $\sigma = Mc/I$ and substituting, we obtain

$$\sigma = \frac{3500 \times 12 \times 5}{\frac{1}{12} \times 4 \times 10^3} = 630 \text{ psi}$$

7-3 Assumptions and Limitations of Flexure Formula

At this point in our use of the flexure formula it is well to consider the assumptions and limitations made in its derivation. They are as follows:

1. A transverse plane perpendicular to the longitudinal axis of the beam remains plane after the beam is bent.

2. The modulus of elasticity in compression is the same as that in tension.

When considered together, assumptions (1) and (2) imply that:

a. The unit stress, unit strain, and total strain are proportional to the distance from the neutral axis.

b. The unit stress does not exceed the proportional limit of the material.

These assumptions disregard the effect of any shearing strains due to the shearing resistance V, and make impossible the use of the flexure formula for curved beams. The difference in length of the fibers between any two transverse sections of a curved beam produces unit strains that are not proportional to the distance from the neutral axis.

3. The beam must deflect normally under elastic bending stresses and not through any local collapse or twisting. This limitation implies the necessity of a beam having an ample width for its depth.

4. The applied loads act in a plane containing the axis of symmetry of each cross section.

This closer scrutiny of the flexure formula should warn the user to view with care the extent of its field of application. To extend its usefulness beyond the scope of this limited field might easily incur disaster. On the other hand, to restrict its usefulness unnecessarily might cause excessive waste.

PROBLEMS

7-1. What is the maximum flexure stress at a section 3 ft from one end of a uniformly loaded simple beam 4 in. wide by 10 in. deep by 12 ft long? $w = 600$ lb/ft.

7-2. What maximum uniform load can a 6- by 12-in. beam withstand if its length between supports is 15 ft and the maximum allowable stress is 1400 psi? *Answer:* 598 lb/ft.

7-3. A wagon axle 2 in. square is loaded as shown in Fig. 7-5. What is the maximum unit stress attained when $P_1 = P_2 = 3600$ lb?

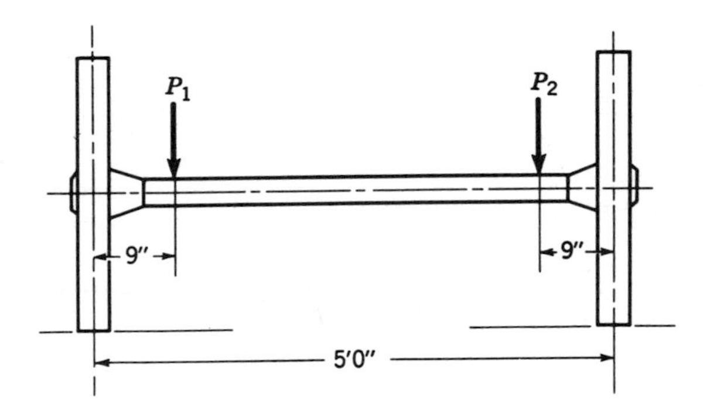

Figure 7-5

7-4. A cast-iron lever (Fig. 7-6) having a $\frac{1}{2}$- by $1\frac{1}{2}$-in. cross section is used to activate a pull rod. How much pull P may be applied before the unit stress will exceed 20,000 psi? *Answer:* 156.4 lb.

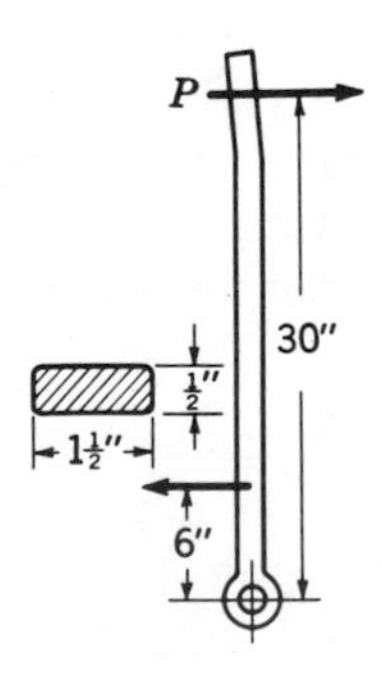

Figure 7-6

7-5. The casting shown in Fig. 7-7 acts as a simple beam and restrains the load P. Is the most dangerously stressed section immediately below P or at section A-A? Prove your answer. *Answers:* $\sigma = 1.35P$, $\sigma = 3.75P$.

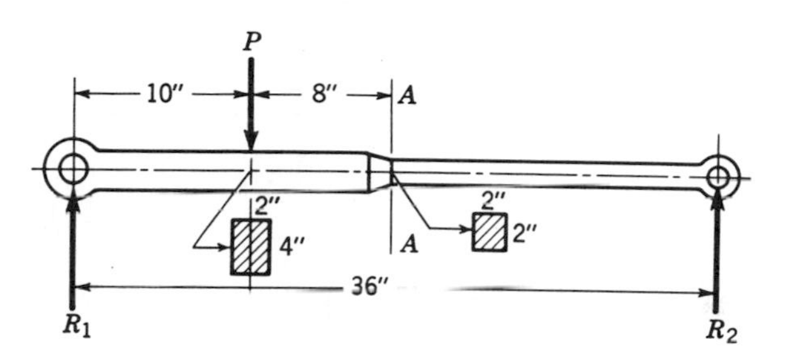

Figure 7-7

7-6a, b, c, d. Determine the maximum flexural stresses at sections A-A and B-B of the beams shown in Fig. 6-13. Use $I/c = 100\text{ in.}^3$.

7-7. Timber sheet piling (Fig. 7-8) is to be used to hold back water from flowing into the foundation pit of a bridge. Assuming complete rigidity at its fixed end, what depth d of water can be held back by the piling whose thickness is 3 in.? Maximum allowable unit stress = 2000 psi. *Hint:* Use typical section 1 ft wide. *Answer:* 6.62 ft.

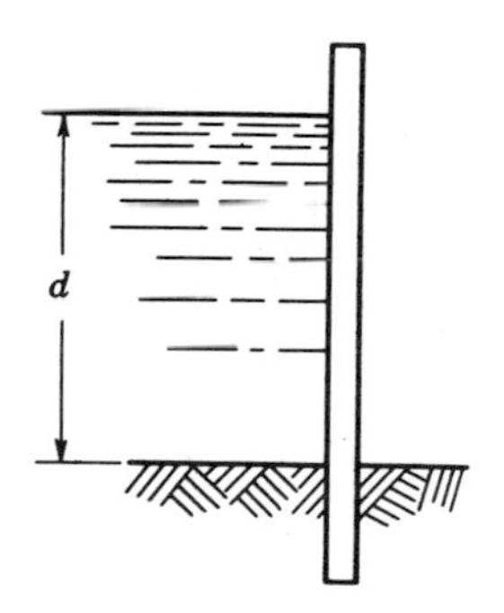

Figure 7-8

7-8. A beam 12 ft long, having a constant cross section 6 in. wide and 12 in. deep, has a flexural tensile stress of 300 psi at a point 2 ft from the left support and 3 in. below the neutral axis. What is the magnitude of the concentrated load located at its center which produces the stress?

7-9. A wood floor with a 15-ft span is to carry 250 lb/ft^2. If 4- by 14-in. beams are to be used, how far apart can they be placed according to the flexure formula if the flexural unit stress is not to exceed 1200 psi? *Answer:* 22.3 in.

7-10. Find the change in length of a fiber extending from A to B in the top surface of the beam shown in Fig. 7-9 when the loads are removed from the beam. $E = 10 \times 10^6$ psi. $b = 6$ in. $d = 8$ in. *Answer:* 0.135 in.

7-11. A section of a loaded beam is shown in Fig. 7-10. If the fibers in layer AA' contract 0.02 in. while the fibers in layer BB' stretch 0.03 in., locate the neutral axis and determine the average bending moment developed over this section. The moment of inertia about the neutral axis is 480 in.^{4}. $E = 10^6$ psi. What is the maximum stress acting on the cross section?

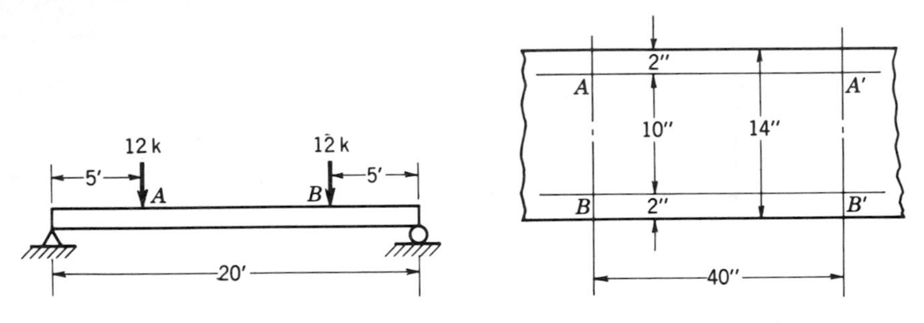

Figure 7-9

Figure 7-10

7-12. If the cross section of a beam shown in Fig. 7-11a were loaded beyond its elastic limit and revealed the stress pattern shown in Fig. 7-11b, what would its resisting bending moment be? *Answer:* $\frac{11}{48}bd^2\sigma_0$.

7-13. A cross section of a beam 6 in. wide by 12 in. deep is subjected to a bending moment of 144,000 in.-lb. Assuming the material of the beam to have

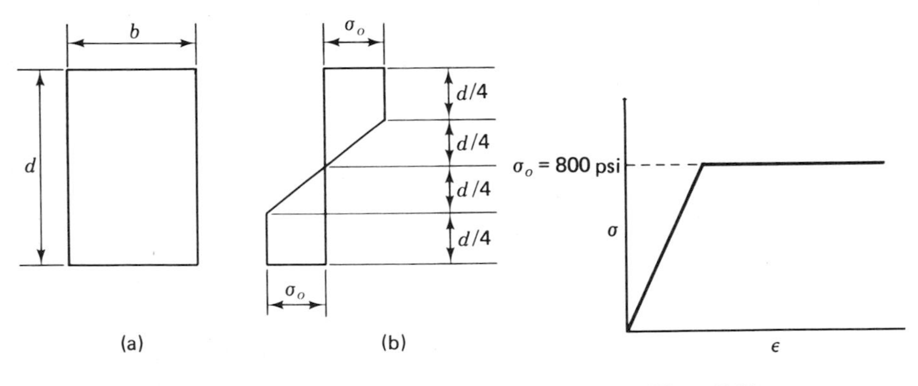

Figure 7-11

Figure 7-12

the ideally elastoplastic stress-strain diagram shown in Fig. 7-12, determine the stress pattern on the cross section required to resist the applied bending moment. *Answer:* σ_0 is constant over top and bottom 1.75 in.

7-14. A timber beam 6 in. wide and 12 in. deep is to be reinforced by $\frac{1}{4}$-in.-thick steel plates. (a) If the plates are 12 in. deep and firmly fixed to both sides of the beam, what is the stress distribution on a cross section if it is subjected to a bending moment of 144,000 in.-lb? $E_{st.} = 30 \times 10^6$ psi. $E_{wd.} = 2 \times 10^6$ psi. Assume the cross sections of the beam to remain plane. (b) If similar plates 6 in. wide are placed on the top and bottom surfaces of the beam, determine the stress distribution on a cross section if it is subjected to the same bending moment.

7-4 Use of Flexure Formula in Design—Section Modulus

To design a beam of uniform symmetrical cross section with regard to its flexural requirements, the value of M is placed equal to the maximum bending moment obtained from the moment diagram, and σ is assigned the value of the maximum allowable flexure stress for the material of which the beam is made. If M in in.-lb is divided by σ in psi, the value obtained is equal to the ratio I/c in in.^{3}. This ratio, called the *section modulus*, reflects that minimum value of resistance required to withstand the imposed bending without exceeding the allowable stress.

If the beam selected to meet this requirement is to be the most economical one, it must provide the necessary section modulus with the least amount of cross-sectional area (or weight). To assist in this selection, tables have been provided (see A.I.S.C. handbook or Appendix Tables) giving the section moduli of all standard beams. The same tables provide other pertinent data, such as weight, centers of gravity, moment of inertia, radii of gyration, and so forth.

Illustrative Problem 2. Determine the cross-sectional dimensions of the beam of Fig. 6-3, based on flexural requirements only, if it is to be made of southern yellow pine having a maximum allowable flexure stress of 1600 psi. The loading- and live-load bending-moment diagrams are redrawn for convenience (Fig. 7-13). Use an American Standard size beam.

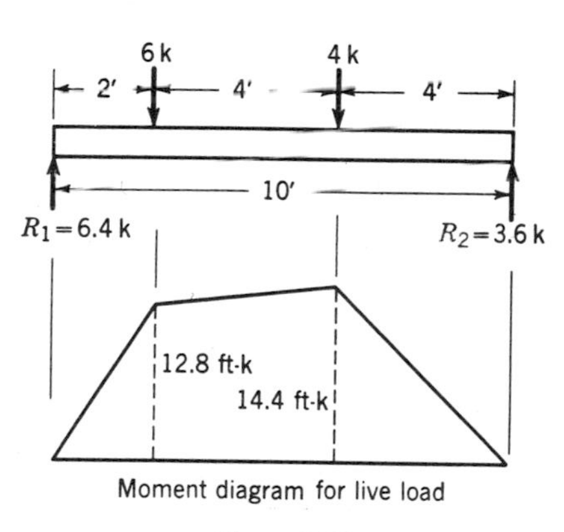

Figure 7-13

SOLUTION. The maximum bending moment for any beam is determined by that greatest algebraic sum of moment occurring at any point on the beam which results from the action of both live and dead load. This maximum moment for our beam occurs at the 4-kip load where the moment due to live load is 14.4 ft-kips (see Fig. 6-14) and that due to dead load (beam assumed 15 lb/ft) is 0.18 ft-kip (say, 0.2 ft-kip). It is upon their sum of 14.6 ft-kips that our flexural design is based.* Substituting in the flexure formula, we obtain

$$\frac{I}{c} = \frac{M}{\sigma} = \frac{14.6 \times 1000 \times 12}{1600} = 109.5 \quad \text{or} \quad 110$$

*The weight of a timber beam is often disregarded in design problems. To ignore the weight of beam in this problem would incur an error of only 1.25 per cent and would not change the size requirements.

But

$$\frac{I}{c} = \frac{\frac{1}{12}bd^3}{d/2} = \frac{bd^2}{6}$$

Thus,

$$\frac{I}{c} = \frac{bd^2}{6} = 110 \quad \text{or} \quad bd^2 = 660$$

Although any cross section whose product of b and d^2 equals 660 would meet the flexural strength requirements, there are, ordinarily, economical and structural limitations that dictate a very definite and more desirable choice.

Timber beams sold at lumber yards are cut and planed to certain standard sizes. (See A.I.S.C., *Steel Handbook*, table on Timber, American Standard sizes, or Table 18 in the Appendix.)* In this *dressing* operation a certain amount of the original cross section as laid out on the tree trunk is lost in the form of sawdust and shavings. This loss of cross section does not reduce the cost to the purchaser, since its price is based upon its *laid out*, or nominal, cross section. For example, a common 3- by 10-in. beam is only $2\frac{5}{8}$ by $9\frac{1}{2}$ in. in cross section when dressed, but is sold on the basis of a 30-in.^{2} cross section. This loss of cross section reduces the value of I and c and, consequently, the strength of the beam.

In the tabulation shown next are given the cross sections of these standard timber beams whose section moduli satisfy with the least excess the minimum required value of 110.

Nominal Size	American Standard Dressed Size	Area of Section (in.^{2})	Section Modulus
3 by 18 in.	$2\frac{5}{8}$ by $17\frac{1}{2}$ in.	45.9	134
4 by 14 in.	$3\frac{5}{8}$ by $13\frac{1}{2}$ in.	48.9	110
6 by 12 in.	$5\frac{1}{2}$ by $11\frac{1}{2}$ in.	63.3	121
8 by 10 in.	$7\frac{1}{2}$ by $9\frac{1}{2}$ in.	71.3	113

Although any one of these four beams would be capable of restraining the imposed load, they would not be equally acceptable on the basis of their cost or structural strength. A designer's axiom is to obtain the greatest strength for the least cost in fulfilling design requirements.

Using this thought as the background for our selection, it would appear that the 3- by 18-in. beam would be the most satisfactory. Its minimum cross-sectional area of 45.9 in.^{2} and its high section modulus would afford what appears to be the greatest economy with more than sufficient strength, assuming no extra charge for the 18-in. depth.† The cross section of this

*Table from *National Lumber Manufacturers Association.*

†Generally, the use of timbers with any dimension above 12 in. calls for an additional charge per board foot (1 board foot = 1 in. thick by 12 in. wide by 12 in. long).

beam may, however, be too narrow for its height if used alone without side support. Common practice generally refrains from using a ratio of depth to width of cross section greater than 4 for beams that are unsupported laterally. *The best tentative selection would, therefore, be the 4- by 14-in. beam, which is only slightly more expensive and has a depth-width ratio of less than 4.* The magnitude of the shearing unit stress must, of course, be determined before this cross section can be finally accepted.

The problem of beam selection, as illustrated in the foregoing paragraph, indicates that the narrower and the deeper a beam can be made without failing by twisting, bearing, or some other method, the stronger it becomes. The reason for this greater strength lies in the fact that the value of the moment of inertia increases as a greater proportion of its area is located farther from the neutral axis. This also accounts for the use of S beams, channels, and other sections (Fig. 7-14) whose flange areas are spread as far

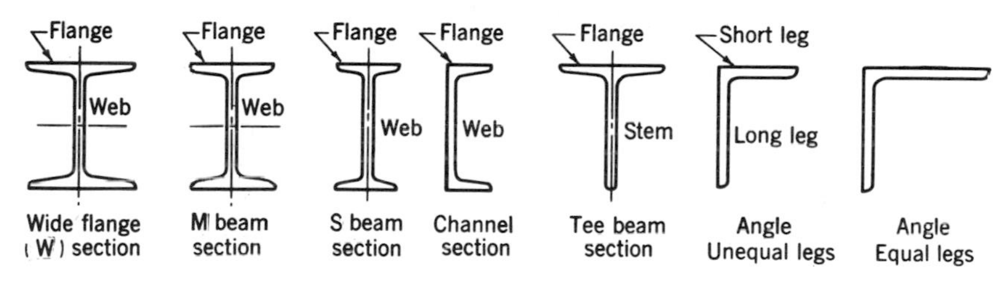

Figure 7-14

from the neutral axis as possible to make the value of the moment of inertia $\int y^2\,dA$ a maximum consistent with other strength requirements, such as local buckling and twisting.

PROBLEMS

Before the beam sections computed for the following problems can be considered as final, they must be checked for excessive shear stresses (§7-13). Where weight of timber beam is to be considered, use weight found in Table 18.

7-15. A simply supported timber beam 20 ft long is to carry a uniformly distributed total load of 600 lb/linear ft. Determine the section modulus required to sustain this load if the maximum allowable unit stress is 1200 psi.

7-16. A cantilever beam 9 ft long is to be made from several 2- by 10-in. nominal-size wooden planks placed vertically and bolted together. If a concentrated load of 1000 lb is to be carried at its free end, determine the number of planks necessary to sustain the load. Maximum allowable fiber stress is 1800 psi. Consider weight of beam. *Answer:* 3.

7-17. A warehouse floor is to be undergirded by beams 15 ft long. If the total load is to be 250 lb/ft^2, what size of American Standard rectangular timber beams would be used, assuming a spacing of 18 in. between beams and a maximum allowable unit stress of 1500 psi? *Answer:* 4 by 14 in.

7-18. What size wood joists, spaced 16 in. center to center on an effective span of 14 ft, is required to carry a total load of 120 lb/ft^2? Allowable unit stress in wood is 1200 psi. *Answer:* 3 by 10 in.

7-19. Determine that American Standard rectangular timber beam required to sustain the loading shown in Fig. 7-15. Assume $\sigma = 1400$ psi. Consider the weight of beam. *Answer:* 6 by 14 in.

7-20. How deep must be the rectangular, steel lever arm included in the scale shown in Fig. 7-16 if its width is $\frac{1}{4}$ in. and its maximum allowable flexure unit stress is 18,000 psi? *Answer:* 0.98 in.

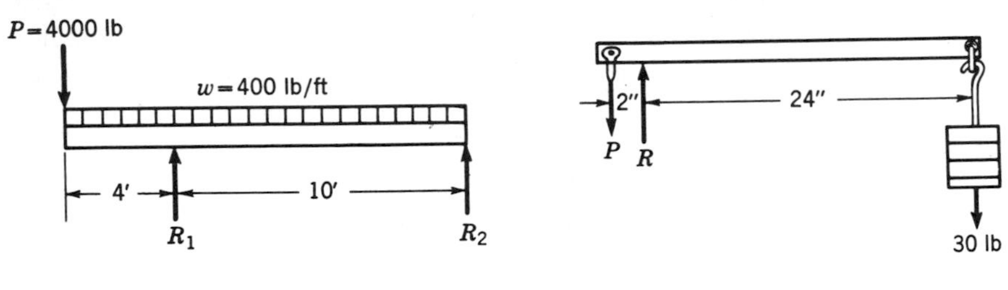

Figure 7-15

Figure 7-16

7-21. If the cast-iron offset arm in Fig. 7-17 is to have a critical cross section whose general dimensions are as shown, what will be the exterior dimensions if $d = 2b$? Use maximum allowable fiber stress of 8000 psi. *Answer:* 4.16 by 8.32 in.

7-22. Find the resisting moment of the cross section shown in Fig. 7-18 if the limiting flexure stress is 4000 psi. *Answer:* 454,000 in.-lb.

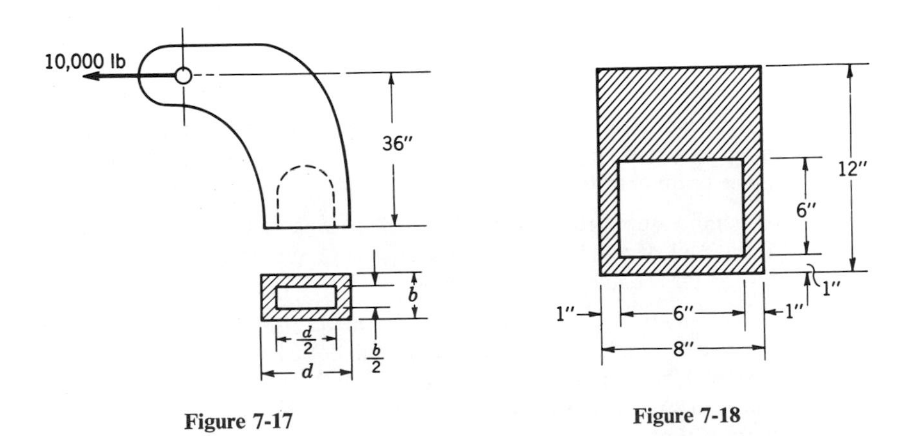

Figure 7-17

Figure 7-18

7-23. To avoid interference in a certain machine, it is necessary to design a bent link as shown in Fig. 7-19. What thickness of rectangular cross section will be needed if its depth is $1\frac{1}{4}$ in. and the allowable tensile stress is 10,000 psi?

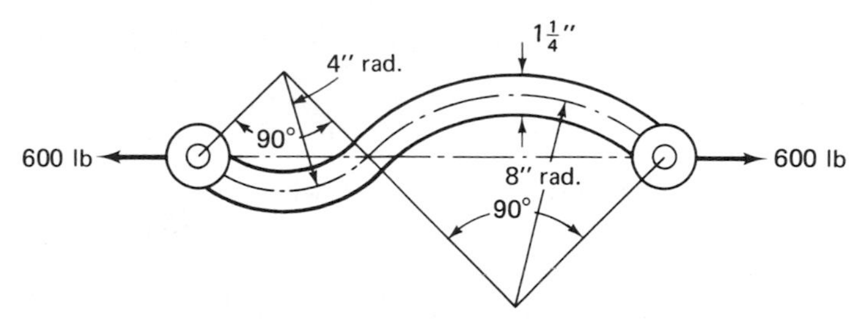

Figure 7-19

7-24. For the bracket shown in Fig. 7-20, determine the minimum depth required for its critical cross section based on flexure if the maximum allowable flexural stress is 10,000 psi. $b = 2$ in. *Answer:* 3.65 in.

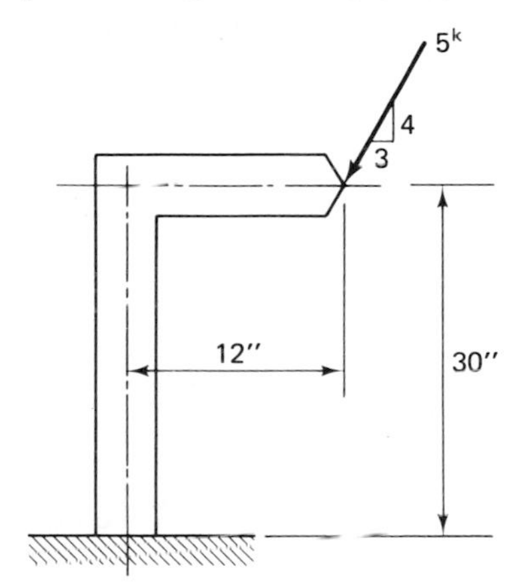

Figure 7-20

USE OF THE FLEXURE FORMULA FOR BEAMS WITH OTHER THAN RECTANGULAR CROSS SECTIONS

The uniform rectangular cross section of the beam used in the derivation of the flexure formula afforded symmetry about both the vertical and horizontal axes. *Because the loads acted in the vertical plane of symmetry no longitudinal twisting could take place, and the unit stress on any horizontal strip y distance from the neutral axis (see Fig. 7-4) was constant over the entire width.* Symmetry about a horizontal axis located the neutral axis at the axis of symmetry (that is, center-of-gravity axis) and caused the rotation about the neutral axis of all plane sections perpendicular to the longitudinal axis.

7-5 Sections Having Two Axes of Symmetry

Since S and W cross sections also have two axes of symmetry, the flexure formula may be extended to include the design of beams having such cross sections. Equation (7.2), developed in the derivation of the flexure formula, presents

$$M_R = \frac{\sigma}{c} \int y^2 \, dA$$

The only difference revealed in this equation when applied to a section such as an S beam is in the variation of dA. This change necessitates only a slightly more involved computation for the moment of inertia of each section considered or its selection from a handbook table. Other than for this added complication, the use of the flexure formula proceeds as with the rectangular sections.

Illustrative Problem 3. Select the most economical S-beam section, based upon flexural requirements, to withstand the loading imposed upon the overhanging beam shown in Fig. 7-21. The beam will be laterally supported. Use an allowable working stress of 20,000 psi. Disregard the weight of the beam.

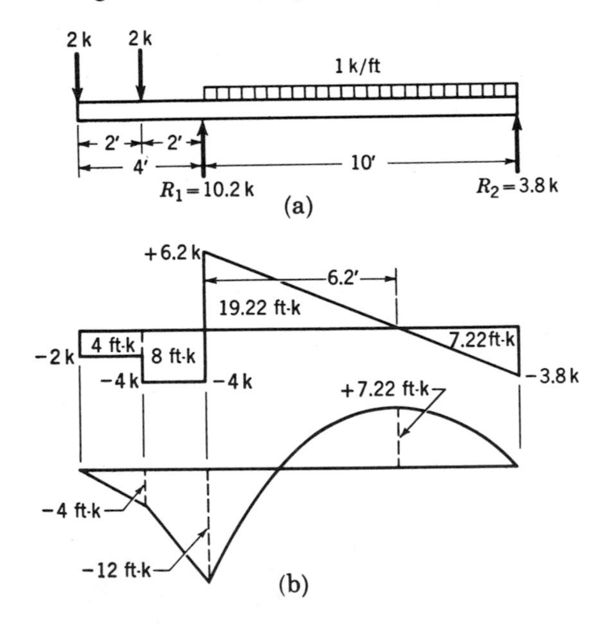

Figure 7-21

SOLUTION. The maximum moment of 12 ft-kips, upon which the flexural portion of the design must be based, is located at the left reaction R_1. Inserting this value with the allowable working stress in the flexure formula, we obtain

$$\frac{M}{\sigma} = \frac{12 \times 1000 \times 12}{20{,}000} = \frac{I}{c} = 7.2 \text{ in.}^3$$

Thus, the S beam required to meet this loading condition most economically must weigh less than any beam having a section modulus greater than

7.2 in.[3] when used in its upright position, that is, resting on its lower flange. Proceeding then to the A.I.S.C. handbook or to Table 13, we note in the column headed by S for the X-X axis that the lightest S beam having a section modulus of 7.2 in.[3] or over is the 6- by $3\frac{3}{8}$-in. nominal-sized S beam weighing 12.5 lb/linear ft.

This is the most economical beam and is recorded as

$$S\,6 \times 12.5$$

Note: The American Institute of Steel Construction permits a maximum allowable tensile and compression stress of $0.66F_y$ on the extreme fibers of laterally supported *compact* rolled shapes and *compact* built-up members having an axis of symmetry in the plane of loading. In order to qualify as a *compact* section, the projecting elements of the compression flange must comply to the width-thickness ratios indicated in the A.I.S.C. specifications. These ratios vary depending on the degree of attachment of the flange to the web, the width-thickness ratio of the outstanding leg and also of the web, and the extent of support provided the outstanding leg. Practically all rolled W shapes of A-36 steel and a large proportion of those shapes having a yield stress of 50 kips/in.[2] meet these provisions.

7-6 Sections Having Vertical Axes of Symmetry—Loaded Vertically

Beams with cross sections having a vertical axis of symmetry, such as the T section in Fig. 7-22, may be designed through the use of the flexure formula, since Eq. (7.2) is still applicable. The position of the neutral axis

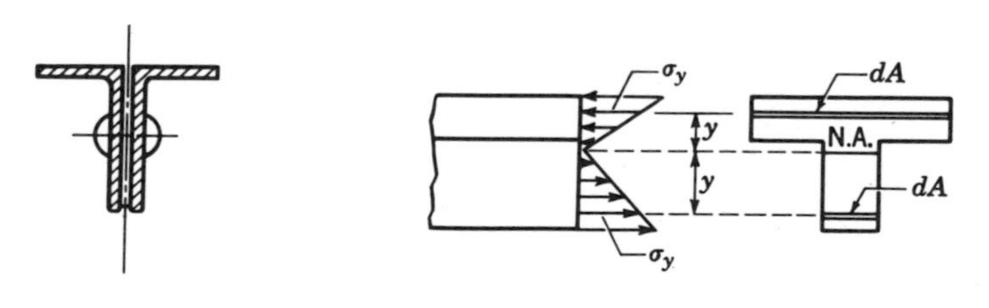

Figure 7-22 **Figure 7-23**

is, however, displaced from its midway position, since symmetry no longer exists about a horizontal axis. To determine the location of the neutral axis on such sections, let us refer to Fig. 7-23. With the resisting compressive forces on the top balancing the resisting tensile forces on the bottom, we may write the equation

$$\int \sigma_y \, dA = 0$$

Multiplying by y/y we obtain

$$\int \frac{\sigma_y}{y} y\, dA = 0$$

Knowing that a product of two values can only equal zero when one of these values also equals zero, we may further stipulate, since σ_y/y is a constant, that

$$\int y\, dA = 0$$

But $\int y\, dA$ is equal to the moment of the cross-sectional area of the beam with respect to the neutral axis. When equal to zero, it indicates that the moments of the areas above and below the neutral axis balance one another. This balance of moments also defines the neutral axis as being a centroidal axis. In other words, *the centroidal axis is coincident with the neutral axis.*

Thus, if a study of the bending stresses were to be made on a T section whose center line of stem was vertical and in the plane of the loads, the location of the centroidal axis parallel to the flange would be required. Once having located the centroidal axis, the moment of inertia of the area about this axis would have to be found for insertion in the flexure formula. It should be noted that the extreme fiber stresses for this type of cross section will not be of the same magnitude.

Illustrative Problem 4. The cast-iron T beam shown in Fig. 7-24 is subjected to two concentrated loads of 4000 and 10,000 lb at its third points. Find

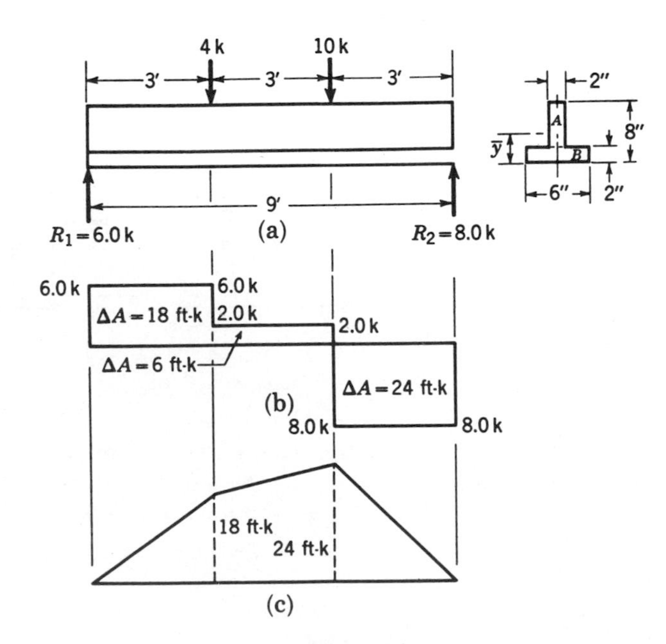

Figure 7-24

the maximum tensile and compressive unit stresses acting on the beam. Disregard the weight of the beam.

SOLUTION. Inasmuch as the use of the flexure formula requires a knowledge of the moment of inertia of the section about its neutral axis, we must first proceed to obtain this value before we can obtain the required stresses.

Location of neutral axis. The moments of the areas A and B, taken about the base of the section, are

SECTION	$\Delta A \quad y$	$\Delta A\, y$
A	12×5	$= 60$ in.^{3}
B	12×1	$= 12$
	$\Sigma\, \Delta A = 24$	$= 72$ in.^{3}

$$\bar{y} = \frac{\Sigma\, \Delta A\, y}{\Sigma\, \Delta A} = \frac{72}{24} = 3 \text{ in.}$$

Moment of inertia about neutral axis

SECTION

$$A \qquad I_{c.g.} = \tfrac{1}{12} \times 2 \times 6^3 = 36 \text{ in.}^4$$
$$Ay^2 = 12 \times 2^2 = 48$$
$$B \qquad I_{c.g.} = \tfrac{1}{12} \times 6 \times 2^3 = 4$$
$$Ay^2 = 12 \times 2^2 = 48$$
$$136 \text{ in.}^4$$

Maximum fiber stresses. The maximum bending moment, as determined from the bending-moment diagram, occurs under the 10-kip load and is equal to 24-ft-kips.

The maximum tensile unit stress at this section occurs at the extreme bottom fiber, located 3 in. below the neutral axis. Its value is equal to

$$\sigma_t = \frac{Mc}{I} = \frac{24 \times 1000 \times 12 \times 3}{136} = 6360 \text{ psi}$$

The maximum compressive unit stress at this maximum-moment section occurs on the extreme top fiber, located 5 in. above the neutral axis. Proceeding with its computation, we obtain

$$\sigma_c = \frac{Mc}{I} = \frac{24 \times 1000 \times 12 \times 5}{136} = 10{,}600 \text{ psi}$$

This stress might also have been obtained from the fact that the flexure stress varies directly with its distance above or below the neutral axis. Knowing that a unit stress of 6360 psi exists 3 in. below the axis, we can find the stress 5 in. above by a similar triangle relationship (Fig. 7-25):

$$\sigma_c = 6360 \times \tfrac{5}{3} = 10{,}600 \text{ psi}$$

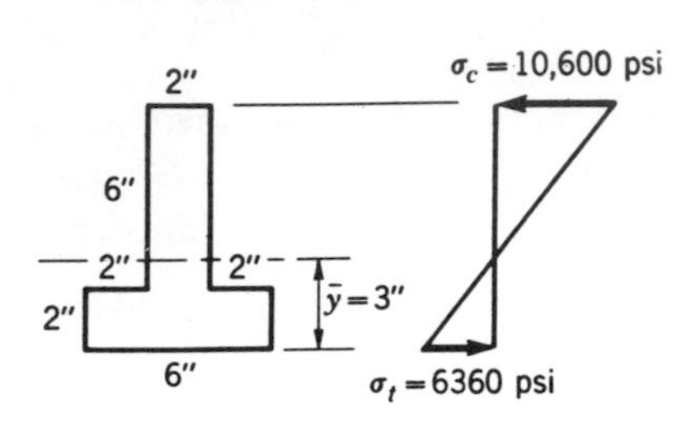

Figure 7-25

The use of T sections in bending is particularly suited for those materials that have different tensile and compressive strengths. If a material, such as cast iron, has a lower ultimate tensile stress than an ultimate compressive stress, the intensity of the more critical tensile stress may be reduced by placing the flange of the T on the tension side. Such sections are frequently used in machine frames.

7-7 Sections Having Horizontal Axis of Symmetry—Loaded Vertically

When a rolled-steel section, such as the channel of Fig. 7-26a, has a horizontal plane of symmetry but is loaded in a plane parallel to its web, care must be taken to locate a loading plane that will prevent longitudinal twisting. Thereafter, the application of the flexure formula can proceed without difficulty.

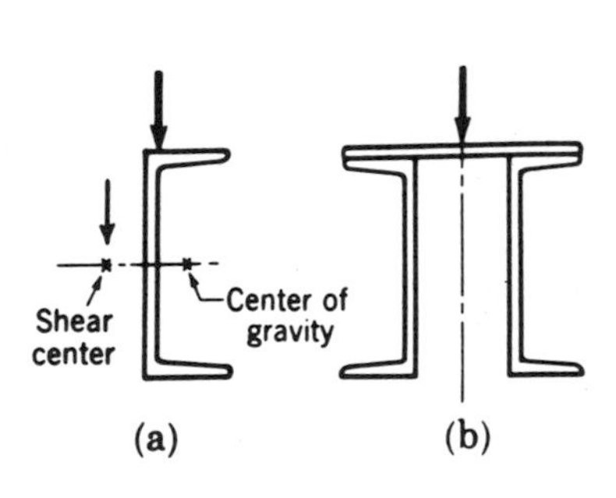

Figure 7-26

To eliminate longitudinal twisting requires determination of the *shear center* of the cross section, the computation of which is discussed in §7-17. It is important to note that it does not correspond to the center of gravity. Frequently, two channels are placed back to back (Fig. 7-26b) to provide the symmetry that will overcome twisting.

7-8 Sections Having No Axes of Symmetry

If a beam has a cross section with no axes of symmetry, the flexure formula can still be used in its design, if it is loaded to prevent twisting.* Further complications arise, however, *since the bending plane will not be in the plane of loading unless it contains one of the principal axes of the section.*† The neutral axis of an unsymmetrical cross section does not coincide with a principal axis unless the above stipulation is maintained.

To illustrate the point, let us consider first a beam with a rectangular cross section acted upon by a moment in the vertical plane of symmetry

*Load must pass through shear center (see §7-17).

†A principal axis is an axis about which the value of I is either a maximum or minimum. See §A-13, Appendix.

(Fig. 7-27a). The tensile and compressive forces *which must produce a restraining moment equal to the applied moment* balance not only about the horizontal axis of symmetry but about the vertical axis as well. Suppose now one quarter of the beam were removed (Fig. 7-27b). Certainly, the center of gravity of the cross section would rise and move to the left. To assume that bending would be about a horizontal axis through this point would be to assume that the stresses varied as indicated in b. The *C-T* resisting couple would under these circumstances then have a horizontal component for which there were no equal applied moment. Such an unbalanced situation could not exist for a body in equilibrium.

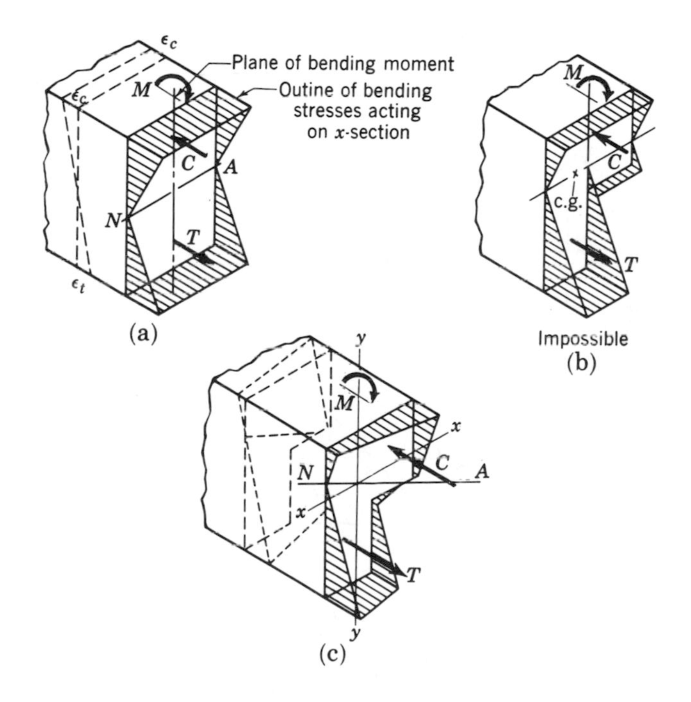

Figure 7-27

The only way a perfect balance could be attained between the applied and resisting moments would be to have the bending axis tilted (Fig. 7-27c). Under this condition, the components of the applied bending moment perpendicular and parallel to the tilted bending axis would exactly balance the corresponding components of the resisting moment.

Because of these complications, the design of a beam with an unsymmetrical cross section proceeds by first resolving the bending moment into components parallel to the two principal axes of the section (see §A-13). *These moments, when considered separately, produce bending and corresponding flexural stresses about their respective principal axes.* If, then, the actions of these component bending moments are combined, bending will occur about some intermediate axis and the resulting flexural stresses are obtained by adding their separate flexural stresses algebraically. That greatest value of resulting flexural stress must not exceed the maximum allowable stress for the material.

Illustrative Problem 5. A cantilever beam, made from a 6- by 6- by $\frac{1}{2}$-in. angle, 6 ft long, is loaded at its free end with a non-twisting vertical load of 500 lb. If the legs of the angle are horizontal and vertical, determine the maximum stress at point *A* (Fig. 7-28) resulting from the imposed load.

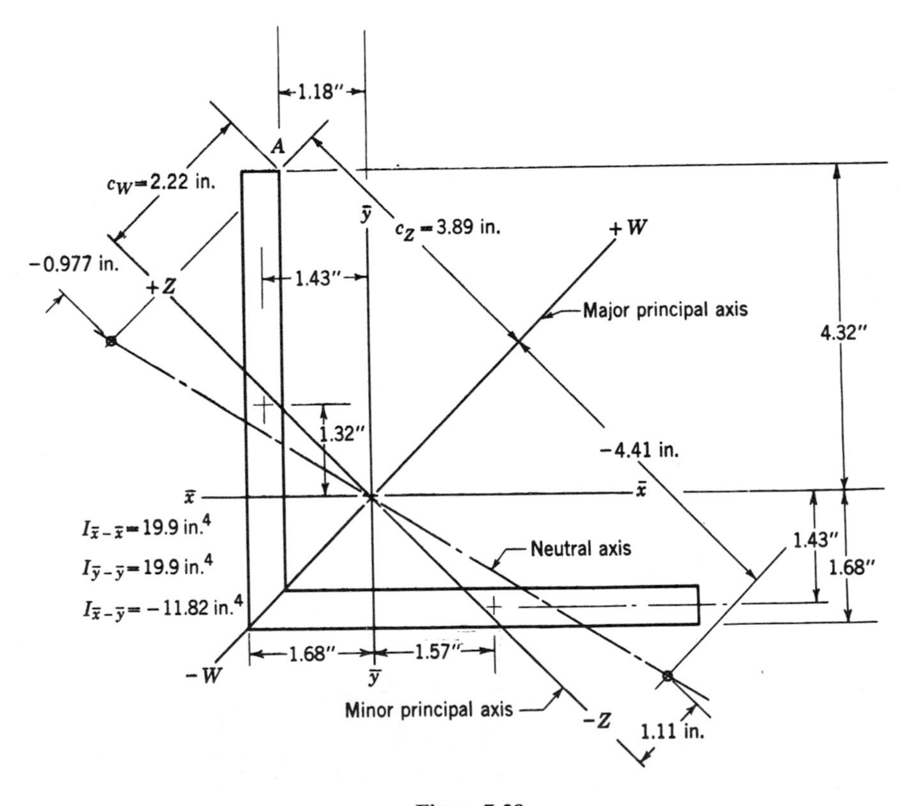

Figure 7-28

Solution. The minor principal axis for the cross section of any standard structural steel angle is commonly known as the *Z-Z* axis and is tilted with respect to each of the legs. When the legs are equal in length, the tilt is 45°. From Table 17 the value of radius of gyration about the *Z-Z* axis for the given 6- by 6- by $\frac{1}{2}$-in. angle is 1.18 in. Its corresponding moment of inertia is equal to Ar^2 or $5.75 \times 1.18^2 = 8.00$ in.^{4}. To obtain the moment of inertia I_W about the major principal axis W-W, which is perpendicular to the minor principal axis Z-Z, use is made of the fact that the sum of moments of inertia on mutually perpendicular planes is a constant.* Thus,

$$I_x + I_y = I_Z + I_W$$

Substituting corresponding values taken directly from Table 17 yields

$$19.9 + 19.9 = 8.0 + I_W$$

from which

$$I_W = 31.8 \text{ in.}^4$$

The maximum bending moment, which equals 3000 ft-lb, is then resolved into components parallel to the principal axis.

$$M_Z = M_W = 3000 \times 12 \times 0.707 = 25{,}450 \text{ in.-lb}$$

The corresponding flexural stresses for point A are therefore

$$\sigma_A = \frac{M_Z \times c_W}{I_Z} + \frac{M_W \times c_Z}{I_W} \tag{7.4}$$

$$\sigma_A = \frac{25{,}450 \times 2.22}{8.0} + \frac{25{,}450 \times 3.89}{31.8}$$

$$\sigma_A = 7070 + 3110$$

$$\sigma_A = 10{,}180 \text{ psi}$$

Locating the neutral axis. The location of the neutral axis of the cross section can be obtained by finding two points at which the sum of the two flexure stresses is equal to zero. The coordinates of these points may be obtained by placing σ_A in Eq. (7.4) equal to zero and solving in turn for values of c'_W and c'_Z that will make the flexural stresses equal to each other. Thus, for $c_Z = 3.89$ in., $c'_W = -0.977$ in., and for $c_W = 1.11$ in., $c'_Z = -4.41$ in.† The neutral axis must pass through these points. As a check, it must also pass through the origin.

7-9 Derivation—General Equation for Nonsymmetrical Bending

An even more general approach to the solution of nonsymmetrical bending problems in which the determination of maximum and minimum moments of inertia is unnecessary is presented as follows.

The bending of a nonsymmetrical cross section with or without an accompanying direct stress provides a stress distribution which is linearly varying. Its intensity variation from point to point on the cross section (Fig. 7-29) may be described by a plane whose equation is

$$\sigma = ax + by + c = z \tag{7.5a}$$

*See development of Eq. A.17 in the Appendix.

†The values of c_Z and c_W used for finding c'_W and c'_Z, respectively, were chosen at random.

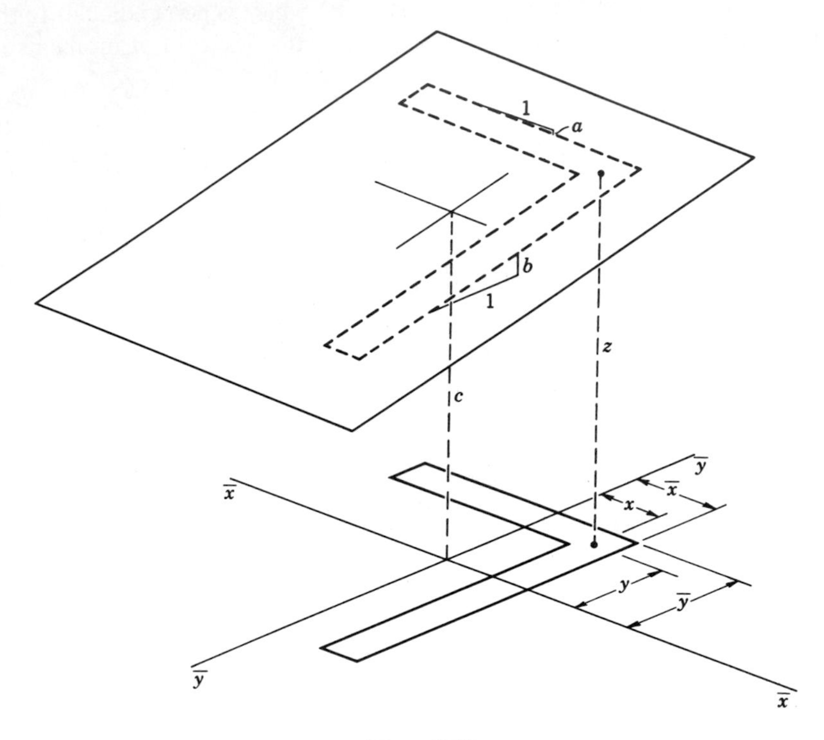

Figure 7-29

where a and b denote rates of stress variation in the x and y directions, and c is a constant direct stress (if any).

But

$$W = \int \sigma \, dA$$

$$M_{xx} = \int \sigma y \, dA$$

and

$$M_{yy} = \int \sigma x \, dA \tag{7.5b}$$

If now Eq. (7.5a) is substituted in Eq. (7.5b), there results

$$W = \int (ax + by + c) \, dA = a \int x \, dA + b \int y \, dA + c \int dA$$

$$M_{xx} = a \int xy \, dA + b \int y^2 \, dA + c \int y \, dA$$

and

$$M_{yy} = a \int x^2 \, dA + b \int xy \, dA + c \int x \, dA \tag{7.5c}$$

In the series of Eq. (7.5c), the values of $\int x\,dA$ and $\int y\,dA$ are equal to zero when the x and y axes of the cross section pass through the centroid. Hence, if x and y are coordinates measured from the centroidal axes,

$$W = c \int dA = cA$$

$$M_{\bar{x}\bar{x}} = aI_{\bar{x}\bar{y}} + bI_{\bar{x}\bar{x}}$$

$$M_{\bar{y}\bar{y}} = aI_{\bar{y}\bar{y}} + bI_{\bar{x}\bar{y}}$$

Solving simultaneously for the value of the constants, we obtain

$$c = \frac{W}{A}, \qquad a = \frac{M_{\bar{y}\bar{y}}I_{\bar{x}\bar{x}} - M_{\bar{x}\bar{x}}I_{\bar{x}\bar{y}}}{I_{\bar{x}\bar{x}}I_{\bar{y}\bar{y}} - I_{\bar{x}\bar{y}}^2}, \qquad b = \frac{M_{\bar{x}\bar{x}}I_{\bar{y}\bar{y}} - M_{\bar{y}\bar{y}}I_{\bar{x}\bar{y}}}{I_{\bar{x}\bar{x}}I_{\bar{y}\bar{y}} - I_{\bar{x}\bar{y}}^2}$$

Substituting these values for a, b, and c into Eq. (7.5a), there results

$$\sigma = x_0\frac{M_{\bar{y}\bar{y}}I_{\bar{x}\bar{x}} - M_{\bar{x}\bar{x}}I_{\bar{x}\bar{y}}}{I_{\bar{x}\bar{x}}I_{\bar{y}\bar{y}} - I_{\bar{x}\bar{y}}^2} + y_0\frac{M_{\bar{x}\bar{x}}I_{\bar{y}\bar{y}} - M_{\bar{y}\bar{y}}I_{\bar{x}\bar{y}}}{I_{\bar{x}\bar{x}}I_{\bar{y}\bar{y}} - I_{\bar{x}\bar{y}}^2} + \frac{W}{A} \tag{7.5d}$$

This is the general equation for flexure plus direct stress when x_0 and y_0 are coordinates measured from the center-of-gravity axes to the point undergoing stress.

The solution of σ_A in the previous problem could have been obtained by substituting directly into Eq. (7.5d). Thus,

$$\sigma_A = (-1.18)\frac{0 \times 19.9 - 36{,}000 \times (-11.82)}{19.9 \times 19.9 - (-11.82)^2} + 4.32\frac{36{,}000 \times 19.9 - 0 \times -11.82}{19.9 \times 19.9 - (-11.82)^2}$$

$$= -1950 + 12{,}100$$

$$\sigma_A = +10{,}150 \text{ psi}$$

Product of inertia calculation:

$$I_{\bar{x}\bar{y}} = 6 \times \tfrac{1}{2} \times (-1.43)(1.32) = -5.65$$

$$5\tfrac{1}{2} \times \tfrac{1}{2} \times (1.57)(-1.43) = \underline{-6.17}$$

$$-11.82$$

PROBLEMS

All beams used in the following problems are assumed to have compact cross sections and sufficient lateral support to develop the maximum allowable flexure stress. They are also subject to review of their shearing stresses, which is considered in §7-13.

7-25a, b, c, d, e, f, g, h. Determine the most economical S beam and W section necessary to sustain the load acting on each beam of Fig. 6-18. Use maximum allowable flexure stress = 20,000 psi.

7-26a, b, c, d, e, f, g, h. Select two channels placed back to back to support the load acting on each beam of Fig. 6-19. Use maximum allowable flexure stress

= 20,000 psi. *Answers:* (a) 8 in.-13.75 lb; (b) 10 in.-20 lb; (c) 10 in.-25 lb; (d) 10 in.-30 lb; (e) 15 in.-40 lb; (f) 8 in.-13.75 lb; (g) 8 in.-13.75 lb; (h) 6 in.-13 lb.

7-27. Select the most economical S beam to sustain the load on the beam of Fig. 7-30. Use $\sigma = 20{,}000$ psi. *Answer:* S 12 × 40.8.

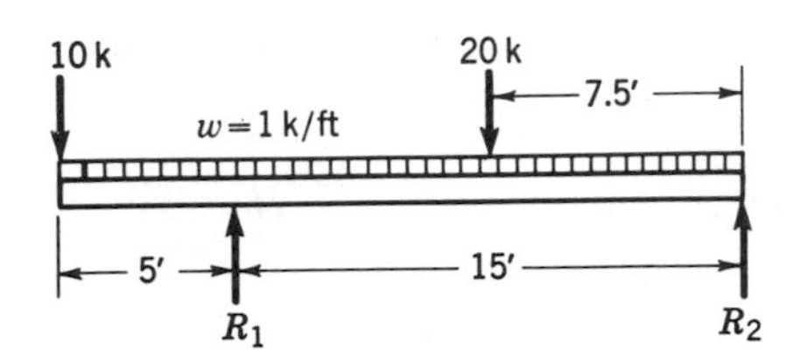

Figure 7-30

7-28. An extra-strong steel pipe having an outside diameter of 4 in. and a wall thickness of 0.318 in. is to be used as a simple beam of 12 ft to support two concentrated loads of 2 kips each, located 4 and 6 ft from the left reaction. Determine the maximum flexure stress developėd by the loads.

7-29. A cast-iron machinery frame having a T section, shown in Fig. 7-31, is to be loaded with force P acting horizontally. What maximum value of P can be applied if the maximum allowable compressive stress is 20,000 psi and the maximum allowable tensile unit stress is 8000 psi? *Answer:* 36,300 lb.

7-30. What width of flange is necessary in the T beam shown in Fig. 7-32 if the maximum tensile unit stress is to be equal to one third the maximum compressive unit stress? The tensile stress is on the lower fibers.

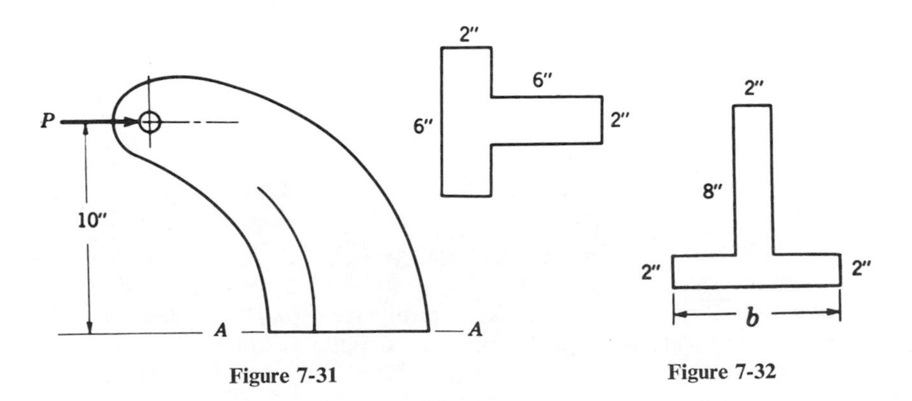

Figure 7-31

Figure 7-32

7-31. What should be the total depth of the cast-iron T section shown in Fig. 7-33 in order to produce simultaneously occurring tensile and compressive stresses of 8000 and 16,000 psi at A and B, respectively. *Answer:* 6 in.

7-32. Two unequal leg angles are to be placed back to back to form a cantilever beam 8 ft long. If a uniformly distributed load of 100 lb/linear ft (includes

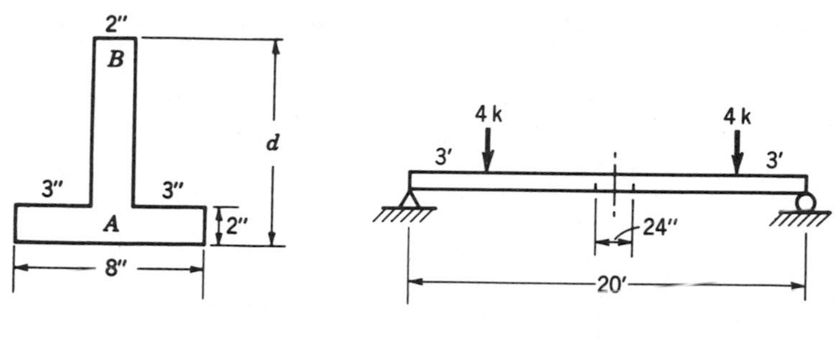

Figure 7-33

Figure 7-34

the weight of the angles) and a concentrated load of 500 lb are located at its free end, what most economical angle section can be arranged to support this load? Use $\sigma = 20{,}000$ psi. *Answer:* Two angles 6 by $3\frac{1}{2}$ by $\frac{1}{4}$ in.

7-33. A steel pipe with a 4-in. outside diameter and a 3.5-in. inside diameter spans 20 ft and carries concentrated loads as shown in Fig. 7-34. If two gage marks are placed in the bottom of the beam, 24 in. apart and symmetrical with the mid-span, what will be the distance between the gage marks after the load is applied? Disregard weight of pipe. $E = 30 \times 10^6$ psi. *Answer:* 24.044 in.

7-34. An overhanging beam (Fig. 7-35) has the cross section of an inverted T with dimensions and properties as indicated. Determine the maximum tensile and compressive unit stress.

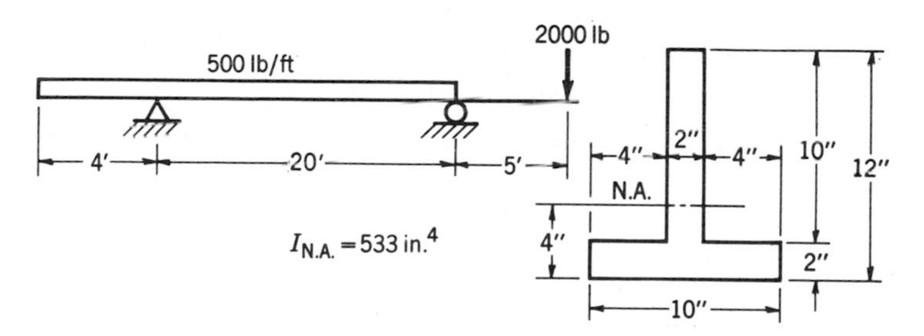

Figure 7-35

7-35. Determine the maximum value of P that can be sustained on the beam of Fig. 7-36 if the allowable tensile and compressive stresses are 8000 and 14,000 psi, respectively. Disregard the weight of beam. $\bar{y} = 3.05$ in. *Answer:* 1940 lb.

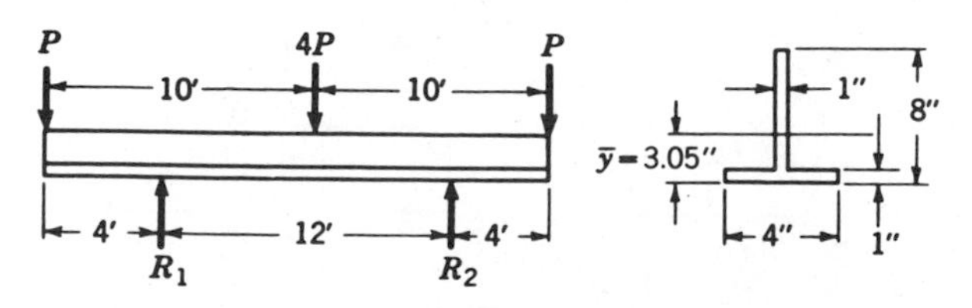

Figure 7-36

7-36. Turn the beam in Problem 7-35 upside down and determine P, using the same allowable stresses.

7-37. The overhanging beam shown in Fig. 7-37 has a cross section of an inverted U with properties indicated. Determine the maximum value of P so that the following extreme fiber stresses are not exceeded: compression, 12,000 psi; tension, 8000 psi. Neglect weight of beam. *Answer:* 6630 lb.

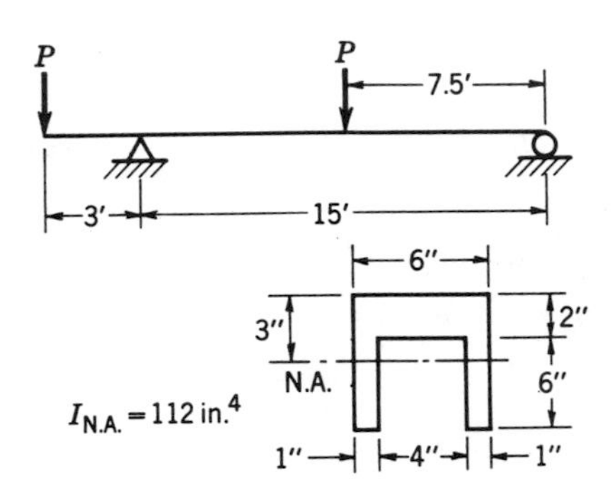

Figure 7-37

7-38. A 12-in. 31.8-lb S beam ($I/c = 36$) simply supported at the ends of a 20-ft span carries two loads of P lb each at points 3 ft from the ends of the beam. Two gage marks were made on the bottom flange of the beam before loading, each mark being 12 in. from midspan. After loading, the gage marks were found to be 24.013 in. apart. Determine the magnitude of the loads P. $E = 30 \times 10^6$ psi. Do not neglect the weight of the beam.

7-39. Determine the maximum flexural stress incurred on the following unsymmetrical sections due to a bending moment of 100,000 in.-lb applied in a vertical plane without longitudinal twisting: (a) An 8- by 8- by $\frac{3}{4}$-in. angle, one leg vertical. (b) An 8- by 6- by $\frac{3}{4}$-in. angle, 6-in. leg horizontal. *Answers:* (a) 10,700 psi; (b) 12,980 psi.

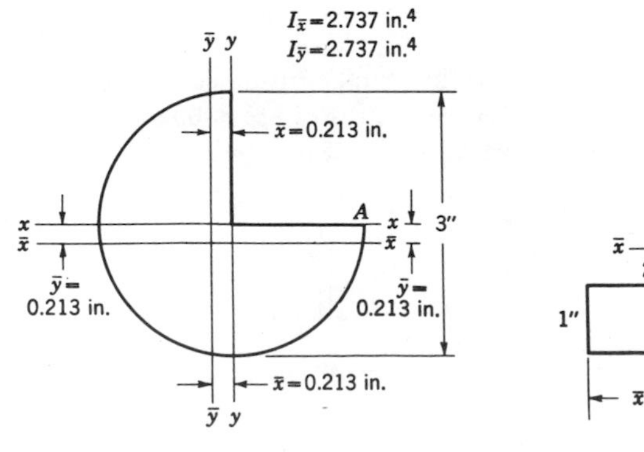

Figure 7-38

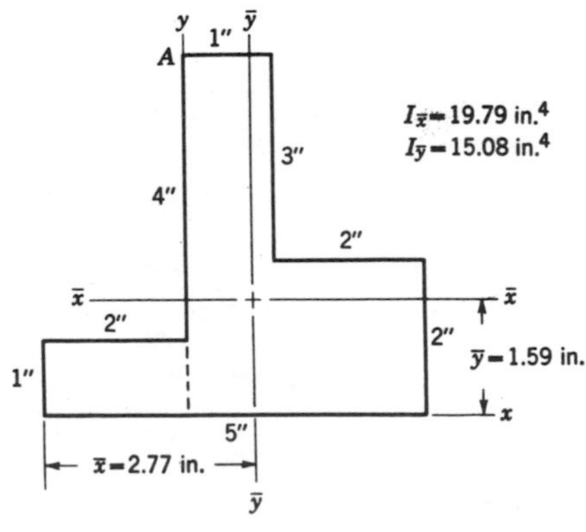

Figure 7-39

7-40. Determine the maximum flexural stress incurred at point A on the unsymmetrical sections shown in Figs. 7-38 and 7-39 due to a bending moment of 50,000 in.-lb applied in a vertical plane without longitudinal twisting. *Answers:* (a) 31,200 psi; (b) 8695 psi.

7-10 Beams of Varying Section Modulus

As indicated in the solutions to previous illustrative problems, the flexural portion of the design of a uniform cross-sectioned beam is governed by the maximum bending moment. On an ordinary beam, this usually occurs on only one section. All the other sections that are of equal size but are acted on by lesser values of bending moment must, consequently, be understressed. The use of material that is understressed makes necessary the carrying of weight in a beam that is not used to its maximum capacity.

This undesirable condition may be eliminated to a large extent by varying the section modulus I/c in the central portion of a beam to conform to the variation or shape of the bending-moment diagram. This would cause the maximum flexural unit stresses in this affected portion of the beam to be equal, since

$$\sigma = \frac{M}{I/c} = \text{constant}$$

This method of reducing the weight of beams is frequently made use of in the fabrication of plate girder beams, heavy truck axles, heavy castings, and so forth (Fig. 7-40). Whether or not the cost of varying the cross section of a beam is always justifiable on the basis of the weight of material saved can be decided only on the individual merits of each case. Certainly the ordinary rolled-steel beams are of constant cross section, because the cost of rolling a beam of variable cross section would not offset the saving in material.

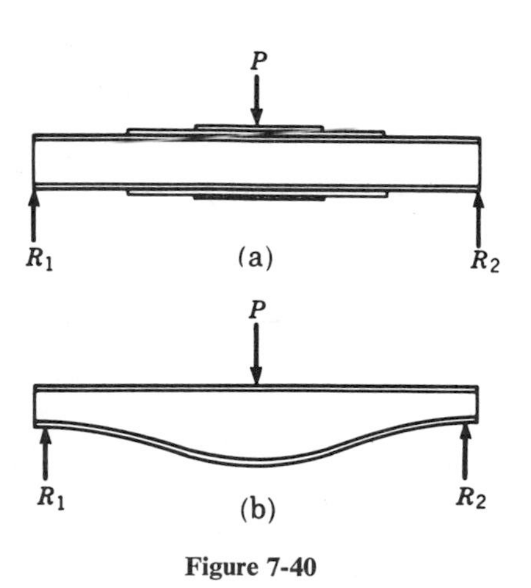

Figure 7-40

Illustrative Problem 6. A uniform load of 6000 lb/ft is to be applied to a cast wide-flange beam 12 ft long, having a variable depth of web. The top and bottom flanges are to be made 6 in. wide by 2 in. thick and are to be held constant throughout the entire length of the beam. What total depth of section will be necessary at the center and quarter points to meet the flexural requirements if the maximum allowable stress of 10,000 psi is to be maintained at each section? Do not consider the weight of the beam.

SOLUTION. The bending moments at the center and quarter points are 108,000 and 81,000 ft-lb, respectively. Their corresponding section moduli required are 129.6 and 97.2 in.^{3}.

By equating these required section moduli to an algebraically expressed section modulus of the proposed cross section in terms of d, a cubic equation can be obtained from which an exact determination of the total depth may be made.

The solution may also be made graphically. On the graph shown in Fig. 7-41 is plotted a curve of available section moduli for various values of d. The values used to obtain the curve were

d (in.)	I/c (in.^{3})
14	148.5
12	115.4
10	85.7
8	58.8

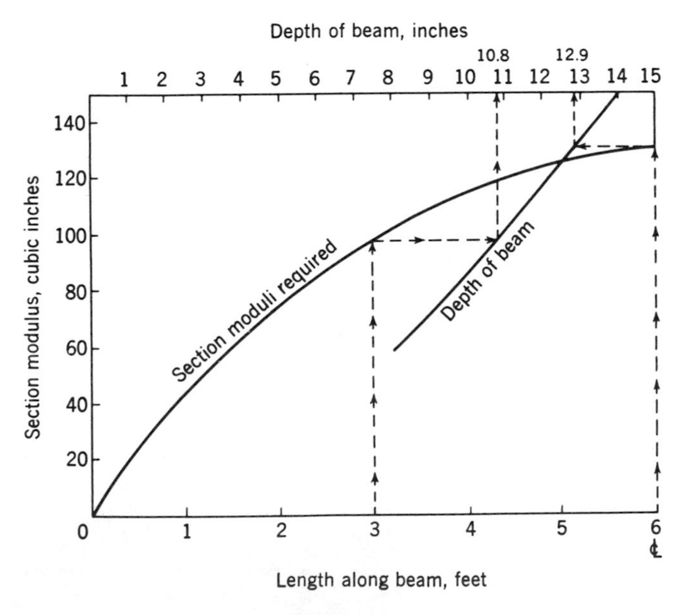

Figure 7-41

An additional curve is drawn giving the variation of M/σ or section moduli required for different sections of the beam. Proceeding horizontally from this curve at the point corresponding to the section under investigation to the available section moduli curve provides the required depth of beam at the point of intersection.

For the center and quarter points, the depths required for flexure are, respectively, 12.9 and 10.8 in.

Note: At the ends of the beam where the bending moment is small, the depth will be controlled by the allowable shear (see §7-12). Both conditions must be investigated thoroughly to determine the permissible depth variations throughout its length.

Illustrative Problem 7. Determine the increase in allowable bending moment obtained by welding $\frac{1}{2}$- by 6-in. cover plates on the top and bottom flanges of an S 15 × 42.9 simple beam. The beam is laterally supported. $\sigma = 20{,}000$ psi.

SOLUTION. The moment of inertia of the section is

S beam		441.8 in.^{4}
Cover plates	$2[\frac{1}{12} \times 6 \times (\frac{1}{2})^3 + 3 \times 7.75^2] =$	360.2
Total moment of inertia		802.0 in.^{4}

The allowable bending moment of the section will, therefore, be

$$M = \frac{\sigma I}{c} = \frac{20{,}000 \times 802}{8.0} = 2{,}000{,}000 \text{ in.-lb}$$

The allowable bending moment of the S beam (previously computed) is 1,180,000 in.-lb.

$$\therefore \quad \text{increase} = \frac{2{,}000{,}000 - 1{,}180{,}000}{1{,}180{,}000} \times 100 = 69.5\%$$

PROBLEMS

In the following problems, use maximum allowable flexure stress for steel = 20,000 psi.

7-41. An S 10 × 25.4 beam, 20 ft long and simply supported at its ends, has 6- by $\frac{1}{2}$-in. cover plates welded to its top and bottom flanges over the middle 10 ft of its length. Determine the magnitude of the two equal concentrated loads that can be placed 7 ft from each support, assuming full development of flexure stress in the cover plates. *Answer:* 8140 lb.

7-42. How thick and how long must the cover plates supplementing the top and bottom flanges of an S 15 × 42.9 beam be to increase its carrying capacity of uniform load by 20 per cent? The cover plates are to be 5.5 in. in width and welded into place. Length of beam equals 12 ft. *Answer:* 0.159 in., 4.90 ft, placed symmetrically with center of beam.

7-43. The flag pole shown in Fig. 7-42 is acted upon by a concentrated load of 200 lb at its free end and by a uniformly distributed load of 20 lb/linear ft over its entire length. Calculate the maximum flexure stress in the flag pole. *Answer:* 53,100 psi.

7-44. A W 16 × 36 ($d = 15.85$ in.) beam is 20 ft long. If one cover plate $\frac{1}{4}$ by 7 in. is placed on each flange, determine the maximum uniform load the beam can carry. Determine the points where the outside cover plate is no longer needed.

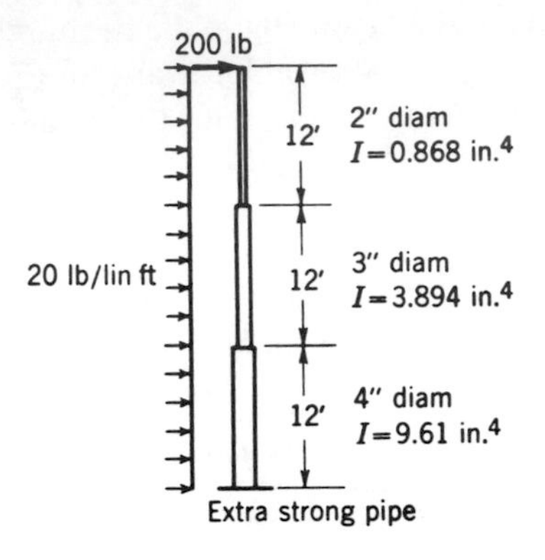

Figure 7-42

I_{x-x} for beam is 446.3 in.^{4}. Consider the cover plates to be welded into place, thus requiring no deduction for rivet holes. *Answer:* 2740 lb/ft.

7-45. A tapered cantilever beam, 10 ft long and 6 in. wide, has a depth which varies linearly from 8 in. at its free end to 20 in. at its fixed end. If the maximum allowable stress is 2000 psi, what would the maximum value be of two identical concentrated loads, one placed at the free end and one at its midpoint? *Answer:* 4440 lb.

7-46. The double overhang beam of Fig. 7-43 is initially loaded as shown with all loads equal to 10^k. (a) Will the section have to be increased if the middle load

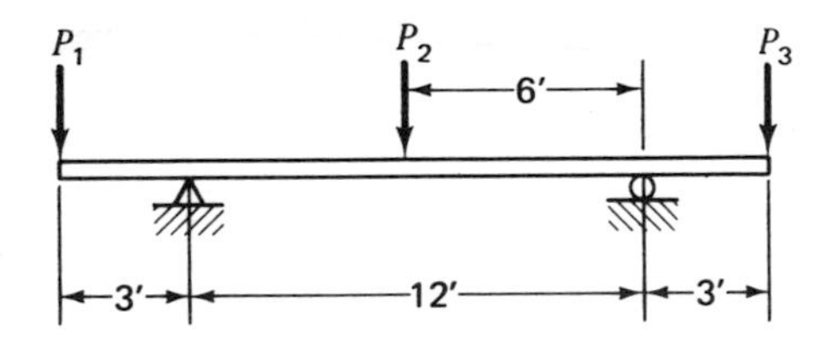

Figure 7-43

is increased to 20^k? If so, what percentage increase in section modulus is required? (b) If the middle load is increased to 30^k, determine the theoretical length and thickness of steel plates, 6 in. wide, that would have to reinforce a W section initially designed for resisting all loads equal to 10^k. Use a maximum allowable stress of 20,000 psi. *Answer:* (b) 2 plates—6 in. by $\frac{1}{4}$ in. theoretical length = 3.2 ft.

7-47. Determine a theoretical expression showing the variation of the cross sections of a cantilever beam carrying a concentrated load P at its free end, for which each section thereof will attain the same maximum allowable stress. (a) Consider the beam first to have a constant depth d. (b) Consider the beam to have a constant width b.

7-48. Design a cantilever beam of constant strength for resisting a uniform load distributed over its entire length. Assume its width to be constant.

7-11 Modulus of Rupture in Bending

An often used, but fictitious, value of ultimate flexure stress is the *modulus of rupture*. It is calculated from the flexure equation by substituting for the value of M the maximum moment attained by the beam during an actual bending test.

To illustrate its calculation, suppose a 12-ft beam having a 4- by 12-in. cross section were to fail under a load of 20,000 lb applied at its midpoint. Its modulus of rupture would then be

$$\sigma_r = \frac{M_u c}{I} = \frac{10{,}000 \times 6 \times 12 \times 6}{\frac{1}{12} \times 4 \times 12^3} = 7500 \text{ psi}$$

Reason for the fallacy. The fallacy developed in the calculation of the modulus of rupture is due to the use of the maximum bending moment. The load producing this maximum moment is almost without exception greater than that required to produce the proportional limit stress. This fact invalidates the use of the flexure equation, since one of the underlying assumptions upon which its derivation is based is that the stress distribution on any section increases linearly with the distance from the neutral axis. This provision is impossible of attainment once the proportional limit is exceeded.

Effect of using the maximum bending-moment. When the stress acting upon the extreme fiber of a beam increases beyond the proportional limit, its rate of stress increase diminishes and the proportion of resisting moment that it previously carried is reduced. This lack of strength increase throws a greater burden on those fibers located closer to the neutral axis. These interior fibers find that they must produce restraining stresses at a rate faster than their ordinary rate. The effect of this increased burden placed on the interior fibers is to *warp* the linear stress distribution (Fig. 7-44) to the extent that at the

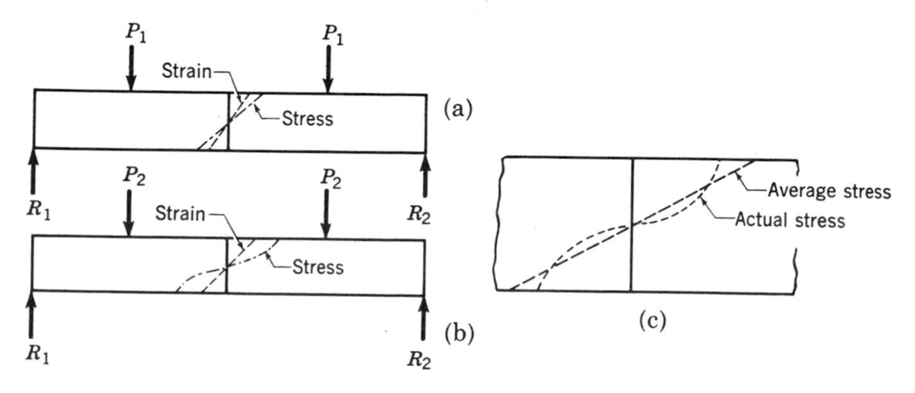

Figure 7-44

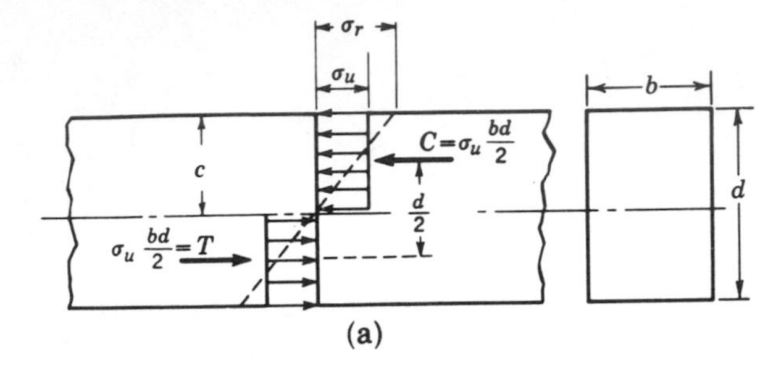

(a)

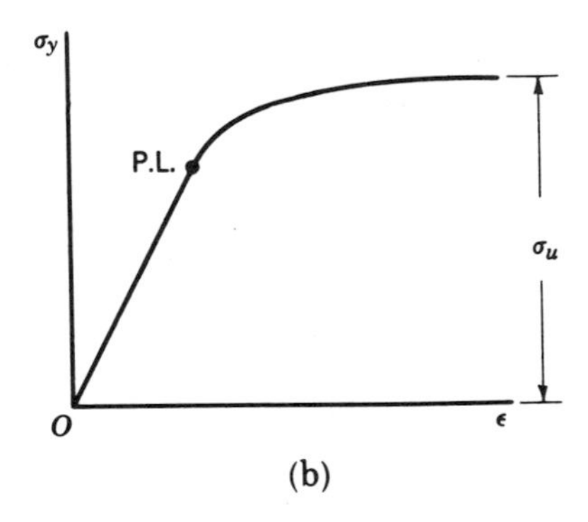

(b)

Figure 7-45

ultimate load it would tend to assume, when the deformation is great, a double rectangle shape (Fig. 7-45a).

The resisting moment corresponding to this stress condition (Fig. 7-45b) acting on a rectangular section is equal to

$$M_u = \sigma_u \times \frac{bd}{2} \times \frac{d}{2}$$

which when multiplied, top and bottom, by $1.5d/2$ becomes

$$M_u = 1.5\frac{\sigma_u}{c} \times \frac{bd^3}{12} = 1.5\frac{\sigma_u I}{c}$$

But $M_u = \sigma_r I/c$. Therefore,

$$\sigma_r = 1.5\sigma_u$$

It is thus apparent that the flexure formula used with the ultimate bending moment produces a fictitious stress which, for rectangular sections, is approximately 1.5 times the more probable ultimate bending stress. It is also apparent that, as the shape of the cross section changes, the factor 1.5 will also change.

Use of the modulus of rupture. The modulus of rupture is used to determine the maximum carrying capacity of beams. It is also used to compare the strengths of similarly shaped sections. When attempts are made to compare sections of dissimilar shape, appreciable error may be involved if the sections are not symmetrical to the neutral axis. Such sections usually attain their tensile and compressive proportional stresses at different times, shifting their neutral axis and making a comparison of strengths impossible.

7-12 Resisting Moment During Yield

Plastic hinge. A beam made of a structural material having a definite yield range similar to that of low-carbon structural steel will develop, on a cross section whose fibers have all commenced yielding, a rectangular stress pattern similar to that developed at its ultimate strength (Fig. 7-45). This moment may be, and often is, considered as its *maximum useful moment* and *for a beam of rectangular cross section* is given as

$$M_p = 1.5\frac{\sigma_y I}{c} = 1.5M_y$$

It is often called the *plastic moment*, for it is developed while the entire section on which it acts is yielding and performing as a *plastic hinge*. For beams of other cross sections, constants other than 1.5, called *shape* or *form factors*, are used to give the plastic moment. For instance, the average shape factor of W structural sections is 1.14.

The shape factor is a measure of one of the sources of reserve strength in a beam. The consideration of this reserve strength forms an important part of limit or plastic design presented in §9-20.

PROBLEMS

7-49. A 2- by 4-in. beam 4 ft long is broken in flexure by a central load of 3000 lb acting on its 2-in. face. What is the modulus of rupture?

7-50. Flashboards are to be placed vertically at the crest of a hydroelectric power dam. They are to be supported by nominal 3-in.-diameter extra-strong pipe ($I_{x-x} = 3.894$ in.^{4}; outside diameter $= 3.5$ in.) placed 3 ft apart. How high should these flashboards be made if they are intended to collapse at the time the water just reaches their top? Modulus of rupture is 70,000 psi. *Answer:* 7.45 ft.

7-51. In the previous problem, what spacing of the 3-in.-diameter pipe would be necessary if the flashboards were to be made 5 ft high?

7-52. A cast-iron flexure test bar is 1.2 in. in diameter and rests on supports placed 18 in. apart. If the central load at the moment of rupture is 2400 lb, determine the modulus of rupture. *Answer:* 63,600 psi.

7-53. Determine the plastic moments of the sections shown in Fig. 7-46. Determine the ratio of these plastic moments with that for a solid rectangular section of width b and depth d.

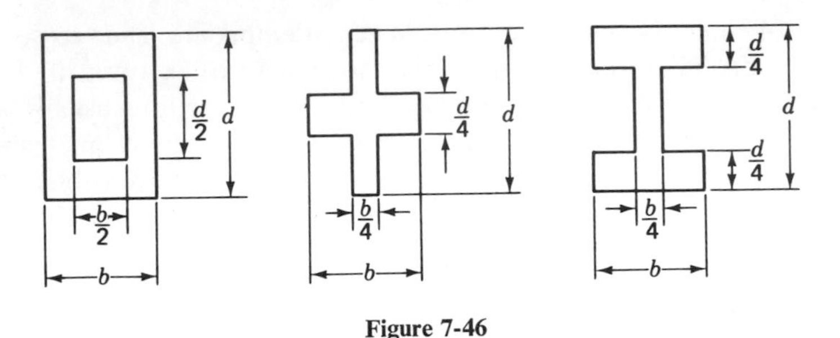

Figure 7-46

7-13 Shearing Stresses in Beams

Rectangular beams. In addition to an internal resisting moment (*C-T* couple) acting on every section of a simple beam, there is generally present an internal shear force, *V*. In the preceding sections of this chapter, no account was taken of the effect of shear in the designs of beams. Because it may be the controlling feature, the design of a beam must be considered incomplete without a thorough investigation of its effects.

If a 4- by 6-in. beam standing on edge were to be sawed longitudinally into 6 boards 1 in. thick and 4 in. wide, and then loaded in its reassembled condition, the beam would bend and cause the individual boards to slide over each other (Fig. 7-47c and d). The evidence of sliding is made apparent from an inspection of end sections *mn* and *op*. Originally, these two sections were perfectly plane (Fig. 7-47a and b). They would have remained so had not the beam been cut. When cutting did take place, the various portions of the beam readjusted their relative positions. Top section m_2o_2, having previously been subjected to compression, elongated when cut from the remainder of the beam. Obviously, the stresses that had restrained the section from sliding when in its original position had been released. Section n_2p_2, having been previously in tension, contracted when severed from the original beam. These sliding adjustments were accompanied by the release of resisting shear stresses present on the sliding

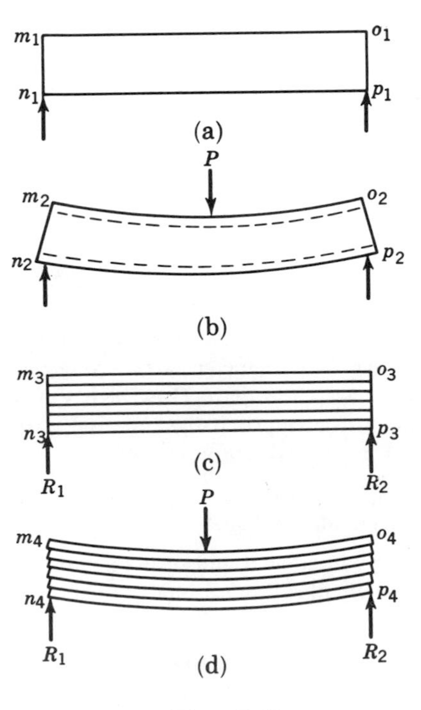

Figure 7-47

planes before cutting took place and caused the end sections m_4n_4 and o_4p_4 to assume a jagged appearance.

The study of the development and variation of these resisting shear stresses acting on horizontal planes resisting failure, as well as the vertical shearing stresses with which they are intimately connected, takes place in the following derivation.

Derivation. Let us consider the simple rectangular beam shown in Fig. 7-48 loaded by a concentrated load P. As the full effect of this load takes

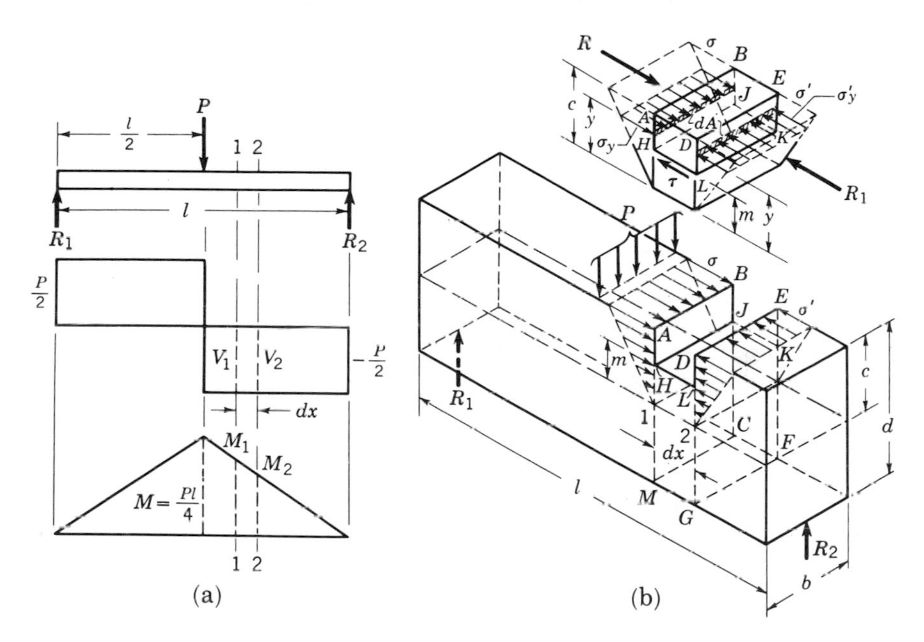

Figure 7-48

place, sections 1 and 2 are acted upon by internal-resisting moments M_1 and M_2 and vertical shearing forces V_1 and V_2, respectively. Endeavoring to analyze the stress condition in the beam, let us cut a free body, a rectangular prism (*ABED-HJKL*), out of the top portion of the beam.

On the exposed sides of this free body act longitudinal compression stresses developed by the internal moments M_1 and M_2. Since M_1 and M_2 are not the same (see moment diagram Fig. 7-48a), the maximum compressive stresses, σ and σ', are also different, σ being larger than σ'. Acting on either side of the free body on the small elementary areas da are the general stresses σ_y and σ'_y situated a distance y from the neutral axis. The product of these stresses times the area on which they act, $\sigma_y\,da$ and $\sigma'_y\,da$, gives the force acting on each elementary area. Since each of the side areas *ABJH* and *DEKL* of the free body contains many such elementary areas, each with its own individual force, the total or resultant forces R and R' acting on these two sides will be equal to the addition of all the individual forces. Hence,

$$R = \int_m^c \sigma_y \, da$$

$$R' = \int_m^c \sigma'_y \, da$$

Since R and R' are unequal, there is obviously an additional force acting on the free body to produce equilibrium. This additional force must act on the only remaining internal plane previously in contact with the original beam (plane *HJKL*). Because this plane is horizontal and the required force must also be horizontal, the stresses induced must be shear stresses.

If the average shear stress on plane *HJKL* equals τ, the total force $\tau b \, dx$ must be equal to the difference in the resultant forces R and R'. Thus, we may write

$$R - R' = \int_m^c \sigma_y \, da - \int_m^c \sigma'_y \, da = \tau b \, dx$$

But, since

$$\frac{\sigma_y}{y} = \frac{\sigma}{c} \quad \text{and} \quad \frac{\sigma'_y}{y} = \frac{\sigma'}{c}$$

we may write

$$\int_m^c \frac{\sigma}{c} y \, da - \int_m^c \frac{\sigma'}{c} y \, da = \tau b \, dx$$

Because the ratios σ/c and σ'/c are included in the force acting on every elementary area, they are constants and may be written outside the integral signs.

$$\frac{\sigma}{c} \int_m^c y \, da - \frac{\sigma'}{c} \int_m^c y \, da = \tau b \, dx$$

or

$$\frac{\sigma - \sigma'}{c} \int_m^c y \, da = \tau b \, dx$$

Substituting $\sigma = M_1 c/I$ and $\sigma' = M_2 c/I$, we obtain

$$\frac{M_1 - M_2}{I} \int_m^c y \, da = \tau b \, dx$$

However, the value of $M_1 - M_2$ is equal to the change of moment over the infinitesimally small distance dx, and is represented by dM. By substituting and rearranging, the equation becomes

$$\tau = \frac{dM}{dx} \times \frac{1}{Ib} \int_m^c y \, da$$

Recalling from Chapter 6 that

$$\frac{dM}{dx} = V$$

and from the study of mechanics that

$$\int_m^c y\,da = \bar{y}\int_m^c da = a\bar{y}\Big]_m^c$$

the moment of the area between m and c about the neutral axis, we may write the equation in its final form:

$$\tau = \frac{V}{Ib}a\bar{y} \tag{7.6}$$

where τ = horizontal shearing unit stress on any very narrow strip passing crosswise through a beam, V = the total internal-shearing force acting on the vertical plane passing through the point where τ is desired (obtained from shear diagram), I = moment of inertia of section of beam passing through the point where τ is desired, b = width of beam at the point where τ is desired, a = area of section located between the closest extreme fiber and the strip on which τ is desired, $\bar{y}$ = distance from the neutral axis to the center of gravity of area a.

A further discussion of this formula may best be directed toward the solution of an illustrative problem.

Illustrative Problem 8. Determine the horizontal shearing unit stress at points 0, 2, 4, and 6 in. above the neutral axis on section *A-A* of the beam shown in Fig. 7-49. The beam is 6 in. wide and 12 in. deep.

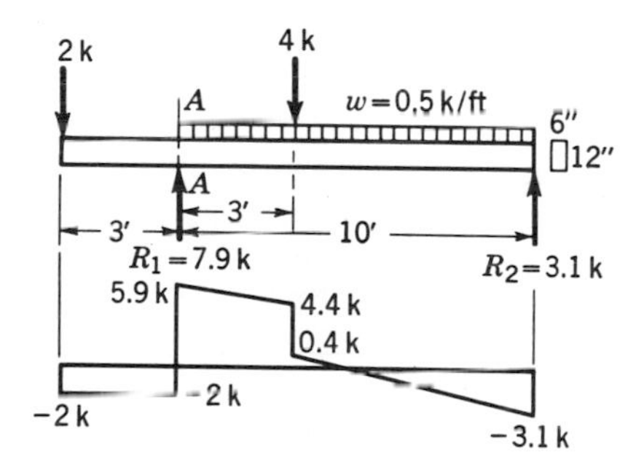

Figure 7-49

SOLUTION. The total shear V acting on section *A-A* (just to the right or R_1) is equal to 5.9 kips.

The moment of inertia of the section about the neutral axis is equal to

$$I = \tfrac{1}{12}bd^3 = \tfrac{1}{12} \times 6 \times 12^3 = 864 \text{ in.}^4$$

The values of $\bar{y}$ and a for each of the four points at which the horizontal shear stress is desired are shown in Fig. 7-50 and are tabulated as follows:

Above Neutral Axis	
0 in.	$a = 36$ in.^{2}
	$\bar{y} = 3$ in.
2 in.	$a = 24$ in.^{2}
	$\bar{y} = 4$ in.
4 in.	$a = 12$ in.^{2}
	$\bar{y} = 5$ in.
6 in.	$a = 0$ in.^{2}
	$\bar{y}$ = meaningless

The values of the horizontal shearing unit stress at each of these four

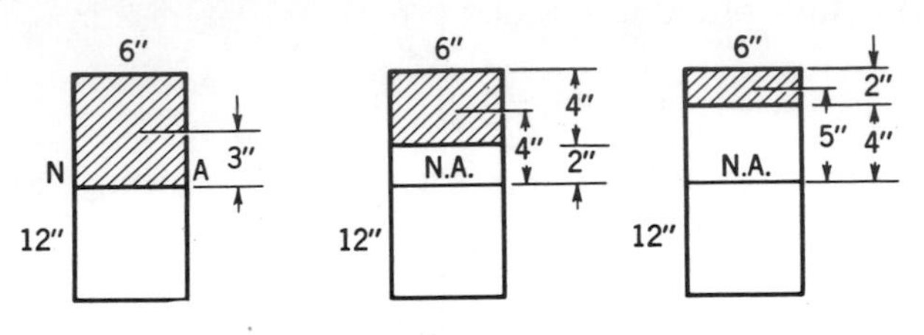

Figure 7-50

points can now be found by substituting the correct value in the shear equation just derived.

Above Neutral Axis

0 in. $\quad \tau = \dfrac{5.9 \times 1000}{864 \times 6} \times 36 \times 3$

$= 123$ psi

2 in. $\quad \tau = \dfrac{5.9 \times 1000}{864 \times 6} \times 24 \times 4$

$= 109.2$ psi

4 in. $\quad \tau = \dfrac{5.9 \times 1000}{864 \times 6} \times 12 \times 5$

$= 68.3$ psi

6 in. $\quad \tau = 0$, since $a = 0$

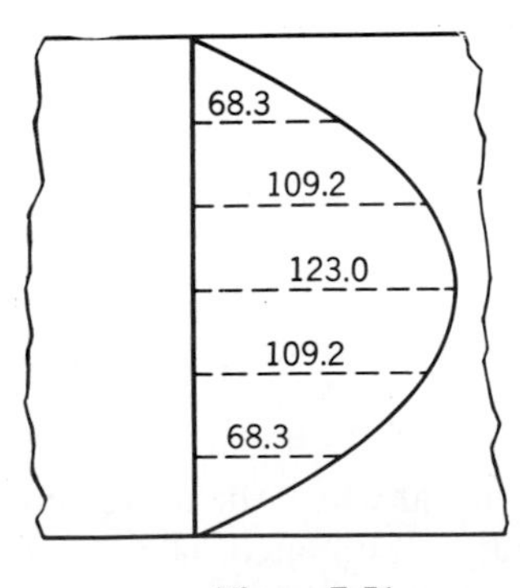

Figure 7-51

Plotting these values on a graph, we obtain a parabolic curve as shown by Fig. 7-51. Had the values of shearing unit stress been obtained for the corresponding points below the neutral axis, we should have found correspondingly equal magnitudes. By completing the curve, it will be noted that the maximum value of horizontal shearing unit stress occurs on the neutral plane, where $a\bar{y}$ is a maximum and the bending stresses are equal to zero.

7-14 Expression for Maximum Shearing Stress in Rectangular Beams

Because of its frequent use in design, an expression for the maximum horizontal shearing stress occurring in *solid* rectangular beams may be derived from Eq. (7.6).

$$\tau_{\max} = \frac{V}{Ib} a\bar{y}$$

$$= \frac{V}{\frac{1}{12}bd^3 b} \times \frac{bd}{2} \times \frac{d}{4} \tag{7.7}$$

$$= \frac{3V}{2bd} = \frac{3}{2}\frac{V}{A}$$

where A is the entire cross-sectional area of the beam.

Thus, the maximum horizontal shearing stress for Illustrative Problem 8 might have been computed as

$$\tau_{\max} = \frac{3}{2}\frac{V}{A}$$

$$= \frac{3}{2}\frac{5.9 \times 1000}{6 \times 12} = 123 \text{ psi}$$

7-15 Relationship Between Vertical and Horizontal Shearing Stresses

In Chapter 5 there was developed a primary law of strength of materials concerning the relationship of the shearing stresses acting at any point on a loaded body. The law stated that "if a plane passing through a point in a body has acting upon it a shearing unit stress, there must be simultaneously induced a shearing unit stress of like magnitude on a plane perpendicular to it." Thus, in our computation of the horizontal shearing stress at any point in a beam, the value obtained applies equally well to the magnitude of the vertical shearing unit stress occurring at the same point.

Because of the similarity in magnitude of the horizontal and vertical shearing unit stresses and because the horizontal as well as the vertical shearing stress varies parabolically with a maximum value of $1.5V/A$ occurring at the neutral surface, it is apparent that the vertical shearing stress given by the equation

$$\tau = \frac{V}{A}$$

where A is the total area of the cross section, is merely an average unit stress acting on a vertical cross section.

7-16 Shear Stresses in I Sections*

Beams made of standard structural shapes that are symmetrical to the plane of loading and whose boundaries are either perpendicular or parallel

*The I sections used in this discussion include all sections with this general shape, viz., S, W, M, etc.

to this same plane have horizontal shear stresses which may be assumed to be uniformly distributed across the width of the beam. The derivation of the shear-stress equation is based upon this assumption which permits its use in the determination of shear stress on S- and W-beam sections.

Inasmuch as the magnitude of the induced shearing stress varies inversely with the thickness or breadth of the section, it is to be expected that the shearing unit stress in the flanges should be small. Conversely, the shearing unit stress in the web should be relatively high. Multiplying the shear stress in the flange and in the web by their respective areas, it is noted that by far the greater percentage of the shearing resistance is developed in the web. In common practice, therefore, the shearing resistance of an S beam is assumed to be taken by the web only. The approximate shearing unit stress computed on this basis will then be equal to

$$\tau_{av} = \frac{V}{td} \tag{7.8}$$

where t = thickness of web and d = depth of S beam.

A comparison of the exact and approximate methods of shear-stress determination on an I-beam section will be made in the following problem.

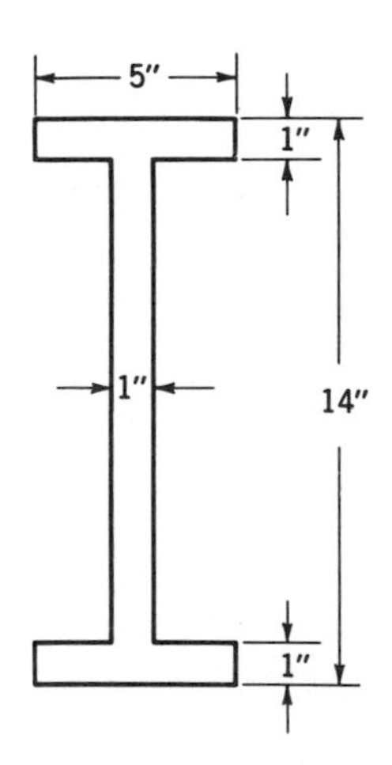

Figure 7-52

Illustrative Problem 9. Determine the shear-stress distribution curve for the I-beam section* shown in Fig. 7-52 by the exact and approximate methods. $V = 20{,}000$ lb.

SOLUTION.

Moment of inertia.

$$I = \tfrac{1}{12} \times 5 \times \overline{14^3} = 1142$$
$$-2 \times \tfrac{1}{12} \times 2 \times \overline{12^3} = -576$$
$$= 566 \text{ in}^4.$$

Values of $a\bar{y}$. Rather than obtain the values of a and $\bar{y}$ separately, it will be easier to determine their product from its equal $\Sigma\,\Delta ay$.

The moments of the areas ($\Sigma\,\Delta ay$) located 0, 2, 4, and 6 in. above and below the neutral axis taken about the same axis are given below. Pertinent cross-sectional areas are shown in Fig. 7-53.

Values of shearing unit stress.

EXACT METHOD

At neutral axis:

$$\tau = \frac{20{,}000}{566 \times 1 \text{ in.}} \times 50.5 = 1785 \text{ psi}$$

*Indicated as an I-beam section because it resembles an I and is not a standard S section.

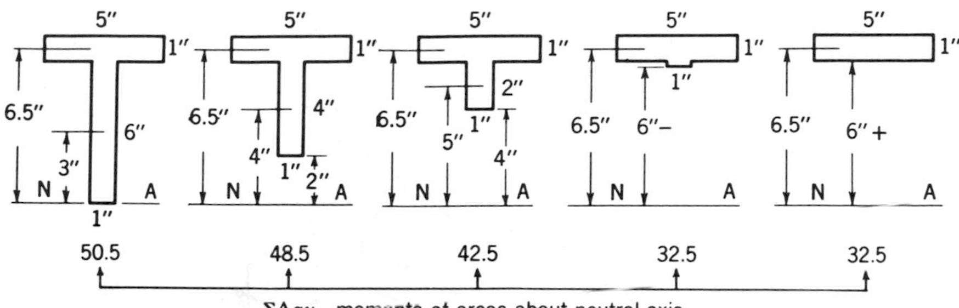

Figure 7-53

At 2 in. above neutral axis:

$$\tau = \frac{20{,}000}{566 \times 1 \text{ in.}} \times 48.5 = 1715 \text{ psi}$$

At 4 in. above neutral axis:

$$\tau = \frac{20{,}000}{566 \times 1 \text{ in.}} \times 42.5 = 1500 \text{ psi}$$

At 6— in. above neutral axis:

$$\tau = \frac{20{,}000}{566 \times 1 \text{ in.}} \times 32.5 = 1150 \text{ psi}$$

At 6+ in. above neutral axis:

$$\tau = \frac{20{,}000}{566 \times 5 \text{ in.}} \times 32.5 = 230 \text{ psi}$$

Approximate Method

$$\tau = \frac{V}{td} = \frac{20{,}000}{1 \text{ in.} \times 14 \text{ in.}} = 1430 \text{ psi}$$

By plotting the values obtained, we may draw the shear-stress distribution curves shown in Fig. 7-54.

The proportion of the shear resistance taken by the flanges, assuming a straight-line variation of stress from the inside to the outside fibers, is

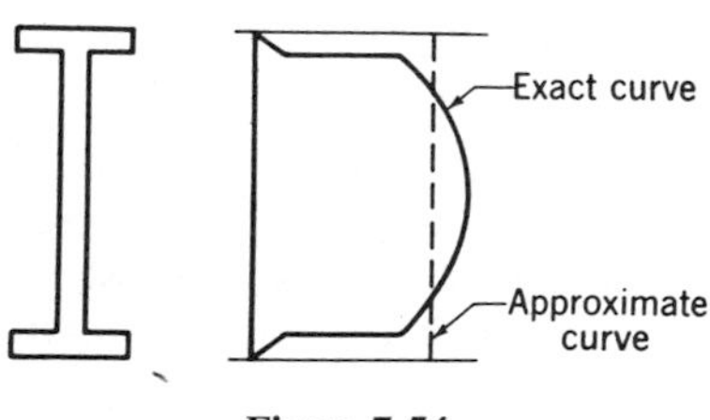

Figure 7-54

$$V_f = \tfrac{1}{2} \times 230 \times 1 \times 5 \times 2 = 1150 \text{ lb}$$

$$\text{per cent shear} = \frac{1150}{20{,}000} \times 100 = 5.74 \text{ per cent}$$

Shear on vertical plane. If the area on which a shear stress is desired lies in a vertical rather than in a horizontal plane, then the shear force acting on that plane is determined, as before, by considering the unbalanced flexural forces

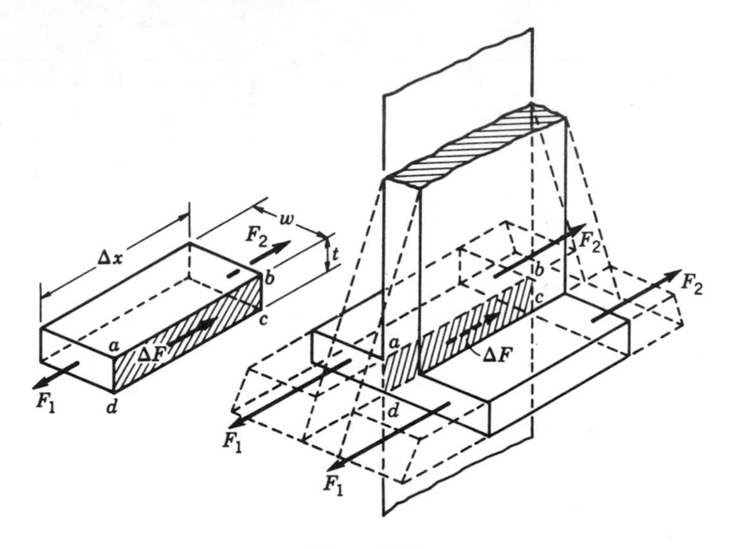

Figure 7-55

which it holds in equilibrium. Thus, in Fig. 7-55 the average shearing stress on the vertical section *abcd* is obtained by equating it to $(F_1 - F_2)/A_{abcd}$ or using $\tau = (V/It)ay$, where ay is equal to the moment of area wt about the neutral axis and t is the thickness *ad*.

PROBLEMS

7-54a, b, c, d. Determine the maximum vertical shearing unit stress in each of the beams shown in Fig. 6-13. Use S beams selected on the basis of allowable flexure stress of 20,000 psi. Use Eq. (7.8).

7-55a, b, c, d, e, f, g, h. Determine the maximum horizontal shearing unit stress for the beams shown in Fig. 6-19. Use American Standard timber sections designed on the basis of 1200 psi flexural stress.

7-56. A timber beam 6 in. wide, 12 in. deep, and 12 ft long is supported at its ends and carries a concentrated load of 10,000 lb at its midpoint. Compute the horizontal shearing unit stresses at 2-in. intervals from the neutral axis on the most critical shear section. *Answer:* On N.A.: 104 psi, 2 in. above; 92.5 psi, 4 in. above; 57.8 psi.

7-57. Determine the average and exact maximum shearing unit stress in a 15-ft steel beam having a cross section as shown in Fig. 7-56 and loaded uniformly with 1000 lb/ft.

7-58. An airplane spar is made as shown in Fig. 7-57. (a) What is the maximum shearing unit stress on this section if V is 2600 lb? (b) What will the average shearing unit stress be on the glued surfaces of the 2- by 2-in. blocks? *Answer:* 228 psi.

7-59. Two 6- by 6-in. beams are bolted together to form one beam 6 in. wide,

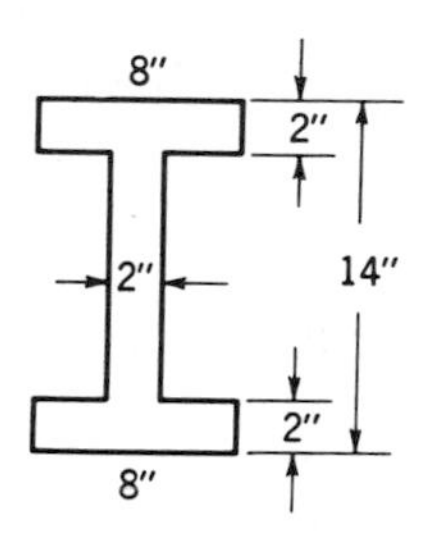

Figure 7-56

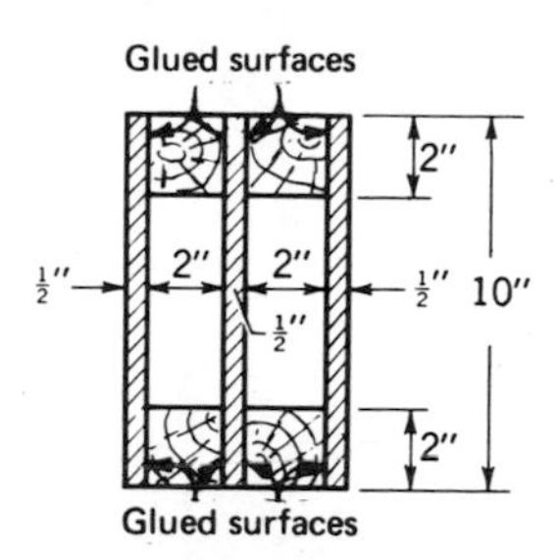

Figure 7-57

12 in. deep, and 10 ft long. If the bolts are placed 4 in. apart in one row and the beam is loaded with a concentrated load at the center to cause a maximum flexural stress of 1800 psi, what is the total shearing force on one bolt?

7-60. If a rectangular cross-sectioned simple beam is loaded at its center with a concentrated load and is b in. in width and d in. in depth, compute the critical depth above which flexure governs and below which shear is the controlling factor. *Answer:* $d = 2l\tau/\sigma$.

7-61. A beam having the cross section shown in Fig. 7-58 is 18 ft long and is freely supported at points 4 ft from either end. It carries a uniformly distributed load of 1000 lb/ft. Find the intensity of the maximum shear stress. *Answer:* 228 psi.

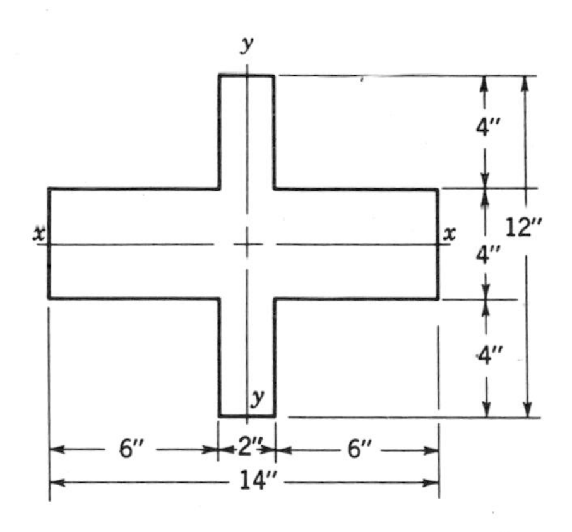

Figure 7-58

7-62. What would be the ratio of length to depth for a beam of rectangular cross section on a simple span supporting a single concentrated load at the center of span, if the maximum shear stress is 8000 psi and maximum flexural stress is 12,000 psi? Length of the beam is measured in feet and depth is measured in inches.

7-17 Shear Center

A prismatic beam subjected to bending will twist about its longitudinal axis (Fig. 7-59) whenever the internally developed shear forces produce an

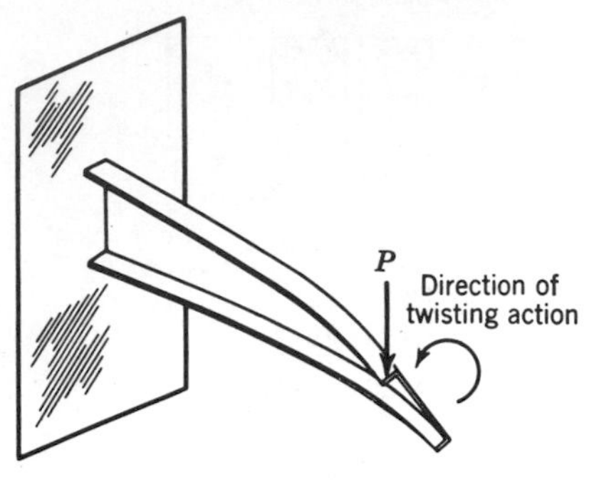

Figure 7-59

unbalanced moment about that axis. To overcome such undesirable twisting action, the resultant of the bending forces should pass through the *shear center*. Where this is done, the twisting moment created by the internal shear forces is balanced by an equal and oppositely directed moment created by the applied forces.

7-18 Analysis of Internal Shear Forces—I Section

In order to firmly establish the basic principles upon which the concept of a shear center rests, let us analyze the development of the internal shear forces in a horizontal beam of *I* cross section acted upon by vertical forces passing through the centroidal axis of its cross sections (Fig. 7-60). Note that

1. The maximum flexural stresses developed by such forces vary in magnitude from section to section in accordance with the variation of the moment diagram.

2. The variation of the flexure stresses acting on two arbitrarily selected cross sections located dx distance apart is proportional to the distance from the neutral axis and is indicated by the dash lines shown on Fig. 7-60c.

Consider now the four free bodies, *E*, *F*, *G*, and *H*, taken from the flanges of the I beam, each Δs in width. Because of the difference in the flexure stresses over the dx distance, the longitudinally directed normal forces in these free bodies differ by the amount dT or dC. To insure equilibrium in the longitudinal direction, these free bodies must have acting on those planes parallel to the vertical plane of symmetry an oppositely directed *shear force* equal to dT or dC. The free bodies are still not balanced, however, because the longitudinal forces are not collinear and produce couples equal to $\frac{1}{2}(dT \times \Delta s)$ or $\frac{1}{2}(dC \times \Delta s)$. To complete the balance of the free body, lateral shear forces V_F are developed in the flanges, the moment of which equals that of the longitudinal forces. It is the development of these laterally directed shear forces that initiates the concept of a shear center.

It is to be noted that each V_F force varies directly with the Δs distance used. It will have a value of zero at the tip of the flange and a maximum value at the junction of the flange and web. The vertical shear force acting on the web varies parabolically, as proved in §7-15 (see Fig. 7-54).

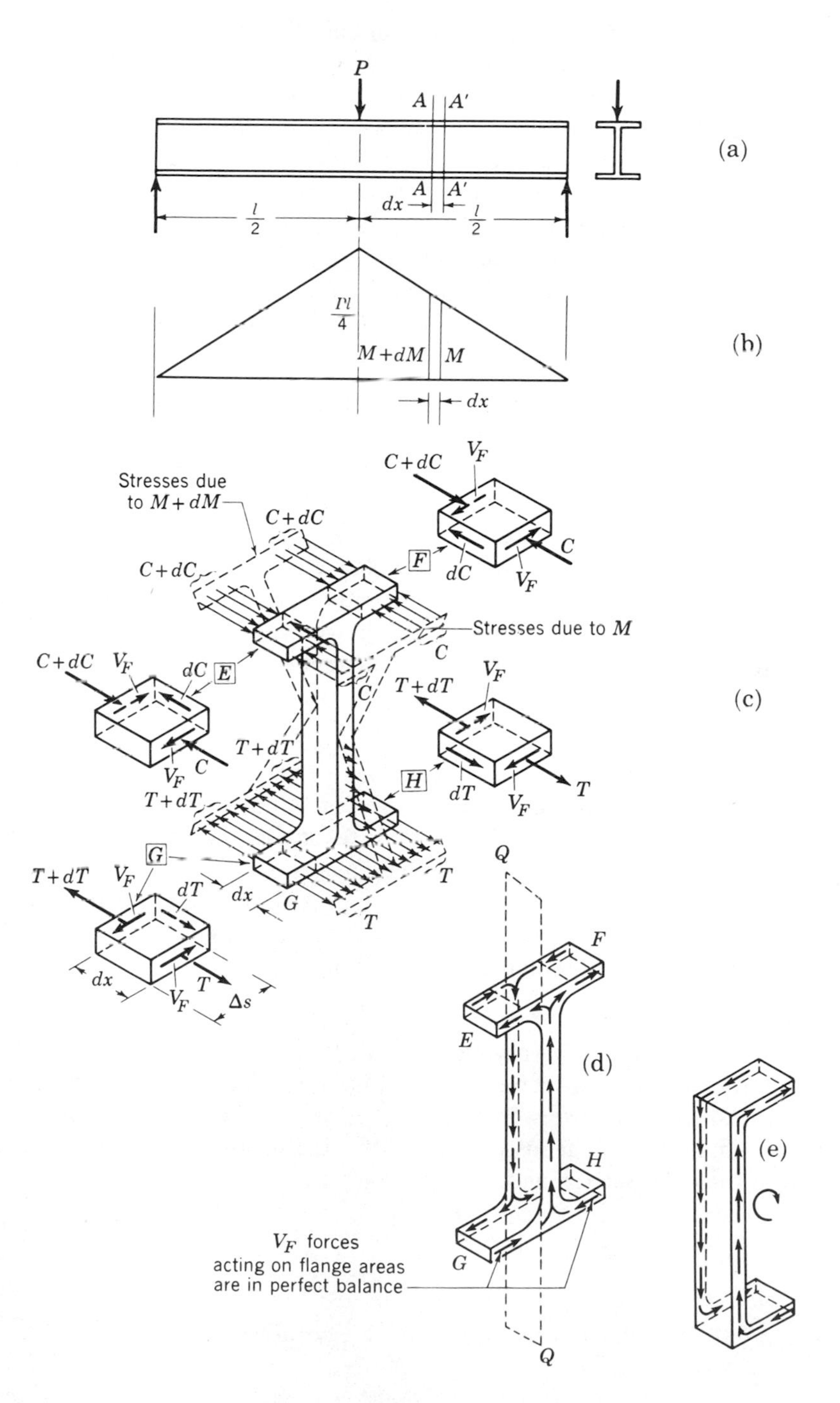
P
A
A'
(a)
l/2
dx
l/2
Pl/4
M+dM
M
(b)
dx
Stresses due to M+dM
C+dC
C+dC
V_F
F
dC
C
Stresses due to M
E
H
G
T+dT
dT
T
(c)
dx
Δs
Q
E
F
(d)
H
G
V_F forces acting on flange areas are in perfect balance
(e)

Figure 7-60

In Fig. 7-60d are shown the location and direction of the shear forces acting on either side of the elementary portion of the I beam. Because the V_F forces on the opposite sides of each flange area act in opposing directions and are equal to each other, perfect balance is obtained in the horizontal direction. Furthermore, the moments of these V_F forces acting on either cross section about any point in their plane are in perfect balance.

To insure the continuance of this perfect balance of shear forces (which is synonymous to saying that we will have no twisting action), the force P acting on the beam must act in the vertical plane of symmetry. To do otherwise would produce a moment about the longitudinal axis which could be equalized only by the development of further shear forces.

7-19 Shear Forces on a Section Unsymmetrical about the Vertical Axis

Suppose now the left side of each flange is removed, changing the I cross section to that of a channel (Fig. 7-60e). Assuming the flexural stresses to remain the same, the shearing forces are not altered and act in the same direction. The shear forces in the flanges, however, because of dissymmetry, are no longer in balance and produce a moment twisting the beam.

7-20 Shear Center—Channel Section

This twisting action can easily be overcome, however, by offsetting the load P from the center line of the web a distance e so that its moment about a point on that center line will equal the moments of the induced shear forces (Fig. 7-61). The balance of the forces involved is shown most clearly on Fig. 7-62, where the force P is resolved into a force and a couple, P_1 and PP_2. Mathematically, the rotational balance is expressed by the equation

$$Pe = 2V_F h \tag{7.9}$$

where V_F is the uniformly varying shear force acting on each flange section and h is the distance between them. See Fig. 7-61. The resolved direct force P_1 is restrained by the two web shears V_w, each equal to $P/2$. The value of e is written most advantageously in terms of the dimension and physical constants of the cross section as follows:

$$e = \frac{2V_F h}{P} \tag{7.10}$$

$$V_F = \tfrac{1}{2}\tau \times b \times t$$

But in §7-13 the value of shearing stress in a beam was found to be

$$\tau = \frac{V}{It} a\bar{y}$$

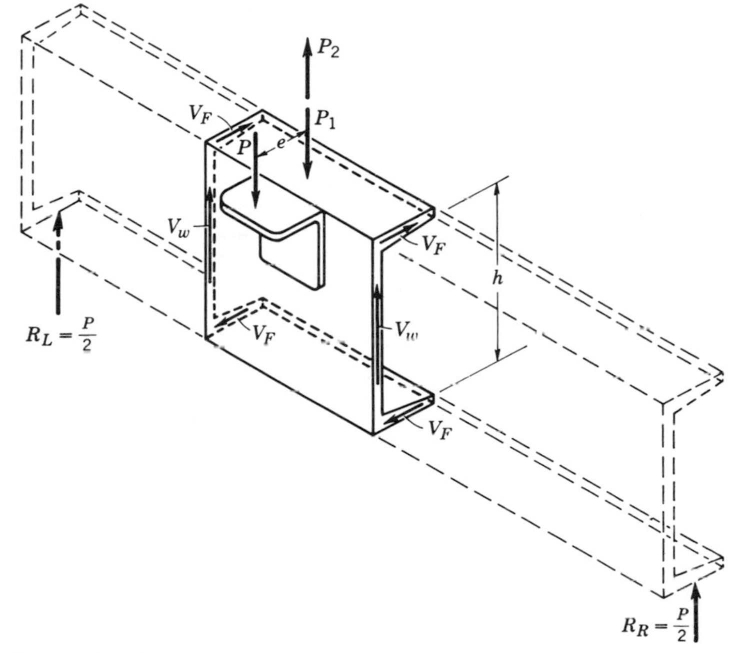

Figure 7-61 Free body of center portion of beam showing how couple Pe is balanced by the two $V_F h$ couples.

where the width of the area under shear has been made equal to t in order to coincide with the presentation now being considered.

Substituting the last two equations into Eq. (7.10), there is obtained

$$e = \frac{[(V/It)a\bar{y}]bth}{P}$$

But $V = P/2$, and $a\bar{y}$ for shear stress at the junction of the flange and web is equal to $bt \times h/2$. Further substitution yields

$$e = \frac{1}{4}\frac{b^2h^2t}{I} \tag{7.11}$$

Were the same beam to be loaded along the horizontal axis of symmetry, balance between the shear forces would be obtained and no twisting would result. The intersection of this axis of symmetry and the vertical line, on which the load would have to be applied to overcome twisting, locates the *shear center*. A load passing through this point on the beam would not produce a tendency to twist.

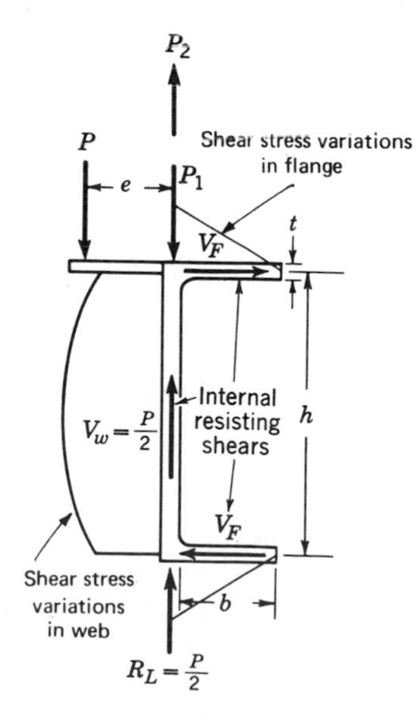

Figure 7-62 End view of Fig.7-61.

The value of I for a channel, if its thickness is small and uniform, may be written as

$$I = 2bt\left(\frac{h}{2}\right)^2 + \frac{t \times h^3}{12} = \frac{th^2}{4}\left(2b + \frac{h}{3}\right) \tag{7.12}$$

Placing this value in Eq. (7.11) gives

$$e = \frac{b^2}{2b + (h/3)} = \frac{b}{2 + (h/3b)} \tag{7.13}$$

It is important to note that Eq. (7.13) is independent of the method of loading or type of beam, i.e., simple cantilever, continuous. It consists of dimensional values only and is a constant for any channel section. It further reveals that e for a channel will never be greater than one half the width of the flange.

7-21 Shear Center of Angle Cross Section

An analysis similar to that used for the channel is employed for determining the shear center of an angle cross section. Let an angle be loaded as a simple beam with its flanges tilted with respect to the horizontal (Fig. 7-63).

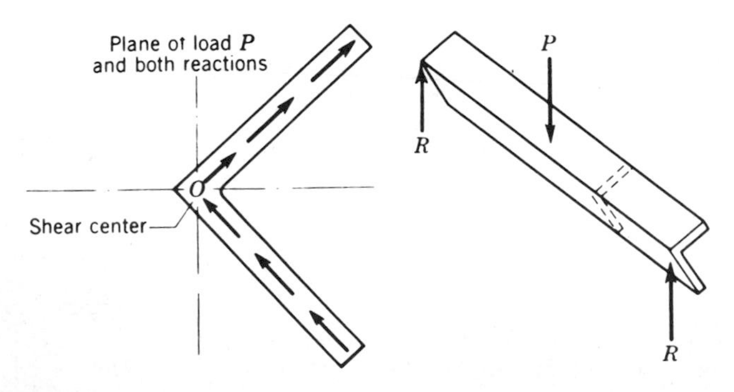

Figure 7-63 All externally applied loads must pass through point O to prevent twisting.

By taking progressively larger free bodies from the legs of the angle, it is noted that the shear stresses vary directly with the distance to the tips and travel along the legs of the cross section from one end to the other. Because the resultant shear forces in the legs intersect at O, their moment about this joint is zero. An applied force acting through point O will also have a zero moment. There being a perfect balance between the moments of the external and internal forces, no twisting will occur. Point O is, therefore, the shear center of the angle.

7-22 Shear Center—Other Rolled Sections

From previous statements made concerning the development of internal shearing forces, it follows that shear on any structural cross section which might be made by bending a single thin plate (i.e., a channel, Z beam, etc.) flows from one end to the other without reversing itself. The flow of shear in a Z section, for example, is continuous, the direction of shear in each flange being the same (Fig. 7-64). If the flanges are of the same length, each will have the same shear force and produce equal but opposite moments about the center of the web. Should the applied force pass through this point, no matter how directed, it could produce no twisting.

Frequently, a beam cross section, consisting of two or more thin rectangular portions placed at right angles to one another, is formed in the shape of T, H, etc., with one axis of symmetry. If loads are applied perpendicular to the plane of symmetry, the internal shear force will be concentrated in the vertical portions of the cross section and directed parallel to their long axis. Any horizontal portion located on the neutral axis where the flexure stresses are zero has an insignificant amount of shear induced in it. The shear center of a T section is, therefore, located at the junction of the two sections

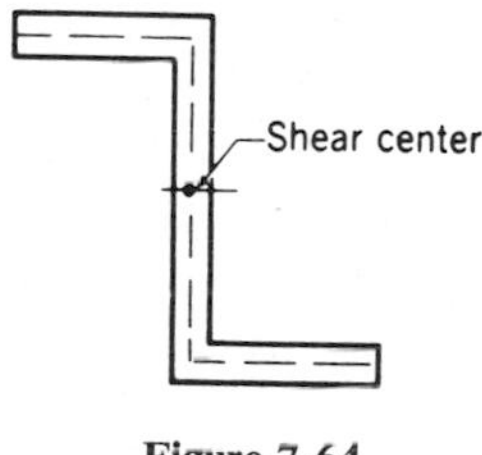

Figure 7-64

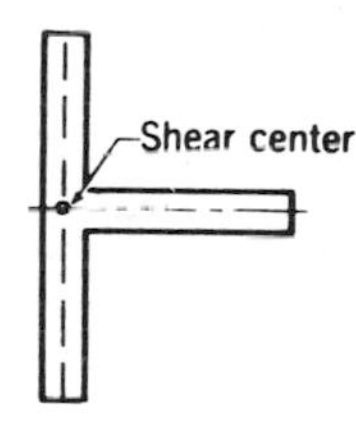

Figure 7-65

(Fig. 7-65). That for the irregular H section, shown in Fig. 7-66, is located at the centroid of the shear forces, where

$$V_1 h_1 = V_2 h_2 \tag{7.14}$$

Because both flanges have the same radius of curvature while being bent, their flexure stresses will reveal the same variation from the neutral axis. Thus, $\sigma_1/c_1 = \sigma_2/c_2$. If now the flexure formulas $\sigma_1 = M_1 c_1/I_1$ and $\sigma_2 = M_2 c_2/I_2$ are substituted, there results

$$\frac{M_1}{I_1} = \frac{M_2}{I_2}$$

where M_1, I_1 and M_2, I_2 refer to the left- and right-hand flanges, respectively (Fig. 7-66).

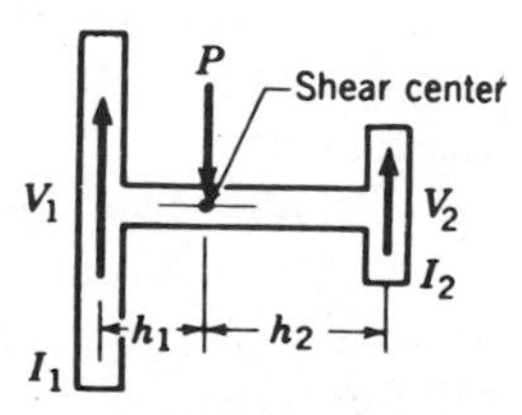

Figure 7-66

If the beam is loaded so as to produce a

constant shear from one end of the beam up to the loading point, covering a distance x_1, $M_1 = V_1 x_1$, and $M_2 = V_2 x_1$, the previous equation may be modified to read

$$\frac{V_1}{I_1} = \frac{V_2}{I_2} \tag{7.15}$$

Substituting V_1 and V_2 with h_1 and h_2 from Eq. (7.14) yields the relationship

$$\frac{h_2}{h_1} = \frac{I_1}{I_2} \tag{7.16}$$

7-23 Shear Center—Solid Sections

The shear stresses which act on solid cross sections often change in both direction and magnitude, causing greater difficulty in locating the shear center. Whereas in the previous discussions of thin-walled and rectangular sections, the shear stress could be assumed to act uniformly, the same assumption applied to wide sections having nonvertical sides could produce considerable error. For instance, to assume a uniform distribution of horizontal shear in a cylindrical beam (Fig. 7-67) would also require the corresponding vertical shear to be uniformly distributed and directed vertically. But this is impossible, for at each end of the horizontal chord CD of this circular cross section located a distance y from its neutral axis, the shear stress is not vertical but is tangent to the surface. To have any component of shear perpendicular to the surface of the cylinder would require the application of a resisting shear on the surface of the cylinder. This is an impossibility and compels the conclusion that the shear stress on a vertical section can neither be uniformly distributed nor directed vertically. Instead, it will form a converging symmetrical pattern (Fig. 7-67) about the vertical axis of symmetry, the shear flowing tangentially at the periphery of the cross section. Because of the nonuniform pattern of the shear on the vertical section, it follows that the horizontal shear on any horizontal plane y distance from the neutral axis will also be nonuniformly distributed.

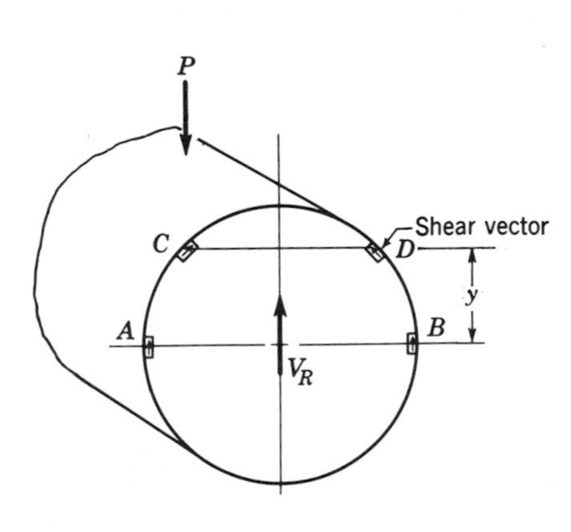

Figure 7-67

It is easy to see after this discussion that to find the shear center of any solid cross section is a difficult task. A complete analysis of this subject is beyond the scope of this text.

If, however, an axis of symmetry does exist, it is evident that the shear center will be in this axis. Furthermore, if two axes of symmetry exist, the shear center will be located at their intersection. The shear center of the circular cross section, previously discussed, will therefore be located at its centroid.

PROBLEMS

7-63a, b, c, d. Locate the shear center of the channel sections shown in Fig. 7-68. Use Eq. (7.11), (7.12), or (7.13) or modification thereof to obtain the most accurate solution possible. *Answers:* (a) 1.00 in.; (b) 0.40 in.; (c) 0.65 in.; (d) 2.88 in.

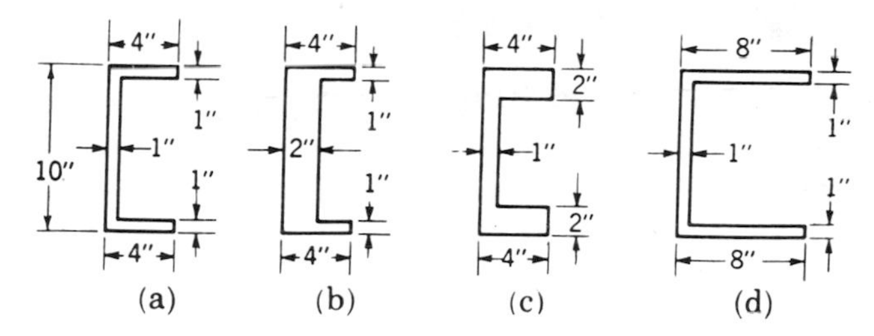

Figure 7-68

7-64. Draw a curve showing the change in location of the shear center as the flanges of the channel shown in Fig. 7-68a vary in length. Use Eq. (7.13).

7-65. Draw a curve showing the change in location of the shear center as the web of the channel shown in Fig. 7-68a changes in thickness. Use Eq. (7.12).

7-66. Determine the shear center of the cross section shown in Fig. 7-69. *Answer:* 10.25 in.

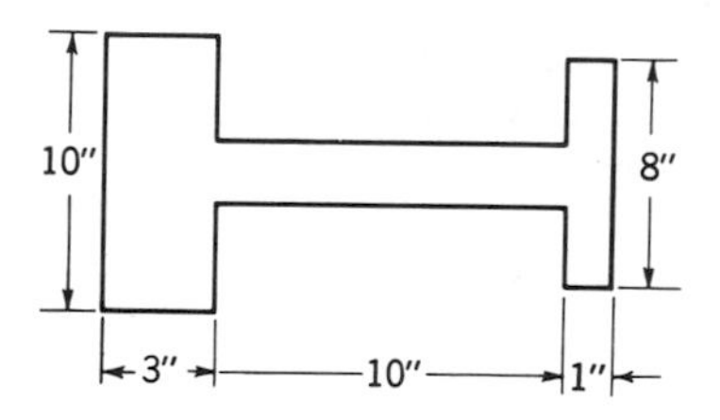

Figure 7-69

8

The Deflections of Statically Determinate Beams

8-1 Reason for Finding Deflections

Quite frequently, the design of a beam is governed by the amount of its permissible deflection. For example, beams that carry plaster ceilings are generally limited to a deflection of $\frac{1}{360}$ of their length. To exceed this value would encourage the formation of cracks. Various parts of accurate metal-shaping equipment, such as milling machines, lathes, boring tools, and so forth, would, if designed solely on the basis of strength, fail to function properly. The resulting deflections would allow distortions of such magnitude as to make precision work an impossibility.

8-2 Fundamental Considerations in Determining Deflections

As previously revealed, a loaded beam must bend. The degree of bending thus incurred varies with the intensity of the internal stresses produced. Because shear and flexure stresses are generally present on every cross section of a beam, both contribute to its deflection. Except for short stubby beams, however, the deflection due to shearing stresses is so small that it is ordinarily neglected. This discussion of the deflection of loaded beams will therefore include only that portion produced by flexure stresses.

8-3 Determination of the Radius of Curvature of a Beam

When a beam deflects, the vertical projection of the neutral plane assumes a curved position, which is known as the *elastic curve*. The radius of curvature at any point on the elastic curve is equal to the radius of that circle whose circumference conforms to the shape of the elastic curve at that point.

In the beam of Fig. 8-1, the length dl represents a minute portion of the elastic curve included in an imaginary circle of which OA and OB are radii.

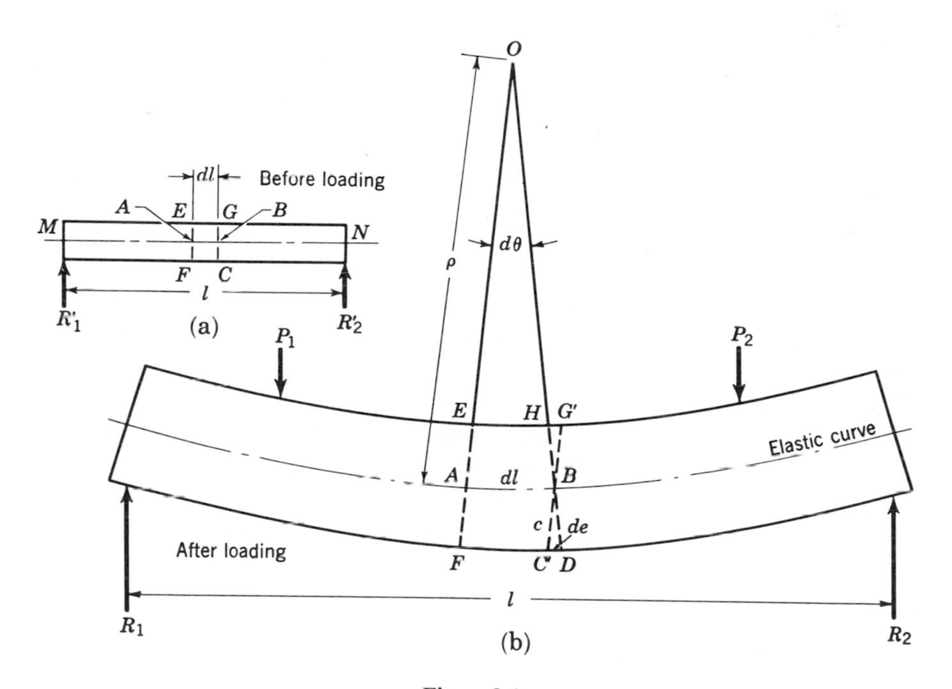

Figure 8-1

Prior to loading, the beam was straight, and cross sections EF and GC were parallel to each other and perpendicular to the longitudinal axis of the beam (Fig. 8-1a). When the loads were applied, section GC rotated with respect to section EF and assumed the position HD. During this rotation, sections EF and HD remained perpendicular to the elastic curve.

Line $G'C'$ drawn parallel to EF locates, therefore, the terminal position of the fibers included in the dl section before bending took place. The small triangles $HG'B$ and $BC'D$ represent the deformation of the fibers that took place during the bending of the beam.

Considering, now, the similar triangles OAB and $BC'D$, we may write the following proportion:

$$\frac{de}{dl} = \frac{BC'}{OA}$$

But de/dl is equal to the unit strain ϵ incurred in fiber FC' while elongating to length FD. Moreover BC' equals c by observation, and OA equals ρ, the radius of curvature at A and B. Inserting these values in the equation above, we obtain

$$\epsilon = \frac{c}{\rho} \tag{8.1}$$

The value of ϵ is also equal to σ/E and from the flexure equation, $c = \sigma I/M$. The substitution of these values in Eq. (8.1) yields

$$\frac{1}{\rho} = \frac{M}{EI} \tag{8.2}$$

where ρ = radius of curvature in inches, M = bending moment at section where ρ is desired in in.-lb, E = modulus of elasticity in psi, I = moment of inertia in in.^{4}.

This equation is known as the *elastic-curve equation expressed in terms of the radius of curvature*. Besides being a basic equation in the development of deflection formulas, it has considerable value in analyzing moment-deflection relationships. For instance, if a beam of uniform cross section (I being constant) made of a homogeneous material (E being constant) were to have a constant value of bending moment M applied to any portion of its length, the value of ρ for such a portion would have to be a constant value; and because a constant value of ρ pertains to the radius of a circular curve, the elastic curve of each constant moment section must be a circular segment (Fig. 8-2). On the other hand, any portion of the member bent into a circular curve would necessarily be subjected to a constant bending moment. For instance, if a steel band-saw blade were bent into a circular curve, as shown in Fig. 8-3, the bending moment at every section would be equal. A special

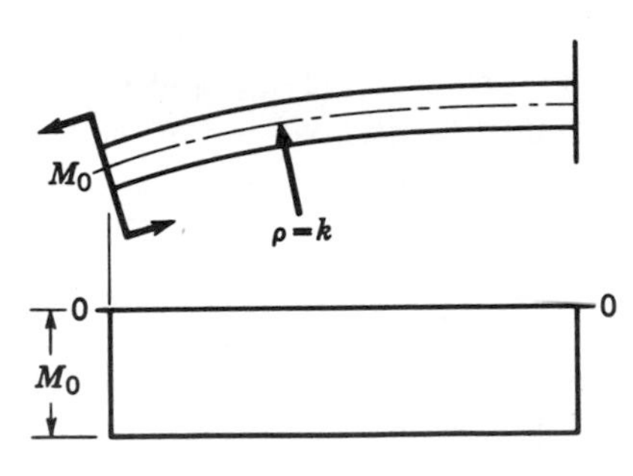

Figure 8-2

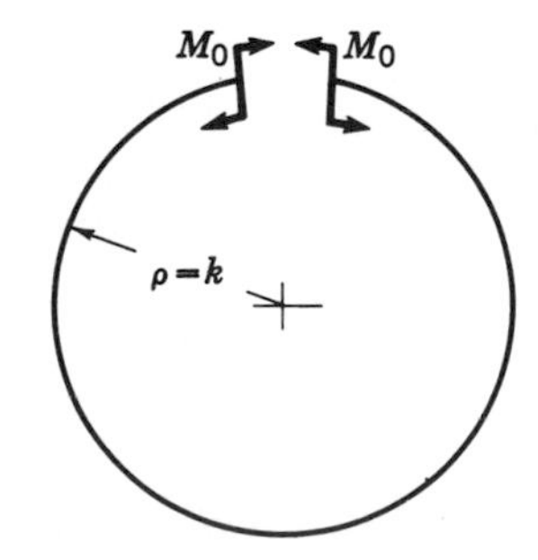

Figure 8-3 The resisting moment at each point on the ring equals M_0.

case to consider in this discussion is that obtained when $M = 0$, i.e., in the development of an *inflection point* (§6-12). Here the value of ρ equals infinity and indicates that the beam is perfectly straight at that point.

PROBLEMS

In the following problems, $E = 30 \times 10^6$ psi for steel.

8-1. A steel band-saw blade $\frac{1}{16}$ in. thick and 1 in. wide moves on the periphery of a circular pulley 30 in. in diameter. What is the maximum flexural stress in the saw blade?

8-2. What value of I would the steel beam in Fig. 8-4 require to maintain its center portion at a radius of 500 ft? Of what importance is the induced maximum flexural stress to the computations of this problem? *Answer:* 48 in.^{4}.

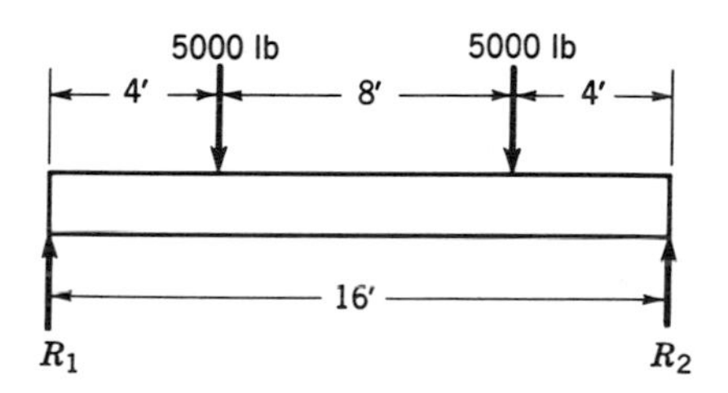

Figure 8-4

8-3. A high-strength steel wire $\frac{1}{8}$ in. in diameter is coiled on a drum 5 ft in diameter. What is the extreme fiber unit stress in this wire?

8-4. A piece of spring steel $\frac{1}{2}$ in. thick, 1 in. wide, and 20 ft long is bent and fastened alongside a piece of circular track having a radius of 20 ft. If the 1-in. side is placed against the track, what is the maximum flexural stress in the steel? *Answer:* 31,250 psi.

8-5. For the cantilever beam shown in Fig. 8-5, find the deflection and slope at A and B in inches and radians, respectively. $I = 50$ in.^{4}, $E = 30 \times 10^6$ psi.

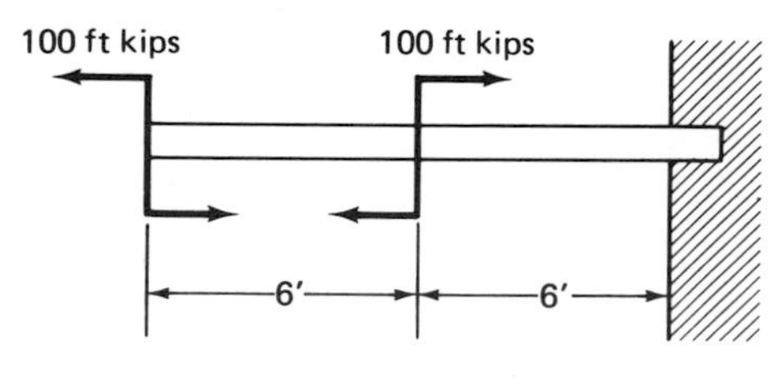

Figure 8-5

8-6. What value of M will be required in the beam of Fig. 8-6 ($l = 6$ ft) to produce an offset at the center of 1 in. $E = 2 \times 10^6$ psi. $I = 50$ in.^{4}. *Answer:* 12.85 ft kips.

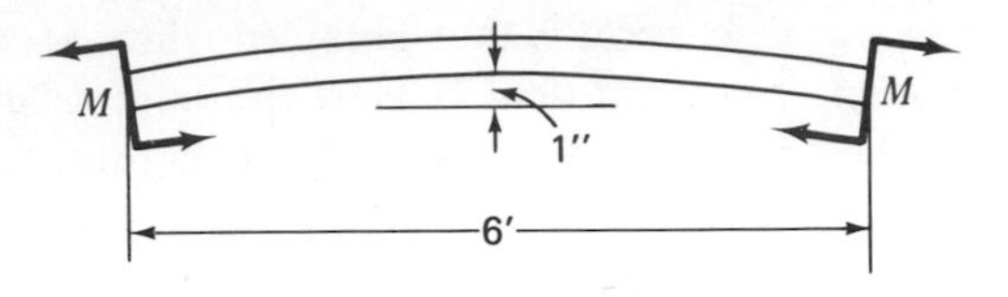

Figure 8-6

8-4 Double-Integration Method of Finding Slopes and Deflections

A classical method for determining slopes and deflections of loaded beams is the double-integration method. Its basic equation is derived from the elastic-curve equation expressed in terms of the radius of curvature.

In a slightly more detailed drawing of Fig. 8-1, shown in Fig. 8-7, dl is

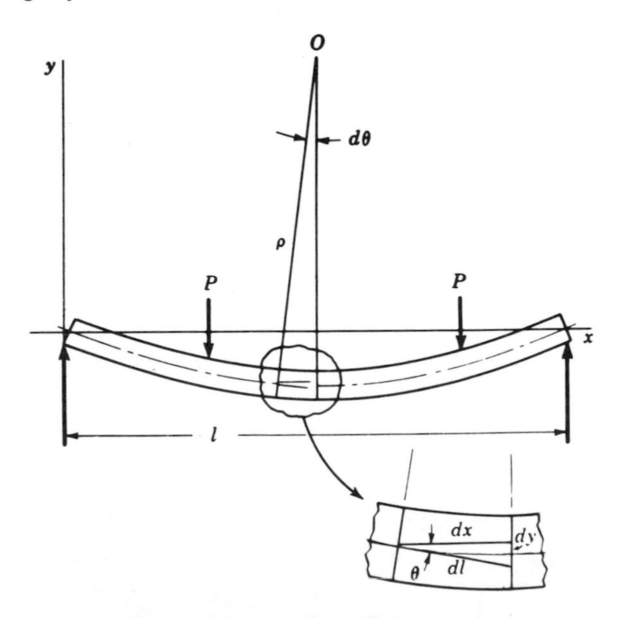

Figure 8-7

approximately equal to dx. Thus, we may write without inducing any appreciable error

$$dx = dl = \rho d\theta, \quad \text{from which} \quad \frac{1}{\rho} = \frac{d\theta}{dx} \tag{8.3}$$

But

$$\theta = \frac{dy}{dx}, \quad \text{making} \quad \frac{d\theta}{dx} = \frac{d^2y}{dx^2} \tag{8.4}$$

By substituting Eq. (8.4) into Eq. (8.3), we obtain

$$\frac{1}{\rho} = \frac{d^2y}{dx^2} \tag{8.5}$$

With further substitution of Eq. (8.5) into the elastic-curve equation expressed in terms of the radius of curvature, we get

$$\frac{M}{EI} = \frac{d^2y}{dx^2} \tag{8.6}$$

This equation is known as the *elastic-curve equation expressed in rectangular coordinates* and forms the basis of the double-integration method. The name of this method is apparent from the fact that two successive integrations of Eq. (8.6) yield the equation of the elastic curve in x and y. From this equation the exact deflection at any point x along that portion of the beam for which M is applicable can be obtained by substituting the value of x and solving for y.

In a similar manner the slope of the elastic curve at any point along this portion of the beam can be obtained from its value of dy/dx produced after one integration.

Illustrative Problem 1. Determine the slope and deflection equations for the simple beam shown in Fig. 8-8 having a concentrated load at its center.

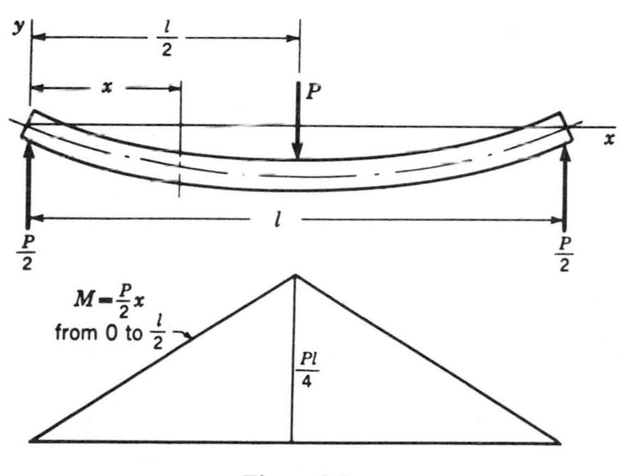

Figure 8-8

SOLUTION. Because the x and y axes were chosen as indicated, d^2y/dx^2 will be positive as x increases. The bending-moment equation for any point on the left-hand side of the beam can be expressed as

$$M = \frac{P}{2}x$$

When substituted into the elastic-curve equation, it gives

$$EI\frac{d^2y}{dx^2} = \frac{Px}{2}$$

Integration of this expression yields

$$EI\frac{dy}{dx} = \frac{Px^2}{4} + C_1$$

To obtain the value of C_1, substitute in the foregoing equation simultaneously known values of dy/dx and x occurring on the left-hand portion of the beam.* The only such point on the left-hand portion of the beam is at $x = l/2$, where $dy/dx = 0$.

Inserting these values in the previous equation and solving for C_1 makes C_1 equal to $-Pl^2/16$.

The dy/dx or *slope equation* for the left-hand portion of the beam is then equal to

$$EI\frac{dy}{dx} = \frac{Px^2}{4} - \frac{Pl^2}{16} \tag{8.7}$$

An additional integration gives

$$EIy = \frac{Px^3}{12} - \frac{Pl^2x}{16} + C_2$$

Using the same portion of the beam, the two simultaneously known values of y and x are found at $x = 0$, where y is also equal to zero. Substituting these values into the above equations makes $C_2 = 0$. Hence, *the deflection equation* for the left portion of the beam is

$$EIy = \frac{Px^3}{12} - \frac{Pl^2x}{16} \tag{8.8}$$

The maximum deflection of the beam occurs at the center where $x = l/2$. Thus, by substituting $l/2$ for x in Eq. (8.8), there results a deflection

$$\Delta = \frac{1}{48}\frac{Pl^3}{EI} \tag{8.9}$$

The deflection at any other point on the left-hand portion of the beam can be determined by inserting its corresponding value of x in terms of l into the elastic-curve equation. Similarly, the slope at any point can be obtained by inserting corresponding values of x into the general slope equation. For instance, the slope at the quarter point may be obtained by inserting the value of $l/4$ for x in Eq. (8.7), resulting in

$$\left(\frac{dy}{dx}\right)_{l/4} = \theta_{l/4} = \frac{1}{EI}\left[\frac{P}{4}\left(\frac{l}{4}\right)^2 - \frac{Pl^2}{16}\right] = -\frac{3}{64}\frac{Pl^2}{EI} \tag{8.10}$$

*Only the left-hand portion of the beam may be used in determining C_1, as the initial d^2y/dx^2 equation was formulated from the equation of moment pertaining to that portion of the beam alone.

PROBLEMS

8-7. Determine by the double-integration method the slope and deflection at the free end of a cantilever beam l ft long loaded uniformly with w lb/ft. Assume E and I constant. *Answers:* $wl^3/6EI$, $wl^4/8EI$.

8-8. Determine by double integration the slope and deflection at the free end of a cantilever beam l ft long with a concentrated load P at the free end. Assume E and I constant. *Answers:* $Pl^2/2EI$, $Pl^3/3EI$.

8-9. Determine by the double-integration method the slope and elastic-curve equations for a simple beam uniformly loaded. Also find the maximum deflection. Assume E and I constant. *Answer:* $\Delta = \frac{5}{384}(wl^4/EI)$.

8-10. In Illustrative Problem 1 determine the exact slope and deflection at a point $\frac{3}{8}l$ from the left support if $P = 10{,}000$ lb, $l = 10$ ft, and the section used is S 10×25.4.

8-11. Develop the procedure involved in using the double-integration method for finding the elastic-curve equation of a simple beam which restrains a concentrated load located off center.

8-12. A cantilever beam (Fig. 8-9) l ft long has a pointer, $l/2$ long, firmly affixed to its midpoint. Determine the vertical distance between the tip of the pointer and the free end of the beam in terms of P, l, E, and I. Use double integration. Use no memorized formulas. *Answer:* $\frac{1}{24}(Pl^3/EI)$.

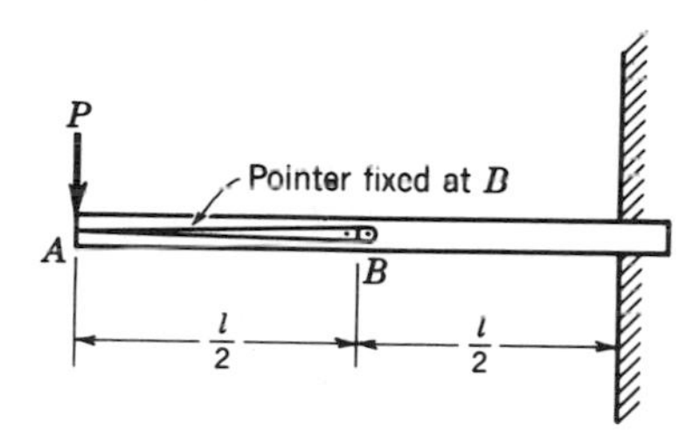

Figure 8-9

8-5 Important Relationships

We have seen the importance of the expressions obtained by the successive integration of the $M/EI = d^2y/dx^2$ equation. Considerable importance is also attached to the expressions obtained by the *successive differentiation* of the same equation. Thus,

$$EI\frac{d^2y}{dx^2} = M$$

$$\text{by one differentiation} = EI\frac{d^3y}{dx^3} = \frac{dM}{dx} = V \qquad (8.11)$$

$$\text{by two differentiations} = EI\frac{d^4y}{dx^4} = \frac{d^2M}{dx^2} = \frac{dV}{dx} = w \qquad (8.12)$$

where w is the intensity of the distributed load at the point x distance from the origin.

The application of these equations to a uniformly loaded (w lb/ft) simply supported beam yields the following expressions.

LOAD EQUATION

$$EI\frac{d^4y}{dx^4} = -w$$

SHEAR EQUATION

$$EI\frac{d^3y}{dx^3} = V = -wx + C_1 = -wx + \frac{wl}{2}$$

MOMENT EQUATION

$$EI\frac{d^2y}{dx^2} = M = -\frac{wx^2}{2} + \frac{wlx}{2} + C_2$$

$$= \frac{wx^2}{2} + \frac{wlx}{2}$$

SLOPE EQUATION

$$EI\frac{dy}{dx} = -\frac{wx^3}{6} + \frac{wlx^2}{4} + C_3$$

$$= \frac{wx^3}{6} + \frac{wlx^2}{4} - \frac{wl^3}{24}$$

DEFLECTION EQUATION

$$EIy = -\frac{wx^4}{24} + \frac{wlx^3}{12} - \frac{wl^3x}{24} + C_4$$

$$= -\frac{wx^4}{24} + \frac{wlx^3}{12} - \frac{wl^3x}{24}$$

PROBLEMS

8-13. Derive the expressions of shear, moment, slope, and deflection for a simply supported beam carrying a uniformly varying load ranging from zero at the left end to w lb/ft at the right end.

8-14. Derive the expressions of shear, moment, slope, and deflection for a cantilever beam loaded over its entire length with w lb/ft.

8-15. Derive the expressions of shear, moment, slope, and deflection for a cantilever beam carrying a uniformly varying load ranging from zero at its free end to w lb/ft at its fixed end.

8-16. Derive the expressions of shear, moment, slope, and deflection for a cantilever beam carrying a uniformly varying load ranging from w lb/ft at its free end to zero at its fixed end.

8-6 Comments on the Double-Integration Method

The double-integration method of obtaining slopes and deflections is academically proficient and is the basis for many of our conventional methods of deflection analysis. However, the usual procedure based on the elastic-curve equation, previously described, can be cumbersome to use, even in dealing with relatively simple problems. It needs to be modified and extended to increase its ease of applicability to more difficult problems. Two methods presently used to increase its usefulness are the step-function and moment-area methods. They will now be presented in that order.

8-7 Step-Function Method*

The step-function method owes its improvement over the double-integration method to the development of a composite bending-moment equation applicable to the entire length of the beam. Each term of this equation expresses the bending-moment contribution offered by one of the loads acting on the beam. Each load is considered in succession from left to right, with the origin always taken at the extreme left end of the beam. If these individual bending-moment contributions—or step functions as they are called—do not apply to the entire length of the beam, they are written in terms of binomials enclosed in pointed brackets. When the value of x selected makes such a binomial negative, it is considered zero and the term provides no contribution to the bending moment at that point. Only when the binomial enclosed in pointed brackets has a positive value does the term have a contributory effect on the total bending moment.

The effect of a couple is indicated by a pointed bracket raised to the zero power, e.g., $\langle x - d\rangle^0$. Thus, when positive, this binomial is always equal to 1.

A typical composite bending-moment equation written in step-function form is that developed for the beam of Fig. 8-10a where x is a variable value of length taken from the origin, i.e., left end.

$$M = R_1 x - P_1\langle x - a\rangle - P_2\langle x - b\rangle - P_3\langle x - c\rangle - w\frac{\langle x - d\rangle^2}{2} + M_0\langle x - d\rangle^0 \tag{8.13}$$

The effect of each term of Eq. (8.13) is indicated graphically in Fig. 8-10c. Those terms which include binomials enclosed in pointed brackets, i.e., step functions, are shown to be additive to the composite bending-moment diagram only when the value of x exceeds or equals the value of the constant included in that binomial.

*Also referred to as the singularity method.

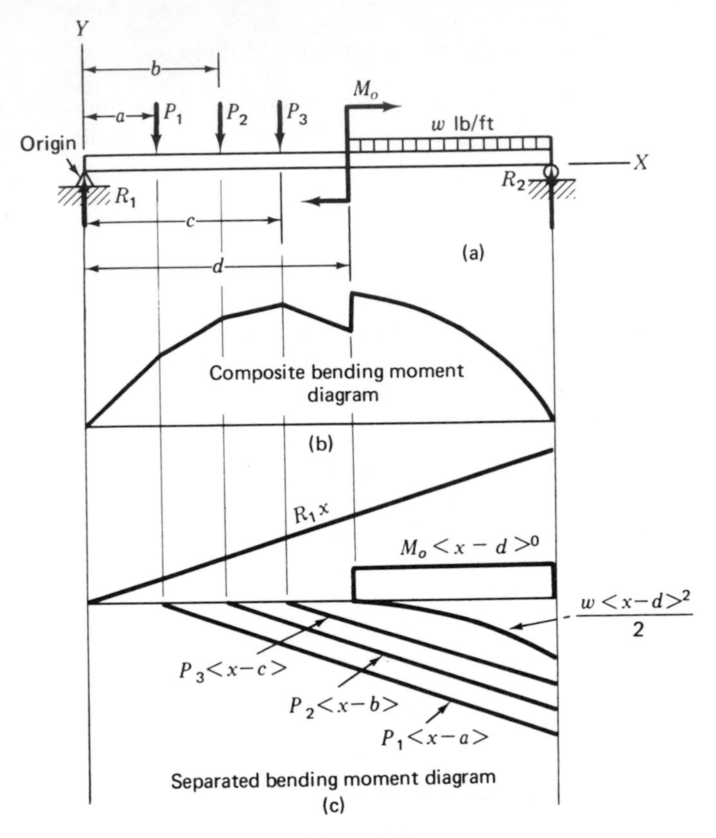

Figure 8-10

To apply the composite bending-moment equation to the determination of slopes and deflections of beams, it is inserted into the double-integration equation and integrated in the usual way. Illustrative Problem 2 will clarify its use.

Illustrative Problem 2. Determine the slope and deflection equations for the simple beam loaded with a concentrated load located off center (Fig. 8-11).

SOLUTION

$$EI\frac{d^2y}{dx^2} = M = \frac{Pb}{l}x - P\langle x - a\rangle$$

Integrating this equation yields

$$EI\frac{dy}{dx} = \frac{Pb}{l}\frac{x^2}{2} - \frac{P\langle x - a\rangle^2}{2} + C_1 \tag{8.14}$$

An additional integration provides

$$EIy = \frac{Pb}{2l}\frac{x^3}{3} - \frac{P}{2}\frac{\langle x - a\rangle^3}{3} + C_1x + C_2 \tag{8.15}$$

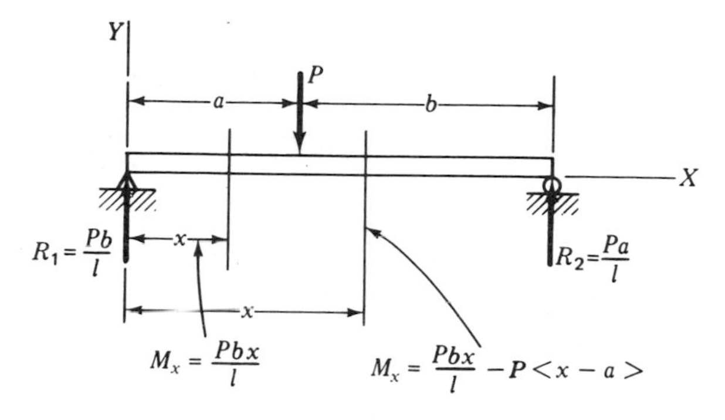

Figure 8-11

To obtain the coefficients of integrations C_1 and C_2, simultaneously occurring values of x and y are inserted in Eq. (8.15). Thus, when $x = 0$ and $y = 0$,

$$C_2 = 0$$

Also, when $x = l$ and $y = 0$

$$C_1 = \frac{Pb^3}{6l} - \frac{Pbl}{6}$$

Inserting these coefficients in Eqs. (8.14) and (8.15), there is obtained

$$EI\frac{dy}{dx} = \frac{Pb}{l}\frac{x^2}{2} - \frac{P\langle x - a\rangle^2}{2} + \frac{Pb^3}{6l} - \frac{Pbl}{6}, \quad \text{slope equation} \quad (8.16)$$

$$EIy = \frac{Pb}{2l}\frac{x^3}{3} - \frac{P}{2}\frac{\langle x - a\rangle^3}{3} + \frac{Pb^3x}{6l} - \frac{Pblx}{6}, \quad \text{deflection equation} \quad (8.17)$$

To obtain the numerical value of a slope or deflection at any point in the beam, the distance to the section desired is inserted in the proper equation. All negative binomials are discarded.

Illustrative Problem 3. Obtain the slope and deflection equations for the cantilever beam loaded as shown in Fig. 8-12a.

SOLUTION. As previously indicated, the origin for x is selected at the extreme left end of the beam. Bending moments using x values taken from this origin will require the calculation of the reaction R and the restraining moment M. Thus, through the summation of vertical forces,

$$R_1 = wa + P$$

Also, by taking moments about the origin

$$M_R = -\frac{wa^2}{2} - Pl$$

Then, to simplify the work entailed in expressing the contribution made

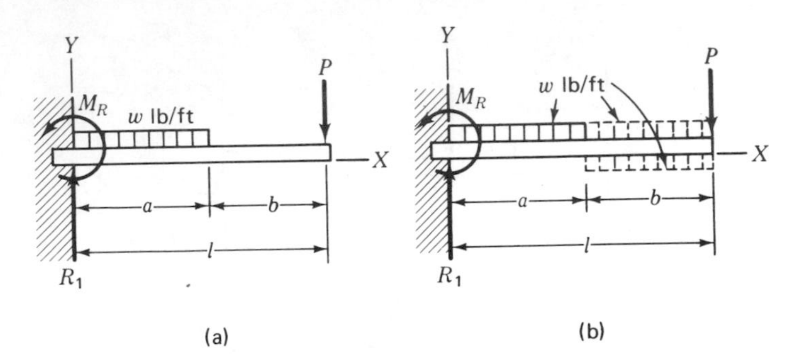

Figure 8-12

by the uniform load, it is permitted to extend to the extreme right end with an equal uniform load counteracting the added load (Fig. 8-12b). The solution then continues with the development of the composite equation:

$$EI\frac{d^2y}{dx} = M_x = \underbrace{-\frac{wa^2}{2} - Pl}_{M_R} + \underbrace{(wa + P)}_{R_1}x - \frac{wx^2}{2} + \frac{w\langle x - a\rangle^2}{2}$$

Integrating this equation twice yields the following equations:

$$EI\frac{dy}{dx} = -\frac{wa^2x}{2} - Plx + (wa + P)\frac{x^2}{2} - \frac{wx^3}{6} + \frac{w\langle x - a\rangle^3}{6} + C_1$$

$$EIy = -\frac{wa^2x^2}{4} - \frac{Plx^2}{2} + (wa + P)\frac{x^3}{6} - \frac{wx^4}{24} + \frac{w\langle x - a\rangle^4}{24} + C_1x + C_2$$

To secure the values of C_1 and C_2,

$$\frac{dy}{dx} = 0 \quad \text{when } x = 0; \quad \therefore C_1 = 0$$

Also

$$y = 0 \quad \text{when } x = 0; \quad \therefore C_2 = 0$$

The slope and deflection equations are therefore, respectively,

$$EI\frac{dy}{dx} = -\frac{wa^2x}{2} - Plx + (wa + P)\frac{x^2}{2} - \frac{wx^3}{6} + \frac{w\langle x - a\rangle^3}{6}$$

$$EIy = -\frac{wa^2x^2}{4} - \frac{Plx^2}{2} + (wa + P)\frac{x^3}{6} - \frac{wx^4}{24} + \frac{w\langle x - a\rangle^4}{6}$$

Illustrative Problem 4. Develop the slope and deflection equations for the beam of Fig. 8-13.

SOLUTION. Here the 1^k/ft uniform load is considered to act over the entire length of the beam. The last 6 ft on the right is then increased by an

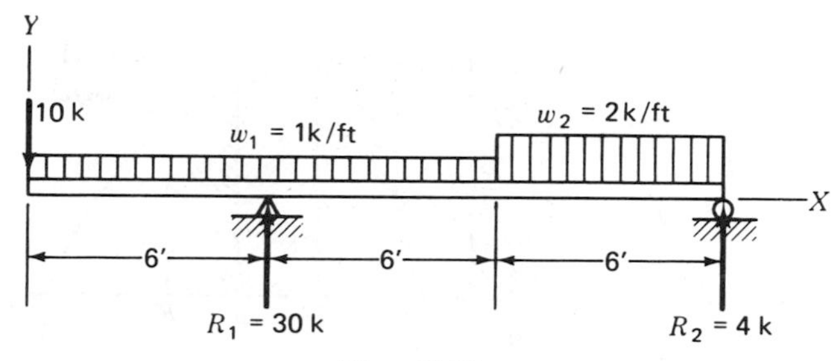

Figure 8-13

additional 1^k/ft to provide the 2^k/ft required. Thus,

$$EI\frac{d^2y}{dx^2} = -10x + 30\langle x - 6\rangle - \frac{x^2}{2} - \frac{\langle x - 12\rangle^2}{2}$$

$$EI\frac{dy}{dx} = \frac{-10x^2}{2} + \frac{30\langle x - 6\rangle^2}{2} - \frac{x^3}{6} - \frac{\langle x - 12\rangle^3}{6} + C_1$$

$$EIy = \frac{-10x^3}{6} + \frac{30\langle x - 6\rangle^3}{6} - \frac{x^4}{24} - \frac{\langle x - 12\rangle^4}{24} + C_1x + C_2$$

To obtain C_1 and C_2, let us use the previous equation. Thus,

$$\text{when } x = 6, \quad y = 0; \quad \therefore C_1 = 69$$

$$\text{when } x = 18, \quad y = 0; \quad \therefore C_2 = 4266$$

Completing the slope and deflection equations, there results

$$EI\frac{dy}{dx} = \frac{-10x^2}{2} + \frac{30\langle x - 6\rangle^2}{2} - \frac{x^3}{6} - \frac{\langle x - 12\rangle^3}{6} + 69$$

$$EIy = \frac{-10x^3}{6} + \frac{30\langle x - 6\rangle^3}{6} - \frac{x^4}{24} - \frac{\langle x - 12\rangle^4}{24} + 69x + 4266$$

8-8 Moment-Area Method of Finding Slopes and Deflections

A simple revision of the elastic-curve equation

$$\frac{M}{EI} = \frac{1}{\rho}$$

forms the basis of the *moment-area method* of determining slopes and deflections. Its simplicity, range of usefulness, and intimate relationship to the bending-moment diagram make it one of the most easily applied.

Derivation. In Fig. 8-14b is shown the elastic curve of a loaded beam whose deflection is grossly exaggerated for clarity. An infinitesimal length *dl* (also shown exaggerated) is laid off on the elastic curve. At its extremities

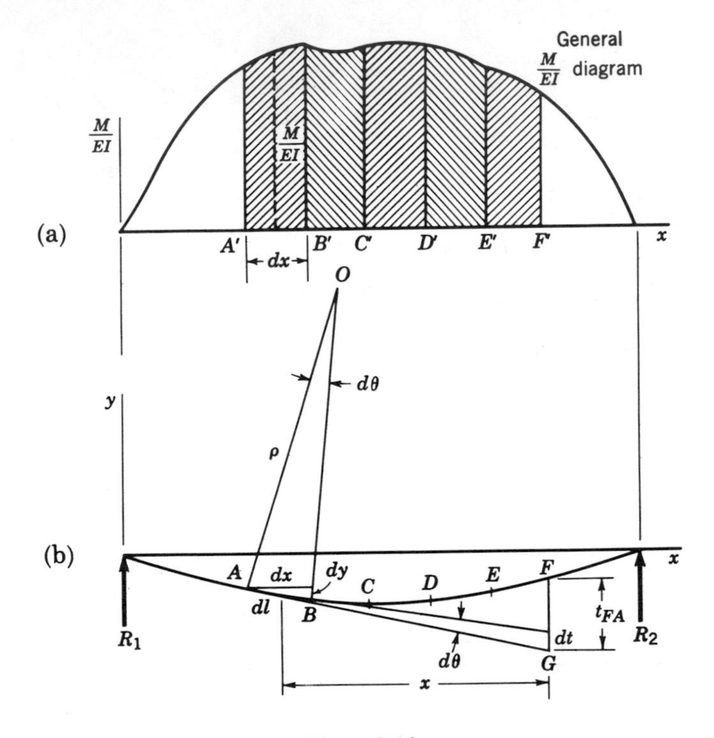

Figure 8-14

A and B are erected perpendiculars that intersect at point O with an angle $d\theta$. Because of the very small distance between points A and B, the value of ρ and the bending moment M for each point may be considered identical.

In an ordinary beam the deflections are relatively small, and the length of the elastic curve is very nearly equal to its horizontal projection. Thus, the value of dl is, for all practical purposes, equal to its horizontal projection dx, and we may write from our knowledge of radian measurement

$$d\theta = \frac{dx}{\rho}$$

By substituting for ρ its value obtained in the elastic-curve equation $1/\rho = M/EI$, we obtain

$$d\theta = \frac{M}{EI}\,dx \tag{8.18}$$

Now, $d\theta$ is also equal to the change in slope from A to B by construction. Therefore, if the change in slope from A to B is equal to $(M/EI)\,dx$, the changes in slope from B to C, C to D, D to E, and E to F are also equal to the same general expression. We may therefore sum up these individual changes in curvature to obtain the total changes in slope from A to F by writing

$$\int_A^F d\theta = \int_A^F \frac{M}{EI}dx \tag{8.19}$$

This expression could, of course, apply equally well to any other two points.

Proposition 1. Moment-area method for finding slopes. Let us now investigate the expression on the right-hand side of the equation. The value of M is the general expression of bending moment for any dx located between A and F and is obtainable from the bending-moment equation applicable to the dx section under consideration. Because of the infinitesimal length of dx, M can be considered a constant value for each such increment.

Equation (8.19) further indicates that each ordinate, M, of the bending-moment diagram is divided by its corresponding value of EI. To simplify the incorporation of this division of M by EI into moment-area problems, computations are based on the so-called *M/EI diagram*. For a homogeneous, uniform cross-sectioned beam, such a diagram would be reduced in height when compared with the bending-moment diagram but would retain exact similarity. Drawn to a new scale EI times the one previously used, however, would make the M/EI diagram identical to the initial moment curve.

Proceeding further, we note that the product of M/EI times the length over which it occurs, dx, is equivalent to the area of a rectangle (Fig. 8-14a) and also to the change in slope over dx. If, therefore, every rectangle of base dx between points A and F were added together, the total area would be equivalent to the change in slope between the same limits. Accordingly, we may state, as Proposition 1 of the moment-area method of beam deflection, the following:

The change in slope expressed in radians between any two points A and B on a loaded beam is equal to the algebraic sum of the M/EI area between those two points.

This proposition may also be expressed for added convenience in terms of finite values of slope and area:

$$\sum_A^F \Delta\theta = \sum_A^F \frac{M}{EI}\Delta x$$

Because most M/EI diagrams consist of rectangles, triangles, and parabolic areas, each of which are easily ascertained, the equation above is very useful.

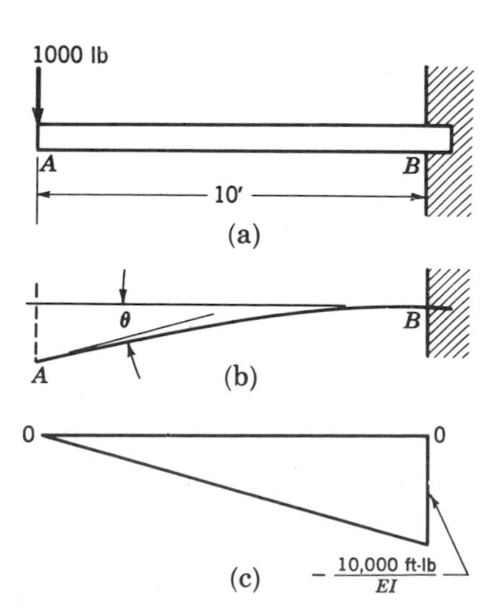

Figure 8-15

Illustrative Problem 5. Determine the slope at the free end of the cantilever beam shown in Fig. 8-15. *Note:* Since the fixed end of a cantilever beam is generally considered horizontal, the terms *slope* and *change in slope* will be synonymous. $E = 2 \times 10^6$ psi and $I = 100$ in.^{4}.

SOLUTION

$$\sum_A^B \Delta\theta = \theta_A = \sum_A^B \frac{M}{EI}\Delta x = -\frac{10{,}000 \times 12 \times 10 \times 12}{2 \times 2 \times 10^6 \times 100} = -36 \times 10^{-3}$$

$$= -0.036 \text{ rad}$$

(minus indicates negative change in slope from A to B)

Proposition 2: Moment-area method for finding deflections. Proposition 2 of the moment-area method involves the determination of *tangential deviations.* By definition, *a tangential deviation is the vertical distance between any point A on the elastic curve of a beam and the tangent drawn from any other point B located on the same curve.* It is designated by the letter t with two subscripts, the first indicating the point of tangential deviation, the second the point from which the tangent is drawn. Several illustrations of tangential deviations are shown in Fig. 8-16.

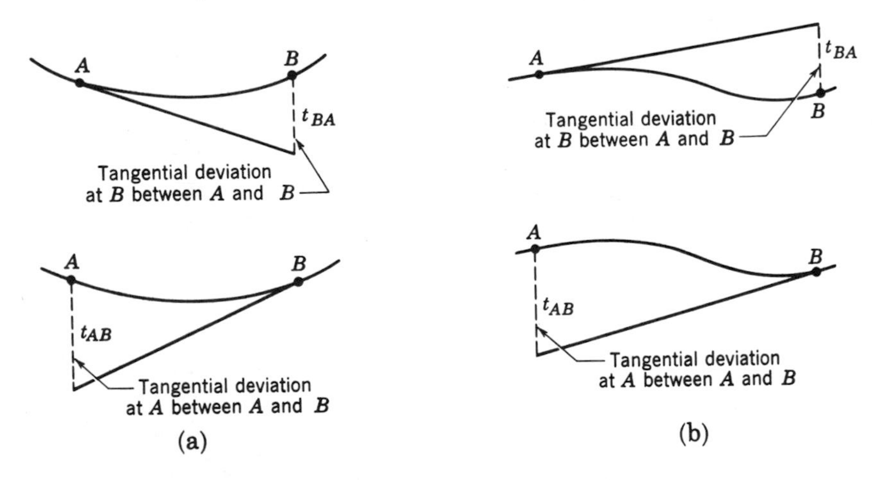

Figure 8-16

Referring once again to Fig. 8-14b, let us draw the tangents to points A and B intercepting a vertical line drawn through point F. The intercept dt, located an average distance x from dx, is a small portion of the total distance FG—a tangential deviation at F between points A and F. Because all practical beams suffer no great sag even when stressed to their maximum allowable limit, the length of the sloping tangents just drawn may be set equal to x without incurring any appreciable amount of error. We may therefore state through the use of radian measurement that

$$dt = x\, d\theta$$

or, by substituting the equation included in Proposition 1,

$$dt = x\frac{M}{EI}dx \tag{8.20}$$

Although this expression for dt was derived using the tangents at A and B, it is not difficult to see that it also pertains to any two adjacent points. Thus, if we were to extend other tangents from points C, D, and E on the elastic curve (Fig. 8-14b), the value of each incremental intercept between points G and F would equal the same general expression given in Eq. (8.20). Furthermore, the sum of these individual intercepts would equal the tangential deviation at F taken between A and F. Expressed in equation form, the tangential deviation at F between A and F equals

$$\int_A^F dt = \int_A^F x\frac{M}{EI}dx = t_{FA} \tag{8.21}$$

By scrutinizing this equation carefully, we again recognize the expression $(M/EI)\,dx$ as the rectangular area used in obtaining Proposition 1. The value of x is the distance between this area and the tangential deviation. The product of the $(M/EI)\,dx$ area and the value x is therefore the moment of an incremental portion of the M/EI diagram taken about the tangential deviation. Thus, the integration of all the products of x and $(M/EI)\,dx$ indicates that the tangential deviation at F between A and F is equal to the sum of the moments of the individual $(M/EI)\,dx$ areas between the limits A and F, taken about FG, the desired tangential deviation. Proposition 2 of the moment-area method is therefore expressed in general form as follows:

The tangential deviation at A between any two points A and B on a loaded beam is equal to the algebraic sum of the moments of the M/EI area between these two points taken about point A, i.e., the point where the tangential deviation is desired.

In the solution of ordinary problems, the M/EI diagrams consist of groupings of triangles, rectangles, parabolas, and so forth, whose centroids are generally known. Under such circumstances, Eq. (8-21) may be simplified to indicate a finite summation:

$$t_{FA} = \sum_A^F \Delta t = \sum_A^F \bar{x}\frac{M}{EI}\Delta x \tag{8.22}$$

8-9 Signs

When moments of M/EI areas are taken, the sign of the M/EI area under consideration shall be ascribed to its moment. The moment of negative areas shall, therefore, be negative, and the moment of positive areas, positive.

It is important to note further that a negative tangential deviation implies a location above the elastic curve, and a positive tangential deviation implies a location below the elastic curve.

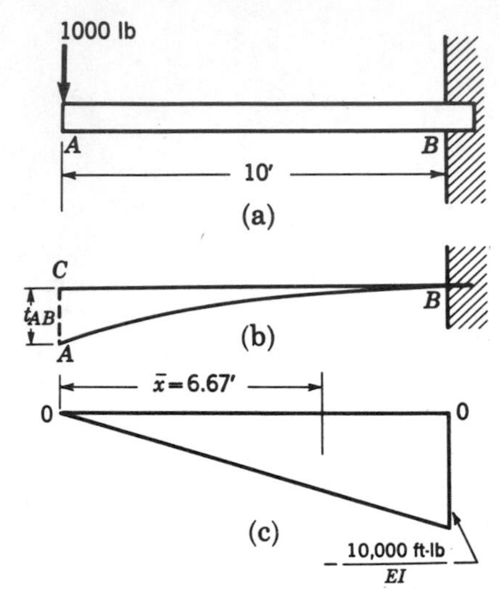

Figure 8-17

Illustrative Problem 6. Determine the tangential deviation at the free end of the cantilever beam taken over its entire length, that is, from A to B (see Fig. 8-17). $E = 2 \times 10^6$ psi and $I = 100$ in.^{4}.

SOLUTION. Taking moments of the M/EI diagram between points A and B about CA, that is, the desired tangential deviation, we obtain

$$t_{AB} = -\frac{10{,}000 \times 12}{2 \times 10^6 \times 100} \times \frac{10 \times 12}{2} \times 6.67 \times 12 = -2.88 \text{ in.}$$

But the tangential deviation at the free end of a cantilever taken over its entire length is also equal to its maximum deflection. Thus, we may write

$$\Delta = -2.88 \text{ in.}$$

WARNING. In our previous illustrative problems no account was taken of the stress developed in the beam. Since the flexure equation was used in the development of the elastic-curve equation, all its assumptions and limitations apply to the moment-area equations. Thus, should the maximum flexural stress exceed the proportional limit stress, the computations of deflection and change in slope would be erroneous.

If the beam in Illustrative Problem 6 were made of yellow pine and had a depth of 8 in., the maximum flexural stress would be

$$\sigma = \frac{Mc}{I} = \frac{10{,}000 \times 12 \times 4}{100} = 4800 \text{ psi}$$

Although this value exceeds the usual allowable flexural stress for this type of wood, it does not exceed the proportional limit stress of approximately 7000 psi.

The calculations of the deflection characteristics of the beam are therefore within the limitations of the previously derived equations and are theoretically correct.

8-10 Further Instructions on Use of Propositions 1 and 2 of Moment Area

Proposition 1 of the moment-area method finds its greatest use in determining the slopes at various points on a beam after some point of known slope has been established. In problems entailing cantilever beams and simple beams with symmetrical loadings, the establishment of a known slope is merely a matter of observation. On unsymmetrically loaded beams, however, the establishment of the point of reference slope takes place, generally, at the ends of the beam, or at the point of maximum deflections where the slope is horizontal. In the former case, the slope is established by dividing the tangential deviation between the two ends of the beam by its length. In the latter case, the tangential deviations from the point of zero slope to the ends of the beam area set equal to each other to determine the location of the point of tangency, or zero slope.

The application of Proposition 2 of the moment-area method to beam-deflection problems calls for nothing more than the judicious use of tangential deviations and similar triangles. Although some deflections may be determined directly from a computed tangential deviation, generally the addition or subtraction of tangential deviations and vertical sides of similar triangles is required. This simple procedure, once mastered, affords solutions of deflections for the most difficult problems.

Several of the more classical problems of beam deflection will now be solved to illustrate the use of the moment-area method.

8-11 Cantilever Beam—Uniform Load

In applying the moment-area propositions to curved M/EI diagrams developed for uniformly loaded beams, care must be taken to locate their proper centers of gravity. See Table 11 on page 586. The position of the center of gravity with respect to the vertex of the parabola is of special importance.

Maximum deflection. The maximum deflection is equal to the tangential deviation at A between A and B (Fig. 8-18). Thus,

$$\Delta = \frac{1}{3} \times -\frac{wl^2}{2EI} \times l \times \frac{3}{4}l = -\frac{1}{8}\frac{wl^4}{EI} \qquad (8.23)$$

Maximum slope. The maximum slope is, of course, located at A. Since the slope at B is zero, it must be equal to the area of the entire M/EI diagram between A and B.

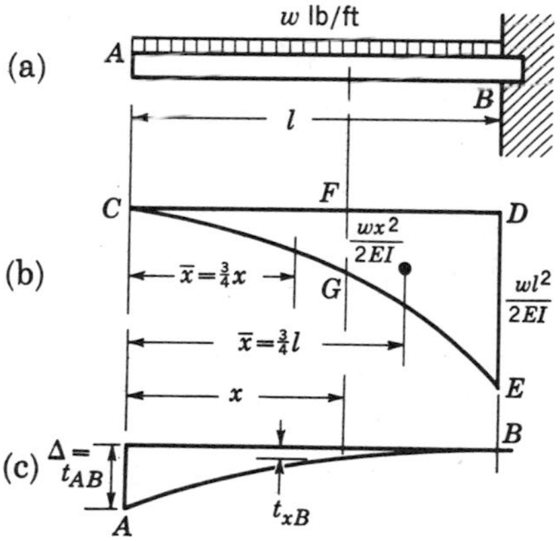

Figure 8-18

$$\theta_A = \frac{1}{3} \times -\frac{wl^2}{2EI} \times l = -\frac{1}{6}\frac{wl^3}{EI} \qquad (8.24)^*$$

*The negative sign for slope in this expression indicates a decreasing change in slope from either A to D or F to D.

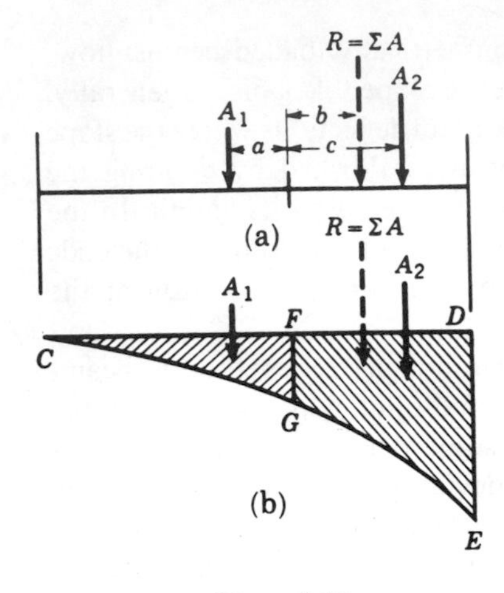

Figure 8-19

Deflection at any point. Finding the deflection at any point x on a uniformly loaded cantilever beam offers some difficulty, since the M/EI area $DEFG$ that must be used in its calculation is not regular. But the moment of this area will be obtained in the process of finding tangential deviation t_{xB} if we subtract the moment of portion CFG, whose center of gravity is known, from the entire parabolic area taken about FG. This procedure is merely an application of the principle, "The moment of a resultant is equal to the algebraic sum of the moments of its components."

Thus, in Fig. 8-19

$$\sum A \times b = -A_1 a + A_2 c$$

or

$$A_2 c = \sum A \times b + A_1 a$$

Proceeding with the calculation of the deflection at point F, any point x distance from the free end where x is less than $\frac{3}{4}l$, we have

$$\begin{aligned} y_x = t_{xB} &= \text{moment of } CDE \text{ about } FG + \text{moment of } CFG \text{ about } FG \\ &= \frac{1}{3}\left(-\frac{wl^2}{2EI}\right) \times l\left(\frac{3}{4}l - x\right) + \frac{1}{3}\left(-\frac{wx^2}{2EI}\right) \times x \times \frac{1}{4}x \qquad (8.25) \\ &= -\frac{1}{8}\frac{wl^4}{EI} + \frac{1}{6}\frac{wl^3x}{EI} - \frac{1}{24}\frac{wx^4}{EI} \end{aligned}$$

A change of signs in the Eq. (8.25) above should be noted when x becomes greater than $\frac{3}{4}l$.

Slope at any point. Since the slope at any point F on the cantilever beam is equal to its change in slope from point B, we may write

$$\theta_x = \frac{1}{3} \times -\frac{wl^2}{2EI} \times l - \frac{1}{3} \times -\frac{wx^2}{2EI} \times x = -\frac{w}{6EI}(l^3 - x^3) \qquad (8.26)^*$$

8-12 Deflection Characteristics of a Simple Beam—Concentrated Load at Center

The simple beam shown in Fig. 8-20a is acted upon by load P at its midpoint. Its bending-moment diagram is shown in Fig. 8-20b. Since the load is symmetrically placed, the maximum deflection occurs at the center of the beam.

*See footnote on p. 293.

Maximum deflection. To determine the magnitude of the maximum deflection requires the construction of a tangent to the elastic curve in such a position that will permit its calculation with the fewest number of tangential deviations and similar triangles.

If, therefore, on the elastic curve of the beam shown in Fig. 8-20c a tangent is drawn at the point of zero slope, that is, maximum deflection, the intercept AE on the vertical line passing through point A will be equal to the maximum deflection DB. Moreover, AE is also the tangential deviation at A between points A and B. The maximum deflection Δ may thus be found by taking the moment of the M/EI diagram between points A and B about the tangential deviation AE, as indicated in Eq. (8.22).

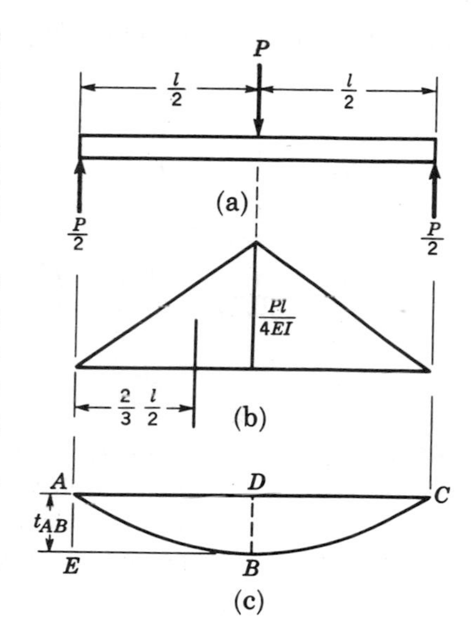

Figure 8-20

$$\Delta = t_{AB} = \sum_A^B \bar{x}\frac{M}{EI}\Delta x$$

Since the center of gravity of a triangle is located a distance equal to one third of its altitude from the base, we may substitute as follows:

$$\Delta = t_{AB} = \frac{2}{3} \times \frac{l}{2} \times \frac{1}{2} \times \frac{Pl}{4EI} \times \frac{l}{2} = \frac{Pl^3}{48EI} \tag{8.27}$$

Deflection at any point on a beam. The deflection of the beam shown in Fig. 8-21a at a point x feet away from the left reaction can be determined in various ways. Perhaps the easiest method of obtaining this deflection is to employ the previous horizontal tangent at B. Thus,

$$y_F = GF = GH - FH$$
$$= AE - FH$$

But both AE and FH are tangential deviations, since they are vertical distances from a point on the elastic curve to a tangent drawn from some other point on the elastic curve. Accordingly, we may write

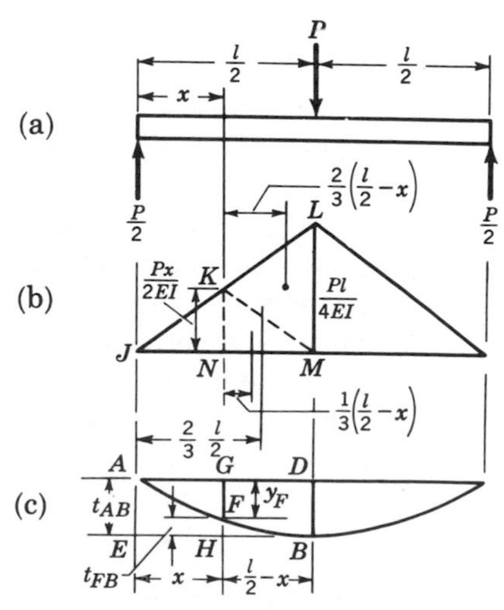

Figure 8-21

$$y_F = t_{AB} - t_{FB} = \text{moment of area } JLM \text{ about } AE$$
$$- \text{moment of area } KLMN \text{ about } FH$$

$$\begin{aligned} y_F &= \frac{Pl^3}{48EI} - \frac{1}{2} \times \frac{Pl}{4EI}\left(\frac{l}{2} - x\right)\frac{2}{3}\left(\frac{l}{2} - x\right) \\ &\quad - \frac{1}{2} \times \frac{Px}{2EI}\left(\frac{l}{2} - x\right)\frac{1}{3}\left(\frac{l}{2} - x\right) \\ &= \frac{Pl^3}{48EI} - \frac{Pl^3}{48EI} + \frac{Pl^2x}{12EI} - \frac{Plx^2}{12EI} - \frac{Pl^2x}{48EI} + \frac{Plx^2}{12EI} - \frac{Px^3}{12EI} \\ &= +\frac{1}{16}\frac{Pl^2x}{EI} - \frac{1}{12}\frac{Px^3}{EI} \end{aligned} \tag{8.28}$$

Maximum slope. The greatest slope on the beam is, of course, at the reaction points. Since all slopes are referred to a horizontal datum, we must determine the *change in slope* from the horizontal tangent at the center to the ends of the beam. By Proposition 1, this is equal to the area JLM.

$$\theta_A = \frac{1}{2} \times \frac{Pl}{4EI} \times \frac{l}{2} = \frac{Pl^2}{16EI} \tag{8.29}$$

Slope at any point. The slope at any point x distance from the left reaction is equal to the area $KLMN$, since its right-hand boundary is located at the point of tangency of the horizontal tangent.

$$\begin{aligned} \theta_x &= \frac{1}{2}\frac{Pl}{4EI}\left(\frac{l}{2} - x\right) + \frac{1}{2}\frac{Px}{2EI}\left(\frac{l}{2} - x\right) \\ &= \frac{Pl^2}{16EI} - \frac{Plx}{8EI} + \frac{Plx}{8EI} - \frac{Px^2}{4EI} = \frac{Pl^2}{16EI} - \frac{Px^2}{4EI} \end{aligned} \tag{8.30}$$

8-13 Consideration of Units in Calculating Slopes and Deflections

When the numerical values of slopes and deflections are calculated from the previously derived equations, it is important that the units of the various terms be carefully employed. Because these terms are not ordinarily expressed in similar units, a correction factor must be employed. Let Eq. (8.27) and (8.29) be used to illustrate the development of the correction factor.

If the units involved in these expressions are P, lb; E, lb/in.^{2}; I, in.^{4}; and l, ft, Eq. (8.29) expressed in unit form will be

$$\theta_A = \frac{Pl^2}{16EI} = \frac{\text{lb} \times \text{ft}^2}{(\text{lb/in.}^2) \times \text{in.}^4} = \frac{\text{ft}^2}{\text{in.}^2}$$

In order to further simplify the expression, either the numerator or denominator must be changed. Choosing to change the numerator to inches, 12 in. must be substituted for each foot. Thus,

$$\theta_A = \frac{(12\text{ in.})^2}{\text{in.}^2} = \frac{144\text{ in.}^2}{\text{in.}^2} = 144$$

The value 144 is the correction factor, which must be used as a multiplier whenever Eq. (8.20) or one similar to it is used with the above units. Because the value of 144 appears with no units, the slope θ_A is a dimensionless term, and is expressed in radians.

Equation (8.27) expressed in unit form is

$$\Delta = \frac{Pl^3}{48EI} = \frac{\text{lb} \times \text{ft}^3}{(\text{lb/in.}^2) \times \text{in.}^4} = \frac{\text{ft}^3}{\text{in.}^2}$$

Desiring the deflection to be calculated in inches, let us substitute 12 in. for each foot term in the numerator. Thus,

$$\Delta = \frac{(12\text{ in.})^3}{\text{in.}^2} = 1728\text{ in.}$$

Therefore, if numerical values bearing the units used above are substituted in this deflection equation, a multiplier of 1728 must be used to obtain the answer in inches.

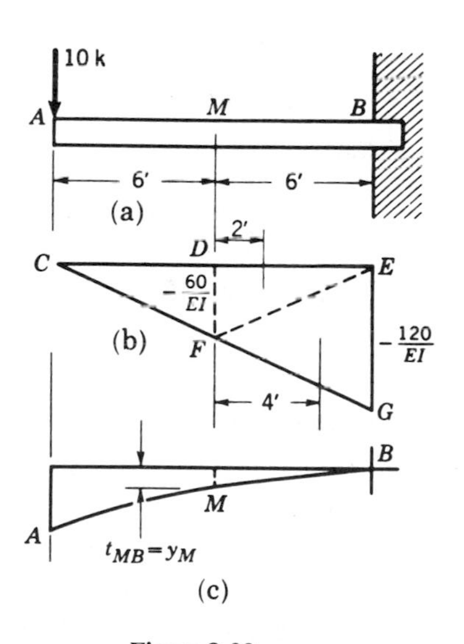

Figure 8-22

Illustrative Problem 7. Find the deflection and slope at the midpoint of a steel cantilever beam 12 ft long, loaded at its free end with a 10-kip load (Fig. 8-22). It is assumed that the maximum flexural stress does not exceed the proportional limit. $E = 30 \times 10^6$ psi and $I = 200$ in.^{4}.

SOLUTION. The slope at point M is equal to the value of the portion of the M/EI diagram between D and E, since this area is by definition the change in slope from point B to point M.

$$\theta_M = \frac{1}{30 \times 10^6 \times 200}\left(\frac{-60 - 120}{2}\right) \times 6 \times 144 \times 1000$$
$$= -0.01296 \text{ rad} \quad \text{or} \quad 0°44'31.2''$$

The deflection at the midpoint is equal to the tangential deviation at M between points M and B. Thus, by taking the moment of the area $DEGF$ (divided into two triangles) about the tangential deviation t_{MB}, the deflection

y_M becomes

$$y_M = \frac{1728}{30 \times 10^6 \times 200}\left(\frac{-60}{2} \times 6 \times 2 - \frac{120}{2} \times 6 \times 4\right) \times 1000$$

$$= -\frac{1728 \times 1800}{30 \times 10^3 \times 200} = -0.518 \text{ in.}$$

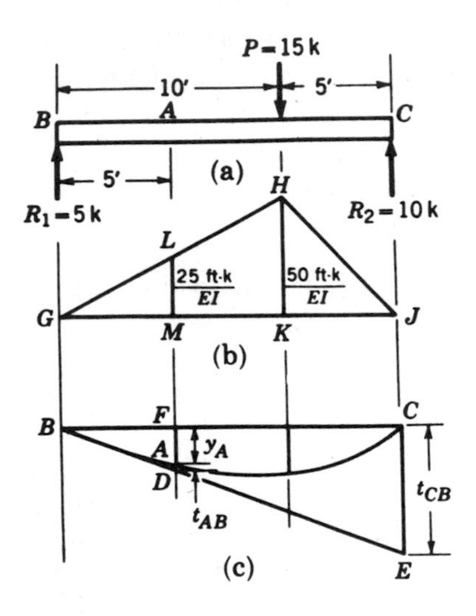

Figure 8-23

Illustrative Problem 8. Find the deflection at point A of the beam shown in Fig. 8-23. Calculate also the position of maximum deflection. $E = 30 \times 10^6$ psi and $I = 250$ in.^{4}.

SOLUTION. Referring to Fig. 8-23c, we may write

$$y_A = FD - AD = \tfrac{5}{15}t_{CB} - t_{AB}$$

Computing t_{CB} as the moment of the entire M/EI diagram about CE, there results

$$t_{CB} = \left(\frac{50 \times 10}{2EI} \times \frac{25}{3} + \frac{50 \times 5}{2EI} \times \frac{10}{3}\right) \times 1728 \times 1000$$

$$= \frac{2500}{EI} \times 1728 \times 1000$$

The value of t_{AB} is now found by taking moments of triangle GLM about AD. Thus,

$$t_{AB} = \left(\frac{25 \times 5}{2EI} \times \frac{5}{3}\right)1728 \times 1000 = \frac{625 \times 1728}{6EI} \times 1000$$

Substituting the values of t_{AB} and t_{CB} in the general equation above, we have

$$y_A = 1728\left(\frac{1}{3} \times \frac{2500}{EI} - \frac{625}{6EI}\right) \times 1000$$

$$= 0.168 \text{ in.}$$

Position of maximum deflection. From experience we know that the point of maximum deflection is located somewhere to the left of load P (Fig. 8-24). Since the slope of the elastic curve at that point must be horizontal, the tangential deviations t_{BH} and t_{CH} located at right and left ends of the beam, respectively, must be equal. Thus, we may write

$$t_{BH} = t_{CH}$$

$$\frac{5x}{EI} \times \frac{x}{2} \times \frac{2x}{3} = \frac{5x}{2EI}(10 - x)\left[5 + \frac{2}{3}(10 - x)\right] + \frac{50}{2EI}(10 - x)\left[5 + \frac{10 - x}{3}\right] + \frac{50}{2EI} \times 5 \times \frac{10}{3}$$

$$10x^3 = 1750x - 275x^2 + 10x^3 + 12{,}500 - 1750x + 50x^2 + 2500$$

from which

$$225x^2 = 15{,}000$$
$$\text{and} \quad x = \pm 8.17 \text{ ft}$$

The negative value of this answer is, of course, meaningless and should be omitted.

The maximum deflection may also be obtained by first differentiating the general elastic-curve equation expressed in x and y developed from a tangent drawn from B. The differential is then set equal to zero and solved for x. The value of x placed back into the elastic-curve equation will provide the maximum deflection.

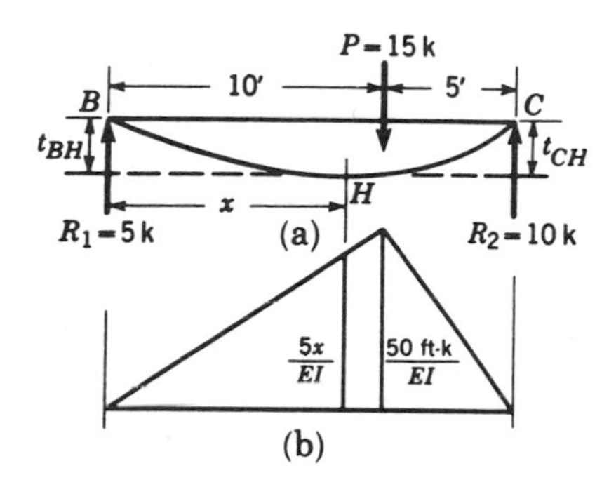

Figure 8-24

8-14 Separated Moment Diagrams

The solution of deflection problems is frequently simplified through the proper use of separated moment diagrams. However, of the infinite number possible there are usually several separated moment diagrams which can be advantageously drawn for the solution of any one problem. Moreover, not every separated moment diagram is equally suited for any one solution. Therefore, familiarity with the construction of all the possible diagrams will save valuable time.

Four different separated moment diagrams are shown for the overhanging beam of Fig. 8-25. The difference in diagrams is readily explained when one remembers that the bending moment at any point on a beam is equal to an algebraic summation of moments *to the left or to the right* of the point. In either diagram, the algebraic sum of the ordinates at any one section is equal to the corresponding ordinate on the composite bending-moment diagram.

Illustrative Problem 9. Determine the deflection and slope of an 18-ft, S 10 × 25.4 simply supported beam (Fig. 8-26a) at a point 6 ft from the left end if it is loaded uniformly with 1000 lb/ft over the right half of its length. $I = 122.1$ in.^{4}, $E = 30{,}000{,}000$ psi.

SOLUTION.

Deflection of point D. The use of a separated moment diagram is desirable (Fig. 8-26c). Let us draw the separated moment curves from the left because of the simple calculations required.

The fundamental equation for y_D obtained from Fig. 8-26b is

$$y_D = \tfrac{6}{18} t_{BA} - t_{DA}$$

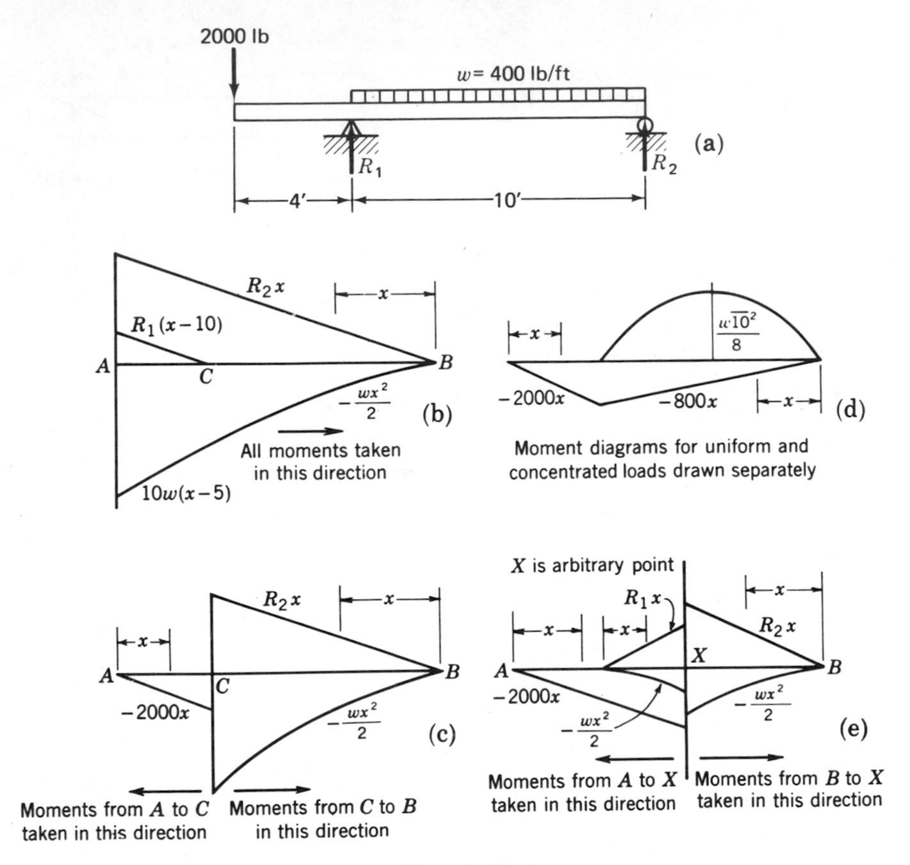

Figure 8-25

The values of t_{BA} and t_{DA} are now determined for insertion into this equation:

$$t_{BA} = \frac{40{,}500}{EI} \times \frac{18}{2} \times 6 - \frac{40{,}500}{EI} \times \frac{9}{3} \times 2.25$$

$$= \frac{2{,}187{,}000}{EI} - \frac{274{,}000}{EI} = \frac{1{,}913{,}000}{EI}$$

$$t_{DA} = \frac{13{,}500}{EI} \times \frac{6}{2} \times 2 = \frac{81{,}000}{EI}$$

The exact value of y_D is therefore

$$y_D = \frac{6}{18}\left(\frac{1{,}913{,}000}{EI}\right) - \frac{81{,}000}{EI} = \frac{556{,}000}{EI}$$

$$= \frac{556{,}000 \times 1728}{30{,}000{,}000 \times 122.1} = 0.262 \text{ in.}$$

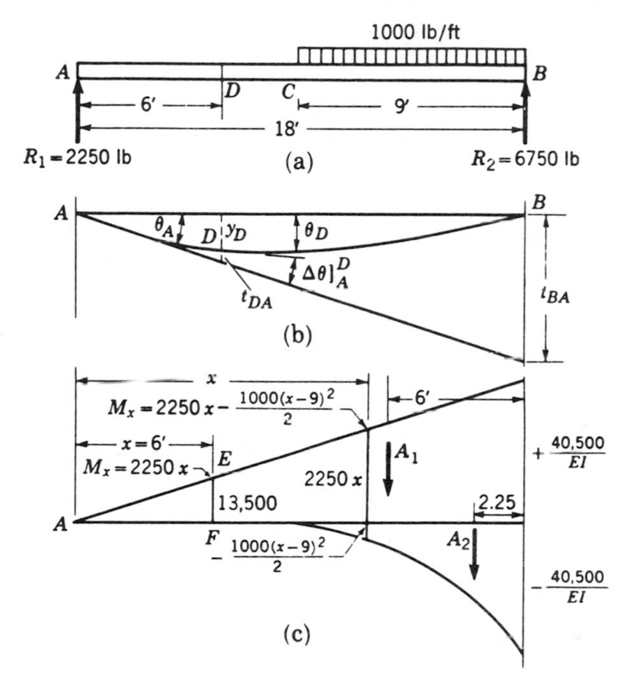

Figure 8-26

Slope at D. The equation for slope at D may be written as

$$\theta_D = \theta_A - [\Delta\theta]_A^D = \frac{t_{BA}}{18} - \text{area } AEF = \frac{1,913,000}{18EI} - \frac{13,500}{EI} \times \frac{6}{2}$$

$$= \frac{65,700}{EI} = \frac{65,700 \times 144}{30,000,000 \times 122.1} = 0.00258 \text{ rad}$$

Illustrative Problem 10. Determine the deflection of the free end of the overhanging beam shown in Fig. 8-27 in terms of E and I.

SOLUTION. The ordinary M/EI diagram, as shown in Fig. 8-27b, would tend to present discouraging complications if used for finding deflections by the moment-area method. But if the M/EI diagram were separated into its several components, as shown in Fig. 8-27c, the computations would be simplified, since each component area is regular and has a known center of gravity.

The M/EI diagram is separated into its component areas by writing the bending-moment equation for a point on the beam x ft to the right of reaction R_1.

$$M_x = -2000(4 + x) + 4800x - \frac{400x^2}{2} \tag{8.31}$$

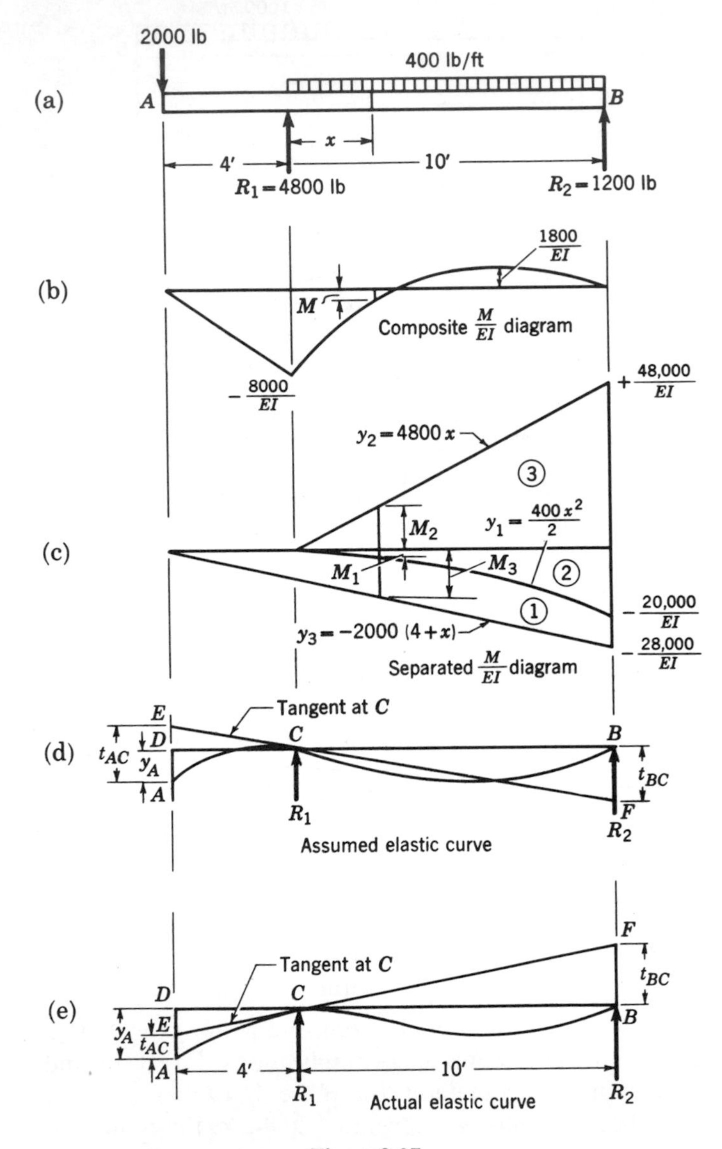

Figure 8-27

If these terms are plotted individually as they appear in Eq. (8.31) divided by EI, the separated M/EI diagram of Fig. 8-27c results. It naturally follows that the sum of the M_1, M_2, and M_3 ordinates of the separated diagram must add up to the M ordinate of the composite diagram.

Proceeding to the computation of the deflection of the free end of the beam, let us draw a tangent to the assumed elastic curve at reaction R_1.

From the geometry of the figure, we may write

$$y_A = t_{AC} - \tfrac{4}{10} t_{BC}$$

The character of the algebraic signs of this equation is doubtful because the slope of the elastic curve (and the tangent) at R_1 was assumed. The correct slope of the tangent can be found by the computation of t_{BC} (see §8-9). If t_{BC} is positive, it must lie below the elastic curve, thereby verifying the assumed position of Fig. 8-27d. If the numerical value of t_{BC} were negative, the shape of the beam would appear as shown in Fig. 8-27e, making

$$y_A = t_{AC} + \tfrac{4}{10} t_{BC} \tag{8.32}$$

The computations follow using the assumed elastic curve (Fig. 8-27d):

$$t_{AC} = -\frac{4 \times 8000}{2EI} \times 2.67 = -\frac{42{,}700}{EI}$$

$$\begin{aligned}
t_{BC} &= -\frac{8000}{EI} \times 10 \times 5 &&= -\frac{400{,}000}{EI} \\
&= -\frac{20{,}000 \times 10 \times 3.33}{2EI} &&= -\frac{333{,}000}{EI} \\
&= -\frac{1}{3} \times 10 \times 20{,}000 \times \frac{10}{4} &&= -\frac{166{,}700}{EI} \\
&= +\frac{48{,}000 \times 10}{2} \times 3.33 &&= +\frac{800{,}000}{EI} \\
t_{BC} &= &&-\frac{99{,}700}{EI}
\end{aligned}$$

Since the algebraic sign of t_{BC} is negative, it must lie above the elastic curve. This invalidates the assumed curve of Fig. 8-27d and substantiates the curve of Fig. 8-27e. Once the correct position of the tangent is ascertained, the values of t_{AC} and t_{BC} can be handled as abstract values to avoid confusion in signs.

Substitution of the numerical values in Eq. (8.32) gives

$$y_A = \frac{42{,}700}{EI} + \frac{4}{10}\left(\frac{99{,}700}{EI}\right) = \frac{82{,}600}{EI}$$

If E is in lb/in.^{2}, I in in.^{4}, and y is desired in inches,

$$y_A = \frac{82{,}600}{EI} \times 1728 = \frac{1.43 \times 10^8}{EI} \text{ in.}$$

Note. Investigation of the uniform and concentrated loads with respect to their contribution to the total deflection of the free end of the beam would reveal that their individual effects, when added, give the total deflection. Moreover, any beam loaded with a complicated system of concentrated and distributed loads can be studied for deflection by considering the various loads or groups of loads separately and adding their individual deflections to give the total deflection at any desired point.

These statements are an outgrowth of the *principle of superposition*, which states that the resultant effect is the sum of the individual effects. For beams this principle holds true as long as the assumptions and limitations of the theory used are not violated and as long as one effect (deflection) does not magnify or interfere with the development of any other effect.

PROBLEMS

Note: Either the step-function or moment-area method may be used in the solution of this set of problems. Use that method designated by the instructor.

8-17. A simple timber beam 9 ft long is acted upon by a concentrated load of 10 kips placed at its midpoint. What is the slope and deflection at a point 3 ft from one of the reactions? $E = 2 \times 10^6$ psi and $I = 300$ in.^{4}.

8-18. Placed at the midpoint of a simple beam 12 ft long is a load of 6 kips. Determine the slope and deflection at one of the quarter points. $E = 30 \times 10^6$ psi and $I = 100$ in.^{4}. *Answers:* 0.001945 rad, 0.085 in.

8-19. A timber cantilever beam 6 ft long is to be so designed that it will deflect exactly 0.3 in. at its free end. Determine the concentrated load required at its free end if $E = 2 \times 10^6$ psi and $I = 500$ in.^{4}.

8-20. A steel cantilever beam 9 ft long carries a concentrated load of 4 kips at its free end. What is the maximum slope and deflection? $I = 100$ in.^{4}. *Answers:* 0.0077 rad, 0.560 in.

8-21. A steel S beam 12 ft long is used as a simple beam carrying a load of 15 kips at one of its quarter points. What is the deflection at the other quarter point? $I = 100$ in.^{4}. *Answer:* 0.136 in.

8-22. A 6-ft steel shaft 3 in. in diameter is supported as a simple beam on adjustable-pillow block bearings. If a load of 4 kips is applied at one of the third points, determine the deflection at the other third point.

8-23. Find the location and magnitude of the maximum deflection of the beam in Problem 8-21. *Answers:* 6.70 ft from the right reaction, 0.218 in.

8-24. Find the location and magnitude of the maximum deflection of the beam of Problem 8-22.

8-25. A cantilever beam 9 ft long has loads of 2 kips applied to its free end and midpoint. What is the deflection and slope at the free end of the beam? $E = 2 \times 10^6$ psi and $I = 300$ in.^{4}. *Answers:* 1.838 in., 0.0243 rad.

8-26. Given the cantilever beam with loads as shown in Fig. 8-28. (a) Find the deflection and slope of the beam at point B, 2.5 ft from the left end. (b) Indicate the direction of both the slope and the deflection at this point, and sketch the deflection curve. E and I are constant. *Answers:* (a) $50/EI$; (b) $136/EI$.

8-27. The beam shown in Fig. 8-29 has a 2-kip force acting on each end as well as a 6-kip force acting at its midpoint. Determine the lengths of overhanging portions a that will make the slope of the beam horizontal at the supports. $EI =$ constant.

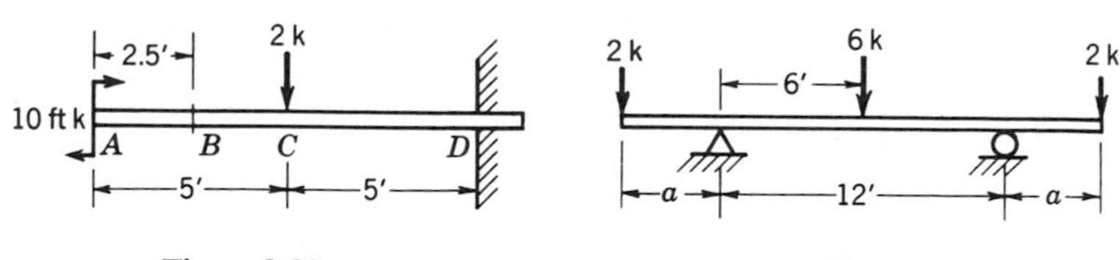

Figure 8-28

Figure 8-29

8-28. Find the amount of load P such that C will be on the same level as the supports A and B (Fig. 8-30). EI = constant. *Answer:* 2.67 kips.

8-29. Find the amount of P which will cause the point B of the beam (Fig. 8-31) to remain on the same level as the supports A and C. EI = constant.

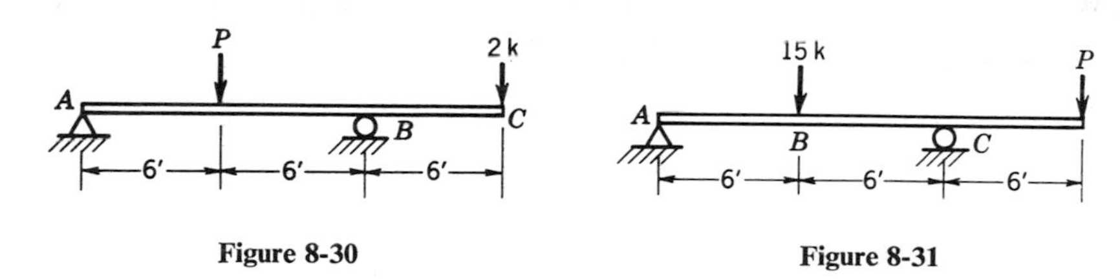

Figure 8-30

Figure 8-31

8-30. The overhanging beam ABC (Fig. 8-32) is 20 ft long, has two simple beam supports 16 ft apart, and is horizontal when not loaded. With the application of the load of 200 lb at the cantilever end of the beam, support B settles 1/4 in. while support A remains at the same elevation. Determine the distance that point C is below its original unloaded position. *Answer:* 0.555 in.

8-31. For the beam shown in Fig. 8-33, determine the deflection at point C in terms of P, L, E, and I. EI = constant.

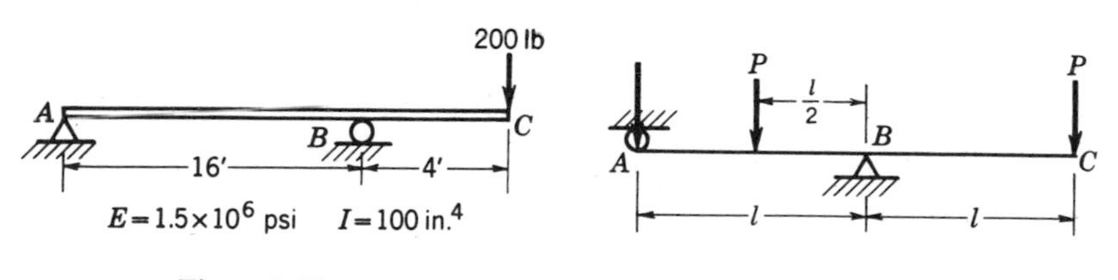

Figure 8-32

Figure 8-33

8-32. Find the deflection at M in inches and the slope at A in degrees of the beam shown in Fig. 8-34. $E = 2 \times 10^6$ psi and $I = 100$ in.^{4}. *Answers:* 0.525 in., 1.48°.

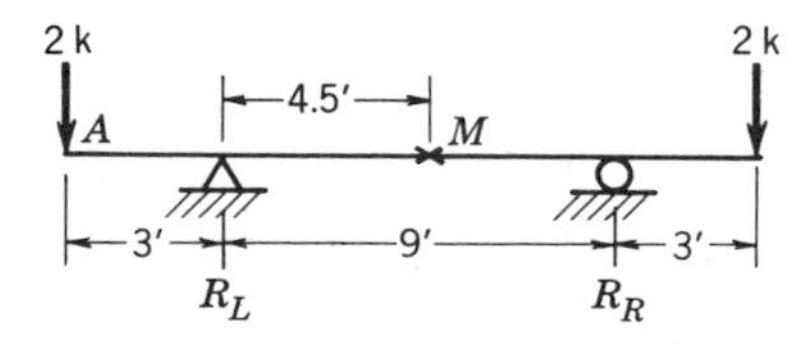

Figure 8-34

8-33. Determine the deflection at the free end of a steel cantilever beam 12 ft long if it is loaded with a uniform load of 1000 lb/ft over the outermost half of its length (Fig. 8-35). $I = 200$ in.^{4}. *Answer:* 0.637 in.

8-34. Determine the deflection at the free end of a steel cantilever beam if it is loaded with a uniform load of 1000 lb/ft over the innermost half of its length (Fig. 8-36). $I = 200$ in.^{4}.

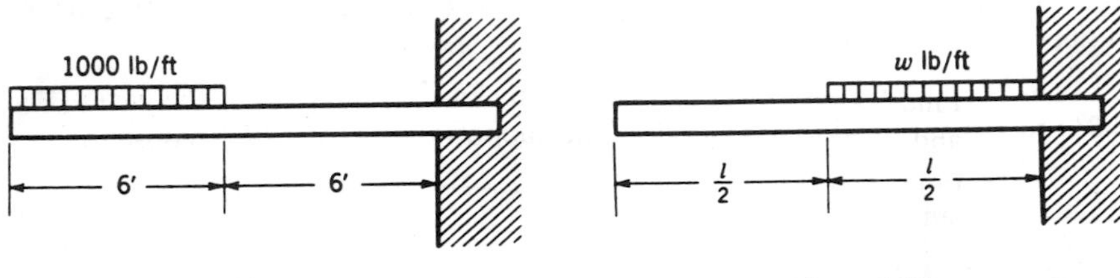

Figure 8-35

Figure 8-36

8-35. Determine the deflection at the quarter points of a uniformly loaded cantilever beam l ft long. Place the answer in terms of w, l, E, and I. *Answers:* $\frac{513}{6144}(wl^4/EI)$, $\frac{27}{2048}(wl^4/EI)$.

8-36. A simple beam is loaded as shown in Fig. 8-37. Find the deflection at the center in terms of w, l, E, and I. *Hint:* Draw the moment diagram in its component parts. *Answer:* $\frac{5}{768}(wl^4/EI)$.

8-37. Determine the deflection at the third points of the beam shown in Fig. 8-38 in terms of w, l, E, and I.

8-38. Find the maximum deflection of the beam shown in Fig. 8-38 in terms of w, l, E, and I. *Answer:* $0.0064(wl^4/EI)$.

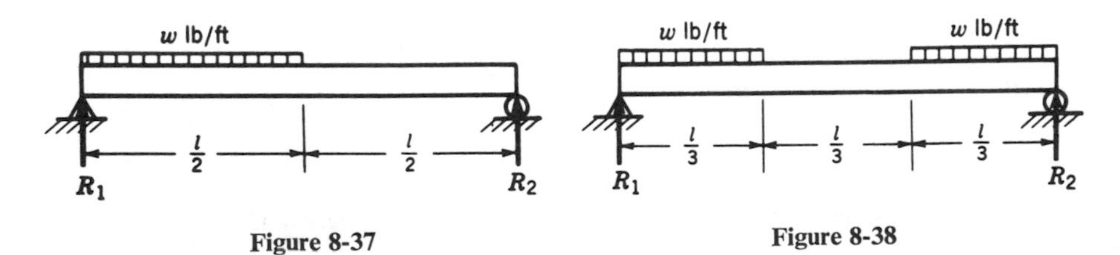

Figure 8-37

Figure 8-38

8-39. Find, by the principle of superposition, the deflection at the free end of the cantilever shown in Fig. 8-39. $E = 10{,}000{,}000$ psi and $I = 200$ in.^{4}. *Answer:* 0.145 in.

8-40. Determine the deflection at point A on the overhanging beam shown in Fig. 8-40. $E = 10{,}000{,}000$ psi and $I = 600$ in.^{4}.

8-41. Determine the deflection at point A on the overhanging beam shown in Fig. 8-41. $E = 30{,}000{,}000$ psi and $I = 500$ in.^{4}. *Answer:* 0.0149 in.

8-42. A couple of M ft-lb is placed at the free end of a cantilever beam l ft in length. Determine the deflection of the free end in terms of M, l, E, and I.

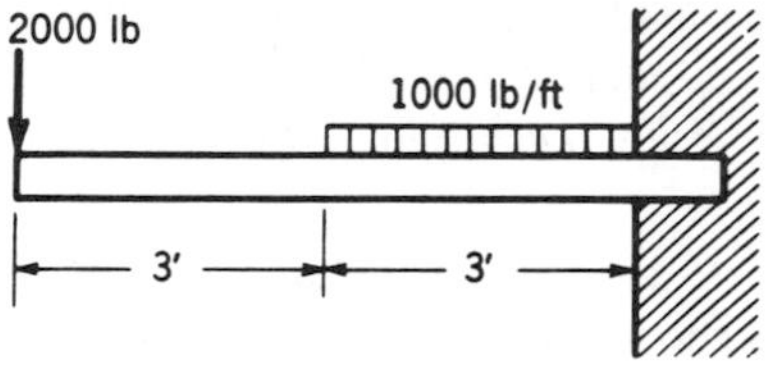

Figure 8-39

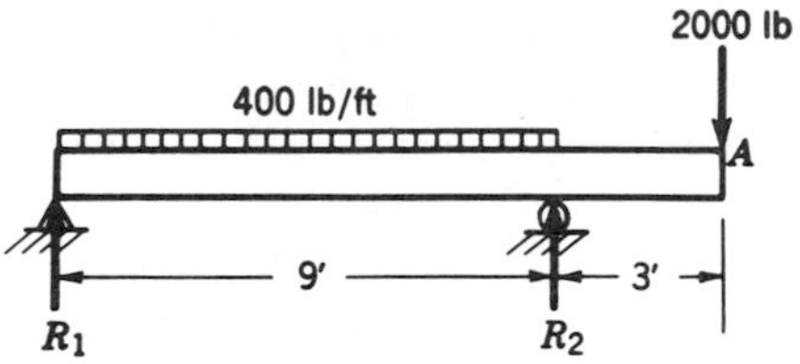

Figure 8-40

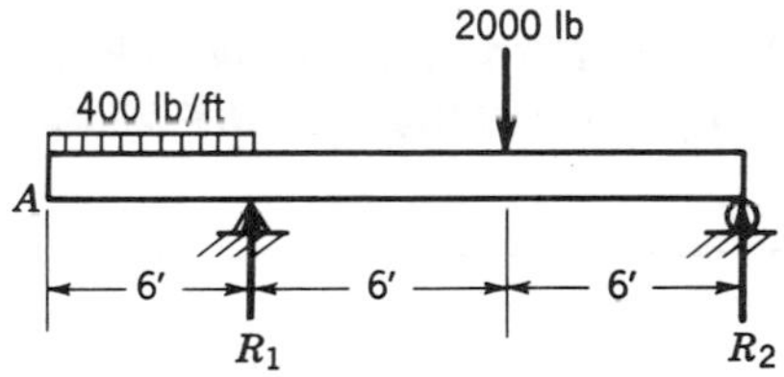

Figure 8-41

8-43. A clockwise moment of M ft-lb is placed at the center of a simple beam. Determine the deflection at $l/4$ from the left reaction, placing the answer in terms of M, l, E, and I. *Answer:* $\frac{3}{384}(Ml^2/EI)$.

8-44. Find the amount of load P acting on the beam in Fig. 8-42 so that C will remain on the same level as supports A and B.

8-45. Find w, in kips per foot, so that B will be at the same elevation as the supports A and C (Fig. 8-43). *Answer:* 10.67 kips/ft.

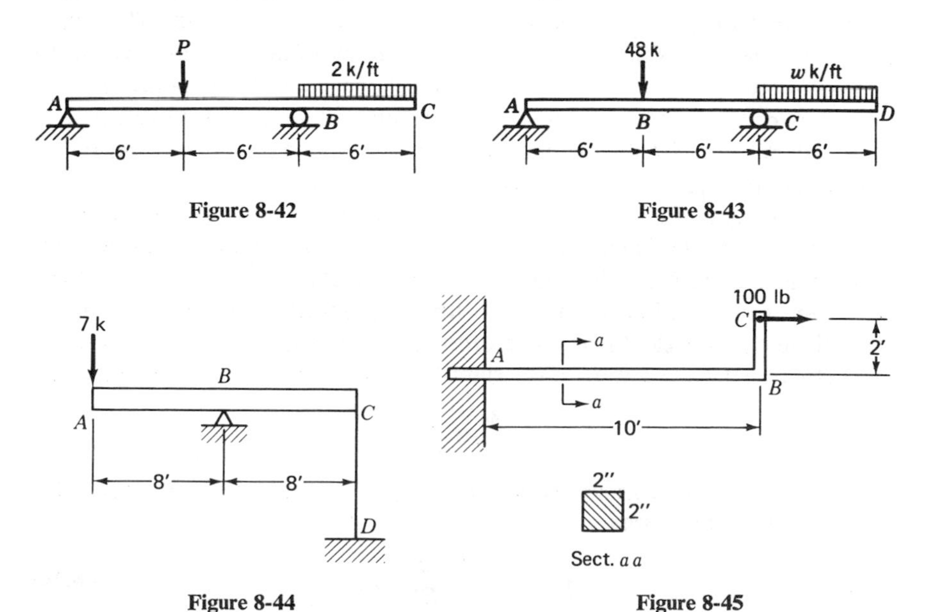

Figure 8-42

Figure 8-43

Figure 8-44

Figure 8-45

8-46. A steel beam with $I = 172.8$ in.^{4} is loaded and supported as shown in Fig. 8-44. A steel cable CD having a cross-sectional area of 1 in.^{2} and a length of 8 ft provides anchorage at the right end of the beam. Determine the deflection of the left end of the beam. $E = 30 \times 10^6$ psi.

8-47. A steel bar with a square cross section 2 by 2 in. is bent and embedded in a support with the 10-ft length horizontal as shown in Fig. 8-45. Determine the vertical deflection of point B due to the application of a horizontal force of 100 lb at point C. $E = 30 \times 10^6$ psi. *Answer:* 0.432 in.

GRAPHICAL DETERMINATION OF SLOPE AND DEFLECTION OF BEAMS

8-15 General Considerations

Occasionally, it is desirable to represent graphically the deflected position of a shaft or beam in order that the clearances of gears, pulleys, or other attached appurtenances may be determined. This desire for graphical representation is usually prompted by the tediousness of conducting the many mathematical solutions necessary to determine an accurate picture of the elastic curve.

The method is based on a graphical representation of the two fundamental equations of moment area given in §8-8. The first of these equations,

$$\int_A^B d\theta = \int_A^B \frac{M}{EI} dx \tag{8.33}$$

indicates that the change in slope between any two points A and B is equal to the *accumulated* sum of the $(M/EI)\,dx$ areas obtained from the M/EI diagram between A and B. If the accumulated sums from point A were plotted as ordinates to a graph whose base and reference line were equal to the length of beam AB, the ordinate at any intermediate point would represent the change in slope between that point and point A.

By changing the position of the reference line to some parallel position crossing the $(M/EI)\,dx$ curve, the new ordinate at any point would present the change of slope from the intersection point to the point in question. Were the point of intersection to coincide with the point of zero slope (i.e., maximum deflection on the beam), the ordinates would represent actual slopes.

Because the integration of the right-hand portion of Eq. (8.33) is equal to

$$\left.\frac{M}{EI} x\right]_A^B$$

which can be plotted as an accumulative curve, the second fundamental equation of moment area

$$\int_A^B dt = \int_A^B \frac{M}{EI} x\, dx \tag{8.34}$$

can also be expressed graphically as an accumulative sum of the $(M/EI)\,dx$ (slope) curve. Were the accumulated sums of incremental $(M/EI)x\,dx$ areas plotted as previously, the ordinate to any point of the resulting graph would be equal to the tangential deviation between that point and point A. Assuming now that the slope curve for the entire beam had been plotted with reference to the zero slope line, the sum of the $(M/EI)x\,dx$ areas on either side of the zero slope point would have had to balance, inasmuch as they represent tangential deviations, and the tangential deviations at the ends of a simple beam created by a horizontal tangent to the elastic curve must be equal.

Therefore, to obtain the zero-slope line for the solution to the type of problem described, one must obtain the average ordinate to the slope curve drawn with respect to the slope at one end.

The deflection curve results from plotting the accumulative sums of the $(M/EI)x\,dx$ areas taken from the slope curve plotted with reference to the zero-slope line.

Illustrative Problem 11. Obtain the deflection curve of the 3-in. diameter steel shaft loaded as shown in Fig. 8-46.

SOLUTION. As the first step toward obtaining the deflection curve of the beam, let us draw an accumulated area curve of the M/EI diagram as we proceed from A to B on the beam. The segmental areas of the M/EI diagram for each of the nine 1-ft intervals are recorded in Fig. 8-46b. If the magnitude of area A_1 is placed as an ordinate at the 1-ft mark of an x axis equal in length to the beam, we have the start of the curve shown in Fig. 8-46c. It should be recalled that area A_1 is the change in slope between points 0 and 1 on the beam. Following the recording of the A_1 ordinate, the magnitude of areas A_1 and A_2 is recorded at point 2. When the other segmental areas are in turn added accumulatively and plotted as ordinates at their respective end points, curve Fig. 8-46c is obtained.

The value of the ordinate at any point on this curve is the difference in slope between that point and the left end of the beam. *To obtain the exact slope of any point on the beam from this curve requires that the point of zero slope or maximum deflection be located.* When area EFG is divided by the length of the beam, an average ordinate is obtained that intersects the slope curve at the point of zero slope. Using HJ, the average ordinate line, as a new axis, any ordinate to the curve now becomes equal to the actual slope at that point. Point K, at the very center of this particular beam, is the point of zero slope.

The deflection curve is obtained by performing a similar accumulated sum operation on the change in slope curve (Fig. 8-46c) just constructed. If these N areas are progressively and algebraically accumulated from point E to F and the intermediate sums plotted as ordinates at their respective points on still another x axis equal to the length of the beam, the actual deflection curve of the beam will result (Fig. 8-46d).

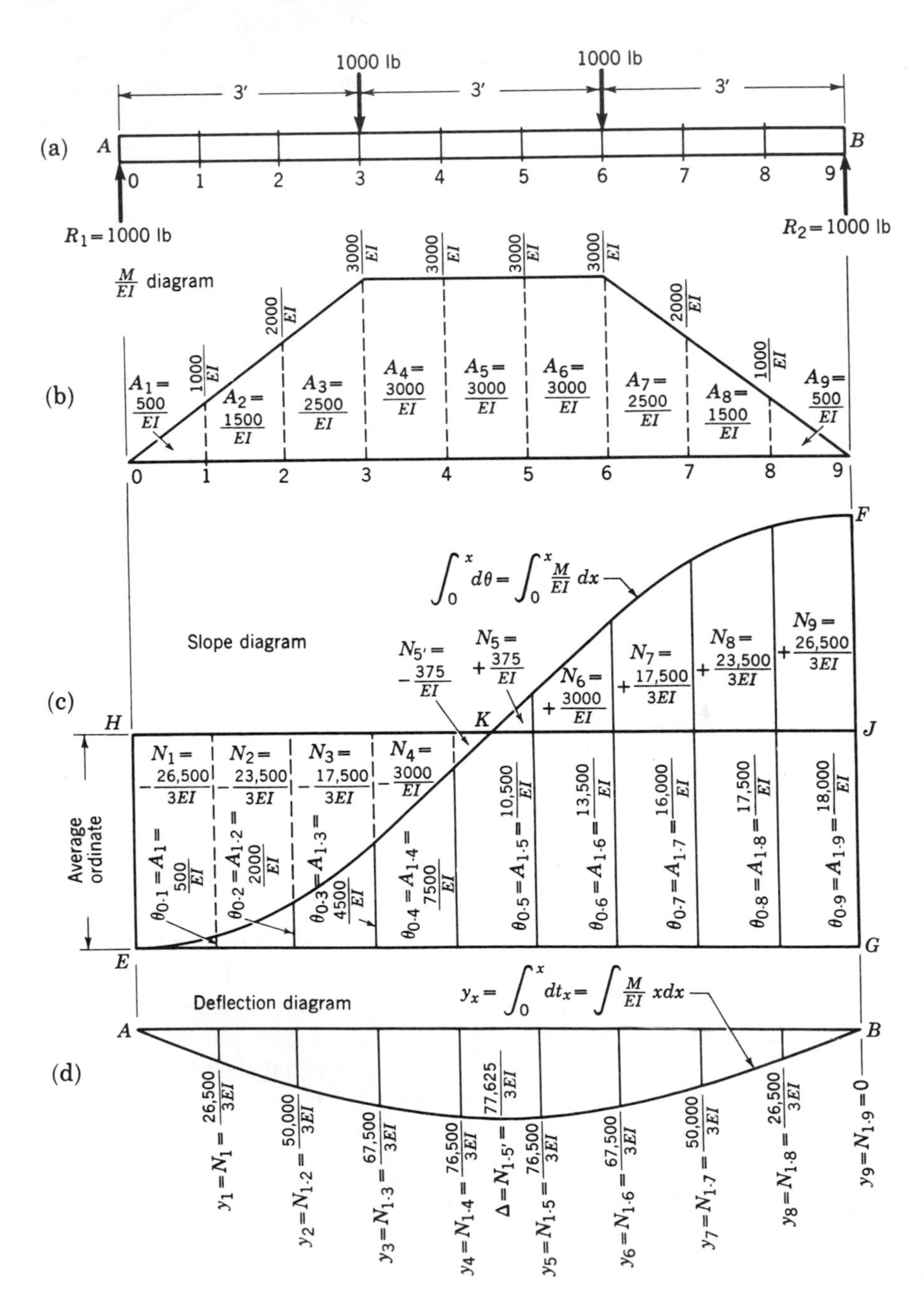
1000 lb
1000 lb
3′
3′
3′
(a)
A
B
0 1 2 3 4 5 6 7 8 9
$R_1 = 1000$ lb
$R_2 = 1000$ lb
$\frac{M}{EI}$ diagram
(b)
$A_1 = \frac{500}{EI}$
$A_2 = \frac{1500}{EI}$
$A_3 = \frac{2500}{EI}$
$A_4 = \frac{3000}{EI}$
$A_5 = \frac{3000}{EI}$
$A_6 = \frac{3000}{EI}$
$A_7 = \frac{2500}{EI}$
$A_8 = \frac{1500}{EI}$
$A_9 = \frac{500}{EI}$
Slope diagram
(c)
$\int_0^x d\theta = \int_0^x \frac{M}{EI}\,dx$
H
J
K
E
G
F
Average ordinate
$N_1 = -\frac{26,500}{3EI}$
$N_2 = -\frac{23,500}{3EI}$
$N_3 = -\frac{17,500}{3EI}$
$N_4 = -\frac{3000}{EI}$
$N_{5'} = -\frac{375}{EI}$
$N_5 = +\frac{375}{EI}$
$N_6 = +\frac{3000}{EI}$
$N_7 = +\frac{17,500}{3EI}$
$N_8 = +\frac{23,500}{3EI}$
$N_9 = +\frac{26,500}{3EI}$
$\theta_{0\cdot1} = A_1 = \frac{500}{EI}$
$\theta_{0\cdot2} = A_{1\cdot2} = \frac{2000}{EI}$
$\theta_{0\cdot3} = A_{1\cdot3} = \frac{4500}{EI}$
$\theta_{0\cdot4} = A_{1\cdot4} = \frac{7500}{EI}$
$\theta_{0\cdot5} = A_{1\cdot5} = \frac{10,500}{EI}$
$\theta_{0\cdot6} = A_{1\cdot6} = \frac{13,500}{EI}$
$\theta_{0\cdot7} = A_{1\cdot7} = \frac{16,000}{EI}$
$\theta_{0\cdot8} = A_{1\cdot8} = \frac{17,500}{EI}$
$\theta_{0\cdot9} = A_{1\cdot9} = \frac{18,000}{EI}$
Deflection diagram
(d)
$y_x = \int_0^x dt_x = \int \frac{M}{EI}\,x\,dx$
$y_1 = N_1 = \frac{26,500}{3EI}$
$y_2 = N_{1\cdot2} = \frac{50,000}{3EI}$
$y_3 = N_{1\cdot3} = \frac{67,500}{3EI}$
$y_4 = N_{1\cdot4} = \frac{76,500}{3EI}$
$\Delta = N_{1\cdot5'} = \frac{77,625}{3EI}$
$y_5 = N_{1\cdot5} = \frac{76,500}{3EI}$
$y_6 = N_{1\cdot6} = \frac{67,500}{3EI}$
$y_7 = N_{1\cdot7} = \frac{50,000}{3EI}$
$y_8 = N_{1\cdot8} = \frac{26,500}{3EI}$
$y_9 = N_{1\cdot9} = 0$

Figure 8-46

The maximum deflection resulting from this progressive summation, corrected for units, is

$$\Delta = \frac{77{,}625}{3EI} \times 1728 \text{ in.} = \frac{25{,}875}{EI} \times 1728 \text{ in.}$$

Theoretically, the maximum value of deflection is computed to be

$$\frac{Pa}{12EI}\left(\frac{3}{4} l^2 - a^2\right)$$

where P = the sum of both concentrated loads, a = the distance from a load to the nearest reaction, l = length of the beam.

Substituting the proper values in this equation, we obtain

$$\Delta = \frac{2000 \times 3}{12EI}\left(\frac{3}{4} \times 9^2 - 3^2\right) = \frac{500}{EI} \times \frac{207}{4} = \frac{103{,}500}{4EI} = \frac{25{,}875}{EI}$$

Assuming the units of E to be in lb/in.^{2} and I to be in in.^{4}, this deflection becomes

$$\Delta = \frac{25{,}875}{EI} \times 1728 \text{ in.}$$

which is exactly equal to that found by graphical integration.

The method of graphical determination of deflection as illustrated in the foregoing problem is applicable to all statically determinate beams under any conditions of loading. Should the areas under the slope curve become difficult to determine theoretically, they can be approximated with sufficient accuracy for most purposes by the use of triangles, rectangles, and trapezoids or by the use of a planimeter.

PROBLEMS

8-48. Draw the deflection curve of the beam shown in Fig. 8-47. Use as large a scale as is possible for accuracy. $E = 2 \times 10^6$ psi and $I = 800$ in.^{4}. *Answer:* $\Delta = 1.48$ in.

8-49. Draw the deflection curve for the beam shown in Fig. 8-48. Determine the deflection under the 3-kip load. $E = 2 \times 10^6$ psi and $I = 500$ in.^{4}.

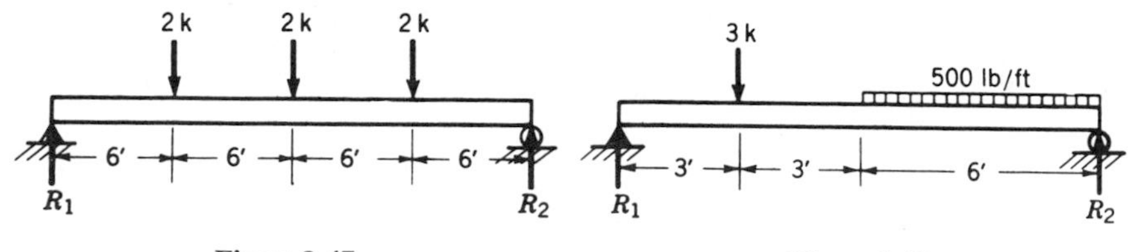

Figure 8-47 **Figure 8-48**

8-50. Obtain from the deflection curve, drawn for the beam shown in Fig. 8-49, the location and magnitude of the maximum deflection. $E = 30 \times 10^6$ psi. *Answers:* 25.6 in. from left reaction, 0.009 in.

8-51. From the slope and deflection curve drawn for the beam shown in Fig. 8-50, determine the location, the magnitude of maximum deflection, and the slope under the 4- and 2-kip loads. $E = 30 \times 10^6$ psi.

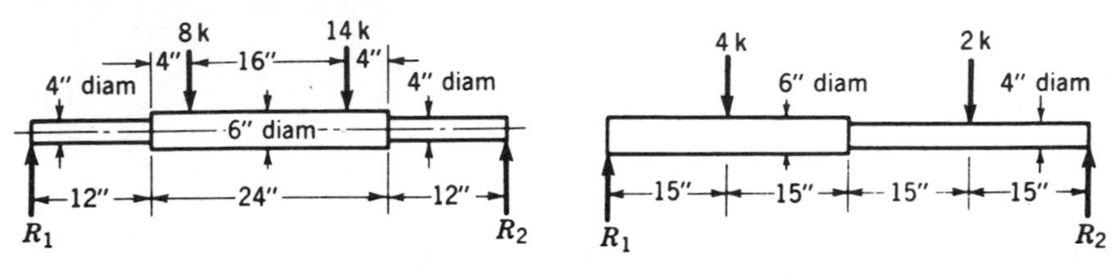

Figure 8-49

Figure 8-50

FINITE-DIFFERENCE METHOD OF OBTAINING DEFLECTIONS OF BEAMS

8-16 General Considerations

The elastic-curve equation $EI(d^2y/dx^2) = M$ expressed in finite form provides the basis of the finite-difference method for obtaining the approximate deflection of beams. The degree of approximation is dependent on the magnitude of the finite increments. With the advent of the computer, however, accuracy no longer presents a problem since the increments can be made as small as necessary to obtain the required refinement.

8-17 Development of Method

To write d^2y/dx^2 in finite form it is necessary to consider it as a rate of change of slope with respect to x and to specify the point at which it will be found. In Fig. 8-51 let B be that point. Now, the slopes between points A and B and points B and C are, respectively,

$$\theta_A^B = \frac{y_{n-1} - y_n}{\Delta x_1}, \qquad \theta_B^C = \frac{y_n - y_{n+1}}{\Delta x_2} \tag{8.35}$$

As the Δx values become smaller, it is apparent that the slopes become more and more equal to that at B. To obtain the rate of change of slope at B over a distance Δx_B, we may write

$$\frac{d^2y}{dx^2} \sim \frac{\theta_A^B - \theta_B^C}{\Delta x_B} = \frac{(y_{n-1} - y_n/\Delta x_1) - (y_n - y_{n+1}/\Delta x_2)}{\Delta x_B} \tag{8.36}$$

where

$$\Delta x_1 = \Delta x_2 = \Delta x_B$$

Since all the Δx values are equal, a simplification of the above expression yields

$$\frac{M}{EI} = \frac{d^2y}{dx^2} \sim \frac{y_{n-1} - 2y_n + y_{n+1}}{\Delta x_2} \tag{8.37}$$

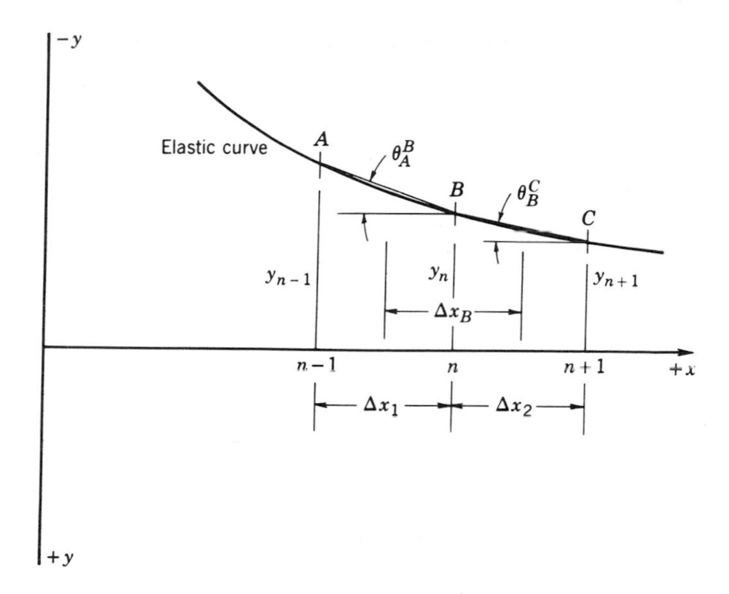

Figure 8-51

This is the fundamental equation of the finite-difference method. In any simple beam of n equal segments the application of this equation to the $n + 1$ division points (two of which are at the ends and have $y = 0$) yields $n - 1$ sets of simultaneous equations. A solution of these simultaneous equations provides the approximate deflections at all division points.

Illustrative Problem 12. Determine the deflection of the beam shown in Fig. 8-52 at the center and at the quarter points.

SOLUTION. Let the beam be divided into four equal segments. The values of the moment M to be inserted in the finite-difference equation will be

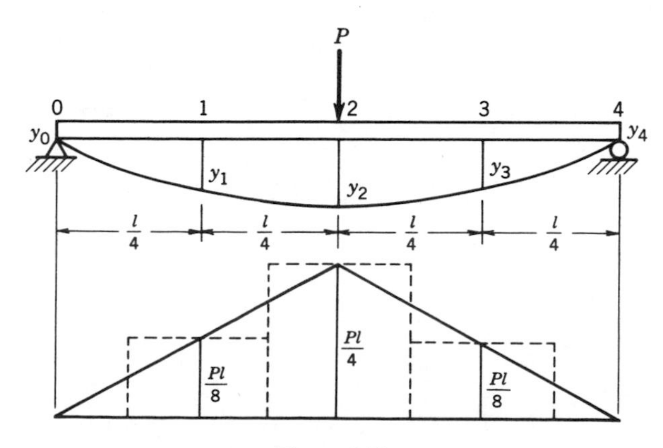

Figure 8-52

those taken at the division points. The equations will then be equal to

for points 0, 1, and 2 $\quad y_0 - 2y_1 + y_2 = \frac{Pl}{8EI}\left(\frac{l}{4}\right)^2$

for points 1, 2, and 3 $\quad y_1 - 2y_2 + y_3 = \frac{Pl}{4EI}\left(\frac{l}{4}\right)^2$

for points 2, 3, and 4 $\quad y_2 - 2y_3 + y_4 = \frac{Pl}{8EI}\left(\frac{l}{4}\right)^2$

But

$$y_0 = y_4 = 0$$

and

$$y_1 = y_3 \qquad \text{by symmetry}$$

The solution to these equations provides

$$y_1 = \frac{1}{64}\frac{Pl^3}{EI}$$

and

$$y_2 = \frac{3}{128}\frac{Pl^3}{EI} \quad \text{or} \quad \frac{1.125}{48}\frac{Pl^3}{EI}$$

The accurate values of y_1 and y_2 obtained using infinitely small subdivisions are

$$y_1 = \frac{0.917}{64}\frac{Pl^3}{EI} \quad \text{and} \quad y_2 = \frac{Pl^3}{48EI}$$

The relative closeness of these values to the finite-difference values obtained using even a coarse subdivision indicates the true worth of the finite-difference method. A finer subdivision will, of course, produce more accurate results, but will also increase the amount of computation necessary, as shown in the following illustrative problem.

Illustrative Problem 13. Determine the deflections at points 1 through 5 of the beam shown in Fig. 8-53. The finite-difference equations are as follows:

$$y_0 - 2y_1 + y_2 = \left(\frac{1}{9}\frac{Pl}{EI}\right)\left(\frac{l}{6}\right)^2 = \frac{1}{324}\frac{Pl^3}{EI}$$

$$y_1 - 2y_2 + y_3 = \left(\frac{2}{9}\frac{Pl}{EI}\right)\left(\frac{l}{6}\right)^2 = \frac{2}{324}\frac{Pl^3}{EI}$$

$$y_2 - 2y_3 + y_4 = \left(\frac{1}{6}\frac{Pl}{EI}\right)\left(\frac{l}{6}\right)^2 = \frac{1}{216}\frac{Pl^3}{EI}$$

$$y_3 - 2y_4 + y_5 = \left(\frac{1}{9}\frac{Pl}{EI}\right)\left(\frac{l}{6}\right)^2 = \frac{1}{324}\frac{Pl^3}{EI}$$

$$y_4 - 2y_5 + y_6 = \left(\frac{1}{18}\frac{Pl}{EI}\right)\left(\frac{l}{6}\right)^2 = \frac{1}{648}\frac{Pl^3}{EI}$$

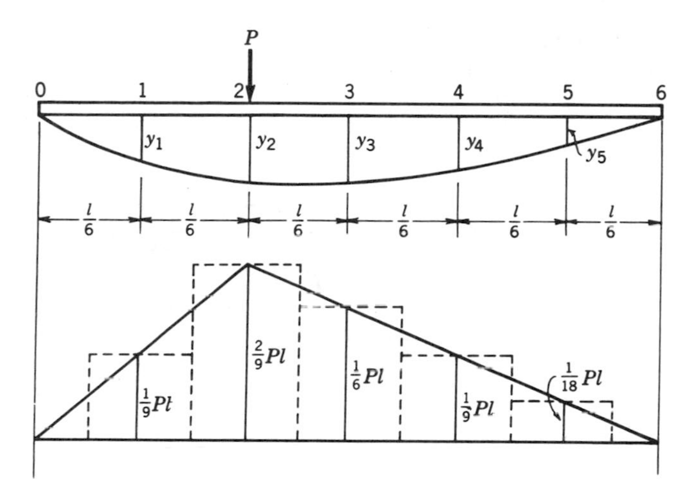

Figure 8-53

The solution to these equations provides

$$y_1 = \frac{-6.67}{648}\frac{Pl^3}{EI}, \qquad y_2 = \frac{-11.33}{648}\frac{Pl^3}{EI}, \qquad y_3 = \frac{-12}{648}\frac{Pl^3}{EI},$$

$$y_4 = \frac{-9.67}{648}\frac{Pl^3}{EI}, \qquad y_5 = \frac{-5.33}{648}\frac{Pl^3}{EI}$$

PROBLEMS

Note: To be done by the finite-difference method. Use segments equal to $l/4$.

8-52. Determine the deflections at the quarter points and at the center of a uniformly loaded simple beam. *Answers:* $\frac{5}{512}(wl^4/EI)$, $\frac{7}{512}(wl^4/EI)$.

8-53a, b, c, d. Determine the deflection at points A and B for the beams shown in Fig. 8-54.

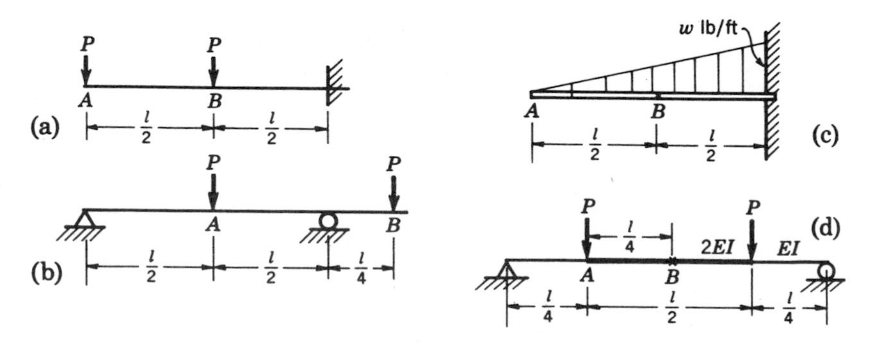

Figure 8-54

8-54. Determine the deflections at the quarter points and the center of the beam for the simple beam shown in Fig. 8-55. *Answers:* $0.00513(wl^4/EI)$, $0.00684(wl^4/EI)$, $0.00464(wl^4/EI)$.

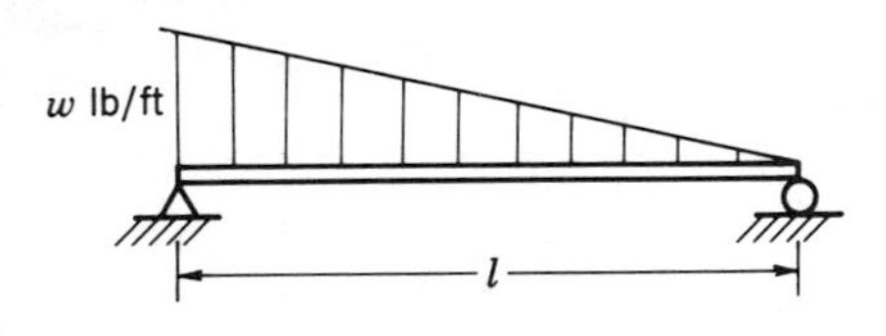

Figure 8-55

9

Statically Indeterminate Beams

9-1 Restraining Moments, Reactions, and Deflections

Problems involving statically indeterminate beams are generally best solved by the calculation of *restraining moments*. These restraining moments are not directly applied, but *are induced whenever the normal deflection of a simply supported beam is hindered.*

For instance, consider a beam having both ends cemented into a stone wall (Fig. 9-1a). The complicated nature of the restraint offered by the walls may be resolved into reactions R_1 and R_2, which maintain vertical balance, and the restraining moments, i.e., couples M_1 and M_2. The induced development of M_1 and M_2 causes a reduction in deflection and imposes upon the beam a statically indeterminate condition. From inspection, the only difference between the loading diagram of the statically indeterminate beam of Fig. 9-1a and the statically determinate beam of Fig. 9-1b is the presence of these two restraining couples.

Restraining moments are also induced at the supports of continuous beams, as shown in Fig. 9-2. Obviously, the elastic curve of each span would change if the beam were to be cut into three simple beams of lengths l_1, l_2, and l_3. The restoration of continuity to the beam could be accomplished only by the restoration of the internal-resisting moments at each interior reaction.

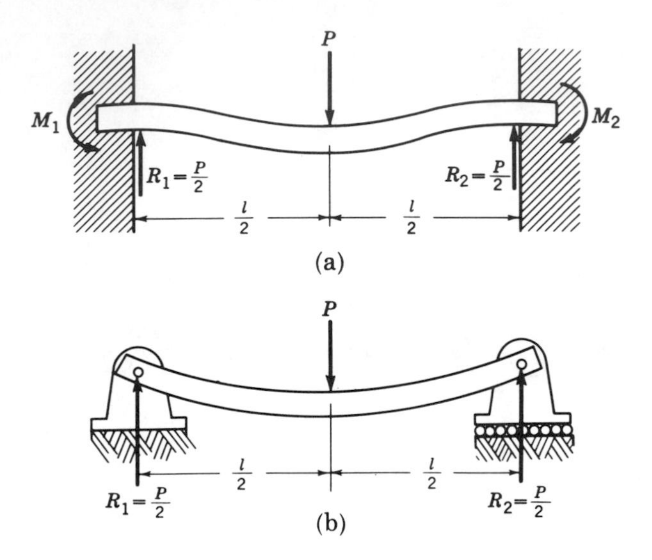

Figure 9-1

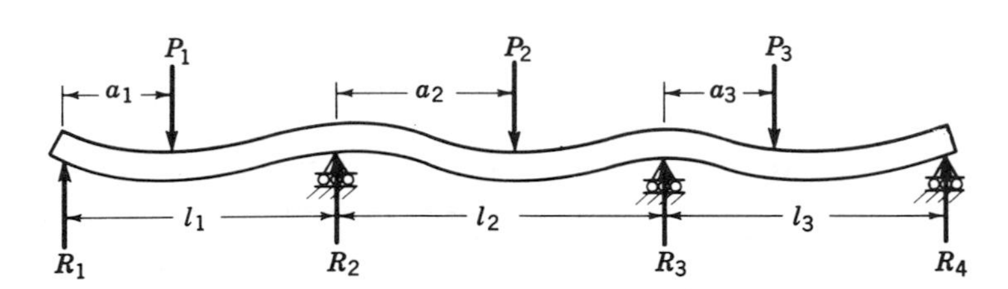

Figure 9-2

9-2 What Happens When End Moments Act Alone on a Beam

One of the effects of applying the counterclockwise couple M_1 to the left end of the beam shown in Fig. 9-3a is to cause the reaction R_2' to act downward. The beam is, therefore, forced to deflect upward, creating negative bending over its entire length. The variation of this bending moment is triangular, as shown in Fig. 9-3b.

If, on the other hand, the same beam were to be acted upon by a clockwise couple M_2 located at its right end, the reaction R_1'' would act down and the bending-moment diagram would again be of triangular shape, zero bending moment occurring this time at R_1'' (Fig. 9-4b).

Should both M_1 and M_2 happen to act simultaneously on the same beam, the bending-moment diagram would by virtue of the law of superposition be a combination of Figs. 9-3b and 9-4b. If, as shown in Fig. 9-5, M_2 is smaller than M_1, the composite-moment diagram is a trapezoid, having

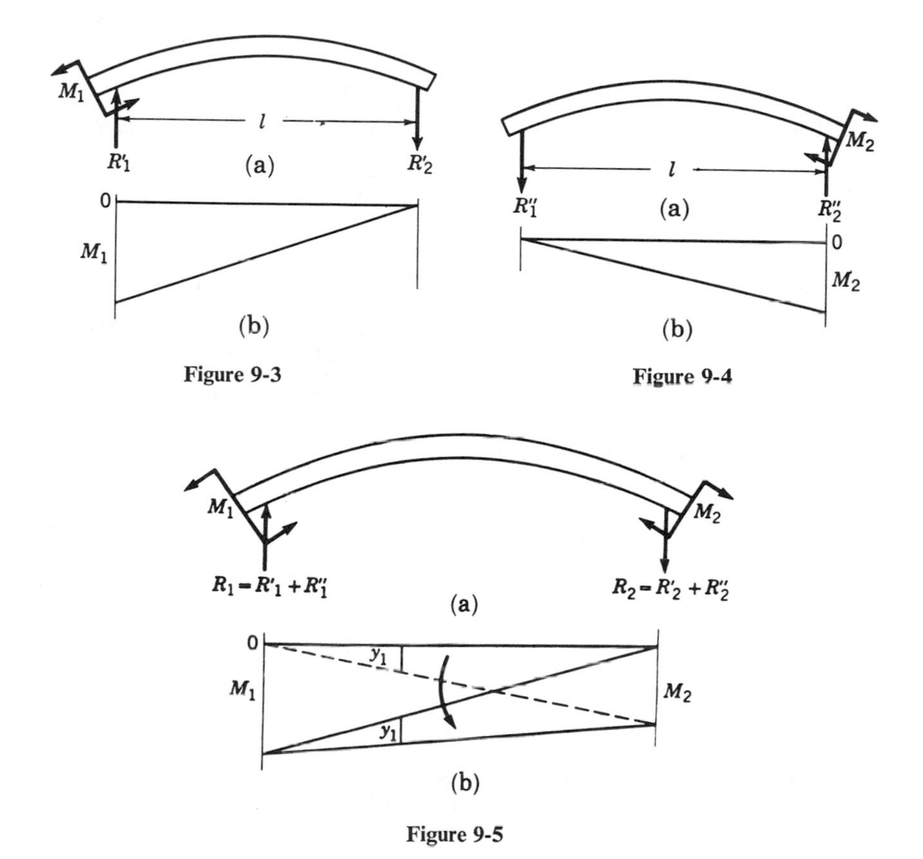

Figure 9-3

Figure 9-4

Figure 9-5

its greatest ordinate at M_1. Note that that portion created by M_2 is not left in an overlapping position with respect to the moment diagram created by M_1, but in a complementary position where the ordinates of each are additive. If M_2 were greater than M_1, the slope of the lower boundary line would be reversed. Having M_1 equal to M_2 would produce a negative rectangular bending-moment diagram.

9-3 What Happens When End Restraints Act on a Loaded Beam

When restraining moments, such as M_1 and M_2, are added in turn to a loaded simple beam, their effect is to rotate the end upon which they act in a direction opposite to that produced by the load. In doing so, they impose upon the bending-moment diagram of the simple beam through the application of the principle of superposition negative bending-moment diagrams of the types shown in Figs. 9-3b and 9-4b. When the magnitudes of restraining moments are such as to rotate the ends of beams on which they act into a horizontal position, they are said to produce ends that are *completely fixed.*

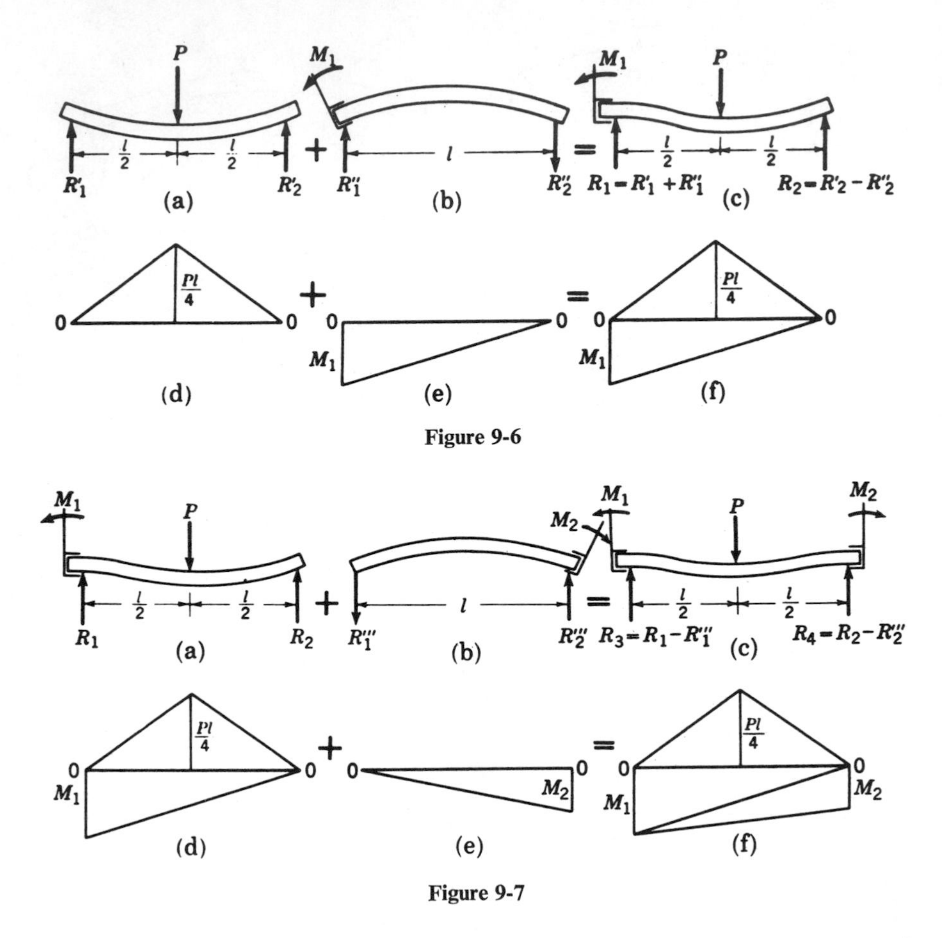

Figure 9-6

Figure 9-7

The addition of these negative diagrams to the bending-moment diagram of a loaded simple beam is shown in Figs. 9-6 and 9-7.

To convey more clearly the rotational effect of a couple, an ordinary *monkey wrench* has been substituted for the more conventional couple representation. The gradual application of a restraining force to the handle of the wrench will

1. Increase the value of the negative bending moment and the reaction at its point of application in exact proportion to the pull or moment applied to the wrench.

2. Decrease the magnitude of the opposite reaction.

3. Decrease and change the location of the maximum deflection of the beam.

4. Impart a reversed curvature to the elastic curve of the beam in the vicinity of its point of application.

If an end restraining moment acting on a beam is incapable of rotating its end to a horizontal position, that end is said to be *partially fixed*. Whether an end moment creates complete or partial fixity, it is represented by an equivalent couple acting in the direction of the restraint.

9-4 Use of the Law of Superposition

As shown in Figs. 9-6 and 9-7, the bending-moment diagram of a loaded beam with end restraints may be compiled by adding thereto the separate bending-moment effects induced by the applied loads and couples. This important principle, made possible by the law of superposition, is used extensively in the study of statically indeterminate beams.

For instance, if a beam with completely fixed ends were acted upon by a uniform load, the bending-moment diagram (Fig. 9-8) would be constructed

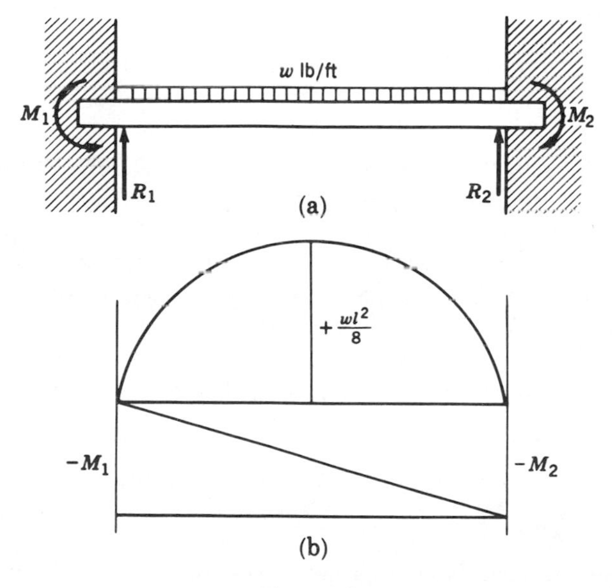

Figure 9-8

by first drawing the positive parabolic-moment diagram for the simple beam and then adding the two equal negative triangular restraining-moment diagrams below the axis. The composite negative portion of the diagram would then be a rectangle, since no diagrams are overlapped. Should one of the ends of the beam be only partially restrained, the bending moment at that point would be reduced with the result that the negative portion of the composite diagram would become a trapezoid.

As a general rule, the drawing of a generalized bending-moment diagram for a statically indeterminate beam proceeds according to the following steps:

1. Assume the beam to be statically determinate by relaxing the end restraints. In case of a continuous beam each span is to be similarly regarded —the relaxation of the restraining moments at each support making each span a simple beam.

2. Draw the bending-moment diagram for the relaxed simple beam either in the combined or separated form, depending on which is more convenient.

3. Remove the loads applied to the simple beam considered above, and reapply all previously relaxed restraining moments.

4. Draw successively the bending-moment diagram for each reapplied restraining moment on the negative side of the axis used in step 2. Avoid any overlapping of the diagrams. The restraining-moment diagrams are of a general character, since the moments themselves are unknown quantities.

5. The positive and negative bending-moment diagrams thus obtained constitute the generalized bending-moment diagram for the given beam when all loads and restraining moments are applied.

APPLICATION OF MOMENT-AREA PRINCIPLES TO STATICALLY INDETERMINATE PROBLEMS

In the previous chapter, the principles of moment area were used exclusively for the determination of slopes and deflections of simple beams. In the solution of statically indeterminate beams, *the principles of moment area may also be used to determine the magnitude of restraining moments.* The following classical solutions indicate their use.

9-5 Beam Fixed* at Both Ends—Concentrated Load at Midspan

Determination of restraining moments. Because of the presence of two unknown end restraints, two unknown reactions, and the applicability of only two equilibrium equations, the beam shown in Fig. 9-9 is statically indeterminate. In accordance with the first principle for drawing the bending-moment diagram of statically indeterminate beams, let us relax the end restraints. In other words, temporarily remove M_1 and M_2. The beam will then be a simple beam having a positive isosceles-triangle bending-moment

*When the term *fixed beam* is used without any qualifying adjective it is understood to mean *completely fixed.*

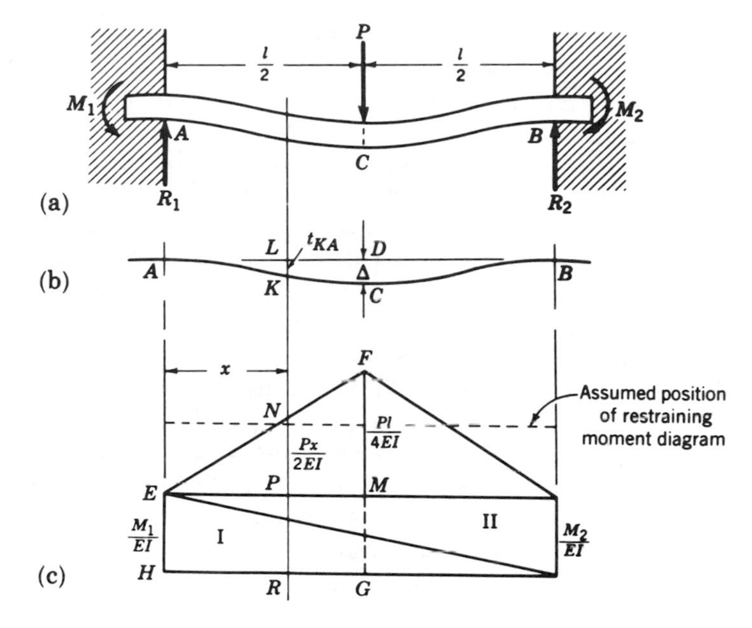

Figure 9-9 E and I are constant values.

diagram with a maximum ordinate of $Pl/4$. Now, remove the load P and reapply in turn the restraining moments M_1 and M_2. Their bending-moment diagrams are represented by triangles I and II, respectively. Because of the symmetrical loading condition imposed upon the beam, M_1 and M_2 are equal, making triangles I and II congruent. Having now completed the drawing of the bending-moment diagram, let us turn our attention to the application of the principles of moment area.

From a perusal of Fig. 9-9b, it is evident that *the change in slope from A to B is equal to zero*, since both ends are completely fixed. Thus, by applying principle 1 of moment area, it follows that *the algebraic sum of the M/EI diagram between points A and B must also equal zero*. Considering M_1 and M_2 to be positive until proven otherwise, the summation equals

$$\frac{Pl}{4EI} \times \frac{l}{2} + \frac{M_1}{EI} \times l = 0, \qquad M_1 = -\frac{Pl}{8} \tag{9.1}$$

and because $M_1 = M_2$,

$$M_2 = -\frac{Pl}{8} \tag{9.2}$$

Algebraic sign of restraining moments. In the computation of M_1 and M_2, their algebraic signs were assumed to be *positive*, even though the bending-moment diagrams representative of their effect were drawn below the axis. Although this may seem contradictory, it is actually a very convenient procedure. Let us see how it works.

Whenever restraining moments such as M_1 and M_2 are assumed positive and become negative upon computation, the assumption is shown to have been wrong. This is a purely algebraic deduction. If, however, the negative sign obtained by calculation corresponds to the standard moment criterion, the restraining moment is actually a negative moment producing tension on the upper fibers of the beam. No further adjustment of sign is necessary. Moreover, if the positively assumed moments M_1 and M_2 are verified as such by a plus sign, the assumption is correct and its sign is again in conformity with the standard moment criterion. *Regardless of the apparent inconsistency of the assumed sign, the computed sign will be in agreement with the actual moment present.* This method is the one to be used in this text.

How much better is this method than that using the opposite approach! Suppose a negative moment were, in its computation, initially assumed negative. The answer would be positive, showing that the initial assumption was correct. And if it were actually a positive moment and initially assumed negative, the answer would be negative in contradiction to the initial assumption. There would be no conformity between the sign of the moment and the moment criterion. Moreover, this swift interchange of algebraic signs, although producing correct results, is also productive of numerous errors and should be avoided.

When computing restraining moments, therefore, it is recommended that the restraining-moment diagrams be drawn below the axis for convenience. In this position, they will not interfere with any positive-moment diagram drawn for the simple beam and will, in 99 cases out of 100, be in their correct general position. The unknown restraining moments should then be assumed positive, since the algebraic sign of the computed value will be the correct one.

The above method pertains only to the determination of restraining moments. Slopes and tangential deviations are computed solely on the basis of algebraic summations after the bending-moment diagram has been established.

Maximum deflections. Because of the symmetrical loading, the maximum deflection occurs at the center of the beam. To determine its method of computation, draw, first, the horizontal line AB. Since line AB is tangent to the beam at A, the maximum deflection is also made equal to the tangential deviation at the center taken between points A and C. Its computation, according to principle 2 of moment area, is equal to the moment of area $EFGH$ taken about DC or EG. But the moment of area $EFGH$ equals the moment of area EFM minus the moment of area $EMGH$; therefore,

$$\begin{aligned}\Delta &= \frac{1}{2} \times \frac{Pl}{4EI} \times \frac{l}{2} \times \frac{1}{3} \times \frac{l}{2} - \frac{Pl}{8EI} \times \frac{l}{2} \times \frac{l}{4} \\ &= \frac{Pl^3}{96EI} - \frac{Pl^3}{64EI} = -\frac{1}{192}\frac{Pl^3}{EI}\end{aligned} \tag{9.3}$$

This maximum deflection is one quarter of the corresponding deflection of a simple beam similarly loaded (see §8-12).

Deflection at any point. Suppose the deflection is desired at any point on the beam of Fig. 9-9, located x distance from the left reaction. By observation, the deflection LK at that point is also equal to the tangential deviation t_{KA}. Thus, taking the algebraic sum of the moments of the M/EI areas between A and K about LK, we have

$$t_{KA} = LK = \frac{Px}{2EI} \times \frac{x}{2} \times \frac{x}{3} - \frac{Pl}{8EI} \times x \times \frac{x}{2} = \frac{Px^3}{12EI} - \frac{Plx^2}{16EI} \tag{9.4}$$

Slope at any point. The slope at any point on the beam of Fig. 9-9 located x distance from the left reaction is equal to the algebraic sum of the M/EI area between the point in question and the point of horizontal tangency at A. Thus,

$$\theta_x = \frac{Px}{2EI} \times \frac{x}{2} - \frac{Pl}{8EI} \times x = \frac{Px^2}{4EI} - \frac{Plx}{8EI} \tag{9.5}$$

9-6 Beam Fixed at One End—Freely Supported at the Other—Concentrated Load at Center

Determining the restraining moment M_1. The analysis of this beam, Fig. 9-10, is initiated as before with the drawing of the M/EI diagram in compliance with the principles previously explained in §9-4. With the removal of the restraining moment M_1, the beam becomes a simple beam whose variation in moment is represented by the positively located isosceles triangle shown in Fig. 9-10c. Removing load P and reapplying an unknown M_1 imposes a uniformly varying moment indicated by line FH placed below the axis line for convenience. The exact location of the line will remain in doubt until the sign and magnitude of M_1 are determined by computation. The simultaneous actions of both P and M_1 result in a combined M/EI diagram consisting of triangles DEF and DFH.

The principles of moment area may again be applied by extending the tangent at A through point B, thus making the tangential deviation at B between points A and B equal to zero. Expressing the tangential deviation at point B in equation form, assuming M_1 to be positive, we get

$$\begin{aligned} t_{BA} &= \text{moment of area } DEF \text{ about vertical through } B \\ &\quad + \text{moment of area } DHF \text{ about vertical through } B = 0 \\ &= \frac{Pl}{4EI} \times \frac{l}{2} \times \frac{l}{2} + \frac{M_1}{EI} \times \frac{l}{2} \times \frac{2}{3} \times l = 0 \end{aligned}$$

from which

$$M_1 = -\tfrac{3}{16}Pl \tag{9.6}$$

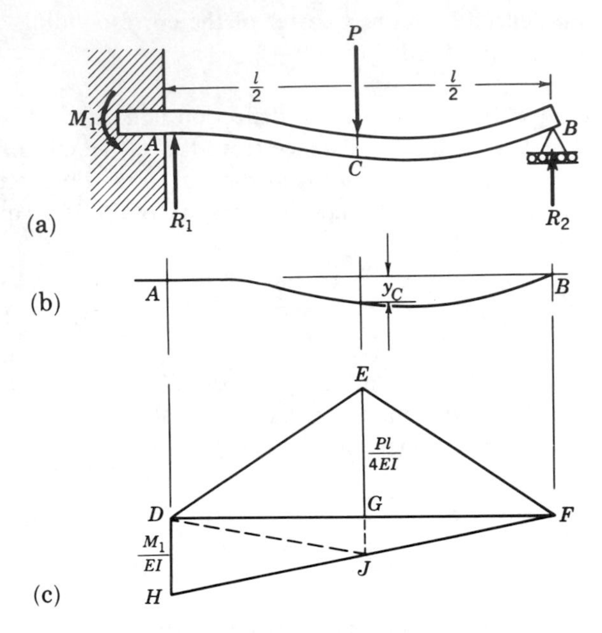

Figure 9-10

The negative sign resulting from this computation reveals that the positive assumption was incorrect. Because of its negative value, it also reveals that the upper fibers of the beam must be in tension and that, as a consequence, M_1 must act counterclockwise. The resulting M/EI diagram is, therefore, shown correctly in Fig. 9-10c with M_1 in a negative position.

Determination of reaction R_1. Because one of the three unknowns acting on the beam has now been found, the beam is no longer statically indeterminate. Reaction R_1 may consequently be determined by taking the algebraic sum of the moments of the forces acting on the free body (Fig. 9-10a) about A.

Note. The restraining moment M_1 acts in the same direction as reaction R_2 about point A.
Therefore,

$$\sum M_A = R_2 l - \frac{Pl}{2} + \frac{3}{16} Pl = 0 \tag{9.7}$$

$$R_2 = \tfrac{5}{16} P$$

The value of R_1 may then be found by setting the algebraic sum of the vertical forces equal to zero.

$$\sum V = R_1 - P + \tfrac{5}{16} P = 0$$

$$R_1 = \tfrac{11}{16} P$$

Having thus determined all the vertical forces acting on the beam, the shear diagram may now be drawn.

Deflection under the load. To obtain the deflection under load P, compute the tangential deviation at C between points C and A. Thus,

$$
\begin{aligned}
y_C = t_{CA} &= \text{moment of area } DEG \text{ about } EJ \\
&\quad - \text{moment of area } DGJH \text{ about } EJ \\
&= \frac{Pl}{4EI} \times \frac{1}{2} \times \frac{l}{2} \times \frac{1}{3} \times \frac{l}{2} - \frac{3}{16}\frac{Pl}{EI} \times \frac{1}{2} \times \frac{l}{2} \times \frac{2}{3} \times \frac{l}{2} \\
&\quad - \frac{3}{32}\frac{Pl}{EI} \times \frac{1}{2} \times \frac{l}{2} \times \frac{1}{3} \times \frac{l}{2} \qquad (9.8) \\
&= \frac{Pl^3}{96EI} - \frac{6}{384}\frac{Pl^3}{EI} - \frac{3}{768}\frac{Pl^3}{EI} = -\frac{7}{768}\frac{Pl^3}{EI}
\end{aligned}
$$

Deflection at any point on the beam. The deflection of any point on the beam of Fig. 9-10 is obtained in the same manner as it is for the midpoint. Because the tangent at A is horizontal, any tangential deviation located between the elastic curve and the tangent is equal to the deflection at that point.

Maximum deflection. Inasmuch as only one restraining moment acts on the beam, the point of maximum deflection is forced off center in a direction away from the restrained end, that is, on the right side of load P (Fig. 9-11).

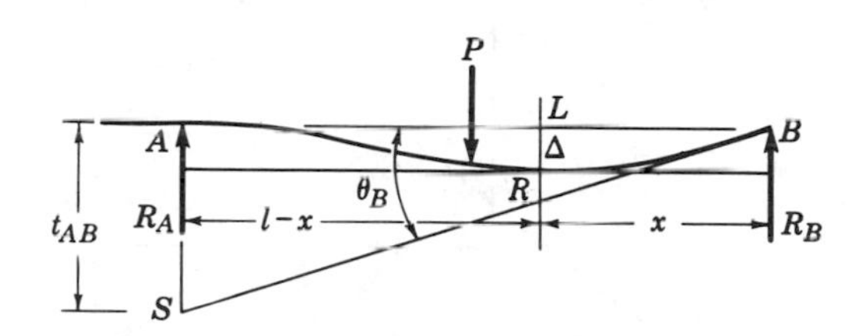

Figure 9-11

Arbitrarily choosing some point D as the point of maximum deflection (or horizontal slope), one must equate the moments of the M/EI areas on either side of LD about the verticals passing through A and B, since the tangential deviations thus obtained will be equal. Because the equation contains only one variable, x, its solution provides the location of the point of maximum deflection ($x = \frac{4}{9}l$, approximately). The location of the maximum deflection may also be found by differentiating the elastic-curve equation of the beam between load P and B and setting it equal to zero.

Slope at any point. To obtain the slope at any point on the beam shown in Fig. 9-10, the algebraic sum of the M/EI area is taken from the horizontally fixed end A to the point in question. Should the end at A be only *partially* *fixed*, the slope at any point would have to be referenced to the zero slope at the point of maximum deflection or to the slope at some point such as *B*, obtained by dividing the tangential deviation t_{AB} (Fig. 9-11) by the length of the beam.

9-7 Composite Bending-Moment Diagram for Restrained Beams

In each of the foregoing discussions the M/EI diagrams (pertains also to the bending-moment diagrams) were drawn in two parts showing the separate effects of the loads and the restraining moments. But by the law of superposition, these two separate effects may be algebraically added to obtain a composite M/EI (or bending-moment) diagram. Thus, the algebraic sum of the ordinates to the M/EI or bending-moment diagrams at any one point on the beam is the actual M/EI, or bending moment at the point.

Referring to Fig. 9-12, the bending-moment diagram for a fixed-ended

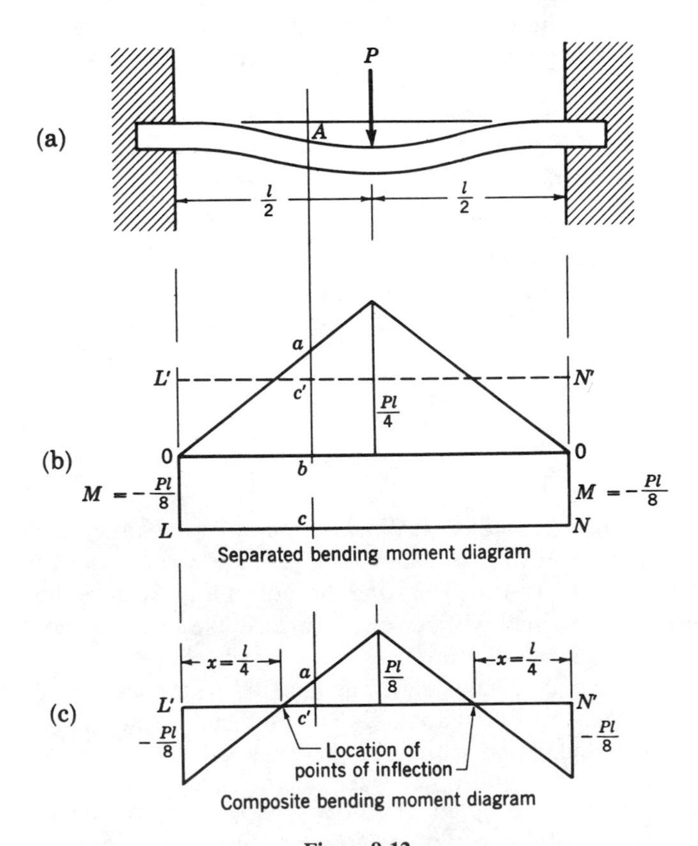

Figure 9-12

beam loaded with a concentrated load at the center, we may perform the algebraic summation at any point A by subtracting the value of bc from ab to obtain the length ac'. The location of c' may also be found geometrically by rotating bc perpendicularly about the zero axis. In fact, this geometric subtraction may be performed for every ordinate in one step by rotating the entire negative-moment diagram about the zero axis, locating LN in a new position $L'N'$. Any positive excess will lie above line $L'N'$. Algebraic summations resulting in negative values will be located below line $L'N'$. Line $L'N'$ is therefore the new zero-moment line, and all magnitudes of M/EI in the composite-moment diagram must be referenced to it.

The points of zero moment or points of inflection determined by the superposition of the negative-moment diagram on the positive-moment diagram are easily computed. From an inspection of the resulting similar triangles, the distance from either end to the closest inflection point is $l/4$.

The composite bending-moment diagram for a beam fixed at only one end with a concentrated load at its center is shown in Fig. 9-13c. A somewhat more favored arrangement places $L'O$ in a horizontal position, as shown in Fig. 9-13d. Both Figs. 9-13c and d are equally correct.

Illustrative Problem 1. Determine the slope and deflection at one of the quarter points of the completely restrained beam shown in Fig. 9-14, loaded with a concentrated load P at the other quarter point. Place the answer in terms of P, l, E, and I.

SOLUTION. The bending moment of the beam after eliminating the effects of the restraining moments M_1 and M_2 is shown as the positive triangle in Fig. 9-14b. The moment diagram for the restraining moments M_1 and M_2 will be a trapezoid, since the beam is not symmetrically loaded. It is placed on the negative side of the axis until such time as M_1 and M_2 (assumed positive) can be definitely ascertained in both magnitude and sign.

Determination of restraining moments. Since the tangents at both A and B are horizontal and collinear, the change in slope from A to B must be zero. Thus, we may set the algebraic sum of the entire M/EI diagram equal to zero:

$$\Delta\theta = \frac{3}{16}\frac{Pl}{EI} \times \frac{l}{2} + \frac{M_1 l}{2EI} + \frac{M_2 l}{2EI} = 0$$

whence

$$M_1 + M_2 = -\tfrac{3}{16} Pl$$

Still another equation is necessary for the solution of the above equation, since two unknowns M_1 and M_2 are involved. This may be obtained by equating the moments of the M/EI areas about the vertical line passing through B, since the tangential deviation at this point will be zero.

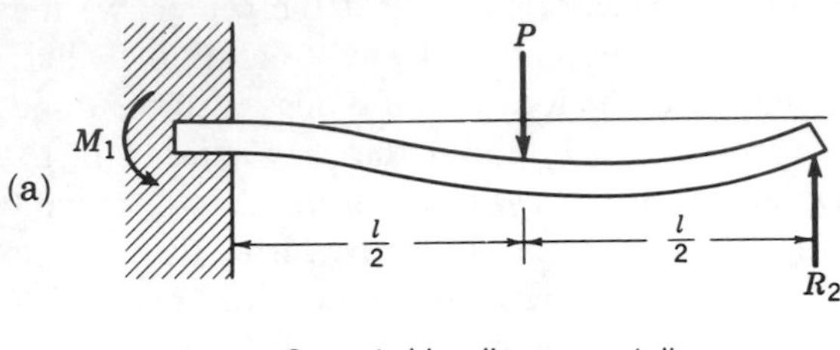

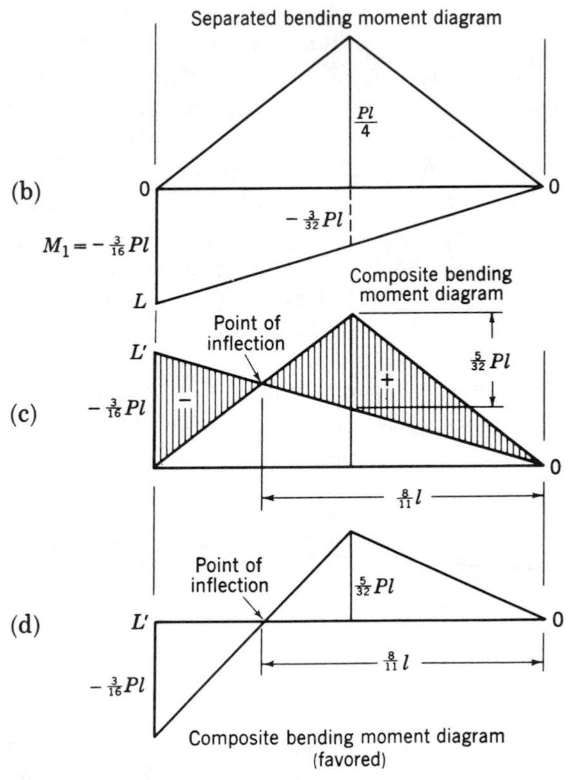

Figure 9-13

$$t_{BA} = \frac{3}{16}\frac{Pl}{EI} \times \frac{l}{4} \times \frac{1}{2} \times \frac{2}{3} \times \frac{l}{4} + \frac{3}{16}\frac{Pl}{EI} \times \frac{3}{4}l \times \frac{1}{2} \times \frac{l}{2}$$

$$+ \frac{M_1}{EI} \times \frac{l}{2} \times \frac{2}{3}l + \frac{M_2}{EI} \times \frac{l}{2} \times \frac{l}{3} = 0$$

whence

$$2M_1 + M_2 = -\tfrac{15}{64}\,Pl$$

Solving for the unknown moments M_1 and M_2 gives

$$M_1 = -\tfrac{3}{64}\,Pl \tag{9.9}$$

$$M_2 = -\tfrac{9}{64}\,Pl \tag{9.10}$$

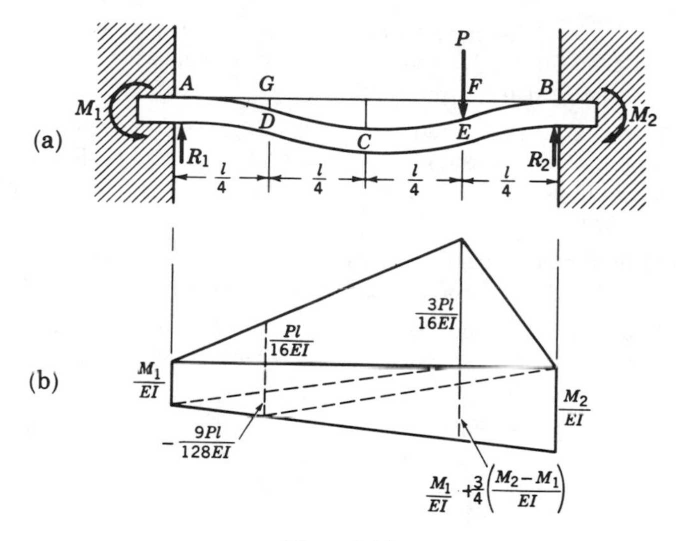

Figure 9-14

Deflection at quarter point D. $y_D = t_{DA} =$ algebraic sum of the moments of M/EI diagram between A and D taken about DG

$$= \frac{1}{2} \times \frac{1}{16}\frac{Pl}{EI} \times \frac{l}{4} \times \frac{1}{3} \times \frac{l}{4} - \frac{1}{2} \times \frac{9}{128}\frac{Pl}{EI} \times \frac{l}{4} \times \frac{1}{3} \times \frac{l}{4}$$

$$- \frac{1}{2} \times \frac{3}{64}\frac{Pl}{EI} \times \frac{l}{4} \times \frac{2}{3} \times \frac{l}{4} \tag{9.11}$$

$$= \left(\frac{Pl^3}{1536} - \frac{3}{4096}Pl^3 - \frac{Pl^3}{1024}\right)\frac{1}{EI} = -\frac{13}{12{,}288}\frac{Pl^3}{EI}$$

Slope at quarter point D. $\theta_D =$ algebraic sum of M/EI diagram between A and D

$$= \frac{1}{2} \times \frac{1}{16}\frac{Pl}{EI} \times \frac{l}{4} - \frac{1}{2} \times \frac{9}{128}\frac{Pl}{EI} \times \frac{l}{4} - \frac{1}{2} \times \frac{3}{64}\frac{Pl}{EI} \times \frac{l}{4}$$

$$= \frac{Pl^2}{128EI} - \frac{9}{1024}\frac{Pl^2}{EI} - \frac{3}{512}\frac{Pl^2}{EI} = -\frac{7}{1024}\frac{Pl^2}{EI} \tag{9.12}$$

PROBLEMS

9-1. A steel S beam 20 ft long ($I = 100$ in.^{4}) is fixed at one end. Located 8 ft from its freely supported end is a 10-ton load. Draw its shear- and bending-moment curves.

9-2. Determine the deflection of the beam of Problem 9-1 directly under the load. *Answer:* 0.903 in.

9-3. Determine the location and magnitude of the maximum deflection of the beam of Problem 9-1.

9-4. An S 10 × 25.4 steel beam, 12 ft long, is fixed at both ends. If it is subjected to a concentrated load of 7500 lb at the center, what will the deflection be directly under the load? *Answer:* 0.0318 in.

9-5. If the fixity at one end of the beam in Problem 9-4 were to be released, what would be the new deflection and slope under the load? Discuss briefly the effect on the bending-moment diagram caused by a gradual relaxation of the one restraining moment.

9-6. Determine the location and magnitude of the maximum deflection of the beam of Problem 9-5. *Answers:* 5.37 ft from simple support, 0.0570 in.

9-7. Draw the shear- and bending-moment diagram for a beam of a length l fixed at one end and freely supported at the other, and loaded at the third points with loads equal to P. *Answer:* $M_{max} = -Pl/3$.

9-8. A 6- by 12-in. timber beam 18 ft long ($E = 1.2 \times 10^6$ psi) is fixed at both ends. It is subjected to a load of 4000 lb at a point 6 ft from one end. Draw the shear- and bending-moment diagram. *Answer:* Maximum moment = 10,680 ft-lb.

9-9. Select an S beam 15 ft long to carry a load of 10,000 lb located 10 ft from its fixed end. The other end is freely supported. Maximum allowable fiber stress is 20,000 psi. *Answer:* S 8 × 23.

9-10. Two horizontal cantilever beams of the same material are supported, as shown in Fig. 9-15, with a cylindrical roller resting on the lower beam and touching the upper beam. I of $BC = 2 \times I$ of AB. Determine V_A, V_C, M_A, and M_C due to the application of loads P and $P/2$ as shown.

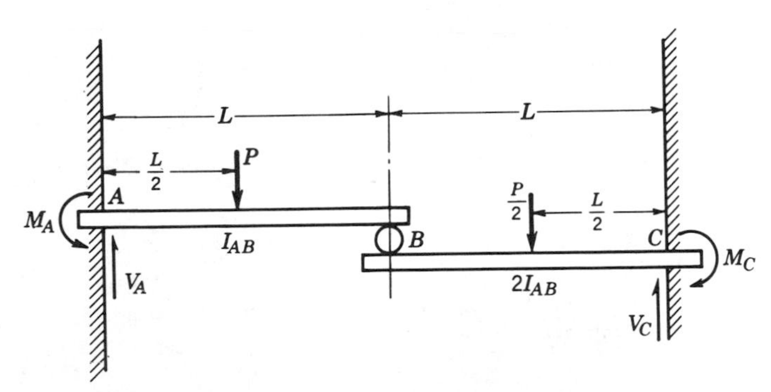

Figure 9-15

9-8 Uniformly Loaded Beam—Fixed at Each End

As indicated in the preceding chapter, uniformly loaded beams offer no difficulty in the applications of the principles of moment area, aside from the determination of the area moments themselves.

Consider the beam shown in Fig. 9-16. As in all statically indeterminate beams, let us first determine the restraining moments M_1 and M_2.

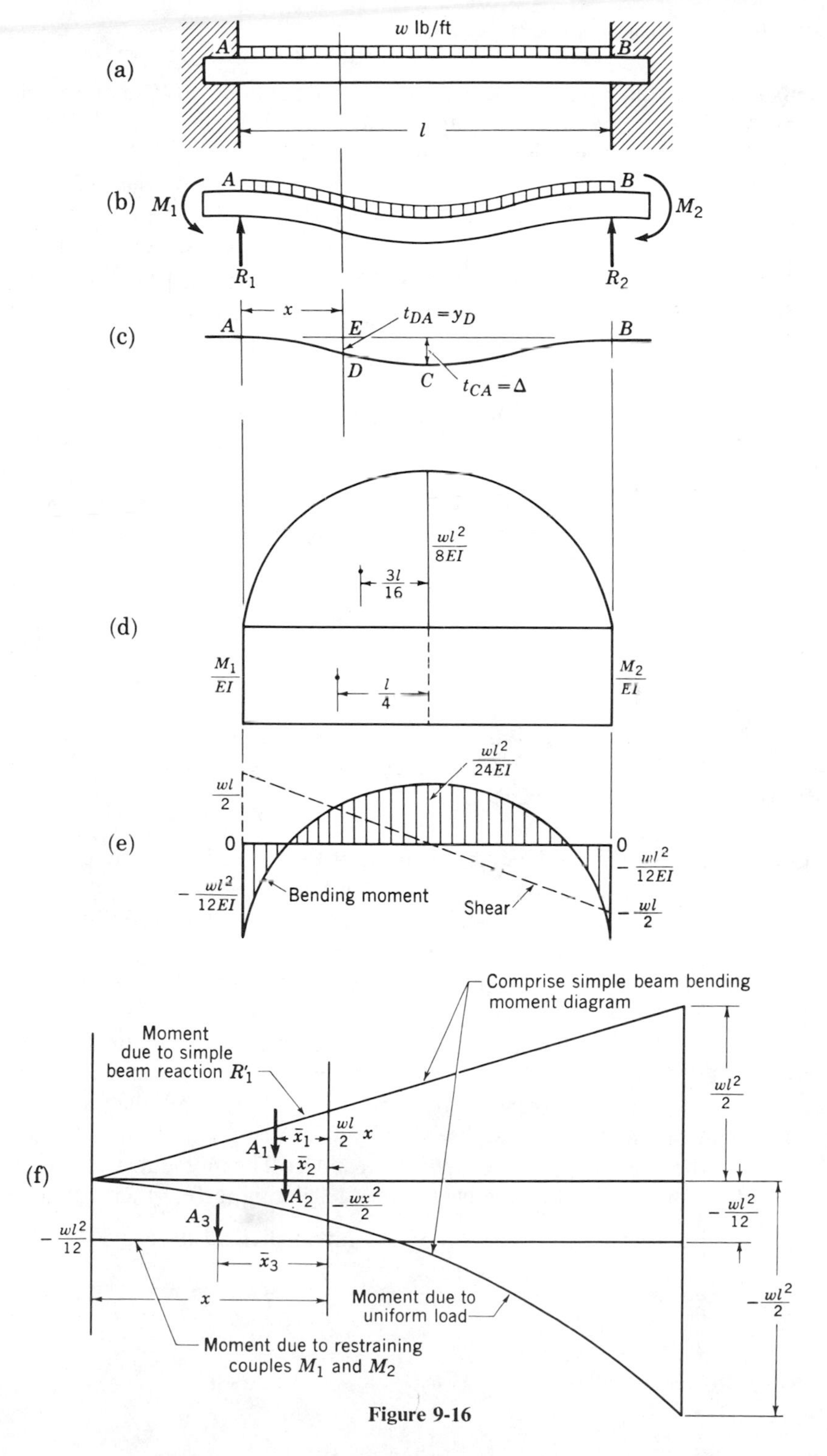

w lb/ft
A
B
(a)
l
(b)
M_1
M_2
R_1
R_2
(c)
x
$t_{DA} = y_D$
E
D
C
$t_{CA} = \Delta$
(d)
$\frac{wl^2}{8EI}$
$\frac{3l}{16}$
$\frac{M_1}{EI}$
$\frac{l}{4}$
$\frac{M_2}{EI}$
(e)
$\frac{wl^2}{24EI}$
$\frac{wl}{2}$
0
$-\frac{wl^2}{12EI}$
Bending moment
Shear
$-\frac{wl}{2}$
(f)
Comprise simple beam bending moment diagram
Moment due to simple beam reaction R'_1
A_1
$\bar{x}_1$
$\frac{wl}{2}x$
$\bar{x}_2$
A_2
$-\frac{wx^2}{2}$
A_3
$-\frac{wl^2}{12}$
$\bar{x}_3$
x
$\frac{wl^2}{2}$
$-\frac{wl^2}{12}$
$-\frac{wl^2}{2}$
Moment due to uniform load
Moment due to restraining couples M_1 and M_2

Figure 9-16

Computation of restraining moments. With only the uniform load acting on the beam, the bending-moment diagram produced is entirely positive, and is shown as the parabolic curve in Fig. 9-16d. Immediately below the axis line is the rectangular bending-moment diagram produced by the action of the equal restraining moments M_1 and M_2. Because R_1 and R_2 are on the same elevation and the tangents at each end are horizontal, the change in slope, or the algebraic sum of the M/EI diagram between the ends, is equal to zero. Thus,

$$\frac{2}{3} \times \frac{wl^2}{8EI} \times l + \frac{M_1}{EI} \times l = 0$$

$$M_1 = -\frac{wl^2}{12} \quad \text{and} \quad M_2 = -\frac{wl^2}{12} \qquad (9.13, 9.14)$$

Maximum deflection. The maximum deflection may be obtained by taking moments of the M/EI diagram between A and C about the vertical line passing through the center of the beam:

$$t_{CA} = \Delta = \frac{2}{3} \times \frac{wl^2}{8EI} \times \frac{l}{2} \times \frac{3}{8} \times \frac{l}{2} - \frac{wl^2}{12EI} \times \frac{l}{2} \times \frac{l}{4}$$

$$\Delta = -\frac{1}{384}\frac{wl^4}{EI} \qquad (9.15)$$

Comparing this deflection with that for the uniformly loaded simple beam of Prob. 8-9 reveals that fixing the ends decreases the deflection to one fifth of the simple beam value.

Deflection at any point. The determination of the deflection at any point D on the beam is simplified a great deal for two reasons: first, because each deflection is a tangential deviation, and, second, tangential deviations can be computed by taking the algebraic sum of the moments of the M/EI areas either to the right or left of their location. The latter simplification is made possible by the horizontal tangents located at each end of the beam.

Although the bending-moment diagram of Fig. 9-16d could be used for the determination of deflection at any point D, the separated diagram shown in Fig. 9-16f will prove to be more desirable. In this diagram, the simple beam moment curve has been divided into two parts, one showing the effect of the uniform load, the other the reaction. The moments were taken to the left for all sections of the beam. The deflection at D is then equal to

$$y_D = t_{DA} = A_1\bar{x}_1 - A_2\bar{x}_2 - A_3\bar{x}_3 \qquad \text{(areas divided by } EI\text{)}$$

$$= \frac{wl}{2EI}x \times \frac{x}{2} \times \frac{x}{3} - \frac{wx^2}{2EI} \times \frac{x}{3} \times \frac{x}{4} - \frac{wl^2}{12EI} \times x \times \frac{x}{2}$$

$$= \frac{2wlx^3 - wx^4 - wl^2x^2}{24EI}$$

9-9 Uniformly Loaded Beam—Fixed at One End—Freely Supported at the Other

Determination of restraining moment. The bending-moment diagram for the separate effects of the uniform load applied to the beam of Fig. 9-17 is shown in part *d*. To obtain the value of the unknown restraining moment M_2, using the application of the principle of tangential deviations, requires the location of a point on the beam where the tangential deviation is known. Such a point is located at the left end where the tangential deviation is zero. Taking the moment of the M/EI areas about the vertical passing through A, assuming M_2 positive, provides the following equation:

$$\sum M = \frac{2}{3} \times \frac{wl^2}{8EI} \times l \times \frac{l}{2} + \frac{M_2}{EI} \times \frac{l}{2} \times \frac{2}{3} l = 0$$

from which

$$M_2 = -\tfrac{1}{8} wl^2 \tag{9.16}$$

The presence of the negative sign reveals that the bending moment is opposite to the assumed positive direction. Moreover, it acts to create the effects of a negative moment, which are tensile stresses on the top and compression stresses on the bottom of the beam. Obviously, the couple to produce these effects at the right-hand end must act clockwise.

From an observation of the combined bending-moment diagram, Fig. 9-17e, it is shown that the maximum moment on the beam occurs at the fixed end. The design of all prismatic beams subjected to uniform load and completely restrained at one end is therefore controlled, insofar as bending moment is concerned, by the moment at the fixed end.

Finding the reactions. Reaction R_1 can be found by taking the moments of the forces acting on the beam about reaction R_2. Thus,

$$R_1 l - \frac{wl^2}{2} + \frac{wl^2}{8} = 0 \tag{9.17}$$

The positive sign given the last term of this equation is in accord with its proven direction and bears no relation to the type of moment involved. The solution of this equation gives

$$R_1 = \tfrac{3}{8} wl \tag{9.18}$$

Reaction R_2 is computed using $\sum V = 0$:

$$\sum V = \tfrac{3}{8} wl - wl + R_2 = 0$$

$$R_2 = \tfrac{5}{8} wl \tag{9.19}$$

The shear diagram for the beam is shown in Fig. 9-17e. From the analysis of this diagram, it is shown that the restraining moment on the right end

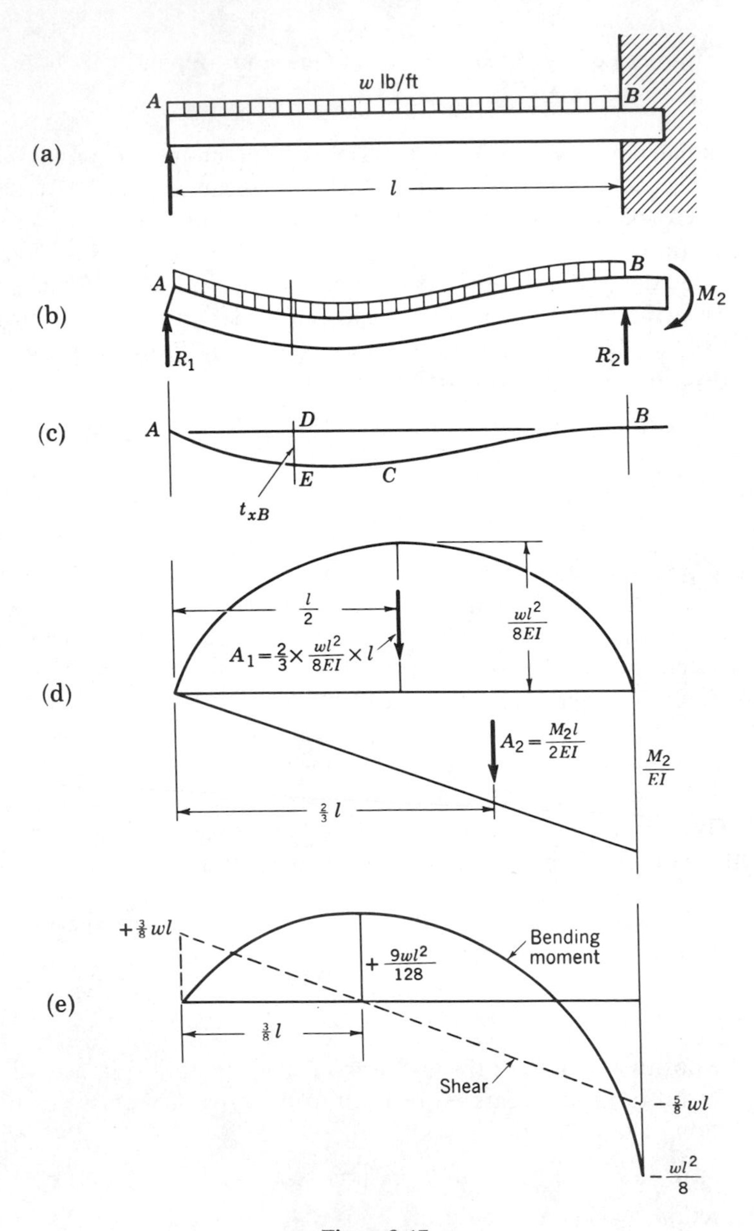

Figure 9-17a-e

of the beam moved the location of the maximum positive moment toward the left end. Its distance from the left reaction is $\frac{3}{8}l$.

Deflection at any point on beam. The deflection at any point x is most advantageously obtained by using the more elaborate separated M/EI diagram

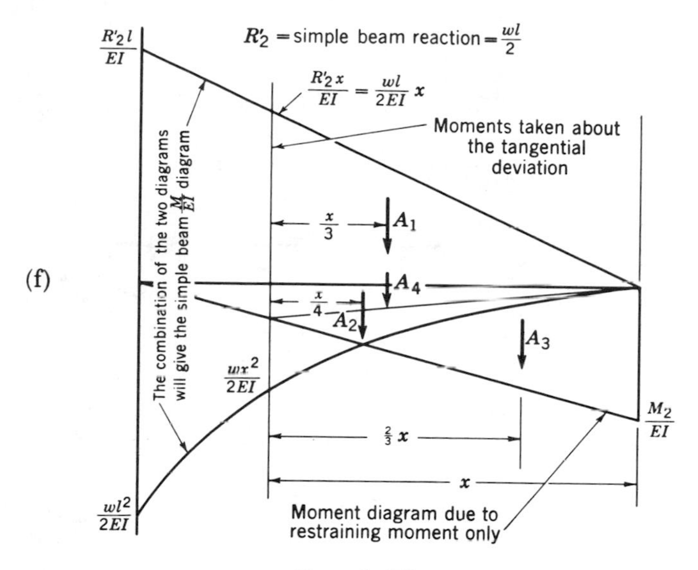

Figure 9-17f

shown in Fig. 9-17f. In this diagram the moment due to the simple beam reaction R'_2 and the uniform load are shown separately. The deflection will then be found from the expression

$$t_{xB} = y_x = A_1\frac{x}{3} - A_2\frac{x}{4} - 2A_3\frac{x}{3} - A_4\frac{x}{3}$$

Illustrative Problem 2. An S 12 × 35 beam 12 ft long and fixed at the right end is loaded as shown in Fig. 9-18a. Determine the reactions and the restraining moment at the right end. $I = 227.0$ in.^{4}.

SOLUTION. A beam, fixed at one end and simply supported at the other, is statically indeterminate to the first degree, meaning that there is one reaction component too many to make possible a solution using the equilibrium equations alone. This requires another equation to unlock the solution for any one or all the reaction components.

A slightly different approach. The one equation required can be determined most easily for the beam in question by using a technique slightly different from that previously recommended. It is dependent upon the use of the moment diagram drawn with *both* applied loads and the restraining moment applied simultaneously. The separated moment diagram (Fig. 9-18c), obtained by taking moments to the left, will then involve the *actual* value of R_A as the *sole redundant*. By taking moments to the left, the use of M_B is avoided in drawing the moment diagram. Because all the reaction components and loads were acting on the beam simultaneously, the resulting moment diagram must present the actual moments acting on the given beam.

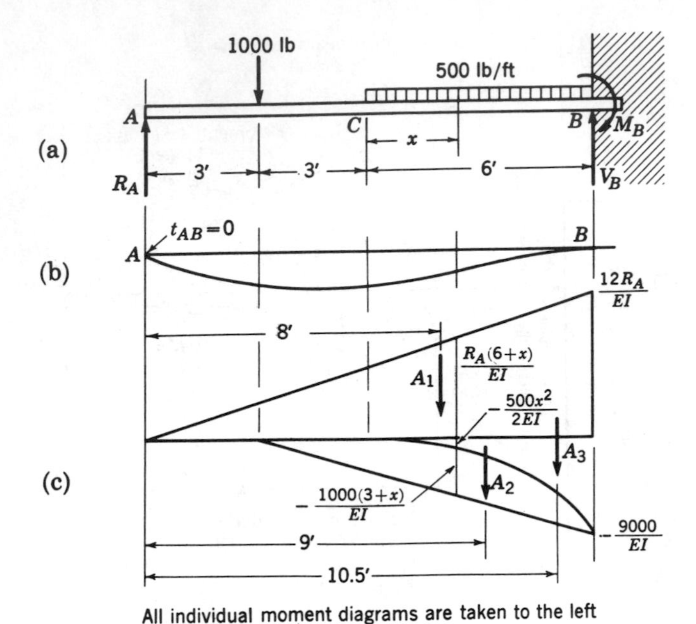

Figure 9-18

Since in Fig. 9-18b $t_{AB} = 0$, the moments of the M/EI diagrams between A and B taken about A must equal zero:

$$8A_1 - 9A_2 - 10.5A_3 = 0$$

$$\frac{12R_A}{EI} \times \frac{12}{2} \times 8 - \frac{9000}{EI} \times \frac{9}{2} \times 9 - \frac{1}{3} \times \frac{9000}{EI} \times 6 \times 10.5 = 0$$

$$576R_A - 364{,}500 - 189{,}000 = 0$$

$$R_A = \frac{553{,}500}{596} = 960 \text{ lb}$$

$$R_B = 4000 - 960 = 3040 \text{ lb}$$

$$\sum M_B = 960 \times 12 - 9000 - 9000 + M_B = 0$$

$$M_B = -6480 \text{ ft-lb}$$

Note: This illustrative problem could have been solved by using the technique recommended before, i.e., by drawing the simple beam and restraining moment diagrams separately. In that case, the number of areas included in the separated moment diagram would have been increased by one (taking moments to left). The student is urged to make this further analysis.

PROBLEMS

9-11. Calculate the position of the points of inflection of a uniformly loaded beam of length l fixed at both ends. *Answer:* $0.2113l$.

9-12. An S 8 × 23 steel beam ($I_{x-x} = 64.2$ in.^{4}) is fixed at both ends and loaded over the center half of its 16-ft length with a uniform load of 2000 lb/ft. Draw its shear and moment curves. *Answer:* $M_1 = 29{,}300$ ft-lb.

9-13. Determine the maximum deflection of the beam of Problem 9-12. *Answer:* 0.248 in.

9-14. Determine the slope and deflection of the beam of Problem 9-12 at one of its quarter points.

9-15. If one end of the beam in Problem 9-12 was simply supported, draw its shear- and bending-moment diagrams.

9-16. A 2- by 8-in. Douglas fir beam (8-in. side vertical) 12 ft long and fixed at both ends is loaded uniformly with 500 lb/ft from its left reaction to a point 4 ft from the right reaction. Using $E = 2{,}000{,}000$ psi and the dressed-size dimensions, find the deflection at the center. *Answer:* 0.325 in.

9-17. Find the deflection and slope of the loaded quarter point of the beam of Problem 9-16. *Answers:* 0.00710 rad, 0.199 in.

9-18. Draw the shear and moment curves for the beam of Problem 9-16 if the left end is simply supported.

9-19. Find the deflection at the right end of the uniform load placed on the beam of Problem 9-16 if the left end is simply supported.

9-10 Partially Restrained Beams

When the restraint at the end of a beam is such as to be incapable of maintaining a horizontal slope, the beam is said to be *partially restrained.* The effect of relaxing a complete restraint is, among other things, to decrease the negative moment applied to the beam at that point. When a single-span beam is relieved of all its end restraint, negative moment no longer exists.

For illustration, let us consider the fixed-ended uniformly loaded beam of Fig. 9-19a. With full restraint, the composite bending-moment diagram is as indicated in Fig. 9-19b with *O-O* as its base line. Suppose, now, the restraint necessary to keep the tangent horizontal at point *B* were to recede slowly in intensity. At some intermediate stage, before complete elimination of the restraint, the zero line *O-O* would move to some new position *O′-D*. The slope at *B* would no longer be horizontal. Upon complete collapse of the end restraint, the zero line would move to *O″-F*. In each of these movements, the positive diagram would remain stationary.

Should the restraint at *A* also pass through a similar recession, the zero line would move through the position *CF* to its final position *EF* upon complete relaxation. With the complete relaxation of both ends, the beam becomes a simple beam, having a bending-moment diagram with line *EF* as a base.

Thus, as restraining moments are applied to the ends of a simple beam and later removed, the only change imparted to its bending-moment diagram is in the position of its zero axis. *The overall height of the combined bending-*

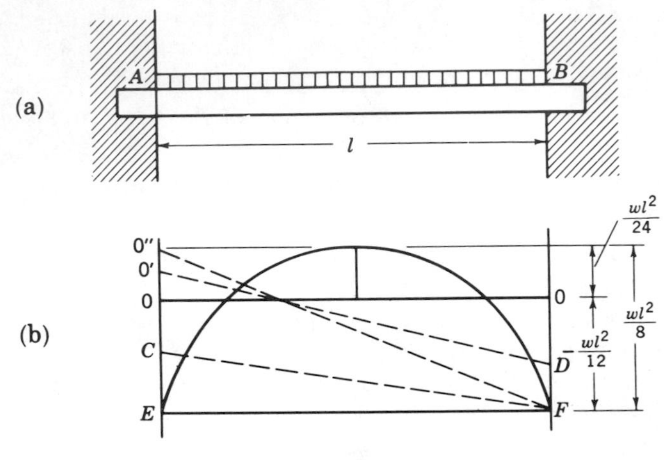

Figure 9-19

moment diagram of any partially supported beam is equal to the maximum moment of the simply supported beam.

Illustrative Problem 3. A beam 12 ft long, loaded with 1000 lb/ft, is supported and partially restrained at both ends (Fig. 9-20a). If the restraint at the left end is $-\frac{1}{24}wl^2$ and that at the right end is $-\frac{1}{16}wl^2$, what is the maximum fiber unit stress if the cross section is 6 in. wide and 12 in. deep?

SOLUTION. Since the maximum bending moment that produces the maximum bending stress is not located at the center of the beam because of unsymmetrical restraints, it is necessary to locate the point of zero shear from the shear diagram. We shall therefore proceed with the determination of the moment and shear diagrams.

$$M_1 = -\tfrac{1}{24}wl^2 = -\tfrac{1}{24} \times 1000 \times 144 = -6000 \text{ ft-lb}$$

$$M_2 = -\tfrac{1}{16}wl^2 = -\tfrac{1}{16} \times 1000 \times 144 = -9000 \text{ ft-lb}$$

$$M_c = \tfrac{1}{8}wl^2 = \tfrac{1}{8} \times 1000 \times 144 = +18{,}000 \text{ ft-lb}$$

The negative signs given to the restraining moments M_1 and M_2 reveal that they impart tensile stresses to the upper fibers of the beam. M_1 must therefore act counterclockwise and M_2, clockwise.

Let us now calculate R_1 by taking moments about R_2.

$$\sum M_{R_2} = -6000 + 12R_1 - 1000 \times 12 \times \tfrac{12}{2} + 9000 = 0$$

$$12R_1 = 69{,}000, \qquad R_1 = 5750 \text{ lb}$$

Reaction R_2 can be found by taking moments about R_1 or by balancing the vertical forces.

$$\sum V = 5750 - 1000 \times 12 + R_2 = 0$$

$$R_2 = 6250 \text{ lb}$$

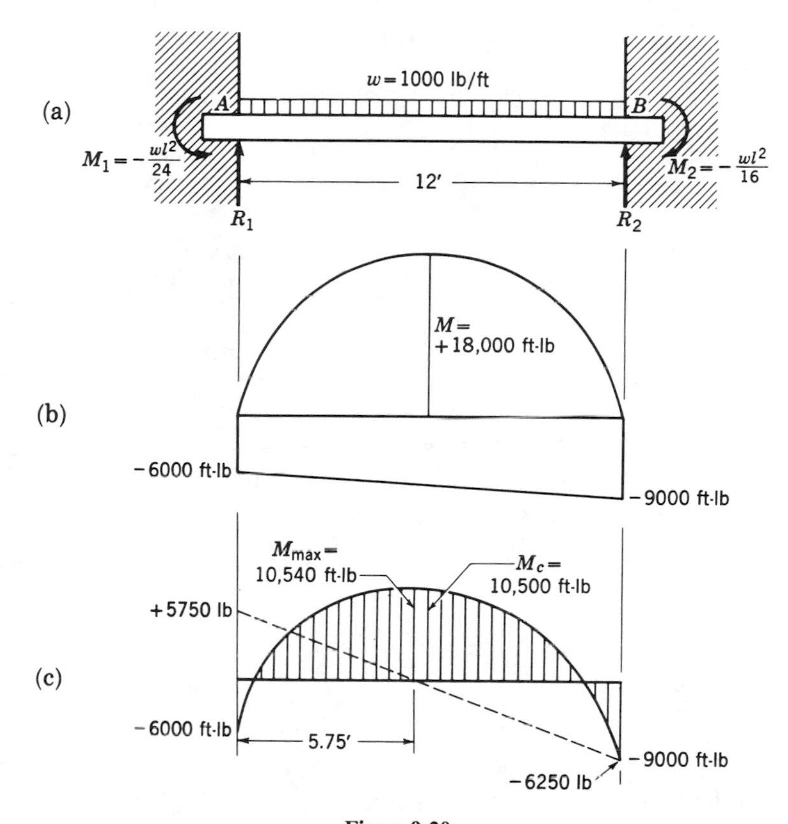

Figure 9-20

The shear diagram may now be drawn, since all the vertical forces have been evaluated.

The point of zero shear may easily be determined by the use of similar triangles.

$$\frac{5750}{6250} = \frac{x}{12 - x}, \qquad x = 5.75 \text{ ft}$$

Taking the moments to the left of the zero shear point gives the maximum positive bending moment M_{max+}.

$$M_{max+} = 5750 \times 5.75 - \frac{1000 \times \overline{5.75^2}}{2} - 6000 = 10{,}540 \text{ ft-lb}$$

Comparing the maximum positive and negative moments on the bending-moment diagram reveals that the maximum positive moment is the largest. The maximum flexural stress is therefore

$$\sigma = \frac{M_{max}c}{I} = \frac{10{,}540 \times 12 \times 6}{\frac{1}{12} \times 6 \times \overline{12^3}} = 875 \text{ psi}$$

PROBLEMS

9-20. The two bronze-pillow block bearings at each end of a 9-ft shaft 4 in. in diameter produce end moments of -9000 and -3000 ft-lb at the left and right ends, respectively. If 5000- and 3000-lb pulley loads act at the left and right third points, respectively, determine the maximum shear and flexure unit stresses produced in the shaft.

9-21. A uniformly loaded beam is initially fixed at each end. If the moment at the left end is reduced to $-wl^2/16$, what will the moment at the right end be? *Answer:* $-\frac{9}{96}wl^2$.

9-22. The ends of a uniformly loaded beam are attached to columns that bend sufficiently to make the moment at each end equal to $-wl^2/16$ when a load is applied. Calculate the moment at the middle of the beam and the position of the points of inflection. *Answers:* $wl^2/16$, $0.147l$, $0.853l$.

9-23. A careful dynamometer check of the reactions of a 30-ft steel bridge span reveals that when loaded with 6000 lb/linear ft, the reactions R_1 and R_2 are, respectively, 72,000 and 108,000 lb. If both end moments are negative, determine the difference between the two and indicate which of the two is the larger. *Answers:* M_2, 540,000 ft-lb.

9-24. A uniformly loaded beam of length l is initially fixed at each end. The left end is then relaxed to make the reaction at that point equal to $\frac{7}{16}wl$. What are the resisting moments at the ends of the beam after this adjustment? What is the slope at the left end? What is the deflection at the center of the beam?

9-25. Draw the shear- and bending-moment diagrams for the beam of Illustrative Problem 3 if the restraining moment M_1 were reversed in direction.

BEAMS ON MULTIPLE SUPPORTS

9-11 Reactions Obtained by Balancing Deflections

When a simply supported beam is bolstered by an additional reaction placed under its midpoint, its center deflection may be reduced to zero. This fact is conveniently employed in the study of statically indeterminate beams and may be illustrated as follows.

In Fig. 9-21a is shown a uniformly loaded simple beam of length l. Its maximum deflection, which occurs at its midpoint, is

$$\Delta = \frac{5}{384}\frac{wl^4}{EI}$$

In Fig. 9-21b the same beam is shown acting upward with a concentrated load R_2 at its midpoint. From §8-12, such a load was shown to produce a maximum displacement of

$$\frac{1}{48}\frac{Pl^3}{EI} \quad \text{or, in our case,} \quad \frac{1}{48}\frac{R_2 l^3}{EI}$$

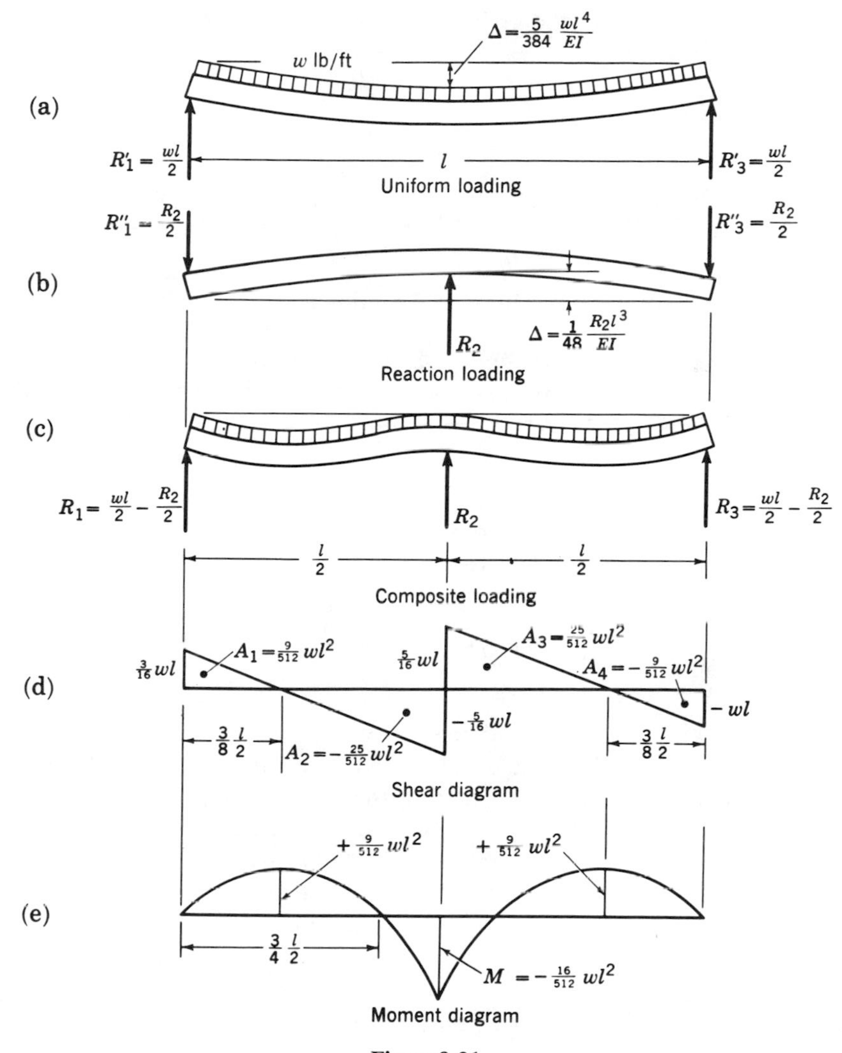

Figure 9-21

Now, if both loads are applied simultaneously, the upward deflection caused by the unknown reaction R_2 can be made equal to the maximum downward deflection caused by the uniform load. This balanced condition will place R_2 at the same elevation as R_1 and R_3, fulfilling the usual requirement of a loaded beam placed on multiple supports. The resulting equation may thereupon be solved for the reaction R_2.

$$\frac{1}{48}\frac{R_2 l^3}{EI} = \frac{5}{384}\frac{wl^4}{EI}$$

$$R_2 = \tfrac{5}{8}wl \tag{9.20}$$

From an analysis of the vertical forces, knowing that $R_1 = R_3$, we obtain

$$2R_1 + \tfrac{5}{8}wl = wl$$

$$R_1 = \tfrac{3}{16}wl \tag{9.21}$$

$$R_3 = \tfrac{3}{16}wl \tag{9.22}$$

The resulting shear diagram is shown in Fig. 9-21d.

9-12 Negative Moments at Interior Reactions

The bending-moment diagram may now be drawn by adding algebraically the shear areas from left to right. Thus, the maximum positive moment in the first span equals

$$M = \frac{1}{2} \times \frac{3}{16}wl \times \frac{3}{8}\frac{l}{2} = \frac{9}{512}wl^2 \tag{9.23}$$

The next shear area to be considered is negative and much larger than the positive area just computed. Its area

$$A_2 = -\tfrac{1}{2} \times \tfrac{5}{16}wl \times \tfrac{5}{16}l = -\tfrac{25}{512}wl^2$$

This, added algebraically to the positive area in A_1, gives

$$M_2 = \tfrac{9}{512}wl^2 - \tfrac{25}{512}wl^2 = -\tfrac{16}{512}wl^2 \tag{9.24}$$

Thus, we note for the first time the presence of a negative moment over the interior reaction. This should not come as a complete surprise, however, since the interior reaction, in assuming the same elevation as R_1 and R_3, caused the center of the beam to convex upward, thereby placing the upper fibers in tension. Since in the analysis of most beams of two or more spans the supports are considered on the same elevation, we may ordinarily expect that a negative moment exists over each interior reaction when each span is loaded.

The successive addition of A_3 and A_4 to the negative moment of R_2 will complete the development of the bending moment diagram.

9-13 Moment-Area Method of Finding Bending Moments at Interior Reactions

It would be amiss to assume that the method used in the foregoing analysis would be as easy to use for all beams placed on three supports. Not all interior reactions are located midway between the two outside reactions. Neither are the loads uniformly distributed or the end moments equal to zero. These complications make advisable another more convenient method of determining interior bending moments and reactions. Such a method is developed from the basic principles of moment area.

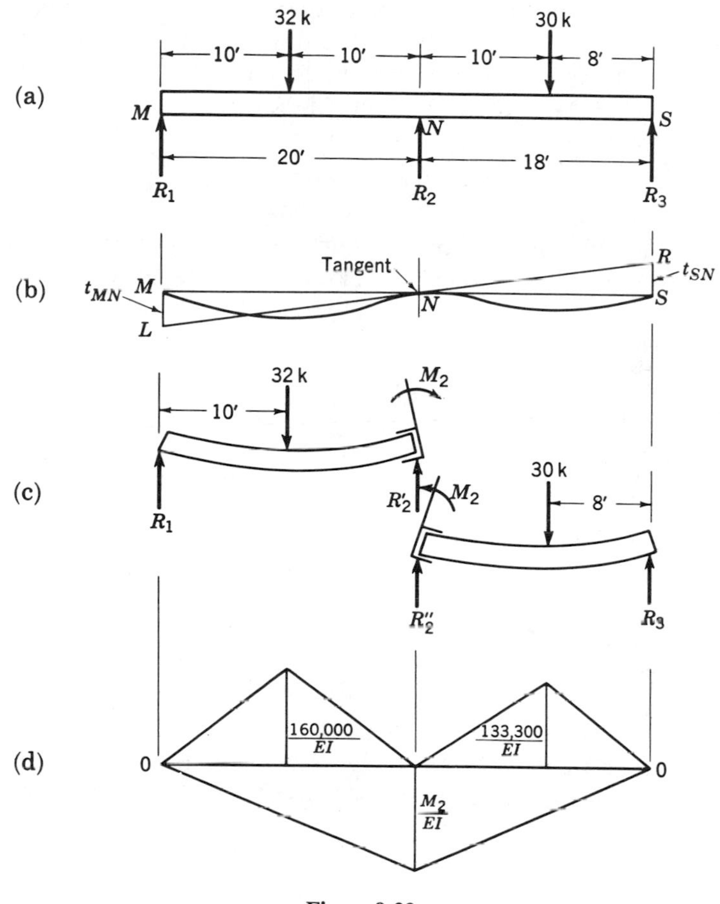

Figure 9-22

In the beam of Fig. 9-22 let it be required to obtain the bending moment over reaction R_2 as well as the magnitude of the three reactions R_1, R_2, and R_3.

To commence our discussion, let us draw a free body of each span length by passing a section through the beam along the action line of R_2. Immediately after being cut, the restraining moment at R_2 is reduced to zero and each span becomes a simple beam. The positive bending-moment diagram (divided by EI to obtain M/EI diagram) for the beam in this separated condition with no restraining moment acting is shown in Fig. 9-22d. If, now, the restraining moments M_2 are restored to the severed ends of the beams in order to obtain the same elastic curve as existed prior to cutting, the additional bending-moment diagrams would be triangular and are placed below the axis for convenience until substantiated by subsequent computation.

Having completed the bending-moment diagram as far as presently possible, let us now turn our attention to the elastic curve of the beam. Assuming the elastic curve at R_2 to be pitched downward to the left, let us draw a tangent at this point extended to cross the action lines of R_1 and R_3, forming similar triangles MNL and RSN. From these triangles we may write the following proportion:

$$\frac{t_{MN}}{20} = -\frac{t_{SN}}{18} \tag{9.25}*$$

This is a very important relationship, because it and others similar to it provide a means for equating the unknowns of two adjacent spans. The value of t_{MN} and t_{SN} may now be found, to be later substituted into Eq. (9.25), M_2 being assumed positive.

$$t_{MN} = \frac{160{,}000}{EI} \times 10 \times 10 + \frac{M_2}{EI} \times 10 \times \frac{40}{3} \tag{9.26}$$

$$t_{SN} = \frac{133{,}300}{EI} \times 4 \times \frac{16}{3} + \frac{133{,}300}{EI} \times 5 \times \frac{34}{3} + \frac{M_2}{EI} \times 9 \times 12 \tag{9.27}$$

Substituting t_{MN} and t_{SN} in Eq. (9.25), we obtain

$$\frac{16{,}000{,}000}{EI} + \frac{400}{3}\frac{M_2}{EI} = -\frac{20}{18}\left(\frac{10{,}390{,}000}{EI} + \frac{108M_2}{EI}\right)$$

$$253.3M_2 = -27{,}520{,}000$$

$$M_2 = -108{,}800 \text{ ft-lb}$$

Proceeding now to the determination of the exterior reactions R_1 and R_3, let us fix our attention upon the two free-body diagrams of Fig. 9-22c. By taking the moments about R_2 of the forces to the left of the cut section, we obtain

$$\sum M'_{R_2} = 20R_1 - 32{,}000 \times 10 + 108{,}800 = 0$$

$$R_1 = \frac{211{,}200}{20} = 10{,}560 \text{ lb}$$

Taking moments to the right of R_2, there results

$$\sum M''_{R_2} = 18R_3 - 30{,}000 \times 10 + 108{,}800 = 0$$

$$R_3 = \frac{191{,}200}{18} = 10{,}610 \text{ lb}$$

In the determination of R_2, it must be remembered that its magnitude is contributed by the weight applied to both spans. Any attempt to obtain its value by moments should take this fact into consideration. Taking moments about R_1 and R_3 to determine R'_2 and R''_2 results in the following:

$$\sum M_{R_1} = 20R'_2 - 32{,}000 \times 10 - 108{,}800 \tag{9.28}$$

$$R'_2 = \frac{428{,}800}{20} = 21{,}440 \text{ lb}$$

*See §8-9 for discussion of signs of tangential deviations.

$$\sum M_{R_3} = 18R_2'' - 30{,}000 \times 8 - 108{,}800 \tag{9.29}$$

$$R_2'' = \frac{348{,}800}{18} = 19{,}390 \text{ lb}$$

$$R_2 = R_2' + R_2'' = 21{,}440 + 91{,}390 = 40{,}830 \text{ lb} \tag{9.30}$$

Check on Vertical Forces

$$\sum V = R_1 + R_2 + R_3 - 32{,}000 - 30{,}000 = 0$$

$$10{,}560 + 40{,}830 + 10{,}610 - 62{,}000 = 0 \tag{9.31}$$

$$0 = 0$$

9-14 Résumé for Solving Beams on Multiple Supports

The solution of all statically indeterminate beams resting on multiple supports follows the general outline given below.

1. Consider each span to be simply supported; that is, remove temporarily all restraining moments over the supports.

2. Draw the bending-moment diagrams for these simple beams acted upon by the applied loads only.

3. Remove the external loads temporarily. Restore the restraining moments which occur over the supports. These restraining moments will be indeterminate in both magnitude and direction.

4. Draw the bending-moment diagrams for the restraining moments below the zero axis for convenience until their location is finally substantiated. Where a restraining moment occurs at each end of a beam, the bending-moment diagram will be a trapezoid. If a restraining moment occurs at but one end, the bending-moment diagram is a triangle. The relative magnitude of the controlling ordinates in the restraining-moment diagrams is unnecessary for a correct solution.

5. Draw an assumed elastic curve of the beam, considering both the applied loads and restraining moments acting.

6. If the beam has no fixed reaction, draw a tangent to the elastic curve at each interior support and obtain relationships of the tangential deviations similar to Eq. (9.25) occurring at the adjacent supports from the similar triangles constructed by the tangent, assuming all restraining moments to be positive.

7. Develop as many tangential deviation relationships as there are unknowns.

8. More convenient moment-area relationships may be found if the beam has fixed ends. These may be found from a known change in slope or a known tangential deviation.

9. The solution of the resulting simultaneous equations will yield the magnitude and direction of each restraining moment.

Illustrative Problem 4. Draw the shear- and bending-moment diagrams for the beam shown in Fig. 9-23. Calculate the necessary *S* beam (A-36) to take the load, the deflection at the center of the left span, and the slope over the

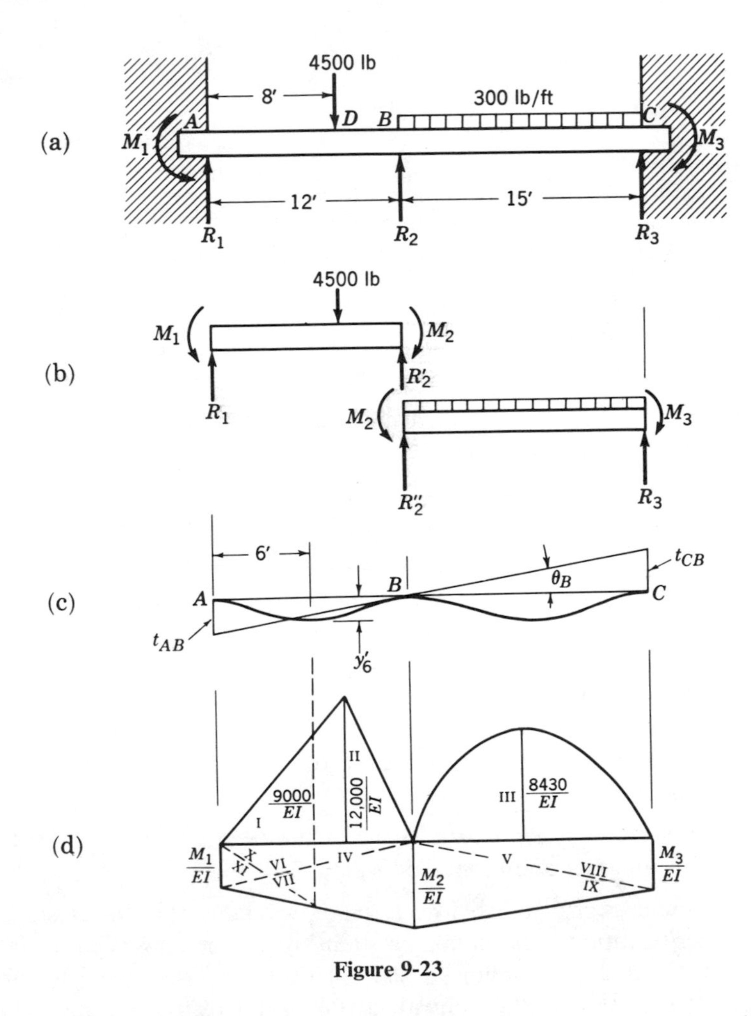

Figure 9-23

middle reaction. The beam is supported laterally over its full length. Use $\sigma = 20{,}000$ psi.

SOLUTION. After removing all restraints from the beam, the bending-moment diagrams for each span are drawn as shown in Fig. 9-23d. Then, considering the restraints alone, their bending-moment diagrams are drawn below the axis, also shown in Fig. 9-23d. All that is known at this stage concerning the restraining-moment diagrams is that they are trapezoidal by virtue of the unknown restraining moment at each end of both beams, and that they have the same moment ordinate at R_2. The bending moment over an interior reaction must be the same for the adjacent spans, since they have a common radius of curvature at this point, Eq. (8.2).

Restraining moments. Having completed the temporary layout of the bending-moment diagram, let us now turn our attention to obtaining moment-area relationships that will provide a solution for the unknown restraining moments. One of the first that might be selected is based on the zero tangential deviation at B between A and B. Expressed in equation form we have

$$\sum M_B \text{ (to left)}^* = 12{,}000 \times \tfrac{4}{2} \times \tfrac{2}{3} \times 4 + 12{,}000 \times \tfrac{8}{2} \times (4 + \tfrac{8}{3})$$
$$+ M_1 \times \tfrac{12}{2} \times \tfrac{2}{3} \times 12 + M_2 \times \tfrac{12}{2} \times \tfrac{1}{3} \times 12 = 0$$

or

$$2M_1 + M_2 = -16{,}000 \tag{9.32}$$

A similar relationship may be found between C and B:

$$\sum M_B \text{ (to right)}^* = 8430 \times 15 \times \tfrac{2}{3} \times \tfrac{15}{2} + M_3 \times \tfrac{15}{2} \times 10$$
$$+ M_2 \times \tfrac{15}{2} \times \tfrac{15}{3} = 0$$

from which

$$2M_3 + M_2 = -16{,}900 \tag{9.33}$$

Since Eqs. (9.32) and (9.33) include three unknown moments, another equation must be found. Noting that a zero slope exists at both A and C, there is no change in slope between these two points. Therefore, the algebraic sum of the M/EI areas between these two points must equal zero.

$$\sum_A^C A^* = 12{,}000 \times 6 + \tfrac{2}{3} \times 15 \times 8430 + \tfrac{1}{2} \times 12(M_1 + M_2)$$
$$+ \tfrac{1}{2} \times 15(M_2 + M_3) = 0 \tag{9.34}$$
$$6M_1 + 13.5M_2 + 7.5M_3 = -156{,}300$$

*The value of EI was purposely omitted from these equations for convenience.

A simultaneous solution of Eqs. (9.32), (9.33), and (9.34) yields the following answers:

$$M_1 = -4675 \text{ ft-lb}, \qquad M_2 = -6650 \text{ ft-lb}, \qquad M_3 = -5125 \text{ ft-lb}$$

Reactions. Referring to the free-body diagrams of Fig. 9-23b, we may obtain the reactions by writing simple equilibrium equations. Thus, R_1 may be obtained by taking moments about point B to the left.

$$\sum M_B \text{(left)} = 12R_1 - 4675 - 4500 \times 4 + 6650 = 0$$
$$R_1 = 1335 \text{ lb} \tag{9.35}$$

Reaction R_3 is obtained by taking moments about the same point to the right.

$$\sum M_B \text{(right)} = 15R_3 - 5125 - 300 \times 15 \times \tfrac{15}{2} + 6650 = 0$$
$$R_3 = 2150 \text{ lb} \tag{9.36}$$

An algebraic sum of all the vertical forces produces R_2:

$$\sum V = 1335 + R_2 + 2150 = 9000$$
$$R_2 = 5515 \text{ lb}$$

The value of R_2 may be checked by finding the sum of R_2' and R_2'' from moment equations of the two free bodies taken about points A and C, respectively.

The shear- and composite bending-moment diagrams are shown in Fig. 9-24.

S *beam required.* A glance at the shear- and bending-moment diagrams reveals that the critical shear and moment values are, respectively, 3165 lb and 6650 ft-lb. To meet the flexural requirements, the section modulus must be

$$\frac{I}{c} = \frac{M}{\sigma} = \frac{6650 \times 12}{20{,}000} = 3.99$$

Use S 5 × 10 beam; $I/c = 4.8 \text{ in.}^3$; web thickness = 0.210 in.; and $I = 12.1 \text{ in.}^4$.

Although this beam has a more than sufficient section modulus, its weight is less than any other capable of taking the load.

A check on the shear strength reveals the stress to be

$$\tau = \frac{V}{td} = \frac{3165}{0.210 \times 5 \text{ in.}} = 3010 \text{ psi}$$

which is much less than the allowable AISC stress of 14,400 psi.

Deflection at midpoint of span AB. Inasmuch as the deflection at any point on span AB is also a tangential deviation between A and the desired

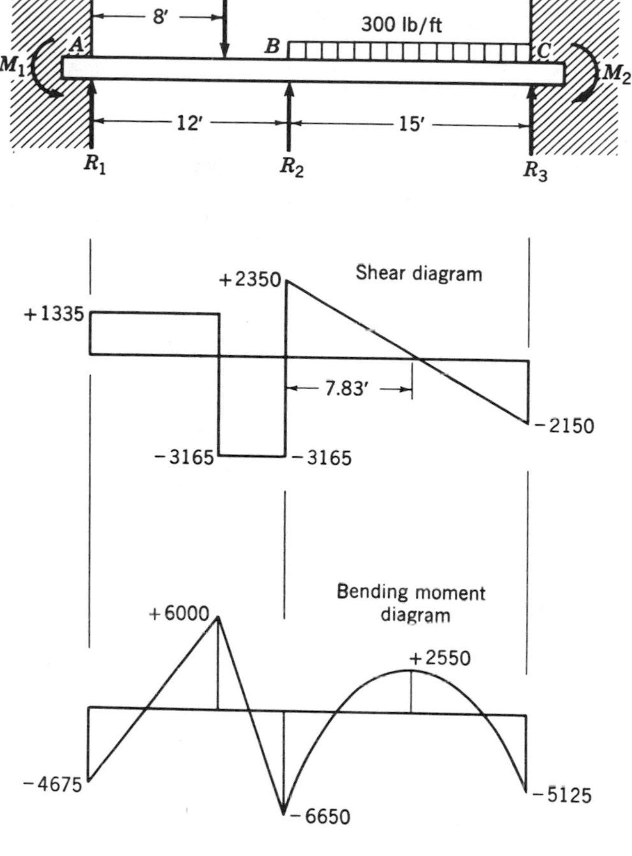

Figure 9-24

point, its value may be computed directly from the separated M/EI diagrams of Fig. 9-23d. Thus,

$$y_6' = t_6' = \frac{9000}{EI} \times \frac{6}{2} \times 2 - \frac{4675}{EI} \times \frac{6}{2} \times 4 - \frac{5660}{EI} \times \frac{6}{2} \times 2$$

$$y_6' = \frac{36{,}060}{EI} = \frac{36{,}060 \times 1728}{30 \times 10^6 \times 12.1} = 0.172 \text{ in.} \tag{9.37}$$

Slope over the middle reaction. The slope at point B is equal to the algebraic sum of the M/EI areas between A and B, since the slope at A is horizontal.

$$\theta_B = \frac{12{,}000}{EI} \times \frac{12}{2} - \frac{4675}{EI} \times \frac{12}{2} - \frac{6650}{EI} \times \frac{12}{2}$$
$$= +\frac{4000}{EI} = +\frac{4000 \times 144}{30 \times 10^6 \times 12.1} = 0.00159 \text{ rad (counterclockwise)} \tag{9.38}$$

Illustrative Problem 5. Draw the shear- and bending-moment diagrams for the continuous beam shown in Fig. 9-25.

SOLUTION. Since no completely fixed ends are present on this continuous beam, it is necessary to solve this problem by drawing tangents to the beam at the two interior reactions and obtaining from the resultant similar triangles the following expressions:

$$-\frac{t_{AB}}{15} = \frac{t_{CB}}{20}, \qquad -\frac{t_{DC}}{9} = \frac{t_{BC}}{20} \tag{9.39, 9.40}$$

The tangential deviations necessary for substitution in Eqs. (9.39) and (9.40) are as follows:

$$t_{AB} = \frac{6670}{EI} \times \frac{5}{2} \times \frac{10}{3} + \frac{6670}{EI} \times \frac{10}{2} \times \frac{25}{3} + \frac{M_2}{EI} \times \frac{15}{2} \times 10 \tag{9.41}$$

$$t_{CB} = \frac{2}{3} \times \frac{15{,}000}{EI} \times 20 \times 10 + \frac{M_2}{EI} \times \frac{20}{2} \times \frac{40}{3} + \frac{M_3}{EI} \times \frac{20}{2} \times \frac{20}{3} \tag{9.42}$$

$$t_{BC} = \frac{2}{3} \times \frac{15{,}000}{EI} \times 20 \times 10 + \frac{M_2}{EI} \times \frac{20}{2} \times \frac{20}{3} + \frac{M_3}{EI} \times \frac{20}{2} \times \frac{40}{3} \tag{9.43}$$

$$t_{DC} = \frac{2000}{EI} \times \frac{3}{2} \times 2 + \frac{2000}{EI} \times \frac{6}{2} \times 5 + M_3 \times \frac{9}{2} \times 6 \tag{9.44}$$

After substitution and simplification, the following simultaneous equations are obtained:

$$10M_2 + 29M_3 = -312{,}000 \tag{9.45}$$

$$35M_2 + 10M_3 = -366{,}200 \tag{9.46}$$

Solving these equations, we obtain

$$M_2 = -8220 \text{ ft-lb}$$
$$M_3 = -7920 \text{ ft-lb}$$

The negative sign indicates that the assumption of positive-restraining moments was incorrect. The restraining bending-moment diagrams are shown correctly in Fig. 9-25c.

Having determined all the restraining moments, we next proceed to the calculation of the reactions and the determination of the shear diagram.

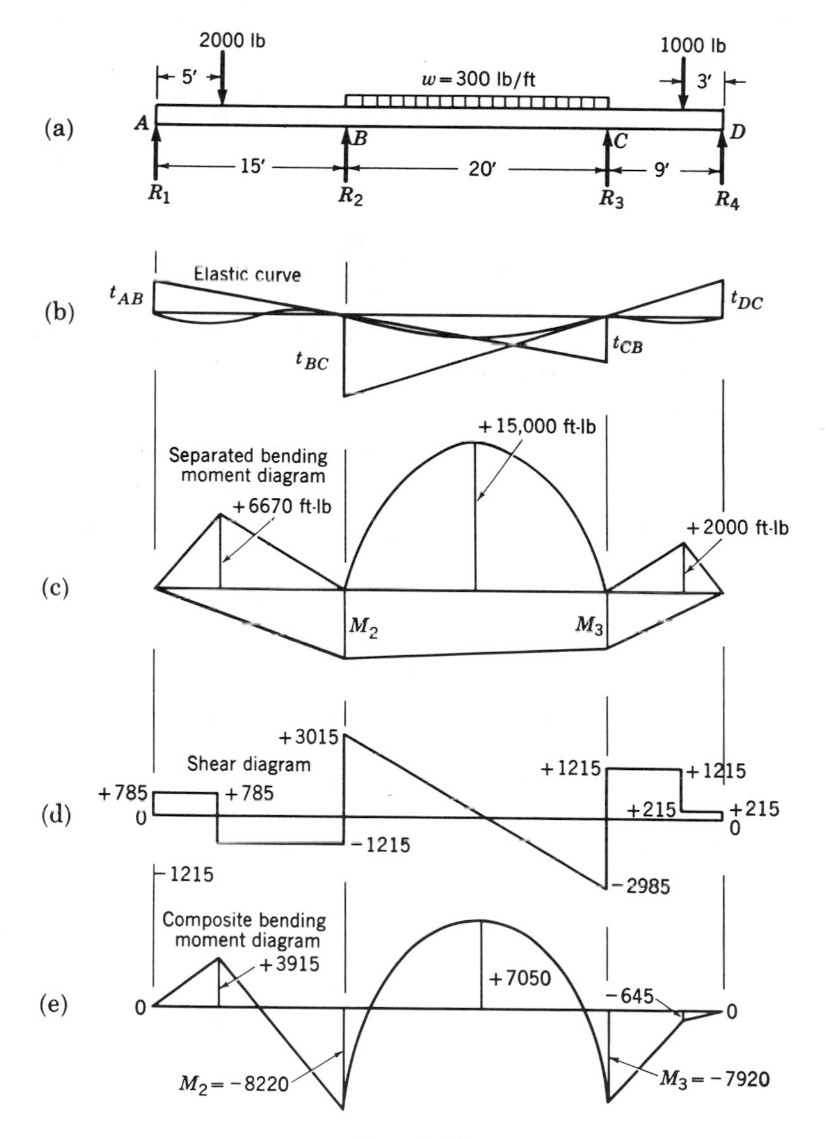

Figure 9-25

To get R_1. Use the first span as a free body.

$$\sum M_{R_2} = R_1 \times 15 - 2000 \times 10 + 8220 = 0$$
$$R_1 = 785 \text{ lb} \tag{9.47}$$

Note: The signs obtained from the solution of moments are used in obtaining the *direction* of the couples. Once the direction is known, the laws of equilibrium are used to obtain other unknown reactions and couples. When writing the equilibrium equation for moments, it is essential that clockwise and counterclockwise moments be given opposite signs.

To get R_2. Use the first two spans as a free body (Fig. 9-26).

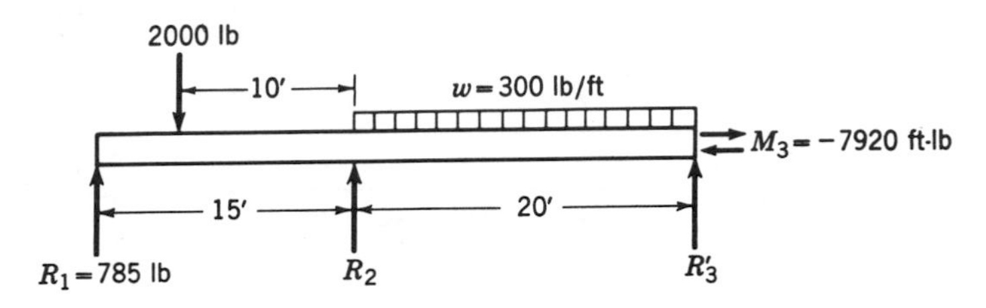

Figure 9-26

$$\sum M_{R_3} = 785 \times 35 - 2000 \times 30 + R_2 \times 20 - 6000 \times 10 + 7920 = 0$$

$$27{,}500 - 60{,}000 + 20R_2 - 60{,}000 + 7920 = 0 \qquad (9.48)$$

$$20R_2 = 84{,}580, \qquad R_2 = 4230 \text{ lb}$$

To get R_4. Use the last span as a free body, with R_4 acting upwards.

$$\sum M_{R_3} = R_4 \times 9 - 1000 \times 6 + 7920 = 0$$
$$R_4 = -215 \text{ lb} \qquad (9.49)$$

which means that the R_4 reaction is acting downward instead of as shown in Fig. 9-25.

To get R_3. Use the last two spans as a free body (Fig. 9-27).

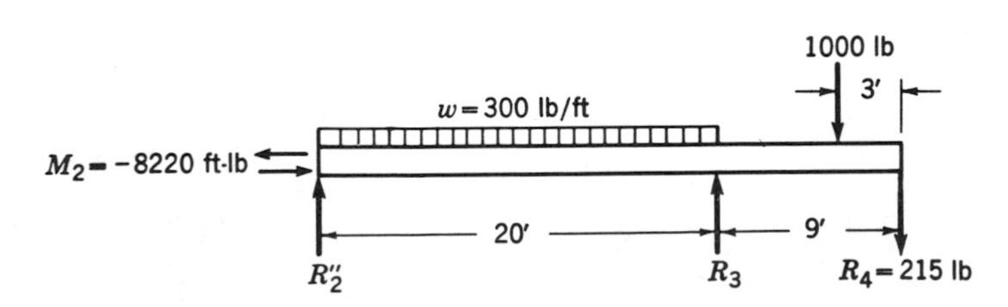

Figure 9-27

$$\sum M_{R_2} = (-215) \times 29 + R_3 \times 20 - 1000 \times 26 - 6000 \times 10 + 8220 = 0 \qquad (9.50)$$

$$R_3 = 4200 \text{ lb}$$

Check $\sum V = 0$

$$\sum \text{reactions} = 785 + 4230 + 4200 - 215 = 9000$$

$$\sum \text{loads} = 2000 + 1000 + 6000 = 9000$$

PROBLEMS

9-26. By the method of balancing deflections, find the magnitude of the roller reaction acting on a uniformly loaded beam fixed at one end and freely supported at the other. *Answer:* $\frac{3}{8}wl$.

9-27a, b, c, d, e, f, g, h, j, k. Draw the shear- and combined bending-moment diagrams for the beams designated in Fig. 9-28. Determine the most economical S beam to withstand the load. Use 20,000 and 13,000 psi as allowable flexure and shear stresses, respectively. Assume beams to be laterally supported. *Answers:* (a) $R_1 = 0.63$ kip, $M_2 = -4.5$ ft-kips; (c) $R_1 = 4.26$ kips, $M_2 = -20.9$ ft-kips; (e) $R_1 = 0.89$ kip, $M_2 = -3.86$ ft-kips; (g) $R_1 = 1.75$ kips, $M_2 = -9.50$ ft-kips; (j) $R_1 = 6.00$ kips, $M_2 = -12.00$ ft-kips.

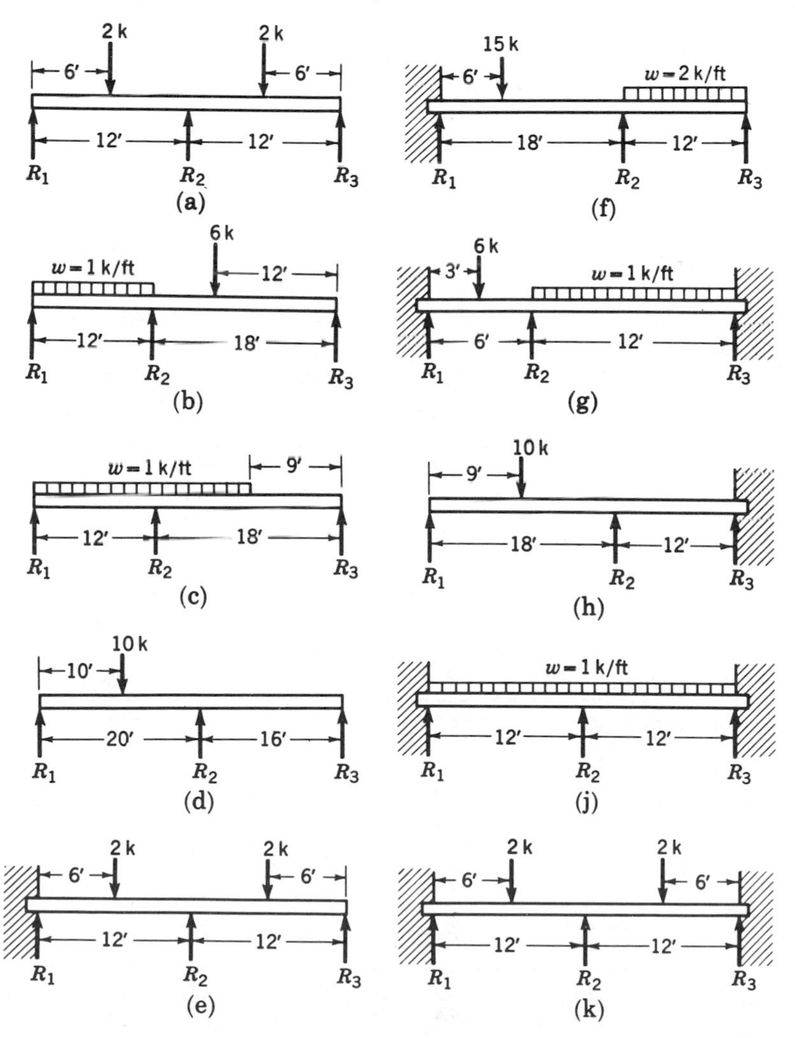

Figure 9-28

9-28a, b, c, d, e, f, g, h. Draw the shear- and combined bending-moment diagrams for the beams designated in Fig. 9-29. Determine the most economical steel, S-beam section to withstand the load. Use 20,000 and 13,000 psi as maximum allowable flexure and shear stresses, respectively. Assume beams to be laterally supported. *Answers:* (a) $R_0 = 8.98$ kips, $R_3 = 13.78$ kips, $M_{max} = 47.9$ ft-kips; (c) $R_0 = 7.06$ kips, $R_3 = 10.70$ kips, $M_{max} = -64.4$ ft-kips; (e) $R_0 = 4.00$ kips, $R_3 = 0.38$ kip, $M_{max} = 40.0$ ft-kips; (g) $R_0 = 6.13$ kips, $R_3 = 3.86$ kips, $M_0 = M_{max} = -32.5$ ft-kips.

9-29. Calculate the position of the points of inflection of the beam of Illustrative Problem 3. *Answer:* 1.15 ft, 10.35 ft.

9-30. Calculate the position of the points of inflection of the beam shown in Fig. 9-22.

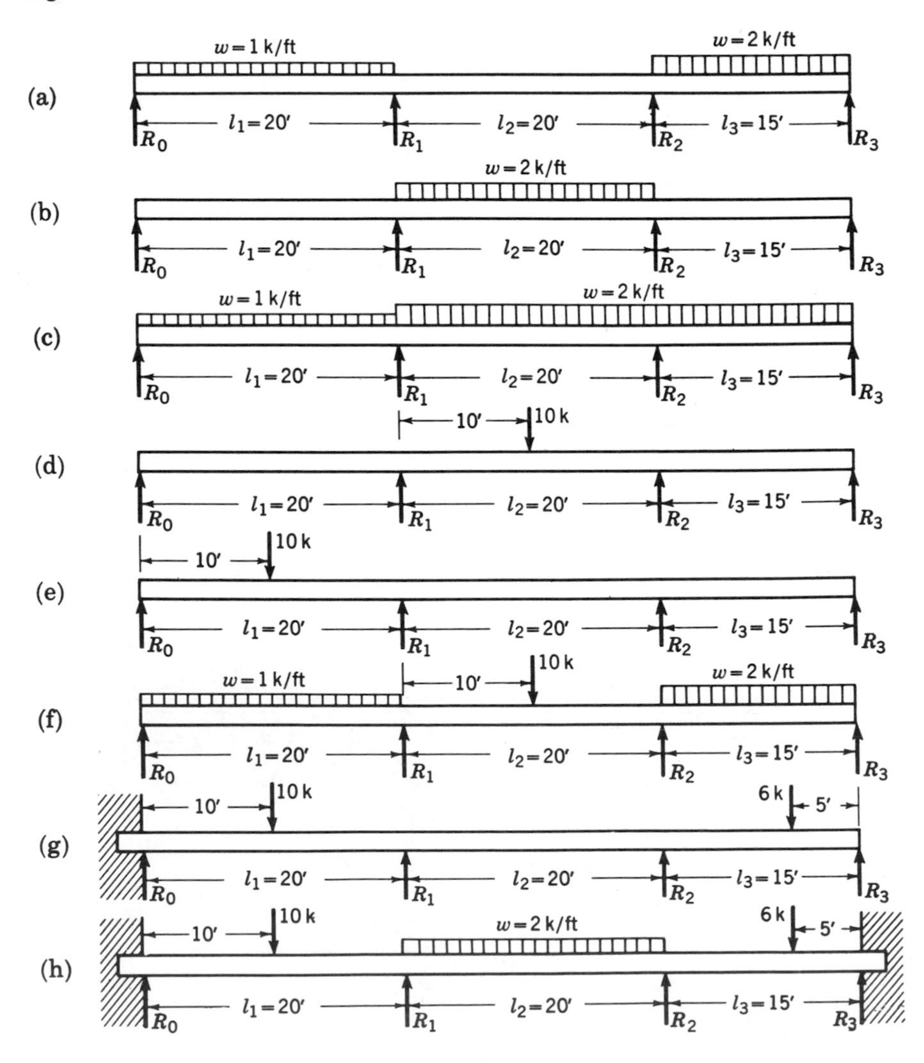

Figure 9-29

9-31. Calculate the slope at the center of the first span of the beam of Illustrative Problem 4 in terms of E and I. *Hint:* Use separated moment diagrams. *Answer:* $576{,}000/EI$ rad.

9-32. Calculate the deflection at the center of span BC of the beam of Illustrative Problem 4 in terms of E and I. *Answer:* $32{,}500/EI$.

9-33. Find the slope of the beam of Problem 9-28g immediately above reaction R_2 in terms of E and I.

9-34. Calculate the position of the inflection points for the beam of Illustrative Problem 4.

9-15 Three-Moment Equations

Because of the frequency with which continuous beams laden with uniform load occur, the relationship of the internal-resisting moments at any three successive reactions and the load applied to the intervening spans is given by the *three-moment equation for uniformly distributed load.* It is expressed as follows:

$$M_1 l_1 + 2M_2(l_1 + l_2) + M_3 l_2 = -\frac{w_1 l_1^3}{4} - \frac{w_2 l_2^3}{4} \tag{9.51}$$

where M_1, M_2, and M_3 are the restraining moments at any three consecutive reactions; l_1 and l_2 are the lengths of the two adjacent intervening spans; and w_1 and w_2 are the loads on spans l_1 and l_2, respectively. The beam must be of constant cross section.

The three-moment equation for concentrated loads

$$\begin{aligned} M_1 l_1 + 2M_2(l_1 + l_2) + M_3 l_2 = &-\sum \frac{P_1 a_1}{l_1}(l_1^2 - a_1^2) \\ &-\sum \frac{P_2 b_2}{l_2}(l_2^2 - b_2^2) \end{aligned} \tag{9.52}$$

bears a great deal of similarity to that for uniform loads. The expression included within the first summation sign is determined for every one of the concentrated loads acting on the left span. P_1 is the general term applied to any one of the concentrated loads acting on this span, and a_1 is the distance of such a load from the No. 1 support.

The second expression is similarly written for every load on the right-hand span, b_2 being the distance of each such load from the No. 3 support. M_1, M_2, and M_3 are, as before, the restraining moments on any three consecutive reactions. Lengths l_1 and l_2 represent the included adjacent span lengths.

Assuming a five-span beam having four intermediate supports, with an unknown restraining moment above each one, four applications of the three-moment equation must be made, that is, for spans 1 and 2, 2 and 3, 3 and 4, and 4 and 5. If the restraining moments at the end supports are equal to zero, the above four equations are sufficient to obtain the four unknown interior restraining moments. The application of any three-moment equation may be made to any two adjacent spans by using the proper subscripts.

Each of the foregoing equations is easily derived by the use of the moment-area method. If used consistently, it may be advisable to memorize them. However, emphasis cannot be too strongly made to study the method by which they are derived. For in so doing all problems of continuous beams are solved with the same approach and require no memorized formula.

MOMENT DISTRIBUTION

9-16 Introduction

The moment-distribution method of determining restraining moments in beams and structural framework was developed by Hardy Cross in 1932. Its speed and simplicity of application have made it one of the best, if not the best, methods of beam and structural analysis available. In the development of its basic principles, all clockwise moments and the angular displacements they produce are considered positive. When incorporated into moment diagrams, however, restraining moments computed by moment distribution must be made to agree in sign with the conventional bending-moment criterion.

9-17 Discussion of Basic Principles of Moment Distribution

In Fig. 9-30b are drawn the bending-moment diagrams of the two spans of a continuous beam, assuming the beam to be *artificially locked in a horizontal position* over support B. It is noted that over this support beam BA requires a clockwise restraining moment of 30 ft-kips to maintain its horizontal position, while BC requires an 18-ft-kip counterclockwise moment.

Actually, there is no outside or artificial restraint holding the beam in that position. The two beams, integrally connected, must provide the restraining moment for each other. If the artificial restraint is removed, joint B must rotate, because beam BC cannot provide the 30-ft-kip moment to keep beam BA horizontal at B. Conversely, beam BA has more than enough moment resistance to keep beam BC horizontal at B. Equilibrium, therefore, can be attained only by rotating joint B in the direction of the unbalanced moment (counterclockwise). In so doing, beam BA reduces its required resisting moment $M_{F_{BA}}$ (see §9-10) and beam BC increases its moment $M_{F_{BA}}$ until they are both equal to each other. The beam is then in equilibrium. By this method of reasoning, the restraining moment over support B should have a value between 18 and 30 ft-kips. The amount of rotation (i.e., the corrective moments) required to bring the beam to its equilibrium position is obtained by the principles of moment distribution.

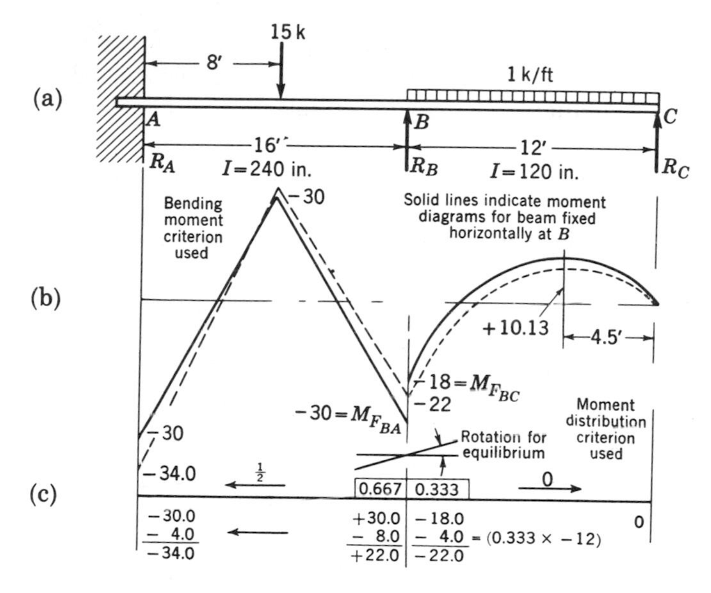

Figure 9-30 Dashed line indicates final moment diagram for balanced beam.

9-18 Development of Basic Principles of Moment Distribution

When a beam fixed at one end and freely supported at the other is depressed an angle θ_a at its freely supported end (Fig. 9-31) by a moment M_{ab}, the value of M_{ab} obtained by moment-area principles is equal to $4EI\theta_a/l$. Closer observation reveals that just *half* the value of the applied moment, or $2EI\theta_a/l$, is transmitted *with the same algebraic sign* to the fixed end. The coefficient of one half is termed *the carry-over factor*.

If the end opposite the applied moment is hinged ($M_{ba} = 0$), the applied moment is equal to $3EI\theta_a/l$. This can be obtained from the moment-area

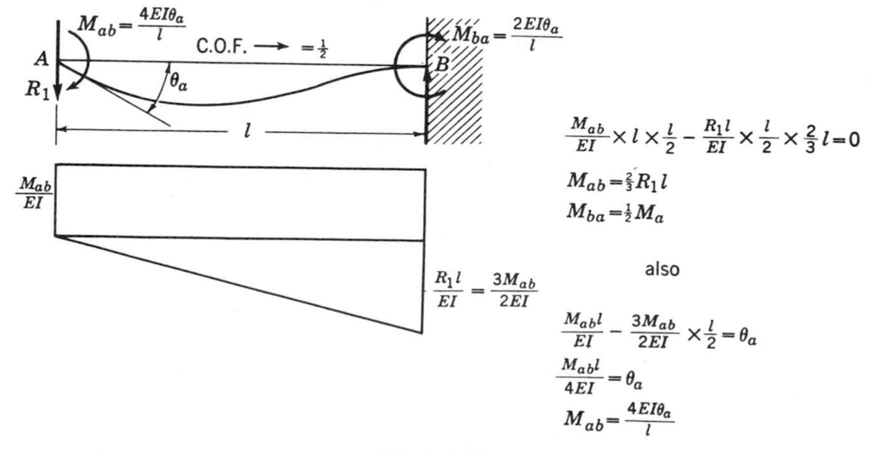

Figure 9-31

principles or by the method of superposition, as shown in Fig. 9-32. The carry-over factor from the left to right end is, of course, equal to zero, and no amount of rotation at the applied moment end will cause a moment at the hinge.

In the typical continuous beam shown in Fig. 9-33, let us consider the beam to be locked in a horizontal position over each support with the exception of E, which is hinged. In this locked, hypothetical condition, assume the

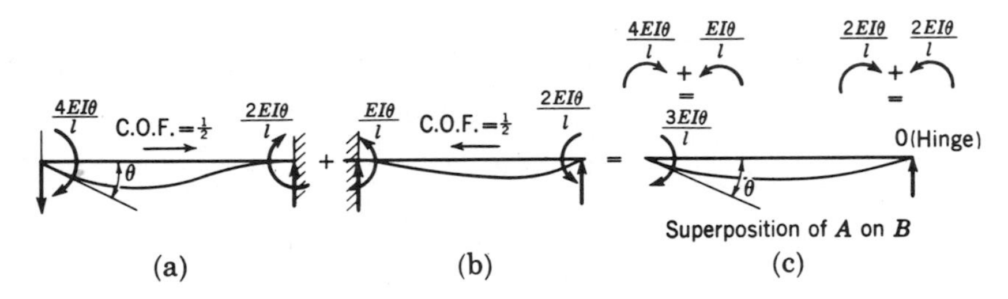

Figure 9-32

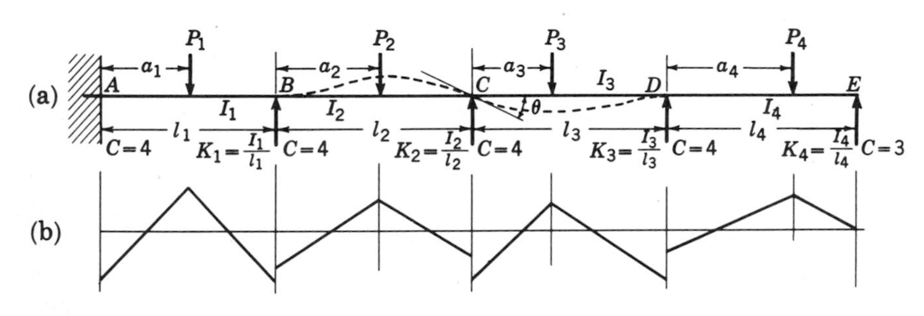

Figure 9-33

fixed-ended moments to be computed and the bending-moment diagrams drawn (Fig. 9-33b). If joint C is unlocked now, it will rotate clockwise into equilibrium. The moments created by rotation will be *superimposed* on the fixed-ended moments present before rotation, making the moments on each side of the joint equal. Stating this equality in general terms,

$$M_{CB} + M_{CD} = 0$$

$$M_{F_{CB}} + \frac{4EI_2\theta_C}{l_2} + M_{F_{CD}} + \frac{4EI_3\theta_C}{l_3} = 0 \tag{9.53}$$

from which

$$E\theta_C = -\frac{M_{F_{BC}} + M_{F_{CD}}}{4K_2 + 4K_3} = -\frac{\sum M_{F_C}}{\sum (CK)_C} \tag{9.54}$$

where

$$K_2 = \frac{I_2}{l_2} \quad \text{and} \quad K_3 = \frac{I_3}{l_3} \tag{9.55}$$

and coefficient $C = 3$ or 4 depending on whether the opposite joint is either hinged or fixed, respectively.

The value of θ_C is the angle of rotation required to bring the joint into equilibrium. Inserting this value in the moment M_{CB} and M_{CD}, we obtain

$$\begin{aligned} M_{CB} &= M_{F_{CB}} + \frac{4EI_2\theta_C}{l_2} = M_{F_{CB}} - \frac{4K_2}{\sum (CK)_C}(\sum M_{F_C}) \\ &= M_{F_{CB}} - r_{CB}(\sum M_{F_C}) \end{aligned} \tag{9.56}$$

and

$$\begin{aligned} M_{CD} &= M_{F_{CD}} + \frac{4EI_3\theta_C}{l_3} = M_{F_{CD}} - \frac{4K_3}{\sum (CK)_C}(\sum M_{F_C}) \\ &= M_{F_{CD}} - r_{CD}(\sum M_{F_C}) \end{aligned} \tag{9.57}$$

These are the moments occurring in the beam at joint C upon the removal of the artificial locking restraint, assuming the joints at B and D are to remain locked. Simultaneous modifications of moment will occur in M_{BC} and M_{DC} equal to one half the balancing moments on the same beam introduced at joint C.

Before operations are moved to the other joints, joint C is locked. Rotation is then permitted at another joint, using similar correction moments, generally stated as $\dfrac{-C_x K_x}{\sum(CK)_x}(\sum M_x)$, where C_x is either 3 or 4 depending on whether the opposite joint is hinged or fixed, respectively. Rotating this joint to a balanced condition may throw the first joint out of equilibrium. Successive balancing and rebalancing will, however, produce a state of equilibrium over the entire beam and permit the removal of all artificial locks.

Let us apply the theory to the beam previously presented in Fig. 9-30. Because the *moment distribution criterion assumes all clockwise restraining moments are positive*, the moments $M_{F_{BA}}$ and $M_{F_{BC}}$ are $+30$ and -18 ft-kips, respectively. Their algebraic sum $(\sum M_{F_B})$ of $+12$ ft-kips represents an unavailable clockwise moment demand or a counterclockwise unbalance.

The distribution factors allocating this unbalance to the two members AB and BC framing into joint B are obtained from the following calculation.

Member	C	K	CK	$\dfrac{CK}{\sum CK} = r$	$-r(\sum M_{F_B})$
BA	4	$\frac{240}{16} = 15$	60	0.667	-8.0
BC	3	$\frac{120}{12} = 10$	30	0.333	-4.0
			$\sum CK = 90$	1.000	-12.0

If now the excess moment of 12 ft-kips is distributed, -8.0 ft-kips to BA and -4.0 ft-kips to BC, joint B will be in balance with 22.0 ft-kips on each entering member. Because joint B rotated and added a negative moment of -8.0 ft-kips to BA, one half that amount of moment, with the same algebraic sign, was induced at joint A. A fixed joint such as A is able to take as much moment as can be imposed upon it. There being no other joints requiring adjustment to equilibrium, the beam is in equilibrium. The restraining moments over the supports are determined by the algebraic addition of the moments at the end of each span.

Illustrative Problem 6. Determine the restraining moment over each support for the continuous beam shown in Fig. 9-34 by moment distribution (I = constant).

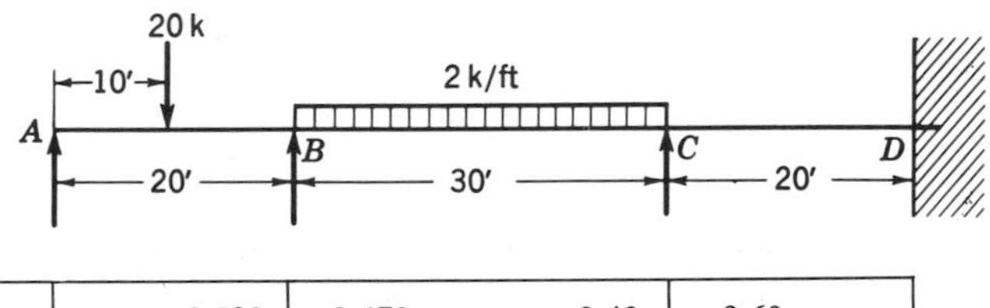

Dist. factors		0.530	0.470	0.40	0.60	
C.O.F.	0 ⟵		0.5 ⟷		0.5 ⟶	
F.E.M.	0.0	+75.0	−150.0	+150.0	0.0	0.0
Joint C balanced			−30.0 ⟵	−60.0	−90.0 ⟶	−45.0
Joint B balanced		+55.6	+49.4 ⟶	+24.7		
Joint C balanced			−4.9 ⟵	−9.9	−14.8 ⟶	−7.4
Joint B balanced		+2.6	+2.3 ⟶	+1.2		
Joint C balanced			−0.3 ⟵	−0.5	−0.7 ⟶	−0.3
Joint B balanced		+0.2	+0.1			
Final restraining moment	0.0	**+133.4**	**−133.4**	**+105.5**	**−105.5**	**−52.7**

Note: Single underscoring lines indicate joint is temporarily balanced; double lines indicate final balance.
All moments in ft. kips.

Figure 9-34

Calculation of Distribution Factors

Joint B

Member	C	K	CK	r
BA	3	$\frac{I}{20}$	$\frac{3I}{20}$	0.530
BC	4	$\frac{I}{30}$	$\frac{4I}{30}$	0.470
				1.000

Joint C

Member	C	K	CK	r
CB	4	$\frac{I}{30}$	$\frac{4I}{30}$	0.40
CD	4	$\frac{I}{20}$	$\frac{4I}{20}$	0.60
				1.000

PROBLEMS

9-35a, b, c, d, e, f, g, h, j, k. Determine the restraining moments over the supports of those beams shown in Fig. 9-28 by moment distribution.

9-36a, b, c, d, e, f, g, h. Determine the restraining moments over the supports of those beams shown in Fig. 9-29 by moment distribution.

PLASTIC ANALYSIS AND DESIGN OF STATICALLY INDETERMINATE BEAMS

9-19 Introduction

The major limiting factor in every elastic design problem is the allowable working stress. Not only does it dictate the most economical section required, but it also provides an ample margin of strength for contingencies. An unnecessarily low allowable stress results in an inexcusable waste of material, while an excessively high stress produces a false and possibly dangerous economy.

The theory of plastic analysis proposes an economical and justifiable procedure to permit the development of the reserve strength in structural materials. This new concept would base analysis and design on the plastic strength rather than on the attainment of the yield point.

9-20 Basic Features of Plastic Analysis and Design

The increase in allowable load permitted by plastic design is derived from a new understanding of *limiting usefulness*. Heretofore, when applied to a beam, this term meant that it lost its usefulness upon the attainment of the yield stress on any one of its sections. The allowable stress was based on this yield stress. It is important to note, however, that even after the stress on a section attains the yield point a considerable amount of elastic strength still remains in the interior fibers (Fig. 9-35); see also §7-12. One of the features of plastic design is the use of this reserve strength in the determination of the allowable load.

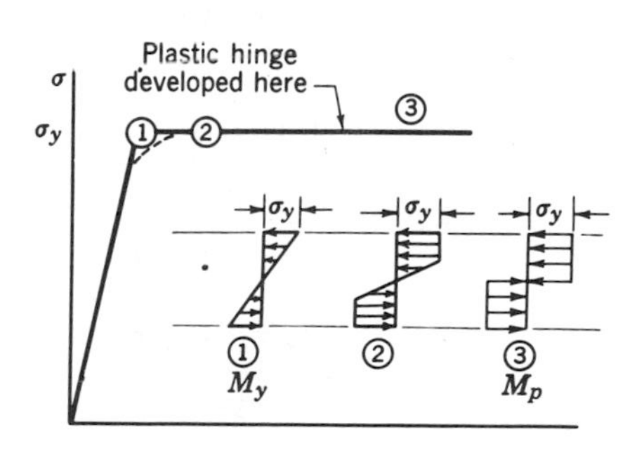

Figure 9-35

The second feature of plastic design applied to indeterminate beams pertains to the use of redistributed moments developed through the formation of *plastic hinges*. These plastic hinges develop at points of maximum moment and when fully formed produce a reasonably constant rotational resistance called the *plastic moment*. Because a plastic moment cannot increase in magnitude and the section on which it operates acts as a hinge, any additional load applied to the beam must be resisted by other portions which have retained some of their elastic restraint. This shifting of resistance under perfectly stable conditions levels off the magnitudes of the maximum moments occurring on an indeterminate beam and permits a new evaluation of its limit of maximum usefulness.

These two features of plastic analysis and design will now be presented in the following illustrative problems.

Illustrative Problem 7. Compare the maximum uniform loads permitted on a low-carbon structural steel beam 6 ft long, and fixed at each end by the elastic and plastic analysis methods, if its cross section is 4 in. wide by 6 in. deep (Fig. 9-36). Use a factor of safety of 2.0 based on a yield point of 30,000 psi. Section modulus equals 24 in.³. Assume elastic limit stress to be equal to yield stress (see Fig. 9-35).

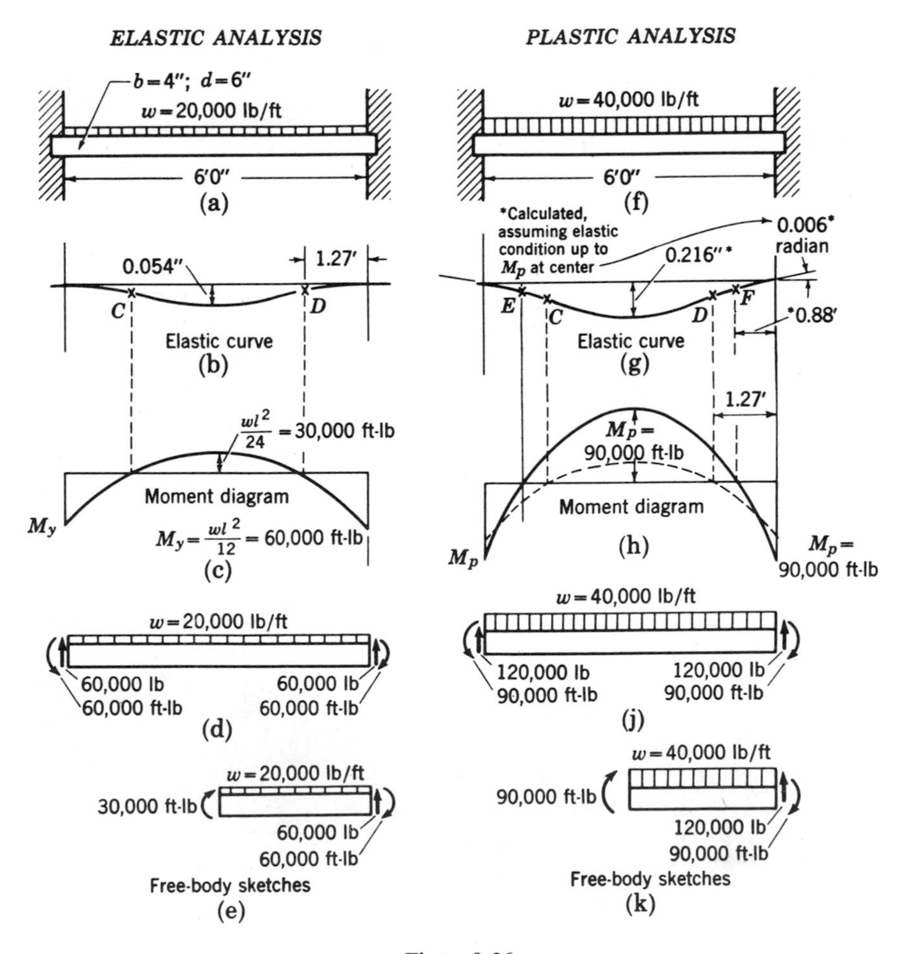

Figure 9-36

SOLUTION.

Elastic design. From §9-8 the end and center moments of a uniformly loaded beam fixed at each end were shown to be $wl^2/12$ and $wl^2/24$, respectively. Using the greater of these two moments in the flexure formula, the maximum uniform load the beam can carry is computed as

$$w = \frac{12 \times 30{,}000 \times 24}{36 \times 12} = 20{,}000 \text{ lb/ft}$$

Its safe load using the factor of safety of 2.0 is equal to 10,000 lb/ft.

Plastic design. The restraining moments at the ends of the beam at the time the load of 20,000 lb/ft is acting are equal to $M_y = wl^2/12$, or 60,000 ft-lb. Suppose now the load were to increase. Immediately the outside fibers of the end sections would yield, and the transformation process from elastic to plastic resistance at that section would start (see Fig. 9-35). This process takes place gradually as more load is applied, and it terminates when the resisting moment attains the value of the constant plastic moment M_p. This, for our rectangular section, is equal to $1.5M_y$, or 90,000 ft-lb (§7-12). Upon the attainment of this moment, each end section rotates as a plastic hinge shifting the points of inflection closer to the ends. Since the moment at these hinges cannot increase, the beam becomes only partially restrained. With the ordinates to the moment diagram increasing with increased load, the moment at the center section must necessarily increase until it, too, attains the value of the plastic moment. At this point the beam will have lost its stability and will have arrived at its maximum useful limit (Fig. 9-36h).

At the time the beam loses stability, the load acting on it computed from the free body of one half the beam is equal to

$$\frac{wl}{2} \times \frac{l}{4} = 2 \times 90{,}000$$

$$w = 40{,}000 \text{ lb/ft}$$

or twice the critical load computed by the elastic analysis.

It should be noted that at no time between the initial yielding at the end sections and the formation of the center hinge was the beam on the verge of collapse. Moreover, the beam deflected less than a corresponding simple beam because of its end restraining moments M_p.

The allowable load permitted on the beam using the plastic analysis with a factor of safety of 2.0 is therefore equal to 20,000 lb/ft. This load will just develop yielding at the end sections. The beam, however, will function as safely as any other.

9-21 Comments on Plastic Design

Plastic design should not be considered as a replacement for elastic design. Actually, only a limited number of structures can be analyzed by this method, although the quantity of materials affected may be large. Obviously, any design calling for exact deflection calculations, or including the effects of fatigue, buckling, brittle fracture, etc., cannot be analyzed by plastic analysis.

Illustrative Problem 8. In the beam of Fig. 9-37 determine the magnitude of P for both elastic and plastic design. Use a factor of safety of 2.0 based on a yield strength of 30,000 psi. Shape factor* = 1.14, and $I/c = 20$ in.³. Use the stress-strain diagram given in Fig. 9-35.

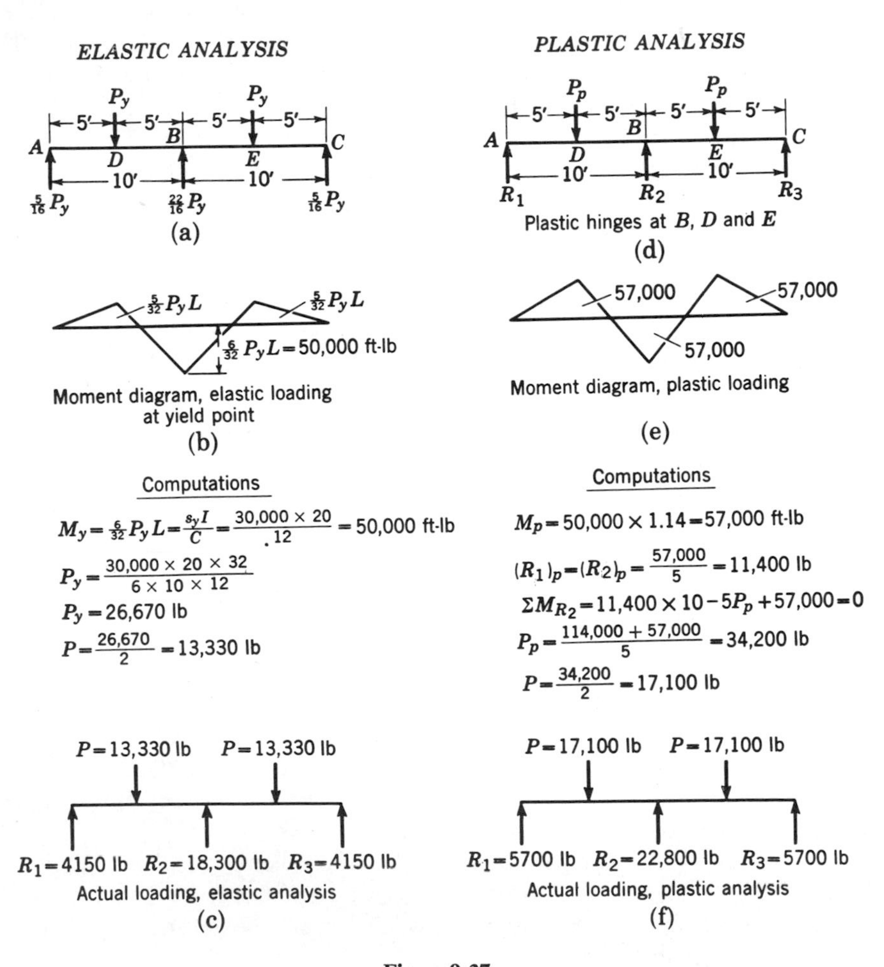

Figure 9-37

SOLUTION.

Elastic analysis. The elastic analysis of this beam limits the load to 13,300 lb, based on the maximum moment on the beam of $\frac{6}{32}PL$.

*Refer to §7-12.

Plastic analysis. Following a procedure similar to that used for Illustrative Problem 7, the loads P are increased until the moments over support R_B and those under each load P attain a value of the plastic moment M_p. The value of P at the occurrence of plastic moments at B, D, and E is determined by a free-body analysis.

Illustrative Problem 9. Calculate the W section required for the continuous beam of Illustrative Problem 8, using the elastic and plastic design methods, if P equals 17,100 lb. Use a factor of safety of 2.0 based on a yield strength of 30,000 psi. Steel has a stress-strain diagram as given in Fig. 9-35. Shape factor equals 1.14. Disregard shear or lateral buckling.

SOLUTION. Note that the plastic design required here is the reverse of the plastic analysis presented in Illustrative Problem 8. The steps comparing the two designs are shown in Fig. 9-38.

ELASTIC DESIGN

Step 1: Calculate the bending moment diagrams, using shear area:

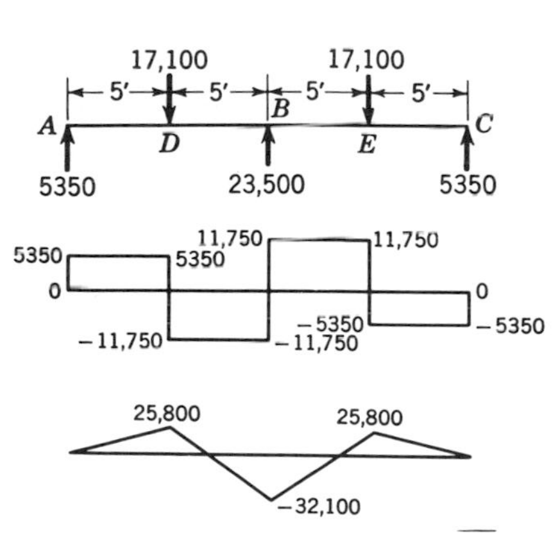

Step 2: Calculate the allowable stress:

$$s=\frac{30{,}000}{2.0}=15{,}000\text{ psi}$$

Step 3: Calculate $\frac{I}{c}$ required:

$$\frac{I}{c}=\frac{32{,}100\times 12}{15{,}000}=25.7\text{ in.}^3$$

Step 4: Select WF beam required:
W 10 × 25

PLASTIC DESIGN

Step 1: Anticipate the location of plastic hinges

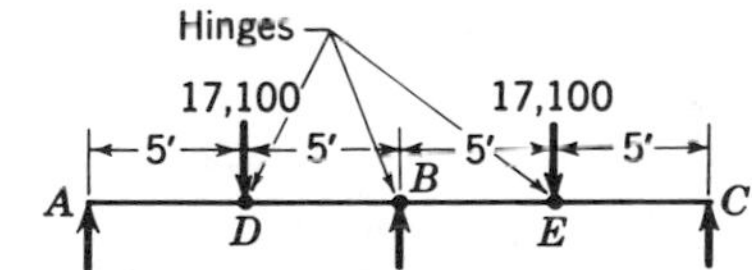

Step 2: Calculate M_p, using moment equation for point D. Apply factor of safety to load.

17,100 × 2.0

5′ 5′ M_p

$R_A = 17{,}100 - \frac{M_p}{10}$

5′ $M_p = M_D$

$R_A = 17{,}100 - \frac{M_p}{10}$

$$M_D=\left(17{,}100-\frac{M_p}{10}\right)5=M_p$$

$$1.5\,M_p=17{,}100\times 5$$

$$M_p=57{,}000\text{ ft-lb}$$

Step 3: Obtain M_y; calculate $\frac{I}{c}$ required:

$$M_y=\frac{57{,}000}{1.14}=50{,}000\text{ ft-lb}$$

$$\frac{I}{c}=\frac{50{,}000\times 12}{30{,}000}=20.0\text{ in.}^3$$

Step 4: Select WF beam required:
W 10 × 21

Figure 9-38

PROBLEMS

Note: Use the stress-strain diagram of Fig. 9-35 in the solution of the following problems. Let yield stress equal 30,000 psi.

9-37. Compare the maximum concentrated load that may be placed at the midpoint of a fixed-ended steel beam, 12 ft long, by the elastic and plastic analysis method, if its cross section is 4 in. wide and 6 in. deep. Use a factor of safety of 2.0 based on the yield stress. *Answers:* 20 kips, 30 kips.

9-38. Do Problem 9-37 using equal concentrated loads at the third points.

9-39. Compute by the elastic and plastic analysis methods the maximum uniform load that may be applied to a W 10 $\times$ 49 continuous beam of three 20-ft spans and fixed at each end. Use a factor of safety of 1.5 based on the yield stress. Shape factor equals 1.14.

9-40. Design a fixed-ended W beam 12 ft long, uniformly loaded with 4000 lb/ft, by the elastic and plastic design methods. Use a factor of safety of 1.5 based on the yield stress. *Answers:* Elastic $I/c = 28.8$ in.^{3}, plastic $I/c = 18.95$ in.^{3}.

9-41. Compare the values of the maximum concentrated loads which act at the centers of a two-span W 10 $\times$ 49 continuous beam (each span equals 20 ft) by the elastic and plastic analysis methods. Use a factor of safety of 2.0 based on the yield stress. Shape factor equals 1.14. *Answers:* 18.2 kips, 23.4 kips.

Note: Many previous problems given in this chapter may be assigned for solution by the method of plastic analysis.

10

Stresses Due to Eccentrically Applied Loads

10-1 A Fundamental Relationship

Whenever an axial load is moved to a position parallel to the longitudinal axis of the member upon which it acts, the cross sections perpendicular to that axis are subjected to a combination of both axial and bending stresses. The reason for the occurrence of this combination of stresses may best be ascertained by a quick review of a fundamental relationship established in the study of elementary mechanics. The illustration concerns the resolution of a force into a force and a couple. A discussion of this relationship follows.

10-2 Resolution of a Force Into a Force and a Couple

If bar AB of Fig. 10-1 were acted upon by force P, its tendency to turn about point A would be measured by the moment of P about A, or Pa. If two opposing forces, P' and P'', equal and parallel to P were now to be added to the same bar, the moment of the force system and its turning effect about point A would still be Pa, because

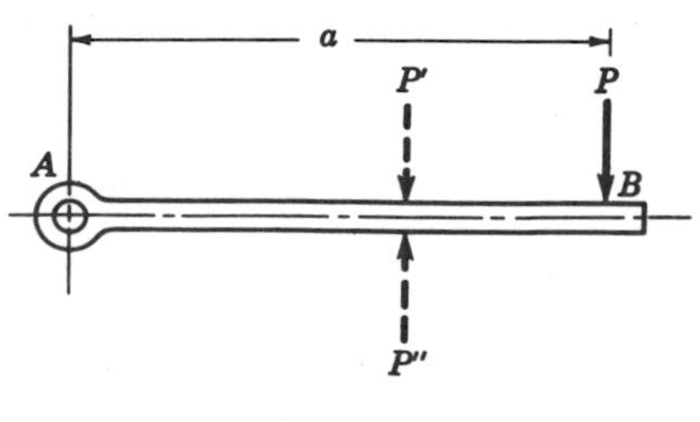

Figure 10-1

the moments of the two added forces would cancel each other. The summation of the vertical forces would also remain the same. Since this resolution allows forces P'' and P to be considered as a couple, and P' as a single force, we may conclude that the *external* effect of a force may be made equal to a force and a couple by adding two forces which are equal, opposite, and parallel to the given forces. Although the point at which the two opposing forces are added is optional, the opportunity of using the remaining single force P' for a specific purpose, such as is indicated in the next paragraph, should not be overlooked. It should be remembered that in this resolution all the forces are externally applied.

10-3 Action of an Eccentric Load on a Short Compression Member

Consider a vertical force P acting in a plane of symmetry of the block shown in Fig. 10-2. It will be noted that the block is subjected to a combination of both compressive and bending strains. Let us now add on the center line of the block two additional *external* forces, P' and P'', that are equal and parallel to P and opposite to each other. The addition of these forces to the block will not, of course, alter its equilibrium condition, since they cancel each other. The three forces comprising couple PP'' and force P' will, therefore, have the same effect as would force P acting alone.

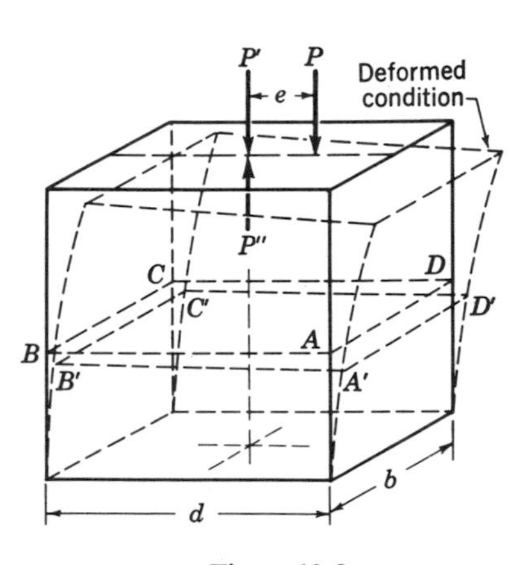

Figure 10-2

But from Chapter 2 we learned that an axial load, such as P', produces only direct (in this case compressive) stress on cross sections perpendicular to its action line. It was also indicated in Chapter 7 that a couple can produce only pure bending stress. Thus, we may further conclude that the addition of forces P' and P'' on the perpendicular passing through the centroid of the area upon which the stress is desired produces the same effect as force P would acting alone, and affords a means of separating the direct from the bending stress.

To illustrate, suppose the block of Fig. 10-2 to have a cross section of 12 by 12 in. If force P equals 14,400 lb and has an eccentricity of 3 in., the moment of couple PP'' will be

$$M_{PP''} = Pe = 3 \times 14{,}400 = 43{,}200 \text{ in-lb}$$

The maximum bending unit stress produced by this bending moment will therefore be

$$\sigma = \frac{Mc}{I} = \frac{Pec}{I} = \frac{43{,}200 \times 6}{\frac{1}{12} \times 12 \times 12^3} = 150.0 \text{ psi}$$

The compressive stress due to force P' will be

$$\sigma_c = \frac{P'}{A} = \frac{14{,}400}{144} = 100 \text{ psi}$$

These two stresses, which act simultaneously, may be algebraically added in accordance with the principle of superposition if their total does not exceed the proportional limit of the material of which the member is made. The maximum and minimum stresses found on the cross section at points A and B, respectively, may be expressed algebraically as follows, using the plus sign to indicate tension and the minus sign to indicate compression.

$$\begin{aligned}\sigma_A &= -\frac{P'}{A} - \frac{Mc}{I} = -\frac{P}{A} - \frac{Pec}{I} \\ &= -\frac{14{,}400}{144} - \frac{43{,}200 \times 6}{\frac{1}{12} \times 12 \times 12^3} \qquad (10.1) \\ &= -100 - 150 = -250 \text{ psi} \qquad \text{(compression)}\end{aligned}$$

$$\begin{aligned}\sigma_B &= -\frac{P'}{A} + \frac{Mc}{I} = -\frac{P}{A} + \frac{Pec}{I} \\ &= -\frac{14{,}400}{144} + \frac{43{,}200 \times 6}{\frac{1}{12} \times 12 \times 12^3} \qquad (10.2) \\ &= -100 + 150 = 50 \text{ psi} \qquad \text{(tension)}\end{aligned}$$

The stress distribution on the cross section is represented graphically in Fig. 10-3.

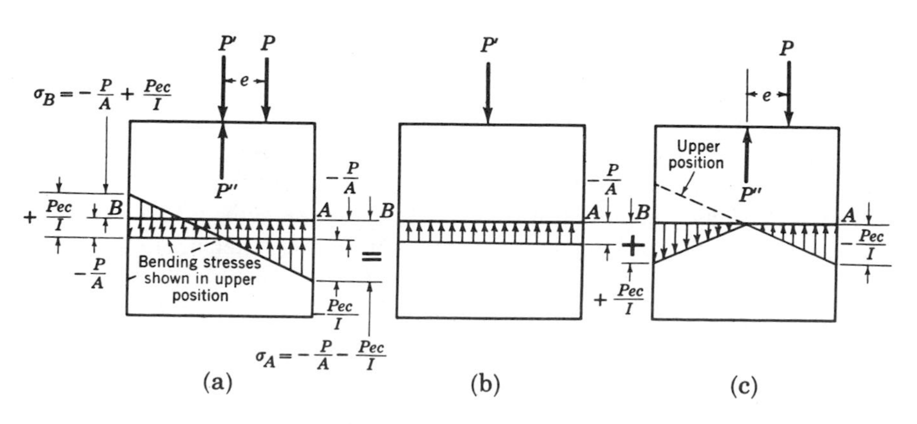

Figure 10-3

Effect of varying the eccentricity. As the value of the eccentricity e (Fig. 10-3) is reduced, the magnitude of the couple PP'' and its accompanying bending unit stress are also reduced. The compressive unit stress P'/A remains the same, however, since the magnitude and location of force P' do not change.

The condition that exists at point B when the direct compression unit stress P'/A equals the tensile bending unit stress Pec/I is of considerable engineering importance, because it determines the limiting position of P necessary to retain a compressive unit stress over the entire cross-sectional area. Any further increase in the value of e would enable the maximum tensile flexural stress at B to exceed the counteracting compressive unit stress, as shown in the problem of the preceding paragraph.

This limiting position of force P is obtained by setting the two stresses at B equal to each other.

$$\frac{P'}{A} = \frac{Pec}{I}$$

Since $P' = P$ and $I = Ar^2$, where r is the radius of gyration of the section about the bending axis, we may write

$$\frac{P}{A} = \frac{Pec}{Ar^2}$$

from which

$$e = \frac{r^2}{c} \tag{10.3}$$

The above equation is general in that it applies to all cross sections. For rectangular sections as used above, the value of e will be

$$e = \frac{1}{c} \times \frac{I}{A} = \frac{1}{c} \times \frac{bd^3}{12} \times \frac{1}{bd} = \frac{d^2}{12c}$$

But $c = d/2$, making

$$e = \frac{d}{6} \tag{10.4}$$

Thus, we are assured that no tensile stress will occur on a cross section of a rectangular block as long as the eccentric load located on an axis of symmetry acts within a distance of $d/6$ to either side of the center. Interpreted in a slightly different way, this statement sometimes reads "Keep the load in the middle third to prevent tension on the outer fibers."

To ascertain the limiting value of eccentricity for a vertical cylinder, one need only substitute the proper values of I, A, and c in Eq. (10.3).

$$e = \frac{1}{c} \times \frac{I}{A} = \frac{1}{d/2} \times \frac{\pi d^4/64}{\pi d^2/4} = \frac{d}{8} \tag{10.5}$$

It follows that to prevent tension in a vertical cylinder, the load must be kept within the *middle fourth*.

The equations for limiting eccentricities find their greatest use in the design of members made of materials that have relatively little strength in tension. Such materials are concrete, brick, cast iron, plaster of Paris, and so forth. In the design of a plain concrete dam, for instance, the resultant force at any rectangular cross section must be kept within the *middle third* to prevent tension.

PROBLEMS

10-1. A short compression member (Fig. 10-4), 12 by 12 in. in cross section, has a load of 80,000 lb applied on one axis of symmetry, 1 in. from the center of the cross section. Determine the stresses on section *A-A*. Show suitable sketches.

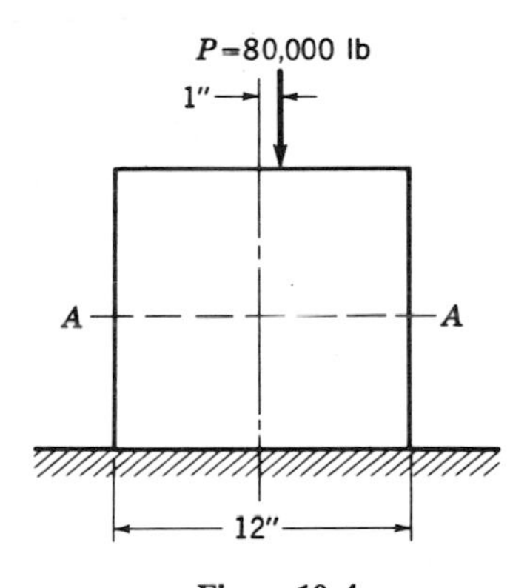

Figure 10-4

10-2. In Problem 10-1, how great an eccentricity should force *P* have to obtain a tensile unit stress of 320 psi on cross section *A-A*? Show sketch of stress distribution. *Answer:* 3.15 in.

10-3. If in Problem 10-1 it is desired to maintain only compressive stress on section *A-A*, what limiting value may the eccentricity have? Show sketch of stress distribution.

10-4. A wooden beam is 12 ft long, 6 in. wide, and 15 in. deep. The beam is hinged at the left end and is simply supported at the right end. There is a uniform load of 1000 lb/ft over the entire length of the beam and a compressive load of 9000 lb applied at the centroid of the right end. Determine the maximum tensile and compressive unit stresses in the beam. Disregard the deflection of the beam. Hinge is located on center line of beam. *Answers:* $\sigma_c = 1057$ psi, $\sigma_t = 857$ psi.

10-5. A rectangular concrete footing 5 ft wide by 6 ft long supports an eccentric load of 400,000 lb (Fig. 10-5) which acts on the longer axis of symmetry but 6 in. off the shorter axis. Determine the stress distribution at the base of footing due to this applied load. *Answer:* Maximum stress is 20,000 lb/ft^2.

10-6. Four bolts, arranged in a rectangular pattern, hold the offset fixture

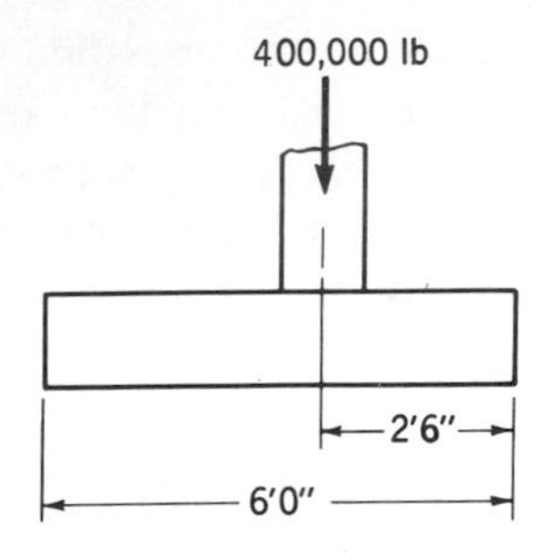

Figure 10-5

of Fig. 10-6. Determine the unit stress on the net area of the bolts if the offset plate, which is acted upon by the eccentric load as shown, is located midway between pairs of resisting bolts. Total bolt area = 1.00 in.^{2}.

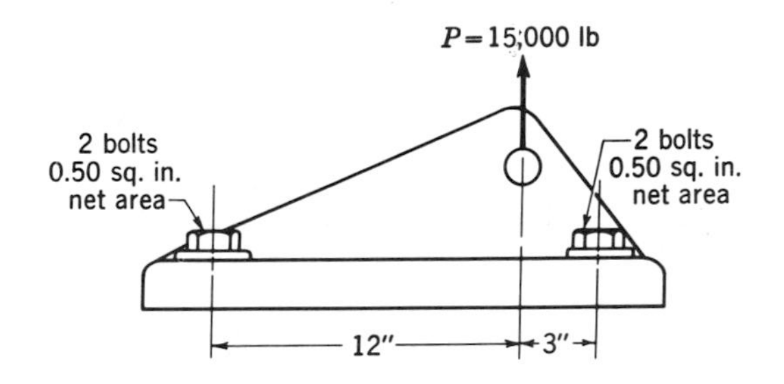

Figure 10-6

10-7. A compression block, whose cross section measures 6 by 12 in., carries two eccentric loads as shown in Fig. 10-7. Find stress at edge *ab* and at edge *cd* of the right section *abcd*. *Answer:* σ_{ab} = 38.4 psi, σ_{cd} = 0 psi.

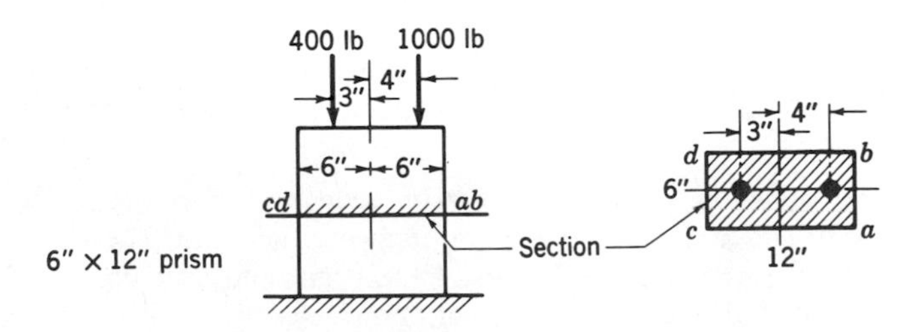

Figure 10-7

10-8. A C frame (cross section 1 by 3 in.) with pin supports at *A* and *B* is subjected to loads as shown in Fig. 10-8. (a) Determine the maximum fiber stress and state where it occurs. (b) What is the maximum fiber stress acting on the vertical section *Y-Y*? The 1-in. dimension is perpendicular to the plane of the paper. *Answers:* (a) 12,800 psi; (b) 8800 psi.

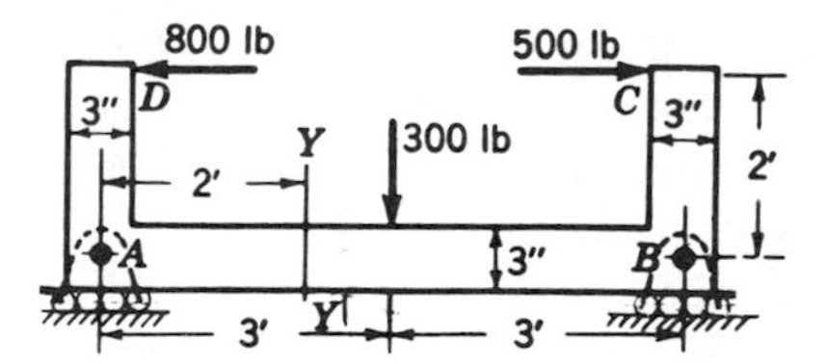

Figure 10-8

10-4 When an Eccentric Load on a Short Compression Member Does Not Lie in a Plane of Symmetry

The more general case of eccentrically applied loads on short compression members occurs when the applied load P does not fall within a plane of symmetry (Fig. 10-9). To determine the stresses imposed on cross section

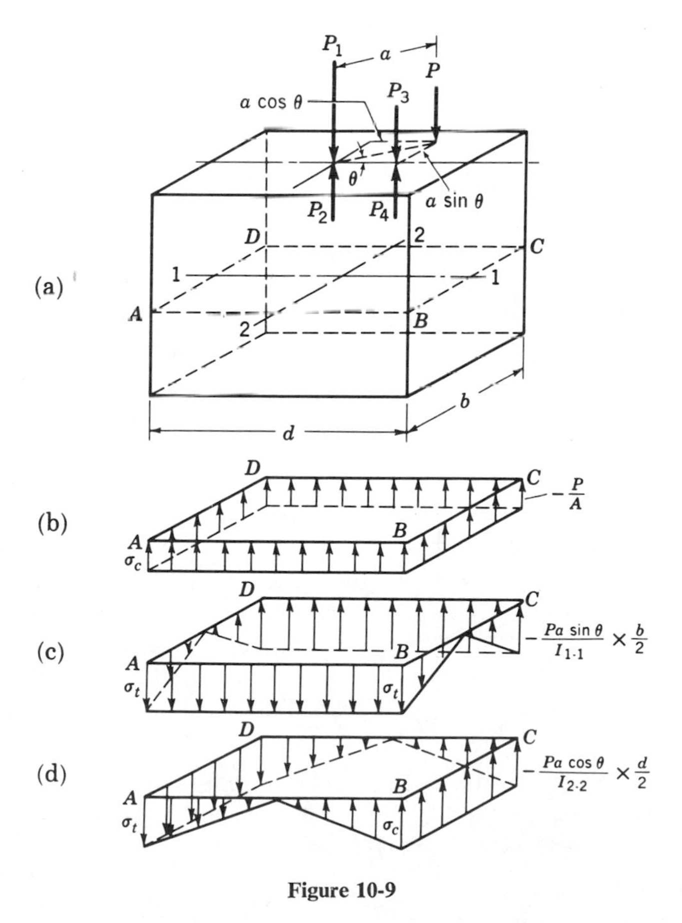

Figure 10-9

ABCD, place two equal, opposite, and parallel forces P_1 and P_2 on the axis line passing through the center of its cross section. The force P_1 will produce direct compression unit stress, and the couple PP_2 bending stress. The plane of the couple is not, however, in either of the two principal bending planes. This difficulty can be overcome by adding two more equal, opposite, and parallel forces P_3 and P_4, placed on axis 1-1 at the foot of a perpendicular dropped from the point of action of force P. The five forces now acting on our member provide the following stresses.

1. Force P_1. A uniformly distributed compression unit stress.
2. Couple $PP_4 = Pa \sin \theta$. A bending unit stress about axis 1-1.
3. Couple $P_2P_3 = Pa \cos \theta$. A bending unit stress about axis 2-2.

When placed in equation form, the stresses at the corners of cross section *ABCD* are as follows:

$$\sigma_A = -\frac{P_1}{A} + \frac{PP_4}{I_{1-1}} \times \frac{b}{2} + \frac{P_2P_3}{I_{2-2}} \times \frac{d}{2} \tag{10.6}$$

$$\sigma_B = -\frac{P_1}{A} + \frac{PP_4}{I_{1-1}} \times \frac{b}{2} - \frac{P_2P_3}{I_{2-2}} \times \frac{d}{2} \tag{10.7}$$

$$\sigma_C = -\frac{P_1}{A} - \frac{PP_4}{I_{1-1}} \times \frac{b}{2} - \frac{P_2P_3}{I_{2-2}} \times \frac{d}{2} \tag{10.8}$$

$$\sigma_D = -\frac{P_1}{A} - \frac{PP_4}{I_{1-1}} \times \frac{b}{2} + \frac{P_2P_3}{I_{2-2}} \times \frac{d}{2} \tag{10.9}$$

From a perusal of the above formulas, it is evident that σ_C is the maximum compressive stress and σ_A is the maximum tensile stress produced on cross section *ABCD*. To obtain the limiting position of P that will mark the beginning of tensile stress at A, the equation of stresses occurring at A must be made equal to zero. When a line is drawn through the locus of points on the cross section, marking the beginning of tensile stress anywhere on the cross section, the area enclosed by this line is known as the *kern*. It is shown as the shaded area in Fig. 10-10.

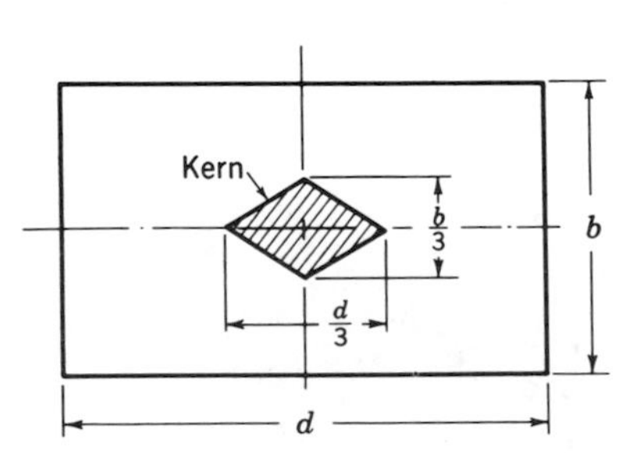

Figure 10-10

Illustrative Problem 1. If in Fig. 10-9 the load $P = 96$ kips, $b = 8$ in., $d = 12$ in., and the eccentricity with respect to both axes is 2 in., determine the stresses at the corners of the plane *ABCD*.

SOLUTION

$$\sigma_A = -\frac{P_1}{A} + \frac{PP_4}{I_{1-1}} \times \frac{b}{2} + \frac{P_2P_3}{I_{2-2}} \times \frac{d}{2}$$

$$\sigma_A = -\frac{96}{96} + \frac{96 \times 2 \times 4}{\frac{1}{12} \times 12 \times 8^3} + \frac{96 \times 2 \times 6}{\frac{1}{12} \times 8 \times 12^3} \tag{10.10}$$

$$= -1.0 + 1.5 + 1.0 = +1.5 \text{ kips/in.}^2$$

$$\sigma_B = -1.0 + 1.5 - 1.0 = -0.5 \text{ kip/in.}^2 \tag{10.11}$$

$$\sigma_C = -1.0 - 1.5 - 1.0 = -3.5 \text{ kips/in.}^2 \tag{10.12}$$

$$\sigma_D = -1.0 - 1.5 + 1.0 = -1.5 \text{ kips/in.}^2 \tag{10.13}$$

These stresses are plotted as ordinates to plane $ABCD$ (Fig. 10-11a). Because each of the three component stress patterns varies linearly from point to point, their combined stress pattern must also vary linearly. Note from Fig. 10-11b that the line FE drawn between the points of zero stress on the periphery is not at right angles to the plane of the couple PP_2. *This is of great significance*. Only if the load acts on one of the principal axes of a section will the plane of bending coincide with the plane of the couple producing the bending. See assumptions of flexure formula in Chapter 7.

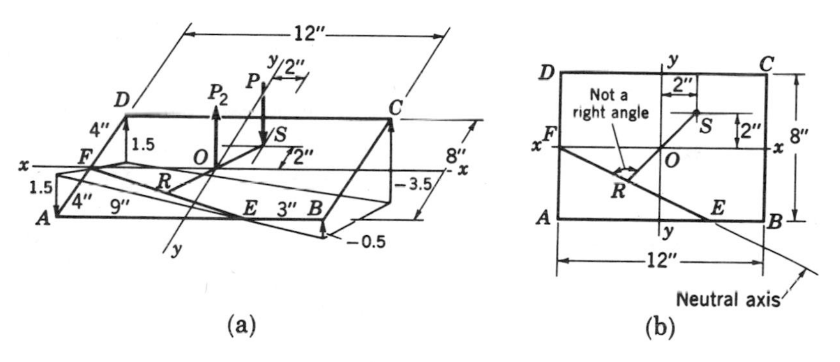

Figure 10-11

PROBLEMS

10-9. If a short compression member has a cross section of 12 by 18 in. and a load of 200,000 lb is placed 2 in. off each axis of symmetry, what stress distribution would be obtained on a cross section perpendicular to its longitudinal axis? The load is parallel to the longitudinal axis of the member. *Answer:* $\sigma_{max} = -2466$ psi.

10-10. Determine the stresses at the corners of cross section $ABCD$ of the compression member in Fig. 10-9, if load P of 100,000 lb acts 2 in. off each axis of symmetry. Cross section equals 12 by 12 in. Show sketch of stress distribution. *Answers:* $\sigma_A = 693$ psi, $\sigma_B = -693$ psi, $\sigma_C = -2080$ psi, $\sigma_D = -693$ psi.

10-11. What would be the theoretical stress distribution on cross section *ABCD* of Fig. 10-9 if load *P* of 100,000 lb acted through one of the corners of the cross section? Cross section equals 12 by 12 in. Show sketch of stress distribution.

10-5 When an Eccentric Load Is Applied to a Nonsymmetrical Cross Section

When an eccentric load is applied to a nonsymmetrical cross section, the load is resolved, as previously suggested, into a force and a couple, the resolution taking place at the centroid of the cross section. Should the plane of the couple not coincide with a principal axis, the procedure explained in §§7-8 and 7-9 should be followed to obtain the flexural stresses. These, combined algebraically with the direct stresses produced by the central force, will give the total stresses acting on the section.

PROBLEMS

10-12. If a couple of 200,000 in-lb is applied on axis *y-y* of the section shown in Fig. 10-12, causing compression on the upper fibers, (a) locate the position of the neutral axis; (b) what will the stresses be at the six corners shown on the cross section? *Answers:* (a) $\theta = 12°55'$; (b) $\sigma_A = -2660$ psi, $\sigma_B = -990$ psi, $\sigma_F = 2040$ psi.

10-13. If a load of 20 kips were applied at point *B* perpendicular to the cross section shown in Fig. 10-12, locate the position of the neutral axis and determine the stresses at points *B* and *F*. *Answers:* $\sigma_B = -2210$ psi, $\sigma_F = +1300$ psi.

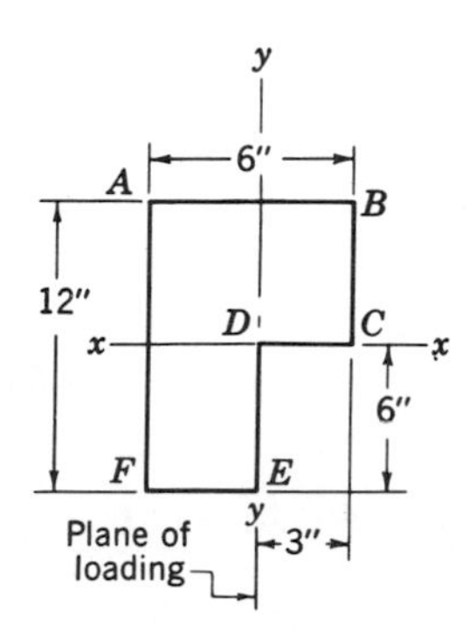

Figure 10-12

10-6 Critical Stress in C Frames

The supporting member or frame for many pieces of industrial equipment is made in the form of a C. Punch and drill presses, riveting and welding machines, cranes, and off-center supports are only some of the more important apparatus that are built with a resemblance to this shape. In each instance forces act on the ends of the C to either spread or compress the opening. Being farthest from the applied loads, those sections opposite the opening generally develop the maximum stresses.

Let us investigate the stresses due to an applied load of 12,000 lb on section *A-A* of Fig. 10-13. To clarify this problem that portion of the C frame above section *A-A* is removed and shown in Fig. 10-14a. The addition of the two equal, opposite, and parallel forces necessary in the resolution of force *P* requires the location of the center of gravity of section *A-A*. Its computation follows, the moments being taken about the inside edge of the section (Fig. 10-14b).

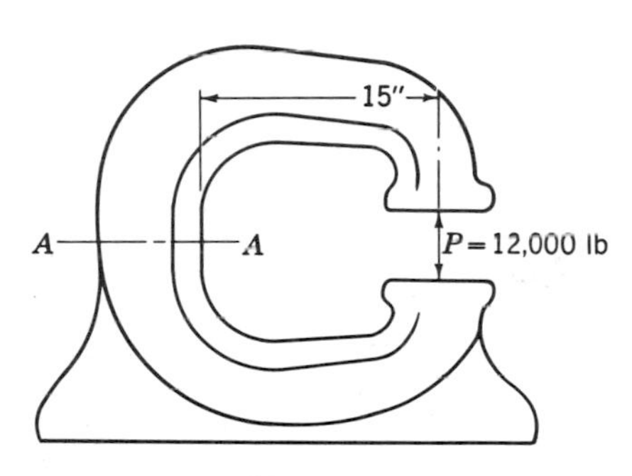

Figure 10-13

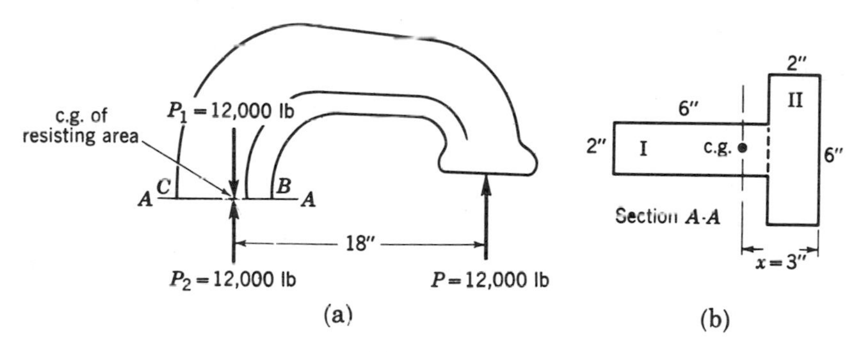

Figure 10-14

	Area (in.²)	x	Ax
I.	12	5	60
II.	12	1	12
	$\sum A = 24$		$\sum Ax = 72$

$$\bar{x} = \frac{\sum Ax}{\sum A} = \frac{72}{24} = 3 \text{ in.}$$

It is now possible to resolve force *P* into a force and a couple by adding the two 12,000-lb forces at the center of gravity of the section. Bearing in mind that each of these three forces is an external force, there are now acting on the free body an external couple PP_1 and a single external force P_2. Although it is not difficult to visualize that the couple produces tensile stress at *B* and compression stress at *C*, some difficulty may be experienced in determining the character of the stress produced by force P_2. This difficulty may be easily overcome, however, by recalling the fact that P_2, as an upward-acting external force, requires a downward-acting internal-resistance stress on section A-A to hold it in equilibrium. Since this stress acts away from section A-A of the free body (Fig. 10-15), it must be a tensile stress.

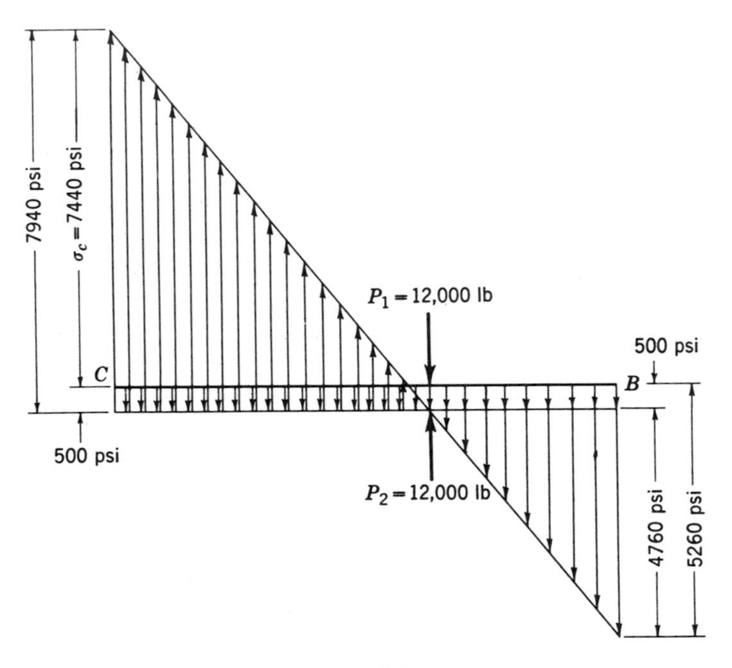

Figure 10-15

Written in equation form, the stresses at B and C are

$$\sigma_B = +\frac{P}{A} + \frac{Pec}{I} \tag{10.14}$$

$$\sigma_C = +\frac{P}{A} - \frac{Pec}{I} \tag{10.15}$$

Before the numerical value of these stresses may be computed, however, the value of I must be obtained about the center-of-gravity axis for the whole section. The required computation follows:

	$I_{c.g.}$	Ad^2	$I_{c.g.} + Ad^2$
I.	$\frac{1}{12} \times 2 \times 6^3 = 36.0$	$12 \times 2^2 = 48.0$	84.0
II.	$\frac{1}{12} \times 6 \times 2^3 = 4.0$	$12 \times 2^2 = 48.0$	52.0
		$\sum (I_{c.g.} + Ad^2) =$	136 in.^{4}

The numerical values of σ_B and σ_C may now be computed, bearing in mind that one is located 3 in., the other, 5 in. away from the neutral axis.

$$\sigma_B = +\frac{12{,}000}{24} + \frac{12{,}000 \times 18 \times 3}{136} = 500 + 4760 = +5260 \text{ psi}$$

$$\sigma_C = +\frac{12{,}000}{24} - \frac{12{,}000 \times 18 \times 5}{136} = +500 - 7940 = -7440 \text{ psi}$$

PROBLEMS

10-14. A metal-shaping machine has a frame as shown in Fig. 10-16. If load *P* has a maximum value of 10,000 lb, find the maximum normal stresses developed on cross section *A-A*. *Answers:* 1990 psi, 2740 psi.

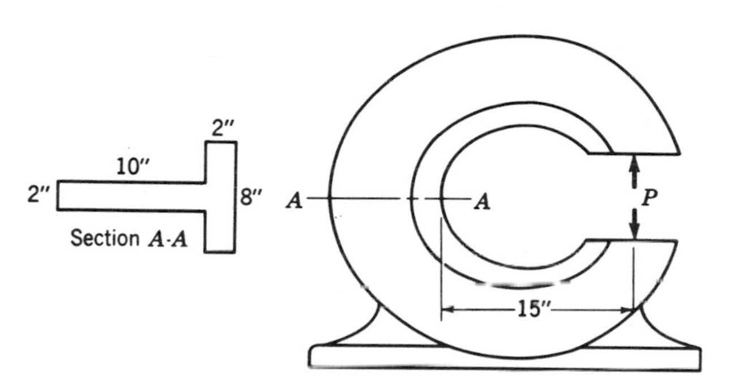

Figure 10-16

10-15. Revise the length of the stem (10 in. length) in the T section of Fig. 10-16 to make the maximum tensile stress on section *A-A* equal to 3000 psi when $P = 8000$ lb.

10-16. The offset bracket shown in Fig. 10-17 is acted upon by a force of 2500 lb. Find the maximum normal stress acting on section *A-A* if its cross section measures $\frac{1}{2}$ in. thick by 3 in. deep. *Answer:* 6670 psi.

10-17. If, in Problem 10-16, the force of 2500 lb were rotated downward through an angle of 45°, what would be the maximum normal stress acting on section *A-A* located 8 in. away from the center of the hole? *Answer:* 16,450 psi.

10-18. In Fig. 10-18 is shown a test specimen for testing the shear strength of glued joints. Assuming no distortion of the joint and a load of 2000 lb, determine the theoretical bending stress imposed on the glued surface under this arrangement. What is the stress distribution on cross section *ABCD*? What arrangement could be more satisfactorily employed in obtaining the shear strength of glued surfaces?

10-19. A lifeboat stanchion (Fig. 10-19) is made from a steel pipe having a 3-in. outside diameter and a $\frac{1}{4}$-in. wall thickness. What load may be applied to the stanchion through an eccentricity of 36 in. if the maximum allowable normal stress is 24,000 psi? *Answer:* 902 lb.

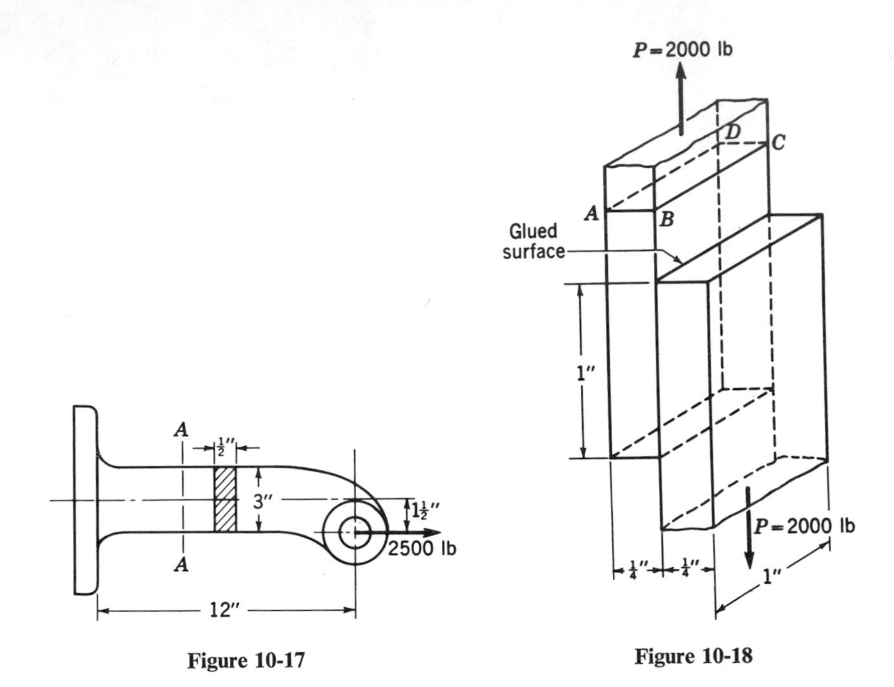

Figure 10-17

Figure 10-18

10-20. A tension member has a 2-in. offset forged into its length. If a cross section at the offset is as indicated in Fig. 10-20, determine the stress distribution acting upon this section if the load is 50,000 lb. Normal stress is uniformly distributed on end sections.

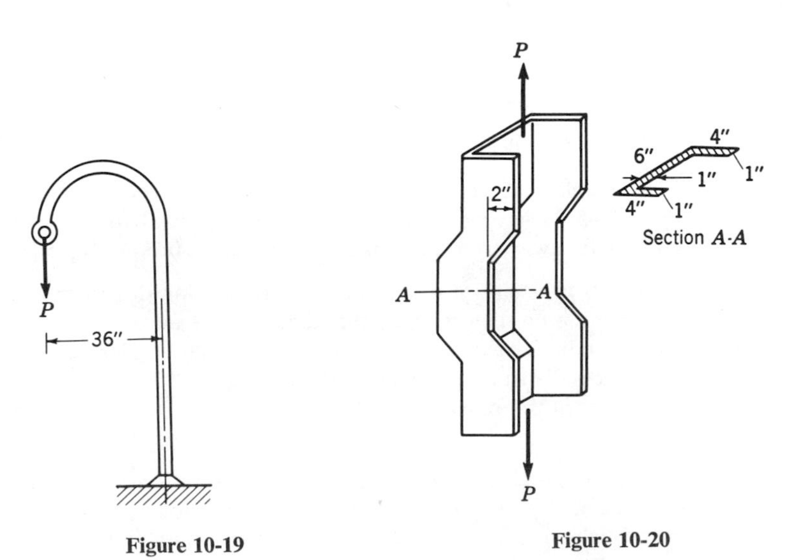

Figure 10-19

Figure 10-20

10-7 Eccentrically Loaded Riveted or Bolted Joints

Whenever a riveted or bolted joint is subjected to an eccentric load, the method used for the determination of the imposed stresses is once again based upon the resolution of the applied force into another force and a couple. Let us illustrate by considering the riveted joint* of Fig. 10-21. Here the loads imposed upon a crane runway girder are carried by two gusset plates placed on either side of a main column. It is desired to find the shearing stresses in the rivets if the load placed on each plate is 10,000 lb and each rivet has a diameter of $\frac{7}{8}$ in.

From our previous discussions, it is evident that the 10,000-lb loads will tend to twist as well as push downward the plate on which each is acting. To separate these two effects such that the single force resulting from the resolution will produce a direct shearing stress $= P/A$, two equal, opposite, and parallel loads need to be added on the vertical center line of each rivet system. Since each plate represents a similar loading condition, let us consider the left plate only.

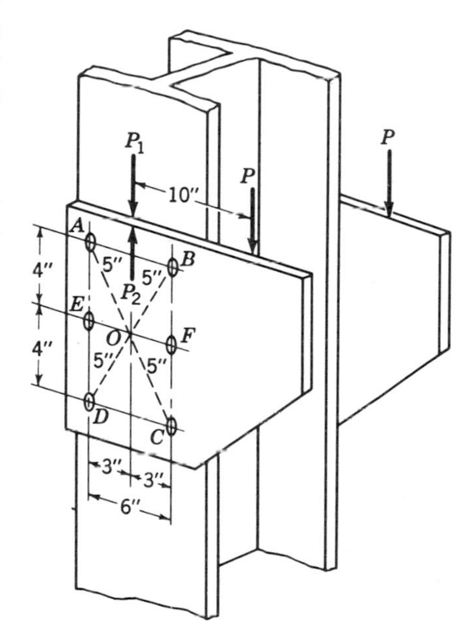

Figure 10-21

Determination of direct-shear stress. Because load P_1 passes through the centroid of the rivet group, it produces a resisting shear of equal magnitude acting upward on each rivet.

$$\tau_D = \frac{P}{A} = \frac{10{,}000}{6 \times \frac{\pi}{4}(\frac{7}{8})^2} = 2780 \text{ psi}$$

Determination of rotational-shear stress. Couple PP_2, as is true of every couple, can only produce rotation, which will always take place about the center of resistance of a rivet group that is assumed to coincide with its centroid. The centroid of the rivet group of Fig. 10-22 is, by virtue of its symmetry, located at point O.

*The discussion presented here concerning riveted joints applies equally well to bolted joints.

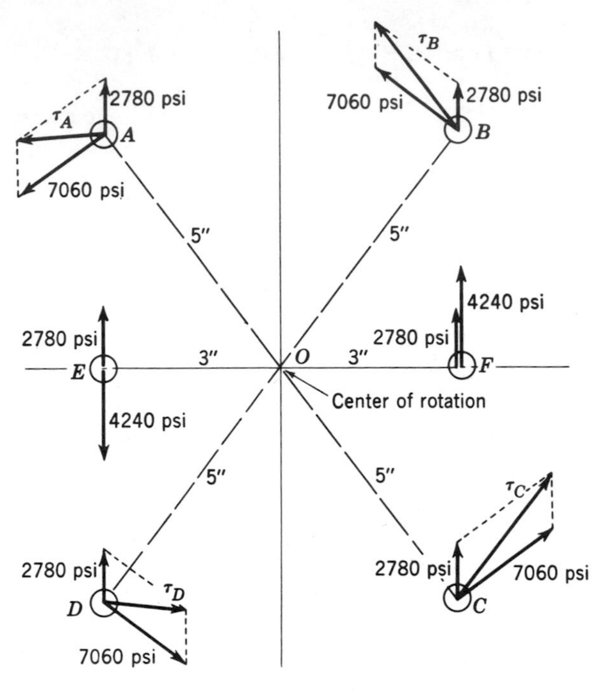

Figure 10-22

The determination of the rotational shearing stresses may now proceed by one of two methods:

Method 1. This method is based on two assumptions:

1. The shearing stress on a body subjected to the action of couples varies with the distance from the center of rotation.

2. The moment of the applied couple PP_2 is equal to the sum of the resisting moments of the rivet forces acting about point O.

Thus, rivets A, B, C, and D have the same rotational shearing unit stress τ_R, since each is located 5 in. away from the center of rotation O. Rivets E and F, located only 3 in. from O, are acted upon only by $\frac{3}{5}\tau_R$. These stresses acting on cross-sectional areas of $\frac{7}{8}$-in.-diameter rivets produce forces whose moments equal the moment of the couple PP_2. Thus, we may write

$$4 \times \tau_R \times 0.60 \times 5 + 2 \times \tfrac{3}{5}\tau_R \times 0.60 \times 3 = 10{,}000 \times 10$$

$$12\tau_R + 2.16\tau_R = 100{,}000$$

$$\tau_R = \frac{100{,}000}{14.16} = 7060 \text{ psi} \qquad \text{(rivets } A, B, C, \text{ and } D\text{)}$$

$$\tfrac{3}{5}\tau_R = 4240 \text{ psi} \qquad \text{(rivets } E \text{ and } F\text{)}$$

Method 2. This method is based upon the application of the torsion formula, its moment of inertia being placed equal to the polar moment of inertia of the rivet group taken about the center of rotation O. Thus,

$$J = \sum (J_{c.g.} + Ad^2)$$

Since, however, the value of the polar moment of each rivet about its own centroidal axis is usually insignificant in contrast to the Ad^2 term, we may write the value of the polar moment equal to

$$J = \sum Ad^2$$

Thus

$$\begin{aligned} J = 4 \times 0.60 \times 5^2 &= 60.0 \text{ in.}^4 \\ 2 \times 0.60 \times 3^2 &= \underline{10.8} \\ & 70.8 \text{ in.}^4 \end{aligned}$$

Inserting this value in the torsion formula, using the distance to the outermost rivet equal to c, we obtain

$$\tau_R = \frac{Pec}{J} = \frac{10{,}000 \times 10 \times 5}{70.8}$$

$$= 7060 \text{ psi} \qquad \text{(rivets } A, B, C, \text{ and } D\text{)}$$

$$\tfrac{3}{5}\tau_R = 4240 \text{ psi} \qquad \text{(rivets } E \text{ and } F\text{)}$$

Note. No variation of stress is considered across the cross-sectional area of a resisting rivet.

A comparison of both methods reveals that either may be used to produce the same result.

Obtaining the resultant stress. In the derivation of the torsion formula, the direction of shear stress at any point was found to be tangent to the circle drawn through that point having its center at the center of rotation of the resisting area. As applied to the present problem, this statement indicates that the torsional stress on the various rivet areas acts tangential to the concentric circle passing through them (Fig. 10-22). To combine these stresses with the upward-acting direct shearing stress, we must obtain their vector sum. In other words,

$$\tau = \tau_D \mathbin{+\!\!\!\!\to} \tau_R \tag{10.16}$$

where the sign $+\!\!\!\to$ infers vector addition.

Thus, the resultant shearing stresses as obtained from the vector diagrams of Fig. 10-22 are as follows:

For Rivets A and D:

$$\tau = 2780 \mathbin{+\!\!\!\!\to} 7060 = 6280 \text{ psi}$$

For Rivets B and C:

$$\tau = 2780 \mathbin{+\!\!\!\!\to} 7060 = 9000 \text{ psi}$$

For Rivet E:

$$\tau = 4240 - 2780 = 1460 \text{ psi}$$

For Rivet F:

$$\tau = 4240 + 2780 = 7020 \text{ psi}$$

PROBLEMS

10-21. Find the maximum shearing unit stress in the bolts of Fig. 10-23. Bolt diameter is $\frac{7}{8}$ in. *Answer:* $\tau_{AB} = 10{,}610$ psi.

10-22. Find the maximum force P permitted on the joint of Fig. 10-24 if the maximum allowable shearing stress on the rivets is 10,000 psi and the rivets are $\frac{7}{8}$ in. in diameter.

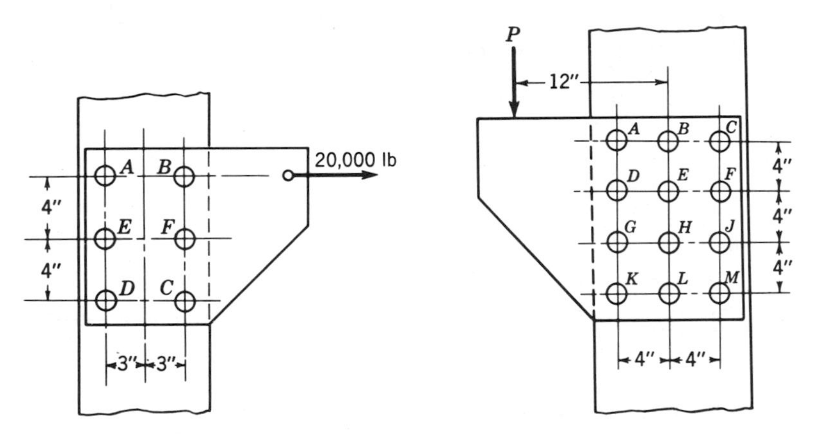

Figure 10-23

Figure 10-24

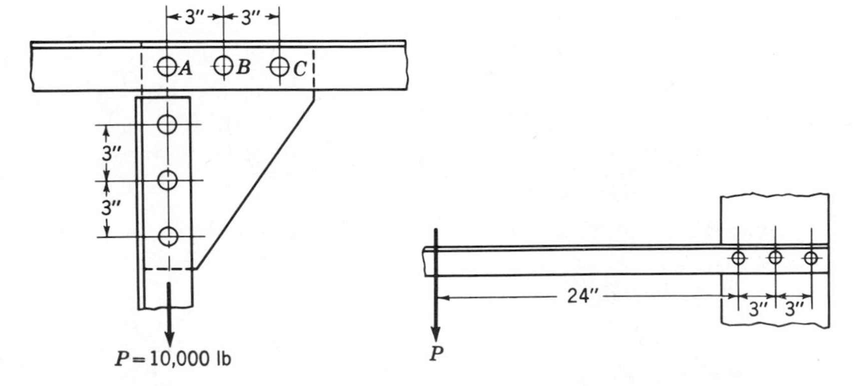

Figure 10-25

Figure 10-26

10-23. Find the shearing unit stress developed in each of the rivets of the horizontal chord member shown in Fig. 10-25. Rivet diameter is $\frac{7}{8}$ in. *Answer:* $\tau_A = 13{,}880$ psi.

10-24. Find the maximum force P that can be permitted on the bracket shown in Fig. 10-26 if the maximum allowable shearing stress on the bolts is 10,000 psi. Bolt diameter is $\frac{7}{8}$ in. Disregard the strength of the angle.

10-25. Find the maximum shearing unit stress in the bolts of Fig. 10-27. Bolt diameter is $\frac{7}{8}$ in. *Answer:* $\tau_A = 2380$ psi.

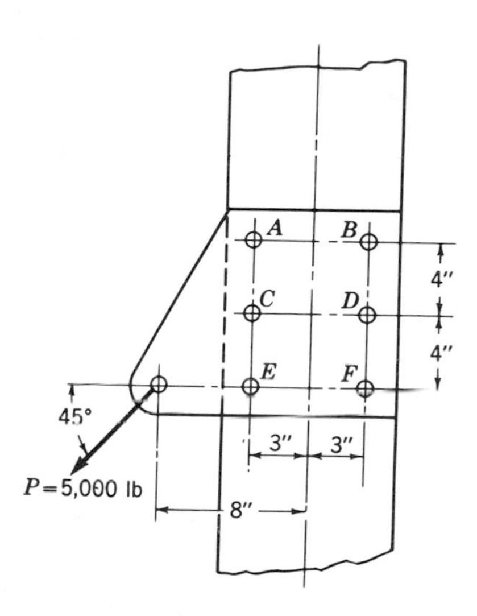

Figure 10-27

10-8 Eccentrically Loaded Welded Joints

The analysis of eccentrically loaded welded joints is similar to the analysis of eccentrically loaded riveted joints. Consider the joint shown in Fig. 10-28. In order to separate the direct and rotational shearing effects, the applied load of 20 kips is resolved into a force and a couple by adding the forces P_1 and P_2 to the joint on a line passing through the centroid of the resisting welds. Thus, the direct stress due to P_1 is equal to

$$\tau_D = \frac{20{,}000}{\frac{1}{2}\text{ in.} \times 0.707 \times 20\text{ in.}} = 2830\text{ psi}$$

The rotational shearing stress due to the couple PP_2 will be maximum at points A, B, C, and D, and is found by modifying the torsion formula, J being equal to the sum of I_{x-x} and I_{y-y} of the throat areas.

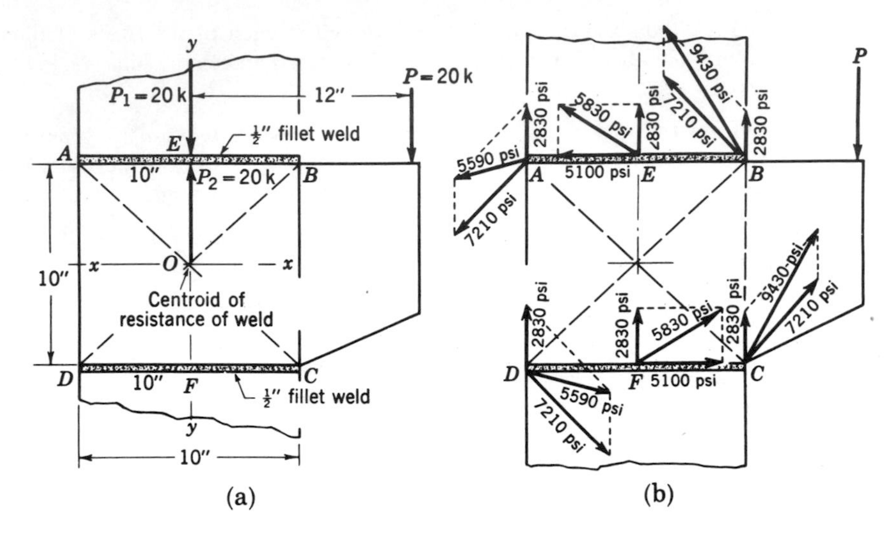

Figure 10-28

$$I_{x-x} = \sum Ad^2$$
$$= (10 \times \tfrac{1}{2} \times 0.707 \times \;\; 5^2)2 = 176.5 \text{ in.}^4$$
$$I_{y-y} = (\tfrac{1}{12} \times \tfrac{1}{2} \times 0.707 \times 10^3)2 = \underline{\;\;58.9\;\;}$$
$$235.4 \text{ in.}^4$$

$$\max \tau_R = \frac{T_R r}{J} = \frac{20{,}000 \times 12 \times 7.07}{235.4} = 7210 \text{ psi}$$

The maximum shear stress at any point is found by taking the vector sum of its direct and rotational shearing stresses. Thus, at points B and C (Fig. 10-28b), where the critical stress occurs, the maximum shear stress equals 9430 psi.

Other more involved eccentrically loaded welded joints, acted upon by a couple lying in the plane of the welds, are solved in the same manner shown above. In such problems it is imperative that the center of gravity of the resisting welds be determined first by considering each straight length of weld as a rectangle having an area equal to its length times its throat distance. The two added forces P_1 and P_2 required for the resolution into a force and a couple are always added on a line passing through the centroid. To calculate the polar moment of inertia, pass x and y axes through the centroid of the resisting welds and compute the I_{x-x} and I_{y-y} of the rectangular areas previously mentioned. The sum of I_{x-x} and I_{y-y} will give the required value of J.

If the load applied to an eccentrically loaded welded joint produces flexural stresses on the weld accompanied by a direct shear (Fig. 10-29), the resultant stress at a point cannot be found by vectorial addition. Reference should be made to Chapter 12 for the solution to such problems. If, however, the flexural stresses produced by the moment occur with direct tensile or

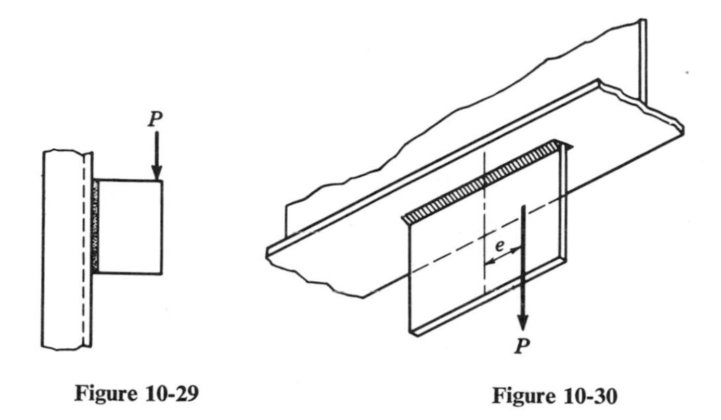

Figure 10-29 **Figure 10-30**

compressive stresses acting in the same direction, such as in Fig. 10-30, a simple algebraic addition provides the total stress at any point.

PROBLEMS

Use A.I.S.C. maximum allowable shearing stress of 21,000 psi on the throat of fillet welds in solving the following problems.

10-26. If, in Fig. 10-28, the top and bottom fillet welds are 8 in. long and have a leg equal to $\frac{5}{16}$ in., what load having an eccentricity of 15 in. can be sustained? *Answer:* 19,200 lb.

10-27. Determine the minimum size of fillet weld required for the weld arrangement shown in Fig. 10-28 if the load is 15 kips and acts with an eccentricity of 12 in.

10-28. In the joint of Fig. 10-31, compute the maximum safe load P that can be applied. *Answer:* 15,300 lb.

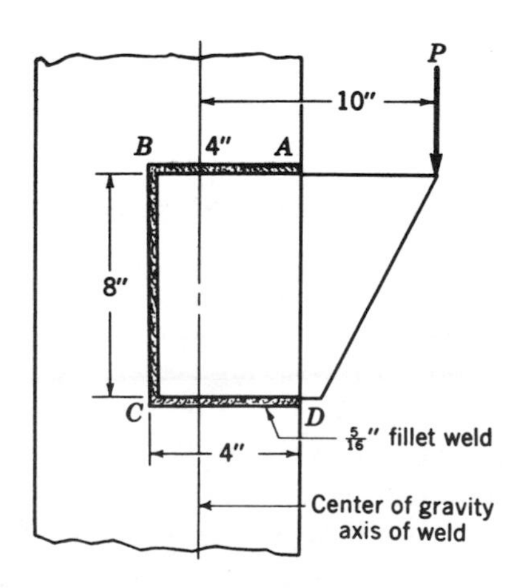

Figure 10-31

10-29. What would be the maximum stress in the weld of Fig. 10-31 if the eccentricity were reduced to 8 in. and the load P were 18 kips?

10-30. What maximum safe load P can the joint of Fig. 10-31 sustain if an additional fillet weld of $\frac{5}{16}$-in. leg is placed from A to D (on opposite side of plate)? $e = 9$ in. *Answer:* 28,600 lb.

10-31. If the weld BC in Fig. 10-31 were placed from A to D, what load P could be safely sustained if its location were not changed? $e = 8$ in. *Answer:* 24,100 lb.

10-9 Stresses in Close-Coiled Helical Springs

The analysis of the resisting stresses of a close-coiled helical spring originates with the resolution of its applied force into a force and couple as it acts on a free body of that spring. Let us consider the free body of the spring in Fig. 10-32b. As the tension in this free body was released by cutting the

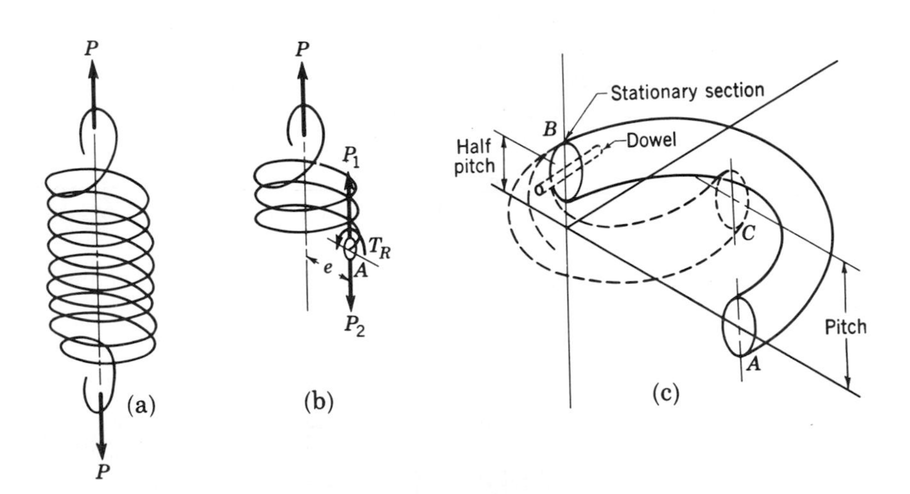

Figure 10-32

spring at section A, the adjoining surfaces must inevitably have slid by each other, indicating the relapse of direct shearing stresses. Then, because the distance between the coils of this spring decreased, there must have been a release of an internal-resisting couple. For illustration, let us consider Fig. 10-32c, a single coil from the same spring. If this coil were now cut at its mid-point by a radial section and pinned together with a dowel placed along the axis of the spring, any movement of one half-section relative to the other would cause rotation of the surfaces at B. Referring again to our free body, Fig. 10-32b, it must now be evident that as the distance between the coils decreased, internal resistance couples that produce rotation must have been released on every section with, of course, their accompanying torsional shear stress. Having thus established the two basic types of stress in a spring, let us determine their magnitude.

Add two forces at the centroid of section A, equal and parallel to P and opposite to each other. Forces P and P_2 constitute an external couple that twists section A. Opposing this action must be a counter couple T_2 acting on the same section, whose action is somewhat similar to the resisting torque developed in a straight cylindrical shaft. This stress will be recalled as being

$$\tau = \frac{T_R c}{J} \quad \text{or} \quad \frac{Pec}{J}$$

where c has been made equal to the radius, i.e., the distance from the center of rotation to the outermost fiber.

In order that this internal couple T_R oppose the couple Pe in Fig. 10-32b, it must act counterclockwise. The resisting torsional shearing unit stress on section A is represented in Fig. 10-33a.

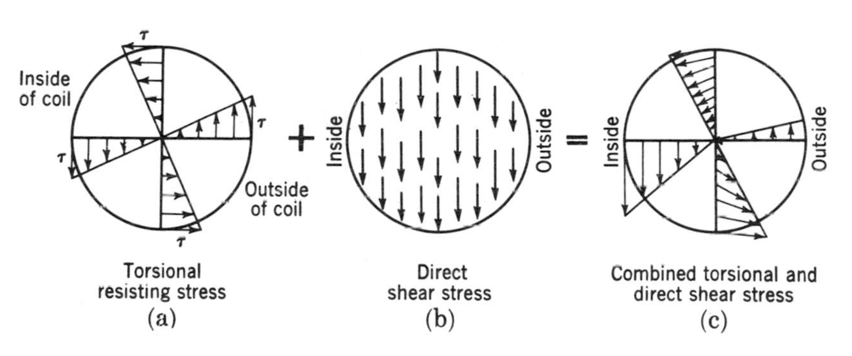

Figure 10-33

The direct stress produced by the upward force P_1 causes the resisting stress to act downward. Its magnitude, equal to $\tau = P/A$, is portrayed in Fig. 10-33b. When combined, the direct and torsional stresses present a picture somewhat as shown in Fig. 10-33c.

Under the action of tensile load P_1, the spring is seen to have its maximum shearing unit stress developed on its inner wall. The same would be true if the load were compressive instead of tensile. Regardless, therefore, of the character of the axial load, the inner wall of the spring is its most critically stressed portion.

Adding the direct and torsional shearing unit stresses vectorially, the maximum shearing unit stress as shown in Fig. 10-33c is obtained:

$$\tau = \frac{P}{A} + \frac{Pec}{J}$$
$$\tau = \frac{P}{A} + \frac{Pec}{Ar^2} = \frac{P}{A}\left(1 + \frac{ec}{r^2}\right) \tag{10.17}$$

The effect of assuming the torsional stress on the cross section of a spring to be identical to that found on the cross section of a straight cylinder is to render the above equation less accurate as the curvature of the spring is increased. To compensate for this difference, a multiplication factor k_1, known as the *Wahl's correction factor*, is inserted in the following equation to produce the theoretical maximum shearing stress.

$$\tau = k_1 \frac{Pec}{J} \tag{10.18}$$

Table 4 shows the variation of the Wahl's correction factor with the spring index, which is a ratio of the mean coil diameter D to the wire diameter d. In Table 4

TABLE 4

WAHL'S CORRECTION VALUES*

D/d	k_1	D/d	k_1	D/d	k_1
2.00	2.06	4.25	1.37	8.00	1.18
2.25	1.87	4.50	1.35	9.00	1.16
2.50	1.75	4.75	1.33	10.00	1.14
2.75	1.65	5.00	1.31	11.00	1.13
3.00	1.58	5.50	1.28	12.00	1.12
3.25	1.52	6.00	1.25	13.00	1.11
3.50	1.48	6.50	1.23	14.00	1.10
3.75	1.44	7.00	1.21	15.00	1.10
4.00	1.40	7.50	1.20	16.00	1.09

*Derived by A. M. Wahl of Westinghouse Electric and Manufacturing Co., Pittsburgh, Pa.

$$m = \frac{D}{d} = \text{spring index} = \frac{\text{mean diameter of coil}}{\text{diameter of wire}}$$

and

$$k_1 = \frac{4m - 1}{4m - 4} + \frac{0.615}{m}$$

10-10 Work Principle

When a *constant* force P is moved a distance Δ in the direction of the force, *the work performed by the force* is $P\Delta$. Suppose, now, that the force started at 0 lb and gradually increased to its full value P after moving through the distance Δ. If the increase in P were uniform, the work performed would only be $\frac{1}{2}P\Delta$.

These introductory remarks also apply to the deformation of any body under the action of a gradually applied load. The amount of external work performed on such a body is also equal to $\frac{1}{2}P\Delta$, where P is the value of the load after producing its maximum deflection.

As this gradually applied force P deforms the body and performs external work equal to $\frac{1}{2}P\Delta$, the resistance forces scattered through the body perform an identical amount of internal work in passing through their respective strains. This balance of external and internal work allows the presentation of the following verbal equation:

work performed by external forces in contact with body
= internal work caused by resistance forces*

10-11 Application of Work Principle to Deflection of Springs

If an axial load P were to be gradually applied to a spring until its maximum axial-deformation value of Δ had been attained, the work performed on the spring would be $\frac{1}{2}P\Delta$. Resisting this deformation are a couple T_R and direct shear force P_R acting on every vertical cross section of the spring. Disregarding the effect of the direct shear force, which is ordinarily very small, the internal work performed on the spring by the couple T_R,† twisting the entire length of the spring through an angle θ, is approximately equal to $\frac{1}{2}T_R\theta$. Thus, we may write the equation

$$\tfrac{1}{2}P\Delta = \tfrac{1}{2}T_R\theta - \tfrac{1}{2}Pe\theta$$

from which

$$\Delta = e\theta \tag{10.19}$$

Substituting in this equation the value of

$$\theta = \frac{Pel}{GJ}$$

(derived in §5-8) and letting $l = 2\pi en$, where n is the number of turns of a close-coiled spring, there results

$$\Delta = \frac{2\pi Pe^3 n}{GJ} \tag{10.20}$$

If the polar moment of inertia $J = \pi d^4/32$ were now to be inserted in

*A more detailed account of external and internal energy is presented in Chapter 14.

†From the study of mechanics, the work produced by a couple T_R rotating through an angle θ, expressed in radians, is $T_R\theta$. In this case, the couple is constant throughout the rotation.

the above expression, the equation would read

$$\Delta = \frac{64Pe^3n}{Gd^4} \tag{10.21}$$

which appears as the deflection formula for close-coiled springs in many of our mechanical-engineering handbooks. The interpretation of each of the items of the above equation, including the units in which each is most frequently used, is as follows:

Δ = spring deflection, in in.
P = axial load on spring, in lb
e = eccentricity of load: the distance from the action line of the load to the center of the cross-sectional area where stress is; developed mean radius in in.
n = number of active coils participating in the deflection
G = shear modulus of elasticity, in lb/in.^{2}
d = diameter of wire, in in.

Illustrative Problem 2. A helical, steel compression spring is to be designed to resist a deflection of 2 in. when subjected to an axial load of 300 lb. If the diameter of the wire is $\frac{1}{4}$ in. and the mean diameter of the coil is $1\frac{1}{4}$ in., find the number of coils required. Allowable shearing unit stress is 80,000 psi.

SOLUTION

$$\Delta = \frac{64Pe^3n}{Gd^4}$$

$$2 = \frac{64 \times 300 \times \overline{0.625^3} \times n}{12 \times 10^6 \times (0.25)^4}$$

$$n = \frac{2 \times 12 \times 10^6 \times (0.25)^4}{64 \times 300 \times (0.625)^3} = 20$$

Such a computation should not be considered complete until the stress is checked. Only if the stress is below the allowable should the previous computation be allowed.

Check on stress.

$$\tau = \frac{P}{A}\left(1 + \frac{ec}{r^2}\right) = \frac{300}{(\pi/4)(1/4)^2}\left(1 + \frac{0.625 \times 0.125}{(1/4)^2(1/8)}\right) = 67{,}400 \text{ psi (approximate)}$$

From Table 4 the Wahl correction factor for a spring having a spring index of 5 is equal to 1.31. The more accurate maximum stress is therefore

$$1.31 \times \frac{300}{\pi} \times 64 \times 10 = 80{,}000 \text{ psi}$$

PROBLEMS

10-32. A helical, steel tension spring is to be stressed to a maximum value of 100,00 psi with a load of 200 lb. If the diameter of the wire is 0.192 in., the outside diameter, $1\frac{1}{2}$ in., and the deflection at full load, 3 in., find the number of coils necessary.

10-33. A phosphor-bronze tension spring having a 0.625-in. outside diameter is to be made from 0.080-in. wire. If the deflection desired is 2 in., how many coils will be necessary if the load applied is 40 lb? $G = 5 \times 10^6$ psi. *Answer:* 7.93.

10-34. Determine the load that must be applied to a compression spring to produce a deflection of 3 in. if the wire diameter is 0.192 in., the outside diameter, 1 in., and the number of coils, 36.6. $G = 12 \times 10^6$ psi.

10-35. A helical tension spring made of steel wire 0.080 in. in diameter has an outside diameter of 0.55 in. Determine the extension of the spring upon the application of an axial load of 40 lb. $n = 10$ coils. *Answer:* 0.677 in.

10-36. Two steel cantilever beams are connected at their ends by a spring as shown in Fig. 10-34. A load of 2000 lb is then applied at the end of the lower beam. Determine the deflection at the end of the upper beam. The spring constant k is the load required to stretch the spring 1 in. $E = 30 \times 10^6$ psi.

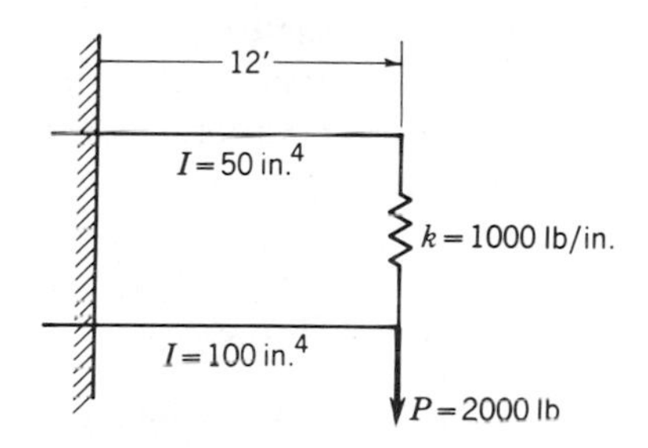

Figure 10-34

10-37. A cantilever beam (Fig. 10-35) is built into a vertical wall at its left end. Under its right end is fitted a steel spring having a spring constant of 9000 lb/in. Calculate the force exerted by the beam upon the spring due to the application of the load P of 4000 lb. Disregard the weight of the beam. I of beam $= 100$ in.^{4}. $E = 30 \times 10^6$ psi. *Answer:* 792 lb.

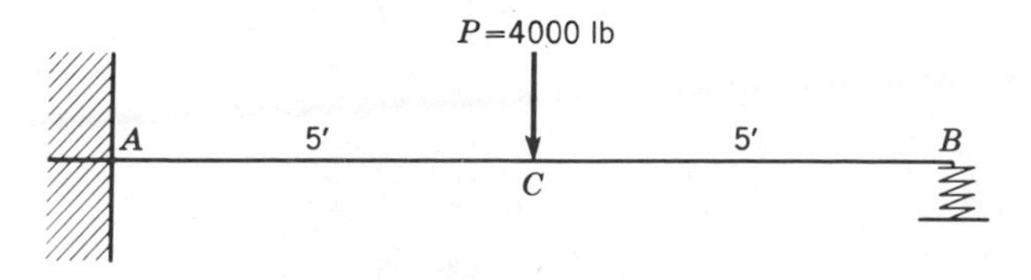

Figure 10-35

10-12 Discussion of Deflection Formula for Springs

The previous deflection formula is approximate for the following reasons:

1. The effect of the deflection due to the direct shear force was not taken into account. Although of little effect for most springs, this feature assumes increased importance as the ratio of wire diameter d to eccentricity e increases. Its omission underestimates the magnitude of the deflection.

2. The length of the spring is not exactly $2\pi en$, but slightly greater, dependent on the overall height of the spring and the number of coils involved.

3. The use of the formula $\theta = Pel/GJ$ applies only to cylindrical shafts subjected to twisting couples. Its use for springs having small radii would result in a smaller computed deflection value than that actually obtained.

4. As the pitch between the coils changes, the value of e also changes.

5. The minimum cross section of the wire is not in the same plane as the axis of the spring. The value of J used in the derivation is slightly smaller than actually exists.

6. The wire diameter may vary slightly from the ordered diameter by the amount of the manufacturer's tolerance. Since the deflection varies with the fourth power of the diameter, any variation in diameter may have an appreciable effect on the deflection.

10-13 Exact Spring Evaluation and Design

Although these discrepancies make the above equation incapable of being used for accurate deflection computations, a fair estimate of its value may be obtained. For a more accurate evaluation of the action of a spring that has already been made, or when required to design a spring to meet certain exacting conditions, it would be well to refer to the various design charts made available by several large spring manufacturers.

11

Columns

11-1 Problem of Column Design

The design of columns presents a problem less possible of exact solution than any other thus far presented. Some of the reasons are

1. There is no definite demarcation point between a column that is relatively short and a compression block that is relatively tall. In some fields of column design, the distinction between the two is arbitrarily assumed with the help of experimental data.

2. Although a column is, for all practical purposes, a straight, homogeneous compression member, it is never made theoretically perfect. Any deviation in its alignment, lack of homogeneity, or presence of internal stresses will act as a source of bending and possible ultimate collapse. Although these defects are of only secondary importance in the design of beams or tension members, they are of primary importance in determining the resistance of a column.

3. The inability to apply a perfectly axial load causes slight eccentricities to be imposed upon the column that may contribute markedly to its bending tendency and possible ultimate collapse.

4. The character and magnitude of the end restraint of ordinary columns may vary greatly. Since the type of end connection and the relative

stiffness of the connected members both contribute to the magnitude of the restraint, it is not difficult to realize the complexity of this phase of the problem alone.

Since an exact quantitative measure of each of these four items would be difficult, if not impossible, of practical attainment, the study of column action and design has resulted, for the most part, in the application of irrational column formulas to existing experimental and theoretical data, notably those obtained from Euler's equation.

11-2 Slenderness Ratio

An item that appears in nearly every column formula is the *slenderness ratio*, which is the ratio of the unsupported length of a column to its least radius of gyration. To evaluate the strength of any column, the l/r ratio should be its maximum value. Should lateral support be offered to one of the principal axes of a column cross section, as for instance, axis y-y of Fig. 11-1, its effect could not be considered in determining the slenderness ratio about the other principal axis x-x. Radii of gyration of all standard sections are found in handbooks under the section entitled *Elements of Sections*. A selected list of standard sections with their common properties, including the radii of gyration, will be found in the Appendix.

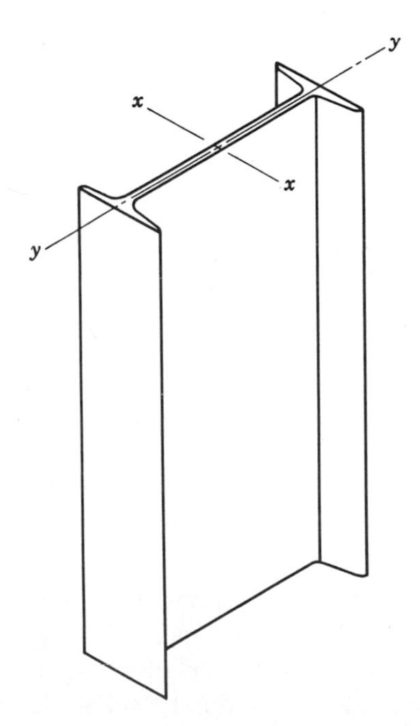

Figure 11-1

Illustrative Problem 1. Determine the slenderness ratio of a W 12 × 65 section, 20 ft long.

SOLUTION. $l = 240$ in., $r_{x-x} = 5.28$ in., and $r_{y-y} = 3.02$ in.

$$\text{slenderness ratio} \left(\frac{l}{r}\right)_{y-y} = \frac{240}{3.02} = 79.4$$

Illustrative Problem 2. If, in the previous problem, lateral support is given to the y-y axis of the W beam at its midpoint, determine the new slenderness ratio of the column.

SOLUTION. $l_{x-x} = 240$ in., $l_{y-y} = 120$ in., $r_{x-x} = 5.28$ in., and $r_{y-y} = 3.02$ in.

$$\left(\frac{l}{r}\right)_{x-x} = \frac{240}{5.28} = 45.4, \qquad \left(\frac{l}{r}\right)_{y-y} = \frac{120}{3.02} = 39.7$$

The critical slenderness ratio, which previously was about axis y-y, has now changed to axis x-x and is equal to 45.4.

Illustrative Problem 3. Determine the slenderness ratio of a 4- by 4- by $\frac{1}{4}$-in. angle, 10 ft long.

SOLUTION. An angle section has a minimum radius of gyration z-z that should be carefully noted (Fig. 11-2). $l = 120$ ft, $r_{z-z} = 0.80$ in., $r_{x-x} = 1.25$ in., and $r_{y-y} = 1.25$ in.

$$\left(\frac{l}{r}\right)_{z-z} = \frac{120}{0.80} = 150$$

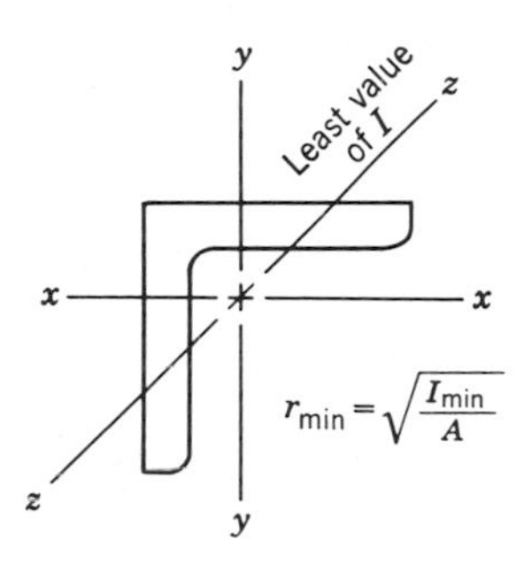

Figure 11-2

PROBLEMS

11-1. Calculate the slenderness ratios of the following sections: (a) An S 8 × 18.4 beam 16 ft long. (b) A 4- by 6-in. timber 10 ft long. (c) A 5- by 3- by $\frac{3}{8}$-in. angle 8 ft long.

11-2. Compare the least radii of gyration of the following steel sections: (a) A 4- by 4- by $\frac{3}{8}$-in. angle. (b) A solid rectangular section 1.7 by 1.7 in. (c) A hollow circular section 4 in. in outside diameter and 3.52 in. in inside diameter. (d) An S 5 × 10 column. In each case, the area is approximately the same. Which of these sections should make the strongest column?

11-3. The symmetrical axes of two 9-in. 13.4-lb channels are placed on the same horizontal axis. What should be the distance between the webs in order to have the radii of gyration the same about the two principal axes? The flanges extend outward. *Answer:* 5.62 in.

11-3 Importance of Radius of Gyration in Column Design

Radii of gyration of column sections bear an important part in the design of columns and serve as measures of their lateral stability. Since the radii of gyration are direct functions of a column cross section, it is through their judicious manipulation that the final column selection is made. Unless a column must meet certain space or directional strength requirements, the radii of gyration about the two principal axes are made as nearly alike as possible. In so doing, the column will not carry an unnecessary and uneconomical amount of material, and it will be equally strong in either direction.

11-4 Experimental Column Curves

When a series of medium-carbon steel columns having hinged ends and varying slenderness ratios are placed under increasing axial compression loads to failure, and the values of l/r and P/A at failure are plotted for each

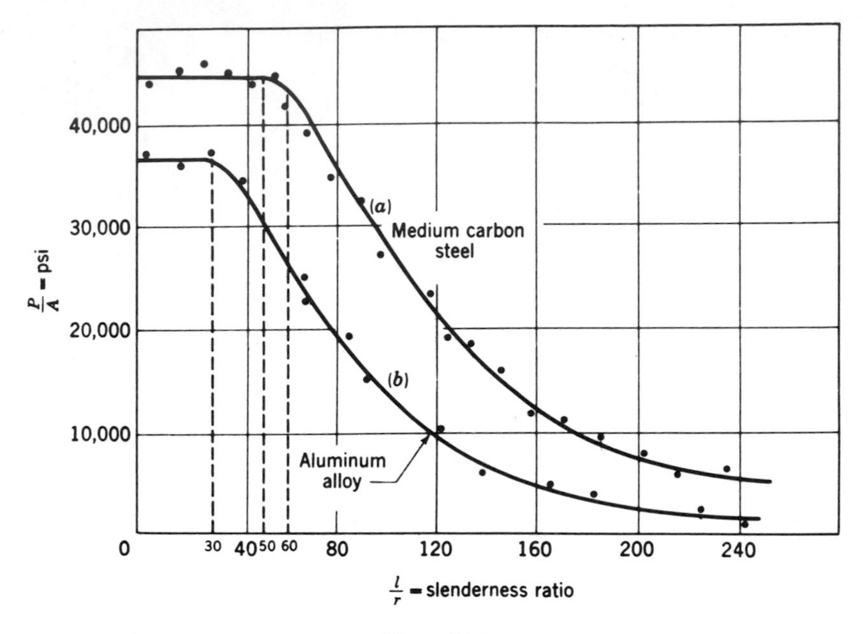

Figure 11-3

column, an average curve similar to that shown in Fig. 11-3a is obtained.

If still another plot were made, using the failure data from a series of aluminum-alloy columns having hinged ends and similar slenderness ratios, it would appear as shown in Fig. 11-3b.

From observation of these two failure curves for columns, it is apparent that for the lower values of l/r the compressive unit stress at failure is fairly constant. This is as we might expect, since the strength of short, stubby compression blocks is not materially affected by even substantial changes in their lengths. For all practical purposes, this limiting value of P/A is the compressive yield strength of the material, the aluminum alloy in this illustration having a lower yield strength than the medium-carbon steel. It is interesting to note at this point that in all previous problems concerning stresses in compression members those members were referred to as *compression blocks*. Their stubbiness prevented any *column action* from interfering with the development and computation of resistance stresses. Column action is indicated by the reduction of the value of P/A (Fig. 11-3) as the value of l/r is increased.

Of significance in Fig. 11-3 is the fact that the curve for the aluminum-alloy is below that for medium-carbon steel. This is primarily due to the lower stiffness of the aluminum alloy. The modulus of elasticity for aluminum alloy is approximately 10,000,000 psi, whereas that for steel is approximately 30,000,000 psi. Although the effect of the modulus of elasticity decreases as the length decreases, its tendency to depress the curve in proportion to the moduli at high values of l/r causes the entire aluminum-alloy curve to fall below that of the steel.

Another point of significance revealed by Fig. 11-3 is the comparative length of the constant P/A portions of both curves. The curve with the greater stiffness (steel) terminates its horizontal portion of the curve at approximately $l/r = 60$. That for the aluminum is terminated at approximately $l/r = 30$. We may expect this variation to persist as materials with varying moduli of elasticity are studied. Wood, as an added example, has a modulus of elasticity of approximately 2,000,000 psi, and has a limiting value of constant P/A of approximately $l/r = 10$. A continued attempt to correlate compression test data will be made after an analysis is made of Euler's formula for tall, slender columns.

COLUMNS WITH HIGH SLENDERNESS RATIOS

11-5 Euler's Formula—General Discussion

Having concerned ourselves to this point primarily with the left-hand portion (low l/r values) of the experimental curves of Fig. 11-3, let us now consider those portions pertaining to columns having high values of l/r. These columns will be referred to as *tall, slender* columns.

A *theoretically perfect* column having an indefinitely high slenderness ratio and loaded with a perfectly axial compressive load will not bend no matter how great the load becomes. It will crush instead. If, however, such a column is purposely bent and is subjected to a compressive load to maintain the deflection, its bending characteristics can be effectively studied in obtaining its buckling load.

Suppose the load maintaining the deflection is gradually decreased. A point is soon attained in the load-removal process when the column becomes perfectly straight. The load still applied to the column upon the attainment of perfect straightness is known as *Euler's critical load*. This load is also that limiting load which the column can carry under practical conditions without buckling. Let us proceed with its derivation.

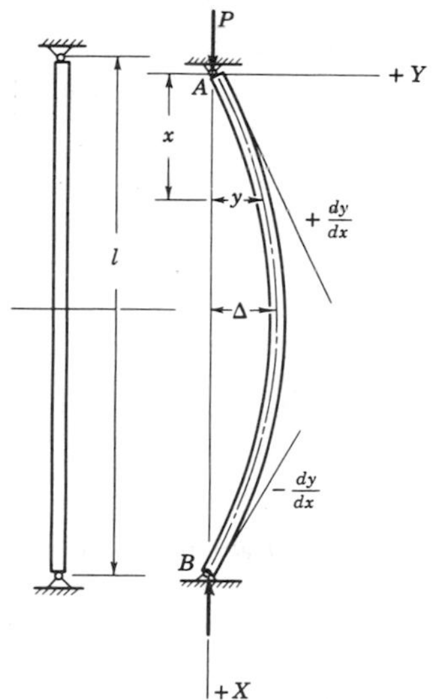

Figure 11-4 The increment $\frac{d}{dx}\left(\frac{dy}{dx}\right)$ must be negative to change the positive slope at A to the negative slope at B.

Derivation. Consider a theoretically perfect column (Fig. 11-4) whose ends are free to rotate about hinges which remain in perfect vertical alignment as load is applied. If the column is now purposely bent and a vertical load P applied to maintain this deflected position, the general elastic-curve equation is assumed to be similar to that derived for a simple beam, Eq. (8.6):

$$EI\frac{d^2y}{dx^2} = M$$

Choosing the positive directions of the x and y axes as downward and to the right, respectively, as indicated on Fig. 11-4, the above equation may be written more specifically as

$$EI\frac{d^2y}{dx^2} = -Py$$

The negative sign results from the fact that the slope dy/dx changes from a positive value at the origin to a negative value at $x = l$, thus making the change of slope d^2y/dx^2 a negative value. Were this equation now to be expanded and multiplied by $2dy$, it would appear as

$$EI\left(2\frac{dy}{dx}\,d\frac{dy}{dx}\right) = -2Py\,dy$$

This expression may then be integrated to give

$$EI\left(\frac{dy}{dx}\right)^2 = -Py^2 + C_1$$

By substituting the value of $dy/dx = 0$ when $y = \Delta$, C_1 becomes $P\Delta^2$, making the equation read

$$EI\left(\frac{dy}{dx}\right)^2 = P(\Delta^2 - y^2) \quad \text{or} \quad \frac{dy}{dx} = \sqrt{\frac{P}{EI}}\sqrt{\Delta^2 - y^2}$$

By separating the variables, the equation may be written

$$\frac{dy}{\sqrt{\Delta^2 - y^2}} = \sqrt{\frac{P}{EI}}\,dx$$

which when integrated gives

$$\sin^{-1}\frac{y}{\Delta} = \sqrt{\frac{P}{EI}}\,x + C_2$$

But $x = 0$ when $y = 0$, making $C_2 = 0$. Thus,

$$\frac{y}{\Delta} = \sin\sqrt{\frac{P}{EI}}\,x \tag{11.1}$$

The Eq. (11.1) above defines the elastic curve of the column as a continuous sine wave whose amplitude equals the maximum deflection Δ. The greatest importance of the equation lies not, however, in defining the deflection curve but in permitting the determination of the critical load.

In determining this load, it is first necessary to note that the projected length of the deflected column is not exactly equal to its total length l. In fact, not until the load is reduced to a value that will just make the column straight will the projected length equal the actual length. The value of the load sustained by the column as it becomes straight is the *Euler critical load.*

In order to obtain the Euler critical load, the column curvature is reduced to the point where Δ is infinitesimally small and the column is just about straight. Under this condition, the left-hand portion of Eq. (11.1) still remains zero, and the value of x is for all intents and purposes equal to l. Thus, Eq. (11.1) becomes

$$\frac{y}{\Delta} = 0 = \sin\sqrt{\frac{P}{EI}}\,l$$

The angle $l\sqrt{P/EI}$ whose sine equals zero must therefore equal π or some multiple of π, depending upon the number of half-sine waves that are included in the column length. It follows that the minimum or critical value of P for a column whose length equals one half a sine wave, i.e., for a column with a bend on one side of the axis and hinged at each end, is found by equating

$$\sqrt{\frac{P}{EI}}\,l = \pi \quad \text{or} \quad P = \frac{\pi^2 EI}{l^2} \tag{11.2}$$

This is known as *Euler's column formula* for tall, slender columns, the variables of which are defined as $E =$ modulus of elasticity of the column material in psi, $I =$ minimum moment of inertia of cross section in in.^{4}, $l =$ unsupported length in inches.

It is important to associate Eq. (11.2) with a column which when loaded will bend into the shape of one half a sine wave. Should the column have a reverse bend and have a shape consisting of two half-sine waves or one complete sine wave, the equation would be modified by setting

$$\sqrt{\frac{P}{EI}}\,l = 2\pi \quad \text{or} \quad P = \frac{4\pi^2 EI}{l^2} \tag{11.2a}$$

In other words, *the strength of a tall, slender column hinged at each end increases by the square of the number of half-sine waves included in its length.*

An important assumption. An important assumption made in the derivation of this formula is that the theoretical column under consideration is so slender that the resistance offered by the direct compressive stresses is negligible when compared to the bending stresses. This assumption would therefore permit the application of the formula only to columns having high slenderness ratios. Fortunately, the agreement of experimental data and the values obtained from the formula for high values of l/r are relatively good. As the slenderness ratios decrease, however, the formula and experimental values diverge (Fig. 11-9).

11-6 Effect of Variables on Euler's Formula

The location of the point of tangency of the Euler and experimental curves is again a function of the relative stiffness of the columns (points S

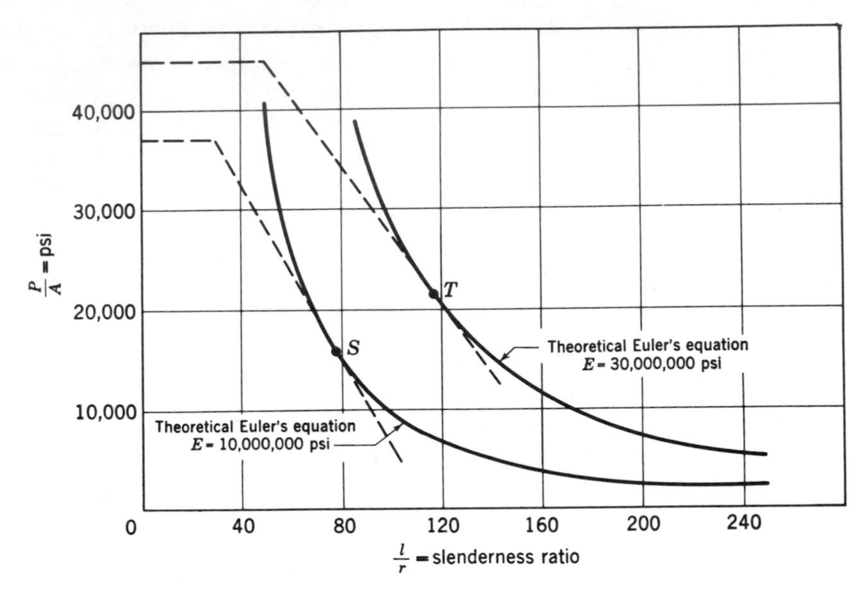

Figure 11-5

and T, Fig. 11-5). As the modulus of elasticity of the column material decreases, the point of tangency moves toward the vertical axis. In fact, Euler's curves move progressively into the corner formed by the P/A and l/r axes as the values of E become smaller and smaller.

11-7 Buckling Action of a Tall, Slender Column

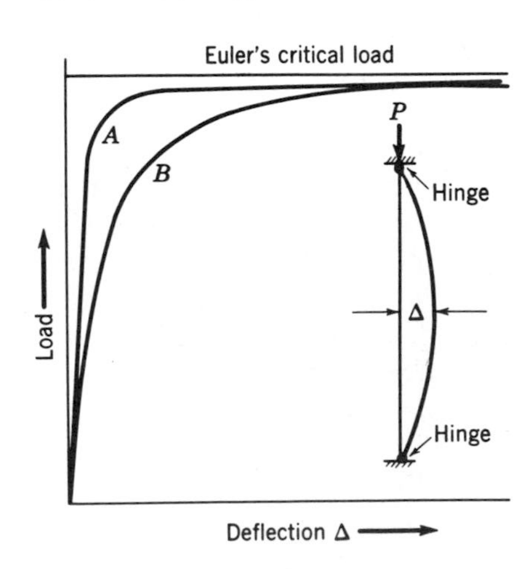

Figure 11-6 Effect of initial eccentricities. A = small initial eccentricity; B = large initial eccentricity.

To bring Euler's formula for tall, slender columns into a more practical realm, let us investigate a tall, slender column of ordinary straightness and uniformity similar in size and shape to that previously used in the theoretical discussion. After carefully placing the column in a vertical position, let us apply an increasing axial load. Immediately upon the application of the load, the column will start to bend (Fig. 11-6, curve A). Although the amount of bend is very small, the effect of the initial eccentricities and residual stresses is in evidence from the very start. As loading continues, deflection will generally increase at a slightly faster rate.

With the approach of Euler's critical load, however, the deflection becomes greater and greater until, at the very moment of its attainment, practically all the bending resistance is gone. The addition of a very small load will at this point cause an excessively large deflection or a possible collapse. Curve B has been drawn on the same diagram to indicate the effect of an increased initial eccentricity of the column or the load on the magnitude of the deflection.

11-8 Inelastic Column Action—the Tangent Modulus*

An examination of Euler's column equation reveals the critical buckling stress of tall, slender columns to be dependent solely on the modulus of elasticity. As slenderness ratios decrease, however, the corresponding P/A stress values at buckling increase and ultimately exceed the proportional limit stress of the material. When this happens, the modulus E begins to decrease slowly. Moreover, below this transition point the normal compressive stress becomes a substantial part of the total stress, causing failure to occur at loads less than those indicated by the Euler equation.

Failure of intermediate and short columns, however, may be predicted from the modified Euler equation developed by F. R. Engesser:

$$\frac{P}{A} = \frac{\pi^2 E_t}{(l/r)^2}. \tag{11.2b}$$

It is based on the use of tangent moduli E_t determined at points between the proportional limit and yield stress. To illustrate, let Fig. 11-7a be the stress-

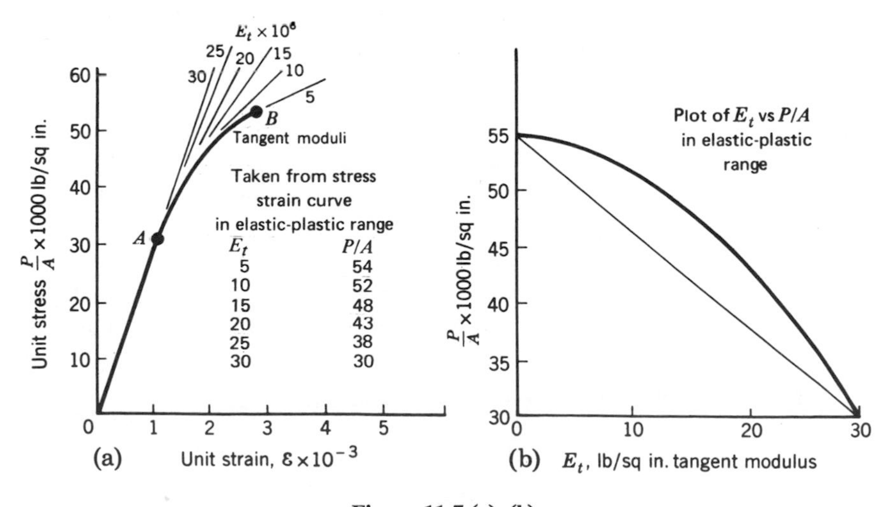

Figure 11-7 (a), (b)

*See §3-4.

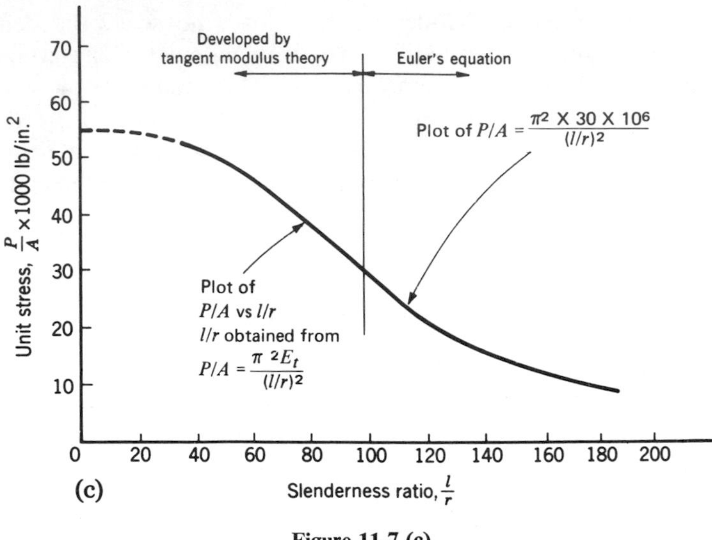

Figure 11-7 (c)

strain diagram of the column material in compression. At several points above the proportional limit the slope of the curve, or, in other words, the tangent modulus, is obtained, These tangent moduli and their corresponding stresses are plotted in Fig. 11-7b. If now corresponding P/A and E_t values are inserted in the modified Euler equation and the values of l/r are computed, the P/A and l/r values, when plotted (Fig. 11-7c), provide an extension to the Euler curve which to a degree represents the failure curve in the intermediate and short column range. This curve is said to have considerable experimental verification.

PROBLEMS

11-4. In Problem 3-1, use the data given to draw a $\sigma - E_t$ curve and the extension to the Euler buckling curve showing the relation of the buckling stress P/A to l/r in the intermediate and short column range. Determine the buckling load for a 6-ft column of this material 6 in. in diameter. Assume pivoted ends. *Answer:* 1,010,000 lb.

11-5. Determine the diameter of a solid Duralumin column 10 ft tall that will fail under an axial load of 2,000,000 lb. Assume hinged ends. Use data of Problem 3-1. *Answer:* 7.80 in.

11-6. Two 3- by 2- by $\frac{1}{4}$-in. angles made of A-36 steel are to be riveted together with the 3-in. legs back to back to form a column with a T section. If both ends are hinged, what length must the column have to fail under an axial load of 4000 lb?

11-7. From the data presented in Problem 3-3 draw a $\sigma - E_t$ curve of this cellulose acetate compound. Draw the complete buckling curve for all values of l/r, assuming pivoted ends.

11-9 Euler's Design Formula

Since the exact magnitude of the eccentricity existing in a column is never known, it is essential that the applied load for a tall column be kept considerably lower than the theoretical Euler load. The general form of Euler's equation used for design purposes is

$$\frac{P}{A} = \frac{\pi^2 E}{f(l/r)^2} \tag{11.3}$$

where I was substituted by its equal Ar^2 and f is a factor of safety.

It is apparent from Eq. (11.3) that every material having a different modulus of elasticity will have a different Euler's design formula. Also, as was intimated in a previous paragraph, the upper limit of l/r, at the point of tangency of Euler's and experimental curves, will also change. For instance, the upper limit of l/r for Euler's equation is approximately 150 for low-carbon steel and 67 for 6061-T6 aluminum alloy. Euler's equation will hold for all higher values of l/r.

Useful from a view point of economy is the fact that low-cost, low-carbon steel will prove as acceptable as high-cost, high-carbon steel for tall, slender columns. This is true because all steel, regardless of its cost, has substantially the same modulus of elasticity.

Illustrative Problem 4. Determine the maximum allowable axial compressive load a 2- by 2- by $\frac{1}{4}$-in. angle made of 6061-T6 structural aluminum can carry if its length is 6 ft.

Euler's equation must be used if the slenderness ratio exceeds 67. Use a factor of safety of 2.25. $A = 0.94$ in.^{2}, r_{x-x} or $r_{y-y} = 0.60$ in., $r_{z-z} = 0.39$ in., and $E = 10{,}000{,}000$ psi.

SOLUTION. In order to determine whether the Euler formula will apply to this column, the value of l/r must first be obtained:

$$\frac{l}{r} = \frac{72}{0.39} = 185$$

Euler's equation does apply. Substituting the proper values, we obtain

$$\frac{P}{0.94} = \frac{9.85 \times 10{,}000{,}000}{2.25 \times (185)^2}$$

$$P = 1200 \text{ lb}$$

Illustrative Problem 5. Select the most economical structural-steel S section 20 ft long to carry an axial compressive load of 50,000 lb with a factor of safety of 2.5. Assume the column to have hinged ends.

SOLUTION. Without the knowledge of a cross-sectional area, the value of the minimum radius of gyration cannot be computed. It also follows that without an l/r ratio the immediate selection of the proper column formula is impossible. We are therefore forced to use the following trial procedure.

Since the column is unusually tall and made of an S section, one axis of which is relatively weak, suspicion dictates a large l/r and the probable use of the Euler equation. Should this assumption prove correct, the maximum allowable stress will be relatively small, since an l/r of 200 provides stress of about 7500 psi with a factor of safety of 1.0. Inasmuch as the stipulated factor of safety is 2.25, let us assume an allowable working stress of 3000 psi. The corresponding required cross-sectional area will be about 16.7 in.^{2}.

From a perusal of Table 13 it is noted that S sections of this approximate area have a radius of gyration of approximately 1.10. The corresponding slenderness ratio of 218 is well within the range of applicability of the Euler equation.

If now an S 20 × 65.4 trial section ($r = 1.21$ in., $I_{min} = 27.9$ in., $A = 19.08$ in.^{2}) is selected, there results upon insertion in Euler's equation

$$2.5 \times 50{,}000 = \frac{9.85 \times 30{,}000{,}000 \times A}{(240 \times 1.21)^2}$$

$$A = 16.7 \text{ in.}^2 \text{ required, available } 19.08 \text{ in.}^2.$$

Although this section will suffice, let us explore further Table 13 for other sections which may be more economical (i.e., of less weight). No lighter beam provides the moment of inertia required. The section selected is therefore the most economical one.

PROBLEMS

Note: The following problems concern columns having hinged ends.

11-8. A steel compression strut 10 ft long is to be made from an equal-leg angle. If the load to be carried is 5000 lb, what most economical angle section should be used, using a factor of safety of 3? *Answer:* $3\frac{1}{2}$ by $3\frac{1}{2}$ by $\frac{1}{4}$ in.

11-9. A certain safety device for a dam supported by two steel angles is designed to collapse at a total specific load of 20,000 lb. If the length of each angle is 12 ft, what angle sections should be used? Keep back-to-back distance less than $\frac{3}{4}$ in.

11-10. A 2024-T3 aluminum-alloy bar having a 1- by 2-in. cross section is to be axially compressed. Determine the critical length at which buckling will take place if $E = 10{,}300{,}000$ psi, and the stress at the beginning of buckling action is 3440 psi. Determine the magnitude of critical stress if the length is 6 ft. *Answers:* 49.6 in., 1640 psi.

11-11. A $2\frac{1}{2}$- by $2\frac{1}{2}$- by $\frac{1}{4}$-in. angle 10 ft long is to be subjected to an axial compressive load. If made of steel, how much may the angle carry, assuming a factor of safety of 2.5? *Answer:* 2340 lb.

11-12. Determine the cross-sectional dimensions of a square timber column, 20 ft long, to take an axial compressive load of 20,000 lb. Use $E = 2{,}000{,}000$ lb/in.^{2}

and a safety factor of 3. The lower limit of l/r for a Euler curve for wood of the above modulus is about 50.

11-13. Determine the axial compressive load a 6-ft length of streamlined aircraft steel tubing can hold if its structural characteristics are as follows: area $= 0.2161$ in.^{2}, $I_{x-x} = 0.0376$ in.^{4}, $I_{y-y} = 0.1494$ in.^{4}, $r_{x-x} = 0.4170$ in., and $r_{y-y} = 0.8313$ in. Use factor of safety of 2.5.

11-10 Effect of End Conditions

In the previous analysis of Euler's equation, each column was assumed to have pivoted ends. This assumption is a rather important one, since a change of end conditions imposed upon such a column may have a marked effect upon its load-carrying capacity. Table 5 indicates the effect of several changes of end condition on the load-carrying capacity of a tall, slender column.

TABLE 5

EFFECT OF VARIOUS END CONDITIONS ON TALL, SLENDER COLUMNS

General Formula: $\dfrac{P}{A} = \dfrac{\pi^2 E}{(kl/r)^2}$

End Condition	k	No. Times Stronger Than Pivot-Ended Column
(a) Both ends fixed	0.5	4
(b) One end fixed, the other pivoted	0.7	2
(c) Both ends pivoted	1.0	1
(d) One end fixed, the other free	2.0	$\frac{1}{4}$
(e) Both ends fixed, one end laterally displaced	1.0	1
(f) One end pivoted, one end fixed and laterally displaced	2.0	$\frac{1}{4}$

In Table 5 and in Figs. 11-8a and b, it is revealed that the load-carrying capacity of a tall, slender column is increased or decreased by a change in end conditions insofar as that change increases or decreases the distance between its hinges or inflection points. As an inflection point is developed and made to travel away from its support toward the midpoint of the column, the load-carrying capacity of the column increases in inverse proportion to the square of the distance between the hinges or inflection points.

Since the column in type (d) would have to be rotated its full length about its base to make a column similar in shape to the type (c) column, its equivalent length is equal to $2l$, and the load reduction to the square of the coefficient of l. Because column strength at low values of l/r is unaffected by the fixity of the ends and is instead dictated by the yield strength of the

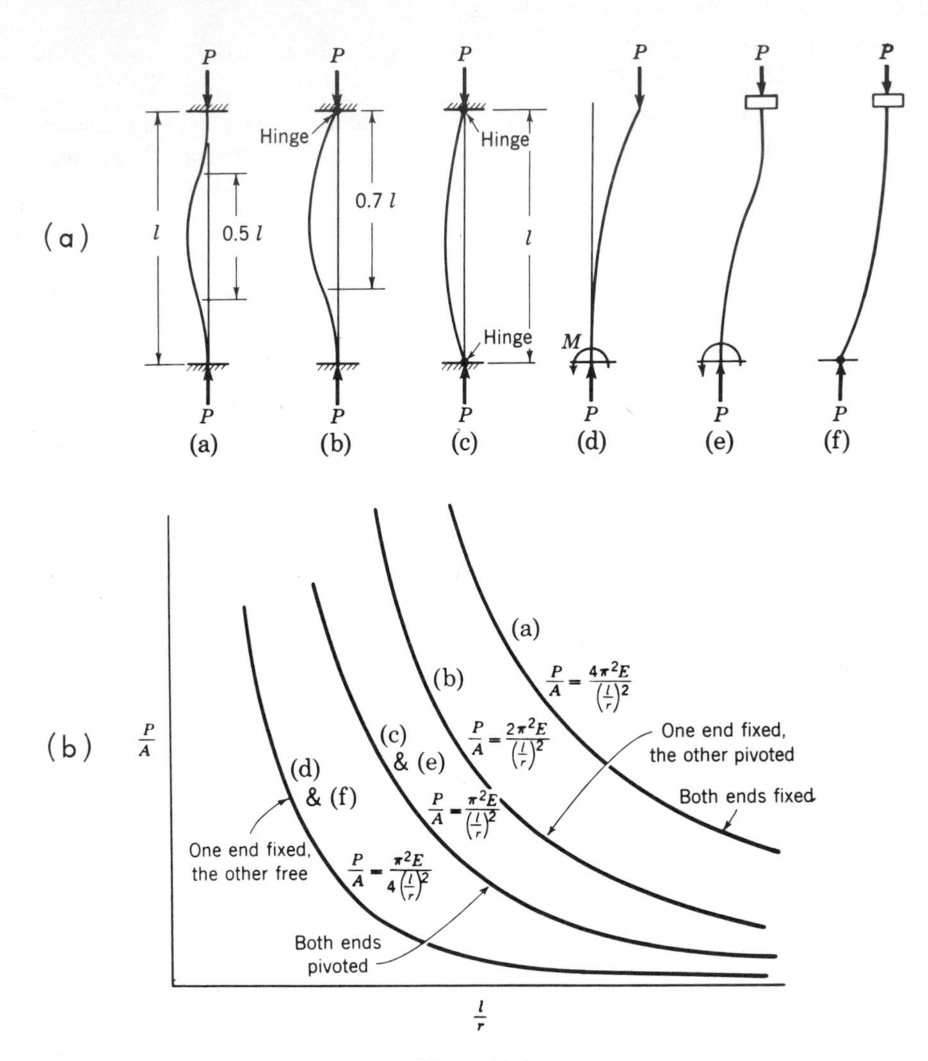

Figure 11-8

material, the point of tangency between the experimental and Euler curves moves toward the right with increased fixity.

PROBLEMS

The following problems are to be solved with the use of modified Euler formulas, assuming that they will satisfy the l/r limitations of the imposed end conditions.

11-14. Determine the allowable compressive load on the column of Problem 11-11, assuming one end fixed and the other pivoted. *Answer:* 4680 lb.

11-15. What angle section should be used for Problem 11-8 if both ends are fixed?

11-16. If one end of each of the angles in Problem 11-9 is fixed and the other free to move, what angle sections must be used to withstand the load most economically? *Answer:* Two 4- by $3\frac{1}{2}$- by $\frac{1}{4}$-in. angles.

11-17. Determine the allowable compressive load on the column of Problem 11-4, assuming both ends fixed.

INTERMEDIATE COLUMN FORMULAS

11-11 Basic Types—Buckling Formulas

Attempts have been made in times past to provide an entirely rational intermediate column formula. None has as yet been forthcoming. The difficulty of anticipating the effect of initial eccentricities seems beyond pure mathematical reasoning. Consequently, the ultimate-strength formulas for axially loaded intermediate columns are based almost entirely on experimental data, modified upon occasion by individual and collective opinion.

In an endeavor to fit an equation to the somewhat scattered experimental points on a P/A, l/r diagram (Fig. 11-9) lying in the intermediate

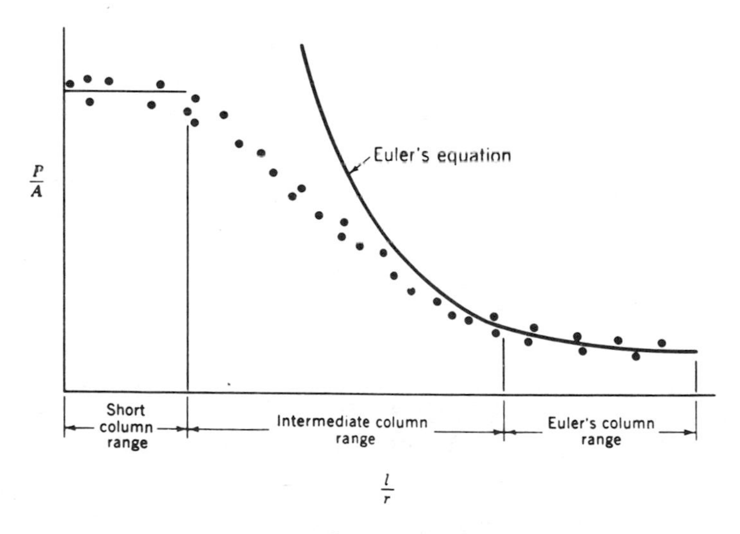

Figure 11-9

column range, several basic types of equations have been suggested. Some disagreement has evolved as to the correct trend of these experimental points. At present, four basic types have emerged: the straight-line, parabolic, Gordon-Rankine, and secant equations. They, together with typical buckling equations for each, are presented in Table 6.

TABLE 6
COLUMN BUCKLING EQUATIONS

1. *Straight-Line Equation*—General Form $\frac{P}{A} = \sigma_y - c\frac{l}{r}$			(11.4)
Low-carbon structural steel	$\frac{P}{A} = 37{,}500 - 125\frac{l}{r}$ maximum = 30,000 psi	upper $\frac{l}{r} = 150$	(11.5)
Aluminum alloy 6061-T6 6062-T6	$\frac{P}{A} = 37{,}600 - 260\frac{l}{r}$ maximum 35,000 psi	upper $\frac{l}{r} = 67$	(11.6)
Duralumin 2024-T3	$\frac{P}{A} = 56{,}600 - 510\frac{l}{r}$ maximum = 46,000 psi	upper $\frac{l}{r} = 73$	(11.7)
Cast iron	$\frac{P}{A} = 34{,}000 - 88\frac{l}{r}$ maximum = 28,300 psi	upper $\frac{l}{r} = 100$	(11.8)
Oak	$\frac{P}{A} = 5400 - 28\frac{l}{r}$	upper $\frac{l}{r} = 64$	(11.9)
2. *Parabolic Equation*—General Form $\frac{P}{A} = \sigma_y - k\left(\frac{l}{r}\right)^2$			(11.10)
Low-carbon structural steel	$\frac{P}{A} = 40{,}000 - 1.35\left(\frac{l}{r}\right)^2$	upper $\frac{l}{r} = 120$	(11.11)
3. *Gordon-Rankine Equation*—General Form $\frac{P}{A} = \frac{\sigma_y}{1 + \phi\left(\frac{l}{r}\right)^2}$			(11.12)
Low-carbon structural steel	$\frac{P}{A} = \frac{50{,}000}{1 + \frac{1}{18{,}000}\left(\frac{l}{r}\right)^2}$		(11.13)
Magnesium AM-C58S	$\frac{P}{A} = \frac{46{,}000}{1 + 0.00071\left(\frac{l}{r}\right)^2}$		(11.14)
4. *Secant Equation*—General Form $\frac{P}{A} = \frac{\sigma_y}{1 + \frac{ec}{r^2}\sec\frac{l}{2r}\sqrt{\frac{P}{AE}}}$			(11.15)
Low-carbon structural Steel	$\frac{P}{A} = \frac{35{,}000}{1 + \frac{ec}{r^2}\sec\frac{fl}{2r}\sqrt{\frac{kP}{AE}}}$		(11.16)

The secant equation is not as much a type equation as are the first three, since it is used as a separate equation under that name. Its complexity and the difficulty accompanying its use have caused it to be employed sparingly in practice. However, it is considered to be the most refined formula that will ever be obtained.

Comment. Of the foregoing column equations, those pertaining to low-carbon structural steel are plotted on Fig. 11-10. It is evident that the spread

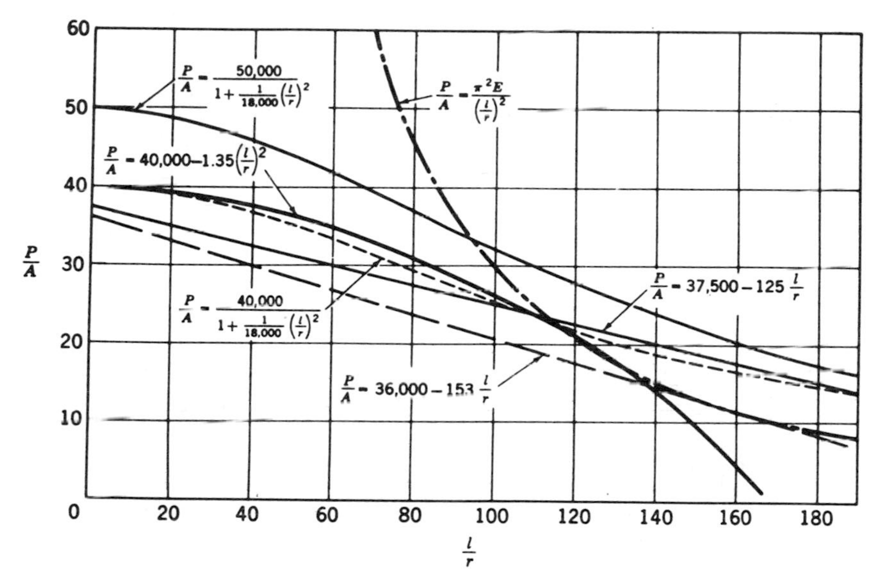

Figure 11-10 Plots of ultimate strength equations for axially loaded low-carbon structural steel columns.

between these graphs is an indication of the scatter of the experimental data and the degree of conservatism used to record their variation in equation form.

It is apparent from Fig. 11-10 that intermediate column formulas may have upper limits of l/r. Some may even have a lower limit given by a minimum l/r or the compressive yield strength of the material. Before using an intermediate column formula, it is essential that the limitations imposed upon it be completely understood.

11-12 Effect of Fixing the Ends of an Intermediate Column

Fixing the ends of an intermediate column produces less improvement in strength than that realized for tall, slender columns (see §11-10).

The added strength given to the intermediate column varies from zero, for that approaching a compression block, to the maximum value already indicated for tall, slender columns. The greatest strength that could accrue to an intermediate column completely fixed at its ends would therefore be computed using its l/r value multiplied by a factor k between 0.5 and 1.0.

But the vast majority of intermediate columns in modern framed steel structures do not have ends which are fixed, but ends which are only *partially fixed*. Therefore, the constant k, which is used to determine the effective l/r, is closer to the value 1.0 than to 0.5. Unless the magnitude of the end restraint can be clearly established, no end restraint is considered to act, and the value of k is taken as 1.0.

Illustrative Problem 6. What axial load can a standard 8-in. S beam 4 ft long, made of 2024-T3 aluminum alloy, carry before it collapses? $I_{x-x} = 57.55$ in.^{4}, $I_{y-y} = 3.73$ in.^{4}, and $A = 5.40$ in.^{2}. It has partially fixed ends ($k = 0.75$).

SOLUTION. The slenderness ratio of the column is as follows:

$$\frac{l}{r} = \frac{0.75 \times 48}{\sqrt{3.73/5.40}} = 0.75 \times 57.8 = 43.2$$

Since this slenderness ratio lies between the limits set for the 2024-T3, straight-line, intermediate column formula, Eq. (11.7), its use is justified.

$$\frac{P}{A} = 56{,}600 - 510 \times 0.75 \times \frac{l}{r}$$

$$P = 5.40(56{,}600 - 510 \times 0.75 \times 57.8) = 186{,}000 \text{ lb}$$

PROBLEMS

All columns referred to in these problems have hinged ends unless specified otherwise.

11-18. What diameter would be required for a low-carbon steel rod 5 ft long if an axial load of 90,000 lb would cause its collapse? Use parabolic formula (11.11) if it is applicable. *Answer:* 2.18 in.

11-19. Determine the critical length of a 4- by 4- by $\frac{3}{8}$-in. steel angle that would fail under a load of 70,000 lb. Use the Gordon-Rankine formula (11.13).

11-20. Solve Problem 11-18, assuming the rod to be made of 2024-T3 aluminum alloy. Use the straight-line formula. *Answer:* 2.89 in.

11-21. Solve Problem 11-19, assuming the angle to be made of AM-C58S extruded magnesium. Use Gordon-Rankine-type formula.

11-22. What maximum axial load can an S 10 × 25.4 steel column 8 ft long withstand? Use the straight-line formula (11.5). *Answer:* 185,400 lb.

11-23. If the above column were partially fixed at each end, what maximum load could be applied? Use the straight-line formula (11.5). $k = 0.75$.

11-13 Intermediate Column Formulas for Design of Compression Members

General Discussion. The equations used in the design of intermediate columns are obtained by dividing buckling equations, such as those of §11-11, by suitable factors of safety. These factors of safety, ranging from approximately 1.6 to 2.5, may be constant values throughout the entire range of applicability of the equation or may be computed from expressions providing variable factors of safety related to the slenderness ratio. If the safety factor is a constant, the design curve will be similar in shape to that representing ultimate strength with approximately the same l/r limits.

The design of an axially loaded column seldom includes, in practice, the effect of end restraints. Even though the base and top of a column may by securely riveted, a sufficient amount of movement is generally possible to warrant the consideration of hinged ends. Although welded column connections seem to provide greater justification for including the effect of fixed ends, it is the exception rather than the rule that they are so treated. The A.I.S.C. code provides for the possibility of semirigid and rigid types of construction however they may be obtained.

All the intermediate column formulas presented in this article are now or have been in common use. To obtain a well-rounded background in column design, it is suggested that those circumscribed equations appearing in this chapter as well as their limitations and supporting equations be committed to memory.

Straight-line equation. *Steel.* One of the most familiar straight-line formulas for designing axially loaded steel columns with hinged ends is obtained by applying a factor of safety of 2.5 to the ultimate-strength equation (11.5). Its final form is

$$\boxed{\frac{P}{A} = 15{,}000 - 50\frac{l}{r}} \qquad (11.17)$$

with the l/r range between 50 and 150.

To permit the design of those columns having lower or higher values of l/r than indicated above, two supplementary adjoining equations are always used for this type of intermediate formula.

Columns having l/r ratios below 50 are designed in a manner similar to that of a compression block, using

$$\boxed{\frac{P}{A} = 12{,}500 \text{ psi}} \qquad (11.18)$$

When the l/r ratio exceeds 150, the Euler-type design formula

$$\boxed{\frac{P}{A} = \frac{6.4E}{(l/r)^2}}^{*} \tag{11.19}$$

is used.

Figure 11-11 records the plots of each of these three equations. Together they constitute a series of equations, hereafter to be called the *straight-line*

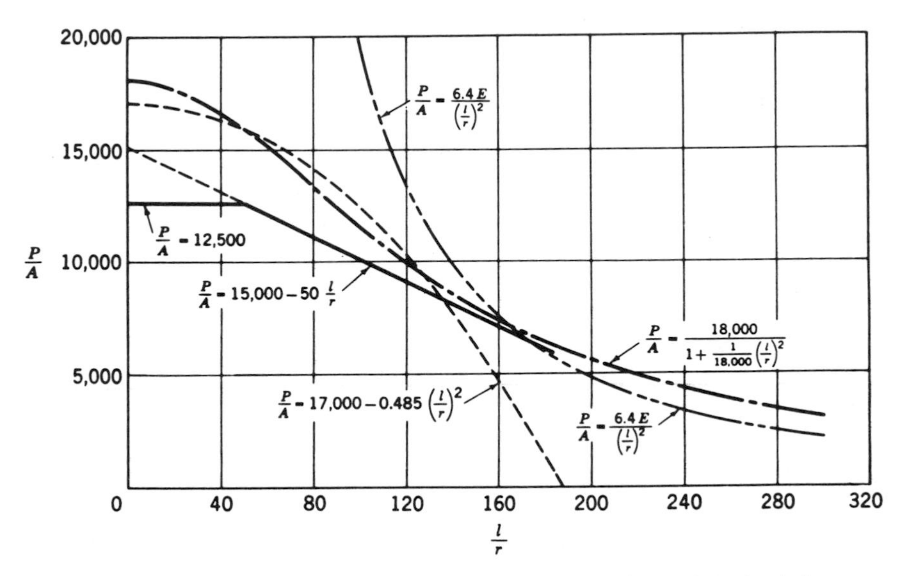

Figure 11-11 Graphs of equations used for designing axially loaded hinge-ended columns of structural steel.

series, that records the maximum allowable stress for each l/r ratio. *It is a very conservative series of column formulas*, as is shown by contrasting the series with the other design equations shown in Fig. 11-11. Since both the straight-line and Euler formulas produce almost identical allowable stresses in the l/r range between 150 and 220, both could be used without noticeable design differences. It is usual, however, to limit the l/r ratio of main compression members to 120 for safety reasons.

Aluminum. Pivot-ended intermediate columns made from structural aluminum alloys 6061-T6 and 6062-T6 and suitable for the construction of buildings are designed using the following straight-line equation:

$$\frac{P}{A} = 20{,}400 - 135\frac{l}{r} \tag{11.20}$$

*Based on an experimentally obtained Euler curve, $P/A = 16E/(l/r)^2$.

Its range of l/r varies from 10 to 67. Below an l/r of 10 the maximum allowable compressive stress is 19,000 psi, and above $l/r = 67$ the Euler equation

$$\frac{P}{A} = \frac{51{,}000{,}000}{(l/r)^2} \qquad (11.21)$$

governs. The factor of safety employed is approximately 1.84.

Other Materials. Straight-line formulas used in the design of axially loaded, pin-ended columns of various other materials are as follows:

CAST IRON

$$\frac{P}{A} = 9000 - 40\frac{l}{r} \qquad (11.22)$$

STRUCTURAL TIMBER

$$\frac{P}{A} = 1000 - 6\frac{l}{r} \qquad (11.23)$$

$$\text{upper } \frac{l}{r} \text{ limit} = 100$$

Gordon-Rankine equation. A formula that has enjoyed widespread use in the design of columns of various materials is the Gordon-Rankine formula (Fig. 11-11). Contributing to its popularity is its ability to give reasonable values of P/A for all values of l/r. At times it has been used over different portions of the l/r range.

As initially conceived, the Gordon-Rankine buckling formula included the effect of superimposed compression and flexure stresses. Because column action, however, cannot be included as a true superimposed effect, it was necessary to include arbitrary constants to cover the experimental points. Hence, it is an irrational equation, as are the other column equations.

To obtain a Gordon-Rankine column design formula from an ultimate-strength formula of the same type, the numerator of the ultimate formula is divided by the factor of safety desired. One such formula that has enjoyed widespread use in the design of axially loaded pin-ended steel columns is

$$\boxed{\frac{P}{A} = \frac{18{,}000}{1 + \dfrac{1}{18{,}000}\left(\dfrac{l}{r}\right)^2}} \qquad (11.24a)$$

This formula is usually specified for use within a lower limit of $l/r = 60$ to an upper limit of 120 when dealing with main compression members.* For bracing and other secondary members, the upper limit is increased to 200.

*This equation may also be used for the design of main columns with l/r ratios between 120 and 200, provided no shock or vibratory loads are present. A further reduction of stress is specified, however, by multiplying Eq. (11.24a) by the factor $[1.6 - (1/200)(l/r)]$.

The design formula for pin-ended columns of type AM-C58S magnesium alloy arranged for a factor of safety of 2.5 is obtained by dividing that value into the ultimate-strength formula given in Table 6.

$$\frac{P}{A} = \frac{18{,}400}{1 + 0.00071(l/r)^2} \tag{11.24b}$$

Maximum P/A stress $= 8620$ psi.

Parabolic equation

General. The parabolic formula proposed by J. B. Johnson for the design of axially loaded columns has become increasingly popular in recent years. Since this formula provides permissible compressive stresses for short as well as intermediate columns, it eliminates the use of a lower l/r limit and the short-column formula. The decision of the American Institute of Steel Construction to employ the parabolic equation in its specification concerning the design of steel columns indicates its widespread acceptance as providing convenient and relatively accurate allowable compressive stresses.

Derivation. Because of the introduction of several new steels, in addition to the retention of those of long standing, the A.I.S.C. has developed a general column equation based upon the individual yield points. Its development is based on the general parabolic equation

$$y = a\left(\frac{l}{r}\right)^2 + b\left(\frac{l}{r}\right) + c$$

which covers the short and intermediate column ranges from $y = \sigma_y$ at $l/r = 0$ to $y = \sigma_y/2$ at $l/r = C_c = \sqrt{2\pi^2 E/\sigma_y}$. See Fig. 11-12. It is asymptotic to the horizontal line $y = \sigma_{yp}$, a, b, and c being constants.

At the y axis where $l/r = 0$, $P/A = c = \sigma_y$.

Thus,

$$\frac{P}{A} = a\left(\frac{l}{r}\right)^2 + b\left(\frac{l}{r}\right) + \sigma_y$$

If now the derivative is found with respect to l/r and set equal to zero, we obtain

$$\frac{d(P/A)}{d(l/r)} = 2a\frac{l}{r} + b = 0$$

Since this relationship must hold for all points of the curve, it follows that $b = 0$ since l/r at the y axis equals zero.

Thus,

$$\frac{P}{A} = a\left(\frac{l}{r}\right)^2 + \sigma_y \tag{11.25a}$$

But from the initial assumption the upper limit of the curve is located at $l/r = C_c$, where $P/A = \sigma_y/2$. If these limiting values are now included,

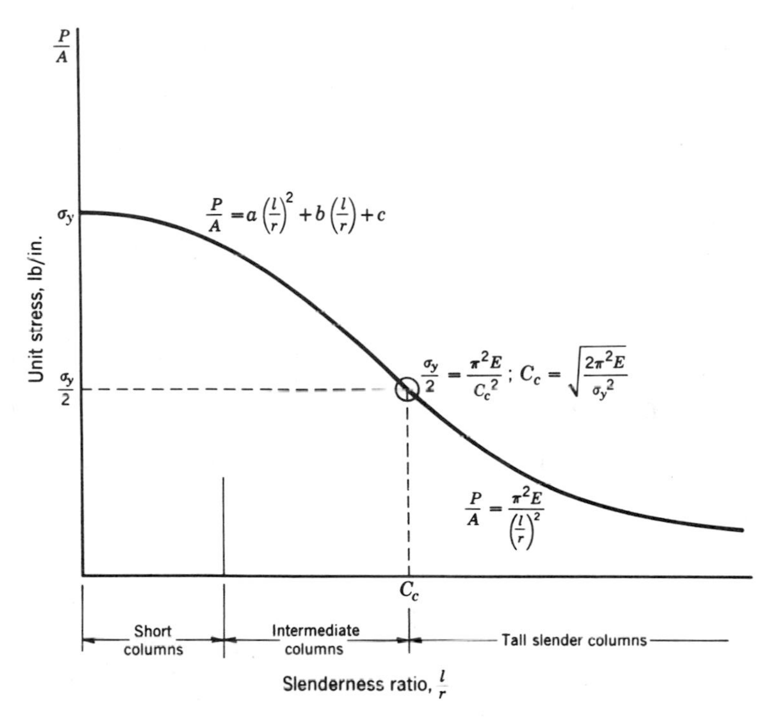

Figure 11-12 Graph showing effect of increased allowable compressive strength and constant modulus of elasticity on straight-line design formulas.

$$\frac{\sigma_y}{2} = a(C_c)^2 + \sigma_y$$

$$a = \frac{-\sigma_y + (\sigma_y/2)}{C_c^2} = -\frac{\sigma_y}{2C_c^2} \tag{11.25b}$$

Inserting this constant into Eq. (11.25a) results in

$$\frac{P}{A} = \frac{-\sigma_y}{2C_c^2}\left(\frac{l}{r}\right)^2 + \sigma_y$$

The addition of a factor of safety and of k, the effective length factor, produces the general design equation

$$\boxed{\frac{P}{A} = \frac{\sigma_y}{\text{F.S.}}\left[1 - \frac{(kl/r)^2}{2C_c^2}\right]} \tag{11.25c}$$

Steel Column Design. The factor of safety applicable to this general A.I.S.C. design equation is obtained from the expression

$$\text{F.S.} = \frac{5}{3} + 3\frac{[k(l/r)]}{8C_c} - \frac{[k(l/r)]^3}{8C_c^3} \tag{11.25d}$$

The values obtainable vary from 1.67 at $l/r = 0$ to 1.92 at $l/r = C_c = \sqrt{(2\pi^2 E/\sigma_y)}$.

If now the values of σ_y of the several available steels are inserted into the general equation, the more specific design equations are obtained for each material. For instance,

A–36:

$$\frac{P}{A} = \frac{36{,}000}{\text{F.S.}}\left[1 - \frac{(l/r)^2}{2 \times 126.1^2}\right] = \frac{36{,}000 - 1.130(l/r)^2}{\text{F.S.}} \tag{11.26a}$$

A-242 (under $\frac{3}{4}$ in. thick):

$$\frac{P}{A} = \frac{50{,}000}{\text{F.S.}}\left[1 - \frac{(l/r)^2}{2 \times 107^2}\right] = \frac{50{,}000 - 2.180(l/r)^2}{\text{F.S.}} \tag{11.26b}$$

When l/r exceeds the value of C_c, the Euler equation obtained using $E = 29{,}000{,}000$ psi and an F.S. = 1.92 is used in the design of columns fabricated from all grades of steel.

$$\frac{P}{A} = \frac{\pi^2 E}{\text{F.S.}(l/r)^2} = \frac{9.85 \times 29{,}000{,}000}{1.92 \times (l/r)^2} = \frac{149{,}000{,}000}{(l/r)^2} \tag{11.27}$$

In the design of axially loaded bracing and secondary members when l/r exceeds 120, the allowable stress P/A is increased by dividing Eqs. (11.26) by the factor

$$1.6 - \frac{l}{200r}$$

Illustrative Problem 7. Select the most economical wide-flange (W) pin-ended steel column, 18 ft long, to sustain an axial load of 200,000 lb. Use the straight-line, Gordon-Rankine, and parabolic (A.I.S.C.-A-36) equations.

SOLUTION. If the column to be designed were to have a round or square section whose minimum radius of gyration could be determined in terms of the area A, the solution would be comparatively simple, entailing but one unknown. Since the difficulty of obtaining r in terms of A for a W section involves considerable time, the determination of a most economical column section is best done by trial.

An assumption is first made of the stress P/A placed on the column. It is generally set at the average of the P/A values computed for the l/r limits of the column equation that is being used. Of course, the closer the assumed stress is to the actual stress in the selected column, the shorter the solution will be.

Let us assume the value of P/A for each of the three equations to be a conservative 10,000 psi, even though the assumption may be poor for the more liberal Gordon-Rankine and parabolic formulas.

The area corresponding to this assumed stress will then be

$$A = \frac{P}{\sigma} = \frac{200{,}000}{10{,}000} = 20 \text{ in.}^2$$

A trial W section must now be found (Table 14) that has approximately this area. To be economical, this trial column should also have a minimum cross-sectional area, or weight per foot, with a maximum value of rigidity expressed by the value of I_{min} or r_{min}. Obviously, therefore, a W column with a square cross section will be more economical than one with a great depth and a narrow flange, since its moments of inertia will be more nearly alike.

Let our first selection be a W 12 × 72 column (area = 21.16 in.^{2} and r_{min} = 3.04 in.). This is a trial section and must now be verified or rejected on the basis of its allowable load and economy.

The slenderness ratio of this column, which shall dictate the proper design formula to be used and the allowable stress to be placed on the column, is

$$\frac{l}{r} = \frac{18 \times 12}{3.04} = 71.0$$

This value of 71.0 lies within the operating range of each of the three intermediate column equations to be used. At this point the solutions using the three equations will vary.

Before continuing the solutions in Fig. 11-13, however, the following comments should be noted. Each step toward the attainment of the most economical column serves to close the gap between the available and required areas. Ordinarily, the most economical, axially loaded, column will be the one whose radii of gyration about its principal axes will have as little variance as possible; this demands the selection of a more or less square column. It is because of space limitations and eccentric loadings that rectangular columns find a great deal of use. The most economical column is obtained when no other column of lesser weight is available.

Let us now proceed to the completion of the problem in Fig. 11-13.

11-14 Straight-Line Equations for Alloyed Materials

The two numerical components of a straight-line equation record the P/A axis intercept and the slope of the line. The intercept value agrees closely with the allowable compressive strength of the column material. The slope value C depends on two items (1) the P/A axis intercept, or the allowable compressive strength, and (2) the position of the Euler design equation corresponding to the modulus of the elasticity of the column material and modified with the same factor of safety as used for the straight-line equation.

If, therefore, the straight-line design formulas for a series of steel columns of varying alloy content were investigated (Fig. 11-14), their P/A intercepts would increase with an increase in allowable compressive strength while the position of the lower extremity of each line would remain relatively fixed, since the modulus of elasticity of all steel alloys is practically the same.

First Trial Column, Common to All-W 12 × 72 $A = 21.16$ sq in. $r_{min} = 3.04$ in. $\frac{l}{r} = 71.0$

Straight Line	Gordon-Rankine	Parabolic
Allowable stress for this $\frac{l}{r}$ (= 71.0) $\frac{P}{A} = 15{,}000 - 50\,(71.0) = 11{,}450$psi Area req. $= \frac{200{,}000}{11{,}450} = 17.45$ in.^{2} Section appears too large. Try smaller section with r_{min} about the same as before. *Second Trial Column.* W 12 × 65 (A = 19.11 in.^{2} $r_{min} = 3.02$ in.) $\frac{l}{r} = \frac{18 \times 12}{3.02} = 71.5$ Allowable stress for this $\frac{l}{r}$ $\frac{P}{A} = 15000 - 50(71.5) = 11{,}430$ psi Area req. $= \frac{200{,}000}{11{,}430} = 17.5$ in.^{2} Ok, but try for smaller weight. *Third Trial Column.* W 10 × 60 ($A = 17.66$ in.^{2} $r_{min} = 2.57$ in.) $\frac{P}{A} = 10{,}800$ psi; area req. = 18.50 in.^{2} NG *Fourth Trial Column.* W 12 × 58 ($A = 17.06$ in.^{2} $r_{min} = 2.51$ in.) NG, in view of W 12 × 65 comparison. ∴ W 12 × 65 most economical.	Allowable stress for this $\frac{l}{r}$ (= 71.0) $\frac{P}{A} = \frac{18{,}000}{1 + \frac{1}{18{,}000}(71)^2} = 14{,}060$ psi Area req. $= \frac{200{,}000}{14{,}060} = 14.22$ in.^{2} Section is too large. Select 10″ W column even though r_{min} is considerably less. *Second Trial Column.* W 10 × 54 (A = 15.88 in.^{2} $r_{min} = 2.56$ in.) $\frac{l}{r} = \frac{18 \times 12}{2.56} = 84.5$ Allowable stress for this $\frac{l}{r}$ $\frac{P}{A} = \frac{18{,}000}{1 + \frac{1}{18{,}000}(84.5)^2} = 12{,}900$ psi Areq req. $= \frac{200{,}000}{12{,}900} = 15.50$ in.^{2} Ok, but try for smaller weight. However, other W columns of lesser weight have insufficient area and r_{min} values less than that used above. ∴ W 10 × 54 most economical.	Allowable stress for this $\frac{l}{r}$ (= 71.0) F.S. $= \frac{5}{3} + \frac{3(71.0)}{8(126.1)} - \frac{(71.0)^3}{8(126.1)^3} = 1.856$ $\frac{P}{A} = \left[1 - \frac{(71.0)^2}{2 \times 126.1^2}\right]\frac{36{,}000}{1.856} = 16{,}300$ psi Area req. $= \frac{200{,}000}{16{,}300} = 12.3$ in.^{2} Section is too large. Try 10″ W column having less area with about same r_{min}. *Second Trial Column.* W 10 × 49 ($A = 14.40$ in.^{2} $r_{min} = 2.54$ in.) $\frac{l}{r} = \frac{18 \times 12}{2.54} = 85.0$ Allowable stress for this $\frac{l}{r}$ F.S. $= \frac{5}{3} + \frac{3 \times 85.0}{8(126.1)} - \frac{(85.0)^3}{8(126.1)^3} = 1.883$ $\frac{P}{A} = \left[1 - \frac{(85.0)^2}{2(126.1)^2}\right]\frac{36{,}000}{1.883} = 14{,}790$ psi Area req. $= \frac{200{,}000}{14{,}790} = 13.52$ in.^{2} No lighter section available. This is a good selection. ∴ W 10 × 49 most economical.

Fig. 11-13. Continuation of Illustrative Problem 7

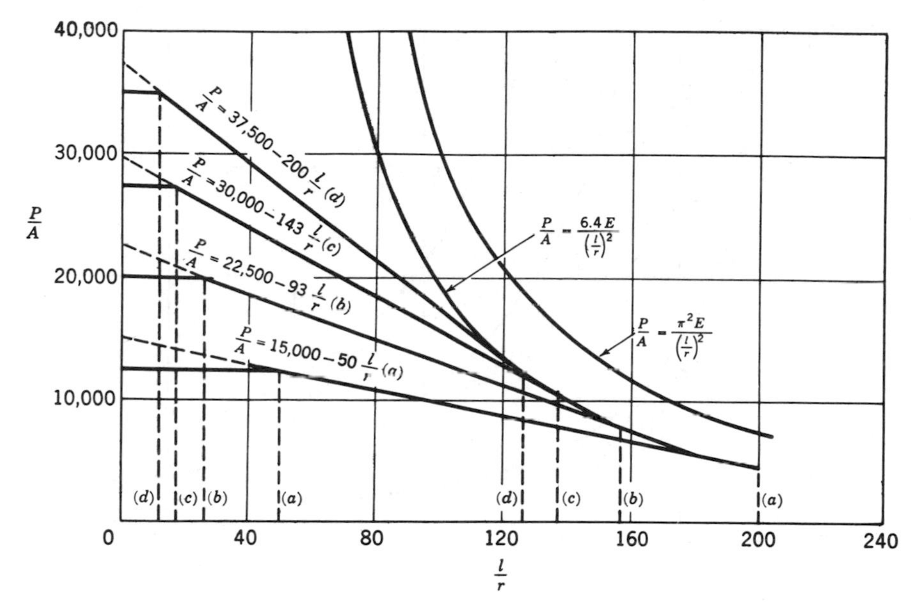

Figure 11-14

We should expect, for example, the straight-line design equations for steel to have (1) an increased value of the P/A intercept σ, (2) an increased slope C, and (3) a decreased upper limit of l/r as the percentage of carbon is increased to approximately 1 per cent, and as other alloying materials are added. The lower l/r limit may also possibly decrease to provide a consistent factor of safety. A perusal of Fig. 11-14 will show how these design formulas vary. Column equation $P/A = 37{,}500 - 200l/r$ would pertain to a steel having a carbon content of about 0.8 per cent, whereas the equation $P/A = 15{,}000 - 50l/r$ refers to a low-carbon steel containing about 0.2 per cent carbon.

A similar variation of relationships exists for columns of other materials.

11-15 Straight-Line Formulas for Columns Having Various End Rigidities

The effect of changing the rigidity of the ends of a column on its column formula for pin ends is shown in Fig. 11-15. Because the rigidity of the ends has no effect on short compression members and an increasing effect as the l/r ratio increases, the straight-line formula may be thought to rotate about its y-axis intersection until it again becomes tangent to the proper Euler curve. As the ends approach a fixed position, therefore, the slope of the straight-line equation decreases and the upper limit increases. If one end is

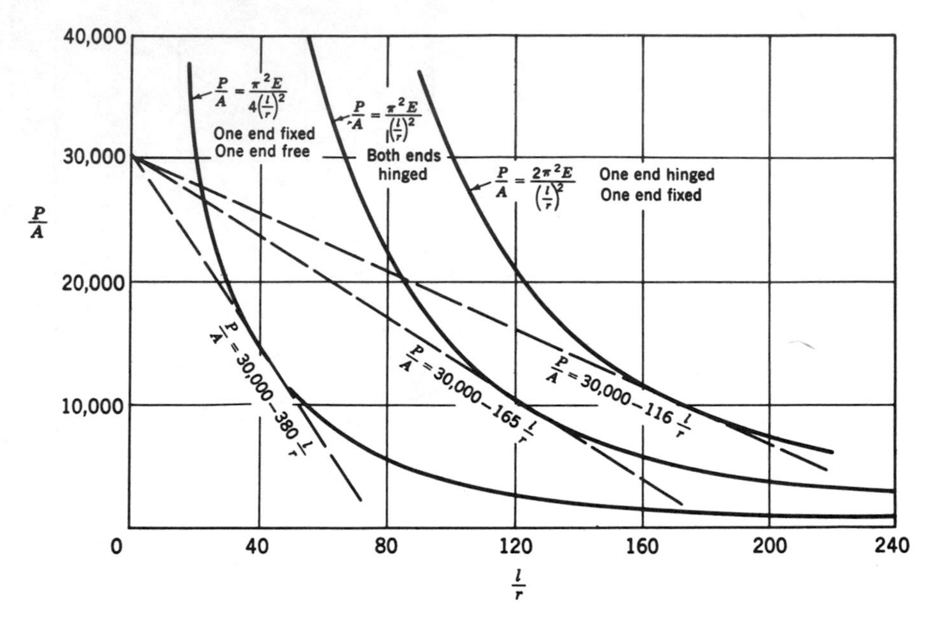

Figure 11-15

free to sway and the other end is fixed, the slope of the equation is increased, and the upper l/r limit is decreased.

Another point of interest to remember in the study of columns is that with decreasing moduli of elasticity, the corresponding Euler curve is placed closer to the origin of the P/A, l/r diagram. Drawing a tangent to such an Euler curve will reveal a smaller range of usefulness for the straight-line equation and a decreased upper l/r limit.

PROBLEMS

11-24. If two 6- by 4- by $\frac{7}{16}$-in. angles, 10 ft long, were placed with their long sides back to back and separated by $\frac{1}{2}$ in., what maximum allowable axial load could be applied? Use A.I.S.C. specifications and A-36 steel.

11-25. What equal-leg angle made of 6061-T6 aluminum alloy should be used to carry an axial load of 20,000 lb, if the length is 10 ft and the straight-line-Euler series of formulas is prescribed? Assume aluminum sections to be similar to those made of steel. *Answer:* 6 by 6 by $\frac{3}{8}$ in.

11-26. What maximum allowable load may be applied to a structural column consisting of two 12- by $\frac{1}{2}$-in. plates and two 10-in. 20-lb channels, 25 ft long? The channels are placed back to back with the plates against the flanges. The tips of the opposing channels are 12 in. apart. Use A.I.S.C. specifications and A-36 steel.

11-27. Determine the most economical A-36 W column to carry an axial load of 150,000 lb, if the length is 12 ft and the A.I.S.C. specifications are prescribed. *Answer:* W 8 × 35.

11-28. Solve Problem 11-27, assuming the length of column to be 22 ft. *Answer:* W 10 × 49.

11-29. Select the most economical W column to carry an axial load of 150,000 lb, if the length is 14 ft and the Gordon-Rankine design formula is prescribed.

11-30. Solve Problem 11-29 using the straight-line formula [Eq. (11.17)]. *Answer:* W 10 × 49.

11-31. Find the lightest A-36 steel S beam, 14 ft long, that can be used as a secondary column under an axial load of 60,000 lb. Use the A.I.S.C. specifications.

11-32. Compare the saving in weight, if any, that would accrue by substituting magnesium for structural steel in a round bar serving as a column to support a load of 2000 lb. $L = 3$ ft. Specific gravity, magnesium $= 1.8$, steel $= 7.80$. Use Gordon-Rankine-type design formulas. *Answer:* 28.9 per cent.

11-33. Determine the maximum allowable axial load that may be placed on a W 10 × 54 column, 16 ft long, using (a) the straight-line formula, Eq. (11.17), (b) the Gordon-Rankine formula, Eq. (11.24a), and (c) the present A.I.S.C. formula, Eq. (11.26a).

11-34. Determine the maximum allowable axial load that may be placed on an S 10 × 25.4 beam, 8 ft long, by the (a) A.I.S.C. formula, Eq. (11.26a), (b) the Gordon-Rankine design formula, Eq. (11.24a), and (c) the straight-line formula, Eq. (11.17). *Answers:* (a) 96,800 lb, (b) 86,000 lb, (c) 74,200 lb.

11-16 Timber Columns

Because solid timber columns are generally rectangular or circular in cross section, their design is based on their l/b ratios, where b is the lesser cross-sectional dimension, rather than on the l/r ratio used heretofore. The design stress for columns having low l/b ratios is obtained from the familiar short column equation

$$\sigma = \frac{P}{A} \tag{11.28}$$

and that for columns having l/b ratios in the intermediate and tall, slender column range from the modified Euler equation

$$\frac{P}{A} = \frac{0.3E}{(l/b)^2} \tag{11.29}$$

It is further stipulated that the transition from one equation to the other take place within the range of l/b from 11 to 18. No timber column shall be used for structural purposes having an l/b greater than 50.

Illustrative Problem 8. Determine the maximum allowable load that may be placed on a pin-ended rectangular timber column of Douglas fir having a cross section of 6 by 8 in. and a length of 6 ft. Maximum allowable stress = 1600 psi. $E = 1.6 \times 10^6$ psi.

SOLUTION

$$\frac{l}{b} = \frac{6 \times 12}{6} = 12$$

$$\frac{P}{A} = \frac{0.3E}{(l/b)^2} = \frac{0.3 \times 1.6 \times 10^6}{12^2} = 3330 \text{ psi}$$

This value exceeds the 1600 psi permissible and is therefore disregarded.

$$\therefore \quad P = \sigma A = 1600 \times 48 = 76{,}800 \text{ lb}$$

Illustrative Problem 9. Determine the dimensions of a square cross-sectioned Douglas fir column 15 ft long to sustain a load of 30,000 lb. $\sigma_{max} = 1600$ psi. $E = 1.6 \times 10^6$ psi.

SOLUTION. Having no cross-sectional dimensions, the slenderness of the column is unknown. A trial-and-error procedure is therefore suggested.

If the $\sigma_{max} = 1600$ psi is assumed to be applicable, a cross-sectional area of $A = 30{,}000/1600 = 18.75$ in.^{2} is required whose side dimension is 4.33 in.

The l/b is then equal to $15 \times 12/4.33 = 41.5$. Thus, l/b being greater than 18 implies that the $\sigma = P/A$ is inapplicable and that the modified Euler equation must be used.

If now a stress of 700 psi is assumed, a cross-sectional area of 42.9 in.^{2} is obtained whose $b = 6.55$ in. The resulting l/b ratio is $180/6.55 = 27.4$. The corresponding allowable stress from the equation is $P/A = (0.3 \times 1.6 \times 10^6/27.4^2) = 640$ psi. The area required for this l/b ratio is $30{,}000/640 = 46.9$, whence $b = 6.85$ in. Obviously, the cross section is just a mite small.

An assumed stress of about 665 psi, found after a few trials, provides a cross section of 6.7 by 6.7 in. whose l/b of 26.8 provides an allowable stress equal to that assumed. The most economical column therefore is one having a cross section of 6.7 by 6.7 in.

PROBLEMS

11-35. Determine the cross-sectional dimensions of a square timber column, 8 ft long, that will carry a load of 75,000 lb. Maximum allowable compressive stress = 1600 psi. $E = 1.6 \times 10^6$ psi. *Answer:* 6.85 by 6.85 in.

11-36. Determine the cross-sectional dimensions of the column of Problem 11-35 if the length were increased to 16 ft.

11-37. Determine the maximum allowable axial load that may be placed on a 4- by 4-in. by 4-ft. pin-ended column of Douglas fir whose maximum allowable compressive stress is 2000 psi. $E = 1.8 \times 10^6$ psi. *Answer:* 32,000 lb.

11-38. What would be the maximum allowable load on the column of Problem 11-37 if the length were increased to 12 ft.

11-17 Eccentrically Loaded Steel Columns—Hinged Ends

Eccentricity about one symmetrical axis. Of the several methods employed in designing eccentrically loaded columns, the one which seems most simple to apply is that included in the A.I.S.C. specifications. It is based on the theory given in §10-3, in which the stress due to an eccentric load placed on a short column is equal to

$$\sigma = \frac{P}{A} + \frac{Pec}{I}$$

The first term indicates the axial stress developed by the eccentric load, and the second term, the magnitude of the superimposed bending stress. In the design of short columns, the value of σ is not allowed to exceed the maximum allowable compressive stress of the material. Longer columns, however, are designed on the basis of a *reduced* working stress obtained from one of the column formulas. Thus, in its application to longer columns, we may write the above formula as

$$A = \frac{P}{\sigma_r} + \frac{Pec}{\sigma_r r^2} \tag{11.30}$$

where σ_r is the reduced working stress and $I = Ar^2$. The value of r is the magnitude of the radius of gyration about the bending axis, not necessarily the minimum value for the cross section.

Let us now consider the relative magnitude of the terms on the right-hand side of the equation as the eccentricity of the applied load varies from zero to a maximum. When the eccentricity is infinitesimally small, the second term is also small and the first term is relatively large. The condition is one approaching axial loading. The stress appearing in the denominator of the first term should therefore approach the reduced stress* σ_c for an axially loaded column. At the same time, the stress appearing in the denominator of the second term could vary considerably from σ_c without materially changing its effect on the total cross-sectional area required for this loading.

When the eccentricity is great, the column is under almost pure bending and the second term $Pec/\sigma_r r^2$ is the much more significant term. The stress more nearly representing this condition is σ_f, the allowable flexural stress in bending. If the first term were still to retain its σ_c stress, it would bear little influence on the area required to resist this loading condition.

In consideration of the stresses produced by the extremes of eccentricity, Eq. (11.30) is rewritten to include the stresses σ_c and σ_f.

$$A = \frac{P}{\sigma_c} + \frac{Pec}{\sigma_f r^2}$$

*Use A.I.S.C. column formulas [Eqs. (11.26)] with their limitations.

This may now be rewritten in the form

$$\frac{P/A}{\sigma_c} + \frac{Pec/Ar^2}{\sigma_f} = 1 \tag{11.31}$$

or in the more condensed form, known as the straight-line interaction formula appearing the in A.I.S.C. specifications, as

$$\frac{f_a}{F_a} + \frac{f_b}{F_b} = 1 \tag{11.32}$$

which indicates that the sum of the two left-hand terms shall not exceed unity for safe design.

The definition of each value is as follows:

f_a = actual axial stress P/A.
f_b = actual bending stress Mc/I.
F_a = maximum allowable column stress, assuming that only axial stress exists.
F_b = maximum allowable bending unit stress, assuming that only bending stress exists. See §7-5.

Failure to include the lateral deflection of the column in the moment arm when computing the moment of the load P has caused this straight-line interaction equation to become less conservative as f_a/F_a exceeds 0.15. To retain the same margin of safety for those loading conditions producing ratios in excess of 0.15, the current bending stress f_b is amplified by the factor

$$\frac{1}{1 - \dfrac{f_a}{F'_e}}$$

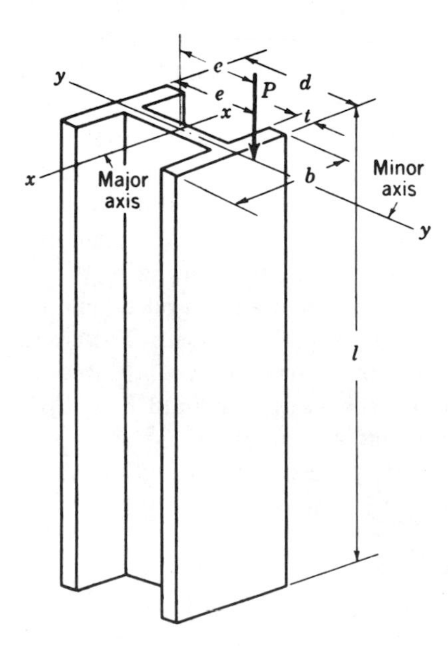

Figure 11-16

where $F'_e = [149{,}000{,}000/(l_b/r_b)^2]$, l_b is the actual unbraced length in the plane of bending, and r_b is the corresponding radius of gyration. The modified interaction equation, used for values of $f_a/F_a > 0.15$ is therefore

$$\frac{f_a}{F_a} + \frac{f_b}{\left(1 - \dfrac{f_a}{F'_e}\right)F_b} \leq 1 \tag{11.33}$$

Illustrative Problem 10. Using A-36 steel, select the most economical W column, 14 ft long, that will sustain an axial load of 200 kips and an eccentric load of 80 kips applied to the minor axis, located 12 in. from the major axis. See Fig. 11-16.

SOLUTION. In order to select a trial section reasonably close to that required, the use of the bending factors (B_x and B_y) found at the bottom of the load tables in the A.I.S.C. handbook is recommended. These factors provide a convenient method of converting bending moments into approximately equivalent direct loads. B_x and B_y are, respectively, equal to the column area divided by the *x-x* and *y-y* section modulus. Since bending in our case takes place about the major axis (*x-x*), we must use B_x. Its range in values for 8- to 12-in. W columns is from 0.333 to 0.177, respectively. Considering our load to be relatively heavy as far as these columns go, let us choose a value of 0.2 as our value of B_x. Thus,

$$\text{bending moment} = M = 80 \times 12 = 960 \text{ in.-kip}$$

$$\text{equivalent direct load} = MB_x = 960 \times 0.2 = 192 \text{ kips}$$

$$\text{approximate direct column load} = 200 + 80 + 192 = 472 \text{ kips}$$

Table 7 was computed from investigations made of several 14-ft columns which, from the load tables found in the A.I.S.C. handbook, were shown to be capable of approximately sustaining the above load.

TABLE 7
INVESTIGATION OF TRIAL COLUMNS
SUBJECTED TO ECCENTRIC LOADING

Section	Allowable Column Load (kips)	Allowable Stress: Axial F_a	Allowable Stress: Bending F_b	Actual Stress: Axial f_a	Actual Stress: Bending f_b	F'_e	Sum Axial and Bending Ratios
W 14 × 103	568	18,670	22,000	9,260	5860	73,010	0.798
W 14 × 95	524	18,750	22,000	10,010	6370	72,500	0.870
W 14 × 87	479	18,750	22,000	10,960	6940	72,370	0.956*
W 14 × 84	441	17,850	22,000	11,330	7340	48,260	1.071
W 14 × 78	409	17,810	22,000	12,220	7920	47,560	1.169

*Use this value.

Typical computation
Investigation of W 14 × 87 column

Area = 25.56 in.^{2} $\qquad r_{\min} = 3.70$ in. $\qquad S = 138.1$ in.^{3}

$$C_c = 126.1 \qquad \frac{l}{r} = \frac{14 \times 12}{3.70} = 45.4$$

$$\text{F.S.} = \frac{5}{3} + \frac{3}{8} \times \frac{45.4}{126.1} - \frac{45.4^3}{8(126.1)^3} = 1.667 + 0.135 - 0.00585$$
$$= 1.796$$

$$F_a = \frac{36,000}{1.796}\left(1 - \frac{45.4^2}{2 \times 126.1^2}\right) = 18,750 \text{ psi}$$

$$f_a = \frac{200 + 80}{25.56} \times 1000 = 10,960 \text{ psi}$$

$$F_b = 22,000 - 0.679 \times \overline{45.4^2} = 20,600 \text{ psi;}$$

$$F_b = \frac{12,000,000}{\dfrac{14 \times 12 \times 14}{14.5 \times 0.688}} = 51,100 \text{ psi}$$

Use $F_b = 22,000$ psi

$$f_b = \frac{960}{138.1} \times 1000 = 6940 \text{ psi}$$

$$F'_e = \frac{149,000,000}{(45.4)^2} = 72,370 \text{ psi}$$

$$\frac{10,960}{18,750} + \frac{6940}{\left(1 - \dfrac{10,960}{72,370}\right)22,000} = 0.584 + 0.372 = 0.956$$

PROBLEMS

Use A.I.S.C. specifications and A-36 steel in the following problems.

11-39. How much axial load can a W 10 × 54 column, 16 ft long, take if it is already withstanding an eccentric load of 40,000 lb placed on the minor axis 10 in. away from the major axis? *Answer:* 98,500 lb.

11-40. Find the greatest unbraced height at which a W 10 × 49 column will safely support a load, at its top, of 80 kips placed on the minor axis 5 in. eccentric to the major axis.

11-41. Determine the most economical W column, 15 ft long, to sustain an axial load of 120,000 lb and an eccentric load of 50,000 lb acting on the major axis located 6 in. from the minor axis. *Answer:* W 12 × 72.

11-42. An axial load of 120,000 lb is being restrained by a W 10 × 54 column 16 ft long. What maximum eccentricity may a 20,000-lb load have when placed on the minor axis without overstressing the column? *Answer:* 19.8 in.

11-43. What change in column section would be required if the length of the column in Problem 11-41 were changed to 20 ft?

11-44. A W 14 × 53 column (main member) has an unbraced height of 24 ft. What load applied on the *y-y* axis at the flange face could the column carry in addition to a concentric load of 54.8 kips? *Answer:* 16.9 kips.

11-45. A W 12 × 53 column carries a top load of 79 kips located on the *y* axis and eccentric 6.02 in. to the *x* axis. The height of the column is 36 ft. It is unbraced with reference to the *x* axis but is braced at a point 12 ft above the base with reference to the *y* axis. How much additional load, concentrically applied, could the column safely carry if additional bracing with reference to the y axis is introduced at 12 ft below the top? *Answer:* 25,400 lb.

11-46. Find the lightest 12-in. W column with an unbraced height of 28 ft to carry a load of 182 kips applied to the top on the minor axis, with an eccentricity of 4 in. with respect to the major axis.

Eccentrically loaded about both symmetrical axes. When a column is eccentrically loaded with respect to both the major and minor axes, the relationship between the sum of the ratios of the applied and allowable stresses includes two bending effects, one about each of the symmetrical axes. When the ratio of axial stresses $f_a/F_a \leq 0.15$,

$$\frac{f_a}{F_a} + \left(\frac{f_b}{F_b}\right)_{x-x} + \left(\frac{f_b}{F_b}\right)_{y-y} \leq 1 \qquad (11.34)$$

When $f_a/F_a > 0.15$,

$$\frac{f_a}{F_a} + \frac{f_{bx}}{\left(1 - \dfrac{f_a}{F'_{ex}}\right)F_{bx}} + \frac{f_{by}}{\left(1 - \dfrac{f_a}{F'_{ey}}\right)F_{by}} \leq 1 \qquad (11.35)$$

Illustrative Problem 11. Design a pin-ended W column made of A-36 steel with an unbraced height of 25 ft, which carries a concentric load of 90 kips applied at the top and an eccentric load of 70 kips (also applied at the top) 3 in. from the major axis and 2 in. from the minor axis. A.I.S.C. specifications.

SOLUTION. Total axial load = 90 + 70 = 160 kips. $M_{xx} = 70 \times 3 = 210$ in.-kips.

$$M_{yy} - 70 \times 2 = 140 \text{ in.-kips}$$

Equivalent axial load ($B_x \sim 0.2$; $B_y \sim 0.6$).

$$P_{eq} = 160 + 0.2 \times 210 + 0.6 \times 140 = 160 + 42 + 84 = 286^k$$

Try W 12 × 72 *section.*

Characteristics of column section:

$r_{min} = 3.04$ in. $r_{max} = 5.31$ in. $A = 21.16$ in.^{2}, $d = 12.25$ in. Flange width = 12.04 in. Flange thickness = 0.671 in.

$$S_{xx} = 97.5 \text{ in.}^3, \qquad S_{yy} = 32.4 \text{ in.}^3$$

$$\left(\frac{l}{r}\right)_{max} = \frac{25 \times 12}{3.04} = 98.7, \qquad \left(\frac{l}{r}\right)_{min} = \frac{25 \times 12}{5.31} = 56.5$$

$$f_a = \frac{160}{21.16} = 7.57 \text{ kips/in.}^2$$

$$\text{F.S.} = \frac{5}{3} + \frac{3}{8} \times \frac{98.7}{126.1} - \frac{98.7^3}{8 \times 126.1^3} = 1.667 + 0.293 - 0.596 = 1.896$$

$$F_a = \frac{36{,}000}{1.896}\left(1 - \frac{98.7^2}{2 \times 126.1^2}\right) = 13{,}200 \text{ psi}$$

$$f_{bx} = \frac{M_{xx}}{S_{xx}} = \frac{210}{97.5} = 2150 \text{ psi}, \qquad f_{by} = \frac{M_{yy}}{S_{yy}} = \frac{140}{32.4} = 4320 \text{ psi}$$

$$F'_{ex} = \frac{149{,}000{,}000}{(98.7)^2} = 15{,}310 \text{ psi}, \qquad F'_{ey} = \frac{149{,}000{,}000}{(56.5)^2} = 46{,}730 \text{ psi}$$

$$F_{bx} = 22{,}000 - 0.679(98.7)^2 = 15{,}400 \text{ psi}$$

$$\text{alt. } F_{bx} = \frac{12{,}000{,}000}{\dfrac{25 \times 12 \times 12.25}{12.04 \times 0.671}} = 26{,}400 \text{ psi}$$

$$F_{by} = 22{,}000 - 0.679(56.5)^2 = 19{,}830 \text{ psi}$$

$$\text{alt. } F_{by} = 26{,}400 \text{ psi}$$

$$\text{Use} \qquad F_b = 22{,}000 \text{ psi}$$

$$\frac{f_a}{F_a} + \frac{f_{bx}}{\left(1 - \dfrac{f_a}{F'_{ex}}\right)F_{bx}} + \frac{f_{by}}{\left(1 - \dfrac{f_a}{F'_{ey}}\right)F_{by}} \leq 1$$

$$\frac{7.57}{13.20} + \frac{2.15}{\left(1 - \dfrac{7.57}{15.31}\right)22} + \frac{4.32}{\left(1 - \dfrac{7.57}{46.73}\right)22} \leq 1$$

$$0.573 + 0.192 + 0.233 < 1$$

$$0.998 < 1.00$$

PROBLEMS

Use A.I.S.C. specifications and A-36 steel in the following problems.

11-47. A W 14 × 219 is used as a column with an unbraced length of 20 ft. The column is to carry a concentric load of 350,000 lb and another load of 150,000 lb located 10 in. from the *x-x* axis. How far to the right or left of the *y-y* axis (but still

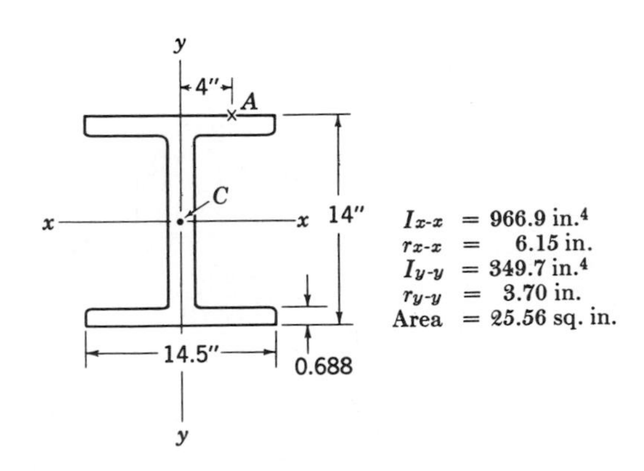

Figure 11-17

10 in. from the x-x axis) may the 150,000-lb load be moved and still comply with A.I.S.C. specifications? *Answer:* 6.60 in.

11-48. Determine the lightest 12-in. W section which will be safe as a column having an unbraced height of 14 ft, and which must provide for a total load at the top of 380 kips, a moment of 128 kip-in. about the major axis, and a moment of 24 kip-in. about the minor axis. *Answer:* W 12 $\times$ 85.

11-49. A W 14 $\times$ 136 column has an unbraced length of 18 ft and carries a concentric load of 300,000 lb. What additional load may be carried at a point 11 in. from the x-x axis and $3\frac{1}{2}$ in. from the y-y axis? *Answer:* 88,000 lb.

11-50. Figure 11-17 shows the cross section of a W 14 $\times$ 87 member used as a column 20 ft long. If there is a load of 50 kips applied at point A, what additional load can be applied at point C?

11-18 Secant Formula

It has already been noted that column design is based mainly on irrational formula because certain factors, such as nonhomogeneity, crookedness, and difficulties of axial loading, produce unavoidable eccentricities. An attempt to evaluate these eccentricities and to include them in a rational column equation has led to the development of the well-known *secant formula*. It is derived as follows: Consider a column with a predetermined unavoidable eccentricity e to be superimposed on the x and y axes as shown in Fig. 11-18. Assuming the column to bend as a beam under the action of load P, its elastic curve equation will be

$$-EI\frac{d^2y}{dx^2} = M = Py$$

Multiplying both sides of the equation by $2dy$ and simplifying, we obtain

$$2\left(\frac{dy}{dx}\right) d\left(\frac{dy}{dx}\right) = -\frac{2P}{EI} y\, dy$$

Then, integrating,

$$\left(\frac{dy}{dx}\right)^2 = -\frac{P}{EI} y^2 + C_1$$

The value of C_1, obtained by substituting the simultaneously occurring values of $dy/dx = 0$ when $y = \Delta$, equals $P\Delta^2/EI$, which, when substituted in the foregoing equation, produces

$$\frac{dy}{dx} = \sqrt{\frac{P}{EI}}\sqrt{\Delta^2 - y^2}$$

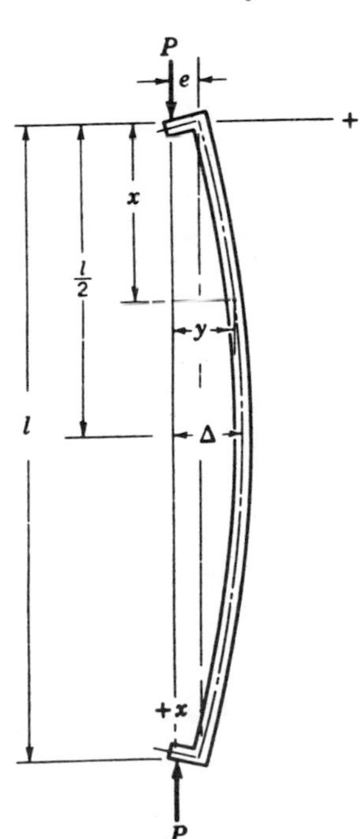

Figure 11-18

Gathering like terms,

$$\frac{dy}{\sqrt{\Delta^2 - y^2}} = \sqrt{\frac{P}{EI}}\,dx$$

Integrating a second time gives

$$\sin^{-1}\frac{y}{\Delta} = \sqrt{\frac{P}{EI}}\,x + C_2$$

But when $x = 0$, $y = e$, making $C_2 = \sin^{-1}(e/\Delta)$. The equation above then reads

$$\sin^{-1}\frac{y}{\Delta} - \sin^{-1}\frac{e}{\Delta} = \sqrt{\frac{P}{EI}}\,x$$

But y is equal to Δ when $x = l/2$ and $\sin^{-1} y/\Delta$ at that point is equal to $\sin^{-1} 1$ or $\pi/2$. Substituting these values makes

$$\frac{e}{\Delta} = \sin\left(\frac{\pi}{2} - \sqrt{\frac{P}{EI}}\frac{l}{2}\right) \quad \text{or} \quad \frac{e}{\Delta} = \cos\frac{l}{2}\sqrt{\frac{P}{EI}}$$

whence

$$e = \Delta\cos\frac{l}{2}\sqrt{\frac{P}{EI}} \quad \text{and} \quad \Delta = e\sec\frac{l}{2}\sqrt{\frac{P}{EI}} \tag{11.36}$$

The maximum bending moment acting on the column is therefore

$$M = P\Delta = Pe\sec\frac{l}{2}\sqrt{\frac{P}{EI}} \tag{11.37}$$

If it is now desired to determine the maximum stress resulting from the application of the load P and its accompanying maximum moment noted above. Let us substitute in

$$\sigma = \frac{P}{A} + \frac{Mc}{I}$$

The maximum fiber stress on the column resulting from the application of a load P will then be

$$\sigma = \frac{P}{A} + \frac{Pec}{Ar^2}\sec\frac{l}{2}\sqrt{\frac{P}{EI}} = \frac{P}{A}\left(1 + \frac{ec}{r^2}\sec\frac{l}{2r}\sqrt{\frac{P}{AE}}\right) \tag{11.38}$$

and

$$\frac{P}{A} = \frac{\sigma}{1 + \dfrac{ec}{r^2}\sec\dfrac{l}{2r}\sqrt{\dfrac{P}{AE}}} \tag{11.39}$$

To correctly use the preceding equations for design, note that (1) the angle $(l/2r)\sqrt{P/AE}$ is expressed in radians; (2) the values of e, c, and r should be in similar linear units—usually inches; and (3) the value of σ should not exceed the proportional limit of the material.

To use Eq. (11.39) in the design of columns, the value of σ is generally taken as the yield strength σ_y of the material, and the value of P, which is the load required to produce the yield stress, is replaced by kP_w, in which k is a factor of safety and P_w is the working load. Furthermore, because the elastic curve is modified by the end conditions applied to the column, the value of l is multiplied by a constant f to give its "effective length." The generalized working formula therefore appears as

$$k\frac{P_w}{A} = \frac{\sigma}{1 + \frac{ec}{r^2}\sec\frac{fl}{2r}\sqrt{\frac{kP_w}{AE}}} \tag{11.40}$$

Substituting $\sigma_y = 32{,}000$ psi, $E = 30 \times 10^6$ psi, $k = 1.6$, and the recommended values of f for hinged ends $= \frac{7}{8}$ and f for riveted ends $= \frac{3}{4}$, the following effective design formulas for structural steel columns are derived.

For hinged ends,

$$\frac{P_w}{A} = \frac{20{,}000}{1 + \frac{ec}{r^2}\sec 0.875\frac{l}{2r}\sqrt{\frac{1.6P_w}{AE}}} \tag{11.41}$$

For riveted or bolted ends,

$$\frac{P_w}{A} = \frac{20{,}000}{1 + \frac{ec}{r^2}\sec 0.75\frac{l}{2r}\sqrt{\frac{1.6P_w}{AE}}} \tag{11.42}$$

Graphs of these equations are shown on Fig. 11-19, using values of $ec/r^2 = 0.25$ and 1.0. The recommended A.I.S.C. equations, Eqs. (11.26a and c), are also

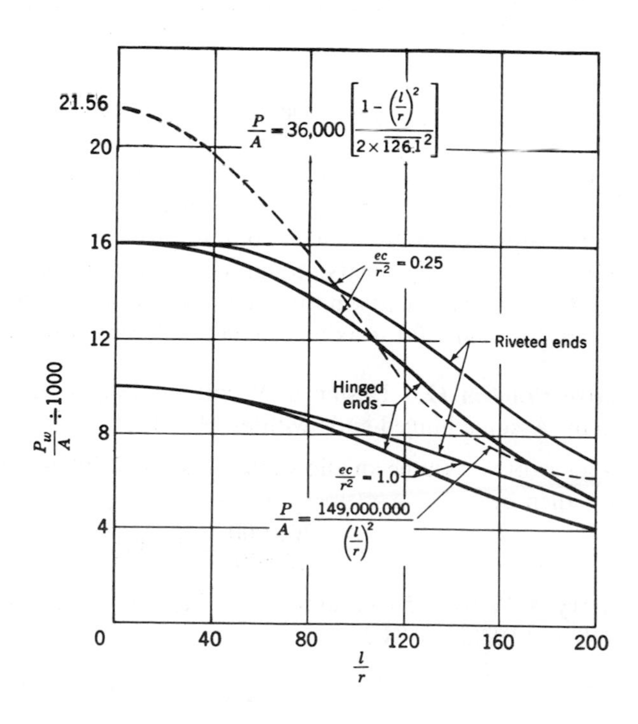

Figure 11-19

shown plotted. Note the close approximation between the A.I.S.C. equations and the secant formula for pin ends, using $ec/r^2 = 0.25$, in the intermediate and tall, slender column ranges. It will also be noted that no l/r limitations are required for the secant formula.

Use of the preceding equations in designing columns is facilitated by first plotting a series of curves showing the variation of P_w/A from l/r for various values of ec/r^2 (Fig. 11-20). From such a family of curves, designs

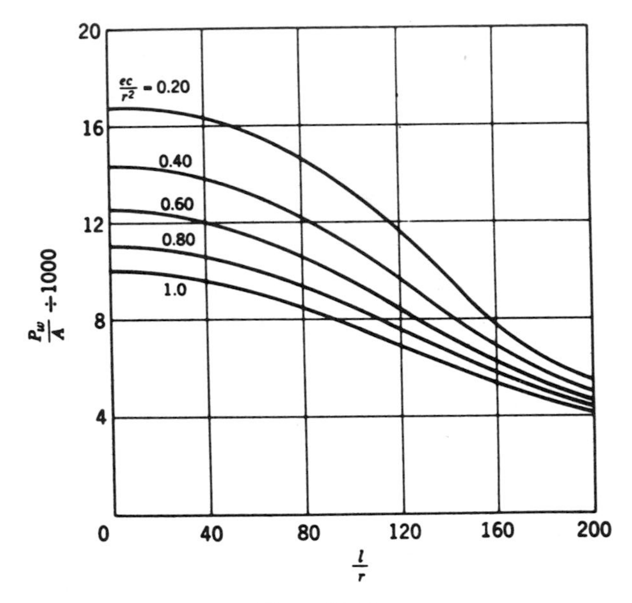

Figure 11-20

can be made not only for columns with inherent natural eccentricity, but also for those with a greater imposed eccentricity.

In the ordinary axially loaded column problem, the value of ec/r^2 must be estimated, depending on its straightness. This is not a simple task and results in further irrationality in the equation. Its value, however, lies generally between 0.10 and 1.0 for axially loaded columns, with 0.25 as the mean.

Illustrative Problem 12. Design a W column 18 ft long to take a load of 200,000 lb. Assume hinged ends with $ec/r^2 = 0.25$.

SOLUTION. Assume an allowable stress of 12,500 psi. The trial required area will then be

$$A = \frac{200{,}000}{12{,}500} = 16 \text{ in.}^2$$

Let us try a W 10 × 54 column: $A = 15.88$ in.^{2}, $r_{\min} = 2.56$ in., and its $l/r = (18 \times 12)/2.56 = 84.3$.

The maximum allowable stress for this value of l/r as given by the secant formula is 13,500 psi (Fig. 11-19). The area required is therefore

$$A = \frac{200{,}000}{13{,}500} = 14.8 \text{ in.}^2$$

This indicates that the section might be too large for economy.

As a second try, consider a W 12 × 53 column:

$$A = 15.59 \text{ in.}^2, \qquad r_{\min} = 2.48$$

$$\text{allowable stress} = 13{,}400 \text{ psi}$$

$$\text{required area} = \frac{200{,}000}{13{,}400} = 14.9 \text{ in.}^2$$

This is the most economical section that can be obtained.

PROBLEMS

11-51. Change the length of the desired column of Illustrative Problem 12 to 14 ft and determine the most economical section required. *Answer:* W 10 × 49.

11-52. If the ends of the column of Illustrative Problem 12 were riveted, what most economical section would be required?

11-53. What maximum axial load can be applied to a W 12 × 79 column, 20 ft long, if $ec/r^2 = 0.25$ and its ends are hinged? *Answer:* 321,000 lb.

11-54. Through what range of length would the W 12 × 53 column of Illustrative Problem 12 remain the most economical column? Use the same load and column characteristics.

11-55. A section of steel column 18 ft long consists of a web 14 by $\frac{3}{8}$ in. and four angles 5 by $3\frac{1}{2}$ by $\frac{7}{16}$ in. placed $14\frac{1}{2}$ in. back to back with long legs outstanding. Determine the maximum load it can carry using the secant formula for hinged ends and $ec/r^2 = 0.40$. *Answer:* 207,000 lb.

11-56. A top chord member of a truss is to be 30 ft long and have applied to it a load of 400,000 lb. What most economical W section should be used if the ends are to be riveted and $ec/r^2 = 0.80$?

11-19 Finite-Difference Method Applied to Tall, Slender Columns

The finite-difference method, already discussed in §8-6 for the determination of deflections in beams, may also be used in the investigation of tall, slender columns. This is due to the fact that the basic bending equation

$$EI\frac{d^2y}{dx^2} = M$$

is applicable to both.

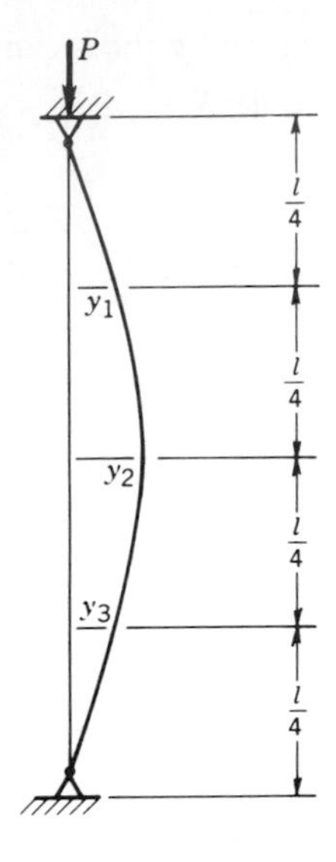

Figure 11-21

Although the finite-difference method provides approximate solutions to the bending equation, the degree of approximation may be reduced to an insignificant amount with little effort through the use of high-speed digital computers.

Let us apply the method of finite differences to the determination of the critical buckling load of a tall, slender column hinged at each end (Fig. 11-21). As previously indicated in the equation $EI(d^2y/dx^2) = M$ may be written in general form as

$$\frac{M}{EI} = \frac{y_{n-1} - 2y_n + y_{n+1}}{\Delta x^2} \tag{11.43}$$

If, now, the length of the column is subdivided into quarters, and the offsets straddling any two successive incremental lengths are inserted into Eq. (11.43), one obtains

$$-2y_1 + y_2 = \frac{Py_1}{EI} \times \frac{l^2}{16}$$

$$y_1 - 2y_2 + y_3 = \frac{Py_2}{EI} \times \frac{l^2}{16}$$

$$y_2 - 2y_3 = \frac{Py_3}{EI} \times \frac{l^2}{16}$$

Note, however, that $y_1 = y_3$. The solution that follows gives

$$P = \frac{9.38EI}{l^2}$$

as against the P for the Euler expression, which equals $9.89EI/l^2$. Of course, as the incremental length Δx decreases, the accuracy of the solution increases.

To determine the critical load for the condition shown in Fig. 11-22, the finite-difference equations become

$$-2y_1 = \frac{P_y}{EI}\frac{l^2}{16}$$

$$y_1 - y_3 = 0$$

$$+2y_3 = \frac{Py_3}{EI}\frac{l^2}{16}$$

whence

$$P = \frac{32EI}{l^2}$$

whereas Euler's equation provides $P = 4\pi^2 EI/l^2$ or $39.48EI/l^2$. The versatility of the finite-difference technique also permits the solution of similar columns with several sinusoidal modes, as well as of those with varying moments of inertia. It may also be extended to apply to columns with fixed ends by

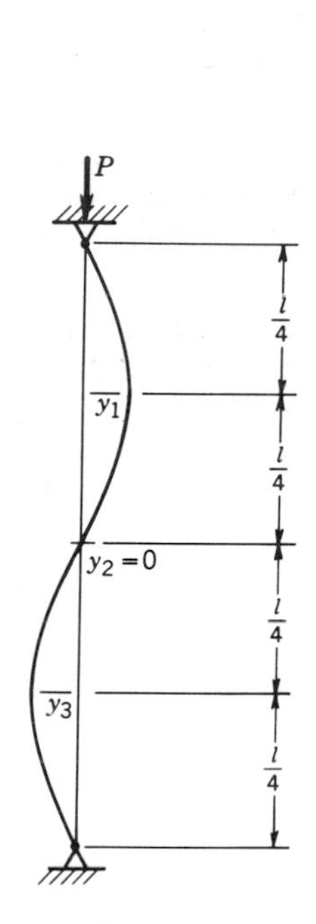

Figure 11-22

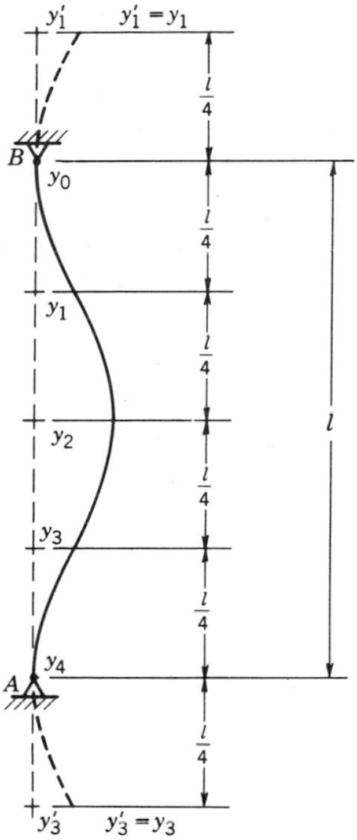

Figure 11-23

extending the elastic curve to provide equal displacements on either side of the support (Fig. 11-23). The equations will then be

$$+y_1' - 0 + y_1 = \frac{Py_1}{EI}\frac{l^2}{16}$$

$$0 - 2y_1 + y_2 = 0$$

$$y_1 - 2y_2 + y_3 = \frac{P(y_2 - y_1)}{EI}\frac{l^2}{16}$$

$$y_2 - 2y_3 = 0$$

$$+y_3 + y_3' = \frac{Py_3}{EI}\frac{l^2}{16}$$

The solution is again $P = 32EI/l^2$ which is an approximation of the theoretically correct Euler load $P = 39.48EI/l^2$.

PROBLEMS

Note: At points of abrupt transition use average value of EI.

11-57. A vertical column hinged at each end has its middle third strengthened to provide an EI twice that of the outer thirds. Using the finite-difference method and $\Delta x = l/3$, what approximate axial load can this column carry? *Answer:* $13.5EI/l^2$.

11-58. A vertical column pivoted at each end has its upper half reinforced to make its value of EI twice as great as its lower half. If it is supported against sidesway at its midpoint, what load can it support? Use $\Delta x = l/4$.

11-59. A tall, slender column which increases its rigidity gradually from a value of EI at the top to a value of $2EI$ at the bottom is hinged at each end. Derive the expression which indicates its load carrying capacity. Use $\Delta x = l/4$. *Answer:* $13.8\ EI/l^2$.

11-60. The upper half of a tall, slender, pin-ended column has an EI twice that of the lower half. By using the finite-difference method and $\Delta x = l/4$, what approximate axial load could this column carry? *Answer:* $12.8\ EI/l^2$.

11-61. Solve Problem 11-60 assuming the lower end to be fixed and the upper end hinged.

12

Combined Stresses and Strains

COMBINED STRESSES

12-1 Introduction

In previous chapters, detailed studies were made of the separate actions of axial, bending, and twisting forces. The question that now arises is: What would happen if two or more of these separate actions were to be applied simultaneously? After all, is not a shaft frequently subjected to bending and twisting forces at the same time? And is not a beam frequently acted upon by a longitudinal compression force while being bent? The development of the fundamental equations involved in resolving the several stresses induced by two or more of these separate actions will be the object of our ensuing study of combined stresses.

12-2 Principal Planes—Principal Stresses

In any stress analysis, a designer must concern himself primarily with the determination of two major items:

1. The types of stresses involved.
2. Their maximum intensity at all critical locations.

Of invaluable assistance to the designer in his analysis is the fact that there occurs, at every point in a solid body, three mutually perpendicular planes, called *principal planes*, on which there is no shearing stress. The normal stresses occurring on these planes are called *principal stresses* and include the maximum and minimum values of the normal stress at that point. The two latter stresses are generally called the *maximum and minimum principal stresses*, the third principal stress being identified as the *intermediate principal stress*, having a magnitude somewhere between the other two. To clarify this point further, it is important that the maximum principal stress be delegated to that principal stress having the greatest numerical value.

The minimum principal stress, however, should be chosen on an algebraic basis as that principal stress having the widest algebraic divergence from the maximum principal stress. For example, if the three principal stresses acting at a point in a body are −3000 psi, +2000 psi, and +1000 psi (the minus sign pertaining to compression, and the plus sign to tension), the maximum principal stress is −3000 psi, the minimum principal stress is +2000 psi, and the +1000 psi is the intermediate principal stress.

The orientation of principal planes may, and generally does, vary from point to point. The relative positions of the planes with respect to each other, however, do not change.

12-3 Finding Stresses at a Point in a Body

In a previous section it was revealed that a unit stress was obtained by dividing a force by the area on which it acted. This is no less true in the study of combined stresses. But because of the large variations of stress generally encountered in a body subjected to combined stresses, the areas considered must necessarily be so small that the average stress for such an area will be, for all practical purposes, its actual stress. In fact, the determination of stresses is generally made *at a point* in a body, the area of which is made visible only by a considerable magnification.

Let us consider the boiler of Fig. 12-1a, and more specifically point M, at which a detailed stress study is to be made. It can be thought of as a dot made by a pencil point that, when expanded, looks like the shaded area of Fig. 12-1b. It is inside this expanded dot that cutting planes, here represented by straight lines, are used to provide a free body from which desired stresses may be computed.* By the judicious employment of the cutting planes, it is generally sufficient to use a free body having only three exposed sides—two sides containing known stresses, such as computed from P/A, Mc/I, Tr/J, and so forth, and the third side, those unknown and perhaps critical stresses that are to be found. For example, the rectangular *free body* shown in Fig.

*The internal pressure force is not shown on the free body $ABCD$, since it is perpendicular to the shell and does not alter the stresses acting on the exposed surfaces. See §12-4.

12-1b, on which the stresses have been computed by the thin-cylinder equations, Eqs. (4.3) and (4.4), must be cut still further to provide a triangular segment *DEF* on which a study might be made of the unknown stresses on its diagonal plane *EF*.

The cutting and recutting of free bodies into smaller and smaller free bodies is justifiable on the basis of equilibrium. If the body from which they were taken was in equilibrium, every portion of that body, no matter how small, was also in equilibrium. The stresses determined from these free bodies through the use of $\sum V = 0$, $\sum H = 0$, and $\sum M = 0$ are therefore necessarily equal to those stresses which were, in effect, in the body before the free bodies were removed.

In the application of the three laws of equilibrium to *free bodies*, it is important to remember that they pertain to forces alone. If stresses are to be included in these equations, they must first be multiplied by the area on which they act.

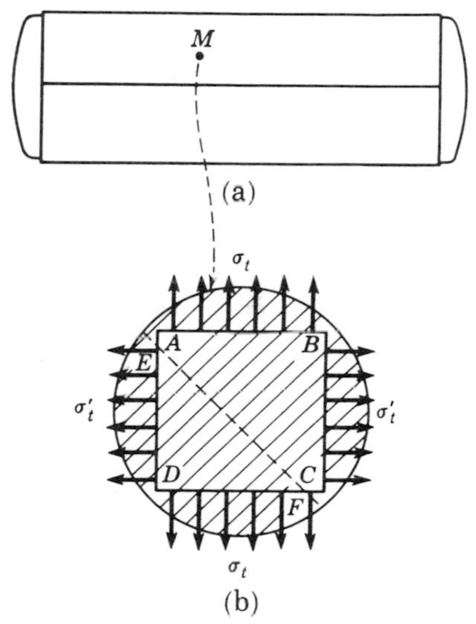

Figure 12-1

12-4 State of Two-Dimensional Stress

When a flat plate is placed under stress by central forces acting on its edges and lying in a plane parallel to its flat surfaces, the plate is said to be subjected to a state of *two-dimensional stress*. Because of the nonexistence of lateral pressures on either of the flat surfaces, they are without shear and must be principal planes on which there is zero principal stress. It follows that the other two principal stresses involved in this action must be perpendicular to each other and lie in the plane of the applied forces.

According to the stress theory basic to this book,* any cubical free body subjected to three-dimensional stress may be solved as a two-dimensional stress problem by progressively projecting the free body with its forces on planes parallel to any two adjacent sides. Viewing the free body in each instance from a position perpendicular to the desired plane of projection, only those forces acting on the periphery of the projected cube are assumed to act. In other words, the forces included within the projected area do not

*Maximum normal stress theory (§12-15).

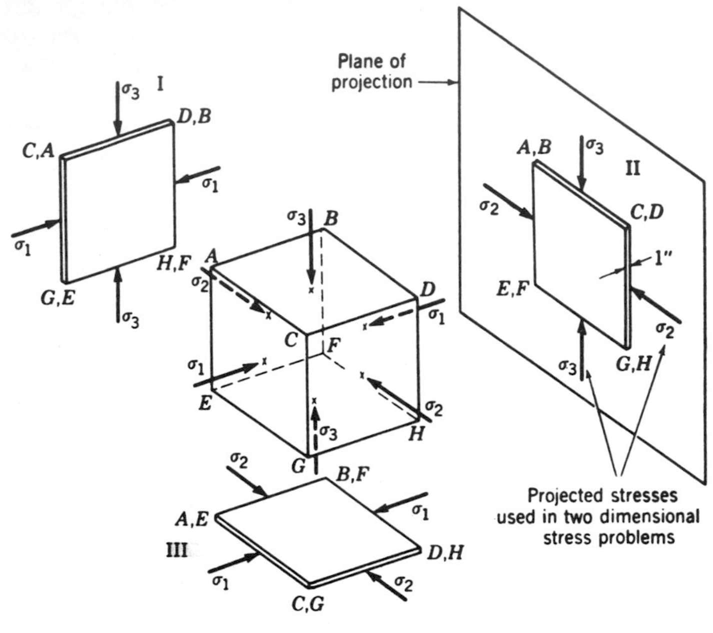

Figure 12-2

affect the stresses on the other mutually perpendicular planes. All projections are assumed to have a unit width in order to make the formation of lateral areas possible. Should, then, the stresses acting on the diagonal plane *ABHG* (Fig. 12-2) be desired, that projection (II) located at the right of the free body would be used in the calculations. Stresses on other diagonal planes—for instance, *ADHE*—would be determined from that projection on whose enclosed area the plane appears as a single line (III).

12-5 Basic Principles Reviewed—Pure Shear

Two basic principles underlying the study of combined stresses have already been advanced in Chapter 5:

1. The first principle concerns the development of mutually perpendicular shear stresses. If a plane passing through a point in a body has acting upon it a shearing unit stress, there is simultaneously induced on a plane perpendicular to it a shearing unit stress of equal magnitude (Fig. 12-3). These shear stresses act in a direction perpendicular to the line of intersection of the planes in which they operate.

If shearing stresses alone act on two perpendicular planes, the stress condition is identified as a state of *pure shear*, and it is under such a stress condition that the second principle is developed.

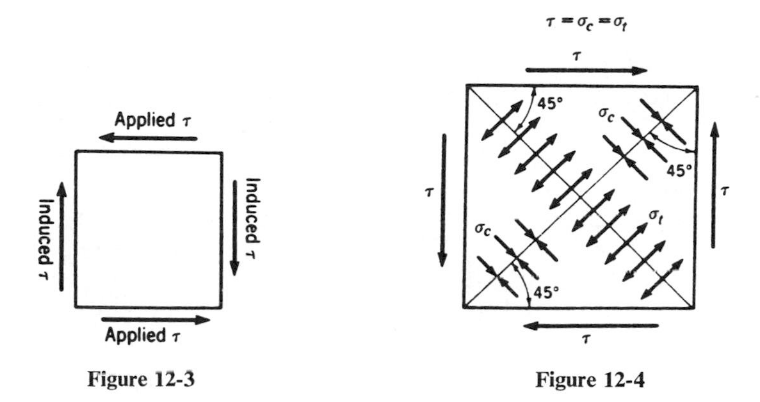

Figure 12-3

Figure 12-4

2. When a state of pure shear exists at a point in a body, normal and shear stresses of varying intensity are developed *on all oblique planes* at that point. On those planes making 45° to the shear planes, the normal stresses are maximum stresses, *acting with no shear*, and are equal in magnitude to the shearing stresses (Fig. 12-4). One maximum normal, or principal, stress is always tension, the other, compression. The planes on which they act are perpendicular to each other.

12-6 Stresses Induced by Principal Stresses

Occasionally a loaded member, such as the boiler of Fig. 12-1, is acted upon by *readily computable principal stresses*. Because these principal stresses induce other stresses which in some instances prove critical to the member being designed, their exact relationship to these stresses becomes an important fundamental in any stress analysis. We shall proceed with its formulation.

Consider the free body shown in Fig. 12-5a, which is stressed in only two directions by principal stresses σ_1 and σ_2, the stress σ_1 being greater than σ_2. Let us now pass a plane through this free body (trace revealed by line AC) which cuts off a triangular segment ABC represented in plane by Fig. 12-5b, and which provides the means for the study of stresses on plane AC.

Because the direction of the equalizing force acting on area AC is unknown, it is shown acting obliquely, having components $\tau \times AC$ and $\sigma_n \times AC$. Balancing now the y components, we may write

$$\sigma_n \times AC = \sigma_1 \times AB \times \cos\theta + \sigma_2 \times CB \times \sin\theta$$

or

$$\sigma_n = \sigma_1 \times \frac{AB}{AC} \cos\theta + \sigma_2 \times \frac{CB}{AC} \times \sin\theta$$

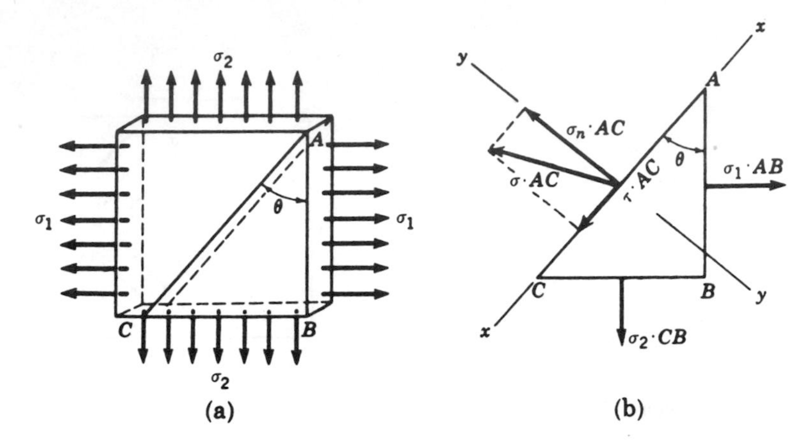

Figure 12-5

But

$$\frac{AB}{AC} = \cos\theta \quad \text{and} \quad \frac{CB}{AC} = \sin\theta$$

$$\therefore \quad \sigma_n = \sigma_1 \cos^2\theta + \sigma_2 \sin^2\theta$$

By substituting $1 - \sin^2\theta$ for $\cos^2\theta$ and $\frac{1}{2}(1 - \cos 2\theta)$ for $\sin^2\theta$, this equation may be changed to read

$$\sigma_n = \frac{\sigma_1 + \sigma_2}{2} + \frac{\sigma_1 - \sigma_2}{2}\cos 2\theta \tag{12.1}$$

Equating also the x components in Fig. 12-5b, we have

$$\tau \times AC = \sigma_1 \times AB \times \sin\theta - \sigma_2 \times CB \times \cos\theta$$

or

$$\tau = \sigma_1 \times \frac{AB}{AC} \times \sin\theta - \sigma_2 \times \frac{CB}{AC} \times \cos\theta$$
$$= \sigma_1 \sin\theta\cos\theta - \sigma_2 \sin\theta\cos\theta$$

But since

$$\sin\theta\cos\theta = \frac{\sin 2\theta}{2}$$

we may write this equation as

$$\tau = \frac{\sigma_1 - \sigma_2}{2}\sin 2\theta \tag{12.2}$$

With the development of Eqs. (12.1) and (12.2), it is revealed that a combination of shearing and normal stresses acts on every plane, making an angle with the principal planes. Each stress varies according to its respective equation, as angle θ is changed. It will be shown in the ensuing paragraphs that the resultant of the shear and normal stresses on any plane can never exceed the intensity of the maximum principal stress.

Maximum Values. The maximum values of σ_n and τ attainable from Eqs. (12.1) and (12.2) are obtained when their trigonometric terms become maximum. For instance, in Eq. (12.1), when $\theta = 0°$, $\cos 2\theta = 1$, its maximum value, making

$$(\sigma_n)_{\max} = \sigma_1$$

Also, the maximum value of τ is obtained when $2\theta = 90°$, making

$$\tau_{\max} = \frac{\sigma_1 - \sigma_2}{2}$$

It is well to note also that the greatest value of shear stress occurs when σ_1 and σ_2 are of opposite sign (i.e., one tension, the other compression).

12-7 Mohr's Circle—for Determining Stresses Induced by Principal Stresses

The use of a Mohr's circle helps to establish firmly in mind the relationships given in Eqs. (12.1) and (12.2). On the coordinate axes of Fig. 12-6, the positive tensile stresses σ_1 and σ_2 are plotted as abscissas, and a circle whose center is on the x axis is drawn through their extremities A and B. The diameter AB of this circle will then be $\sigma_1 - \sigma_2$ and its radius equal to $\frac{1}{2}(\sigma_1 - \sigma_2)$.

If a radius OC is now drawn at a counterclockwise angle 2θ to the positive direction of the x axis, the ordinate DC to the end of the radius will equal $\frac{1}{2}(\sigma_1 - \sigma_2) \sin 2\theta$ or, as we have previously shown in Eq. (12.2), the shearing unit stress τ. The abscissa DE to this point is

$$\frac{\sigma_1 + \sigma_2}{2} + \frac{\sigma_1 - \sigma_2}{2} \cos 2\theta$$

which, as revealed from Eq. (12.1), is equal to the normal unit stress σ_n.

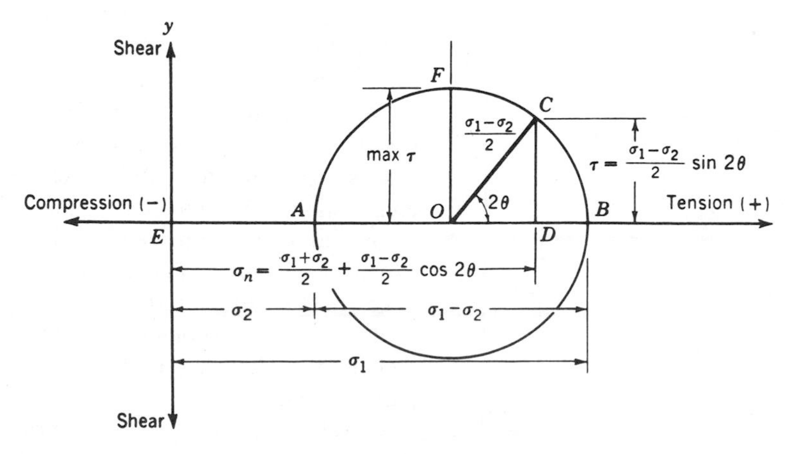

Figure 12-6

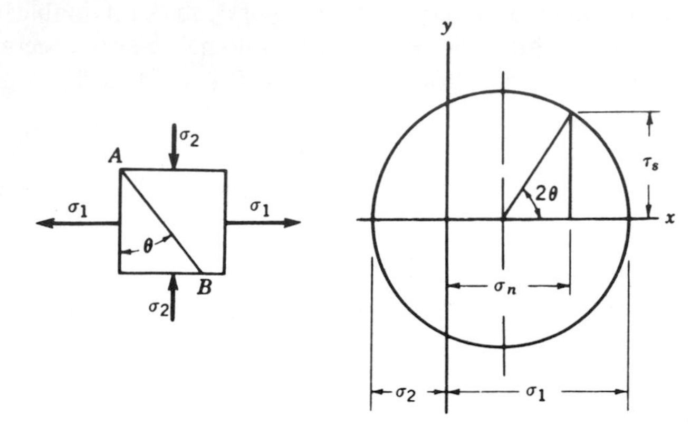

Figure 12-7

It cannot be emphasized strongly enough, however, that in order to correlate the values given by the Mohr's circle to those on the free body, the value of θ must always be the angle between the plane of the maximum principal stress and the plane on which the stresses are desired. That this must be so is shown as θ decreases to zero. The value of σ_n, the normal stress acting on the diagonal plane, will then become equal to σ_1 and the shear stress will equal zero.

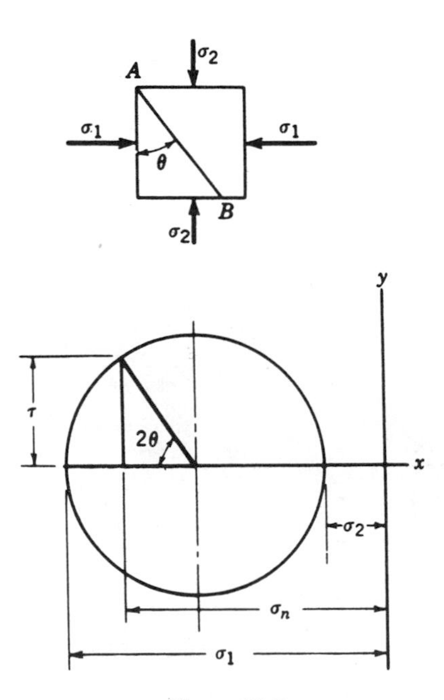

Figure 12-8

In the event one or both of the principal stresses σ_1 and σ_2 are compressive, their values are plotted to the left (on the negative side) of the origin of the coordinate axes. The circle is always drawn through the extremities of σ_1 and σ_2 regardless of on which side of the origin they may have to be placed; i.e., in Fig. 12-7, σ_1 is a tensile stress and σ_2 is a compressive stress.

Both σ_1 and σ_2 are shown as compressive stresses in Fig. 12-8. It is of interest to note that Mohr's circle becomes smaller in diameter as the values of σ_1 and σ_2 approach each other. Whenever both principal stresses are alike in both magnitude and direction, Mohr's circle becomes a point and renders the shearing unit stress on all planes equal to zero.

Whenever one of the principal stresses σ_1 and σ_2 equals zero, Mohr's circle is tangent to the y axis and a state of axial stress results. Such a condition was previously discussed in §2-3.

Maximum shearing stress. As radius OC of Fig. 12-9 is rotated about O, the ordinate of point C passes through a maximum value equal to $\frac{1}{2}(\sigma_1 - \sigma_2)$ when angle 2θ equals 90 and 270°, or when θ is equal to 45 and 135°. Since

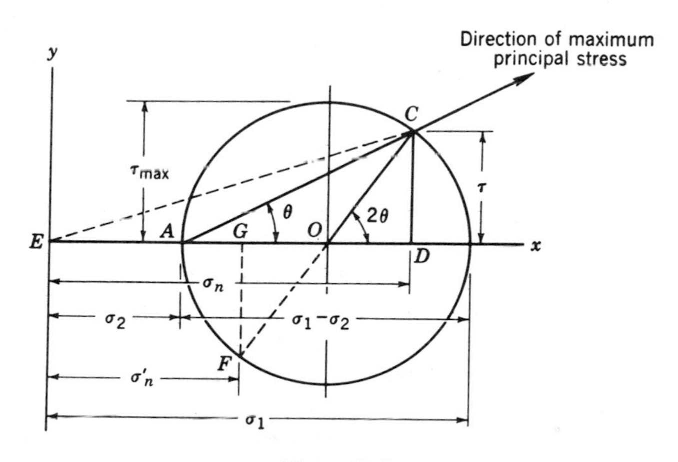

Figure 12-9

both of these shear stresses are equal and occur on mutually perpendicular planes, further substantiation is introduced of basic principle 2 on the inducement of equal shear stresses given in §12-5.

As previously mentioned, three principal stresses occur at every point in a stressed body, even though one may be zero. To obtain the maximum shear stress at such a point, it is necessary to select from the three principal stresses those two which have the greatest algebraic difference. For instance, an element subjected to $\sigma_1 = 800$ psi, $\sigma_2 = 200$ psi, and $\sigma_3 = 0$ will have its maximum shearing stress developed by σ_1 and σ_3.

$$\tau_{\max} = \frac{\sigma_1 - \sigma_3}{2} = \frac{800 - 0}{2} = 400 \text{ psi}$$

The graphical solution for this problem is shown in Fig. 12-10a. The solutions for two similar problems are shown in Fig. 12-10b and c.

Maximum normal stress. Also of great significance and vital concern to the designer is the fact that the maximum principal stress is the greatest normal stress that can act at a point. This relieves his anxiety and helps expedite the problem of endeavoring to determine the most critical condition acting on a body. Of course, several critical points may have to be investigated, but as the principal stresses are found at each such point and the most critical

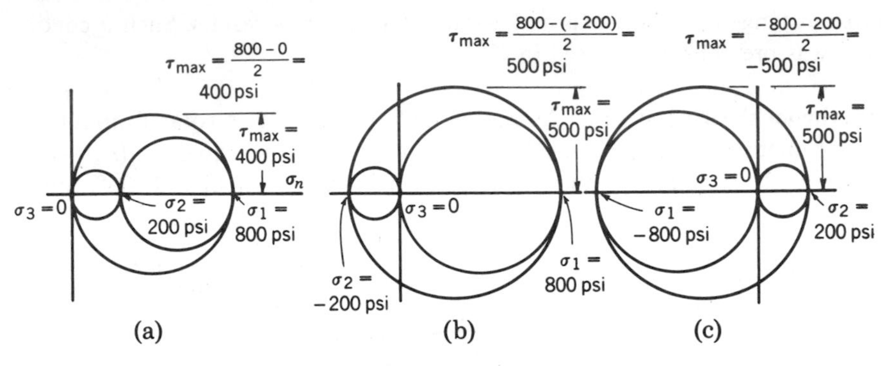

Figure 12-10

condition is ascertained by comparison, no further concern about greater normal stresses acting on the body is necessary.

The justification for the above statements is based on the fact that the abscissa to point C, shown in Fig. 12-9, can never be greater than σ_1.

Sum of normal stresses on mutually perpendicular planes. Another interesting relationship deduced from the Mohr's circle is that *the sum of the normal stresses on any two mutually perpendicular planes remains constant for any angle θ*. Referring again to Fig. 12-9, we note that

$$ED + EG = 2OE$$

Substituting the appropriate values of stress in this equation gives the substantiating statement

$$\sigma_n + \sigma_n' = \sigma_1 + \sigma_2 = (\sigma_n)_x + (\sigma_n')_x \tag{12.3}$$

where $(\sigma_n)_x$ and $(\sigma_n')_x$ are the normal stresses on any two mutually perpendicular planes.

Illustrative Problem 1. A boiler 60 in. in diameter is subjected to a steam pressure of 250 psi. If its walls are $\frac{1}{2}$ in. thick, calculate the maximum normal and shearing unit stresses that occur in the shell.

SOLUTION. From the thin-wall cylinder formulas derived in §4-2, we may determine longitudinal and transverse stresses.

Transverse stress:

$$\sigma_t = \frac{RD}{2t} = \frac{250 \times 60}{2 \times \frac{1}{2}} = 15{,}000 \text{ psi}$$

Longitudinal stress:

$$\sigma_t' = \frac{RD}{4t} = \frac{250 \times 60}{4 \times \frac{1}{2}} = 7500 \text{ psi}$$

These stresses are principal stresses because they occur on planes unaccompanied by shear. They are shown in Fig. 12-11, acting on a small rectangular portion of the cylinder wall.

The third principal stress is produced by the internal pressure acting on the wall. It gradually decreases to zero as the outside of the wall is approached. On cubical free bodies A and B (lateral areas equal one unit of area) obtained from the inside and outside of the wall (Fig. 12-12) are shown the three principal stresses (or forces), the third stress being at full intensity on body A and zero on body B.

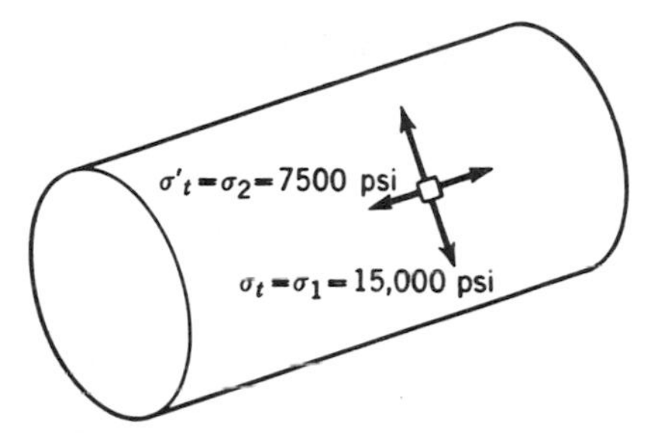

Figure 12-11

Having now ascertained the principal stresses, the computation of the maximum shear stress follows, using Eq. (12.2). Because compressive stresses are considered as negative tensile forces, Eq. (12.2), as applied to free body A, will produce a greater maximum shearing unit stress than the two-dimen-

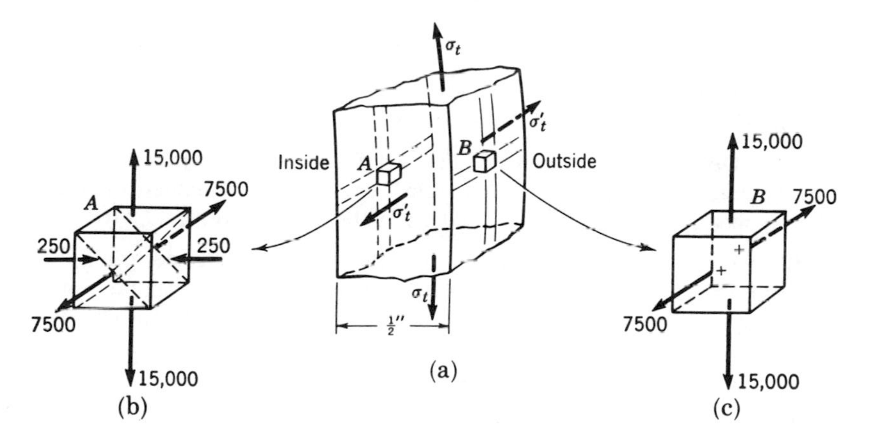

Figure 12-12

sional system at free body B. Substituting in this equation the maximum principal stress of 15,000 psi and the minimum principal stress of 250 psi, there results

$$\tau_{\max} = \frac{\sigma_1 - \sigma_2}{2} = \frac{15{,}000 - (-250)}{2} = 7625 \text{ psi}$$

The traces of the planes on which the maximum shearing unit stress acts are shown as dash lines in Fig. 12-12b. Both shearing planes are perpendicular to the plane containing the 15,000- and 250-lb forces and are located at a 45° angle to the maximum principal stress.

PROBLEMS

12-1. A minute free body taken from a stressed member is shown in Fig. 12-13. Determine the shearing and normal unit stresses on a plane making 30° to the plane of the maximum principal stress. *Answers:* 5630 psi, 6750 psi.

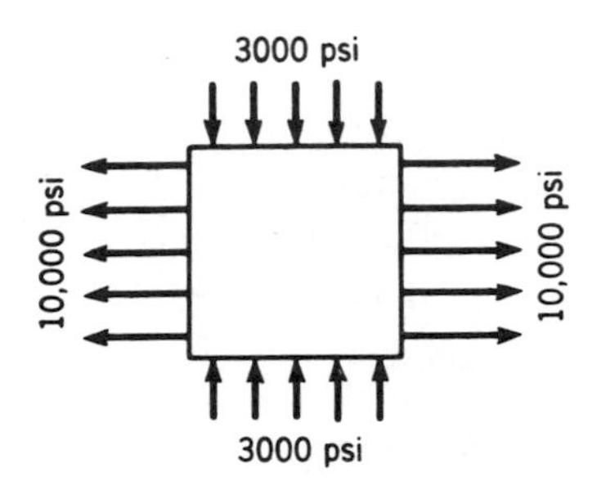

Figure 12-13

12-2. Determine the shearing and normal unit stress in Problem 12-1 if the minor principal stress of −3000 psi were changed to a tensile unit stress of +8000 psi.

12-3. If the angle used in Problem 12-1 were to be changed to 60°, what would the shearing and normal unit stresses be? *Answers:* 5630 psi, 250 psi.

12-4. Determine the maximum diameter for a sphere subjected to an internal gas pressure of 80 psi, fabricated from $\frac{1}{2}$-in. plates and limited in size by a maximum shearing unit stress of 12,000 psi in the shell. *Answer:* 49.8 ft.

12-5. A cylindrical tank 48 in. in diameter is to be subjected to an internal pressure of 150 psi. If the plates are $\frac{3}{8}$ in. thick, what are the maximum normal and maximum shearing unit stresses developed in the shell of the tank?

12-6. Determine theoretically the maximum shear stress for the following states of stress occurring on a cubical element. Substantiate answer by graphical solution.

(a) $\sigma_x = +2000$ $\quad \sigma_y = +5000$ $\quad \sigma_z = -3000$
(b) $\sigma_x = -3000$ $\quad \sigma_y = +1000$ $\quad \sigma_z = 0$
(c) $\sigma_x = +4000$ $\quad \sigma_y = +4000$ $\quad \sigma_z = +4000$
(d) $\sigma_x = -4000$ $\quad \sigma_y = -4000$ $\quad \sigma_z = 0$

Answers: (a) 4000 psi; (b) 2000 psi; (c) 0; (d) 2000 psi.

12-7. Draw Mohr's circle for the following states of stress occurring on a square (two-dimensional) element.

(a) Only tensile stress σ_x.
(b) Equal tensile stresses σ_x and σ_y.
(c) One tensile stress σ_x and one compressive stress σ_y.
(d) Only shear stress acting on each face.

SIMULTANEOUS APPLICATION OF A TENSION (OR COMPRESSION) AND SHEARING LOAD

12-8 Preliminary Considerations

In contrast to the previous problem where principal stresses were found directly from well-known stress equations, such as Eqs. (4.3) and (4.4), the more general problem concerns the determination of the magnitude and location of the principal stresses in structural members to which two or more different types of loading (that is, axial, bending, twisting, and so forth) are applied simultaneously. In such a case, the critical stress induced by each type of loading is determined separately and combined through convenient stress relationships to find the principal stresses. We shall endeavor to obtain these relationships.

One combined stress problem frequently encountered in engineering practice concerns the tightening of an engine stud bolt (Fig. 12-14a). In the

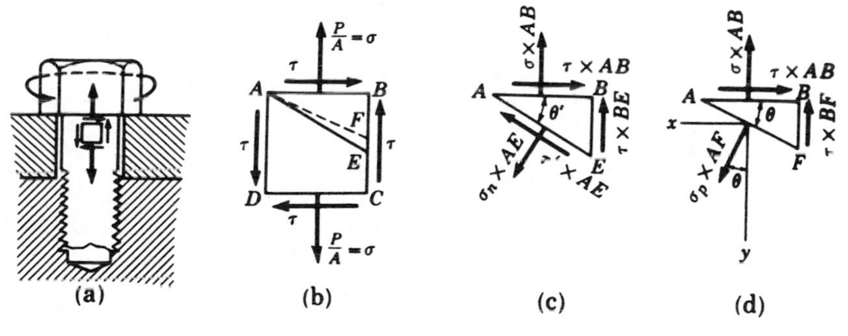

Figure 12-14

final tightening operation, the bolt is twisted and placed in tension to hold the cylinder head to the cylinder block. A free body taken from a point just below the head of the bolt appears in Fig. 12-14b. Our problem is now to find the principal stresses, since they are the maximum and minimum normal stresses and, as seen from §12-6, produce the maximum shearing unit stress.

12-9 Finding the Maximum Principal Stress

Selecting the orientation of the free body as shown in Fig. 12-14a, the shearing and normal (tensile) stresses, τ and σ, are obtained from

$$\sigma = \frac{P}{A} \quad \text{and} \quad \tau = \frac{Tr}{J}$$

Since principal stresses, however, act on their planes unaccompanied by shearing unit stress, it is evident that the normal unit stress found above cannot be a principal stress. For this determination, we must select a particular plane AF (Fig. 12-14d), a principal plane, on which only the maximum normal stress is assumed to act. Note the difference between this plane and the general plane AE shown in Fig. 12-14c, on which both a normal and shearing unit stress act simultaneously.

Equating the x and y components of the forces shown acting on this free body, we obtain the following expressions:

$$\sum F_x = \tau \times AB - \sigma_p \times AF \times \sin\theta = 0 \qquad (12.4)$$

$$\sum F_y = \sigma \times AB + \tau \times BF - \sigma_p \times AF \cos\theta = 0 \qquad (12.5)$$

These equations may be simplified by dividing them by AF and noting that $AB/AF = \cos\theta$ and $BF/AF = \sin\theta$. Thus,

$$\tau \times \frac{AB}{AF} = \sigma_p \sin\theta$$

from which

$$\tau \cos\theta = \sigma_p \sin\theta \qquad (12.6)$$

and

$$\sigma \times \frac{AB}{AF} + \tau \times \frac{BF}{AF} = \sigma_p \cos\theta$$

from which

$$\sigma \times \cos\theta + \tau \times \sin\theta = \sigma_p \cos\theta \qquad (12.7)$$

By rearranging Eq. (12.7) so that the cosine terms are on the left-hand side and the sine term on the right, there is obtained

$$(\sigma_p - \sigma)\cos\theta = \tau \sin\theta \qquad (12.8)$$

Equation (12.6) may now be divided into Eq. (12.8) to give

$$\frac{\sigma_p - \sigma}{\tau} = \frac{\tau}{\sigma_p}$$

from which

$$\sigma_p^2 - \sigma \times \sigma_p - \tau^2 = 0$$

and

$$\sigma_p = \tfrac{1}{2}\sigma \pm \tfrac{1}{2}\sqrt{\sigma^2 + 4\tau^2} \qquad (12.9a)$$

This equation may also be written as

$$\sigma_p = \frac{\sigma}{2} \pm \sqrt{\left(\frac{\sigma}{2}\right)^2 + \tau^2} \qquad (12.9b)$$

The two values obtained for σ_p from this equation are the maximum and minimum principal stresses. There must therefore be two values of θ that satisfy Eqs. (12.4) and (12.5). *Because the value under the radical will* *always exceed the magnitude of $\frac{1}{2}\sigma$, the maximum and minimum principal stresses will be of opposite types—that is, one a tensile stress, the other compressive.*

12-10 Finding the Planes of Principal Stress

The planes on which these principal stresses act are found by solving Eq. (12.6) for σ_p and substituting it in Eq. (12.7). The latter equation will then provide the value of θ in terms of the known stresses σ and τ.

$$\sigma_p = \tau \times \frac{\cos\theta}{\sin\theta}$$

$$\sigma\cos\theta + \tau\sin\theta = \tau \times \frac{\cos\theta}{\sin\theta} \times \cos\theta$$

$$\sigma\sin\theta\cos\theta = \tau(\cos^2\theta - \sin^2\theta)$$

From our knowledge of double-angle trigonometric functions we obtain

$$\sigma \times \frac{\sin 2\theta}{2} = \tau\cos 2\theta$$

from which

$$\tan 2\theta = \frac{2\tau}{\sigma} \tag{12.10}$$

Again recalling our trigonometry, it is remembered that $\tan\theta$ has identical values of sign and magnitude for all pairs of angles separated by 180°. For instance, the values of tan 2θ for angles $2\theta = 30°$ and for 210° are the same. Thus, if

$$\tan 2\theta = \tan 30° = \tan 210°$$

the angles that satisfy the equation are $\theta = 15°$ and $\theta = 105°$.

We may therefore state that two values of θ satisfy Eq. (12.10) and that they are 90° apart. Furthermore, these two values of θ define the planes on which the maximum and minimum principal stresses act.

Illustrative Problem 2. A steel shaft (Fig. 12-15), 2 in. in diameter, is acted upon by an axial tensile force of 50,000 lb and a twisting moment of 30,000 in.-lb. Calculate the maximum and minimum principal stresses and the planes on which they act.

SOLUTION.

Maximum and minimum principal stresses. Let us first select a free body from the surface of the shaft, having a radial depth of unity (Fig. 12-15b).

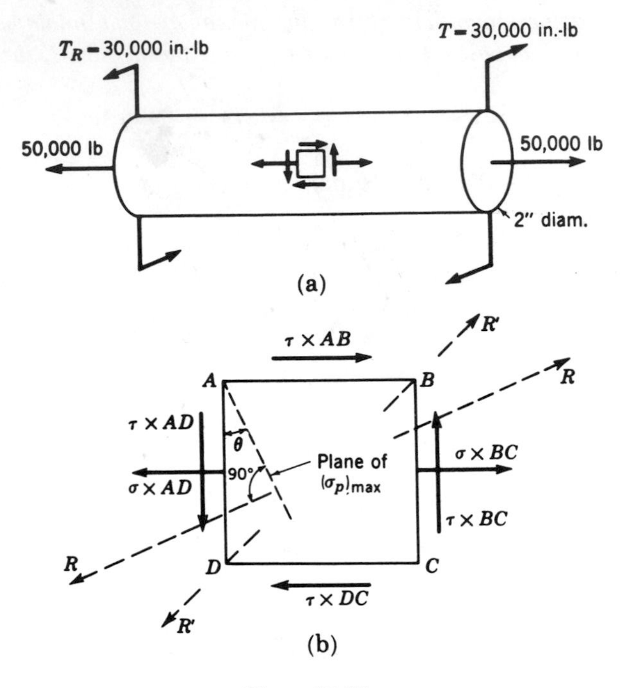

Figure 12-15

The values of σ and τ are then computed as

$$\sigma = \frac{P}{A} = \frac{50{,}000}{\pi/4 \times 2^2} = 15{,}900 \text{ psi}$$

$$\tau = \frac{Tr}{J} = \frac{30{,}000 \times 1}{\pi/32 \times 2^4} = 19{,}100 \text{ psi}$$

Inserting these values in Eq. (12.9a), there are obtained the maximum and minimum principal stresses:

$$\begin{aligned}\sigma_p &= \tfrac{1}{2}\sigma \pm \tfrac{1}{2}\sqrt{\sigma^2 + 4\tau^2} \\ &= \frac{15{,}900}{2} \pm \frac{1}{2}\sqrt{15{,}900^2 + 4 \times 19{,}100^2} \\ &= 7950 \pm \frac{41{,}300}{2}\end{aligned}$$

$$(\sigma_p)_{\max} = 28{,}600 \text{ psi}$$

$$(\sigma_p)_{\min} = -12{,}700 \text{ psi}$$

These stresses and the planes in which they act are shown in Fig. 12-16.

Principal planes. It is well to recall at this point that the angle θ in the previous derivation is the angle between the principal plane and the plane of

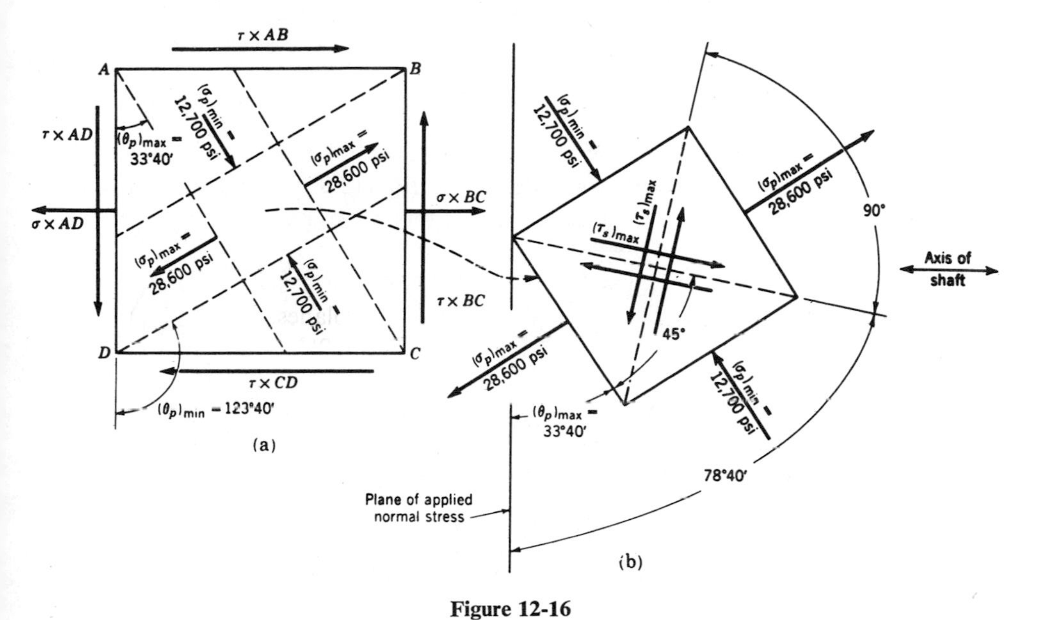

Figure 12-16

the applied normal stress. However, two sets of planes may be drawn at an angle θ to the plane of applied normal stress, and it becomes necessary to further qualify which θ is correct.

An easy method that may be employed in the determination of the correct value of θ is to find first the approximate location and magnitude of the resultant R' of both pairs of colliding shear forces (see Fig. 12-15). A further approximate combination of R' with the applied normal force, $\sigma \times BC$, directs the resultant R in a direction approximately perpendicular to the plane of the maximum principal stress. It is this plane that is angle θ (magnitude below 90°) from the plane of the applied normal stress.

The value of θ may now be determined from Eq. (12.10).

$$\tan 2\theta = \frac{2\tau}{\sigma} = \frac{2 \times 19{,}100}{15{,}900} = 2.4$$

$$2\theta = 67°20' \quad \text{or} \quad 247°20'$$

$$\theta = 33°40' \quad \text{or} \quad 123°40'$$

$$(\theta_p)_{max} = 33°40', \qquad (\theta_p)_{min} = 123°40'$$

12-11 Maximum Shearing Stress

As explained in §12-6 the maximum shearing unit stress produced by the application of two tensile principal stresses σ_1 and σ_2 acting at a single point in a body is equal to

$$\tau_{max} = \frac{\sigma_1 - \sigma_2}{2} = \frac{(\sigma_p)_{max} - (\sigma_p)_{min}}{2}$$

and occurs on planes located 45° from the principal planes.

If the two principal stresses as derived in Eq. (12.9b) are inserted in the above equation of maximum shearing unit stress, there results

$$\begin{aligned} \tau_{max} &= \frac{1}{2}\left[\frac{1}{2}\sigma + \sqrt{\left(\frac{\sigma}{2}\right)^2 + \tau^2} - \left(+\frac{1}{2}\sigma - \sqrt{\left(\frac{\sigma}{2}\right)^2 + \tau^2}\right)\right] \\ &= \sqrt{\left(\frac{\sigma}{2}\right)^2 + \tau^2} \end{aligned} \tag{12.11}$$

Illustrative Problem 3. Determine in Illustrative Problem 2 the maximum shearing unit stress and its location with respect to the plane of the applied normal stress σ.

SOLUTION. The maximum shearing stress is equal to

$$\tau_{max} = \sqrt{\left(\frac{\sigma}{2}\right)^2 + \tau^2} = \sqrt{\left(\frac{15{,}900}{2}\right)^2 + 19{,}100^2} = 20{,}650 \text{ psi}$$

It may also have been computed directly from the principal stresses as follows:

$$\tau_{max} = \tfrac{1}{2}[28{,}600 - (-12{,}700)] = \tfrac{1}{2} \times 41{,}300 = 20{,}650 \text{ psi}$$

The location of the plane of maximum shearing stress with respect to the plane of the applied normal stress is

$$\theta_s = 33°40' + 45°00' = 78°40'$$

or

$$\theta_s = 33°40' + 135°00' = 168°40'$$

PROBLEMS

12-8. A shaft 2 in. in diameter is subjected to a torque of 600 ft-lb and a bending moment of 500 ft-lb. Determine the maximum normal and shearing unit stresses in the shaft. *Answers:* 9790 psi, 5970 psi.

12-9. The solid, circular steel shaft of a vertical hydraulic turbine is 30 in. in diameter. If the axial load supported by the shaft is 1,500,000 lb and the torque produced by the water wheel is 3,500,000 ft-lb, what maximum normal and shearing stresses are produced?

12-10. If the torque in Problem 12-9 were not given, what maximum value could it have withstood without exceeding a shearing stress of 10,000 psi, or a maximum principal stress of 15,000 psi? *Answer:* 4,380,000 ft-lb.

12-11. A hollow, steel hanger rod, $1\frac{1}{2}$ in. in outside diameter and 1 in. in inside diameter and 12 ft long, is subjected to a static load of 12,000 lb. If the rod is twisted 5°, what maximum normal and shearing stresses are incurred? *Answers:* 14,300 psi, 8200 psi.

12-12. A stud bolt $\frac{3}{4}$ in. in diameter is to be tightened with a 50-lb force applied at the end of a 12-in. monkey wrench. If the tensile unit stress in the bolt at the time of the development of maximum torque is 10,000 psi, what are the maximum, principal, and shearing unit stresses in the body of the bolt? Locate the angles on which they act with respect to the cross section.

12-13. The 2-in.-diameter steel shaft of Fig. 12-17 is acted upon by force P of 400 lb. Determine the maximum and minimum principal stress, the maximum shearing unit stress, and the direction in which each acts. *Answers:* 10,600 psi, −400 psi, 5500 psi, 10°54′, 100°54′, 55°54′.

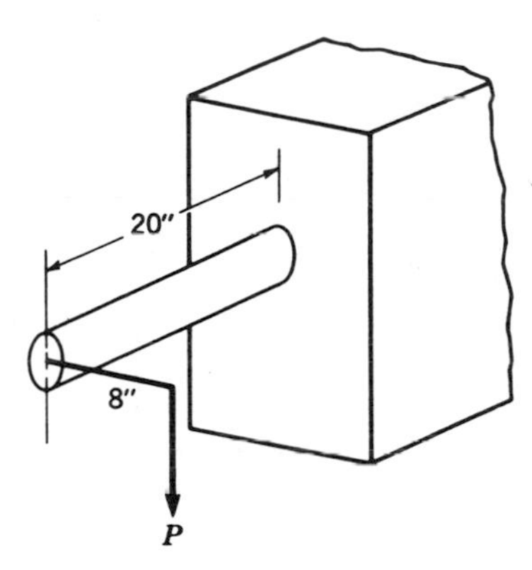

Figure 12-17

12-14. Determine and show on a sketch the maximum and minimum principal stresses and the maximum shearing stress developed in the shaft of Fig. 12-18. Determine the planes on which they act and indicate clearly on your sketch. *Answers:* 9660 psi, −1660 psi, 5660 psi, 22.5°.

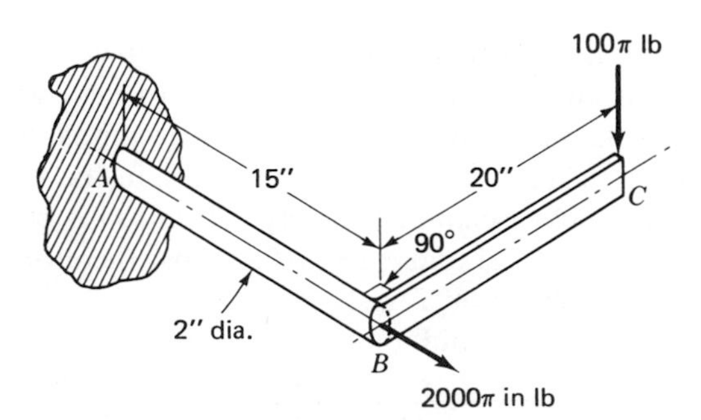

Figure 12-18

12-15. A solid steel shaft 2 in. in diameter is subjected to a horizontal compressive force of 5000 lb acting along the axis, a torque of 5000 in.-lb, and a bending moment of 5000 in.-lb which causes tension in the top fibers of the shaft. Determine the maximum tensile stress, maximum compressive stress, and maximum shearing stress in the shaft due to this combined loading and indicate the points where these stresses occur.

12-16. A 2-in.-diameter solid circular bar (Fig. 12-19) is loaded at the free end by a vertical force of 1000 lb and a torque T. An experimental analysis determines the maximum principal stress at point A to be 16,170 psi. Assuming linearly elastic behavior, determine the amount of torque T that must have been applied. *Answer:* 6000 in.-lb.

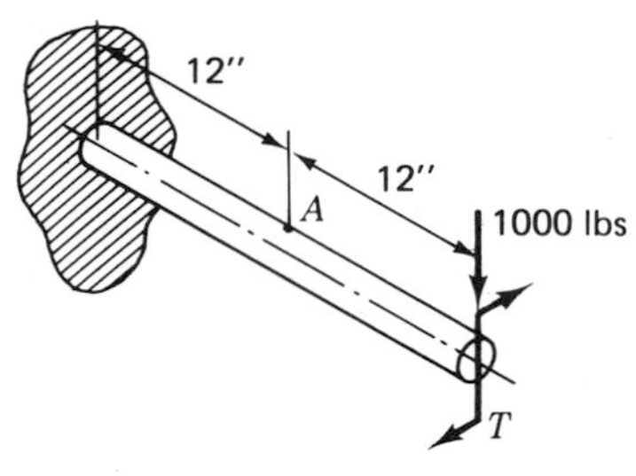

Figure 12-19

12-12 Use of Mohr's Circle for Determining Principal Stresses

From the computed normal and shearing unit stresses σ and τ acting at any point in a body, the corresponding principal stresses can also be obtained through the use of a Mohr-circle diagram. Consider the solution of Illustrative Problem 2, the computed stresses of which are shown on the element of Fig. 12-20a.

The diagram is developed, as previously shown in Chapter 2, by plotting the normal stress as an abscissa, the tensile stress to the right and the compressive stress to the left, and the shear stress accompanying the normal stress as an ordinate at the end of the normal stress, in the positive direction if its couple rotates clockwise and negatively if rotation is counterclockwise. The values of $\sigma = 15{,}900$ psi and $\tau = -19{,}000$ psi therefore determine the location of point B as shown in Fig. 12-20b.

Now locate C at the midpoint of OA. Line CB is then equal to $\sqrt{(\sigma/2)^2 + \tau^2}$. Then using CB as a radius, draw a circle with C as a center. If, now, we investigate the value of OD, we note that it is equal to the maximum principal stress, since

$$OD = OC + CD = \frac{\sigma}{2} + \sqrt{\left(\frac{\sigma}{2}\right)^2 + \tau^2} = (\sigma_p)_{\max} = 28{,}600 \text{ psi}$$

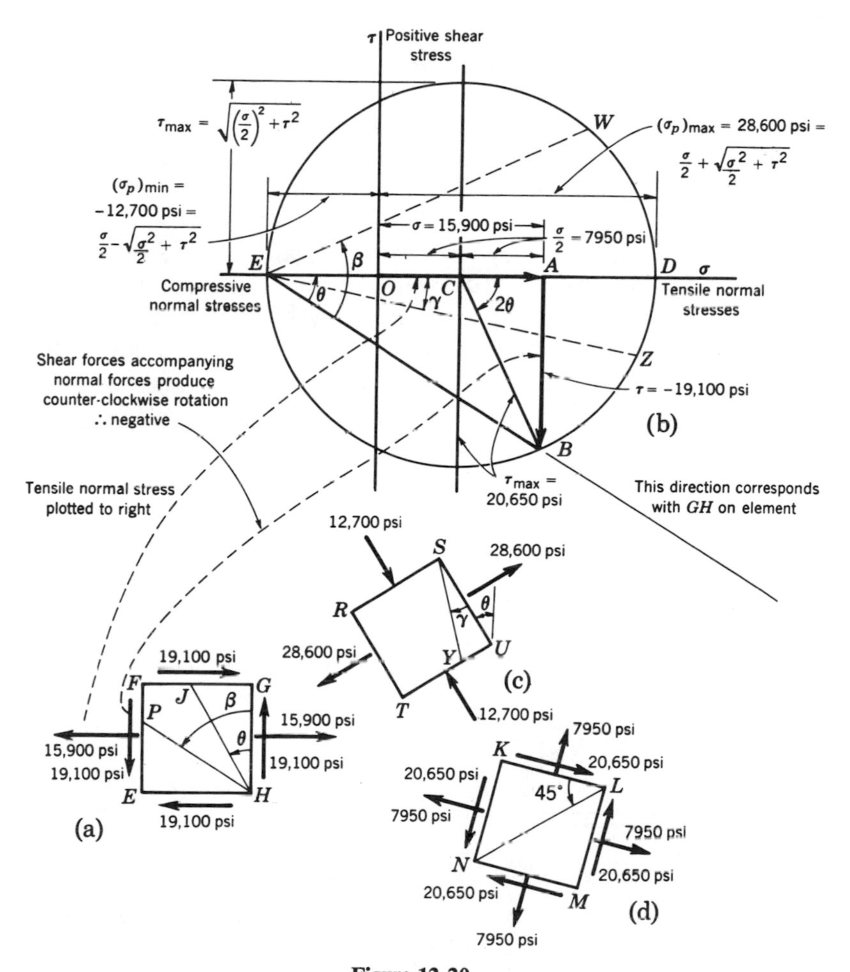

Figure 12-20

This equation was derived in §12-9. In like manner OE is shown to be equal to the minimum principal stress.

$$OE = OC - CE = \frac{\sigma}{2} - \sqrt{\left(\frac{\sigma}{2}\right)^2 + \tau^2} = (\sigma_p)_{\min} = -12{,}700 \text{ psi}$$

Because the maximum ordinate of a point on the circle is equal to the radius of the circle, its value is equal to the maximum shear stress.

$$\tau_{\max} = \sqrt{\left(\frac{\sigma}{2}\right)^2 + \tau^2} = 20{,}650 \text{ psi}$$

Furthermore, in Eq. (12.10) it was shown that the angle of inclination of the plane of principal stresses to the plane of the computed normal stress was given by the relationship

$$\tan 2\theta = \frac{2\tau}{\sigma}$$

If this is written as $\tan 2\theta = \tau/(\sigma/2)$, it reveals that angle BCA is equal to 2θ. Because angle BCA is a central angle cutting the arc BD, included angle BED, which cuts the same arc, is equal to one half of 2θ, or θ. This is the angle that the maximum principal stress plane makes with the plane of the given normal stress. It is also the angle between these two stresses.

If then we consider line EB on Fig. 12-20b to correspond to the plane GH of the element (Fig. 12-20a) and rotate both counterclockwise through an angle θ, they will coincide respectively with line ED, representing the magnitudes of the principal stress, and line HJ, its actual plane of action on the element. Figure 12-20c represents the orientation of the element required

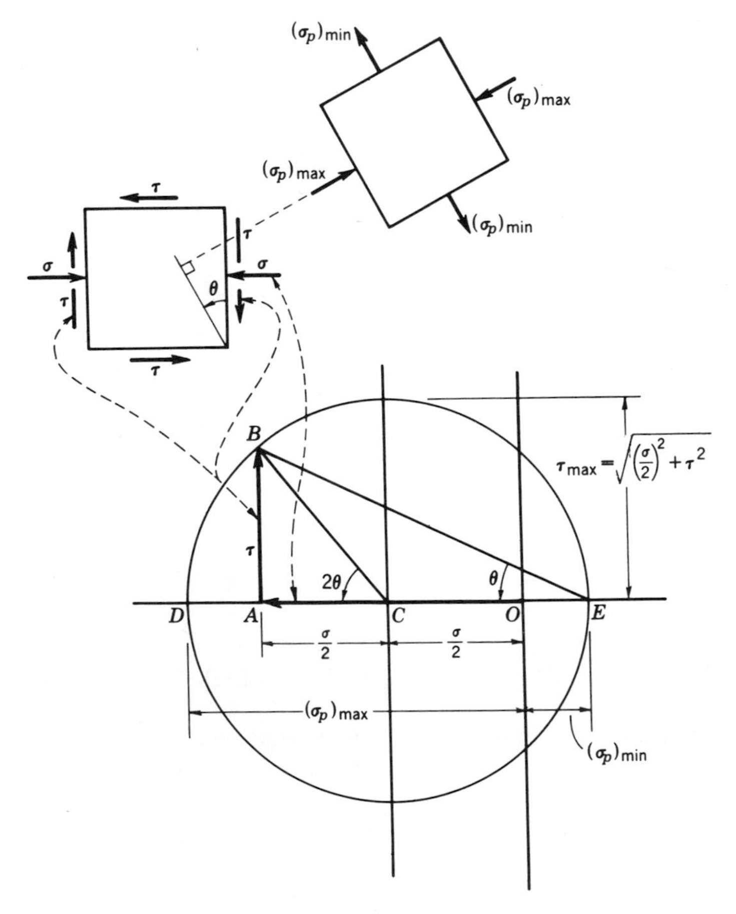

Figure 12-21

for the development of the principal stresses. Note that the maximum principal stress planes are rotated 90° from the minimum principal stress planes, on which act the 12,700-psi stress, given by line OE on the Mohr diagram.

The orientation of the element required to produce the maximum shear stresses (45°) to principal stresses is shown in Fig. 12-20d.

The procedure, as outlined in the paragraphs above, follows exactly any combined stress problem consisting of shear stress and one normal stress. See Fig. 12-21 for solution of an element subjected to compression and shear stress.

To Find the Stresses on Any Plane. The stresses acting on any plane within the element may be found by orienting that plane with respect to either the plane of the maximum principal stress JH (Fig. 12-20a) or the plane of the given normal stress GH. For instance, the stresses acting on any random plane PH rotated through an angle β from GH, or a similarly rotated angle $\beta - \theta$ from JH, are determined from the coordinates to point W located on the Mohr's circle. The point W is located at the end of line EW rotated angle β from EB, the line corresponding to the given stress plane GH, or angle $\beta - \theta$ from ED, the line corresponding to the principal stress plane. Note that the rotation within the element and on the Mohr circle is in the same direction. In a similar manner the stresses acting on plane SY (Fig. 12-20c) rotated through an angle γ *clockwise* from the principal plane SU are obtained by drawing EZ at a *clockwise* angle γ from ED on the Mohr circle. The negative shear and positive normal stress corresponding to the coordinates of point Z act on the face SY.

PROBLEMS

12-17. If a free body of a loaded member were stressed as shown in Fig. 12-22, with $\tau = 6000$ psi and $\sigma_n = 10{,}000$ psi, what would the maximum and minimum normal and maximum shearing unit stresses be? What would the normal and shearing unit stresses be on a plane making 30° to the plane of $(\sigma_p)_{max}$? Use mathematical analysis. *Answers:* 12,800 psi, −2800 psi, 7800 psi, 6750 psi, 8900 psi.

12-18. Solve Problem 12-17, assuming stress σ_n to be reversed. Use mathematical analysis. *Answers:* −12,800 psi, +2800 psi, 7800 psi, 6750 psi, −8900 psi.

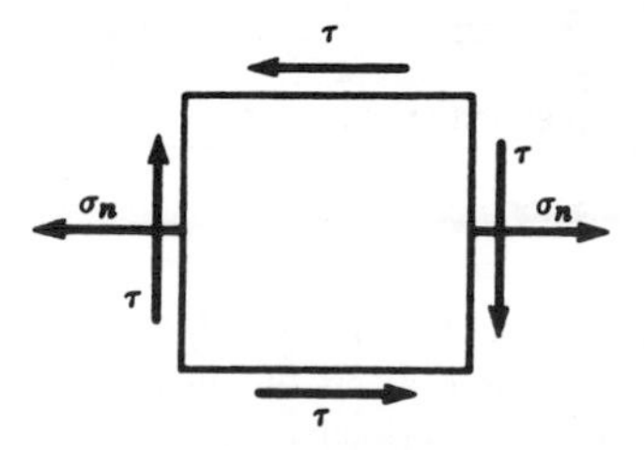

Figure 12-22

12-19. Solve Problem 12-17 using Mohr's circle.

12-20. Solve Problem 12-18 using Mohr's circle.

12-21. Solve Problem 12-13 using Mohr's circle.

12-22. Solve Problem 12-8 using Mohr's circle. *Answers:* 9790 psi, 5970 psi.

12-23. Find the maximum normal shear stresses and directions of each for the following states of stress.

(a) $\sigma_x = +10{,}000$ psi $\quad \tau_{xy} = +6000$ psi
(b) $\sigma_x = -10{,}000$ psi $\quad \tau_{xy} = +6000$ psi
(c) $\sigma_x = +10{,}000$ psi $\quad \tau_{xy} = -6000$ psi

Determine the magnitude of the normal and shear stress acting on a surface measured 45° counterclockwise from the right vertical face.

12-24. Solve Problem 12-14 using Mohr's circle.

12-25. Solve Problem 12-15 using Mohr's circle.

12-13 Combined Stresses in Beams

When a beam is loaded as shown in Fig. 12-23, flexural and shear stresses are developed throughout the beam in varying intensity according to the formulas $\sigma = Mc/I$ and $\tau = (V/Ib)a\bar{y}$. The principal stresses resulting therefrom are also of varying magnitude and direction as shown. Further study reveals that below the neutral axis the maximum principal stresses are in tension. Above the same axis, they are in compression. When a line such as ABA' is drawn tangent at all points to either the tensile or compressive

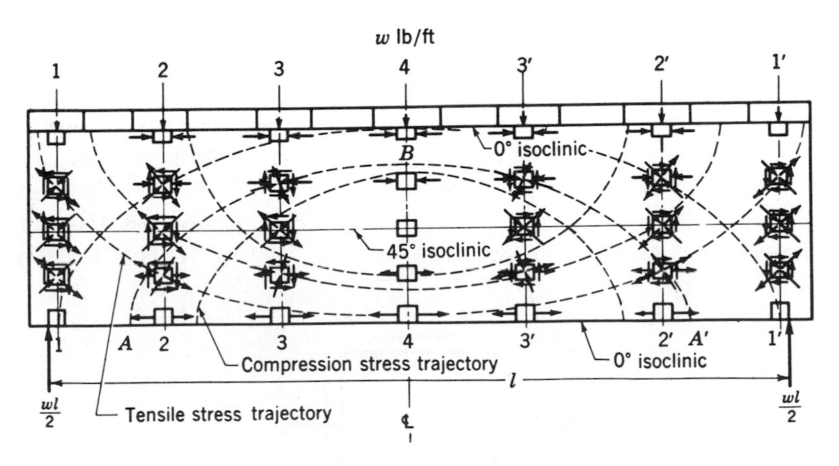

Figure 12-23 Variation of flexural and shearing stresses throughout a beam, disregarding localized effects of uniform load and concentrated reactions. Also shown are the principal stress directions and compressive stress trajectories.

principal stresses, it forms a *stress trajectory*. Many such stress trajectories can be drawn. If stress trajectories for the compressive principal stresses are drawn superimposed on corresponding stress trajectories for the tensile principal stresses, a system of orthogonal trajectories (intersections at right angles) is formed.

Note that the direction of the principal stresses on the neutral axis is the same, i.e., 45°. Note also that the directions of the maximum principal stresses on the top and bottom surfaces are at 0°. Lines connecting points of these or other *equal* directions of principal stress are called *isoclinics*.

Another set of lines of invaluable assistance to the stress analyst is that composed of *isochromatics*. These lines obtained photographically from a two-dimensional translucent model placed in a circular polarized light field while under stress reveal equal differences of maximum and minimum principal stress. When stress trajectories are superimposed on isochromatics, sufficient data are obtained to compute the magnitude and direction of all normal and shearing unit stresses occurring in the model and its prototype.

12-14 Finding Internal Stresses by Photoelasticity

A basic method for determining internal stresses by photoelasticity is the *graphical integration method*. It is based on the equilibrium equation of an element located between two pairs of orthogonally located stress trajectories (Fig. 12-24). Because the element is bordered by stress trajectories, the stresses acting on the periphery are normally directed. Using an average rather than the actual stress distribution, the equilibrium equation in the x direction is as follows:

$$
\begin{aligned}
-\sigma_1 \Delta s_2 \frac{\cos \Delta\phi}{2} + \left(\sigma_1 + \frac{\partial \sigma_1}{\partial s_1}\Delta s_1\right)\Delta s_2' \cos\left(\Delta\theta + \frac{\Delta\phi}{2}\right) \\
+ \sigma_2 \Delta s_1 \sin\frac{\Delta\theta}{2} - \left(\sigma_2 + \frac{\partial \sigma_2}{\partial s_2}\Delta s_2\right)\Delta s_1' \sin\left(\Delta\phi + \frac{\Delta\theta}{2}\right) = 0
\end{aligned}
\tag{12.12}
$$

This equation is simplified by the use of two assumptions. The first considers the stress trajectories to be circular, in which case $\Delta s_2' = \Delta s_2 [1 + (\Delta s_1/\rho_2)]$. The second considers the angles through which the stress trajectories rotate to be so small as to make the cosines equal to unity and the sines of angles equal to the angle in radians.

Applying these assumptions, Eq. (12.12) reduces to

$$\frac{\partial \sigma_1}{\partial s_1} + \frac{\sigma_1 - \sigma_2}{\rho_2} = 0 \tag{12.13}$$

A similar expression is obtained from the summation in the y direction:

$$\frac{\partial \sigma_2}{\partial s_2} + \frac{\sigma_1 - \sigma_2}{\rho_1} = 0 \tag{12.14}$$

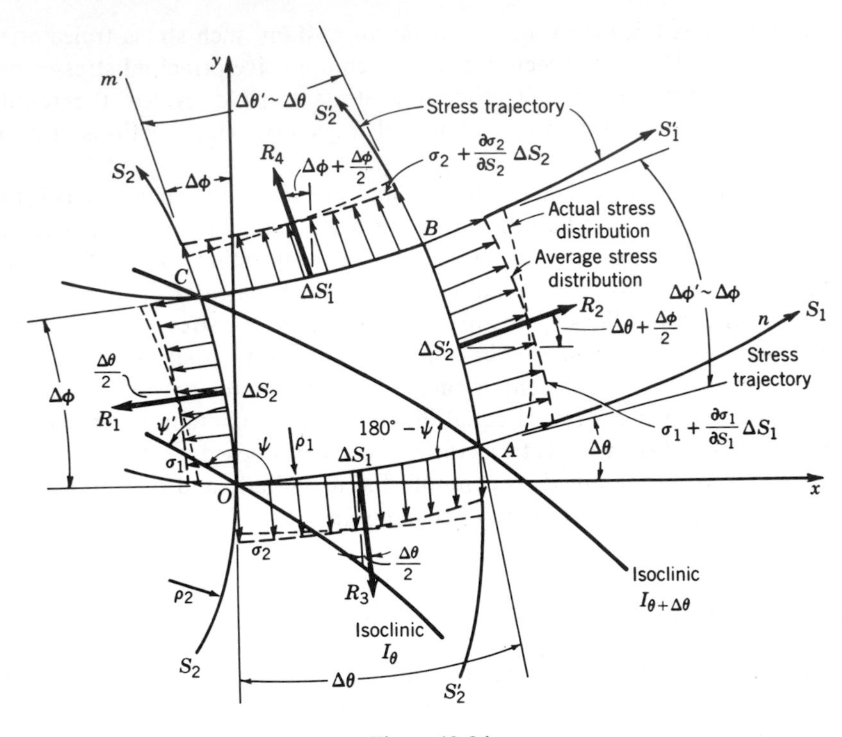

Figure 12-24

Equations (12.13) and (12.14) are frequently referred to as the Lamé-Maxwell equations of equilibrium.

If the variation of stresses is desired along the S_1 and S_2 stress trajectories, Eq. (12.13) and (12.14) are integrated respectively as follows:

$$\int_0^n d\sigma_1 = \int_0^n \frac{\sigma_1 - \sigma_2}{\rho_2}\, ds_1 \quad \text{and} \quad \int_0^m d\sigma_2 = \int_0^m \frac{\sigma_1 - \sigma_2}{\rho_2}\, ds_2$$

from which

$$(\sigma_1)_n = (\sigma_1)_0 - \int_0^n \frac{\sigma_1 - \sigma_2}{\rho_2}\, ds_1 \quad \text{and} \quad (\sigma_2)_m = (\sigma_2)_0 - \int_0^m \frac{\sigma_1 - \sigma_2}{\rho_1}\, ds_2 \tag{12.15}$$

Placed in finite form, these expressions become

$$(\sigma_1)_n = (\sigma_1)_0 - \sum_0^n \frac{\sigma_1 - \sigma_2}{\rho_2}\Delta s_1 \quad \text{and} \quad (\sigma_2)_m = (\sigma_2)_0 - \sum_0^m \frac{\sigma_1 - \sigma_2}{\rho_1}\Delta s_2 \tag{12.16}$$

The direct use of these equations in a step-by-step integration process requires a slight modification due to the difficulty of evaluating exact values of the radius ρ. This is accomplished through the use of the triangle OCA

formed by the stress trajectories and isoclinic $I_{\theta+\Delta\theta}$ from which

$$\frac{OC}{OA} = \tan(180° - \psi) = -\tan\psi$$

or

$$\frac{OA}{OC} = -\cot\psi$$

But

$$OC = \rho_2 \, \Delta\phi \quad \text{and} \quad \frac{1}{\rho_2} = \frac{\Delta\phi}{OC}$$

Multiplying both sides by OA yields

$$\frac{OA}{\rho_2} = \frac{OA}{OC}\Delta\phi \quad \text{or} \quad \frac{OA}{\rho_2} = -\cot\psi \, \Delta\phi \tag{12.17}$$

Similarly, it may be shown that

$$\frac{OC}{\rho_2} = \cot\psi' \, \Delta\phi \tag{12.18}$$

Inserting expressions (12.17) and (12.18) in Eq. (12.16) provides

$$(\sigma_1)_n = (\sigma_1)_0 + \sum_0^n (\sigma_1 - \sigma_2) \cot\psi \, \Delta\phi \tag{12.19}$$

$$(\sigma_2)_m = (\sigma_2)_0 - \sum_0^m (\sigma_1 - \sigma_2) \cot\psi' \, \Delta\phi \tag{12.20}$$

The graphical integration process using Eq. (12.19) or (12.20) proceeds along a specified stress trajectory, Eq. (12.19) for σ_1 and Eq. (12.20) for σ_2. A plot is made of the $\sigma_1 - \sigma_2$ values (isochromatics) occurring at the isoclinic lines along the length of the trajectory together with the corresponding cot ψ values. A further plot of $(\sigma_1 - \sigma_2)$ cot ψ areas as ordinates and $\Delta\phi$ values as abscissas (Fig. 12-25) provides areas equal to the incremental stress increases over the length of the integrated stress trajectory.

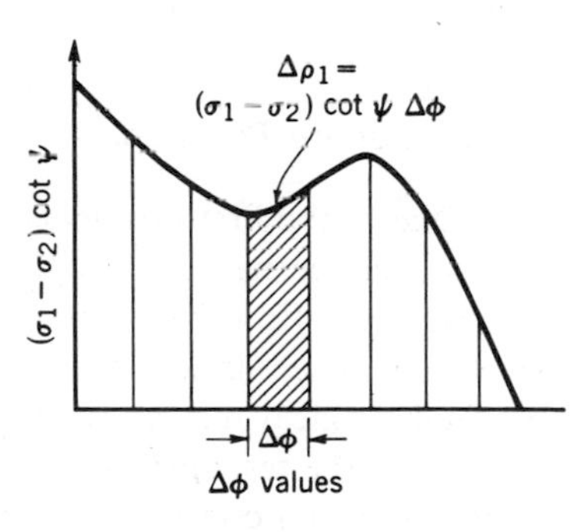

Figure 12-25

PROBLEMS

12-26. Draw the stress trajectories for a cantilever beam of rectangular cross section which has a concentrated load acting at its free end. Assume no stress concentrations due to the load and its reaction.

12-27. Draw the stress trajectories for a cantilever beam of rectangular cross section which has a uniform load acting over its entire length.

12-28. In Fig. 12-26, determine the maximum normal and shearing unit stresses on the section 3 ft to the right of R_1 at 0, 2, 4, and 6 in. below the neutral axis. Determine the direction of each stress with respect to the horizontal.

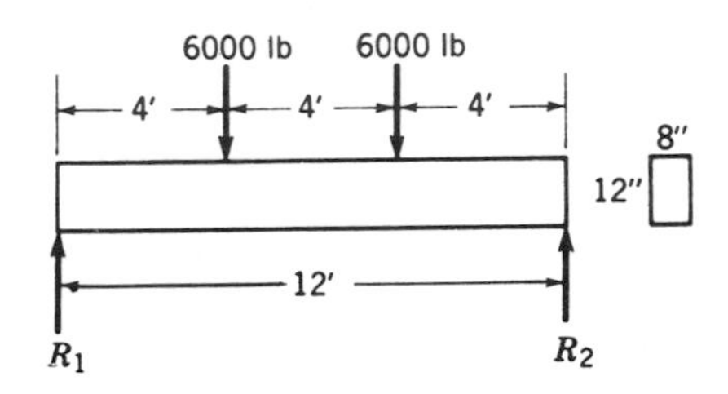

Figure 12-26

12-29. A rectangular beam 4 by 10 in. in cross section (10-in. dimension vertical) is subjected to a positive bending moment of 43,000 in.-lb and a vertical shear of 3580 lb at a short distance from the left end of the beam. The beam is simply supported at the ends. Determine the magnitude and orientation of the principal stresses for a point in the cross section 2 in. below the centroidal axis of the beam. *Answers:* $(\sigma_p)_{\max} = 301$ psi, $(\sigma_p)_{\min} = -43$ psi, $\theta_p = 20.6°$.

12-30. A timber cantilever beam, 25 in. long, with a 2- by 12-in. cross section (12-in. vertical dimension), has a concentrated load of 800 lb at its free end. Determine the magnitude and orientation (with the horizontal) of the principal stresses for a point 2 in. above the neutral axis on the section 10 in. from the free end. *Answers:* $(\sigma_p)_{\max} = 80.3$ psi, $(\sigma_p)_{\min} = -24.7$ psi, $\theta_p = 29°$.

12-15 Failure Theories

One of the first problems encountered in the study of the mechanics of materials concerned the design of a tension member. For this problem there was used an allowable stress based on the yield or ultimate strength of a tensile specimen made of a similar material. This procedure followed a seemingly logical reasoning, because the action of the tension member was like that of the test specimen. Suppose, however, the member to be designed is subjected to a complicated combined stress system, the action of which is different from that of a tensile test. Can the yield or ultimate strength of the tension test still be employed as a basis for securing the maximum allowable design stress? Can the attainment of a certain maximum tensile stress be the only possible method of failure, and the only basis for its design? Would the development of the other stresses, shear and compression, or the resulting strains produce earlier failure? Or of even more significance, does the tensile-test specimen which is used to determine the maximum allowable working stress in tension actually fail in tension?

The complete answers to these questions are still being sought. Various theories, based on experimental data, have been proposed. To date, no one theory has successfully explained the cause of failure in all materials. Some theories work well with brittle materials, others with ductile materials. None is perfect. Let us consider five of the more prominent theories that have been proposed, in the order of their development.

Maximum normal stress theory. The maximum normal stress theory, generally attributed to Lamé and Rankine, is based on the assumption that failure will occur when the maximum normal stress at any point attains the value of the limiting stress obtained by a simple tensile or compression test, regardless of the other stresses acting at the point. Failure may be considered to take place by excessive yielding or by fracture.

The dependency of this theory upon the tensile test is one reason for its inability to predict accurately the failure of ductile materials subjected to combined stress. Observation of a ductile tensile specimen during and after a test reveals that shear stresses contribute heavily to its failure. Test specimens of brittle materials, however, show little or no shear deformation, and the test data obtained from them permit the application of the maximum normal stress theory with satisfactory accuracy.

The failure of an element subjected to axial stresses σ_1 and σ_2 may be shown graphically under the maximum normal stress theory by using a plot of σ_1/σ values as abscissas and σ_2/σ values as ordinates, σ being either the ultimate- or yield-strength value in tension or compression, dependent on what mode of failure will take place. Failure will occur whenever $\sigma_1/\sigma = \pm 1$ and $\sigma_2/\sigma = \pm 1$. This condition is represented by the square plot shown as curve A in Fig. 12-27. According to this theory, any state of biaxial stress represented by a point within this square will not produce failure.

Maximum normal strain theory. The maximum normal strain theory, which was championed principally by Poncelet and St. Venant in the middle of the nineteenth century, is based on the assumption that failure occurs when the maximum normal strain at any point attains the value of the limiting strain obtained by a simple tension or compression test regardless of the stresses or other strains acting at the point. For an element acted on by principal stresses σ_1 and σ_2, it is expressed mathematically by

$$\epsilon_1 \pm \mu\epsilon_2 = \epsilon \quad \text{or} \quad \epsilon_2 \pm \mu\epsilon_1 = \epsilon$$

where ϵ is the strain attained at the proportional limit of a simple tension or compression test. Substituting stresses for the strains,

$$\frac{\sigma_1}{E} \pm \mu\frac{\sigma_2}{E} = \frac{\sigma}{E} \quad \text{and} \quad \frac{\sigma_2}{E} \pm \mu\frac{\sigma_1}{E} = \frac{\sigma}{E}$$

These may be reduced further to

$$\frac{\sigma_1}{\sigma} = 1 \pm \mu\frac{\sigma_2}{\sigma} \quad \text{and} \quad \frac{\sigma_2}{\sigma} = 1 \pm \mu\frac{\sigma_1}{\sigma}$$

Plotted on σ_1/σ and σ_2/σ axes, the graphs of these equations result in a

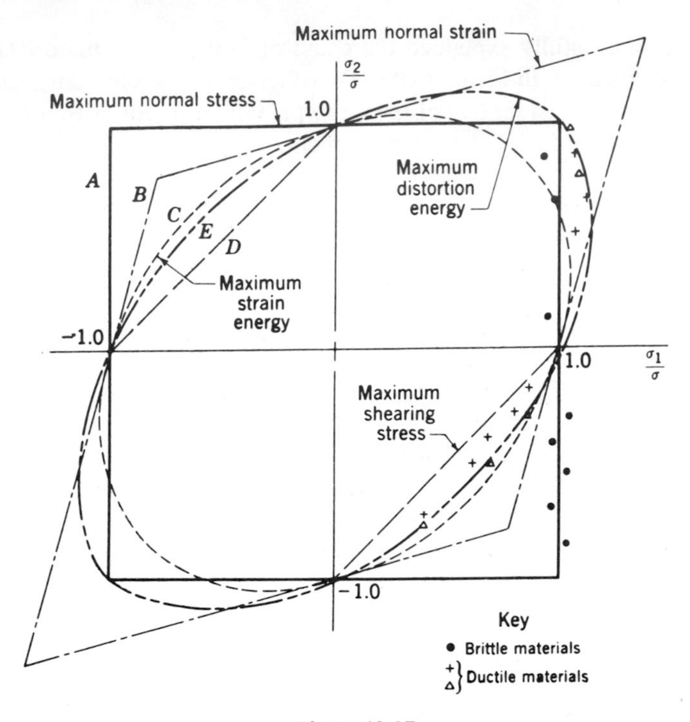

Figure 12-27

diamond-shaped figure (curve B, Fig. 12-27). The figure forms a boundary of limiting stress conditions outside of which failure will result.

Maximum strain energy theory. In an earlier chapter it was shown that when a load acts on an axial member there is stored in each unit volume an amount of strain energy equal to $\sigma^2/2E$, where σ is a stress below the proportional limit created by the load. The maximum strain energy theory, proposed by Maxwell, assumes that elastic failure takes place in a material whenever the strain energy per unit volume at any point exceeds the energy per unit volume absorbed by a tension or compression test specimen of similar material when stressed up to its proportional limit.

An element subjected to the same biaxial state of stress indicated previously produces a unit maximum energy absorption equal to

$$
\begin{aligned}
U &= \frac{1}{2}\sigma_1\epsilon_x + \frac{1}{2}\sigma_2\epsilon_y \\
&= \frac{1}{2}\sigma_1(\epsilon_1 - \mu\epsilon_2) + \frac{1}{2}\sigma_2(\epsilon_2 - \mu\epsilon_1) \\
&= \frac{1}{2}\sigma_1\left(\frac{\sigma_1}{E} - \mu\frac{\sigma_2}{E}\right) + \frac{1}{2}\sigma_2\left(\frac{\sigma_2}{E} - \mu\frac{\sigma_1}{E}\right) \\
&= \frac{1}{2}\frac{\sigma_1^2}{E} + \frac{1}{2}\frac{\sigma_2^2}{E} - \mu\frac{\sigma_1\sigma_2}{E}
\end{aligned}
$$

According to the maximum strain energy theory, this energy produces failure when equal to that obtained in a tension or compression test. Thus,

$$\frac{\sigma_1^2}{2E} + \frac{\sigma_2^2}{2E} - \mu\frac{\sigma_1\sigma_2}{E} = \frac{\sigma^2}{2E}$$

or

$$\sigma_1^2 + \sigma_2^2 - 2\mu\sigma_1\sigma_2 = \sigma^2$$

Dividing through by σ^2 gives

$$\left(\frac{\sigma_1}{\sigma}\right)^2 + \left(\frac{\sigma_2}{\sigma}\right)^2 - 2\mu\left(\frac{\sigma_1}{\sigma}\right)\left(\frac{\sigma_2}{\sigma}\right) - 1$$

If the proportional limit stresses in tension and compression are equal, this equation will form, when plotted graphically, an ellipse, shown as curve C on Fig. 12-27. According to this theory, to remain within the ellipse guarantees elastic action.

Maximum shear stress theory. Proposed by J. J. Guest at the beginning of the twentieth century, the maximum shear stress theory assumes that failure in a material will occur only when the maximum shear stress at any point reaches the value of the shear stress obtained at failure in a simple tension test of the same material. This condition expressed mathematically occurs when

$$\tau_{\max} = \tfrac{1}{2}\sigma$$

In an element under a biaxial state of stress (Fig. 12-28a), where the principal stresses are σ_1, σ_2, and 0, the maximum shear occurs on different planes, depending on which projection of the element will give the maximum difference of the principal stresses. When σ_1 and σ_2 have like signs, the greatest difference will be between the greater of the σ_1 and σ_2 stresses and the zero stress, and will act on planes 45° to them. (See Fig. 12-28c.) Expressed mathematically, the relationship of these stresses at failure is presented by the following equation.

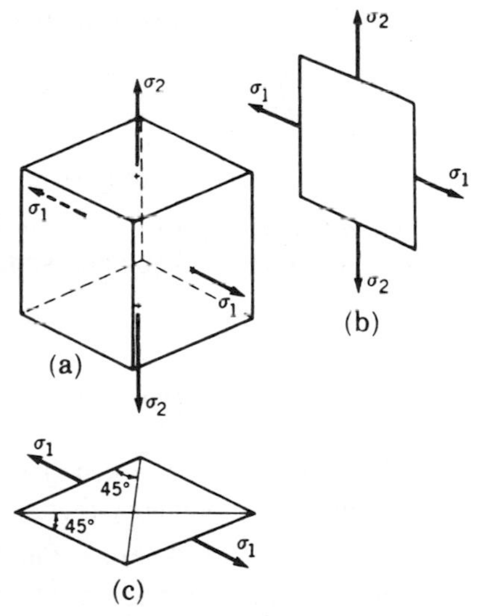

Figure 12-28

$$\pm\frac{\sigma_1 - 0}{2} = \pm\frac{\sigma_1}{2} = \frac{\sigma}{2} \quad \text{or} \quad \sigma_1 = \pm\sigma$$

where σ is the ultimate- or yield-strength value in tension or compression, dependent upon what mode of failure takes place. In a similar manner when σ_2 is greater than σ_1,

$$\sigma_2 = \pm\sigma$$

Thus, when the signs of σ_1 and σ_2 are the same, this theory corresponds to the maximum normal stress theory.

When the signs of σ_1 and σ_2 are different, however, the greatest shear stress equals one half their arithmetic sum (Fig. 12-28b, reverse sign of σ_2 on side projection). Failure is then represented by the equation

$$\tfrac{1}{2}(\pm\sigma_1 \pm \sigma_2) = \tau_{\max} = \tfrac{1}{2}\sigma$$

$$\pm(\sigma_1 + \sigma_2) = \sigma$$

$$\frac{\sigma_1}{\sigma} + \frac{\sigma_2}{\sigma} = \pm 1$$

The graphs of this equation and those above enclose the area shown in Fig. 12-27 by D.

Maximum distortion energy theory. This theory, formulated independently by Von Mises and Hencky, and later supplemented by Huber, assumes that failure under combined stresses occurs when the amount of strain energy per unit volume due to shear (U_c) equals the strain energy per unit volume due to shear produced at the proportional limit of the same material subjected to a simple tension or compression test (U_t). Thus, the equality of energies

$$U_c = U_t$$

in a general state of stress becomes

$$\frac{1}{12G}[(\sigma_1 - \sigma_2)^2 + (\sigma_1 - \sigma_3)^2 + (\sigma_2 - \sigma_3)^2] = \frac{\sigma^2}{6G}$$

or

$$(\sigma_1 - \sigma_2)^2 + (\sigma_1 - \sigma_3)^2 + (\sigma_2 - \sigma_3)^2 = 2\sigma^2$$

in which σ is the proportional limit stress, and σ_1, σ_2, and σ_3 are the principal stresses. When $\sigma_3 = 0$ (biaxial state of stress), this equation becomes

$$\sigma_1^2 - \sigma_1\sigma_2 + \sigma_2^2 = \sigma^2$$

or

$$\left(\frac{\sigma_1}{\sigma}\right)^2 - \left(\frac{\sigma_1}{\sigma}\right)\left(\frac{\sigma_2}{\sigma}\right) + \left(\frac{\sigma_2}{\sigma}\right)^2 = 1$$

A marked similarity is noted between this equation and that derived for the maximum strain energy theory. It is indicated as curve E on Fig. 12-27.

Comparison of Theories. On Fig. 12-27 where the graphs of the five theories are shown superimposed on the same set of axes, the typical results of combined stress tests of various materials are also plotted. Note that the maximum normal stress theory appears to be most correct for brittle materials and the maximum distortion energy theory for ductile materials.

12-16 Machine Shafting

Most shafts are subjected to the combined action of twisting and bending forces. The tension and compression stresses produced by bending, applied simultaneously to the shearing stresses produced by twisting, develop the critical normal and shearing stresses expressed by formulas developed in §12-9. No stresses, with the possible exception of stresses produced at keyways, shrink fits, etc., occur elsewhere in the shaft to exceed them. In order to insure the safety of the shaft, these stresses, or the strain or strain energy that they produce, must be maintained at a safe level. The safe level is customarily determined as some percentage of the controlling feature of one of the five failure theories presented in §12-15.

The design of a typical shaft will now be presented. Each of the five failure theories will be used to determine its controlling diameter.

Illustrative Problem 4. In Fig. 12-29 a solid, circular, steel shaft, 80 in. long, has affixed to it a 30- and an 18-in. pulley and a 14-in. PD spur gear in the position shown. The belt tensions at each pulley are assumed parallel, with $T_2 = 0.3T_1$, and $T_4 = 0.3T_3$. If the driven pulley A is rotated clockwise at 200 rpm by a 20-hp motor and has 16 and 4 hp tapped off at B and C, respectively, what will the diameter of the shaft have to be if based on developing one third of its elastic strength of 42,000 psi for each of the five previously mentioned failure theories? $\mu = 0.30$; $\tau = 0.40\sigma_t$.

SOLUTION. The torques supplied at points A, B, and C are as follows [Eq. (5.6)]:

$$\text{hp} = \frac{2\pi Tn}{12 \times 33{,}000}$$

$$T_A = \frac{12 \times 33{,}000 \times 20}{6.28 \times 200} = 6300 \text{ in.-lb}$$

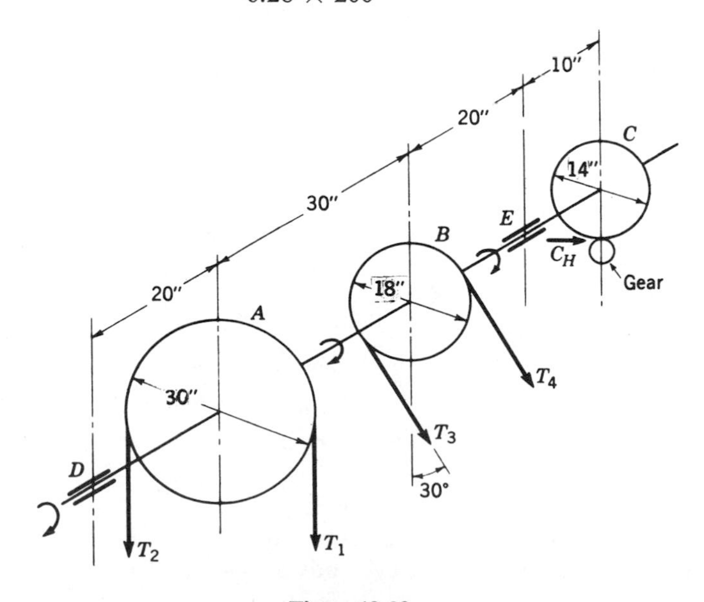

Figure 12-29

Because the torques B and C are respectively proportional to the horsepower tapped off, we may write

$$T_B = 6300 \times \tfrac{16}{20} = 5040 \text{ in.-lb}$$

$$T_C = 6300 \times \tfrac{4}{20} = 1260 \text{ in.-lb}$$

The belt tensions on pulley A equal

$$15.0(T_1 - 0.3T_1) = 6300$$

$$T_1 = 600 \text{ lb}, \qquad T_2 = 180 \text{ lb}$$

The belt tensions on pulley B equal

$$9.0(T_3 - 0.3T_3) = 5040$$

$$T_3 = 800 \text{ lb}, \qquad T_4 = 240 \text{ lb}$$

The horizontal thrust on the gear at C is found from the expression

$$7.0 \times C_H = 1260, \qquad C_H = 180 \text{ lb}$$

The effect of these applied loads in the horizontal and vertical planes is now determined. When the bending moments in each of these planes are found, those providing the greatest resultant effect are used in the determination of the proper size of the shaft.

$$T_1 + T_2 = 600 + 180 = 780 \text{ lb}$$

$$T_3 + T_4 = 800 + 240 = 1040 \text{ lb}$$

$$\text{horizontal component} = 1040 \times 0.5 = 520 \text{ lb}$$

$$\text{vertical component} = 1040 \times 0.866 = 900 \text{ lb}$$

Reactions are:

$$\begin{aligned} \sum M_{D_V} &= 20 \times 780 = 15{,}600 \\ &= 50 \times 900 = \underline{45{,}000} \\ 70E_V &= 60{,}600 \\ E_V &= 866 \text{ lb} \\ D_V &= 814 \text{ lb} \end{aligned}$$

$$\begin{aligned} \sum M_{D_H} &= 50 \times 520 = 26{,}000 \\ & 80 \times 180 = \underline{14{,}400} \\ 70E_H &= 40{,}400 \\ E_H &= 577 \text{ lb} \\ D_H &= 123 \text{ lb} \end{aligned}$$

The critical stress condition, from observation of the two bending-moment diagrams (Fig. 12-30), is evidently at point B. The vector sum of the

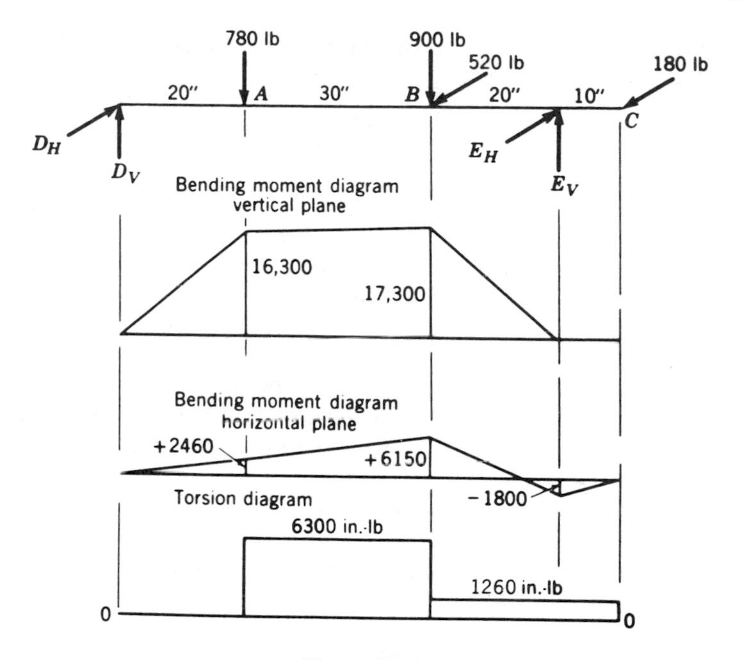

Figure 12-30

bending moments is their resultant and is found by

$$M_R = \sqrt{M_V^2 + M_H^2} = \sqrt{17{,}300^2 + 6150^2} = 18{,}360 \text{ in.-lb}$$

The solution has progressed to the point where the diameter must now be found by each of the five separate failure theories.

Maximum normal stress theory. According to this theory, elastic failure occurs when the maximum normal stress $(\sigma_p)_{\max}$ attains the value of the limiting elastic strength σ_t of the tensile test.

$$\sigma_t = (\sigma_p)_{\max} = \frac{\sigma}{2} + \frac{1}{2}\sqrt{\sigma^2 + 4\tau^2}$$

but

$$\sigma = \frac{Mc}{I}, \quad \tau = \frac{Tr}{J}, \quad I = \frac{\pi d^4}{64}, \quad \text{and} \quad J = \frac{\pi d^4}{32}$$

Therefore,

$$\sigma_t = \frac{16}{\pi d^3}(M + \sqrt{M^2 + T^2})$$

$$\frac{42{,}000}{3} = \frac{16}{3.14 \times d^3}(18{,}360 + \sqrt{18{,}360^2 + 6300^2})$$

$$d = 2.39 \text{ in.}$$

Maximum shear stress theory. According to the maximum shear stress theory, elastic failure occurs when the shearing elastic limit $\tau_{\max} =$

$\frac{1}{2}\sqrt{\sigma^2 + 4\tau^2}$. If $\tau_{max} = 0.4\sigma_t$ and substitutions are made for σ and τ in terms of M and T, then

$$0.4\sigma_t = \frac{16}{\pi d^3}\sqrt{M^2 + T^2}$$

$$5600 = \frac{16}{\pi d^3}\sqrt{\overline{18{,}360}^2 + \overline{6300}^2}$$

$$d^3 = \frac{16 \times 19{,}400}{3.14 \times 5600} = 17.65$$

$$d = 2.61 \text{ in.}$$

Maximum normal strain theory. From the discussion of this theory in §12-15, elastic failure will occur if the maximum normal strain at any point attains the value of the limiting strain obtained by a simple tension test.

$$\epsilon = \epsilon_1 - \mu\epsilon_2 = \frac{\sigma_1}{E} - \mu\frac{\sigma_2}{E}$$

If, now, the values of maximum and minimum principal stresses σ_1 and σ_2 are substituted,

$$E\epsilon = \frac{\sigma}{2} + \frac{1}{2}\sqrt{\sigma^2 + 4\tau^2} - \mu\left(\frac{\sigma}{2} - \frac{1}{2}\sqrt{\sigma^2 + 4\tau^2}\right)$$

$$= \frac{(1 - \mu)}{2}\sigma + \frac{1 + \mu}{2}\sqrt{\sigma^2 + 4\tau^2}$$

A further substitution of Mc/I and Tr/J for σ and τ, respectively, gives

$$E\epsilon = \frac{32}{\pi d^3}(0.35M + 0.65\sqrt{M^2 + T^2})$$

The value of ϵ is the strain that produces elastic failure in the tension test: $42{,}000/E$ psi. If substituted in the above equation, $E\epsilon$ becomes equal to the limiting stress of 42,000 psi. The allowable stress is 42,000/3 or 14,000 psi. Therefore,

$$d^3 = \frac{32}{14{,}000 \times 3.14}(0.35 \times 18{,}360 + 0.65\sqrt{\overline{18{,}360}^2 + \overline{6300}^2})$$

$$d = 2.41 \text{ in.}$$

Maximum strain energy theory. Failure, according to this theory, is attained when the strain energy of any point under combined stress equals the strain energy of the tensile test at the elastic limit: $\sigma_t^2/2E$. This is equal to $\overline{42{,}000}^2/(2 \times 30 \times 10^6)$ or 29.4 in.^{3}. If the elastic limit stress is now divided by the safety factor 3, this value becomes 3.27 in.^{3}. Thus, the design according to this theory states that

$$\frac{(\sigma_t/3)^2}{2E} = 3.27 = \frac{1}{2E}(\sigma_1^2 + \sigma_2^2 - 2\mu\sigma_1\sigma_2)$$

Inserting the equivalents of σ_1 and σ_2 in terms of the initially computed normal and shearing stresses provides

$$2E \times 3.27 = \left(\frac{\sigma}{2} + \frac{1}{2}\sqrt{\sigma^2 + 4\tau^2}\right)^2 + \left(\frac{\sigma}{2} - \frac{1}{2}\sqrt{\sigma^2 + 4\tau^2}\right)^2$$
$$-2\mu\left(\frac{\sigma}{2} + \frac{1}{2}\sqrt{\sigma^2 + 4\tau^2}\right)\left(\frac{\sigma}{2} - \frac{1}{2}\sqrt{\sigma^2 + 4\tau^2}\right)$$

$$6.54E = \frac{\sigma^2}{2} + \frac{1}{2}(\sigma^2 + 4\tau^2) - \frac{\mu\sigma^2}{2} + \frac{\mu}{2}(\sigma^2 + 4\tau^2)$$
$$= \frac{(1-\mu)}{2}\sigma^2 + \frac{(1+\mu)}{2}(\sigma^2 + 4\tau^2)$$

To further simplify the expression for use in designing shafts, let Mc/I and Tr/J be substituted for σ and τ, respectively. There follows

$$\sqrt{6.54E} = \frac{32}{\pi d^3}\sqrt{0.35M^2 + 0.65(M^2 + T^2)}$$

Now, inserting the numerical values of M and T, let us find d.

$$d^3 = \frac{32 \times 19{,}200}{14{,}000 \times \pi}, \qquad d = 2.41 \text{ in.}$$

Maximum distortion energy theory. Placing the fundamental equation of this theory

$$(\sigma_t)^2 = \sigma_1^2 + \sigma_2^2 - \sigma_1\sigma_2 = (\sigma_p)^2_{\max} + (\sigma_p)^2_{\min} - (\sigma_p)_{\max}(\sigma_p)_{\min}$$

in terms of the initially calculated shear and normal stress yields

$$\sigma_t^2 = \sigma^2 + 3\tau^2 = M^2\left(\frac{32}{\pi d^3}\right)^2 + 3T^2\left(\frac{16}{\pi d^3}\right)^2$$

or

$$\sigma_t = \frac{32}{\pi d^3}\sqrt{M^2 + \tfrac{3}{4}T^2}$$

Substituting numerical values with the factor of safety of 3 gives

$$d^3 = \frac{32}{\pi \times 14{,}000}\sqrt{18{,}360^2 + \tfrac{3}{4}6300^2}$$
$$d = 2.40 \text{ in.}$$

Note. The closeness of the answers to this problem indicates a condition of stress represented on Fig. 12-27 by the near intersection of all five failure theories. Change in the relative magnitudes of maximum bending moment and torque, as well as in the allowable normal and shear stresses, would unquestionably have caused greater differences in diameter.

PROBLEMS

12-31. Two pulleys 15 and 12 in. in diameter are placed at the third points of a 6-ft shaft operated at 200 rpm by a 20-hp motor. If one pulley is driven and the other is the driver, what diameter of shaft is necessary if elastic failure occurs in shear at 24,000 psi, and in normal stress at 48,000 psi? Assume the tension on one side of each pulley to be 0.3 of the other. Factor of safety = 3.0. $\mu = 0.3$. Use all five theories presented. *Answer:* 3.10 in.

12-32. Determine the diameter of the shaft shown in Fig. 12-31 if elastic failure occurs at shearing and normal unit stresses of 24,000 and 40,000 psi, respectively. Use factor of safety of 4.0. $\mu = 0.3$. Use all five failure theories presented. *Answer:* 1.388 in.

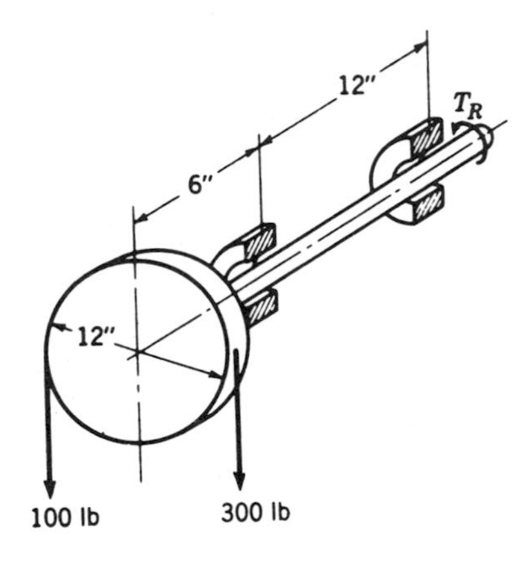

Figure 12-31

12-33. Determine the diameter of the solid circular shaft required to operate the pulley system of Fig. 12-32. Design is governed by a maximum allowable shearing unit stress of 8000 psi. *Answer:* 1.91 in.

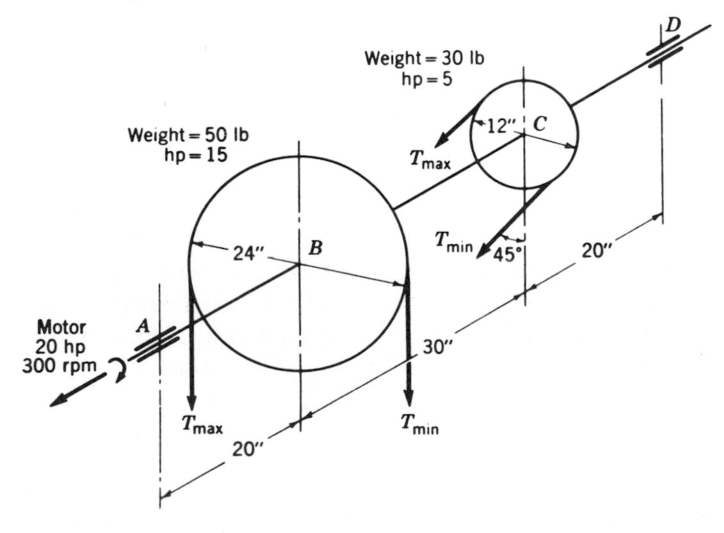

Figure 12-32

12-34. Determine the diameter of the shaft shown in Fig. 12-33 if elastic failure occurs at shearing and normal unit stresses of 24,000 and 42,000 psi, respectively. Use factor of safety of 3.0. $\mu = 0.3$. Use all five theories presented.

12-35. What would be the diameter of the shaft of Problem 12-34 if pulley

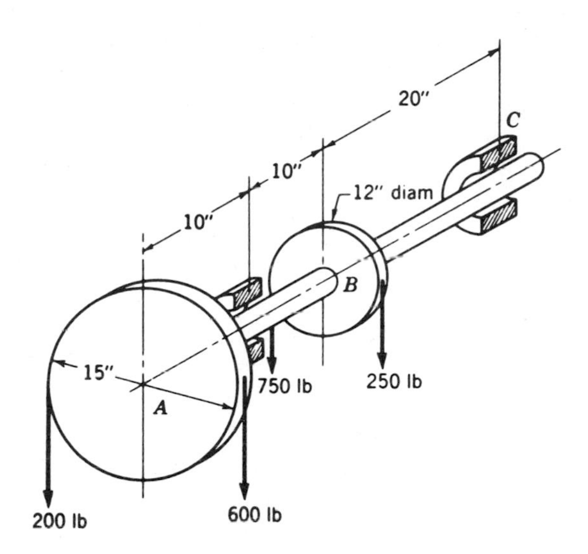

Figure 12-33

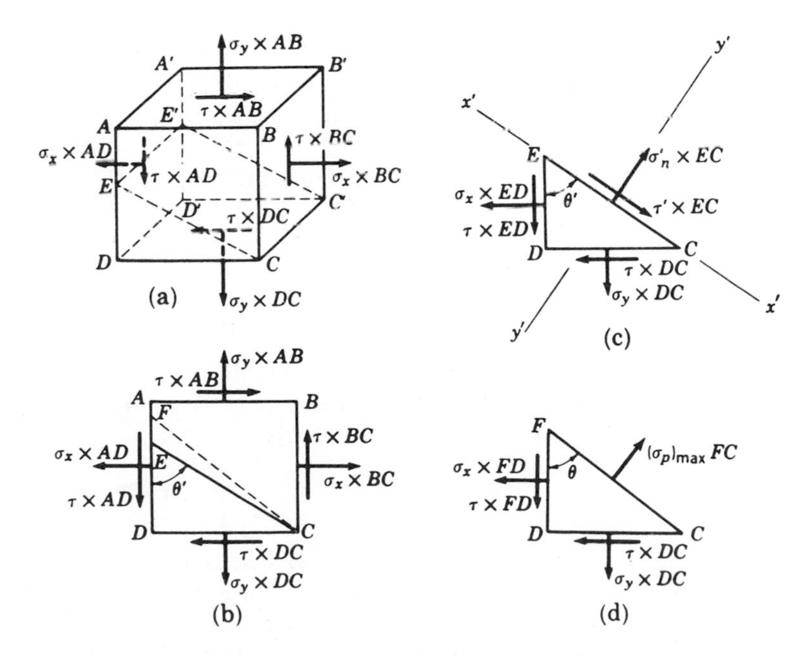

Figure 12-34

B were removed and a vertical load of 4000 lb placed midway between the bearings? The torque produced at A is to be dissipated at C. Consider all design conditions to be the same. Use all five theories presented. *Answer:* 2.67 in.

12-17 The Most General Two-Dimensional Stress System

When an element of a body has acting upon it a coplanar system of shear and normal forces, and when projected has stresses resulting from these forces acting on each of its sides, it is subjected to the most general two-dimensional stress system (see Fig. 12-34a and b). To investigate the possibility of a more critical stress condition existing within the element, a still smaller free body is secured by passing the diagonal plane $ECC'E'$ through it perpendicular to its front face. Using the projected free body (Fig. 12-34c), equilibrium in the y'-y' direction is given by the following expression:

$$\sigma_n' \times EC = \sigma_x \times ED \times \cos\theta' + \sigma_y \times DC \sin\theta' + \tau \times ED \times \sin\theta' + \tau \times DC \times \cos\theta'$$

Dividing each side of the equation by EC and setting $ED/EC = \cos\theta'$ and $DC/EC = \sin\theta'$, there results

$$\sigma_n' = \sigma_x \cos^2\theta' + \sigma_y \sin^2\theta' + \tau \sin 2\theta' \tag{12.21}$$

Similarly, the value of τ' is obtained by equating the components parallel to the x'-x' axis.

$$\tau' \times EC = \sigma_x \times ED \times \sin\theta' - \sigma_y \times DC \times \cos\theta' - \tau \times ED \times \cos\theta' + \tau \times DC \times \sin\theta'$$

$$\begin{aligned}\tau' &= \sigma_x \sin\theta' \cos\theta' - \sigma_y \sin\theta' \cos\theta' - \tau \cos^2\theta' + \tau \sin^2\theta' \\ &= \frac{(\sigma_x - \sigma_y)}{2} \sin 2\theta' - \tau \cos 2\theta'\end{aligned} \tag{12.22}$$

Planes of maximum normal and shear stress. Were the two general equations of normal and shear stress, Eqs. (12.21) and (12.22), now differentiated with respect to θ and set equal to zero, the planes of maximum and minimum intensities would be defined. Those planes defining the maximum and minimum normal stress are indicated by the derivative

$$\tan 2\theta = \frac{2\tau}{\sigma_x - \sigma_y} \tag{12.23}$$

This expression is satisfied by two values of 2θ between 0 and 360°, which are 180° apart. Therefore, the two values of θ indicating the planes of *the maximum and minimum normal stresses are separated by 90°.*

The planes of maximum shear stress, obtained by differentiating Eq.

(12.22), are defined by the expression

$$\tan 2\theta = -\frac{\sigma_x - \sigma_y}{2\tau} \qquad (12.24)$$

This expression, also, is satisfied by two values of 2θ between 0 and 360°, which are 180° apart. The two values of θ which are 90° apart, however, define planes of maximum shear stress.

Were the general shear equation, Eq. (12.22), set equal to zero and solved for θ, the plane of zero shear stress would be obtained. The result of this operation gives

$$\tan 2\theta = \frac{2\tau}{\sigma_x - \sigma_y} \qquad (12.25)$$

which defines the same angles as those for the maximum and minimum normal stress.

It must therefore be concluded, in verification of our previous statement, that *the maximum and minimum normal stresses act on planes where the shearing stress is zero* (Fig. 12-34d).

Maximum and minimum normal stresses. The maximum and minimum normal stresses are obtained by inserting the value θ of the plane on which they occur, Eq. (12.23), in the general normal stress equation, Eq. (12.21). These principal stresses are therefore

$$(\sigma_p)_{\substack{\max \\ \min}} = \frac{\sigma_x + \sigma_y}{2} \pm \sqrt{\left(\frac{\sigma_x - \sigma_y}{2}\right)^2 + \tau^2} \qquad (12.26)$$

Maximum shearing stress. Similarly, the maximum shearing stress is obtained by substituting the value θ of the plane on which that stress occurs, Eq. (12.24), in the general shearing stress equation, Eq. (12.22). There results

$$\tau_{\max} = \pm\sqrt{\left(\frac{\sigma_x - \sigma_y}{2}\right)^2 + \tau^2} \qquad (12.27)$$

The $\pm$ sign in front of the radical indicates that there are two equal maximum shearing stresses which occur in a general two-dimensional stress system. As indicated by Eq. (12.24), these maximum stresses act on planes 90° apart.

Directional considerations. By substituting each of the two angles derived from Eq. (12.23) into Eq. (12.21), an accurate analytical determination can be made to connect the proper principal plane with the maximum or minimum normal stress. A more simple determination, however, may be made by sketching to an approximate scale the resultants of the shear and normal forces on one side of the two diagonals passing through the element. The greater total resultant will be the maximum principal stress, and the plane on which it acts will be the maximum principal plane. See discussion at end of §12-10.

12-18 Mohr's Circle for General Two-Dimensional Stress System

The procedure to be followed in drawing Mohr's circle for the general two-dimensional stress system includes those steps used in drawing previous circle diagrams. The normal stresses σ_x and σ_y are plotted first (Fig. 12-35b),

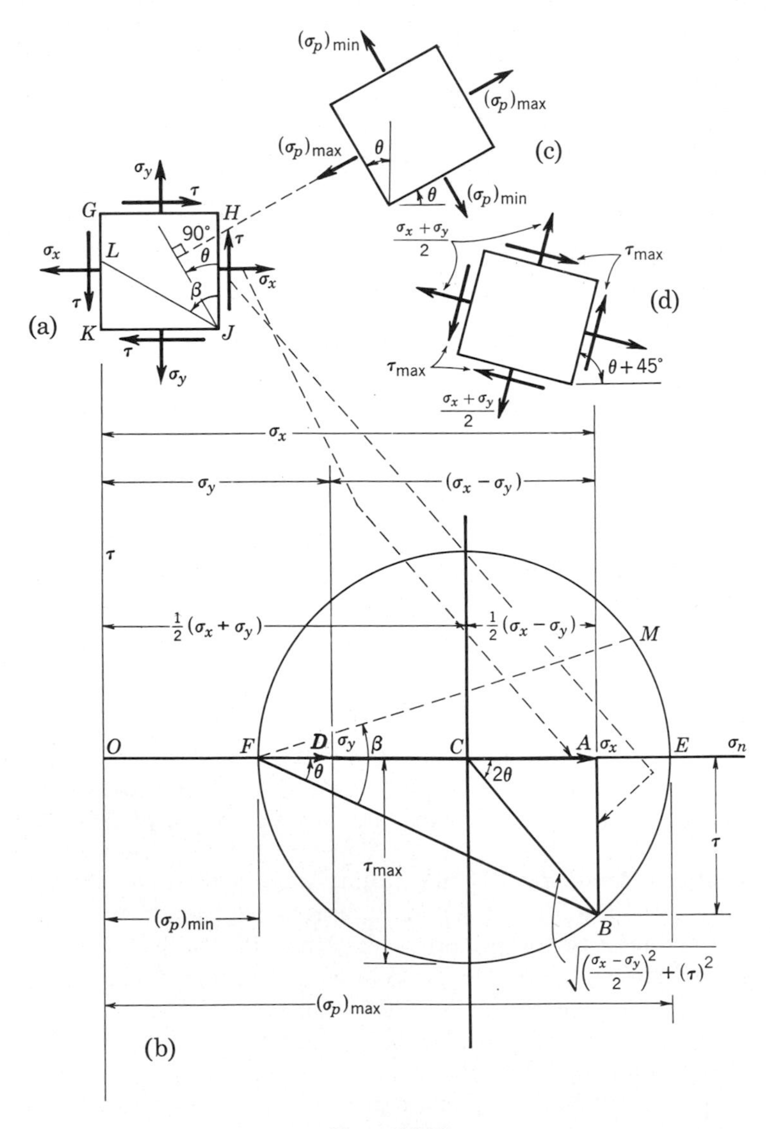

Figure 12-35

care being taken to plot the tensile stresses on the right of the origin and the compression stresses on the left.

The shear stress τ is plotted perpendicular to the x axis at the extremity of σ_x, downward (negatively) in conformity with the sign of the shear couple acting on the planes as σ_x. Then selecting a center on the x axis $\frac{1}{2}(\sigma_x + \sigma_y)$ away from the origin, draw a circle passing through the free end of the shear stress ordinate τ.

The principal stresses, maximum shear stress, and the directions of the planes on which each acts are determined in accordance with the discussion given in §12-12.

Should the shear and normal stress be desired on any plane JL at an angle β measured counterclockwise from plane HJ shown on Fig. 12-35a, the same angle measured counterclockwise from the line FB will provide the line FM whose intersection with the circle at point M provides coordinate shear and normal stresses acting on the plane JL.

12-19 General Triaxial State of Stress

The most general stress condition that may be developed on any element of a body occurs when normal and component shear stresses act on each of its faces (Fig. 12-36a). Because of the three-dimensional character of this

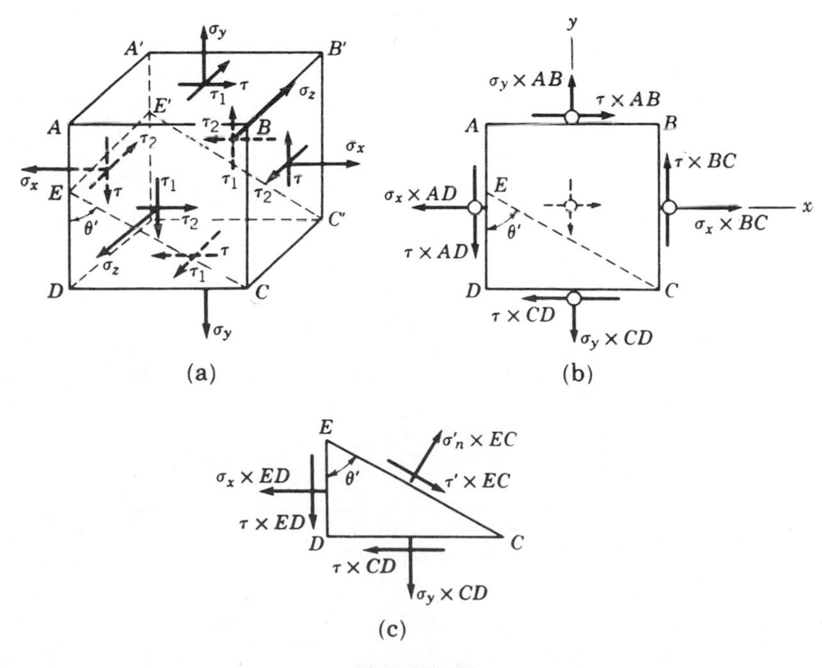

Figure 12-36

stress system, any investigation of a more critical stress condition existing within the element involves complications heretofore unencountered. Consider the small free body (Fig. 12-36c) obtained by passing the diagonal plane $ECC'E'$ through the element perpendicular to its front face. When projected into the XY plane, several of its forces (Fig. 12-36b) are reduced to zero effect. However, these forces can still produce stresses on plane $ECC'E'$, a truth which is revealed by projecting the element into other planes.

Referring momentarily to the elementary body, note that the shear forces created by τ_1 and τ_2 on the front face are balanced by like forces on the rear face and may be eliminated from equilibrium considerations. The force created by τ_2 on the left face, however, has no balancing force except on plane $ECC'E'$ because the right face was removed in obtaining the wedge-shaped free body. A similar condition also exists relative to the shear force developed by τ_1 on the bottom face. It is evident, therefore, that to find the total stress condition their combined effect must be added to the stresses found acting on plane $ECC'E'$ from the two-dimensional analysis. Furthermore, in order to obtain the maximum effect of all the active stresses, plane $ECC'E'$ must be rotated not only about line CC' but also about line EC.

Because of these complications, the determination of the principal stresses within the element due to the general stress system is beyond the scope of this text.

PROBLEMS

12-36. A horizontal cylindrical tank is subjected to an internal pressure which causes a longitudinal tensile stress $\sigma_x = 5000$ psi and a circumferential tensile stress $\sigma_y = 10{,}000$ psi at a point on the external surface. An external torque applied to the cylindrical tank causes a shearing unit stress $\tau = -1500$ psi at the point. Determine the principal stresses (magnitude and sign) and draw a small sketch showing how the principal stresses are inclined with the horizontal (i.e., the x) axis. State the angle in terms of its natural function. *Answers:* $(\sigma_p)_{\max} = 10{,}420$ psi, $(\sigma_p)_{\min} = 4580$ psi, $(\theta_p)_{\min} = +15°30'$.

12-37. In each of the sketches of Fig. 12-37 an element is shown with normal and shearing unit stresses acting on each peripheral face. Determine for each element the magnitude, character, and orientation of the principal and maximum

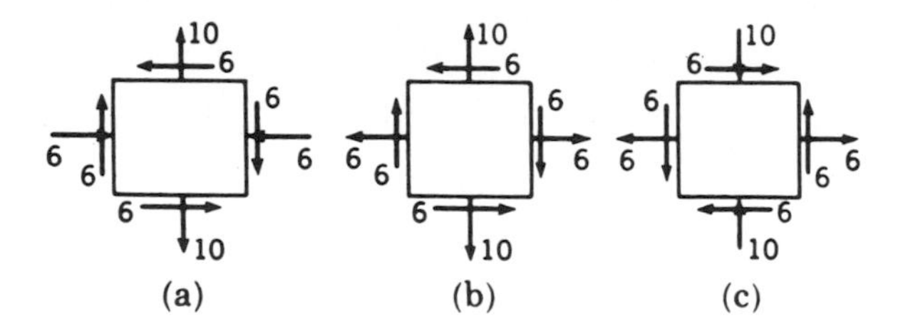

Figure 12-37

shear stresses. Use Mohr's circle for each solution. Find the normal and shear stress on a plane rotated counterclockwise 30° from the right-hand vertical face. Also on the plane rotated clockwise 45° from the same face. *Answers:*

(a) $(\sigma_p)_{\max} = 12.0$, $(\sigma_p)_{\min} = -8.0$, $(\tau)_{\max} = 10.0$, $\theta = \tan^{-1} - \frac{6}{18}$;

(b) $(\sigma_p)_{\max} = 14.3$, $(\sigma_p)_{\min} = 1.7$, $(\tau)_{\max} = 6.3$, $(\theta_p)_{\max} = 54.1°$;

(c) $(\sigma_p)_{\max} = -12.0$, $(\sigma_p)_{\min} = +8.0$, $(\tau)_{\max} = 10.0$, $\theta = \tan^{-1} + \frac{6}{18}$.

12-38. A pipe, 10-in. O.D., $9\frac{1}{2}$-in. I.D., and 6 ft long, has its ends capped. When subjected to an internal pressure of 50.0 psi and a torque of 60,000 in.-lb, what will be the maximum and minimum principal stress and maximum shearing stress appearing on a section 3 ft from one end?

12-39. A thin-walled aluminum tube 6 in. square and 4 ft long is to serve as a cantilever beam while being subjected to an internal pressure of 500 psi (ends capped). If the tube is also loaded axially with a horizontal compressive force of 10,000 lb and a vertical force of 4000 lb at its free end acting downward 24 in. from the vertical plane of symmetry, what is the maximum and minimum principal stress and maximum shearing stress acting at the vertical and horizontal axes of symmetry of the cross section 2 ft from the free end? Thickness of the tube is $\frac{1}{4}$ in. The dimensions of the cross section are taken from the median lines of the sides. Eliminate effect of corners.

COMBINED STRAINS

12-20 Importance of Study of Strains

As revealed in the preceding sections, the theoretical determination of combined stresses and their use in design is, without question, of great importance. Yet the accuracy of their determination is dependent on the degree to which the equations used and the assumptions made in their derivation

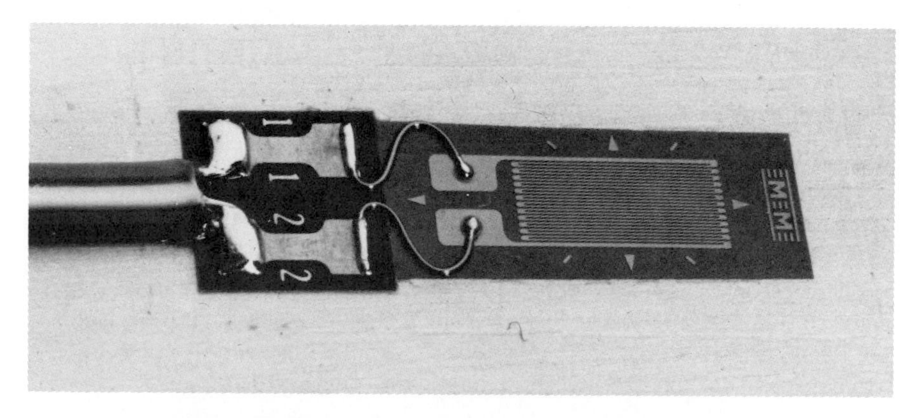

Figure 12-38 Strain gauge for experimental stress analysis. (Courtesy of Micro-Measurements. A Division of Vishay Intertechnology, Inc.)

depict the actual conditions present. It often occurs that the calculated and existing stresses vary considerably. What is even more serious, a theoretical analysis may prove to be very difficult, if not altogether impossible.

Whether theory must be correlated with actual conditions present, or a stress analysis required where a theoretical solution is unobtainable, recourse may be taken to an *experimental* stress analysis based on indirect measurements of strain. These strain measurements are obtained by the correlated measurements of resistance changes of fine wire or semiconductor strain gages (Fig. 12-38) cemented to the surface of the body at the point where the stresses are desired. From a suitable number of such strain measurements, the complete combined stress picture at that point may be obtained. The relationships of strain and stress required to make such a study possible are presented in the ensuing paragraphs.

12-21 Two-Dimensional Strain Relationships

Linear strains. Consider the infinitesimal surface element $ABCD$ shown in Fig. 12-39a to be drawn at a point on an unstressed body and having inscribed upon it line DE, oriented at any angle θ with respect to line DC,

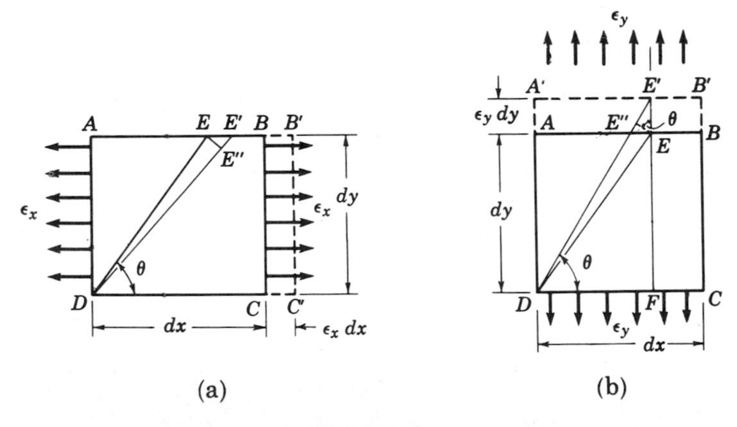

Figure 12-39

an arbitrary datum. If the member is loaded and a unit tensile strain ϵ_x applied to the element, line DE will move to line DE', assuming DA to remain stationary. The total elongation in AE would then equal EE', and if DE'' is made equal to DE, $E'E''$ will be the total elongation in DE. The unit strain applied to line DE will therefore equal

$$\epsilon_\theta = \frac{E'E''}{DE''} = \frac{EE' \cos\theta}{DE} = \frac{EE' \cos\theta}{AE/\cos\theta} = \frac{EE'}{AE}\cos^2\theta = \epsilon_x \cos^2\theta \qquad (12.28)$$

If, in like manner, a unit tensile strain ϵ_y is applied in the y direction (Fig. 12-39b), the unit strain applied to line DE will equal

$$\epsilon_\theta = \frac{E'E''}{DE''} = \frac{EE' \sin\theta}{DE} = \frac{EE' \sin\theta}{EF/\sin\theta} = \frac{EE'}{EF}\sin^2\theta = \epsilon_y \sin^2\theta \qquad (12.29)$$

Considering both tensile strains ϵ_x and ϵ_y to be applied simultaneously, the strain on line DE will be

$$\epsilon_\theta = \epsilon_x \cos^2\theta + \epsilon_y \sin^2\theta \qquad (12.30a)$$

$$= \frac{\epsilon_x + \epsilon_y}{2} + \frac{\epsilon_x - \epsilon_y}{2}\cos 2\theta \qquad (12.30b)$$

It is important to note that ϵ_x and ϵ_y are principal strains and that Eq. (12.30b) has been obtained for the strain of *any* line DE oriented at an angle θ with CD the plane of minimum principal strain.

Shear strains. Let us now consider the magnitude of the shear strain produced by the two principal strains ϵ_x and ϵ_y.

In Fig. 12-40a is shown the infinitesimal element $ABCD$ upon which has been drawn the rectangle $EFGH$. If now the element is subjected to a tensile

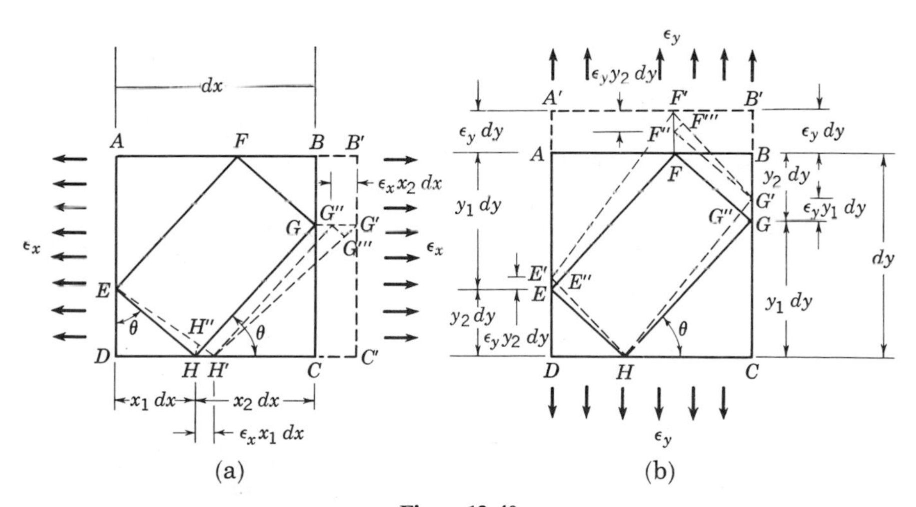

Figure 12-40

strain ϵ_x, the resulting shear strain on a plane making θ to plane CD of the element will be indicated by the angular change in the angles of the rectangle $EFGH$ (§2-12). Let us consider the angle EHG. After being subjected to the horizontal tensile strain of ϵ_x, the angle is increased by the angles $G''H'G'$ and $H'EH$. The sum of these two angles is equal to the shear strain associated with angle θ.

Expressed mathematically, this shear strain is equal to

$$\begin{aligned}\gamma_\theta &= \text{angle } G''H'G' + \text{angle } H'EH \\ &= \frac{G''G'''}{G''H'} + \frac{HH''}{EH} \\ &= \frac{\epsilon_x x_2\, dx \sin\theta}{x_2\, dx/\cos\theta} + \frac{\epsilon_x x_1\, dx \cos\theta}{x_1\, dx/\sin\theta} \\ &= \epsilon_x \sin\theta\cos\theta + \epsilon_x \sin\theta\cos\theta \\ &= \epsilon_x \sin 2\theta\end{aligned} \tag{12.31}$$

In like manner, the shear strain γ_θ due to an applied tensile strain in the y direction (Fig. 12-40b) is equal to

$$\begin{aligned}\gamma_\theta &= \frac{EE''}{HE} + \frac{GG''}{HG} \\ &= \frac{\epsilon_y y_2\, dy \sin\theta}{y_2\, dy/\cos\theta} + \frac{\epsilon_y y_1\, dy \cos\theta}{y_1\, dy/\sin\theta} \\ &= \epsilon_y \sin\theta\cos\theta + \epsilon_y \sin\theta\cos\theta \\ &= \epsilon_y \sin 2\theta\end{aligned} \tag{12.32}$$

Because the tensile strain in the x direction enlarges angle EHG and the tensile strain in the y direction decreases the angle, their combined effect will create a subtraction of the shear strains. Assuming that enlarging the angle produces positive strain,

$$\gamma_\theta = (\epsilon_x - \epsilon_y) \sin 2\theta \tag{12.33}$$

An analysis of Eq. (12.33) reveals that if both tensile principal strains were equal to each other, no shear strain or angular distortion could be developed within its field of action. It is also to be noted that if the principal strains were of different character, one of tension, the other of compression, their corresponding effects would be additive. The maximum shear strain would, however, be found occurring on an angle of 45° to the principal strains.

12-22 Mohr's Circle of Strain

A comparison of Eqs. (12.30b) and (12.33) with those for shear and normal stress developed on diagonal planes by maximum and minimum principal stresses,

$$\sigma_n = \frac{\sigma_1 + \sigma_2}{2} + \frac{\sigma_1 - \sigma_2}{2}\cos 2\theta = \frac{\sigma_x + \sigma_y}{2} + \frac{\sigma_x - \sigma_y}{2}\cos 2\theta \tag{12.1}$$

$$\tau = \frac{\sigma_1 - \sigma_2}{2}\sin 2\theta = \frac{\sigma_x - \sigma_y}{2}\sin 2\theta \tag{12.2}$$

reveals a marked degree of mathematical similarity. The one major difference is the omission of the 2 in the denominator of the shear-strain equation.

In order to make the right-hand sides of both shear equations similar, let the shear-strain equation be divided by 2; thus,

$$\frac{\gamma_\theta}{2} = \frac{\epsilon_x - \epsilon_y}{2} \sin 2\theta \tag{12.34}$$

Now because similar trigonometric relationships exist for σ_n and ϵ_θ, and τ and $\gamma_\theta/2$, a Mohr's circle of strain similar to Mohr's circle of stress may be drawn on rectangular axes ϵ and $\gamma/2$ (Fig. 12-41), having a radius equal to

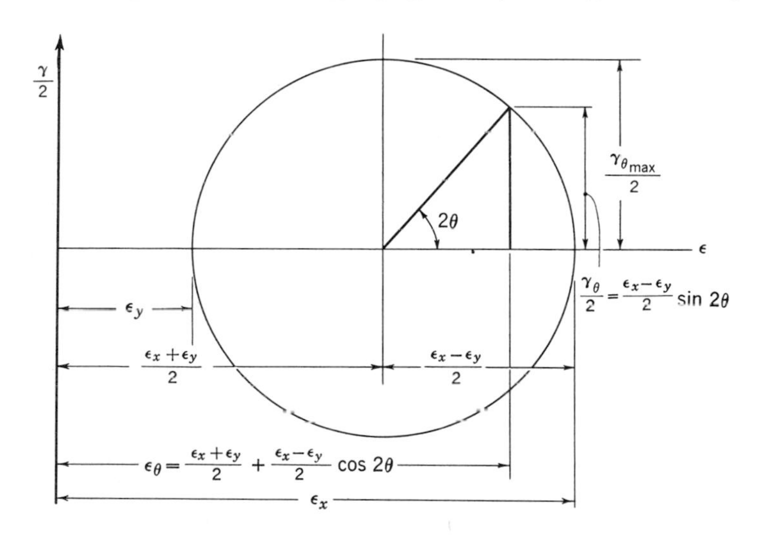

Figure 12-41

$(\epsilon_x - \epsilon_y)/2$ and its center located at $(\epsilon_x + \epsilon_y)/2$. To find the shear and normal strain on any plane making an angle θ with the direction of the minimum principal strain, in this case ϵ_y, a radius is drawn making an angle 2θ to the positive end of the normal strain axis measured counterclockwise. The coordinates to the intersection of the radius and circle equal the normal and shear strains desired.

Illustrative Problem 5. The principal strains at a certain point on a stressed body (Fig. 12-42a) are $\epsilon_x = 0.00060$ in./in. and $\epsilon_y = -0.00040$ in./in. Determine the shear and normal strain on a plane making $\theta = 30°$ with the direction of minimum principal strain. Determine the maximum shear strain.

SOLUTION. On the rectangular axes ϵ and $\gamma/2$ (Fig. 12-42b), plot the value of $\epsilon_x = 0.00060$ and $\epsilon_y = -0.00040$. Find the midpoint of these two values, which equals $(\epsilon_x + \epsilon_y)/2 = 0.00010$, and plot on the ϵ axis. This is the center of the strain circle whose radius is equal to $(\epsilon_x - \epsilon_y)/2 = 0.00050$. After drawing the circle, lay off radius at $2\theta = 60°$. The intercept of the radius with the circle reveals $\epsilon_{30°} = 0.00035$ and $\gamma_{30°} = 0.000866$.

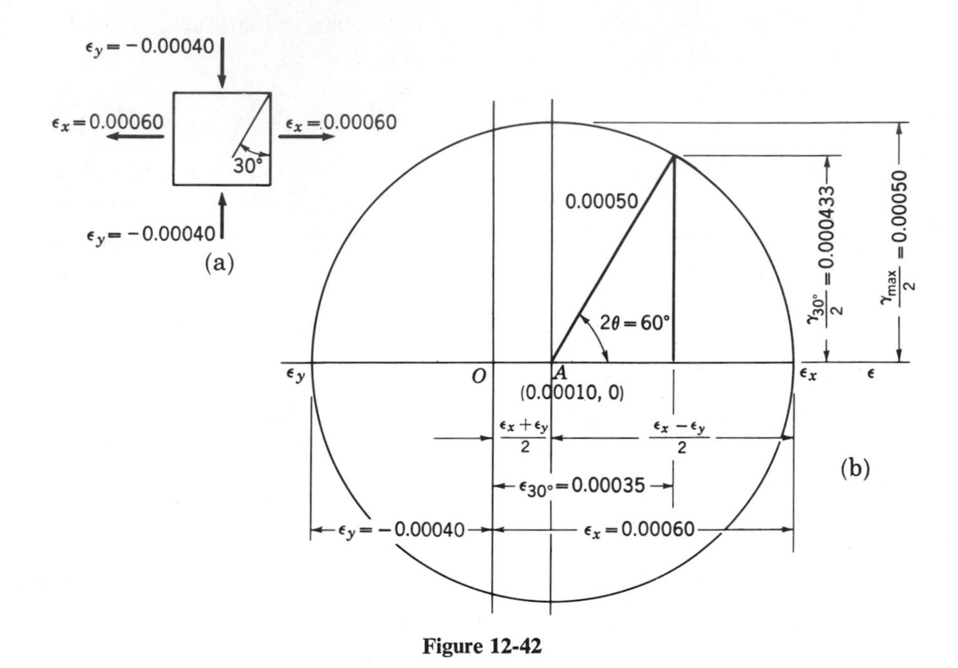

Figure 12-42

The maximum shear strain, which is equal to the diameter of the circle, is equal to 0.001000.

PROBLEMS

12-40. Two mutually perpendicular strains ϵ_x and ϵ_y equal to 0.000600 and −0.000200 occur with no accompanying shear strain. Calculate the strain condition existing on a plane making an angle of 60° with the plane on which the 0.000600 strain acts. *Answers:* $\epsilon_{60°} = 0$, $\gamma_{60°} = 0.000693$.

12-41. The maximum shear strain at a point in a body is equal to 0.000800. If the maximum principal strain is equal to 0.000600, what is the minimum principal strain? *Answer:* 0.000200.

12-42. A square element has the following normal strains acting on its surfaces.

(a) $\epsilon_x = +0.00060 \quad \epsilon_y = -0.00040$
(b) $\epsilon_x = -0.00040 \quad \epsilon_y = +0.00060$
(c) $\epsilon_x = -0.00060 \quad \epsilon_y = -0.00060$
(d) $\epsilon_x = +0.00060 \quad \epsilon_y = +0.00060$

Find the maximum shear strain and the normal and shear strain on a plane rotated 30° counterclockwise from the right-hand vertical face.

12-23 General Two-Dimensional Strain Condition

The most general strain condition that may act on any surface element of a body includes both shear and normal strains on all four faces. Because of the shear strains, the normal strains are not principal strains. The principal strains must therefore be produced at a different orientation from that given.

Development of linear strain. Consider the element shown in Fig. 12-43c acted upon by shear and normal strains. To obtain the linear and angular unit deformation ϵ_θ and γ_θ on any line DE under this general strain condition, those deformations for the biaxial strain condition (Fig. 12-43a) previously considered are superimposed and algebraically added to those produced for the shear condition (Fig. 12-43b), as yet undetermined.

The unit elongation for line DE located on the element subjected to shear as shown in Figs. 12-43b and 12-44a is equal to

$$\epsilon_\theta = \frac{E'E''}{DE''} = \frac{EE' \cos\theta}{DE} = \frac{AA' \cos\theta}{AD/\sin\theta} = \gamma \sin\theta \cos\theta \qquad (12.35a)$$

$$\epsilon_\theta = \gamma \sin\theta \cos\theta = \frac{\gamma}{2} \sin 2\theta \qquad (12.35b)$$

The net total unit elongation for line DE for the general strain condition of Fig. 12-43c is therefore equal to

$$\epsilon_\theta = \epsilon_x \cos^2\theta + \epsilon_y \sin^2\theta + \gamma \sin\theta \cos\theta \qquad (12.36a)$$

$$\epsilon_\theta = \frac{\epsilon_x + \epsilon_y}{2} + \frac{\epsilon_x - \epsilon_y}{2} \cos 2\theta + \frac{\gamma}{2} \sin 2\theta \qquad (12.36b)$$

Development of shear strain. The shear strain associated with angle θ of the same element subjected to shear (Fig. 12-44b) is equal to

$$\begin{aligned}
\gamma_\theta &= -(\alpha - \beta) \\
&= -\left(\frac{E'E''}{E'H} - \frac{GG''}{GH}\right) \\
&= -\left(\frac{EE' \cos\theta}{ED/\cos\theta} - \frac{GG' \sin\theta}{GC/\sin\theta}\right) \qquad (12.37) \\
&= -(\gamma \cos^2\theta - \gamma \sin^2\theta) \\
&= -\gamma(\cos^2\theta - \sin^2\theta) = -\gamma \cos 2\theta
\end{aligned}$$

Adding this strain algebraically to the shear strain obtained for the same element subjected to biaxial strain gives the net shear strain due to the general strain condition:

$$\gamma_\theta = (\epsilon_x - \epsilon_y) \sin 2\theta - \gamma \cos 2\theta \qquad (12.38)$$

Maximum linear and shear strain. The maximum strains occurring in an element subjected to a general strain condition occur, as did the principal

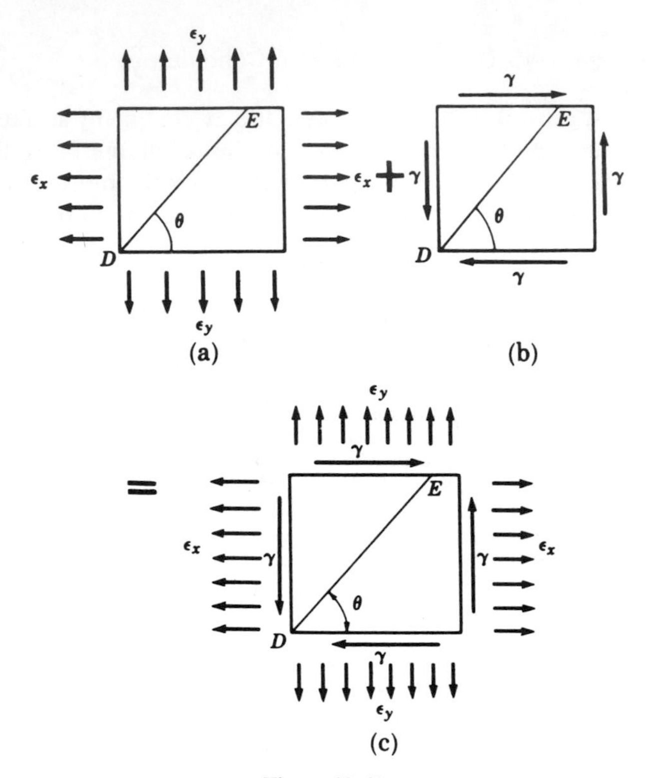

Figure 12-43

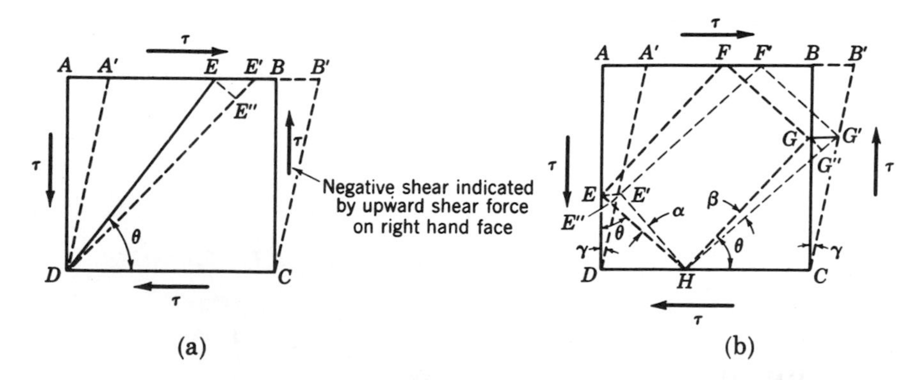

Figure 12-44

stresses, on specific planes, obtained by differentiating the expressions for ϵ_θ and γ_θ with respect to θ. Thus, Eq. (12.36b) becomes, after differentiating,

$$\frac{d\epsilon_\theta}{d\theta} = -\frac{\epsilon_x - \epsilon_y}{2} 2 \sin 2\theta + \frac{\gamma}{2} 2 \cos 2\theta$$

Set equal to zero and solved for tan 2θ, this equation gives

$$\tan 2\theta_p = \frac{\gamma}{\epsilon_x - \epsilon_y} \tag{12.39}$$

The two angles θ_p which satisfy this expression identify the planes on which the maximum or minimum linear strains, otherwise known as the *principal strains*, occur. If the values of sin 2θ and cos 2θ, derived from the expression for tan 2θ, are inserted in the general strain equation (12.36), these principal strains become

$$(\epsilon_p)_{\substack{\max \\ \min}} = \frac{\epsilon_x + \epsilon_y}{2} \pm \sqrt{\left(\frac{\epsilon_x - \epsilon_y}{2}\right)^2 + \left(\frac{\gamma}{2}\right)^2} \tag{12.40}$$

It is of interest to note that the directions of the principal strains correspond to those of the principal stresses.

In a similar manner, the maximum shear strain is found to equal

$$\gamma_{\max} = \sqrt{\left(\frac{\epsilon_x - \epsilon_y}{2}\right)^2 + \left(\frac{\gamma}{2}\right)^2} \tag{12.41}$$

and acts on planes 45° to the planes of principal strain.

Mohr's circle for general strain condition. The expressions for ϵ_θ and γ_θ, as given in Eqs. (12.36) and (12.38), may be represented by a Mohr's circle of strain similar to that explained in the previous article. Thus, for the element subjected to the general strain condition shown in Fig. 12-45a, the values of ϵ_x and ϵ_y are laid off with algebraic magnitude on the ϵ axis, and the $\gamma/2$ value is plotted perpendicular to the ϵ axis at the extremity of ϵ_x downward (negatively) in conformity with the negative sign of the counterclockwise shear-strain couple acting on the same planes as ϵ_x. The value of $(\epsilon_x + \epsilon_y)/2$ is then located (point A) and a circle scribed using AD as a radius. The line OB will then be the maximum principal strain, OC the minimum principal strain, and AE the maximum shear strain.

The previously discussed diagram can be considered as that resulting from the following illustrative problem.

Illustrative Problem 6. At a point on the surface of a stressed body mutually perpendicular normal strains ϵ_x and ϵ_y found to equal 0.00040 and −0.00020, respectively, occur with a negative* shear strain (shown in Fig. 12-45a) of 0.00080. Determine the magnitude of the principal strains and the maximum shear strain, and the angles on which they act.

SOLUTION. Supplementary to the previous discussion, there are plotted on the $\gamma/2$ and ϵ axes, first, the values of ϵ_x (0.00040, 0) and ϵ_y (−0.00020, 0). Then, noting the counterclockwise (negative) rotation of the shear-strain

*A negative shear strain is produced by a shear-strain couple which rotates counterclockwise. Clockwise rotation produces positive shear strain.

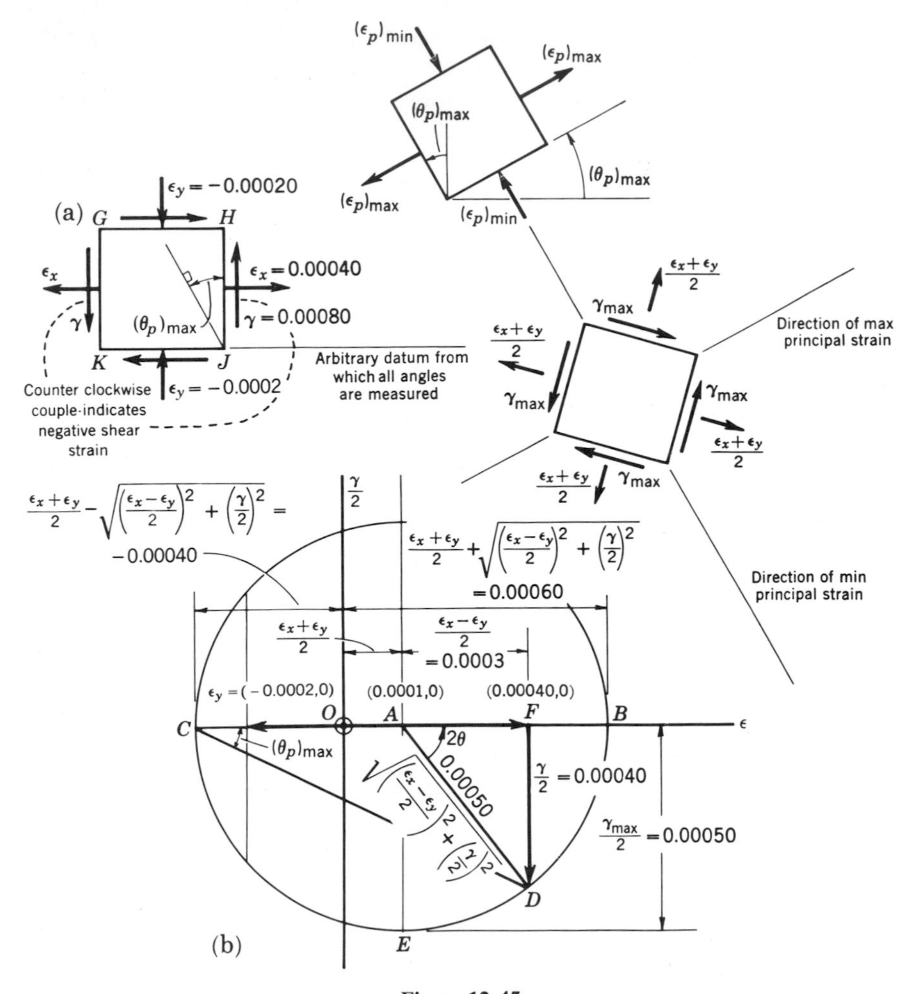

Figure 12-45

couple acting with ϵ_x, the value of $\gamma/2 = -0.00040$ is erected at the extremity of ϵ_x downward (line FD). The midpoint between ϵ_x and ϵ_y, which is equal to $(\epsilon_x + \epsilon_y)/2$, is then plotted at point A (0.00010, 0) and used as the center of a circle whose radius is AD (0.00050). The maximum principal strain is then equal to $OB = 0.00060$ and the minimum principal strain equal to OC or -0.00040. Line AE, the maximum ordinate, is equal to one half the maximum shear strain or 0.00050. The maximum shear strain is therefore equal to 0.00100. Angle DAF is equal to $2(\theta_p)_{max}$, having its tangent equal to

$$\frac{\gamma}{\epsilon_x - \epsilon_y} \quad \text{or} \quad \frac{0.00080}{0.00060} = 1.33$$

The plane of the maximum principal strain is obtained by rotating line *JH* counterclockwise (Fig. 12-45a) through angle *DCA*, the angle located between line *CD* on the Mohr circle representing ϵ_x and γ, and line *OB* representing the maximum principal strain. By geometry, angle *DCA* is equal to $(\theta_p)_{max}$, the angle between the planes perpendicular to the maximum principal strain and ϵ_x. The minimum principal strain occurs at right angles to the maximum principal strain with the maximum shear strain occurring at angles of 45° with each of them.

PROBLEMS

12-43. Two mutually perpendicular strains ϵ_x and ϵ_y are equal to 0.000600 and −0.000200. If they are accompanied by a positive shear strain of 0.000800, determine the maximum and minimum principal strains, the maximum shear strain, and the angles at which each acts relative to the direction of ϵ_x. *Answers:* $(\epsilon_p)_{max} = 0.000766$, $(\epsilon_p)_{min} = -0.000366$, $\gamma_{max} = 0.001133$, $(\theta_p)_{max} = 22.5°$.

12-44. In Fig. 12-43c, ϵ_x, ϵ_y, and γ are respectively equal to 0.000800, 0.000200, and 0.000800. Determine the maximum and minimum principal strains, the maximum shear strain, and the angles at which each acts relative to the direction of ϵ_x.

12-45. In a plane state of strain, the strain components are $\epsilon_x = 500 \times 10^{-6}$, $\epsilon_y = 3500 \times 10^{-6}$, and $\gamma_{xy} = 4000 \times 10^{-6}$. (a) Determine the magnitude and direction of the principal strains. (b) Determine the angles θ measured counterclockwise from the positive x axis of those directions for which the normal strain vanishes. *Answers:* (a) $(\epsilon_p)_{max} = 4500 \times 10^{-6}$, $(\epsilon_p)_{min} = -500 \times 10^{-6}$, $\gamma_{max} = 5000 \times 10^{-6}$.

12-46. Solve Problem 12-44 using ϵ_x, ϵ_y, and γ equal to −0.000800, +0.000200, and 0.000800, respectively. *Answers:* $\epsilon_{max} = -0.000940$, $\epsilon_{min} = 0.000340$, $\gamma_{max} = 0.001280$, $(\theta_p)_{max} = 160°40'$.

12-47. Solve Problem 12-44 using ϵ_x, ϵ_y, and γ equal to 0.000800, −0.000200, and −0.000800, respectively.

MATRIX AND TENSOR ANALYSIS APPLIED TO COMBINED STRESS AND STRAIN

12-24 Introduction

As has been previously shown in this chapter, the determination of the maximum principal stress or strain proceeds from a general state of stress or strain, so oriented that its component values can be obtained by simple, well-established formulas. As the orientation of the element is changed, its component values also change. The mathematical relationships governing these variations reveal maximum and minimum values at specific orientations. The object of *matrix* or *tensor theory* is to obtain the laws of variation of these components, so that specific orientations for which desired stresses or strains occur may be evaluated.

12-25 Fundamentals

A matrix is a rectangular array of values, m rows in height and n columns in width. These values may be of numerical, trigonometric, algebraic, etc., origin—in fact, any kind of values that have a relationship to each other. As an example, a 3×3 square matrix is

$$A = \begin{pmatrix} a_{11} & a_{12} & a_{13} \\ a_{21} & a_{22} & a_{23} \\ a_{31} & a_{32} & a_{33} \end{pmatrix}$$

The parentheses enclosing the array of terms distinguish it from a determinant.

A tensor is a special kind of matrix of physical significance in which each term bears a *rotational* relationship to the other. Tensors are also classified as of zero, first, or second order. If the terms of a tensor are related to a vector quantity, such as force, it is a first-order tensor. If the tensor is related to the combined effect of two vectors, so that their combined unit can be considered to be the product of the units of the separate vectors, e.g., $\text{lb/in.}^2 = \text{lb} \times 1/\text{in.}^2$, then this is a vector of the second order. If a tensor consists of terms which bear no rotational relationship to each other, it is a tensor of zero order. Such a tensor represents a quantity independent of axial orientation, such as temperature, pressure, work, etc.

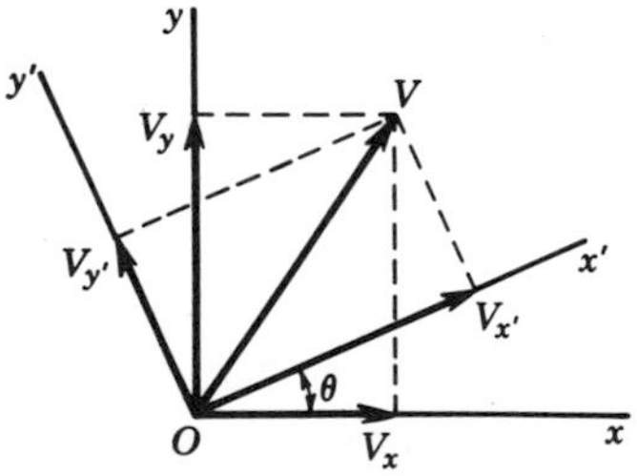

Figure 12-46

First-Order Tensor. Consider a vector V in the xy plane (Fig. 12-46). If its components on the x' and y' axes are desired in terms of those obtained on the x and y axes, there results

$$V'_x = V_x \cos\theta + V_y \sin\theta$$
$$V'_y = -V_x \sin\theta + V_y \cos\theta$$

In tensor notation these equations may be written

$$\begin{pmatrix} V'_x \\ V'_y \end{pmatrix} = \begin{pmatrix} \cos\theta & \sin\theta \\ -\sin\theta & \cos\theta \end{pmatrix} \begin{pmatrix} V_x \\ V_y \end{pmatrix} \tag{12.42}$$

The multiplication of two tensor quantities proceeds by multiplying successively the terms in each corresponding row and column. This is possible only when the number of columns in the first tensor equals the number of rows of the second.

The term $\begin{pmatrix} \cos\theta & \sin\theta \\ -\sin\theta & \cos\theta \end{pmatrix}$, abbreviated as R, is called the rotation matrix and would be the same for any set of rectangular coordinate axes rotated through an angle θ about the origin O. R may also be written as a direction cosine matrix (Fig. 12-47) in which

$$R = \begin{pmatrix} \cos\theta_{x'x} & \cos\theta_{x'y} \\ \cos\theta_{y'x} & \cos\theta_{y'y} \end{pmatrix}$$

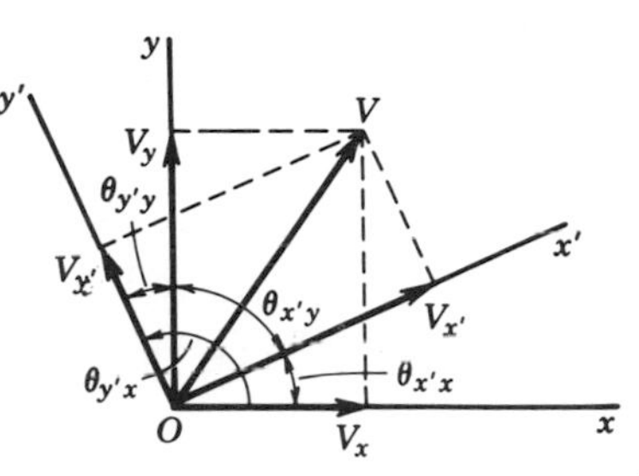

Figure 12-47

The relationship between the two sets of components may be simply expressed as

$$V' = RV \tag{12.43}$$

Interestingly enough, a similar transformation law may be written for the components of alternate sets of three-dimensional vectors (Fig. 12-48), both emanating from the same origin:

$$\begin{pmatrix} V'_x \\ V'_y \\ V'_z \end{pmatrix} = \begin{pmatrix} \cos\theta_{x'x} & \cos\theta_{x'y} & \cos\theta_{x'z} \\ \cos\theta_{y'x} & \cos\theta_{y'y} & \cos\theta_{y'z} \\ \cos\theta_{z'x} & \cos\theta_{z'y} & \cos\theta_{z'z} \end{pmatrix} \begin{pmatrix} V_x \\ V_y \\ V_z \end{pmatrix} \tag{12.44a}$$

or

$$V' = RV \tag{12.44b}$$

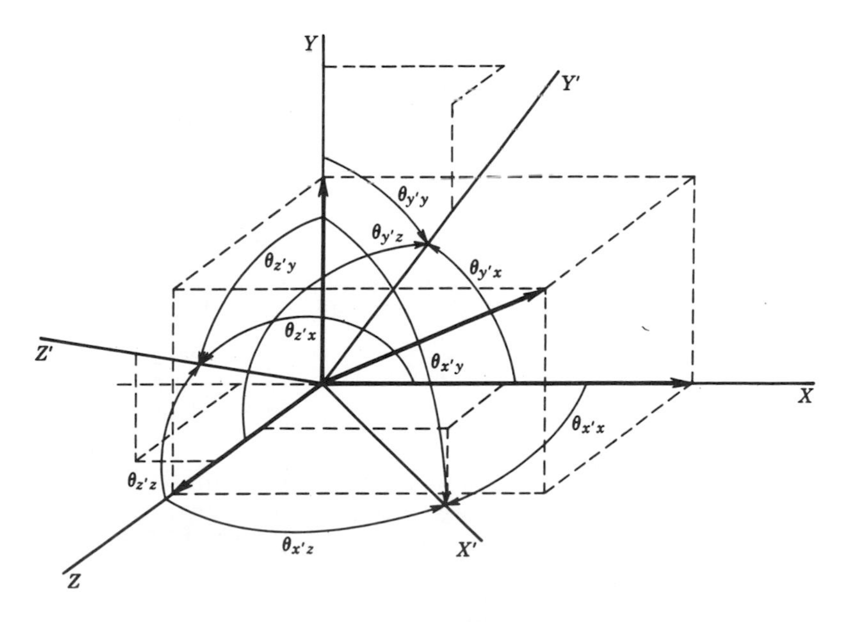

Figure 12-48

The transpose of a tensor A is designated as A^* and is obtained by interchanging the rows and columns of A. Thus, if

$$A = \begin{pmatrix} a_{11} & a_{12} & a_{13} \\ a_{21} & a_{22} & a_{23} \end{pmatrix}, \qquad A^* = \begin{pmatrix} a_{11} & a_{21} \\ a_{12} & a_{22} \\ a_{13} & a_{23} \end{pmatrix} \tag{12.45}$$

Second-Order Tensor. A second-order tensor is the simple algebraic product of a vector and the transpose of another. Thus, for vectors $U = (u_x u_y u_z)$ and $V = (v_x v_y v_z)$ their second-order tensor is $T = U^* V$ and is obtained by arranging the individual vector tensors as

$$\begin{pmatrix} u_x \\ u_y \\ u_z \end{pmatrix} (v_x v_y v_z) \tag{12.46}$$

and multiplying successively the term in each row of the first tensor by the term in each column of the second. Thus,

$$T = \begin{pmatrix} u_x v_x & u_x v_y & u_x v_z \\ u_y v_x & u_y v_y & u_y v_z \\ u_z v_x & u_z v_y & u_z v_z \end{pmatrix} \tag{12.47}$$

Let us now proceed to find the rotational transformation law for the second-order tensor from the first-order tensor transformation equations $U' = RU$ and $V' = RV$. From these equations we obtain

$$U'^* = RU^*$$

$$V'^* = RV^*$$

It is also simple to show that $V' = VR^*$, so that

$$U'^* V' = RU^* VR^*$$

from which

$$T' = RTR^* \tag{12.48}$$

In this expression, as well as in that for first-order tensors, $V' = RV$, means are available for expressing relationships between the components of any force or vector relative to any set of mutually perpendicular axes rotated about their origin.

12-26 Stress Tensor

If a cubical element subjected to a general state of stress is rotated into different orientations, the stresses which act on its faces change. At any one orientation these stresses may be presented as a second-order tensor:

$$T = \begin{pmatrix} \sigma_{xx} & \tau_{xy} & \tau_{xz} \\ \tau_{yx} & \sigma_{yy} & \tau_{yz} \\ \tau_{zx} & \tau_{zy} & \sigma_{zz} \end{pmatrix} \tag{12.49}$$

A different orientation would provide a T' tensor with all stresses modified but related through the equation $T' = RTR^*$.

12-27 Strain Tensor

In a similar manner the strains acting upon any cubical element can be presented for any one orientation as

$$\eta = \begin{pmatrix} \epsilon_{xx} & \frac{1}{2}\gamma_{xy} & \frac{1}{2}\gamma_{xz} \\ \frac{1}{2}\gamma_{yx} & \epsilon_{yy} & \frac{1}{2}\gamma_{yz} \\ \frac{1}{2}\gamma_{zx} & \frac{1}{2}\gamma_{zy} & \epsilon_{zz} \end{pmatrix} \tag{12.50}$$

12-28 Properties of a Tensor

Thus far we have described the tensor and noted its directional properties. Two direct inferences can be made from these properties which are of considerable importance in engineering analyses. These are given below without proof.

a. *Diagonalization of a Symmetrical Tensor.*† In our previous discussion of combined stresses and strains we found one orientation of the reference element for which the shearing values become zero and the three normal values are maximum, minimum, or intermediate. The corresponding tensors appear as

$$T_1 = \begin{pmatrix} (\sigma_x)_{\max} & 0 & 0 \\ 0 & (\sigma_y)_{\max} & 0 \\ 0 & 0 & (\sigma_z)_{\max} \end{pmatrix}, \qquad \eta = \begin{pmatrix} (\epsilon_x)_{\max} & 0 & 0 \\ 0 & (\epsilon_y)_{\max} & 0 \\ 0 & 0 & (\epsilon_z)_{\max} \end{pmatrix} \tag{12.51}$$

The axes which provide this condition are, of course, the principal axes.

b. *Invariants of the Tensor.* As tensors are rotated, three combinations of their elements never change. These are the invariants of the tensor and are as follows:

1. The sum of the terms of the main diagonal is a constant regardless of rotation. For example, in the second-order stress tensors, Eqs. (12.49) and (12.50),

$$I_1 = \sigma_{xx} + \sigma_{yy} + \sigma_{zz} = \sigma_{x'x'} + \sigma_{y'y'} + \sigma_{z'z'} \tag{12.52}$$

and

$$I_1 = \epsilon_{xx} + \epsilon_{yy} + \epsilon_{zz} = \epsilon_{x'x'} + \epsilon_{y'y'} + \epsilon_{z'z'} \tag{12.53}$$

†A symmetrical tensor is a square matrix in which the element bears the relationship $a_{ij} = a_{ji}$, where i and j are the row and column subscripts, respectively.

2. The sum of the principal two row minors is a constant, regardless of rotation.

$$I_2 = \begin{vmatrix} \sigma_{xx} & \tau_{xy} \\ \tau_{yx} & \sigma_{yy} \end{vmatrix} + \begin{vmatrix} \sigma_{yy} & \tau_{yz} \\ \tau_{zy} & \sigma_{zz} \end{vmatrix} + \begin{vmatrix} \sigma_{xx} & \tau_{xz} \\ \tau_{zx} & \sigma_{zz} \end{vmatrix} = \begin{vmatrix} \sigma_{x'x'} & \tau_{x'y'} \\ \tau_{y'x'} & \sigma_{y'y'} \end{vmatrix} + \begin{vmatrix} \sigma_{y'y'} & \tau_{y'z'} \\ \tau_{z'y'} & \sigma_{z'z'} \end{vmatrix} + \begin{vmatrix} \sigma_{x'x'} & \tau_{x'z'} \\ \tau_{z'x'} & \sigma_{z'z'} \end{vmatrix} \tag{12.54}$$

3. The determinant of the matrix is a constant, regardless of rotation.

$$I_3 = \begin{vmatrix} \sigma_{xx} & \tau_{xy} & \tau_{xz} \\ \tau_{yx} & \sigma_{yy} & \tau_{yz} \\ \tau_{zx} & \tau_{zy} & \sigma_{zz} \end{vmatrix} = \begin{vmatrix} \sigma_{x'x'} & \tau_{x'y'} & \tau_{x'z'} \\ \tau_{y'x'} & \sigma_{y'y'} & \tau_{y'z'} \\ \tau_{z'x'} & \tau_{z'y'} & \sigma_{z'z'} \end{vmatrix} \tag{12.55}$$

Thus, it is shown that as a cubical stress (or strain) element takes different orientations within a loaded member, the stresses (or strains) acting on its faces have a definite relationship to one another dependent on the angular rotations between them. The shorthand representation of the tensor provides a convenient and effective method of identifying and handling the general stress (or strain) pattern without the danger of overlooking any one element. In a 2×2 tensor, there are only two invariants.

$$I_1 = \sigma_{xx} + \sigma_{yy} = \sigma_{x'x'} + \sigma_{y'y'} \tag{12.56}$$

and

$$I_2 = \begin{vmatrix} \sigma_{xx} & \tau_{xy} \\ \tau_{yx} & \sigma_{yy} \end{vmatrix} = \begin{vmatrix} \sigma_{x'x'} & \tau_{x'y'} \\ \tau_{y'x'} & \sigma_{y'y'} \end{vmatrix} \tag{12.57}$$

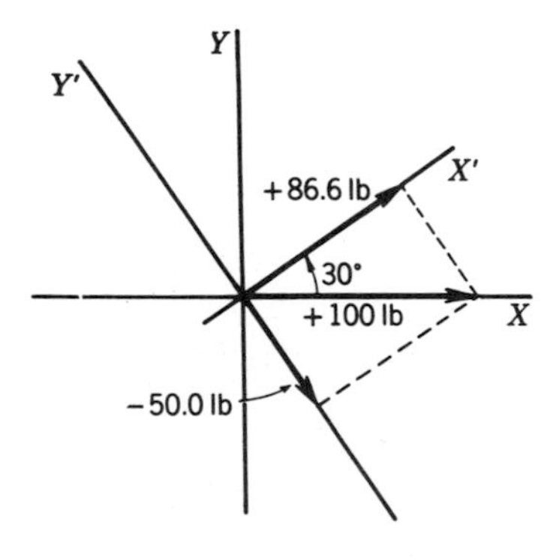

Figure 12-49a

Illustrative Problem 7. Determine the components of a force of 100 lb acting positively along the x axis, with reference to the x' and y' axes rotated 30° counterclockwise from the respective x and y axes (Fig. 12-49a).

SOLUTION. Components related to two sets of mutually perpendicular axes separated by an angle θ are interrelated by the Eq. (12.43).

$$V' = RV$$

or

$$\begin{pmatrix} V'_x \\ V'_y \end{pmatrix} = \begin{pmatrix} \cos\theta_{x'x} & \cos\theta_{x'y} \\ \cos\theta_{y'x} & \cos\theta_{y'y} \end{pmatrix} \begin{pmatrix} V_x \\ V_y \end{pmatrix}$$

Inserting the values of the angle obtained from Fig. 12-49 in the direction cosine matrix,

$$\begin{pmatrix} \cos 30^\circ & \cos 60^\circ \\ \cos 120^\circ & \cos 30^\circ \end{pmatrix} = \begin{pmatrix} 0.866 & 0.500 \\ -0.500 & 0.866 \end{pmatrix}$$

Using this expression in the full equation, we have

$$\begin{pmatrix} V'_x \\ V'_y \end{pmatrix} = \begin{pmatrix} 0.866 & 0.500 \\ -0.500 & 0.866 \end{pmatrix}\begin{pmatrix} 100 \\ 0 \end{pmatrix} = \begin{pmatrix} 0.866 \times 100 + 0.500 \times 0 \\ -0.500 \times 100 + 0.866 \times 0 \end{pmatrix}$$

$$= \begin{pmatrix} 86.6 \\ -50.0 \end{pmatrix}$$

Illustrative Problem 8. Determine the stress components on the faces of the two-dimensional square element $A'B'C'D'$ rotated on angle 30° clockwise with respect to the element $ABCD$ shown in Fig. 12-49b.

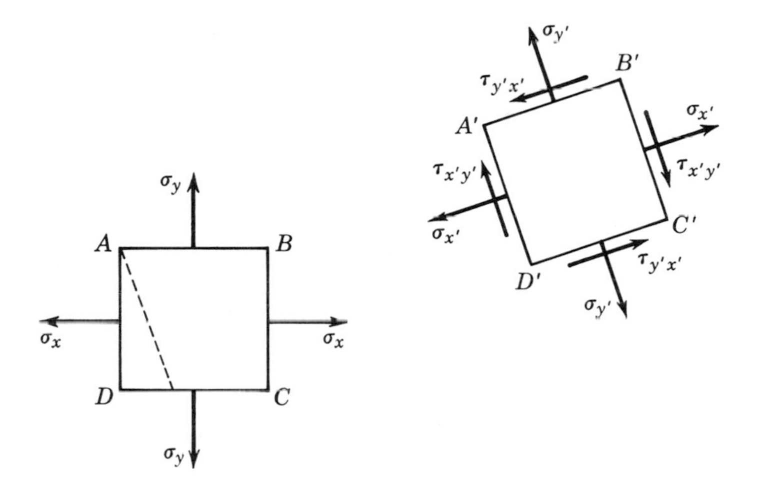

Figure 12-49b

SOLUTION. Because this problem concerns tensors of the second order and is of a two-dimensional character,

$$T' = RTR^*$$

$$= \begin{pmatrix} \cos 30^\circ & \sin 30^\circ \\ -\sin 30^\circ & \cos 30^\circ \end{pmatrix}\begin{pmatrix} \sigma_x & 0 \\ 0 & \sigma_y \end{pmatrix}\begin{pmatrix} \cos 30^\circ & -\sin 30^\circ \\ \sin 30^\circ & \cos 30^\circ \end{pmatrix}$$

Multiplying the first two terms using the numerical values of the angles, we obtain

$$T' = \begin{pmatrix} 0.866\sigma_x & 0.500\sigma_y \\ -0.500\sigma_x & 0.866\sigma_y \end{pmatrix}\begin{pmatrix} \cos 30^\circ & -\sin 30^\circ \\ \sin 30^\circ & \cos 30^\circ \end{pmatrix}$$

Completing the multiplication of the tensors,

$$T' = \begin{pmatrix} 0.750\sigma_x + 0.250\sigma_y & -0.433\sigma_x + 0.433\sigma_y \\ -0.433\sigma_x + 0.433\sigma_y & +0.250\sigma_x + 0.750\sigma_y \end{pmatrix} = \begin{pmatrix} \sigma_{x'} & \tau_{x'y'} \\ \tau_{y'x} & \sigma_{y'} \end{pmatrix}$$

It is to be noted that the resulting tensor reveals the effect of σ_x and σ_y on the stresses in the rotated position. If σ_x and σ_y were equal to each other, say 1.0, the $\sigma_{x'}$ and $\sigma_{y'}$ terms would also equal 1.0, and the shear terms would equal zero.

Illustrative Problem 9. What are the components of stress acting on the three-dimensional stress element of Fig. 12-49c if it is rotated 30° clockwise about the y axis?

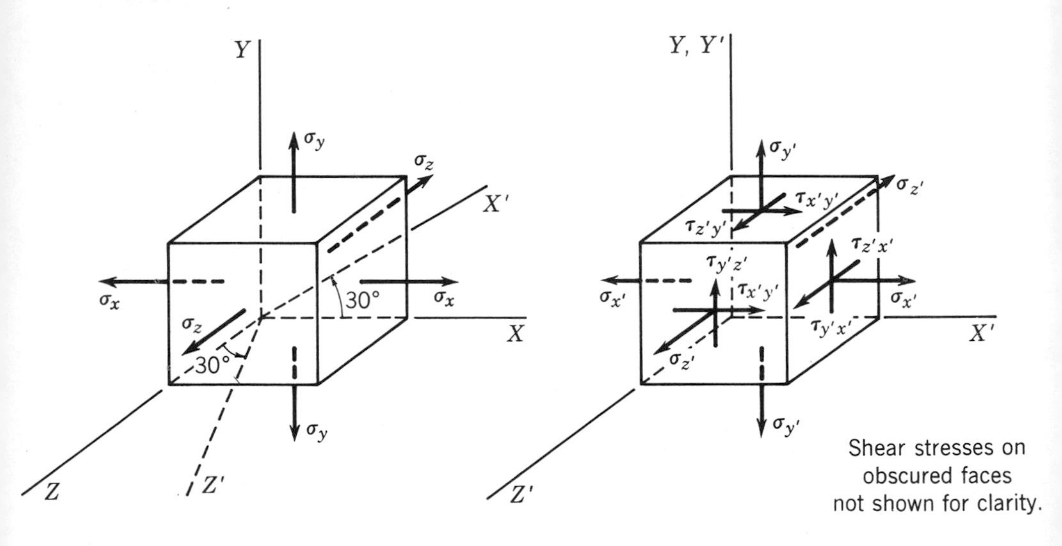

Figure 12-49c

SOLUTION. This problem also concerns a second-order tensor and requires the use of the transformation law $T' = RTR^*$. Applied to a three-dimensional stress situation,

$$R = \begin{pmatrix} \cos\theta_{x'x} & \cos\theta_{x'y} & \cos\theta_{x'z} \\ \cos\theta_{y'x} & \cos\theta_{y'y} & \cos\theta_{y'z} \\ \cos\theta_{z'x} & \cos\theta_{z'y} & \cos\theta_{z'z} \end{pmatrix} = \begin{pmatrix} \cos 30^\circ & \cos 90^\circ & \cos 120^\circ \\ \cos 90^\circ & \cos 0^\circ & \cos 90^\circ \\ \cos 60^\circ & \cos 90^\circ & \cos 30^\circ \end{pmatrix}$$

Substituting in the transformation law,

$$T' = RTR^*$$
$$= \begin{pmatrix} 0.866 & 0 & -0.500 \\ 0 & 1 & 0 \\ 0.500 & 0 & 0.866 \end{pmatrix} \begin{pmatrix} \sigma_x & 0 & 0 \\ 0 & \sigma_y & 0 \\ 0 & 0 & \sigma_z \end{pmatrix} \begin{pmatrix} 0.866 & 0 & 0.500 \\ 0 & 1 & 0 \\ -0.500 & 0 & 0.866 \end{pmatrix}$$

Multiplying the first two tensors, we obtain

$$T' = \begin{pmatrix} 0.866\sigma_x & 0 & -0.500\sigma_z \\ 0 & \sigma_y & 0 \\ 0.500\sigma_x & 0 & 0.866\sigma_z \end{pmatrix} \begin{pmatrix} 0.866 & 0 & 0.500 \\ 0 & 1 & 0 \\ -0.500 & 0 & 0.866 \end{pmatrix}$$

Thus, completing the multiplication,

$$T' = \begin{pmatrix} 0.750\sigma_x + 0.250\sigma_z & 0 & 0.433\sigma_x - 0.433\sigma_z \\ 0 & \sigma_y & 0 \\ 0.433\sigma_x - 0.433\sigma_z & 0 & 0.250\sigma_x + 0.750\sigma_z \end{pmatrix}$$
$$= \begin{pmatrix} \sigma_{x'} & \tau_{x'y'} & \tau_{x'z'} \\ \tau_{y'x'} & \sigma_{y'} & \tau_{y'z'} \\ \tau_{z'x'} & \tau_{z'y'} & \sigma_{z'} \end{pmatrix} \tag{12.58}$$

Corresponding terms in the tensors of Eq. (12.58) reveal that if the initial element had $\sigma_x = \sigma_y = \sigma_z = 1.0$ the normal stresses in the rotated position would also equal 1.0, and the shear stresses would equal zero. This would describe a hydrostatic stress situation.

PROBLEMS

12-48. Show that R^*R for first- and second-order tensors develops into a tensor whose main diagonal consists of terms equal to 1 and all other terms equal zero.

12-49. The components of force V are $V_x = +500$ lb and $V_y = +300$. What are the components related to axes x' and y' rotated 45° counterclockwise?

12-50. If the components in Problem 12-49 are $V_x = 500$ and $V_y = -300$, what will the components be related to the same x' and y' axes?

12-51. Normal stresses of 100 psi act on the x and y axes of a cubical element located at the intersection of the x, y, and z axis frame. Find the components acting on the x', y', and z' axis frame which is located by counterclockwise rotation about the z axis through an angle of 45° from the x axis. Use the relationship for second-order tensors, $T' = RTR^*$.

12-52. Normal stresses of 100 psi act on the x and y axes of a cubical element located at the intersection of the x, y, z axis frame. If this orthogonal frame of axis is rotated about the origin so that the $+x$ axis (now $-x'$) is located at equal angles from the $+x$, $+y$, and $+z$ axis, find the components of the two stresses on the x', y', and z' axes. Use the relationship $T' = RTR^*$.

12-53. For the two-dimensional system σ_x, σ_y, τ_{xy}, and τ_{yx} prove that $T' =$

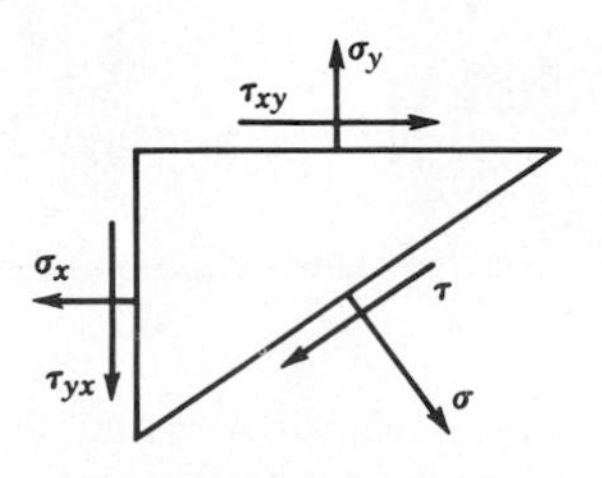

Figure 12-50

*RTR** by considering the equilibrium of forces on the faces of the element in Fig. 12-50 and taking T to be a tensor of the second order.

EXPERIMENTAL DETERMINATION OF STRESS

12-29 Brittle-Lacquer Method

A widely used method of determining the presence of critical stresses on a body subjected to load employs the use of brittle lacquer sprayed on its surface. If the lacquer is applied before loading and permitted to dry thoroughly, it becomes brittle and relatively weak in tension. When the load is applied, the lacquer cracks wherever the strain attains its threshold value; i.e., the strain causes cracking. The cracks are always perpendicular to the direction of the principal stresses (Fig. 12-51).

The quantitative measure of the stress or strain produced at the time the lacquer cracks is obtained by spraying the lacquer on a cantilever calibration beam at the same time and under the same conditions as the test body. At the time of the test the calibration beam is depressed a known amount at its free end; this causes the lacquer to crack over a portion of its total length, starting at its fixed end, where the bending moment and stress will be greatest,

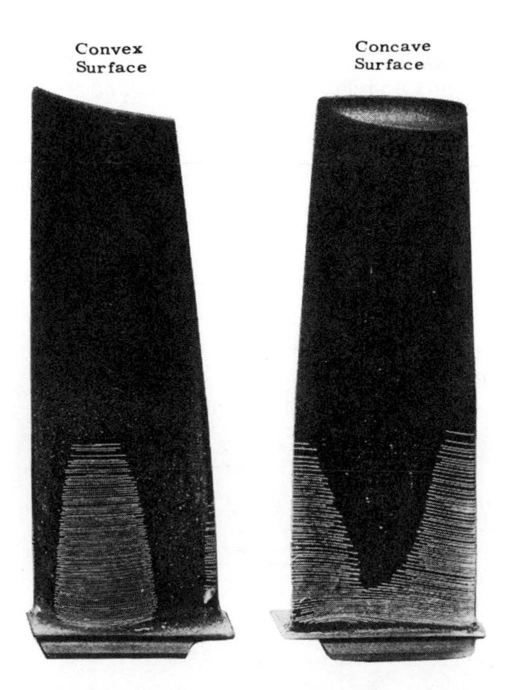

Figure 12-51 Stresscoat all-temp patterns on a turbine blade. Brought out by application of Statiflux electrostatic powder method. (Courtesy of Magnaflux Corporation, Chicago, Illinois.)

and extending to a point where the strain is just sufficient to produce cracking, i.e., the threshold value. This strain and its corresponding stress may be computed by using the equations $\Delta = Pl^3/3EI$, $\sigma = Mc/I$, and $\sigma = \epsilon E$.

Thus, during a load test the progress of the cracking in the lacquer is carefully noted. The threshold value occurring at the edge of the cracked zone indicates the same strain that developed at the threshold value on the calibrating beam. If the value of E is the same for calibrating beam and test body, their stresses will also be the same. Otherwise, a simple calculation for stress in the test body can be made, knowing its value of modulus.

A similar approach may be made in the determination of excessive compressive stresses by applying and drying the lacquer after the test specimen has been loaded. The release of the load will cause the lacquer to crack, revealing compressive strains.

12-30 Experimental Determination of Stresses by Electric Resistance Gages—Strain Rosette

The purpose of a strain gage is to measure the deformation that occurs over the test length of the gage, from which the unit strain may be calculated. Of the many kinds of strain gages now available, one of the most refined and widely applicable is the electric resistance gage. It consists of a grid of very fine wire (Fig. 12-38) which is bonded to the surface of a member at a point where a stress study is desired. When the member is subjected to load, the gage deforms exactly the same amount as the surface to which it is attached. Because wire, when deformed longitudinally, changes its resistance to the flow of electricity, and because the deformation per inch and the corresponding resistance change can easily be calibrated, the strain at the point in question may be obtained by measuring the change in resistance of the gage through the use of a Wheatstone bridge.

Should a resistance gage be bonded lengthwise to a tensile test bar which is thereafter loaded axially, the unit stress in the bar would be equal to the product of its longitudinal unit strain and modulus of elasticity. If, however, the same gage were to be placed in a biaxial state of strain in the direction of one of the principal strains (Fig. 12-52), the product of strain reading and modulus would not be equal to the principal stress. Its stress would be influenced by the action of the lateral strain, as indicated previously in Eq. (3.9).

$$(\sigma_p)_{\max} = \frac{(\epsilon_x + \mu\epsilon_y)E}{1 - \mu^2} = \frac{E[(\epsilon_p)_{\max} + (\mu\epsilon_p)_{\min}]}{1 - \mu^2} \tag{12.59a}$$

$$(\sigma_p)_{\min} = \frac{(\epsilon_y + \mu\epsilon_x)E}{1 - \mu^2} = \frac{E[(\epsilon_p)_{\min} + (\mu\epsilon_p)_{\max}]}{1 - \mu^2} \tag{12.59b}$$

Thus, to obtain the principal stresses occurring in a biaxial state of strain, two strain gages would have to be placed at the point in question, in the

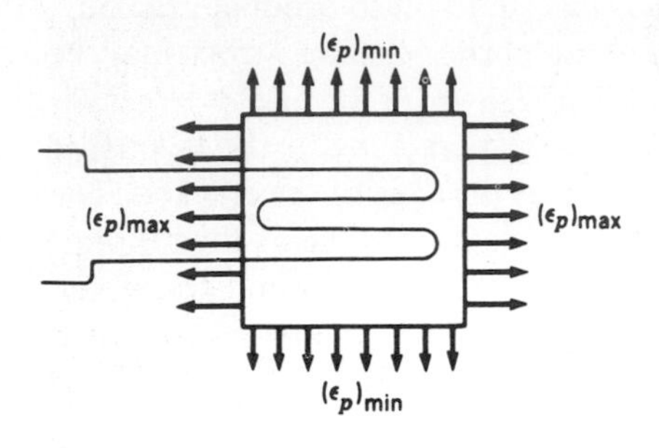

Figure 12-52

direction of the principal strains. The simultaneously obtained strain readings introduced into Eqs. (12.59a and b) give the principal stresses acting at the point.

Strain rosette. Ordinarily, however, the directions of the principal strains are not discernible on the surface of a body subjected to combined stress. Either an alternative means must first be employed to determine their orientation—such as the use of a brittle coating (see §12-29)—or, as is commonly done, a more general use of strain gages is instituted, for which no prior knowledge of strain direction is required.

This more general method is based on the threefold use of Eq. (12.36a), at the designated point. Thus, the equations

UNKNOWNS

$$\epsilon_{\theta 1} = \epsilon_x \cos^2\theta_1 + \epsilon_y \sin^2\theta_1 + \gamma \sin\theta_1 \cos\theta_1 \tag{12.60}$$

$$\epsilon_{\theta 2} = \epsilon_x \cos^2\theta_2 + \epsilon_y \sin^2\theta_2 + \gamma \sin\theta_2 \cos\theta_2 \tag{12.61}$$

$$\epsilon_{\theta 3} = \epsilon_x \cos^2\theta_3 + \epsilon_y \sin^2\theta_3 + \gamma \sin\theta_3 \cos\theta_3 \tag{12.62}$$

MEASURABLE

indicate for such a point the linear strains along three different lines making angles of θ_1, θ_2, and θ_3 *with some arbitrary axis*. The values of $\epsilon_{\theta 1}$, $\epsilon_{\theta 2}$, and $\epsilon_{\theta 3}$ are linear strains measured by means of three resistance gages superimposed on one another. These gages comprise a unit called a *strain rosette* (Fig. 12-53). Occasionally a fourth gage is added to serve as a check on the measurements obtained by the other three.

Because the values of ϵ_θ and θ are measurable quantities, the three common unknowns ϵ_x, ϵ_y, and γ may be obtained by the solution of these three simultaneous equations. The values of ϵ_x, ϵ_y, and γ, in turn, may be used either in a Mohr's circle of strains or in Eqs. (12.39) and (12.40) to obtain the principal strains and their orientation. Once these strains are obtained, the principal stresses are computed from Eqs. (12.59a and b).

To simplify the calculation of the values of ϵ_x, ϵ_y, and γ, the gages of the strain rosette are set at predetermined angles to each other, the trigonometric values of which are well known. The angles most frequently used are 0, 45, and 90°. The name assigned to the combination of gages set at these angles is the *rectangular strain rosette* (Fig. 12-54).

Figure 12-53 Strain rosettes being used in experimental stress analysis. (Courtesy of Baldwin-Lima-Hamilton Corp.)

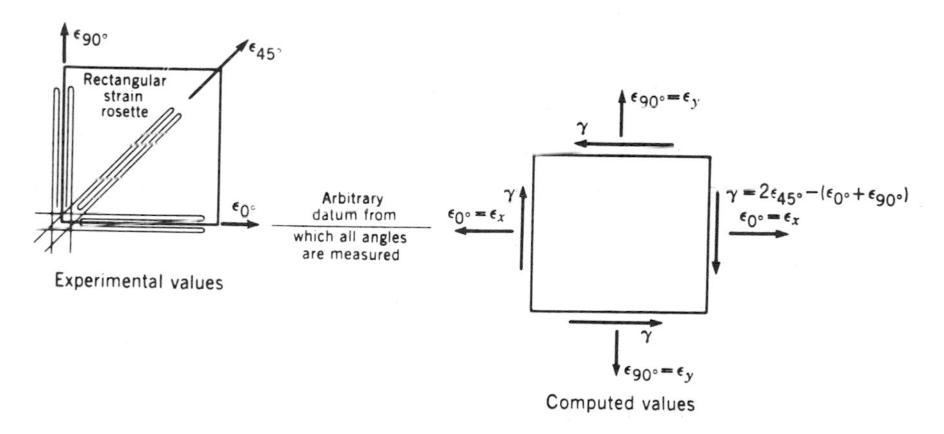

Figure 12-54

Inserting these angles in Eqs. (12.60), (12.61), and (12.62) results in

$$\epsilon_{0°} = \epsilon_x, \qquad \epsilon_{90°} = \epsilon_y \tag{12.63}$$

$$\epsilon_{45°} = \frac{\epsilon_x}{2} + \frac{\epsilon_y}{2} + \frac{\gamma}{2}$$

from which

$$\gamma = 2\epsilon_{45^\circ} - (\epsilon_x + \epsilon_y)$$

and

$$\gamma = 2\epsilon_{45^\circ} - (\epsilon_{0^\circ} + \epsilon_{90^\circ}) \tag{12.64}$$

Another strain rosette used employs the angles of 0, 60, and 120°. This is the *equiangular strain rosette*. Should these values be substituted into Eqs. (12.60), (12.61), and (12.62) and properly simplified, the result would be

$$\epsilon_x = \epsilon_{0^\circ} \tag{12.65}$$

$$\epsilon_y = \frac{1}{3}(2\epsilon_{60^\circ} + 2\epsilon_{120^\circ} - \epsilon_{0^\circ}) \tag{12.66}$$

$$\gamma = \frac{2}{\sqrt{3}}(\epsilon_{60^\circ} - \epsilon_{120^\circ}) \tag{12.67}$$

The use of a strain rosette may best be illustrated by a practical problem.

Illustrative Problem 10. The principal stresses are desired at a certain critical location on a web of a steel crankshaft. The strain readings of its three gages are $\epsilon_{0^\circ} = 0.00040$, $\epsilon_{45^\circ} = 0.00050$, and $\epsilon_{90^\circ} = -0.00020$. $\mu = 0.30$. $E = 30 \times 10^6$ psi.

SOLUTION. From the simplified linear strain equations, there are obtained from the strain readings

$$\epsilon_x = \epsilon_{0^\circ} = 0.00040$$

$$\epsilon_y = \epsilon_{90^\circ} = -0.00020$$

$$\begin{aligned}\gamma &= 2\epsilon_{45^\circ} - (\epsilon_{0^\circ} + \epsilon_{90^\circ})\\ &= 2 \times 0.00050 - (0.00040 - 0.00020)\\ &= +0.00080 \qquad \text{(acts upward on left-hand face of element)}\end{aligned}$$

To obtain the principal strain relative to this condition, use may be made of a Mohr's circle of strains (shown in Fig. 12-45) or direct computation may be made using Eq. (12.40). Using the latter method,

$$\begin{aligned}(\epsilon_p)_{\max} &= \frac{0.00040 - 0.00020}{2} + \sqrt{\left(\frac{0.00040 + 0.00020}{2}\right)^2 + \left(\frac{0.00080}{2}\right)^2}\\ &= 0.00060\end{aligned}$$

$$\begin{aligned}(\epsilon_p)_{\min} &= \frac{0.00040 - 0.00020}{2} - \sqrt{\left(\frac{0.00040 + 0.00020}{2}\right)^2 + \left(\frac{0.00080}{2}\right)^2}\\ &= -0.00040\end{aligned}$$

The principal stresses may now be obtained using Eqs. (12.59a and b)

$$\begin{aligned}(\sigma_p)_{\max} &= \frac{E}{1 - (0.30)^2} \times [0.00060 + 0.3(-0.00040)]\\ &= \frac{30 \times 10^6}{0.91} \times 0.00048 = 15{,}820 \text{ psi}\end{aligned}$$

$$(\sigma_p)_{min} = \frac{E}{1 - (0.30)^2}[-0.00040 + 0.3(0.00060)]$$
$$= 7260 \text{ psi}$$

The direction of the maximum principal strain (stress) is given by Eq. (12.39).

$$\tan 2\theta = \frac{\gamma}{\epsilon_x - \epsilon_y} = \frac{0.00080}{0.00040 + 0.00020} = 1.33$$
$$2\theta = 49°, \qquad \theta = 24.5°$$

PROBLEMS

12-54. The strain readings obtained on the 0, 45, and 90° gages of a rectangular strain rosette are, respectively, 0.000300, 0.000400, and 0.000100. Determine the principal stresses, the maximum shear stress, and the angles on which each acts relative to the direction of the 0° gage. $E = 30 \times 10^6$ psi. $\mu = 0.30$. *Answers:* $(\sigma_p)_{max} = 13{,}750$ psi, $(\sigma_p)_{min} = 3400$ psi, $\tau_{max} = 5180$ psi, $(\theta_p)_{max} = 31°43'$.

12-55. Solve Problem 12-54 using gage readings of −0.000300, 0.000400, and +0.000100.

12-56. Solve Problem 12-54 using gage readings of −0.000300, −0.000400, and +0.000100. *Answers:* $(\sigma_p)_{max} = -12{,}600$ psi, $(\sigma_p)_{min} = +1030$ psi, $\tau_{max} = 8320$ psi, $(\theta_p)_{max} = 28°9'$.

12-57. An equiangular strain rosette placed on the surface of a stressed machine part reveals strains along the $\epsilon_{0°}$, $\epsilon_{60°}$, and $\epsilon_{120°}$ directions equal to 0.000600, 0.000200, and −0.000400, respectively. If $E = 10.6 \times 10^6$ psi and $\mu = 0.33$, determine the principal stresses and their directions relative to the assumed 0° direction. *Answers:* $(\sigma_p)_{max} = 6740$ psi, $(\sigma_p)_{min} = -2530$ psi, $\tau_{max} = 4635$ psi, $(\theta_p)_{max} = 18°18'$.

12-58. Solve Problem 12-57 using gage readings of −0.000600, 0.000200, and −0.000400.

12-59. Solve Problem 12-57 using gage readings of 0.000600, −0.000200, and 0.000400. *Answers:* $(\sigma_p)_{max} = 8070$ psi, $(\sigma_p)_{min} = 428$ psi, $\tau_{max} = 3830$ psi, $(\theta_p)_{max} = -23°5'$.

12-60. Will two strain gages placed parallel and perpendicular to the longitudinal axis of a thin-walled cylinder give readings in the ratio of 1:2 when the cylinder is under pressure? Give reason for your answer.

12-61. The strain readings obtained on the 0, 45, and 90° gages of a rectangular strain rosette are, respectively, −0.000300, −0.000400, and +0.000100. Determine the principal stresses, the maximum shear stress, and the angles on which each acts. $E = 30 \times 10^6$ psi. $\mu = 0.30$. *Answers:* $(\sigma_p)_{max} = -11{,}580$ psi, $(\sigma_p)_{min} = 3000$ psi, $\tau_{max} = 7290$ psi.

12-62. In a plane state of stress ($\sigma_z = \tau_{zx} = \tau_{zy} = 0$), a strain rosette in the xy plane measures the following strains: $\epsilon_{0°} = 0.002$, $\epsilon_{45°} = 0.001$, $\epsilon_{90°} = -0.0005$. (a) Determine the principal strains and directions. (b) Assuming that the material is isotropic linearly elastic and that $\sigma_x = 20{,}000$ psi, $\sigma_y = 0$, determine the modulus of elasticity E, Poisson's ratio μ, and the shearing stress τ_{xy}.

12-31 Grid Method

One of the simplest methods of detecting surface strains on the surface of a member is through the use of a grid applied to the surface (Fig. 12-55).

Figure 12-55 Commercial test using grid analysis. Elastomer model of tenons cast into rim of turbine disc. It was desired to determine the possible development of tensile stresses at the base of the tenons when three successive tenons were loaded with similar couples. A superimposed exposure was made at zero and full load. No tensile deformation was noted at the base of the tenon.
(Courtesy of Curtiss-Wright Corp., Wood-Ridge, N.J.)

The grid which deforms with the surface upon application of load is analyzed to determine the amount of normal end shear strain. For example, if one square grid element $A_1B_1C_1D_1$ of Fig. 12-56a remains a square or assumes a rectangular shape $A_2B_2C_2D_2$ after loading, it follows that only normal stresses could have acted on its sides, and the average strain is equal to

$$\epsilon_x = \frac{(x_2' - x_2) - (x_1' - x_1)}{x_2 - x_1} = \frac{u_2 - u_1}{\Delta x} \tag{12.68}$$

$$\epsilon_y = \frac{(y_2' - y_2) - (y_1' - y_1)}{y_2 - y_1} = \frac{v_2 - v_1}{\Delta y} \tag{12.69}$$

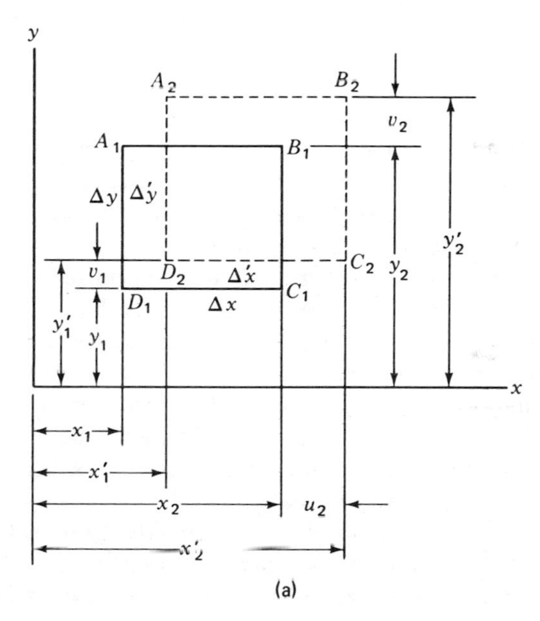

(a)

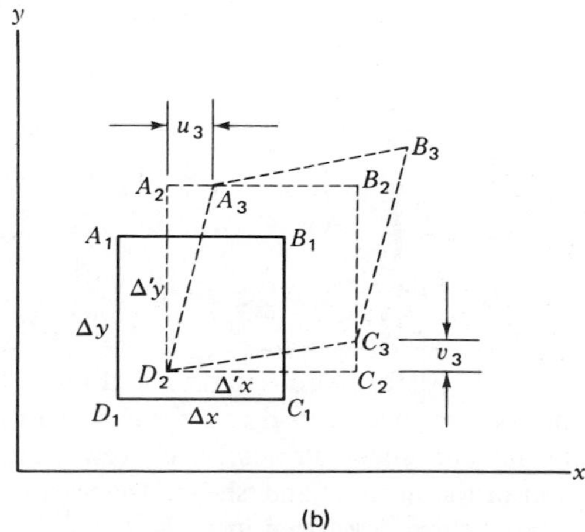

(b)

Figure 12-56

If the distorted square element $A_2B_2C_2D_2$ is further deformed into a rhombus $A_3B_3C_3D_2$ (Fig. 12-56b) retaining the same length of sides, only shear stresses could have produced the subsequent deformation. The average shear strain is equal to the change in the right angle at D:

$$\gamma = \sphericalangle A_2D_2A_3 + \sphericalangle C_3D_2C_2 = \frac{v_3}{\Delta x'} + \frac{u_3}{\Delta y'} \tag{12.70}$$

Should these normal and shear strains be superimposed and develop simultaneously, element $A_1B_1C_1D_1$ would deform into slope $A_3B_3C_3D_2$ with no intermediate stage. To determine the normal and shear strains contributing to its deformation would require a reverse procedure to that used for its step-by-step development; i.e., draw $A_2B_2C_2D_2$ and then determine the strains.

The accuracy of the grid method is dependent upon the grid-line spacing. The smaller the specimen, and the smaller the discontinuity to be studied on that specimen, the smaller the grid-line spacing, and the finer the lines must be. Grids of 100 lines/in. are not uncommon. On large specimens, grids of up to $\frac{1}{2}$ to 1 in. are frequently used.

Various methods are used for applying the grid to a specimen. Photographic printing is probably the most common method. Commercially prepared grids on thin, easily stretchable films cemented to the surface, as well as etching, scribing, and embossing, are others.

The grid method is more widely used for problems entailing large displacement and large strain. For instance, grids are frequently used for studies of models made from rubber, or rubber-like materials, care being taken to keep the deformations relatively small.

When greater accuracy is desired using the grid technique, a more closely spaced grid and a microscope for measuring deformations are required. As the grid spacing becomes infinitesimally small, the calculus must be used and the previously indicated equations of normal and shear strain become

$$\epsilon_x = \lim_{\Delta x \to 0} \frac{u_2 - u_1}{\Delta x} = \frac{du}{dx} \tag{12.71}$$

$$\epsilon_y = \lim_{\Delta y \to 0} \frac{v_2 - v_1}{\Delta y} = \frac{dv}{dy} \tag{12.72}$$

$$\gamma_y = \lim_{\substack{\Delta x' \to 0 \\ \Delta y' \to 0}} \frac{v_3}{\Delta x'} + \frac{u_3}{\Delta y'} = \frac{dv}{dx} + \frac{du}{dy} \tag{12.73}$$

Equations expressing deformations in the x and y direction (i.e., u and v) can be differentiated to produce normal and shearing unit strains.

Once the normal and shear strains have been determined using the grid, the calculation of the normal and shear stresses may be made using the stress/strain relationships developed in § 12-23.

PROBLEMS

12-63. A grid element $A_1B_1C_1D_1$ in Fig. 12-57 is distorted into the position $A_3B_3C_3D_3$. Determine the shear and normal strains at D.

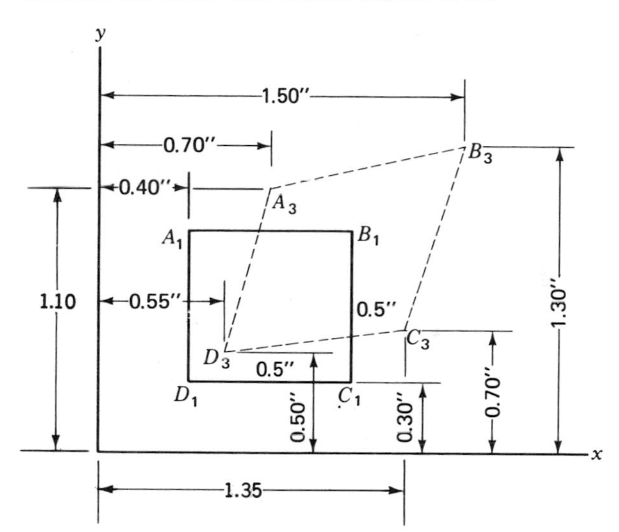

Figure 12-57

12-64. A model of a turbine blade is to be made of an elastomer and tested in torsion to determine whether tension develops under this type of loading at its point of attachment to the turbine housing. What procedure would you follow in making this determination from a grid analysis?

12-65. A flat plate $ABCD$ is deformed into position $AB'CD$ (Fig. 12-58), sides AD and DC remaining fixed. If the normal strains ϵ_x and ϵ_y at any point x and y on the surface arc given as

$$\epsilon_x = \frac{exy}{L^3} \quad \text{and} \quad \epsilon_y = \frac{exy}{L^3}$$

(a) Find the total displacement in the x and y direction.

(b) Show that $\gamma_{xy} = er^2/L^3$, where $r^2 = \sqrt{x^2 + y^2}$.

(c) Determine the principal strains and directions at point B.

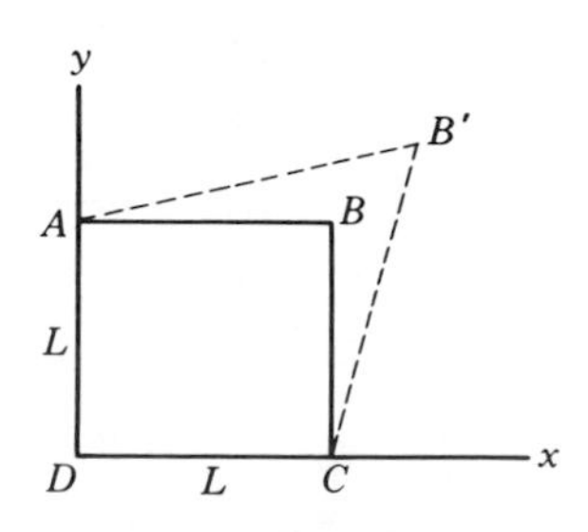

Figure 12-58

12-66. The grid element $ABCD$ (Fig. 12-59) is deformed into position $AB'CD$ so that straight lines parallel to the x axis remain straight parallel to the x axis and are uniformly stretched. (a) Show that at any point having coordinates x and y $\epsilon_x = ye/L^2$, $\epsilon_y = 0$, and $\gamma_{xy} = xe/L^2$. (b) Determine the principal strains at point B.

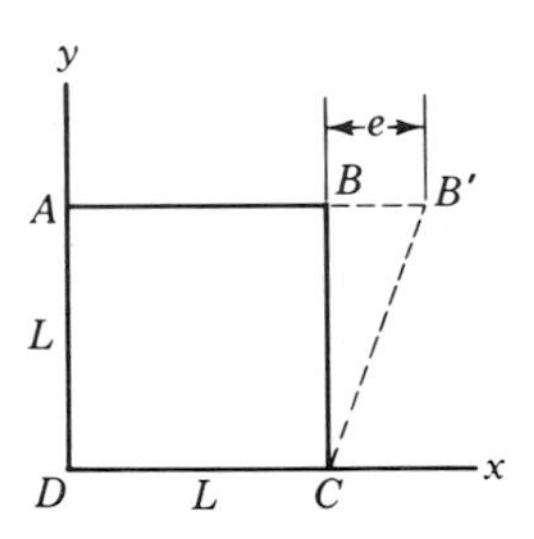

Figure 12-59

12-67. A rectangular grid element $ABCD$ is deformed so that its lower side falls along the parabola $y = kx^3$, and no point in the element displaces in the x direction $AB'C'D$ (Fig. 12-60). Find the shearing strain anywhere along the bottom edge of the element.

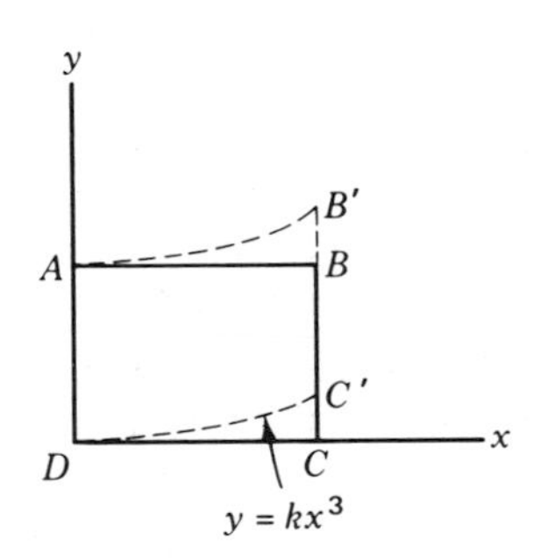

Figure 12-60

13

Fatigue Strength—Stress Concentrations

13-1 Phenomena of Fatigue

Perhaps our first introduction to the subject of the fatigue strength of materials took place when, in order to remove a nail from the surface of a piece of wood, we wiggled the nail back and forth until it broke. Although at the time the term *fatigue* may have had no significance to us other than a purely physical one, we are now to apply it to the *gradual reduction of strength in a material subjected to repeated stresses.*

In our wiggling operation we may have noticed that a greater sweep to and fro produced a quicker failure. Expressed in engineering terms, rupture is hastened by increasing the range of the applied stresses. Furthermore, if our cycle sweeps through alternating tensile and compressive stresses of equal magnitude, the range of stress will have attained its maximum destructiveness.

Such a balanced stress reversal is called a *completely reversed cycle of stress*. It may be contrasted with an *incomplete cycle of stress*, whose extremes of magnitude are not alike and which may prove to be more critical since the range of stress must decrease if fatigue failure is to be avoided.

Many members employed in structures and machines, *such as crankshafts*, *axles*, *turbine disks*, and so forth, are subjected to complete or incomplete cycles of stress. Some of the magnitudes of the stress reversals may be small and of little significance to the life of structural or machine parts. Others may be large and cause the design of members to be made dependent upon a contemplated reduction of strength with use.

13-2 σ-N Curves—Endurance Limit

Of invaluable assistance to the designer in such problems is the σ-N curve, σ indicating unit stresses in lb/in.^{2} plotted as ordinates, and N signifying the number of completely reversed cycles of stress plotted as abscissas (Fig. 13-1). To gather information from one material for the plotting of such

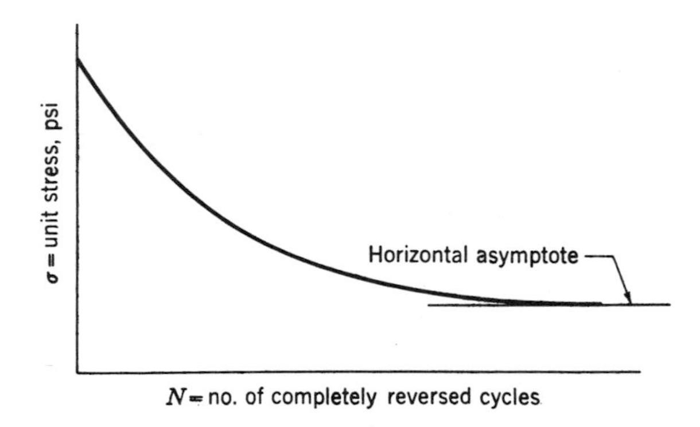

Figure 13-1

a curve requires a series of fatigue tests, each employing a different unit stress. Each test is run to failure, the number of cycles required for rupture being equal to N and the corresponding maximum unit stress being equal to σ. When these values are plotted, they locate one point on the curve. A sufficient number of similar tests must be performed to provide the points necessary for a well-defined graph.

The σ-N curve of Fig. 13-1, which is typical for many materials, shows that as the unit stress is decreased, the number of cycles required for failure is increased. This increase in N is rather rapid for the lower values of σ, making the σ-N curve asymptotic to a horizontal line. Should a completely reversed stress be imposed upon a specimen of a magnitude less than that of the asymptote, it would not fail regardless of the number of stress cycles to which it was subjected. That maximum unit stress which may be applied to a material through an indefinite number of completely reversed stress cycles without failure is called its *endurance limit*.

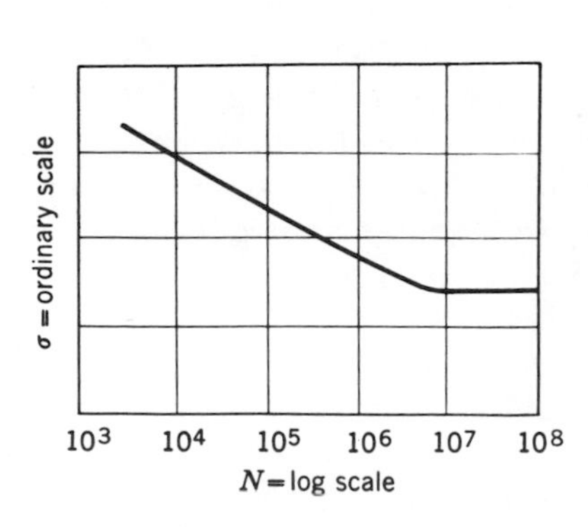

Figure 13-2

Because of the numerous tests required, the construction of σ-N curve is necessarily time consuming. It is

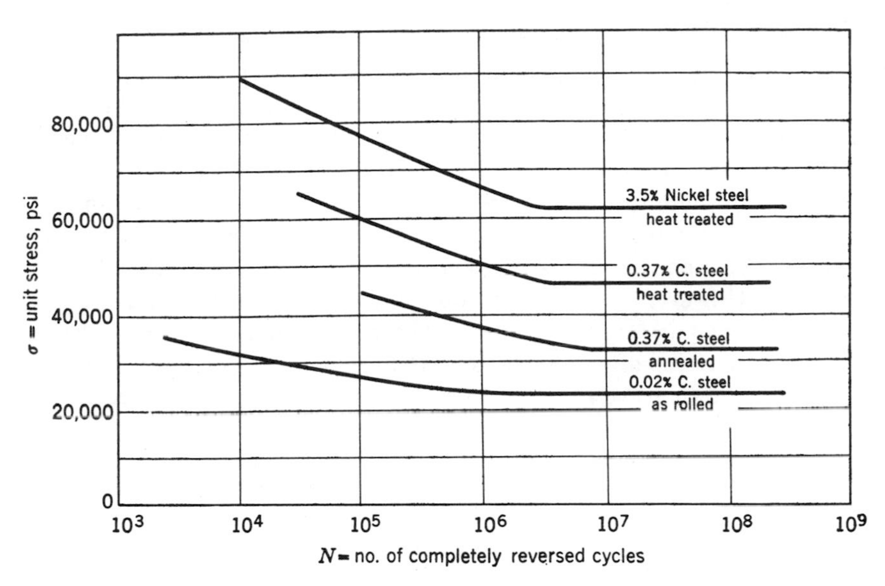

Figure 13-3

TABLE 8

FLEXURAL ENDURANCE-LIMIT STRESSES

Material	Flexural Endurance Limit* (psi)
Plain carbon steel:	
0.25% carbon, as rolled	26,000
0.50% carbon, normalized	33,000
0.75% carbon, annealed	38,000
1.25% carbon, normalized	50,000
Alloy steels:	
Nickel steel, not quenched	50,000
Nickel steel, quenched	65,000
Chrome-nickel steel, not quenched	50,000
Chrome-nickel steel, quenched	68,000
Cast iron	11,000
Aluminum, rolled	10,500
Duralumin, rolled	18,000
Magnesium forgings	16,000
Bronze, 95-5 annealed	23,000
Bronze, cold-rolled	22,500
Brass, 70-30, cold-rolled	17,500
Copper, annealed	10,000
Copper, cold-rolled	16,000
Beryllium copper, heat-treated	45,000
Douglas fir	3,900
Spruce	1,700

*Flexural endurance limits are approximately one-half the ultimate tensile strength.

seldom that one can obtain definite assurance of an endurance limit with less than 10,000,000 cycles. Should the presence of the endurance limit still be questioned, this number of stress cycles would have to be further increased. To include the widely scattered data of a σ-N curve on one graph sheet, a semilogarithmic, or a log-log, plot is used (Fig. 13-2). Such plots indicate the presence of an endurance limit more clearly than an ordinary Cartesian plot.

In Fig. 13-3 are drawn the σ-N curves of a few common metals. Those curves representing ferrous materials generally show a clearly defined endurance limit. However, some nonferrous materials, notably those containing aluminum or magnesium, have not as yet been shown to have a clearly defined endurance limit even after 500,000,000 complete stress reversals. It is reasonable to assume that even these materials will reveal an endurance limit if sufficient time is taken for such a study. Table 8 gives the flexural endurance-limit stresses for a number of commonly used materials.

13-3 Other Methods of Producing Fatigue Failure

A fatigue failure is not only possible by the action of repeated flexural stresses, but by repeated torsional and direct stresses as well. In fact, combinations of the above methods are common in the design of machine parts and frequently prove to be of primary significance. The endurance-limit ratio in torsion to that in bending has an average value of about 0.56. The fatigue strength for axially applied stress is approximately equal to that for bending.

Illustrative Problem 1. A rod that activates the cantilever arm of a valve mechanism pushes and pulls with the same intensity of force at a rate of 120 complete cycles per minute. If the maximum activating force is 500 lb and the rod made of unquenched chrome-nickel steel is 12 in. long, what diameter must the rod have if operation is to be on the basis of an 8-hr day for a minimum of 10 years? Use a factor of safety of 2.

SOLUTION. Number of complete cycles of stress anticipated equals

$$120 \times 60 \times 8 \times 365 \times 10 = 210{,}000{,}000$$

Since this number exceeds the 10,000,000 cycles generally required for an endurance limit, the design of the valve mechanism must be based on the 50,000-psi endurance limit. See Table 8.

The working stress is therefore equal to

$$\frac{50{,}000}{2} = 25{,}000 \text{ psi}$$

$$\text{area required} = \frac{500}{25{,}000} = 0.02 \text{ in.}^2$$

$$\text{diameter required} = \sqrt{\frac{4 \times 0.02}{\pi}} = 0.16 \text{ in.}$$

PROBLEMS

13-1. A 2-in.-diameter circular shaft made of 0.25 per cent carbon steel is 6 ft long and is freely supported in bearings placed at each end. If several million completely reversed cycles of stress are applied at its midpoint, what maximum value of load may be applied without rupture? Use factor of safety of 2.

13-2. A beryllium-copper spring, 0.25-in. wire diameter, is to be subjected to millions of complete cycles of torsional stress. What maximum load may be fluctuated if the coil diameter is 1.5 in., the number of coils is 10, and the factor of safety is 2? *Answer:* 56.0 lb.

13-3. The axle of a freight car is expected to perform its intended function 8 hr a day for 10 years. If during this period the loading is as shown in Fig. 13-4, and the speed of the freight car averages 20 mph, calculate the diameter of the axles. The axle is made of normalized 0.50 per cent carbon steel. Use a factor of safety of 2.5. *Answer:* 4.40 in.

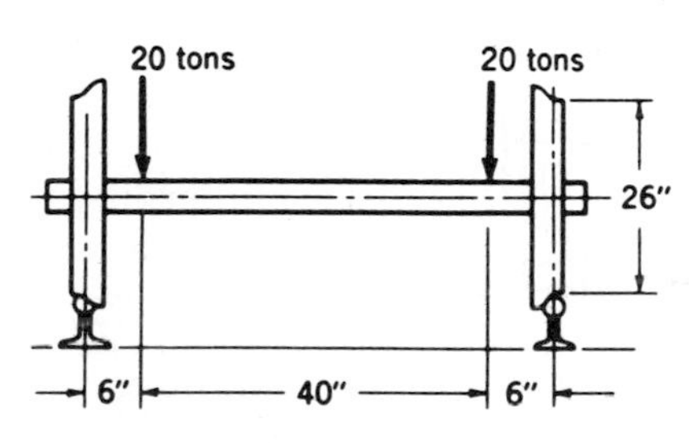

Figure 13-4

13-4 Why Materials Fail in Fatigue

Fatigue failures are the result of the poor transmission of stress developed by discontinuities both in the body and in the design of the working part. The discussion of internal and external discontinuities follows.

Internal discontinuities. No natural or synthetic material is ever homogeneous. If you were to peer into a microscope at the surface of a polished and etched piece of metal, you would see that it consisted of a tightly packed and irregular granular structure (Fig. 13-5). A plastic, which might be considered more uniform, is still not homogeneous because of the large irregular molecules that make up its structure. Wood is an even more irregular material, consisting of a highly complex and nonuniform cell structure.

Because of their nonhomogeneity, regular geometrically shaped stress-bearing members do not have imposed upon their cross sections the theoretical stress patterns we assume to be present. Take, for example, the ordinary uniform cross-sectioned tension member. If loaded axially, the stress theoretically applied to each cross section is P/A. This, however, is an average

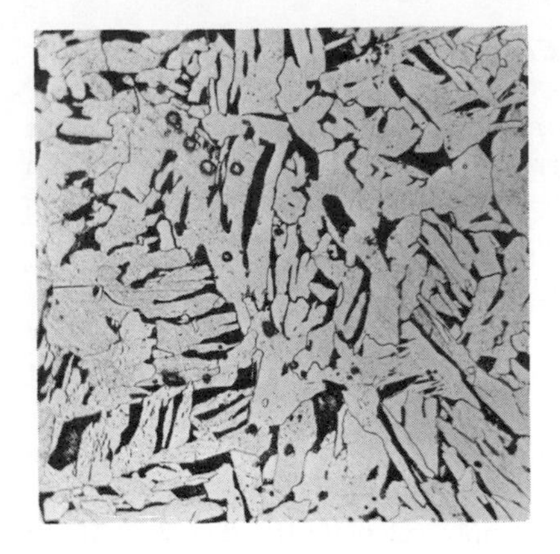

Figure 13-5 Photomicrograph of prestressed structural steel surface (200 ×). (Courtesy of Oliver E. Olsen)

stress, perfectly adapted for ordinary engineering design. On the other hand, if we were to investigate this stress across the various minute components of the cross section, we would find a distribution of stress pictorically represented by Fig. 13-6.

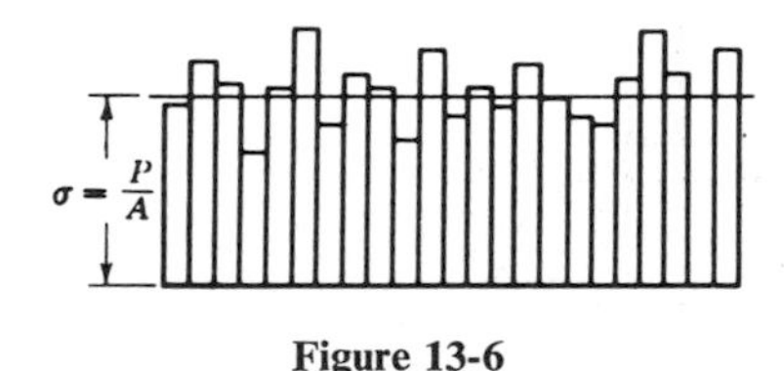

Figure 13-6

Those points more highly stressed than the others may pass through their proportional limit and suffer some mechanical weakening even though held from straining further by the adjoining understressed material. If this weakening action takes place in a metal, it usually takes place across the grains, along *slip planes*. When a slip plane is developed, it forms a source of incipient failure. Repeated stresses gradually cause the slip plane of adjoining grains to form a line perpendicular to the direction of the principal stress or parallel to the direction of the maximum shear stress, whichever is the more critical. With a decrease in the resisting area occurring with the development of each slip plane, failure is hastened.

This type of failure is caused by an internal discontinuity and is possible in every engineering material.

External discontinuity. Whenever the resisting portion of a stressed member has its cross section suddenly changed, another type of structural discontinuity is developed that is even more serious than that previously described. We shall call such a discontinuity an *external discontinuity*. The more common examples of this type of discontinuity are screw threads, rivet holes, keyways, re-entrant corners, and so forth, although even the peculiar shapes of many engineering members have discontinuities that may result in fatigue failure.

The seriousness accorded this type of discontinuity is apparent when it is discovered that at points of abrupt cross-sectional change the stress is materially increased, oftentimes to three or four times that obtained with the ordinary stress formulas derived in the earlier portions of this book. It is at such stress concentrations that most fatigue failures take place. Should such a stress concentration be unavoidable, the designer should correctly evaluate its magnitude and reduce the design stresses accordingly. The corrected axial, torsional, and bending formulas giving stresses at stress concentrations are

$$\sigma = k_1 \frac{P}{A}, \qquad \tau = k_2 \frac{Tr}{J}, \qquad \sigma = k_3 \frac{Mc}{I} \tag{13.1}$$

where k_1, k_2, and k_3 are stress concentration factors.

Ductile materials, when subjected to static loads, are not as seriously affected by stress concentrations as are more brittle materials. This is due to the fact that ductile materials yield sufficiently to reduce the potentially high-stress concentration. When subjected to repeated loads, however, ductile materials fare only slightly better.

Theoretical stress concentration factors caused by various types of design discontinuities are recorded in Table 9.

TABLE 9

STRESS CONCENTRATION FACTORS*

I

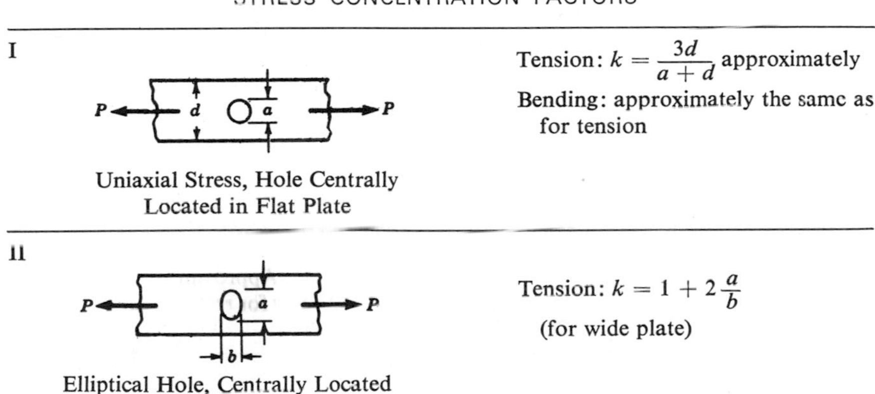

Uniaxial Stress, Hole Centrally Located in Flat Plate

Tension: $k = \frac{3d}{a + d}$ approximately

Bending: approximately the same as for tension

II

Elliptical Hole, Centrally Located in Flat Plate

Tension: $k = 1 + 2\frac{a}{b}$

(for wide plate)

III(a)

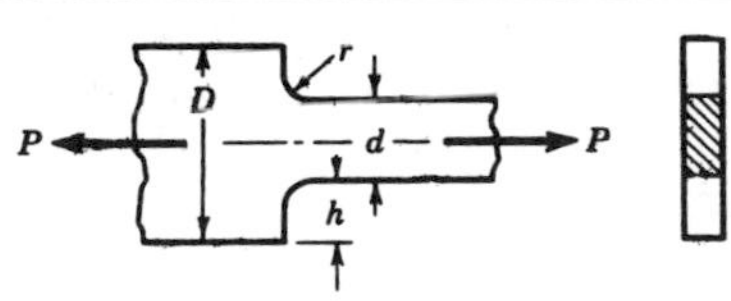

Rectangular Bar, Reduced Section, Square Shoulder

Tension:

$\frac{D}{d} \backslash \frac{r}{d}$	$\frac{1}{16}$	$\frac{1}{8}$	$\frac{3}{16}$	$\frac{1}{4}$	$\frac{3}{8}$	$\frac{1}{2}$
3.0	3.10	2.66	2.28	2.00	1.76	1.60
2.0	2.66	2.29	1.97	1.74	1.52	1.41
1.25	1.96	1.84	1.66	1.50	1.32	1.26

*Compiled from the following sources:

R. J. Roark, *Formulas for Stress and Strain*, New York: McGraw-Hill Book Company, 1938, pp. 298–311.

H. F. Moore and P. E. Henward, "The Strength of Screw Threads Under Repeated Tension," *Univ. Ill. Expt. Sta. Bull. 264*, 1934.

R. R. Moore, "Effect of Grooves, Threads and Corrosion upon the Fatigue of Metals," *Proc. A.S.T.M.*, **26** (1926), Pt. II, 255.

TABLE 9 (*Cont.*)

III(*b*)

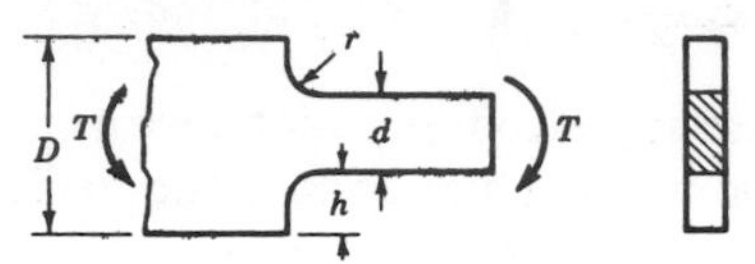

Bending:

$\frac{h}{r} \backslash \frac{r}{d}$	0.05	0.10	0.20	0.25	0.50	1.0
0.5	1.61	1.49	1.39	1.34	1.22	1.07
1.5	2.00	1.73	1.50	1.39	1.22	1.08
3.5		1.76	1.54	1.40	1.23	1.10

IV

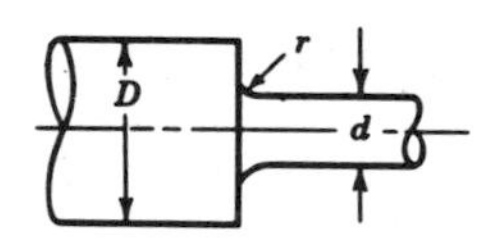

Circular Shaft, Reduced Section, Square Shoulders

Tension: Approximately the same as tension for the rectangular bar of case III

Bending: Approximately the same as bending for rectangular bar of case III

Torsion:

$\frac{D}{d} \backslash \frac{r}{d}$	0.005	0.01	0.02	0.03	0.06
2.00		3.00	2.25	2.00	1.65
1.20	3.00	2.50	2.00	1.75	1.50
1.09	2.20	1.88	1.53	1.40	1.20

V

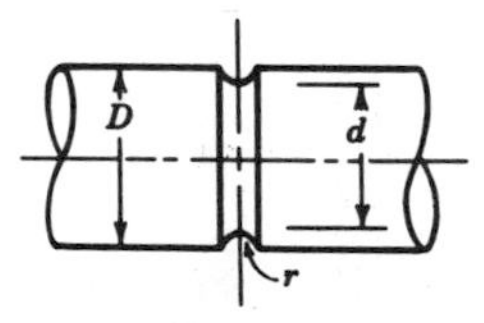

Semicircular Notch in Circular Shaft

Tension: Approximately the same as for the elliptical hole in the plate of case II, tension

Bending: Approximately the same as for the elliptical hole in the plate of case II, tension

Torsion: $k = \dfrac{2D}{D + 2r}$

VI

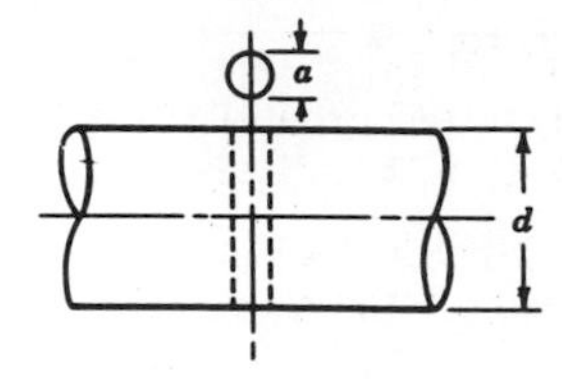

Radial Hole in Circular Shaft

Bending:

$\frac{a}{d} \backslash d$	2	3
0.0625	1.84	1.65
0.125	1.50	1.70
0.250	1.37	1.46

Torsion: for $\frac{a}{b}$ very small, $k = 4$

TABLE 9 (*Cont.*)

VII Screw Threads	
Tension: $D = 0.375$ in., 0.30% C steel	
Whitworth thread	1.76
U.S. standard	2.84
Rolled	2.15
Bending: $D = 0.375$ in., $h = 0.0271$ in. 24 threads per in., 0.43% C steel	
Continuous thread	1.38
Single notch	2.42

Illustrative Problem 2. A long tension member having a cross section $\frac{1}{2}$ in. thick by 4 in. wide has a 1-in.-diameter hole bored through its longitudinal axis perpendicular to its 4 in-dimension. Determine the maximum tensile stress at the sides of the hole if the load applied is 30,000 lb. Stress concentration factor is 2.

SOLUTION. The cross section of the bar through the hole is $A = \frac{1}{2}(4 - 1) = 1.5$ in.^{2}.

The average stress on this cross section is

$$\sigma = \frac{P}{A} = \frac{30{,}000}{1.5} = 20{,}000 \text{ psi}$$

However, due to a stress concentration factor of 2 at the side of the hole, the actual stress at this point will be

$$\sigma = k_1 \frac{P}{A} = 2 \times \frac{30{,}000}{1.5} = 40{,}000 \text{ psi}$$

It follows that the actual stress at the outside edge of the cross section will be less than 20,000 psi, since the average stress on the cross section must remain the same.

PROBLEMS

13-4. A Steatite insulator $\frac{3}{4}$ by $1\frac{1}{2}$ by 6 in. long is to be made with a hole placed 1 in. from either end for connection purposes. If the hole were made elliptical with its major axis of $\frac{3}{8}$ in. coincidental with the longitudinal axes of the insulator and its minor axis equal to $\frac{1}{4}$ in., what load could be safely applied through millions of axial stress reversals? Endurance limit under imposed conditions of direct stress is 1000 psi. Use stress concentration factor of 2.

13-5. What load could be sustained by the insulator of Problem 13-4 if the hole were $\frac{1}{4}$ in. in diameter? *Answer:* 182 lb.

13-6. The rolled Duralumin shaft shown in Fig. 13-7 is subject to millions of complete cycles of flexural load applied at its midpoint. If the stress concentration factor is 2, determine the maximum load that may be applied. Reactions at ends. *Answer:* 147 lb.

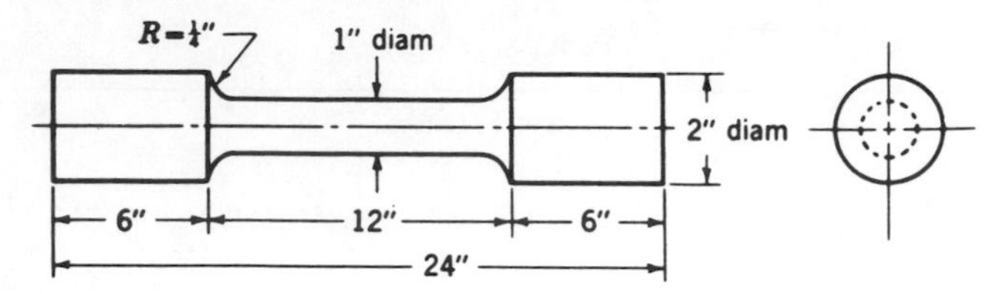

Figure 13-7

13-7. A plaster of paris specimen (Fig. 13-8) is loaded at the third points. If rupture takes place in the threaded portion when both loads equal 50 lb, what stress concentration is present? Root diameter is 1.25 in. Modulus of rupture of plaster of paris is 1100 psi. *Answer:* 1.41

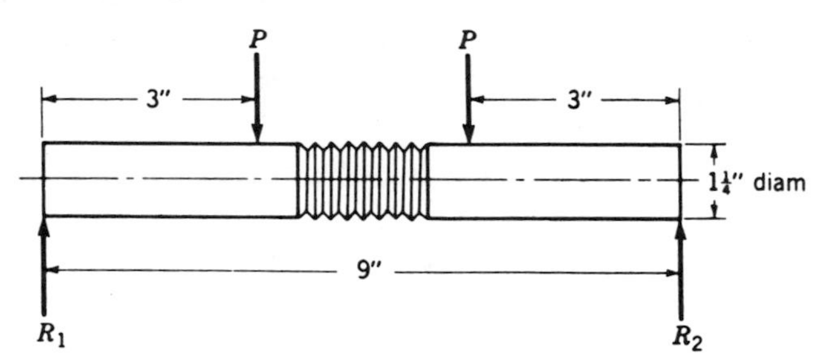

Figure 13-8

13-8. What maximum load P may be applied to the 95-5 annealed-bronze shaft shown in Fig. 13-9, if it is to fluctuate through many millions of reversed cycles? Use a stress concentration factor of 2.

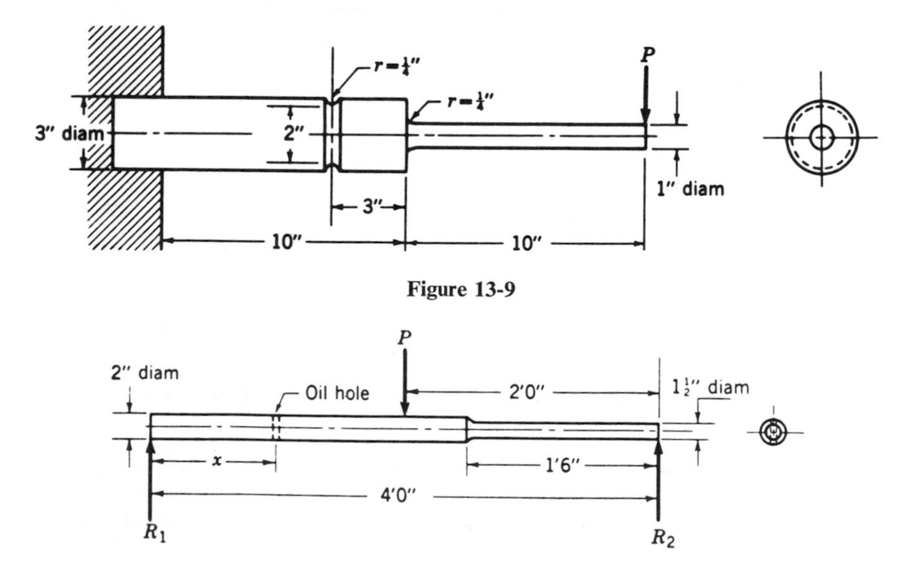

Figure 13-9

Figure 13-10

13-9. If the stress concentration at the oil hole (Fig. 13-10) equals 3, how far must it be placed from R_1 to create a stress equal to that obtained at the fillet? *Answer:* 21.4 in.

13-5 Methods of Obtaining Stress Concentration Factors

Various methods are employed to determine the stress concentration factor at an external discontinuity. Among the methods employed are

1. Strain-gage method.
2. Plaster-of-paris method.
3. Brittle-lacquer method.
4. Photoelastic method.
5. Fatigue method.
6. Mathematical method.

Strain-gage method. By methods discussed in Chapter 12, the strains occurring at any point on a stressed surface can be used to obtain the *average* normal stress occurring in any direction. The smaller the gage length of the strain recorder, the closer will be the agreement between the actual and average stress.

The electric strain gage, manufactured in lengths as small as $\frac{1}{8}$ in., is conveniently employed for stress concentration determinations at fillets and other sharp changes in surface contour. The *stress concentration factor* equals the experimental normal stress (σ_a) acting in any desired direction divided by the theoretical stress (σ_t) acting in the same direction obtained by the simple strength theory and developed by the same load ($k = \sigma_a/\sigma_t$). It indicates the relative speed of stress increase at the stress concentration over that of a similarly located point having no stress concentration.

Plaster-of-paris method. Plaster of paris, if properly hardened, has a fairly straight stress-strain graph when stressed in tension. When subjected to load, it will, as most other brittle materials, fail along the planes of maximum normal tensile stress. Thus, when a model of a load-bearing member made from plaster of paris is stressed to its breaking point, the failure stress is equal to the ultimate unit stress on the tensile stress-strain graph.

By comparing the experimental or actual stress with the theoretical stress produced by the same load, the stress concentration factor may be found.

Brittle-lacquer method. This method, discussed in detail in §12-29, is based on the brittleness of a surface coated with lacquer. Whenever the surface strain exceeds the ultimate tensile unit strain of the lacquer, it cracks. Correlating the stress obtained at the threshold value of the calibrating beam to the point on the mechanical part where lacquer failure begins affords a simple means of ascertaining the stress at that location caused by the applied load.

Comparing once again the experimental and theoretical stress at the same point provides a simple means of determining the stress concentration factor.

At the present time, industry uses the brittle-lacquer method more extensively for the determination of the direction of the principal stresses rather than for the computation of their magnitude. Electric strain gages, oriented accurately by the cracks in the lacquer, are employed for the actual stress determination. The greater accuracy of the electric strain gage has prompted this arrangement.

Photoelastic method. When a doubly refracting, isotropic, two-dimensional body, such as a glass, celluloid, or cast-Bakelite plate, is stressed and placed in a polariscope (Fig. 13-11), dark lines passing through the body are

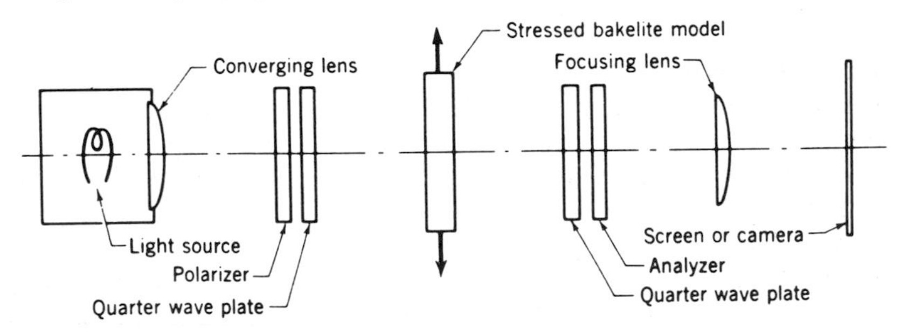

Figure 13-11

viewed on the projecting screen (Fig. 13-12) and are proportional to the difference of the principal stresses. Under the same loading condition, the change in thickness of the plate, which is proportional to the sum of the principal stresses, is determined. A solution of the two resulting simultaneous equations, including the terms $\sigma_1 + \sigma_2$ and $\sigma_1 - \sigma_2$, may then be made to obtain the principal stresses.* A detailed discussion of the optical problems involved is beyond the scope of this book. Knowing the actual stresses and those obtained by the simple strength formulas, the computation of the stress concentration factor is readily obtained.

Fatigue method. When σ-N curves are obtained for two series of specimens similar in every detail with the exception of a stress concentration, the ratio of the endurance limits approximates the stress concentration factor. Because

*Other methods can be used to afford a solution of the σ_1 and σ_2 stresses, as found in the $\sigma_1 - \sigma_2$ equation, which do not require a determination of the change of model thickness. The reader is referred to any textbook on photoelasticity for further explanation.

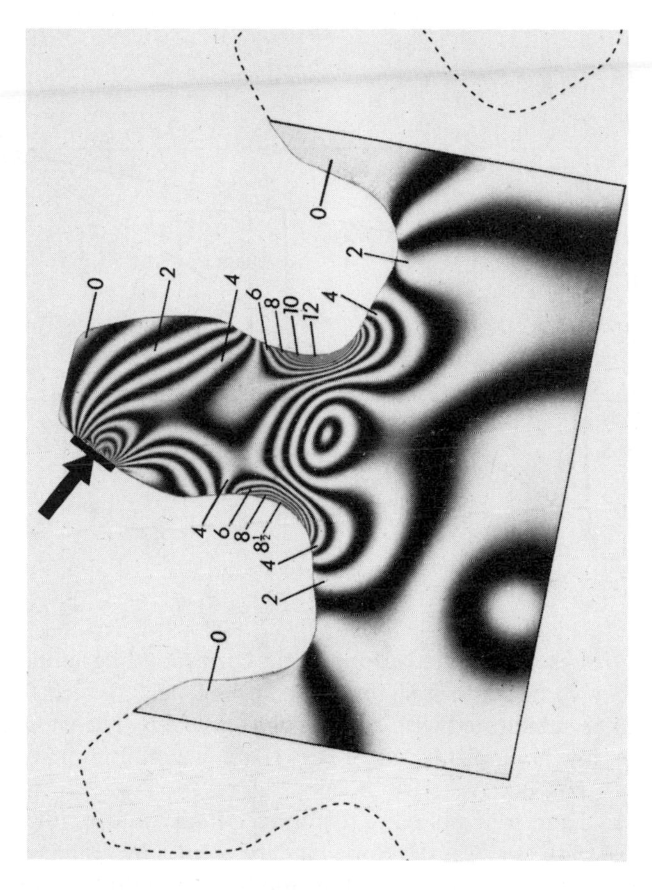

Figure 13-12

of the yielding of the material at the stress concentration, the value of k is less for more ductile materials.The value of k obtained is of real importance, however, for members subjected to repeated stress.

Mathematical method. Stresses at certain external discontinuities, such as circular and elliptical holes, may be determined mathematically. Comparing the exact theoretical unit stress with that obtained from the application of the simple theory provides the stress concentration factor.

13-6 Incomplete Stress Reversals

On many occasions mechanical parts are subjected to incomplete stress reversals. Under such conditions the endurance limit will be higher as the range of stress decreases. The limiting value is, of course, the static strength.

Of invaluable assistance in ascertaining the endurance limit for various cycles of stress is a diagram of the type shown in Fig. 13-13. It is constructed by plotting along a convenient straight line, such as *COB*, the values of the minimum stresses used in various incomplete stress cycles. If, then, the maxi-

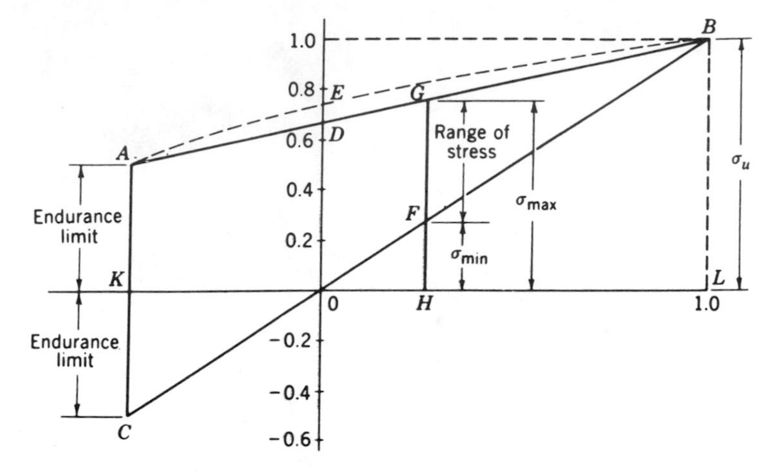

Figure 13-13

mum unit stresses are plotted along their corresponding minimum stress ordinates, they conform to a slightly curved line *AEB*, which for practical purposes can be substituted with the straight line *ADB*. The endurance limit, therefore, for any incomplete cycle of stress having a minimum normal tensile stress equal to *HF*, is *GH*.

Such a diagram employing endurance-limit values for completely reversed stress of $0.33\sigma_u$ was developed independently by Goodman and J. B. Johnson and is commonly known as the Goodman-Johnson diagram. Greater accuracy is obtained, however, if the actual endurance limit is used.

PROBLEMS

13-10. A cantilever beam is subjected to millions of incomplete load reversals. If it is to be made from a 1-in.-diameter steel (0.25 per cent C) rod, 12 in. long, and has a minimum downward load of 30 lb, what maximum downward load can it take if a factor of safety of 2 is used?

13-11. A 1-in.-diameter quenched nickel-steel rod is twisted through an indefinite number of incomplete loading cycles. If the torque in one direction is 1500 in.-lb, what can be the maximum torque in the opposite direction without rupture occurring? Ultimate shearing unit stress is 75,000 psi. *Answer:* 10,220 in.-lb.

13-12. A simple, gray cast-iron beam, $\frac{1}{2}$ in. thick by $1\frac{1}{2}$ in. wide by 24 in. long, is subjected to an incomplete load reversal at its center having as its maximum and minimum downward values 500 and 200 lb, respectively. Will it last indefinitely if the $1\frac{1}{2}$-in. dimension is kept vertical? Prove your answer. Ultimate flexural unit stress is 25,000 psi. *Answer:* Yes.

Hint: Construct a Goodman-Johnson diagram using ultimate stress and the endurance limit for completely reversed cycles of stress. Determine the endurance limits for the incomplete cycles of stress from the conditions of the problem.

13-7 Corrosion Fatigue

The corrosion of metals proceeds more rapidly for stressed than for unstressed materials. McAdam, in his exhaustive study of the effects of stress on corrosion, found that corrosion without stress reduced the fatigue strength of metals as much as 30 per cent. This reduction of strength was caused by the pits and irregularities formed on the surface. When corrosion and stress were applied simultaneously, the results were much more serious than when either acted alone.

Since the experimental σ-N curves are obtained using finely polished specimens, it would be necessary to consider a reduced endurance limit should parts be designed to perform in a corrosive atmosphere. A reduction of over 50 per cent in the endurance limit of steel may be expected when the repeated stress is developed in the presence of salt water.

14

The Applications of Energy Relationships

14-1 Introduction

Energy, by definition, is the ability or capacity of doing work. Strong men reveal their energy by the weights they lift, i.e., the work they perform. *Work* is always associated with the displacement of force and *is defined as the product of the distance moved and the component of the applied force in the direction of motion.*

Sometimes the force varies. For instance, when a body is pushed, the magnitude of force acting on the body varies in proportion to the resistance encountered, and the work it performs on the body is the product of its average magnitude and the displacement it produces. This work is classified as *external work*. Because energy can neither be created nor destroyed, the resistive forces acting within the body passing through internal deformations must produce an equivalent amount of *internal work*. If an equivalent amount of internal work is not developed, unrestrained motion or instability will result.

The more dynamic form of energy, such as that accruing to a falling weight, produces stresses of a much greater magnitude upon impact than those produced by the same weight applied gradually. The kinetic energy of the weight at impact is equated to the total internal energy developed to determine the resulting stress.

The succeeding sections will discuss in detail the external and internal energies resulting from gradually applied and impact loads, and their use in determining deformations and stresses.

ENERGY DEVELOPED BY GRADUALLY APPLIED LOADS

14-2 External Energy

When a load P is gradually applied to an elastic body, it produces a gradually increasing deformation which attains a maximum value at the time of full load application (see Fig. 14-1). The product of the average load

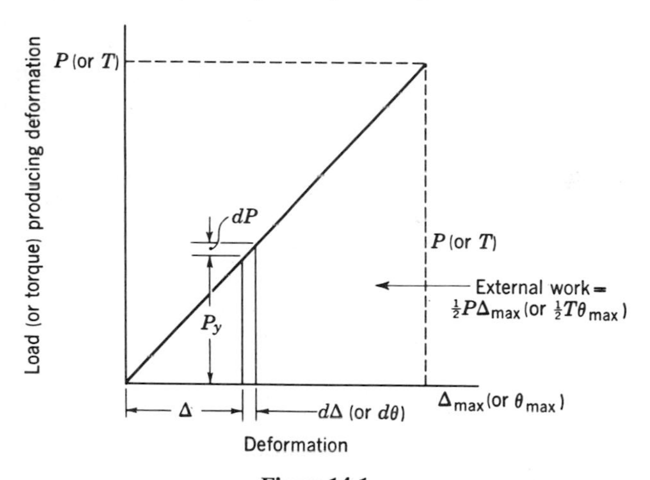

Figure 14-1

($P/2$) applied or transferred to the body and the deformation Δ_{max} produced by the full load P are the measure of the external energy *put into* the body. Derived mathematically,

$$\text{external energy} = \int_0^{\Delta_{max}} P_y \, d\Delta$$

But if the load varies directly with the deformation,

$$P_y = \frac{P}{\Delta_{max}} \Delta$$

$$\text{external energy} = \int_0^{\Delta_{max}} \frac{P}{\Delta_{max}} \Delta \, d\Delta = \frac{P}{\Delta_{max}} \left[\frac{\Delta^2}{2}\right]^{\Delta_{max}} = \frac{P\Delta_{max}}{2} \qquad (14.1)$$

The value of $\frac{1}{2}P\Delta_{max}$ should be recognized as the area under the load deformation curve up to full load.

In a similar fashion, the work performed by a gradually applied couple T acting on an elastic body is equal to $\frac{1}{2}T\theta_{max}$ (Fig. 14-1).

14-3 Internal Energy

Whereas those forces producing external energy are *externally applied*, those producing internal energy are *internally applied.* Internally applied forces are those resisting forces (viz., *C-T* couples, T_R, etc.), which are induced by the action of the external forces. When developed by gradually applied external forces, the internal forces must, of course, also be gradually applied. The internal energy developed must therefore also equal an average force or couple times its maximum internal deformation.

Internal or resisting forces are of four different kinds: axial, bending, torsional, and shear. Each produces its own type of deformation and magnitude of internal energy. They will now be considered separately.

Axial. An axial load P produces an equal and opposite resisting load P_R on every cross section of body (Fig. 14-2). Every incremental length dx of

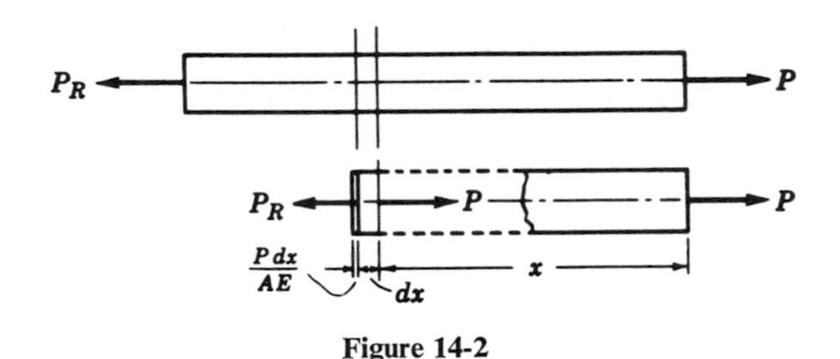

Figure 14-2

the body is subjected to the same deformation $P\,dx/AE$ [Eq. (2.10)], assuming A and E to remain constant. When P is gradually applied, each incremental length resists with an amount of internal energy equal to $(P/2)(P\,dx/AE)$. If the total energy is now obtained for all the incremental lengths, we find that the

$$\begin{matrix}\text{total internal energy due}\\ \text{to an axially applied load}\end{matrix} = \int_0^l \frac{1}{2}\frac{P_R^2\,dx}{AE} = \frac{1}{2}\frac{P_R^2 l}{AE} = \frac{1}{2}\frac{P^2 l}{AE} \quad (14.2)$$

Bending. The *C-T* couples produced on the cross sections of a member subjected to bending can produce only rotational deformation. These deformations produce relative rotations of the opposite sides of each segment in the beam equal to $d\theta$ (Fig. 14-3). From Chapter 8 it was learned that $d\theta = (M/EI)\,dx$. If now a bending moment M is gradually applied to a segment of a beam whose internal *C-T* couple or M_R twists that segment through an angle $d\theta$, there is developed within that segment an internal energy equal to

$$\frac{M_R}{2} \times \frac{M_R\,dx}{EI}$$

If the energy from every such segment is added together, there results

$$\begin{matrix}\text{total internal energy}\\ \text{due to bending}\end{matrix} = \int_0^l \frac{M_R^2\,dx}{2EI} = \int_0^l \frac{M^2\,dx}{2EI} \quad (14.3)$$

Torsional. When a shaft is twisted about its longitudinal axis by a couple T, every cross section of that shaft has acting on it an equalizing couple

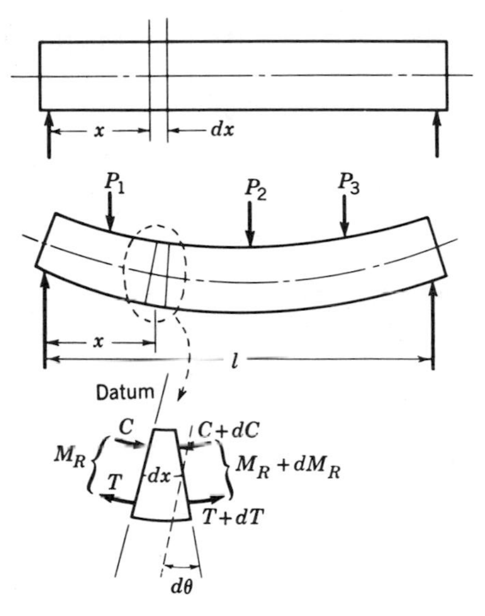

Figure 14-3

T_R (Chapter 5). These couples rotate about the longitudinal axis. The relative rotation of one side of a segment dx in length with respect to its other side is equal to $(T_R/GJ)\,dx$ (Fig. 14-4). Adding the energies of all the disk-like segments included in length l yields

$$\text{total internal energy due to torsion} = \int_0^l \frac{T_R^2\,dx}{2GJ} = \int_0^l \frac{T^2\,dx}{2GJ} \tag{14.4}$$

Shear. When two adjacent parallel surfaces dx distance apart are subjected to shearing forces, the shear strain is equal to de_s/dx. Assuming the shear

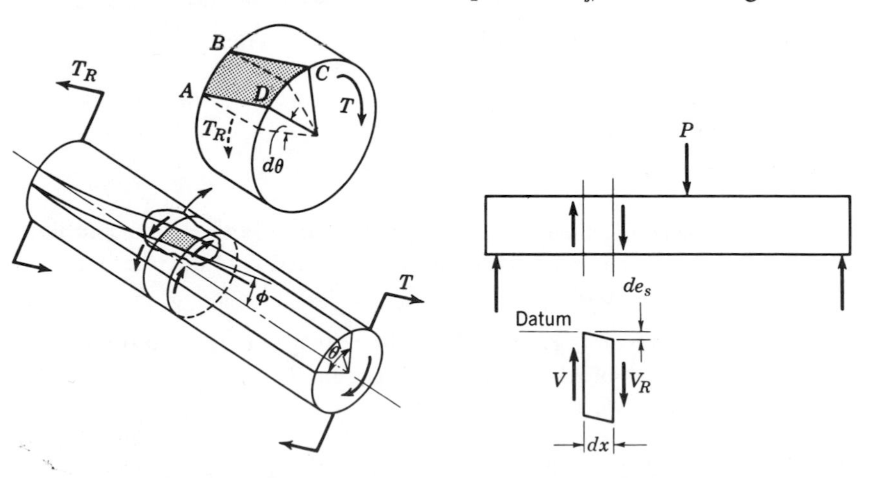

Figure 14-4

Figure 14-5

stress to be uniformly distributed, the total deformation is then equal to

$$de_s = \gamma\, dx = \frac{\tau}{G}\, dx = \frac{V_R}{A_s G}\, dx$$

The internal energy exerted by the element in resisting shear is equal to $V_R/2 \times (V_R\, dx/A_s G)$ (Fig. 14-5). The total internal shear energy exerted by all the shearing forces is then equal to

$$\begin{array}{r}\text{total internal energy}\\ \text{due to shear}\end{array} = \int_0^l \frac{V_R^2}{2A_s G}\, dx = \int_0^l \frac{V^2\, dx}{2A_s G} \tag{14.5}$$

14-4 Equivalence of External and Internal Energy

Whenever an elastic body is subjected to a system of forces and couples producing internal energies of the four types discussed in the previous section, the law of conservation of energy is used to produce highly important relationships between these forces and their displacements. This law, which states that energy can neither be created nor destroyed, makes the external energy equal to the internal energy. Therefore, discounting the effects of friction and hysteresis, the following general statement drawn from the preceding derivations may be written for any elastic body subjected to a general force system.

$$\text{external energy} = \text{internal energy}$$

$$\Sigma\frac{P\Delta}{2} + \Sigma\frac{T\theta}{2} = \Sigma\int\frac{M^2\, dx}{2EI} + \Sigma\int\frac{P^2\, dx}{2AE} + \Sigma\int\frac{T^2\, dx}{2GJ} + \Sigma\int\frac{V^2\, dx}{2AG} \tag{14.6}$$

Of course, not all bodies are subjected to all kinds of external and internal energy. Also, under certain conditions of loading some terms may be relatively small. In such cases those terms in Eq. (14.6) not applicable or inconsequential to the analysis may be deleted.

A simple illustration will reveal its usefulness.

Illustrative Problem 1. Determine the maximum deflection of a cantilever beam loaded at its free end with a concentrated load. Omit internal shear energy as being negligible.

SOLUTION. Assuming the origin of the x and y axes to be at the free end of the beam, the value of M_x at any point is equal to $-Px$. Writing the equation of internal and external energy gives

$$\frac{P\Delta}{2} = \int_0^l \frac{M^2\, dx}{2EI} = \int_0^l \frac{(-Px)^2\, dx}{2EI} = \frac{P^2 l^3}{6EI}$$

from which

$$\Delta = \frac{Pl^3}{3EI}$$

14-5 Disadvantages of Using Full Energy Equation

If the foregoing cantilever beam had an additional load applied to its midpoint, the energy equation (omitting internal shear energy) would be written as

$$\frac{1}{2}P_1\Delta_1 + \frac{1}{2}P_2\Delta_2 = \int_0^{l/2} \frac{M^2\,dx}{2EI} + \int_{l/2}^{l} \frac{M^2\,dx}{2EI}$$

To obtain the values of Δ_1 and Δ_2 would require another relationship between them. The difficulty involved in obtaining this relationship and the simplicity with which either deflection may be obtained by other methods have resulted in the almost complete abandonment of the full energy equation in determining deflections. The virtual-work method, based on an energy equation of a dummy or virtual load, the superposition method, or that based on Castigliano's theorem, have superseded the use of the full energy expression. Only the latter method will be discussed here.

14-6 Castigliano's Theorem

Where an elastic body is deformed under the action of a general force system, Castigliano's theorem states that the derivative of the strain energy with respect to any one of the forces or couples included in the force system is equal to the displacement of that force or couple. The proof of this statement follows:

Proof. Let there be applied to the beam shown in Fig. 14-6 a differential load dP_1 creating an internal strain energy equal to dU. If loads P_1 and P_2 are then added to the beam, developing y_1 and y_2 displacements, respectively, the additional internal energy developed in the beam will be equal to

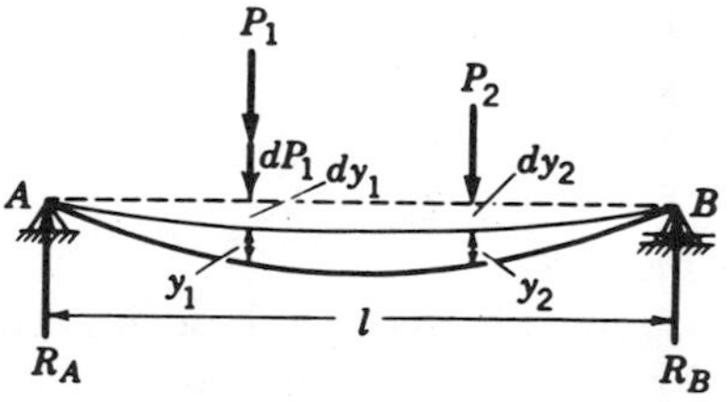

Figure 14-6

$$U_1 = \frac{P_1 y_1}{2} + y_1\,dP_1 + \frac{P_2 y_2}{2}$$

Had only P_1 and P_2 been added (dP_1 omitted), the internal energy of the beam would have equaled

$$U_2 = \frac{P_1 y_1}{2} + \frac{P_2 y_2}{2}$$

Taking the difference of these two energies, there results

$$dU = U_1 - U_2 = \frac{P_1 y_1}{2} + y_1\, dP_1 + \frac{P_2 y_2}{2} - \frac{P_1 y_1}{2} - \frac{P_2 y_2}{2}$$

$$dU = y_1\, dP_1 \tag{14.7}$$

$$\frac{dU}{dP_1} = y_1$$

Had a differential dP_2 been used at the location of P_2, a similar equation could have been written:

$$\frac{dU}{dP_2} = y_2 \tag{14.8}$$

Upon further analysis, this equation states that the differentiation* of the entire internal energy of an elastic body produced by some external force system with respect to any one of the forces in that system is equal to the elastic displacement of that force in the direction of its action line. In a similar manner, had a torque T been acting in place of either of the forces, and dT been used as its differential, the result would have been

$$\frac{dU}{dT} = \theta \tag{14.9}$$

where θ is the rotational displacement of T.

The versatility of these equations will be demonstrated by the use of some simple problems.

In the case of an axial loaded tensile bar, the internal energy equals $P^2l/2AE$ (§14.3). The derivative of this expression with respect to P is equal to Pl/AE, or the elongation of the bar due to P as derived in Chapter 2.

The internal energy of a shaft twisted by a torque T is equal to $T^2l/2GJ$. Its derivative with respect to T is equal to Tl/GJ, or θ, the angle of twist caused by T (see Chapter 5).

The maximum deflection of a cantilever beam loaded at its free end with a load P is equal, disregarding shear, to the derivative of its internal flexural energy:

$$\Delta_P = \frac{dU}{dP} = \frac{d}{dP}\int_0^l \frac{(-Px)^2\, dx}{2EI} = \frac{d}{dP}\left(\frac{P^2 l^3}{6EI}\right) = \frac{Pl^3}{3EI}$$

In problems such as these, it is generally much more convenient to perform the differentiation first, then the integration.

$$\Delta_P = \frac{dU}{dP} = \frac{d}{dP}\int_0^l \frac{M^2\, dx}{2EI} = \int_0^l \frac{M(dM/dP)\, dx}{EI} = \int_0^l \frac{(-Px)(-x)\, dx}{EI} = \frac{Pl^3}{3EI}$$

*If two or more forces in the force system are variable, the displacement of any one of them is obtained by taking the partial differential of the total energy with respect to that force.

When the deflection desired occurs at a point where there is no load, a dummy load or couple is placed at that point. The strain energy equation is then written including the dummy load or couple and differentiated with respect to it. This load or couple is then reduced to zero and the integration performed to obtain the answer.

Illustrative Problem 2. Determine by Castigliano's theorem the deflection at the free end of a uniformly loaded cantilever beam. Consider flexural energy alone (Fig. 14-7).

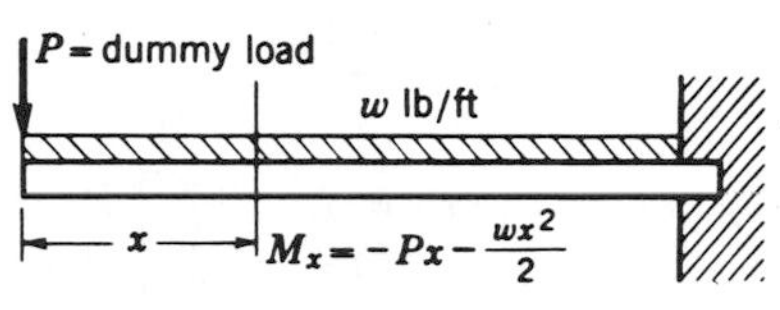

Figure 14-7

SOLUTION

$$M_x = -Px - \frac{wx^2}{2}$$

$$U = \int_0^l \frac{M^2\,dx}{2EI} = \int_0^l \frac{\left(-Px - \frac{wx^2}{2}\right)^2 dx}{2EI}$$

$$\Delta_P = \frac{dU}{dP} = \int_0^l \frac{\left(-Px - \frac{wx^2}{2}\right)(-x)\,dx}{EI}$$

If, now, P is reduced to zero,

$$\Delta_P = \int_0^l \frac{(wx^3/2)\,dx}{EI} = \left[\frac{wx^4}{8EI}\right]_0^l$$

$$\Delta_P = \frac{wl^4}{8EI}$$

Illustrative Problem 3. Determine by Castigliano's theorem the vertical displacement of point C of the member having an I cross section shown in Fig. 14-8. Use all energies. $A = 10$ in.^{2}, $I = 100$ in.^{4}, $E = 30 \times 10^6$ psi, $G = 12 \times 10^6$ psi, area of web $A_s = 5$ in.^{2}.

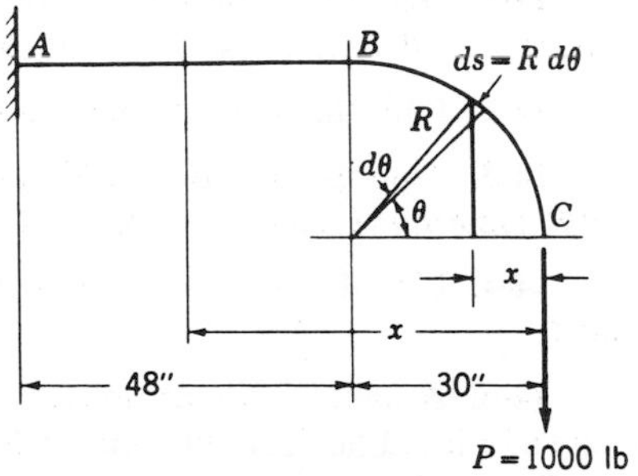

Figure 14-8

SOLUTION. Because of the necessity of differentiating with respect to the load, let us consider the load as P.

Forces Acting on AB	Forces Acting on BC
Moment $= Px$	Moment $= PR(1 - \cos\theta)$
Shear $= P$	Shear $= P\sin\theta$
Axial $= 0$	Axial $= P\cos\theta$

The signs of these effects are all considered positive, inasmuch as their contributory displacements are all similarly directed, i.e., downward.

$$U = \int_{30}^{78} \frac{M^2\,dx}{2EI} + \int_{30}^{78} \frac{V^2\,dx}{2A_sG} + \int_{30}^{78} \frac{P^2\,dx}{2AE} + \int_{0}^{\pi/2} \frac{M^2\,ds}{2EI} + \int_{0}^{\pi/2} \frac{V^2\,ds}{2A_sG} + \int_{0}^{\pi/2} \frac{P^2\,ds}{2AE}$$

$$U = \int_{30}^{78} \frac{(Px)^2\,dx}{2EI} + \int_{30}^{78} \frac{P^2\,dx}{2A_sG} + 0 + \int_{0}^{\pi/2} \frac{[PR(1-\cos\theta)]^2 R\,d\theta}{2EI} + \int_{0}^{\pi/2} \frac{(P\sin\theta)^2 R\,d\theta}{2A_sG} + \int_{0}^{\pi/2} \frac{(P\cos\theta)^2 R\,d\theta}{2AE}$$

$$\Delta_c = \frac{dU}{dP} = \int_{30}^{78} \frac{2(Px)(x)\,dx}{2EI} + \int_{30}^{78} \frac{2P\,dx}{2A_sG} + \int_{0}^{\pi/2} \frac{2[PR(1-\cos\theta)][R(1-\cos\theta)]R\,d\theta}{2EI} + \int_{0}^{\pi/2} \frac{2(P\sin\theta)(\sin\theta)R\,d\theta}{2A_sG} + \int_{0}^{\pi/2} \frac{2(P\cos\theta)(\cos\theta)R\,d\theta}{2AE}$$

After integrating, inserting the limits, and setting $P = 1000$ lb,

$$\Delta_c = 0.0537 \text{ in.}$$

PROBLEMS

Note: Use Castigliano's theorem. Consider only flexural energy except where noted.

14-1. Determine the vertical displacement at the midpoint of a simple beam uniformly loaded. *Answer:* $5wl^4/384EI$.

14-2. Find the slope at the end of the beam of Problem 14-1.

14-3. A simple beam is loaded at one of its third points by a concentrated load P. Determine the displacement at the other third point. *Answer:* $7/486(Pl^3/EI)$.

14-4. Find the slope at the end of the beam of Problem 14-3 farthest from the load P.

14-5. A steel I beam having the shape of a quarter circle (radius $= R$) has its fixed end placed horizontally and its load P acting vertically downward on its free end. Free end is below fixed support. Determine the vertical displacement of the load in terms of R and P. Use all energies. $A = 10$ in.^{2}, $A_s = 5$ in.^{2}, $I = 10$ in.^{4}. *Answer:* $(1.18PR^3 + 1.57PR)10^{-8}$.

14-6. Determine the displacement and slope of the free end of the quarter-circle beam of Problem 14-5 if its fixed end is placed vertically and is above the applied load, which acts vertically downward at its free end. Use all energies.

14-7 Solution of Statically Indeterminate Members by Castigliano's Theorem

Any statically indeterminate member may be solved by means of Castigliano's theorem by the simple method of

1. Replacing a sufficient number of redundant forces with *variable, movable* forces to make the member or structure statically determinate. Stability must, of course, be maintained.

2. Writing the expression for internal energy over the *entire* structure including the effect of all the movable, variable redundant forces.

3. Taking the partial differential of the expression with respect to the force (or forces) whose magnitude is desired. Generally, there will be as many differentiations as there are redundant forces.

4. Setting the differential(s) equal to zero, inasmuch as a fixed point cannot move.

5. Solving the equations so obtained to determine the redundant(s).

These steps will best be illustrated by the following problems.

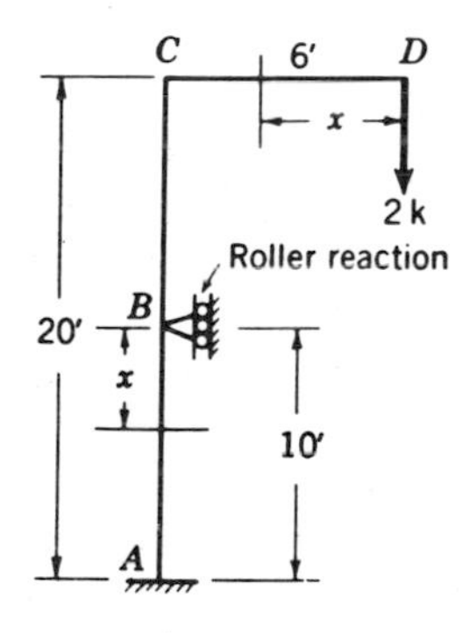

Figure 14-9

Illustrative Problem 4. Find the reactions at A and B for the bracket shown in Fig. 14-9 by Castigliano's theorem using flexural energy alone. E and I are constant.

SOLUTION. Replace the reaction at B with a variable movable unknown R_B. Then, as

$$M_{CD} = -2x, \qquad M_{CB} = -12, \qquad M_{AB} = -12 + R_B x$$

$$U = \int_0^6 \frac{(-2x)^2\,dx}{2EI} + \int_0^{10} \frac{(-12)^2\,dx}{2EI} + \int_0^{10} \frac{(-12 + R_B x)^2\,dx}{2EI}$$

$$\Delta_B = \frac{dU}{dR_B} = 0 + 0 + \int_0^{10} \frac{(-12 + R_B x)(x)\,dx}{2EI}$$

$$\Delta_B = 0 = \frac{1}{2EI}\left[\frac{-12x^2}{2} + \frac{R_B x^3}{3}\right]_0^{10}$$

$$R_B = 1.8 \text{ kips}$$

from which $H_A = 1.8$ kips, $V_A = 2.0$ kips, and $M_A = +6.0$ ft-kips.

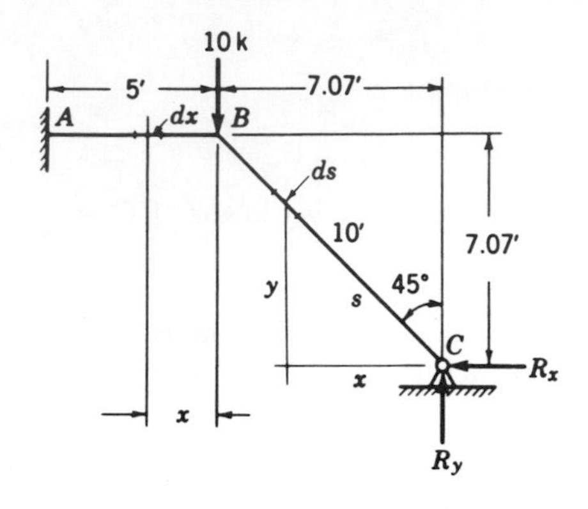

Figure 14-10

Illustrative Problem 5. Determine the components of the reaction at C acting on the beam shown in Fig. 14-10 by Castigliano's theorem. Use only flexural energy.

SOLUTION. Assume the beam to be disengaged from the hinge at C and supplied with two variable component forces R_x and R_y at that same point. Under this loading

$$M_{AB} = R_y(x + 7.07) - 7.07R_x - 10x$$

$$M_{BC} = 0.707sR_y - 0.707sR_x$$

The expression for energy is then equal to

$$U = \int_0^5 \frac{[R_y(x + 7.07) - 7.07R_x - 10x]^2}{2EI} dx + \int_0^{10} \frac{(0.707sR_y - 0.707sR_x)^2}{2EI} ds$$

from which

$$\frac{\partial U}{\partial R_x} = \int_0^5 \frac{[R_y(x + 7.07) - 7.07R_x - 10x]}{EI}(-7.07)\, dx + \int_0^{10} \frac{[0.707s(R_y - R_x)](-0.707s)\, ds}{EI} = 0$$

Now integrating and simplifying, we obtain

$$\left[-7.07R_y\frac{x^2}{2} - 50R_y x + 50R_x x + 70.7\frac{x^2}{2}\right]_0^5 + \left[-0.5\frac{s^3}{3}(R_y - R_x)\right]_0^{10} = 0$$

from which

$$+20.20R_y - 16.67R_x = +35.35 \qquad \text{(A)}$$

Now, taking the partial differential of U with respect to R_y and setting it equal to zero, there follows

$$\frac{\partial U}{\partial R_y} = \int_0^5 \frac{[R_y(x + 7.07) - 7.07R_x - 10x](x + 7.07)}{EI} dx + \int_0^{10} \frac{[0.707s(R_y - R_x)](0.707s)}{EI} ds = 0$$

$$\left[\frac{R_y x^3}{3} + 14.14R_y\frac{x^2}{2} + 50R_y x - 7.07R_x\frac{x^2}{2} - 50R_x x - \frac{10x^3}{3} - 70.7\frac{x^2}{2}\right]_0^5 + \left[0.5\frac{s^3}{3}(R_y - R_x)\right]_0^{10} = 0$$

$$25.41R_y - 20.20R_x - 52.0 = 0 \qquad \text{(B)}$$

Then by solving Eqs. (A) and (B) simultaneously,

$$R_x = 10.0 \text{ kips} \quad \text{and} \quad R_y = 10.0 \text{ kips}$$

PROBLEMS

14-7. Determine by means of Castigliano's theorem the restraining moment on a uniformly loaded beam of span l, fixed at one end and simply supported at the other. *Answer:* $-wl^2/8$.

14-8. Determine the reactions acting on the frame shown in Fig. 14-11, using Castigliano's theorem. *Answers:* $A_H = 6.67$ kips, $A_V = 16.67$ kips.

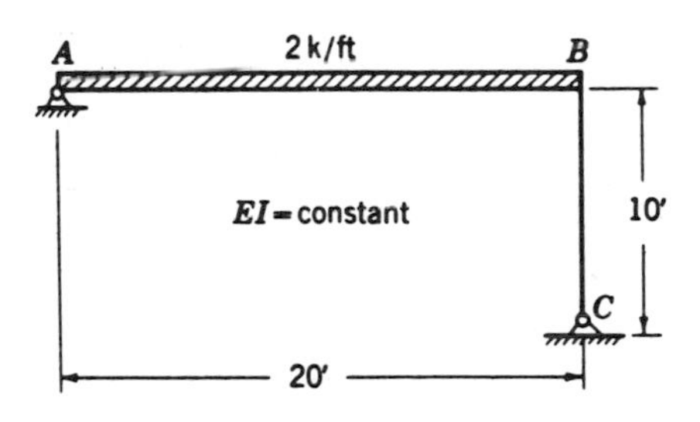

Figure 14-11

14-9. If the support at A shown in Fig. 14-11 were fixed, find the reactions acting on the frame using Castigliano's theorem. *Answers:* $A_H = 4.00$ kips, $A_V = 22.0$ kips, $M_A = -80.0$ ft-kips.

14-10. By applying Castigliano's theorem, find reactions R_C and R_D for the eccentrically loaded pole shown in Fig. 14-12a. Use flexure energy alone. *Answers:* $R_C = 8.57$ kips, $R_D = 11.43$ kips.

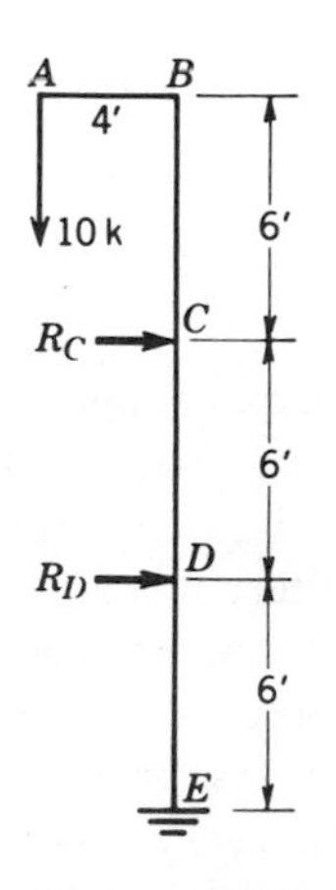

Figure 14-12a

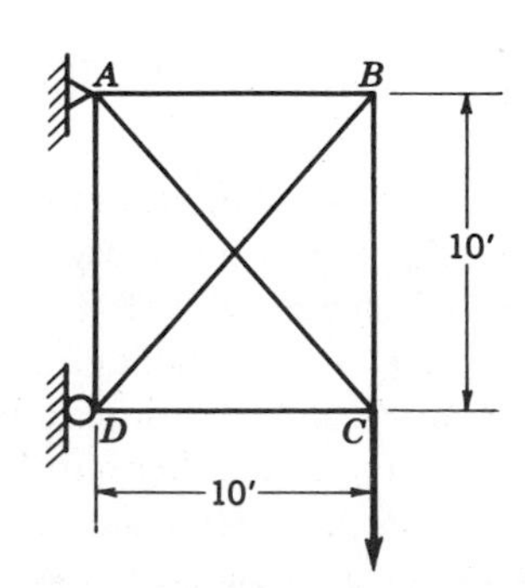

Figure 14-12b

14-11. Find the forces of the statically indeterminate frame shown in Fig. 14-12b by Castigliano's theorem. Consider one of the diagonals as a redundant member. All members have an area equal to 10 in.^{2}.

FAILURE DUE TO INSTABILITY

14-8 Introduction

It has been stated previously that a load acting on an elastic body produces external work which must be matched by an equal amount of internal work if equilibrium is to be maintained. If, on the other hand, the body attains its maximum capacity of internal resistive energy with less than full load, any addition of load beyond this critical state will cause instability.

This broad principle is, in a very real sense, a second criterion to be applied to the design of engineering members. The first, based on stress, has already been discussed in the previous chapters. The second, *stability*, can be equally important and becomes critical in the design of slender members and flexible assemblies. The following sections show the principles of energy applied to problems where stability is critical.

14-9 Instability Applied to Spring Assemblies

In the assembly shown in Fig. 14-13 a spring is used to keep two pin-ended rigid column segments in equilibrium under the applied load P. The stress on the column cross section is assumed to be so small as to make the geometric displacement or stability of the assembly the sole criterion for design. As load P increases, the external and internal energies developed in the assembly are as follows:

$$\text{external energy} = \frac{P\lambda}{2} \tag{14.10}$$

$$\text{internal energy} = \frac{k\Delta^2}{2} \tag{14.11}$$

(regarding normal energy in the column as negligible)

Significant in the analysis of stability is the comparison of the increase of internal energy with that of external energy. If the external energy (i.e., work input) for any small deformation exceeds the internal energy (i.e., energy of resistance), deformation continues and a general condition of instability is present. If, on the other hand, the internal energy exceeds the external energy, no continued deformation can occur and stability prevails. A critical condition occurs when the internal and external energies are equal.

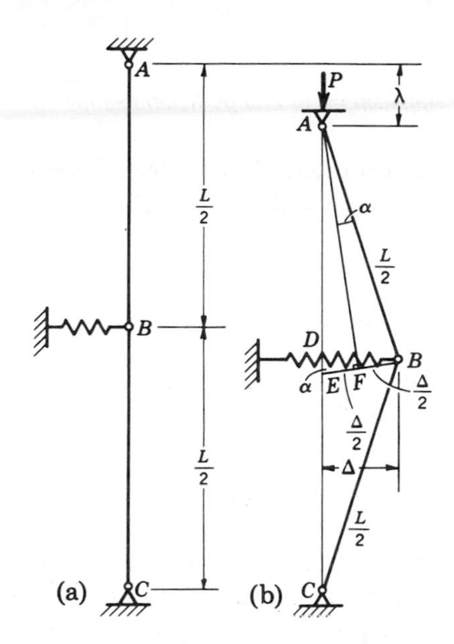

Figure 14-13

Expressing these concepts mathematically, we obtain

$$\frac{P\lambda}{2} > \frac{k\Delta^2}{2} \quad \text{for instability} \tag{14.12a}$$

$$\frac{P\lambda}{2} < \frac{k\Delta^2}{2} \quad \text{for stability} \tag{14.12b}$$

$$\frac{P\lambda}{2} = \frac{k\Delta^2}{2} \quad \text{for the critical condition when } P = \text{critical load } P_{cr}, \text{ limiting load between stability and instability} \tag{14.12c}$$

To establish an expression for P_{cr} free from deformation values, let us find the relationship of λ to Δ using the similar triangles ABF and DBE (Fig. 14-14b), assuming the displacements to be small.

$$\frac{\Delta/2}{L/2} = \frac{\lambda/2}{\Delta}, \qquad \lambda = \frac{2\Delta^2}{L} \tag{14.13}$$

The critical load P_{cr} is obtained by substituting Eq. (14.11) into Eq. (14.13c), eliminating λ.

$$\frac{P_{cr}}{2} \times \frac{2\Delta^2}{L} = \frac{k\Delta^2}{2}, \qquad P_{cr} = \frac{kL}{2} \tag{14.14}$$

Illustrative Problem 6. Find the critical load for the assembly shown in Fig. 14-14, consisting of a rigid vertical bar loaded axially with load P and a torsional resisting spring at the base.

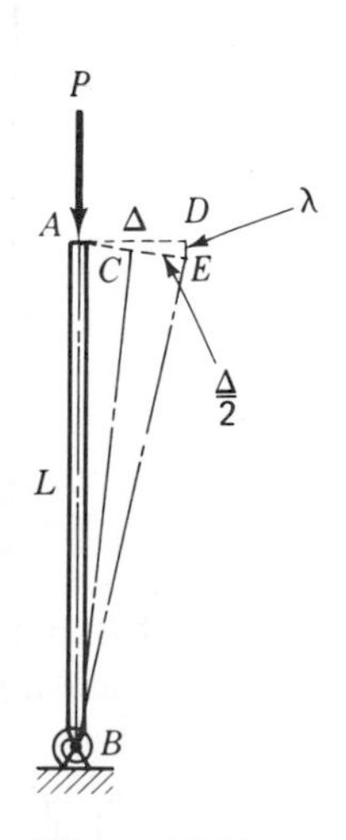

Figure 14-14

SOLUTION. Assuming a small lateral displacement and the similar triangles ADE and BCE,

$$\frac{\Delta/2}{L} = \frac{\lambda}{\Delta}, \qquad \lambda = \frac{\Delta^2}{2L}$$

Equating the critical external and internal energies and recalling that P_{cr} remains constant during the displacement λ,

$$P\lambda = \frac{k\theta^2}{2}$$

Inserting the previously computed value of λ and $\theta = \Delta/L$, we obtain

$$P_{cr} \cdot \frac{\Delta^2}{2L} = \frac{k\Delta^2}{2L^2}; \qquad \text{thus } P_{cr} = \frac{k}{L}$$

14-10 Instability—Slender Columns

External and internal energies may also be equated to determine the critical loads producing instability in tall, slender columns.

In the hinge-ended slender column shown in Fig. 14-15, it is assumed

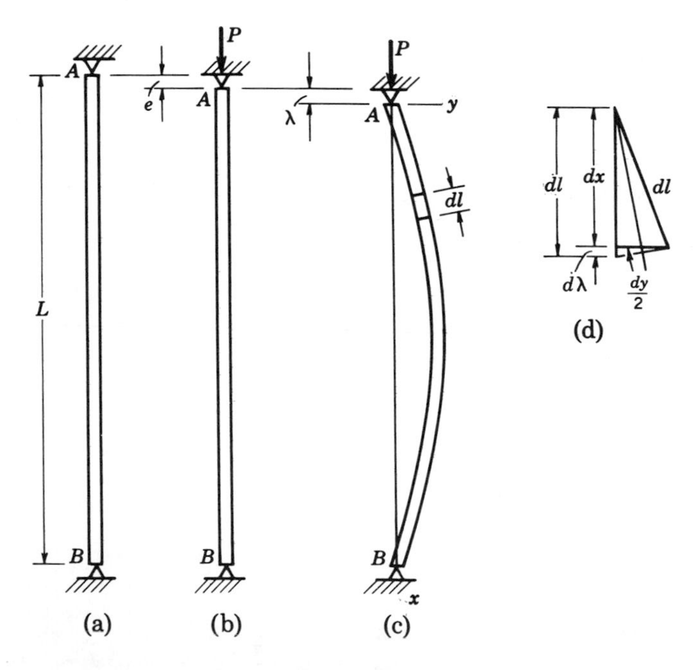

Figure 14-15

that load P first compresses the column a value of e, and then is assisted by means of a temporarily applied lateral force to deflect the column into the bent position shown. The value of λ through which the constant force P moves is obtained by accumulating the differences between the inclined dl lengths and their vertical projections.

$$\frac{d\lambda}{dy} = \frac{dy/2}{dx}$$

$$d\lambda = \frac{1}{2}\frac{dy}{dx}dy$$

If the right side is now multiplied by dx/dx, there results

$$d\lambda = \frac{1}{2}\left(\frac{dy}{dx}\right)^2 dx$$

or

$$\lambda = \int_0^L \frac{1}{2}\left(\frac{dy}{dx}\right)^2 dx \tag{14.15}$$

Due to the movement of the constant load P through λ, the external energy will then be equal to

$$U_E = P\lambda = P\int_0^L \frac{1}{2}\left(\frac{dy}{dx}\right)^2 dx \tag{14.16}$$

The internal energy of a bent member is equal essentially to that developed by the flexural stresses—and is, according to §14-3,

$$U_f = \int \frac{M^2\,dx}{2EI} = \int_0^L \frac{[EI(d^2y/dx^2)]^2\,dx}{2EI} = \frac{EI}{2}\int_0^L \left(\frac{d^2y}{dx^2}\right)^2 dx \tag{14.17}$$

If we assume the deflected curve of the column to have the shape of a sine wave,

$$y = A\sin\frac{\pi x}{L}$$

its first and second differentials will be

$$\frac{dy}{dx} = A\frac{\pi}{L}\cos\frac{\pi x}{L} \tag{14.18}$$

and

$$\frac{d^2y}{dx^2} = -A\frac{\pi^2}{L^2}\sin\frac{\pi x}{L} \tag{14.19}$$

Substituting Eqs. (14.18) and (14.19) into Eqs. (14.16) and (14.17) and equating to obtain the critical load, we have

$$P_{cr} = \frac{U_f}{\lambda} = \frac{\frac{EI}{2}\int_0^L \left(-A\frac{\pi^2}{L^2}\sin\frac{\pi x}{L}\right)^2 dx}{\frac{1}{2}\int_0^L \left(A\frac{\pi}{L}\cos\frac{\pi x}{L}\right)^2 dx} = \frac{\pi^2 EI}{L^2} \times \frac{\int_0^L \sin^2\frac{\pi x}{L}\,dx}{\int_0^L \cos^2\frac{\pi x}{L}\,dx} \tag{14.20}$$

With each of the integrals having the value of $L/2$, there results

$$P_{cr} = \frac{\pi^2 EI}{L^2} \tag{14.21}$$

This limiting or critical load that a tall, slender column can withstand is the same as given by the well-known Euler column formula, previously derived from a stress point of view. This stability study not only verified its authenticity but convincingly proved that once the critical load of an ordinary tall, slender column is obtained there is no chance for the column to remain straight.

14-11 Instability—Slender Column Supported Laterally by Springs

Should the tall, slender column of the previous section be supported laterally by springs placed along its entire length (Fig. 14-16), its critical load P_{cr} can be obtained in the same manner using external and internal energies.

In this problem the external energy $P\lambda$ applied to the column is used to produce the internal resistive energy in the springs as well as in the column. Thus,

$$P_{cr} = U_c + U_s \tag{14.22}$$

and since

$$\lambda = \frac{1}{2}\int_0^L \left(\frac{dy}{dx}\right)^2 dx$$

$$U_c = \frac{EI}{2}\int_0^L \left(\frac{d^2y}{dx^2}\right)^2 dx$$

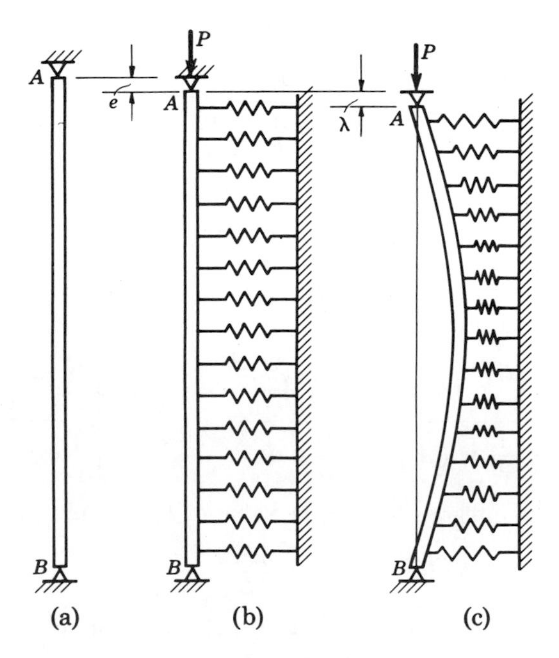

Figure 14-16

and

$$U_s = \frac{1}{2}\int_0^L ky^2\,dx$$

$$P_{cr} = \frac{\dfrac{EI}{2}\displaystyle\int_0^L \left(\frac{d^2y}{dx^2}\right)^2 dx + \frac{k}{2}\int_0^L y^2\,dx}{\dfrac{1}{2}\displaystyle\int_0^L \left(\frac{dy}{dx}\right)^2 dx} \tag{14.23}$$

Assuming the same sine wave as in the previous problem,

$$P_{cr} = \frac{EI\displaystyle\int_0^L \frac{A^2\pi^4}{L^4}\sin^2\frac{\pi x}{L}\,dx + k\int_0^L A^2\sin^2\frac{\pi x}{L}\,dx}{\displaystyle\int_0^L \frac{A^2\pi^2}{L^2}\cos^2\frac{\pi x}{L}\,dx} \tag{14.24}$$

$$P_{cr} = \frac{\pi^2 EI}{L^2} + \frac{kA^2\dfrac{\pi}{2}}{\dfrac{A^2\pi^2}{L^2}\dfrac{\pi}{2}} = \frac{\pi^2 EI}{L^2} + \frac{kL^2}{\pi^2} \tag{14.25}$$

The second term reflects the contribution offered by the springs in sustaining the column load. As L becomes longer it is apparent that the effect of these springs increases rapidly.

PROBLEMS

14-12. A rigid column, l ft long and hinged at the bottom, has two similar springs attached to the top in the same vertical plane (Fig. 14-17). Determine the critical load P_{cr} applied to the slightly rotated column that will just cause failure of stability. P remains constant during the deflection. *Answer:* $2kL$.

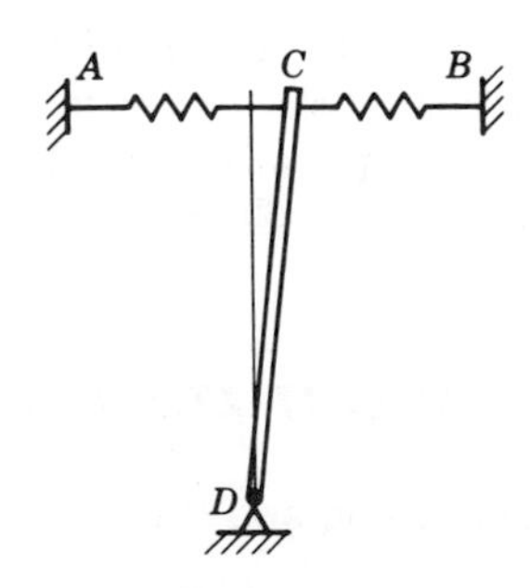

Figure 14-17

14-13. If the column of Problem 14-12 had the two springs attached at its midpoint, what would the critical load be?

14-14. In Fig. 14-18 the middle third of the rigid bar assembly remains vertical while subjected to load P. P is assumed to increase linearly with the deflection. Find the critical load. *Answer:* $\frac{2}{3}kL$.

14-15. Find the critical load for the rigid bar assembly shown in Fig. 14-19. P varies linearly with the deflection.

14-16. Find the critical load for the rigid bar assembly shown in Fig. 14-20. P varies linearly with the deflection. *Answer:* $\frac{4}{9}kL$.

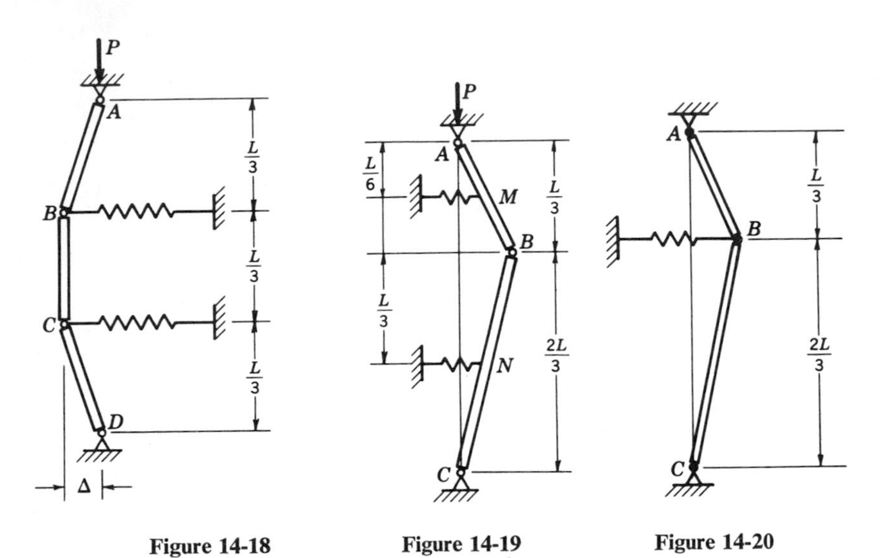

Figure 14-18 Figure 14-19 Figure 14-20

14-17. A tall, slender column is fixed at its lower end and free to sway in one vertical plane at its top. If two sets of equal-strength springs were to be placed in this same vertical plane, one on each side of the column to resist swaying, what critical load applied at the top would just produce instability? P remains constant during deflection. The curve of the column is sinusoidal.

14-18. The middle half of a tall, slender column hinged at each end is supported against swaying by a continuous system of equal-strength springs. What is the critical load applied at the top that will just produce instability? P remains constant during deflection. Assume column curve to be sinusoidal. *Answer:* $(\pi^2 EI/L^2) + (0.409kL^2/\pi^2)$.

IMPACT LOADS

14-12 Difference Between Static and Impact Loads

When a blacksmith forges a red-hot piece of steel, he subjects the steel to repeated blows from his heavy hammer. These blows, delivered with tremendous force, cause large deformations which form the material into its desired shape. It is evident that the faster the blacksmith can propel the hammer, the greater the energy he possesses and the greater will be the resultant deformation. The force which is produced by the hammer *by virtue of its sudden change of velocity is called an impact load* as differentiated from the force of the dead weight of the hammer, which is called its *static load.*

14-13 Stress-Strain Diagram—Static and Impact Loading

With the introduction of the static-load stress-strain diagram in Chapter 3, it was shown that for many materials the unit stress is proportional to the unit strain below the proportional limit (Fig. 14-21). All the basic design

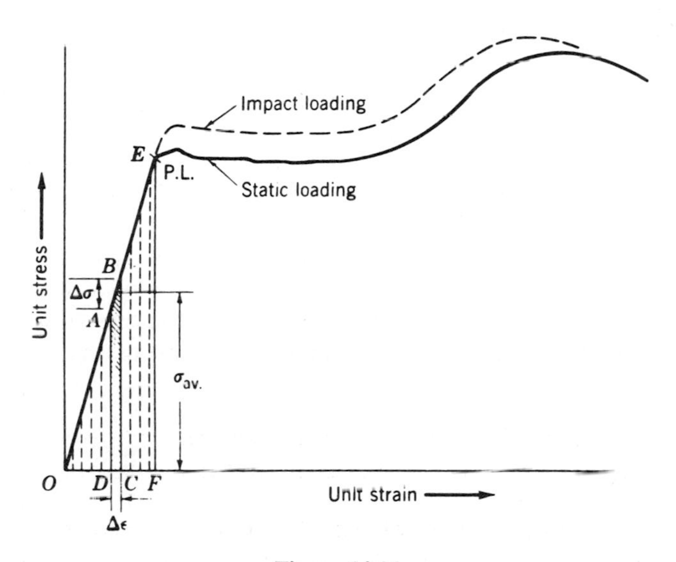

Figure 14-21

relationships employing static loads are dependent upon the proportionality of stress and strain. Through the use of modern strain gages, it has been shown that this important relationship is approximately equal for either static or impact loads. It is thus possible to use the proportionality of stress and strain as given by the static-load stress-strain diagram in formulating the principles used in designing members subjected to impact loading.

Further study of Fig. 14-21 shows that, although the stress-strain diagram for impact loading will have a higher yield point and ultimate unit stress than that recorded for a static loading, the amount of elongation is appreciably less. The moduli of resilience and toughness for both static and impact tests are sufficiently similar, however, for interchangeable use in either type of design (see §§3-8 and 3-9).

14-14 When Kinetic Energy Is Delivered to a Resisting Member

When a moving load is suddenly stopped by a resisting member, its kinetic energy is dissipated into all those bodies taking part in the collision. These bodies include not only the members that take the full force of the impact, but every supporting, loading, or restraining member no matter how lightly stressed. Even the surrounding air acts as a means of dissipating the heat energy generated by the impact. The uncertainty of the nature and magnitude of the dissipation of energy makes the design of structural and machine members less accurate than those subjected to static loads only.

Assuming, however, that the net energy of an impact load applied to a certain member could be accurately found, its diffusion into the volume of this member would be of next concern. If the member were of uniform cross section and the applied energy were spread uniformly throughout its interior it would be capable of sustaining the maximum amount of energy for a member of that volume. This ideal situation is complicated in practice with the inclusion of irregular cross sections, holes, screw threads, and many other forms of stress concentrations. Since the stress increases more rapidly at a stress concentration than elsewhere in the member, it arrives at the ceiling value of permissible energy absorption before points of lower stress. This condition curtails the amount of energy absorption that the other portions of material included in the member are capable of withstanding.

14-15 Calculation of Impact Stresses in Tension or Compression Member of Uniform Cross Section

Consider the steel rod *AB* of Fig. 14-22 to have a uniform cross section *a*, a length *l*, and to be suspended from some stationary point *A*. If weight *W* were now allowed to fall onto the flange at *B* from a height *h*, its kinetic energy at the time of impact would be *Wh*. However, owing to the elongation *e*, the weight will have moved a distance of $h + e$ before coming to rest. The total energy derived from the fall is, therefore, $W(h + e)$.

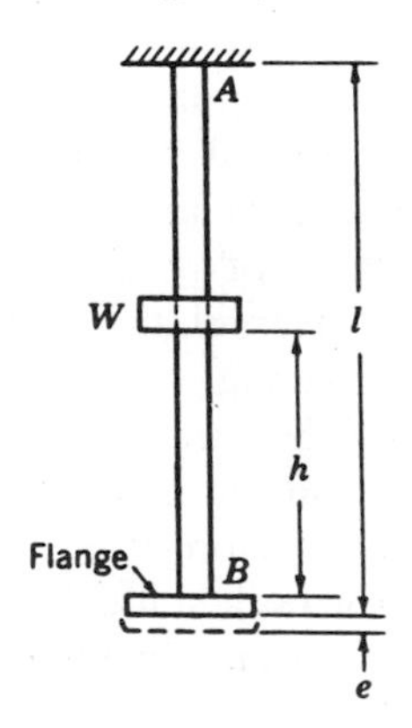

Figure 14-22

Although some energy will have been absorbed by the weight and the air during impact, let us suppose a complete transfer into the rod. Assume also the maximum stress incurred to be below the elastic limit and to be acting on each cubic inch of material in the rod. The kinetic energy applied by the weight will then be equal to the total internal energy absorbed by the rod as given by the following relationship.

$$U = W(h + e) = \frac{1}{2}\frac{\sigma^2}{E}al \tag{14.26}$$

Solving for σ, we obtain

$$\sigma = \sqrt{\frac{2UE}{al}} = \sqrt{\frac{2WE(h + e)}{al}} \tag{14.27}$$

If, under the conditions of impact, e is insignificant when compared to h, it may be eliminated from Eq. (14.27). This permits the equation to be solved, since e is dependent on σ.

Should the deformation e be appreciable, a solution is obtainable by substituting for W/a the static stress σ_q and for E/l its equal σ/e. Equation (14.27) thus becomes

$$\sigma = 2\sigma_q\left(\frac{h}{e} + 1\right) \tag{14.28}$$

Recalling the proportionality between unit stress and unit strain below the proportional limit, we may write

$$\frac{\sigma}{\epsilon} = \frac{\sigma_q}{\epsilon_q}$$

in which ϵ_q is the static strain caused by the weight W. Dividing each side by l gives

$$\frac{\sigma}{\epsilon l} = \frac{\sigma_q}{\epsilon_q l}$$

which is equal to

$$\frac{\sigma}{e} = \frac{\sigma_q}{e_q} \quad \text{or} \quad e = \frac{\sigma e_q}{\sigma_q}$$

Substituting this value into Eq. (14.28) and rearranging gives

$$\sigma^2 - 2\sigma\sigma_q + \sigma_q^2 = \sigma_q^2 + \frac{2\sigma_q^2 h}{e_q}$$

from which

$$\sigma = \sigma_q + \sigma_q\sqrt{1 + \frac{2h}{e_q}} \tag{14.29}$$

Equation (14.29) reveals that the stress developed by a falling weight acting on a uniform cross-sectioned tension or compression member may greatly exceed that obtained by the same weight when gradually applied.

To obtain the deformation of a tension or compression member when acted upon by an axially applied energy load, the values of $\sigma = Ee/l$ and $\sigma_q = Ee_q/l$ are substituted in Eq. (14.29). Thus,

$$e = e_q + e_q\sqrt{1 + \frac{2h}{e_q}} \tag{14.30}$$

14-16 Axially Applied Impact Loads on Tension or Compression Members of Nonuniform Cross Section

Let us compare the permissible energy loads of the two tension members of Fig. 14-23 whose static strengths are equal.

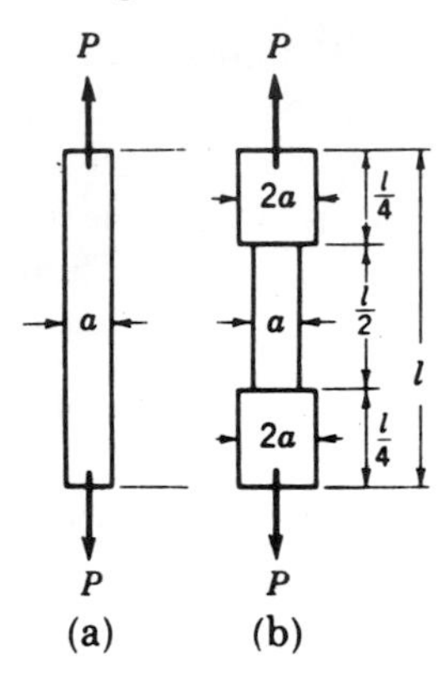

Figure 14-23

The uniform cross-sectional bar of Fig. 14-23a will, upon the application of an impact stress σ to any cross section, have an imposed energy load of $\frac{1}{2}(\sigma^2/E)al$.

In the member of Fig. 14-23b, however, the stress in the upper and lower sections will be equal to one-half that in the middle section. If the stress in the middle section is σ, the total energy in the member will be

$$
\begin{aligned}
U &= \frac{1}{2}\frac{\sigma^2}{E}a\frac{l}{2} + 2 \times \frac{1}{2}\frac{\sigma^2}{4E}2a\frac{l}{4} \\
&= \frac{\sigma^2}{4E}al + \frac{\sigma^2}{8E}al = \frac{3}{8}\frac{\sigma^2}{E}al
\end{aligned} \tag{14.31}
$$

which is less than that for the uniform cross-sectioned bar of lesser volume.

The reduction of the permissible energy load in the larger member can be traced to the fact that in the enlarged ends the stress decreases to $\sigma/2$ as the area increases to $2a$. The energy-per-unit volume, being proportional to σ^2, decreases to one fourth that value in the middle portion, whereas its volume only increases twice.

The foregoing discussion reveals the startling fact that members subjected to impact loading must be designed not only on the basis of static loading but on energy loading as well. It is also shown that those members possessing maximum impact resistance have each unit of volume equally stressed. Adding material at certain cross sections may weaken rather than increase the resistance of a member to impact loads. Conversely, the removal of material from an axially loaded member may increase its impact resistance. An excellent example of such an increase in impact strength is revealed in the reduction of the shank diameter of threaded bolts. The larger diameter threads cause severe stress concentrations. By making the shanks smaller, the stress more nearly approximates the maximum stress at the concentration, thereby increasing the energy resistance of the bolt.

PROBLEMS

14-19. A weight of 60 lb drops 12 in. onto the top of a cast-iron cylinder 6 in. in diameter by 12 in. high. If the loading is axial, what compressive stress is induced?

14-20. What would the compressive unit stress be in Problem 14-19 if the cylinder were 4 in. in diameter by 6 in. high? *Answer:* 16,900 psi.

14-21. Which of the two bodies shown in Fig. 14-24 will absorb the most energy when acted upon by the same energy load and limited to a maximum stress equal to σ?

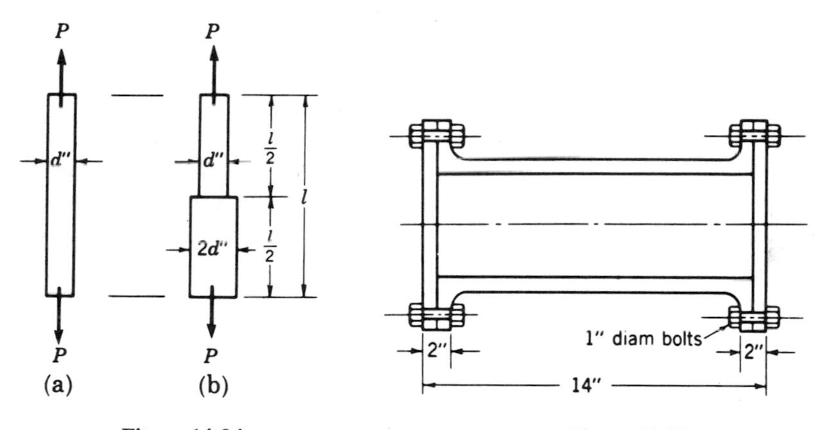

Figure 14-24

Figure 14-25

14-22. An internal explosion in the chamber shown in Fig. 14-25 resulted in the rupture of the short 1-in.-diameter (2-in. active length) bolts. Would the use of longer bolts (14-in. active length), passing through both plates, be more resistant to a similar energy load? Prove your answer. *Answer:* The smaller bolt will have 2.64 times more stress than the larger bolt.

14-23. A load of 600 lb falls 2 in. on a stop at the end of a vertical steel rod 12 ft long and 1 in.^{2} in cross section. Compute the stress and elongation produced in the bar.

14-24. Find the greatest height from which the weight in Problem 14-23 can fall without exceeding an allowable working unit stress of 20,000 psi in the bar. *Answer:* 1.6 in.

14-25. How long would a 1-in.-diameter steel rod have to be to absorb the energy load of a 50-lb weight falling 4 ft and to attain a maximum tensile stress of 20,000 psi? Suppose the rod were bronze?

14-26. A steel rod must elongate 0.010 in. when subjected to a falling weight of 50 lb. If its diameter is $\frac{3}{4}$ in. and the weight is dropped the full length of the rod, how long must it be? *Answer:* 3.64 in.

14-17 Effect of Impact Loads on Beams

Because the energy of resistance in any body depends on the stress applied to it, it follows that the resisting energy of a beam depends on the moment diagram of the imposed loads. This means that the distribution of load as well as the type of beam used makes a considerable difference in the resistance of the beam to impact loads.

It should also be noted that because the moment and stress generally vary along the length of a beam, it is not possible to expect as high an impact resistance in a beam as it would be in a tension member of the same dimensions.

Consider a simple beam having a rectangular cross section on which a concentrated load W falls on its midpoint from a height h. The external energy will be equal to $U = W(h + \Delta)$, where Δ is the maximum deflection of the beam. This maximum deflection will be equal to $\Delta = \frac{1}{48}Pl^3/EI$ (Chapter 8), P being the equivalent static load required to produce it. The maximum moment produced by load P is, of course, equal to $Pl/4$. Placing these expressions in $\frac{1}{2}P\Delta$, the external energy developed by the equivalent static load, there results

$$\begin{aligned} U = \frac{1}{2}P\Delta &= \frac{1}{2} \times \frac{4M}{l} \times \frac{1}{48}\frac{Pl^3}{EI} \\ &= \frac{1}{6}\frac{\sigma^2}{c^2}\frac{Il}{E} \end{aligned} \tag{14.32}$$

Inserting the value of $I = \frac{1}{12}bd^3$ for a rectangular beam, there is obtained

$$U = \frac{1}{18}\frac{\sigma^2}{E}al = \frac{1}{9}\left(\frac{\sigma^2}{2E}\right)al \tag{14.33}$$

A comparison of Eqs. (14.26) and (14.33) reveals the relative ineffectiveness of a constant cross-sectioned beam in developing internal flexural energy when contrasted to a tension member of similar size. When it is realized that the maximum stress is developed theoretically at only two points in the beam, and all others will have $\sigma^2/2E$ values less than they are capable of carrying, it is not surprising that the resistance offered is so low.

In order to improve the energy-carrying capacity of beams, I sections are used so that a greater volume of the material will be subjected to a higher stress. Beams are also made with varying section moduli approximately proportional to the moment diagram of the applied loads.

PROBLEMS

14-27. Derive the expression for the energy that a circular cross-sectioned simple beam can withstand when subjected to an impact load applied at its midpoint. *Answer:* $U = \frac{1}{12}(\sigma^2/2E)al$.

14-28. Compare Eq. (14.33) with the energy developed in a cantilever beam of similar size subjected to an impact load at its free end.

15

Miscellaneous Topics

CURVED BEAMS

15-1 Introduction

In contrast to an ordinary straight beam, the theory of which is presented in §7-2, a curved beam when subjected to bending, produces flexural stresses which do not vary directly with the distance from the neutral axis (Fig. 15-1). Neither does the neutral axis coincide with the centroidal axis. As will be shown by the theory presented in the succeeding paragraphs, the magnitudes of these divergences are dependent on the ratio of the depth of the beam to its degree of curvature.

15-2 Nonlinear Stress Variation Explained

Let Fig. 15-1 represent an infinitesimal portion of a curved beam located between the two radii AO and CO. Consider it also to have a constant cross section which is symmetrical about the plane of bending (or vertical axis). As such, the horizontal centroidal axes of all the intervening sections would form a plane curve.

Suppose, now, the couples M are applied to the beam in the plane of bending. The radial sections AB and CD, which are assumed to remain plane under stress, will rotate through an angle $d\alpha$ with respect to each other. Sec-

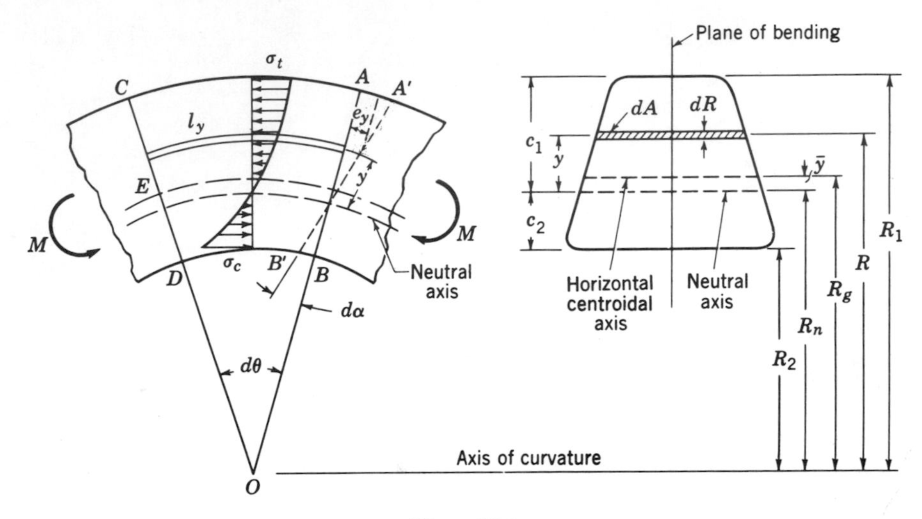

Figure 15-1

tion AB will move to its new position $A'B'$ producing deformations which vary linearly. The unit strains, however, cannot vary linearly since the fibers between sections AB and CD vary in length. Because of this nonlinear variation of unit strain, it follows that the unit stresses must also vary nonlinearly, since unit strain is proportional to unit stress (E is constant) (Fig. 15-1).

15-3 Curved-Beam Theory

Expressed mathematically, the unit strain for any fiber y distance from the neutral axis is equal to

$$\epsilon_y = \frac{e_y}{l_y} = \frac{y\,d\alpha}{(R_n + y)\,d\theta}$$

where y is positive above the neutral axis and negative below. Assuming Hooke's law to hold, the stress on this fiber is equal to

$$\sigma_y = E\epsilon_y = \frac{E\,d\alpha}{d\theta} \times \frac{y}{R_n + y} \tag{15.1}$$

An analysis of this equation reveals that for any one section, all its terms with the exception of y are constants. The ratio $y/(R_n + y)$ indicates, therefore, the variable nature of the stress.

Since the normal forces on any section must equal zero to insure equilibrium,

$$\int_{R_2}^{R_1} \sigma_y\,dA = \frac{E\,d\alpha}{d\theta} \int_{R_2}^{R_1} \frac{y}{R_n + y}\,dA = 0$$

Because the term $E\,d\alpha/d\theta$ is constant, it must necessarily follow that

$$\int_{R_2}^{R_1} \frac{y}{R_n + y}\,dA = 0 \tag{15.2}$$

Placing $y = R - R_n$, there is obtained

$$\int_{R_2}^{R_1} \frac{R - R_n}{R}\,dA = 0, \quad \text{from which} \quad R_n = \frac{A}{\int_{R_2}^{R_1} \frac{dA}{R}}$$

and

$$\bar{y} = R_g - \frac{A}{\int_{R_2}^{R_1} \frac{dA}{R}} \tag{15.3}$$

It is thus revealed that in bending a curved beam the neutral axis shifts away from the horizontal centroidal axis and toward the axis of curvature of the beam. The integral appearing in the equation is evaluated by placing dA in terms of dR.

It is also true that the resisting moment of the normal forces acting on the section (*C-T* couple) must equal the applied moment. This statement may be expressed by

$$\int_{R_2}^{R_1} y\sigma_y\,dA = \frac{E\,d\alpha}{d\theta} \int_{R_2}^{R_1} \frac{y^2\,dA}{R_n + y} = M \tag{15.4}$$

The expression within the integral may be simplified as follows:

$$\int_{R_2}^{R_1} \frac{y^2\,dA}{R_n + y} = \int_{R_2}^{R_1} y - R_n\left(\frac{y}{R_n + y}\right) dA = \int_{R_2}^{R_1} y\,dA - R_n \int_{R_2}^{R_1} \frac{y}{R_n + y}\,dA$$

From Eq. (15.2) previously derived, the second integral was shown to equal zero, thus making

$$\int_{R_2}^{R_1} \frac{y^2\,dA}{R_n + y} = \int_{R_2}^{R_1} y\,dA = \bar{y}A$$

Inserting this simplified expression in Eq. (15.4),

$$\frac{M}{\bar{y}A} = \frac{E\,d\alpha}{d\theta}$$

in which case, Eq. (15.1) may be written

$$\sigma_y = \frac{M}{\bar{y}A} \times \frac{y}{R_n + y} \tag{15.5}$$

This is a general expression for flexural stress in a curved beam. If the values of $y = c_1$ and $y = c_2$ are inserted in the above formula to obtain the outside fiber stresses, there are obtained

$$\sigma_1 = \frac{Mc_1}{\bar{y}AR_1} \tag{15.6}$$

and

$$\sigma_2 = \frac{Mc_2}{\bar{y}AR_2} \tag{15.7}$$

The value of the greater of the two stresses, σ_2, will be located on the inside of the beam. It should be remembered that these stresses are developed by the application of pure bending only.

Illustrative Problem 1. A critically stressed rectangular cross section of a steel crane hook (Fig. 15-2) has a width of 1 in. and a depth of 4 in. If R_1

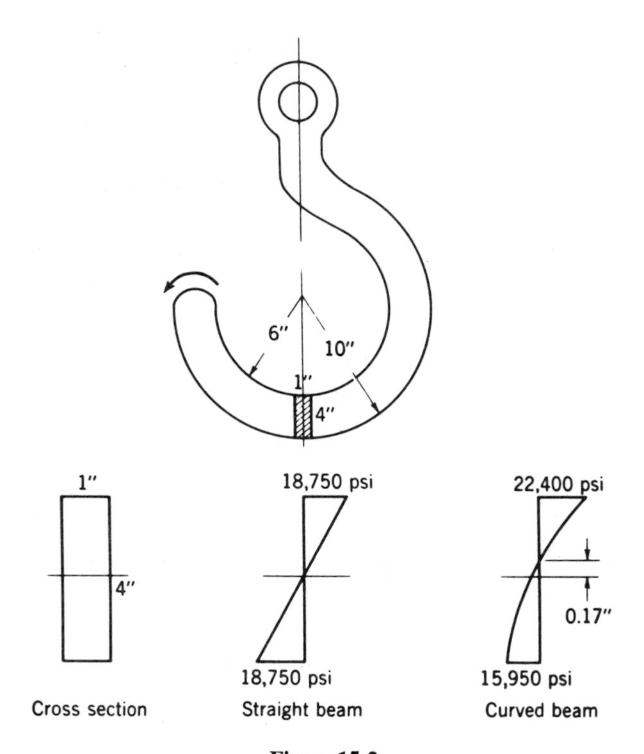

Figure 15-2

and R_2 at this section are respectively 10 and 6 in., and $M = 50{,}000$ in.-lb, determine its maximum tensile and compressive stresses. Compare these stresses to those obtained with the straight-beam formula.

SOLUTION. The position of the neutral axis must first be located.

$$\bar{y} = R_g - \frac{A}{\int_{R_2}^{R_1} \frac{dA}{R}} = 8 \text{ in.} - \frac{4}{\int_{R_2}^{R_1} \frac{1 \times dR}{R}} = 8 - \frac{4}{\ln R \Big]_{R_2}^{R_1}}$$

$$= 8 - \frac{4}{2.3026 \log(10/6)} = 8 - \frac{4}{2.3026 \times (1.000 - 0.788)}$$

$$= 8 - 7.83 = 0.17 \text{ in.}$$

The maximum fiber stresses are now found from Eqs. (15.6) and (15.7):

$$\sigma_1 = \frac{50{,}000 \times 2.17}{0.17 \times 4.0 \times 10} = 15{,}950 \text{ psi}$$

$$\sigma_2 = \frac{50{,}000 \times 1.83}{0.17 \times 4.0 \times 6} = 22{,}400 \text{ psi}$$

Straight-beam comparison:

$$\sigma = \frac{50{,}000 \times 2.0}{\frac{1}{12} \times 1 \times 4^3} = 18{,}750 \text{ psi}$$

Comments. It is observed from Illustrative Problem 1 that the stress on the inner radius of a curved beam increases as its initial curvature increases.

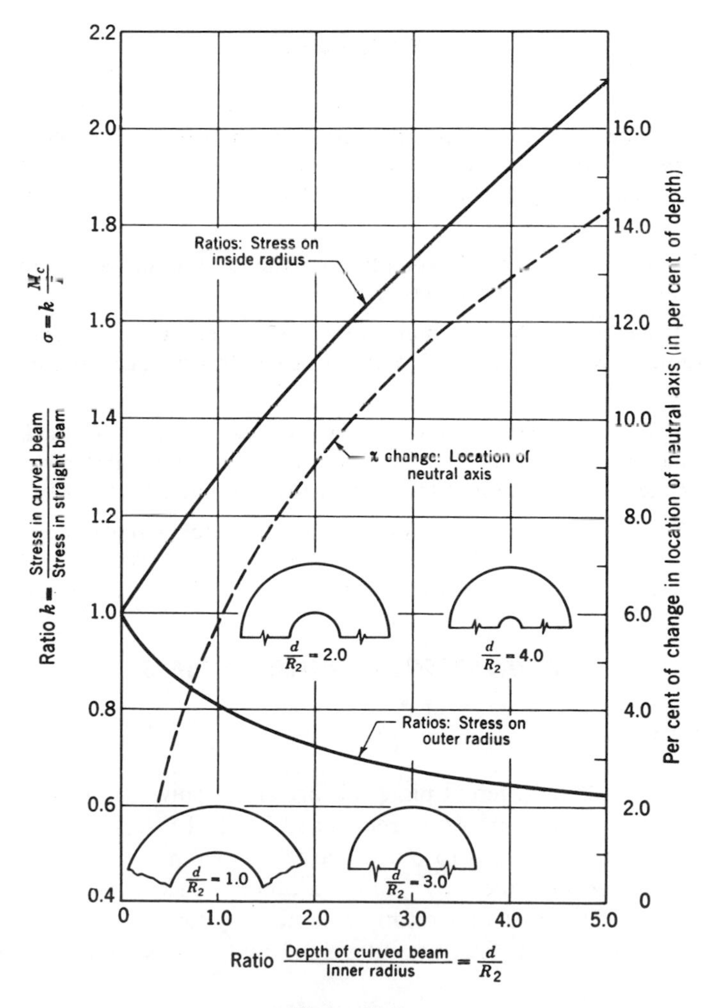

Figure 15-3

Contrariwise, the stress on the outer radius decreases. The variation of these stresses to the ratio of d/R_2 is shown in Fig. 15-3. The percentage of change in position of the neutral axis as the curvature increases is also shown.

It becomes apparent as a study is made of Fig. 15-3 that a beam, as commonly used, could have a considerable curvature before the ordinary flexure formula need be discarded. It is only as a relatively deep beam is bent into a C shape or a portion thereof that one need pay particular attention to curved-beam theory.

It should be carefully noted that these curves apply to rectangular beams subjected to bending moment only. Should curved beams of other cross sections be subjected to bending, the theory would have to be applied specifically to them by a proper use of Eq. (15.3).

PROBLEMS

15-1. In Problem 10-14, the stresses incurred were obtained in using the straight-beam flexure theory. Apply the curved-beam theory to obtain the maximum normal stresses acting on a cross section, if the inside radius of the T section is 6 in. *Hint:* For each rectangular section set $\int dA/R = \int b\, dR/R = b \log_e R$. *Answers:* -2072 psi, $+2868$ psi.

15-2. Derive the general expression for the location of the neutral axis of a curved beam of circular cross section.

15-3. Determine the maximum normal stress occurring on section *A-A* of the C frame shown in Fig. 10-13 if the inside radius is equal to 8 in. *Answers:* -5920 psi, $+6410$ psi.

15-4. A curved beam has a rectangular cross section 1 in. thick and 2 in. deep with an inside radius of 3 in. Show the variation of normal stress over a cross section of the curved portion of the beam if the moment is 5000 in.-lb.

15-5. A 2-in.-diameter steel rod is to be bent into a circular shape. To what mean radius must it be bent for a maximum stress of 20,000 psi to develop on a cross section when subjected to a moment of 11,820 in.-lb. *Answer:* 3.0 in.

NONHOMOGENEOUS BEAMS

15-4 Introduction

Beams made of different materials are commonly used in practice. For instance, timber beams are often reinforced by steel plates. Concrete beams are reinforced by steel rods to overcome their weakness in tension. Plastic materials are strengthened by gluing alumium sheets to their surfaces. The addition of these stronger supplementary materials to beams made of cheaper and weaker materials provides a flexibility in design which may result in appreciable economy. The theory relative to the flexural design of non-homogeneous beams will now be presented.

15-5 Composite Beams—Theoretical Design Principles

Let us consider the composite beam of Fig. 15-4, which consists of a timber beam reinforced with steel plates firmly fixed to its sides. When subjected to load P, every longitudinal element of the composite beam, dx

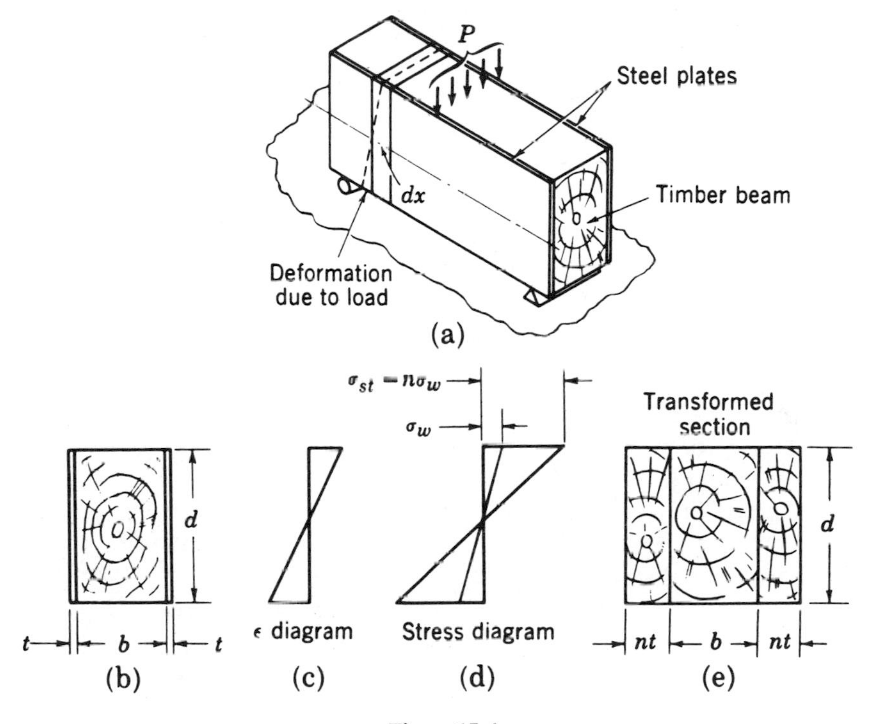

Figure 15-4

in length, is deformed in direct proportion to the distance from the neutral axis. The linear variation of this deformation produced without regard to the materials encountered implies that the unit strains across these materials will also vary in like manner. The flexure stresses, however, while varying linearly within each material ($\sigma = \epsilon E$), do not vary at the same rate because of the difference of their moduli of elasticity. (See Fig. 15-4d.) If $n = E_{st}/E_{wd}$, the stress in the higher modulus material will be n times the other.

Because of the dissimilar stress pattern on the cross section of the composite beam, the flexure formula cannot be used directly (see assumption,

§7-3). A modification of the cross section, however, will make its use possible. Let us investigate this interesting possibility.

The resisting moment of the entire cross section is obviously the sum of the resisting moments of the steel and wood. These individual resisting moments (see derivation of flexure formula) are

$$(M_R)_{wd.} = \frac{\sigma_{wd.}I_{wd.}}{c} \quad \text{and} \quad (M_R)_{st.} = \frac{\sigma_{st.}I_{st.}}{c}$$

where I_{wd} and I_{st} are the moments of inertia of the wood and steel cross section, respectively. Their total equals

$$M = M_R = \frac{\sigma_{wd.}I_{wd.}}{c} + \frac{\sigma_{st.}I_{st.}}{c}$$

But $\sigma_{st} = n\sigma_{wd}$ which makes

$$M = \frac{\sigma_{wd.}I_{wd.}}{c} + \frac{n\sigma_{wd.}I_{st.}}{c}$$

or

$$M = \frac{\sigma_{wd.}}{c}(I_{wd.} + nI_{st.})$$

This equation is now a combined flexural formula providing the resisting moment of a *transformed cross section* whose moment of inertia is $I_{wd} + nI_{st}$ and whose maximum stress is σ_{wd}. The section may be considered as that obtained by expanding the steel section laterally n times on either side of the given wood section. By that transformation it will act as though it were made entirely of wood (Fig. 15-4e).

Suppose, however, the reinforcing plates were placed on its top and bottom surfaces. Would the beam be stronger? Let us investigate.

Once again, the strains are proportional to the distance from the neutral axis, regardless of the material involved. At the interfaces the strains ϵ_2 in both materials are the same. The corresponding stresses will be $\epsilon_2 E_1$ and $\epsilon_2 E_2$, and if $E_1/E_2 = n$, the stress in the higher modulus material will be n times the other.

Because of the discontinuity in stress across the composite cross section, it must be modified to permit the use of the flexure formula. Let us investigate.

The C and T resisting forces developed on the cross section (Fig. 15-5) are the forces of two couples whose moments $C_1(d + t)$* and $C_2(\frac{2}{3}d)$, or $T_1(d + t)$ and $T_2(\frac{2}{3}d)$, equal the total resisting moment of the cross section. Their sum must remain the same for the modified cross section to insure equilibrium.

*The moment arm $(d + t)$ is approximate due to the variation of stress across the thickness of the steel plates. It is sufficiently accurate, however, for most engineering problems when the plates are relatively thin, as is the case in this illustration.

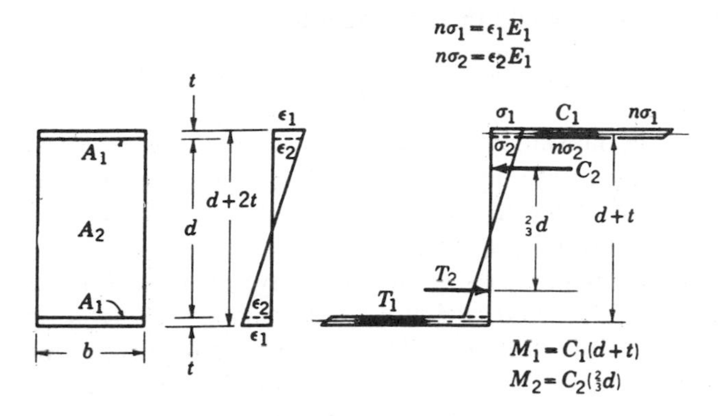

Figure 15-5

Let us use only the moment $C_1(d+t)$ or, more specifically, its factor C_1 in modifying our cross section, since the moment arm $(d+t)$ remains constant. The value of C_1, as given by Fig. 15-5, is equal to

$$C_1 = T_1 = \left(\frac{n\sigma_1 + n\sigma_2}{2}\right)A_1 \tag{15.8}$$

This may be modified by making

$$C_1 = T_1 = \left(\frac{\sigma_1 + \sigma_2}{2}\right)nA_1 \tag{15.9}$$

If we considered the steel sections to be expanded laterally n times, their areas would equal nA_1 and their centers of gravity would still be $(d+t)$ apart (Fig. 15-6). Holding C_1 constant, *the stresses on this transformed area would then be reduced n times to make its stress variation similar to that of the wood section.* The straight-line stress variation assumed in the derivation of the flexure formula would now be attained, permitting its use for the transformed section. As mentioned in the previous analysis, the transformed section will act as though it were made entirely of wood. The resisting moment

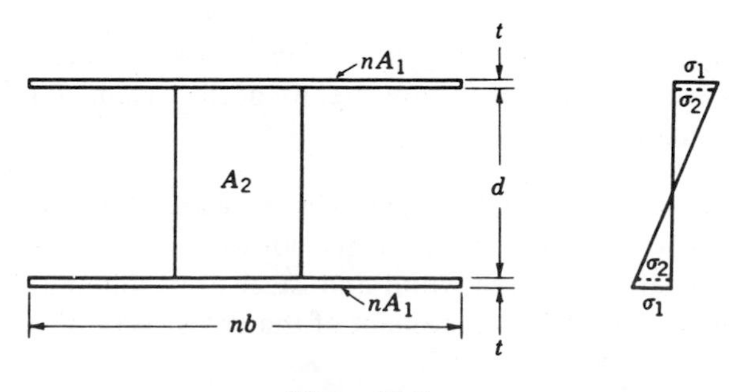

Figure 15-6

of the cross section would then be equal to

$$M = C_2 \times \frac{2}{3} d + C_1(d + t) = \frac{\sigma_1 I_T}{c_T} \tag{15.10}$$

where σ_1 = the maximum flexural stress in the transformed section, I_T = the moment of inertia of the transformed section about its neutral axis, and c_T = the distance from the neutral axis to the extreme fiber of the transformed section, i.e., $(d + 2t)/2$.

In a similar manner, the section may have been transformed by decreasing the width of the wood section n times, making the rate of stress increase similar to that of the steel plate. The beam would then act as though it were made entirely of steel and would have a transformed section as shown in Fig. 15-7. Its limiting stress would be $n\sigma_1$ and its moment of inertia would be that of the shaded area taken about the neutral axis.

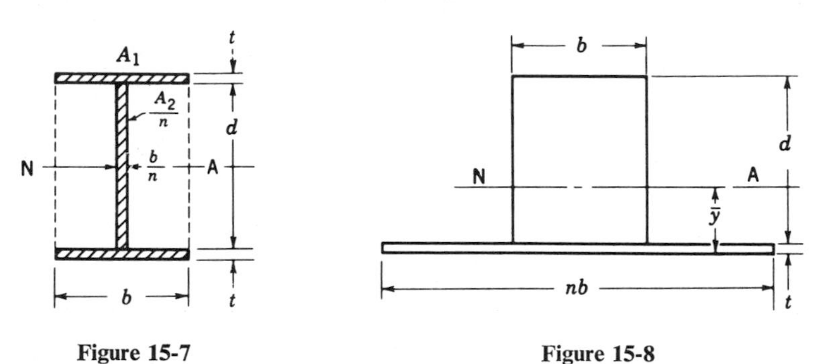

Figure 15-7

Figure 15-8

The foregoing analysis may also be applied to members reinforced on the bottom (or top) only. The actual stress on the reinforcing plate may be reduced to vary with that of the wood by expanding it laterally by the ratio of their moduli (Fig. 15-8). Or the wood may be decreased in width by the same ratio as indicated in the previous paragraph. In any event, the flexure formula would apply to the transformed section. The center-of-gravity axis of the transformed section would have to be found, of course, and the moment of inertia taken about that axis.

Shear investigation. The investigation of critical shear stresses proceeds in the same manner as that for homogeneous beams, using Eq. (7.6) and the transformed section.

Illustrative Problem 2. A timber beam 6 in. wide and 12 in. deep is reinforced by affixing a 6 by $\frac{1}{4}$-in. steel plate to its bottom face (Fig. 15-9). If the allowable flexure stresses in the wood and steel are 1400 and 20,000 psi, respectively, what is the resisting moment of the cross section? $E_{st.} = 2{,}000{,}000$ psi, $E_{wd.} = 30{,}000{,}000$ psi, $n = 15$.

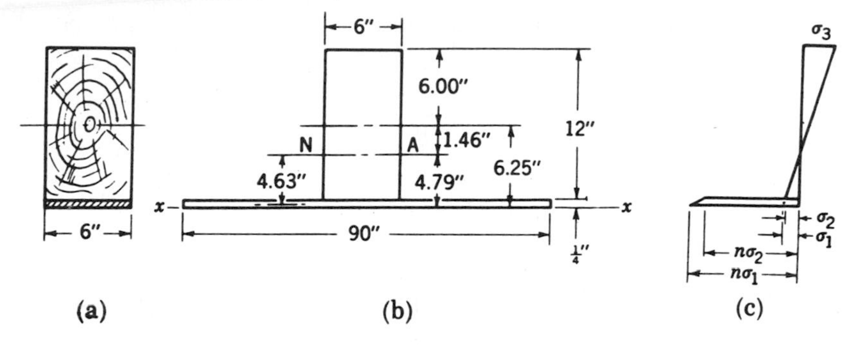

Figure 15-9

SOLUTION. The center of gravity of the transformed section shown in Fig. 15-9b is first obtained by taking moments about the x-x axis:

$$\begin{array}{lllll} & A & y & & Ay \\ 6 \times 12 = & 72.0 & \times\ 6.25 & = & 450.0 \\ 90 \times \frac{1}{4} = & 22.5 & \times\ 0.125 & = & 2.8 \\ & \overline{94.5} & & & \overline{452.8} \end{array}$$

$$\bar{y} = \frac{452.8}{94.5} = 4.79 \text{ in.}$$

The moment of inertia about the center-of-gravity (neutral) axis is then equal to

$$I_{\text{wd.}} = \tfrac{1}{12} \times 6 \times (12)^3 = 864$$

$$(Ad^2)_{\text{wd.}} = 72 \times \overline{1.46^2} = 154$$

$$I_{\text{st.}} = \frac{1}{12} \times 90 \times \left(\frac{1}{4}\right)^3 = 0.117 \qquad \text{(negligible)}$$

$$(Ad^2)_{\text{st.}} = 22.5 \times \overline{4.63^2} = \frac{482}{1500} \text{ in.}^4$$

To determine the resisting moment, the allowable stresses must be checked to ascertain perchance whether the attainment of one stress will cause the other to be exceeded. Suppose the maximum allowable stress of 20,000 psi ($n\sigma_1$) were to be attained in the steel. The corresponding σ_1 stress will then be 1333 psi. By similar triangles, the maximum stress in the timber (σ_3) is equal to 1333 × 7.46/4.79 or 2080 psi. This exceeds the allowable stress of 1400 psi in the wood. The stress of 20,000 psi in the steel, therefore, is impossible to attain without overstressing the wood. If, on the other hand, the maximum stress in the wood (σ_3) is set at its allowable value of 1400 psi, the corresponding maximum stress in the steel will be 1400 × 4.79/7.46 × 15 = 13,500 psi. Although neither material is overstressed, the stress in the steel is far below its maximum allowable value. The maximum resisting moment will be

$$M_R = \frac{\sigma_{\text{wd.}} I_T}{c_T} = \frac{1400 \times 1500}{7.46} = 281{,}000 \text{ in.-lb}$$

PROBLEMS

15-6. A beam consists of a timber 6 ft by 12 in. with a 12- by $\frac{1}{4}$-in. steel plate fastened to each of the 12-in. sides. What is the resisting moment of this section about the short axis under the following assumptions? Steel and timber act together without slipping. $E_{\text{st.}} = 30 \times 10^6$ psi, $E_{\text{wd.}} = 1{,}200{,}000$ psi, maximum allowable stress in steel = 18,000 psi, in timber = 1000 psi. *Answer:* 319,000 in.-lb.

15-7. A simply supported timber beam, 10 ft long, has a cross section 4 in. wide by 10 in. deep. If $\frac{1}{4}$- by 3-in. steel plates, also 10 ft long, were fastened to the top and bottom faces, what maximum flexural stresses would be developed in the timber and the steel due to a load of 5000 lb applied at the center of the span? $E_{\text{wd.}} = 1.5 \times 10^6$ psi, $E_{\text{st.}} = 30 \times 10^6$ psi. *Answers:* $\sigma_{\text{wd.}} = 668$ psi, $\sigma_{\text{st.}} = 14{,}050$ psi.

15-8. What maximum uniform load could be applied to the beam of Problem 15-7 if the maximum allowable stresses for the steel and timber are 20,000 and 1500 psi, respectively? Moduli of elasticity are the same.

15-9. A sheet of plastic laminate $\frac{3}{8}$ in. thick, 12 in. wide, and 36 in. long is to be strengthened by bonding sheets of Duralumin 0.05 in. thick to its top and bottom surfaces. What maximum concentrated load may be placed at its midsection if the sheet is simply supported at its ends and the maximum allowable stresses for the laminate and Duralumin are 2500 and 10,000 psi, respectively? $E_{\text{L}} = 2{,}000{,}000$ psi, $E_{\text{D}} = 10{,}500{,}000$ psi. Consider flexure alone. *Answer:* 301 lb.

15-10. A timber beam 6 in. wide by 12 in. deep has recessed into its upper face a steel strip $\frac{1}{2}$ in. thick by 2 in. wide (Fig. 15-10). If the beam is 10 ft long and is simply supported at each end, what maximum uniform load can it carry? Allowable stresses in timber and steel are 1200 and 20,000 psi, respectively. $E_{\text{st.}} = 2 \times 10^6$

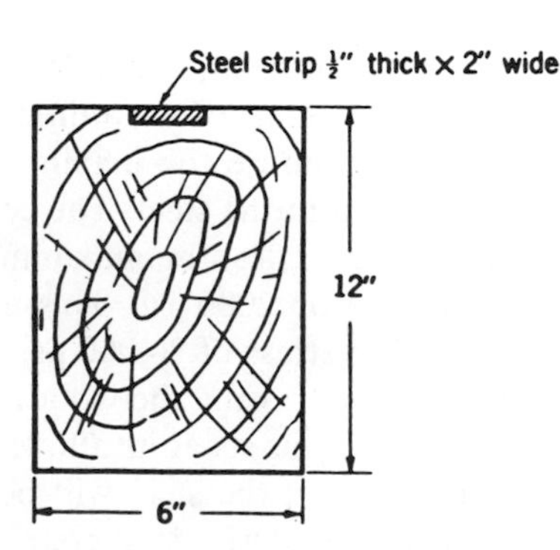

Figure 15-10

psi and $E_{st.} = 30 \times 10^6$ psi. What uniform load could it have carried if no steel had been used and the cross section had been the full 6 by 12 in. deep?

15-11. A rectangular cross section timber beam having a width of 6 in. and a depth of 10 in. is used as a cantilever, 8 ft long. If steel plates, 6 by $\frac{1}{4}$ in. in cross section are placed on the top and bottom faces, and the maximum allowable stress in the wood is 1000 psi, what is the maximum unit stress in the steel and the permissible uniform load on the beam? $E_{st.} = 30 \times 10^6$ psi, $E_{wd.} = 2 \times 10^6$ psi. *Answers:* 15,750 psi, 845 lb/ft.

15-12. Compute the needed thickness of a pair of steel side plates 10 in. in depth in order to reinforce an 8- by 12-in. (nominal) timber beam to carry an applied moment of 50 kip-ft. Assume the timber to have dressed dimensions $7\frac{1}{2}$ by $11\frac{1}{2}$ in. Assume allowable flexure strength of timber to be 1600 psi, and that for steel to be 18,000 psi. $n = 20$. Each plate to be centered on the $11\frac{1}{2}$-in. side. Disregard effect of bolt hole.

REINFORCED CONCRETE BEAMS*

15-6 Introduction

A concrete beam is reinforced with steel rods to overcome its weakness in tension. Its flexural resistance is determined by using a transformed section obtained by expanding laterally the area of the steel rods by the the ratio of the moduli of steel to concrete ($n = E_{st}/E_c$) and by *disregarding the strength of the tensile concrete*, Fig. 15-11. If the compressive concrete and tensile steel are stressed to their maximum allowable values *simultaneously*, the design of the beam is said to be *balanced* and its resisting moment (see Fig. 15-11) is equal to that developed by either stress.

$$M_R = \tfrac{1}{2}f_c bkdjd = \tfrac{1}{2}f_c kjbd^2 = Kbd^2 \qquad (15.11)$$

$$= nA_s\frac{f_s}{n}jd = A_s f_s jd \qquad (15.12)$$

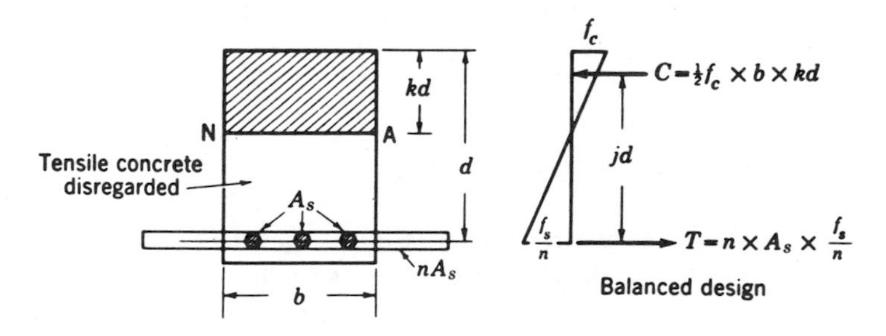

Figure 15-11

*Design provisions and nomenclature presented in accordance with ACI Standard "Building Code Requirements for Reinforced Concrete" (ACI 318–71)—Working Stress Design, except where noted.

Because of the simultaneous attainment of maximum allowable stresses in a balanced beam, it is also most economical and used whenever possible.

The design of a balanced beam is greatly simplified through the use of pertinent design data given in Table 10 for various values of f_c and f_s.

TABLE 10

CHARACTERISTICS FOR USE IN DESIGN OF REINFORCED CONCRETE BEAMS

f_c'*	n	f_c	f_s	k	j	p	K
5000	7	2250	20,000	0.440	0.853	0.0247	423
			18,000	0.467	0.844	0.0292	440
4000	8	1800	20,000	0.418	0.861	0.0188	324
			18,000	0.445	0.852	0.0222	341
3000	9	1350	20,000	0.378	0.874	0.0127	223
			18,000	0.403	0.866	0.0151	236
2500	10	1125	20,000	0.360	0.880	0.0101	178
			18,000	0.385	0.872	0.0120	189
2500	12	1125	20,000	0.403	0.866	0.0113	196
			18,000	0.429	0.857	0.0134	207
2000	15	900	20,000	0.403	0.866	0.0091	157
			18,000	0.428	0.857	0.0107	165

*f_c' = ultimate strength of concrete in 28 days, $f_c = 0.45 f_c'$, $K = \frac{1}{2} f_c jk$, $p = A_s/bd$.

With the values of K and M (or M_R) known, the dimensions of a rectangular beam are easily obtained from Eq. (15.11). The value of p may also be used instead of Eq. (15.12) to obtain the steel area required.

Simply supported slabs, reinforced in one direction only, are also designed using the above procedure. A 1-ft width of slab is used as the design unit, it being typical of every other 1-ft width.

15-7 Unbalanced Flexural Design

Under extraordinary circumstances one may be forced to use an unbalanced design in which the allowable stresses for the compressive concrete and tensile steel are *not* attained simultaneously. In such instances, Eqs. (15.11) and (15.12) cannot be used, and new values similar to k and j, as well as the actually impressed f_c and f_s stresses must be introduced to obtain a corrected resisting moment equation. Moreover, all investigations of existing or already designed concrete beams must be made on the assumption that their design is unbalanced.

An unbalanced condition encountered in either design or investigation usually involves the determination of the *critical stress*. Because the maximum allowable stresses in concrete and steel are not attained simultaneously, one of these materials must be stressed to its maximum critical value, the other to a value *less* than the allowable. When it is determined what these stresses are, the resisting moment may be obtained by finding the new values of k and j and introducing them into formulas similar to Eqs. (15.11) and (15.12). Of invaluable aid in the determination of the value of k is its relationship with p and n. It is derived as follows, using Fig. 15-12.

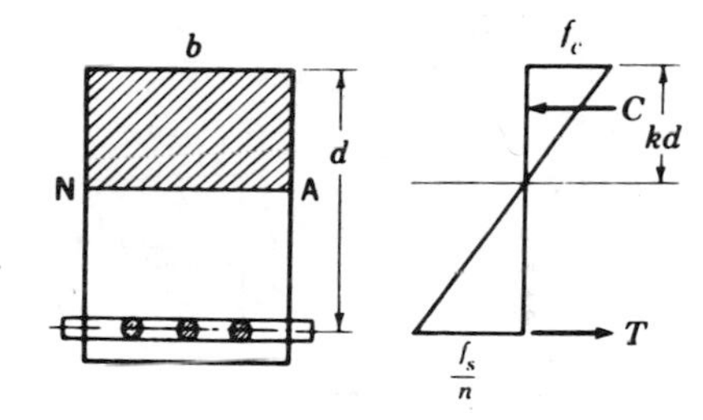

Figure 15-12

Knowing the neutral axis to be the center-of-gravity axis, regardless of the magnitude of stresses f_c and f_s we may write

$$bkd\frac{kd}{2} = nA_s(d - kd)$$

Inserting $A_s = pbd$ and simplifying,

$$\frac{k^2}{2}bd^2 = npbd^2 - npkbd^2$$

and

$$k^2 = 2np - 2npk$$

from which

$$k = \sqrt{2np + (np)^2} - np \tag{15.13}$$

15-8 Shear Investigation

In §7-14 it was shown that the vertical shear stress at any point in a beam develops an equal horizontal shear stress at the same point. The combined effect of these two shearing stresses produces an equal maximum tensile stress on a 45° plane to these shears. The effect of this tensile stress, which occurs in addition to the flexural tensile stress, must be curtailed, if excessive, with the use of steel reinforcement placed across the lines of potential cracking.

The shearing unit stress acting on a vertical section is assumed to be uniformly distributed over the area *bd*.

$$v = \frac{V}{bd} \tag{15.14}$$

The value of shear stress obtained by Eq. (15.14) is also equal to the diagonal tensile stress and is restricted to a value of $1.1\sqrt{f'_c}$ at a distance d from the supports of a beam whose web is unreinforced in shear.

15-9 Bond Investigation

Of equal importance to the shear stress in the design of reinforced concrete beams is the bond stress which occurs between the concrete and the steel bars. It must not be permitted to exceed a maximum allowable value. The magnitude of the bond stress may be derived as follows.

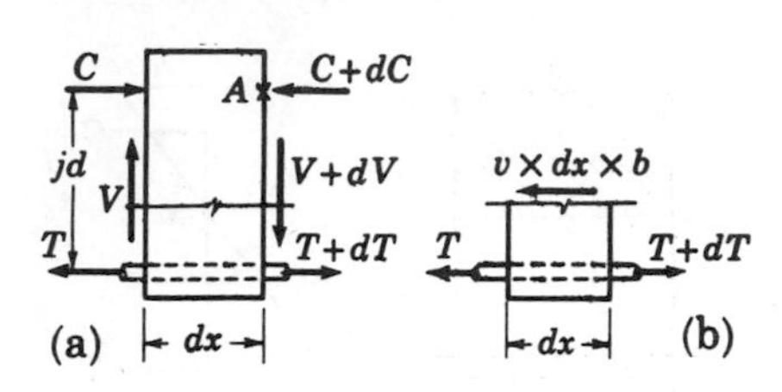

Figure 15-13

In Fig. 15-13b the unit bond stress occurring on the reinforcing rods is equal to the differential excess load dT divided by the peripheral area ($\Sigma o \times dx$) of the rods.

$$u = \frac{dT}{\Sigma o \times dx} \tag{15.15}$$

If now we let Fig. 15-13a represent a section of a loaded reinforced concrete beam, dx in length and b in width, and take moments about A, there is obtained

$$dT\, jd = V\, dx$$

Inserting the value of dT from this relationship into Eq. (15.15), there results

$$u = \frac{V}{\Sigma ojd} \tag{15.16}$$

The bond stress is limited by code to a value, for most cases,* equal to $(4.8\sqrt{f'_c})/D$ but not exceeding 500 psi, where D is the nominal diameter of the bar in inches.

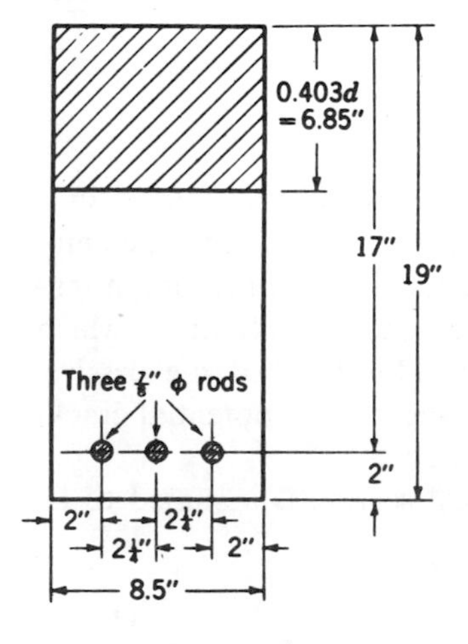

Figure 15-14

Illustrative Problem 3. A simply supported reinforced concrete beam (Fig. 15-14) having a length of 12 ft has applied to it a uniformly distributed live load of 2000 lb/ft. If it is to have a rectangular cross section in which $b = d/2$, what will its dimensions be if the maximum allowable f_s and f_c equal 18,000 and 1125 psi, respectively? $n = 10$. Check shear and bond.

The maximum design load is equal to

2000 lb/ft	live load
170	dead load (assumed)
2170 lb/ft	total weight of beam

*Top bars in horizontal beams are limited to a bond stress of $(3.4\sqrt{f'_c})/D$ but not exceeding 350 psi.

This load will produce a maximum moment of

$$M_{\max} = \frac{wl^2}{8} = \frac{2170 \times 144}{8} = 39{,}000 \text{ ft-lb}$$

SOLUTION. From Table 10 those design factors corresponding to the given f_s and f_c stresses are as follows:

$$k = 0.385, \qquad j = 0.872, \qquad p = 0.0120, \qquad K = 189$$

Inserting the values of $M_{\max}$ and K into Eq. (15.11), there is obtained

$$bd^2 = \frac{M}{K} = \frac{39{,}000 \times 12}{189} = 2480 \text{ in.}^3$$

The corresponding b and d will then be

$$d = 17.05 \text{ in.}$$

$$b = 8.52 \text{ in.}$$

Using a $1\frac{1}{2}$-in. cover of concrete and $\frac{7}{8}$-in. round bars, the practical overall dimension of the beam will be 8.5 in. wide by 19.0 in. deep (see Fig. 15-14), making j equal to 17.06 in.

Check on assumed weight. The weight of this beam per foot (based on concrete = 150 lb/ft³) will then be $(8.5 \times 19.0 \times 12)/1728 \times 150 = 169$ lb/ft, which checks the assumed weight within a sufficient degree of accuracy.

Steel required

$$A_s = \frac{M}{f_s jd} = \frac{39{,}000 \times 12}{18{,}000 \times 0.872 \times 17.06} = 1.75 \text{ in.}^2 \text{ required}$$

A good selection might then be 3 — $\frac{7}{8}$ in. round bars at 0.60 in.² = 1.80 in.².

Shear investigation

$$v = \frac{V}{bd} = \frac{2170 \times 6 - 2170 \times (17.06/12)}{8.50 \times 17.06} = 68.4 \text{ psi}$$

Because this value exceeds the allowable shearing unit stress of 55 psi, or $1.1\sqrt{2500}$, shear reinforcement will be required.

Bond investigation

$$u = \frac{V}{\Sigma ojd} = \frac{2170 \times 6}{3 \times 2.75 \times 0.872 \times 17.06} = 106.0 \text{ psi}$$

This value does not exceed the 264 psi $[(4.8\sqrt{2500})/0.875]$ permitted, indicating that the steel perimeter is satisfactory for the required bond resistance.

Illustrative Problem 4. Determine the maximum stresses in the concrete and steel of the rectangular reinforced concrete beam shown in Fig. 15-15 if the maximum bending moment is 800,000 in.-lb. $n = 12$.

SOLUTION. This is obviously an investigative problem. Since the stresses in either the concrete or steel are unknown, the neutral axis must be located using Eq. (15.13).

$$k = \sqrt{2np + (np)^2} - np$$

$$p = \frac{A_s}{bd} = \frac{3.0}{14 \times 20} = 0.0107$$

$$k = \sqrt{2 \times 0.0107 \times 12 + (12 \times 0.0107)^2} - 12 \times 0.0107$$

$$= 0.525 - 0.129 = 0.396$$

$$kd = 0.396 \times 20 = 7.92 \text{ in.}$$

$$jd = d - \frac{kd}{3} = 20 - \frac{7.92}{3} = 17.36 \text{ in.}$$

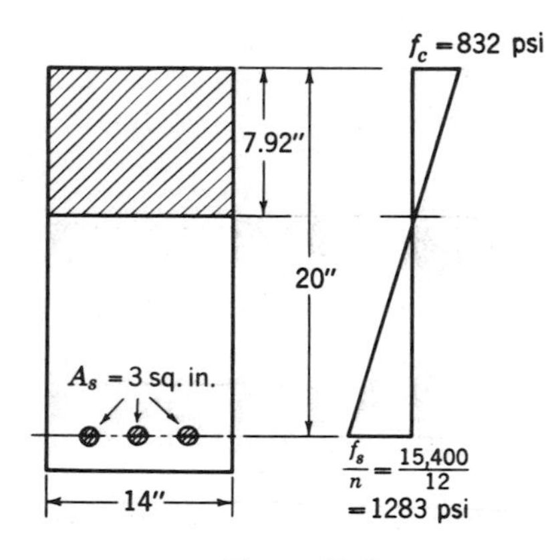

Figure 15-15

If now the moment of the compressive force in the concrete is taken about the center of the steel,

$$M = \tfrac{1}{2} f_c b \times kd \times jd$$

$$800,000 = \tfrac{1}{2} \times f_c \times 14 \times 7.92 \times 17.36$$

$$f_c = 832 \text{ psi}$$

The corresponding stress in the steel is

$$f_s = \frac{M}{A_s jd} = \frac{800,000}{3.0 \times 17.36} = 15,400 \text{ psi}$$

Illustrative Problem 5. Suppose the beam of Illustrative Problem 4 is to be subjected to an increased bending moment of 1,300,000 in.-lb. (a) What maximum stresses would then be induced in the concrete and steel? (b) If these stresses exceed the allowables of 20,000 psi in the steel and 1125 psi in the concrete ($n = 12$), what additional reinforcement must be added to make the beam safe? The dimensions of the beam are not to be altered.

SOLUTION. (a) Because the transformed section is the same as in Illustrative Problem 4, the location of the neutral axis will not change. Moreover, the stresses in the steel and concrete will be proportional to the bending moment, since j, k, b, and d will not change.

$$f_s = 15,400 \times \frac{1,300,000}{800,000} = 25,000 \text{ psi}$$

$$f_c = 832 \times \frac{1,300,000}{800,000} = 1350 \text{ psi}$$

(b) The above stresses, being in excess of the allowables, must be reduced by the addition of steel. This conclusion is evident when it is realized that the location of the T and C resultant forces on the maximum moment section does not change, but their magnitude does. In order for the T resultant to be counteracted without exceeding the 20,000-psi limiting stress, its area must be increased. In the upper or compression portion of the beam, however, it is impossible to increase the resultant C force beyond $\frac{1}{2} \times 1125 \times 7.92 \times 14.0 = 62{,}400$ lb without exceeding the maximum allowable concrete stress of 1125 psi.

Since there is no way to increase the compressive concrete area, compression steel must be added to supplement the concrete. This steel will be stressed to a value n times the stress on that concrete corresponding to its level in the beam.

The solution to this problem proceeds in three steps (Fig. 15-16a):

1. The determination of the bending moment the beam can carry assuming its design to be balanced.

2. The determination of the amount of tensile and compression steel required to produce the additional moment required.

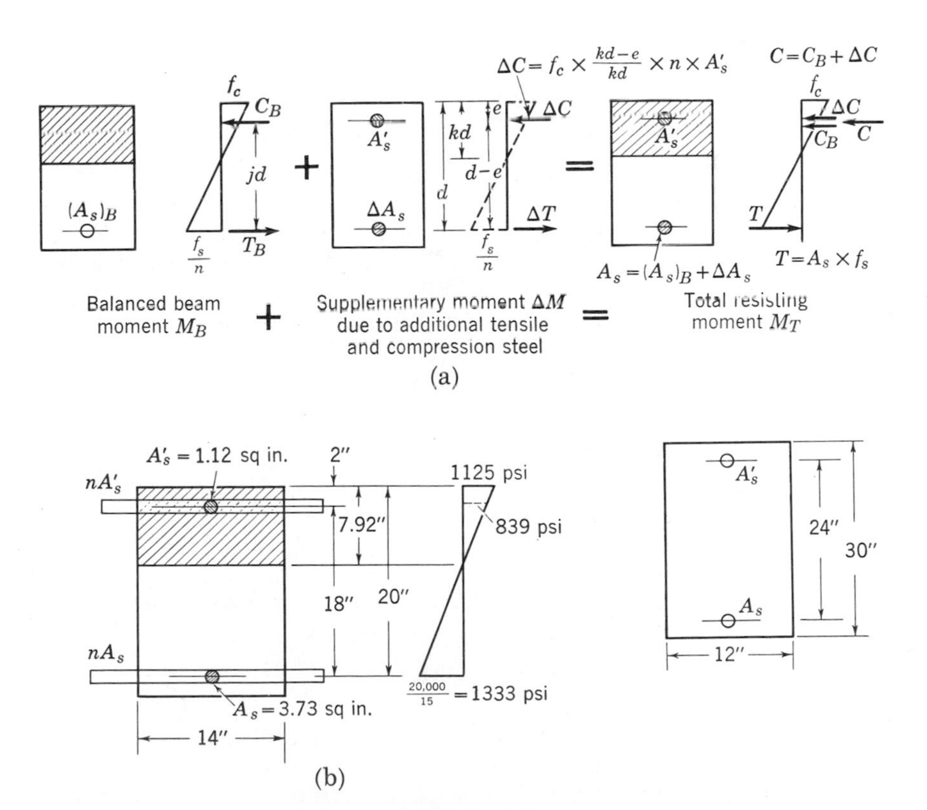

Figure 15-16

3. The algebraic summation of the steel and concrete areas to provide the cross section required to resist the total moment.

The value of K for the balanced design of this beam is equal to 196 (see Table 10), making the corresponding resisting moment

$$M_B = 196 \times 14 \times 20^2 = 1{,}097{,}000 \text{ in.-lb}$$

Resisting this moment will be a tensile steel area of

$$(A_s)_B = 0.0113 \times 14 \times 20 = 3.17 \text{ in.}^2$$

The supplementary moment (ΔC, ΔT couple) will therefore be

$$\Delta M = M_T - M_B = 1{,}300{,}000 - 1{,}097{,}000 = 203{,}000 \text{ in.-lb}$$

Proceeding now to the reinforcement in the compression concrete, let us assume this steel to be placed 2 in. from the top surface. Assuming the maximum allowable stresses in steel and concrete to be maintained, the concrete stress at this level will be

$$f_c'' = 1125 \times \frac{7.92 - 2.00}{7.92} = 839 \text{ psi}$$

The steel stress in the compression steel, therefore, cannot exceed

$$f_s'' = 839 \times 12 = 10{,}080 \text{ psi}$$

The area of compression steel required at this stress is thus

$$A_s' = \frac{\Delta C}{nf_c''} = \frac{\Delta M}{(kd - e)f_s''}$$

$$= \frac{203{,}000}{18 \times 10{,}080} = 1.12 \text{ in.}^2$$

The tensile steel area required to support the extra moment is

$$\Delta A_s = \frac{\Delta T}{f_s} = \frac{\Delta M}{(d - e)f_s}$$

$$= \frac{203{,}000}{(20 - 2) \times 20{,}000} = 0.56 \text{ in.}^2$$

making the total area of tensile steel required

$$A_s = (A_s)_B + \Delta A_s = 3.17 + 0.56 = 3.73 \text{ in.}^2$$

The cross section of the strengthened beam is shown in Fig. 15-16b.

PROBLEMS

15-13. Find the dimensions of a rectangular reinforced concrete beam if the maximum moment is 60,000 ft-lb. $E_c = 2 \times 10^6$ psi, $f_c = 900$ psi, $f_s = 20{,}000$ psi, $n = 15$. Find area of steel. Make $b = \frac{1}{2}d$. *Answers:* $b = 10.5$ in., $d = 21.0$ in., $A_s = 1.98$ in.^{2}.

15-14. Solve Problem 15-13 using $f_c = 1688$ psi, $f_s = 20{,}000$ psi, and $n = 8$.

15-15. Solve Problem 15-13 using $f_c = 1125$ psi, $f_s = 20{,}000$ psi, and $n = 12$.

15-16. The width of a reinforced concrete beam is 8 in. and the depth to the center of steel is 10 in. The beam is 10 ft long and is loaded with 650 lb/ft. Area of steel = 1.0 in.². Calculate the unit stress in the concrete and the steel. $n = 15$. *Answers:* $f_c = 737$ psi, $f_s = 13{,}100$ psi.

15-17. Calculate the areas of the compressive and tensile steel for the beam shown in Fig. 15-17. Bending moment = 180,000 ft-lb. Allowable stresses: concrete = 900 psi, steel = 18,000 psi, and $n = 15$. *Answers:* $A'_s = 2.98$ in.², $A_s = 5.12$ in.².

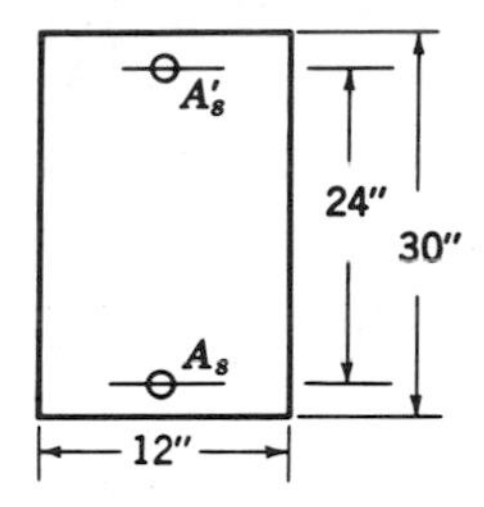

Figure 15-17

15-18. A reinforced concrete beam subjected to a bending moment of 1,500,000 in.-lb is restricted to 12 in. in width and 24-in. total depth. Design section for $f_s = 1000$ psi, $f_s = 18{,}000$ psi, and $n = 12$. Keep steel 2 in. away from nearest surface. *Answers:* $A'_s = 2.60$ in.², $A_s = 4.30$ in.².

15-19. Design a simply supported reinforced concrete slab having a span of 8 ft and carrying a superimposed load of 500 lb/ft². The maximum allowable stresses in the steel and concrete are 18,000 and 650 psi, respectively. $n = 15$. Check shear and bond. *Answers:* $d = 7.0$ in., $A_s = 0.517$ in.²/foot width of slab.

15-20. Design a simply supported reinforced concrete slab on a 12-ft span to carry a live load of 500 lb/ft². Assume allowable stresses of 800 psi in the concrete, 18,000 psi in the steel, and take n as 12. Disregard bond and shear.

15-21. Design a cinder-concrete floor slab on a simple span of 6 ft to carry a live load of 250 lb/ft². using the following data: allowable compressive stress in cinder concrete, 200 psi, allowable tensile stress in steel reinforcement, 18,000 psi; $n = 30$. Assume weight of cinder concrete as 105 lb/ft³, and use $\frac{3}{4}$-in. cover under reinforcement.

PRE-STRESSED CONCRETE

15-10 Pre-tensioning; Post-tensioning

To further develop the bending or tensile resistance of reinforced concrete, its reinforcing steel may be placed in tension, i.e., pre-stressed, before the design loads are applied. This, in effect, produces deformations in the

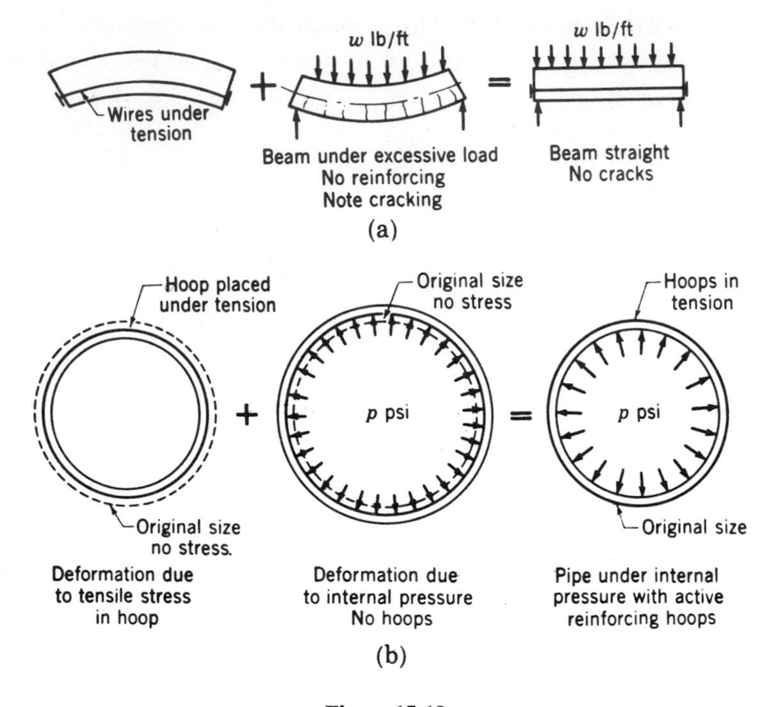

Figure 15-18

concrete of opposite character to those subsequently introduced by the applied forces. In this way, undesirable tensile stresses are avoided. See Fig. 15-18 as applied to beams and pipes.

Pre-stressing may be done in one of two ways, by either *pre-tensioning* or by *post-tensioning*.

Pre-tensioning employs the use of a stressing bed to place the reinforcement in tension before the concrete is poured around it. After the concrete is cured sufficiently to provide good bond, the force applied by the stressing bed is released. With the release, the steel wires (rods are not generally used) attempt to contract to a zero stress condition but are prevented from doing so by the bond and compressive resistance of the concrete. The development of a compressive stress prior to the application of tension increases the range of tensile stress that may safely be applied and consequently its load-carrying capacity.

Post-tensioning, on the other hand, employs wrapped or sheathed bond-free reinforcement which is placed in tension after the concrete has been cured. Because the steel is not bonded to the concrete, the tensile force applied to it is developed by pushing against the concrete. This compressive force counteracts the effects of the tensile stresses produced in the concrete by the applied loads, thereby increasing the resistance of the member to tension or bending. After the tensioning has been completed, the reinforcing wires are grouted to the beam to promote bond and provide resistance to corrosion.

15-11 Internal Deformations Due to Pre-stressing

When the load developed by a straining frame (Fig. 15-19) is transferred to the concrete of a pre-tensioned member, the concrete deforms. This causes the reinforcement to shorten with a consequent reduction in its stress. More-

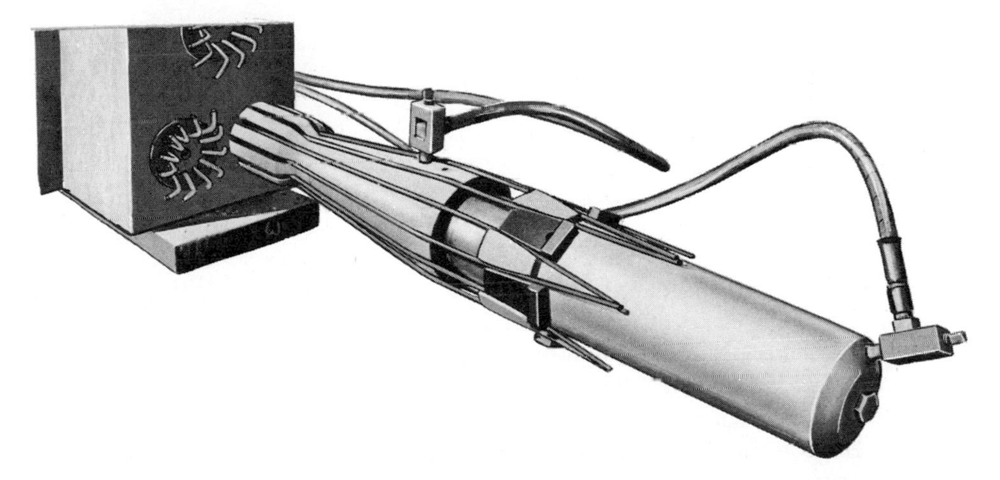

Figure 15-19 Freyssinet pre-stressing jack. (Courtesy of Freyssinet Co.)

over, concrete also shrinks as it hardens and flows plastically when it is placed in compression. Even the steel will creep, i.e., deform inelastically, when subjected to continuous high stress. These effects must be anticipated in determining the final desired stress in the steel. The approximate loss of pre-tension stress in the steel amounts to 20 to 25 per cent.

In the post-tensioning procedure, the effect of shrinkage is greatly reduced and that of initial concrete deformation is eliminated in the determination of the final stress in the steel. Plastic flow and creep, however, must be considered, even though their effects may be small.

15-12 Considerations in Designing Pre-stressed Beams

In designing a pre-stressed beam, the eccentric pre-stress load is generally applied to develop compressive stress over the entire cross section, varying from zero at the top fibers to a maximum on the bottom. The flexural stress pattern produced by the design loads is then superimposed on that caused by the pre-stressing load. It is common practice to limit the magnitude of

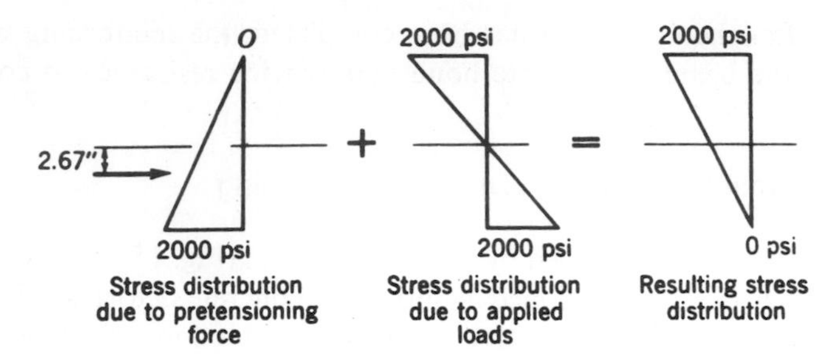

Figure 15-20

the design load to a value which will reduce the intensity of the pre-stress compression on the lower fibers of the concrete to either zero or some small fractional part of its ultimate tensile strength (Fig. 15-20). The stress on the upper fibers, correspondingly, will be equal to the sum of the compression stresses due to the pre-stressing and design loads.

Because the entire concrete cross section of the beam is placed under effective resisting stresses, its gross moment of inertia may be used in computing the safe design load. It is of interest to note, too, that no cracks will form in the lower part of the beam under the application of this load.

In order to produce maximum economy in a pre-stressed concrete member, the steel and concrete used must be of high strength. The reinforcement is generally of high carbon or alloy steel with a maximum allowable stress up to 200,000 psi. The required concrete strength at 28 days is from 4000 to 6000 psi.

Let us now consider the actual design of a pre-stressed concrete beam.

Illustrative Problem 6. A rectangular concrete beam 12 in. wide, 16 in. deep, and 20 ft long is to be pre-tensioned before assuming its design load. What uniform load per foot may it carry if the allowable $f_c = 2000$ psi and the allowable $f_s = 150{,}000$ psi? Allow 20 per cent for the combined effects of shrinkage, compression, and creep.

SOLUTION.

Pre-tensioning analysis. In order to produce a uniformly varying compressive stress over the cross section ranging from zero at the top fibers to a maximum at the bottom, the pre-tension load P_p applied to the end cross section must provide

$$\frac{P_p}{A} - \frac{P_p ec}{I} = 0$$

at the top. From this equation we obtain

$$\frac{P_p}{A}\left(1 - \frac{ec}{r^2}\right) = 0$$

or

$$e = \frac{r^2}{c} = \frac{I}{Ac}$$

Because

$$I = \tfrac{1}{12} \times 12 \times 16^3 = 4096 \quad \text{and} \quad A = 12 \times 16 = 192 \text{ in.}^2$$

$$e = \frac{4096}{192 \times 8} = 2.67 \text{ in.}$$

The value of P_p, the pre-tensioning force, is then computed using the 2000-psi limiting stress.

$$\frac{P_p}{192} + \frac{P_p \times 2.67 \times 8}{4096} = 2000, \qquad P_p = \frac{2000}{0.01042} = 192{,}000 \text{ lb}$$

If a uniform load is to be applied to the beam so as to reduce the 2000-psi compressive stress to zero, there must be applied a flexural stress pattern having a 2000-psi tensile stress on the bottom fibers. The allowable uniform load may then be computed as follows.

$$M = \frac{w \times 20^2}{8} \times 12 = \frac{2000 \times 4096}{8}$$

$$w = 1705 \text{ lb/ft}$$

less	200 lb/ft	(weight of concrete beam)
	1505 lb/ft	(allowable uniform load)

Area of steel required. The area of steel required may be obtained by dividing the required pre-tensioning force by the reduced allowable stress.

$$A = \frac{P_p}{f_s k_s} = \frac{192{,}000}{150{,}000 \times 0.80} = 1.60 \text{ in.}^2$$

It should be noted that the required pre-tensioning force will not change because of the imposed bending loads. This may be verified by considering the maximum compressive stress on the lower fibers caused by pre-tensioning to be counteracted by the equal and opposite maximum tensile stress caused by the applied loads.

$$\frac{P_p}{A} + \frac{P_p ec}{I} = \frac{Mc}{I}$$

$$\frac{P_p}{A}\left(1 + \frac{ec}{r^2}\right) = \frac{Mc}{Ar^2}$$

$$P_p = \frac{M}{(r^2/c) + e} = \frac{M}{2e}$$

But

$$M = \frac{1705 \times 20^2}{8} \times 12 = 1{,}023{,}000 \text{ in.-lb}$$

Therefore,

$$P_p = \frac{1{,}023{,}000}{2 \times 2.67} = 192{,}000 \text{ lb}$$

Extreme fiber stresses. The algebraic sum of the stresses (see Fig. 15-20) imposed by the pre-tensioning and applied loads reveals the maximum stresses at the top and bottom of the beam to be 2000 and 0 psi, respectively, with no tension appearing on the cross section.

PROBLEMS

15-22. A rectangular concrete beam 10 in. wide, 20 in. deep, and 20 ft long is to be pre-tensioned before assuming its design load. What uniform live load may it carry if the allowable concrete and steel stresses are 2000 and 150,000 psi, respectively? Allow 20 per cent for the combined effects of shrinkage, compression, and creep. *Answer:* 2020 lb/ft.

15-23. Design the beam of Problem 15-22 assuming the reinforcement to be post-tensioned.

15-24. A water pipe of 5-ft inside diameter is intended to operate under a pressure of 50 psi. The pipe is cast of concrete 3 in. thick and then wound with $\frac{1}{4}$-in. wire under tension. If the maximum allowable unit stress in the wire is 30,000 psi, what spacing of wire must be used so that the concrete will be under a compressive stress of 100 psi when the pipe is under pressure? *Answer:* 0.818 in. on centers.

15-25. A 3-in.-thick water pipe of 5-ft inside diameter is pre-stressed to give compression in the concrete by wrapping the concrete with $\frac{1}{4}$-in. steel wire at 2-in. centers under tension of 2500 lb. $n = 10$. (a) Compute the internal pressure intensity which would reduce compression in the concrete to 30 psi. (b) What will be the unit stress in the wire? *Answers:* (a) 41.7 psi; (b) 54,866 psi.

Appendix

Center of Gravity and Moment of Inertia

CENTER OF GRAVITY

A-1 General Principles

The method of locating the center of gravity of a mass or an area is based upon the method of determining the resultants of parallel force systems.

Should a mass or an area be divided into a large number of small, equal units, each of which is represented by one of a system of parallel vectors acting at its centroid, the resultant of the vector system would act through the center of gravity of the total mass area. As found in Chapter 1, the resultant of a force or vector system has an effect equal to its component forces or vectors. Thus, when we find the center of gravity of a mass or of an area, we determine that point at which the entire mass or area can be considered to be concentrated.

A-2 Center of Gravity of Composite Areas

Suppose the center of gravity of the area shown in Fig. A-1a, is desired. After placing the area in the xz plane, as shown in Fig. A-1b, subdivide the area into 1- by 1-in. squares and locate at the center of each square a vector of 1 in.^{2} acting perpendicular to the xz plane. The system of parallel vectors

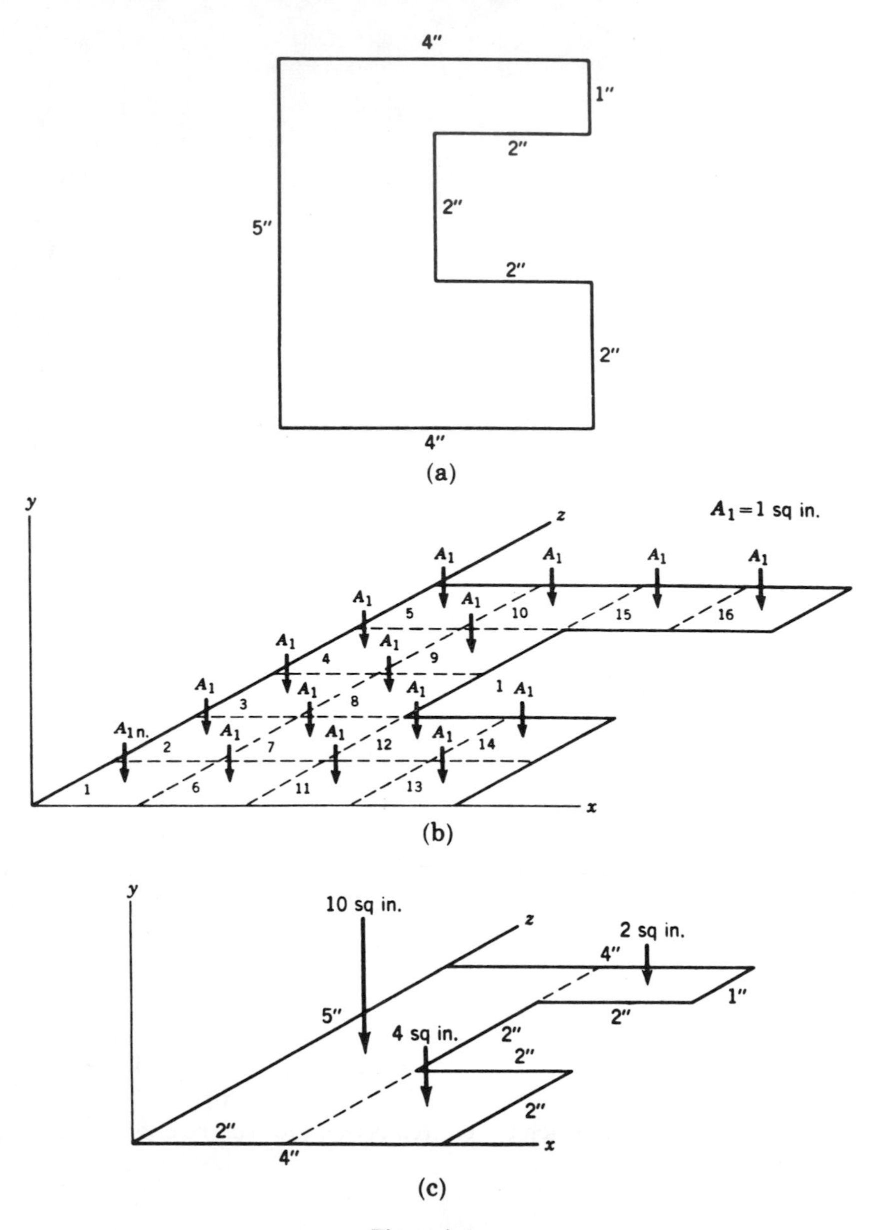

Figure A-1

thereby developed can be concentrated into a resultant, the location of which will act at the center of gravity of the figure.

The magnitude of the resultant is equal to the algebraic sum of the component vectors—that is, 16 in.^{2}. To determine its location—that is, the center of gravity—we shall make use of the fact that *the moment of the resultant of a force or vector system must equal the moment of its component forces or vectors about any axis.* Taking moments, therefore, about the z axis of the vectors and their resultant and setting them equal to one another gives us an equation whose solution will provide the perpendicular distance of the resultant from that axis. Since this distance is parallel to the x axis, it is called $\bar{x}$. A similar equation taken about the x axis provides the $\bar{z}$ which, together with the $\bar{x}$, fixes the location of the center of gravity.

A simplification of the moment equation may be made by grouping certain vectors, the location of whose resultants is obvious. For instance, the squares from 1 to 10 form a rectangle, whose resultant vector of 10 in.^{2} will act at its own center of gravity. Squares 11 to 14 and 15 and 16 can also be grouped to give a more simplified vector system, shown in Fig. A-1c. The more simple the diagram can be made, the easier the solution.

Taking the moment of the vectors shown in Fig. A-1c about the z axis, gives the following equation:

	Moments of Vectors		Moment of Resultant
Areas 1 to 10	$10 \times 1 =$	10 in.^{3}	$16 \times \bar{x}$
Areas 11 to 14	$4 \times 3 =$	12 in.^{3}	
Areas 15 to 16	$2 \times 3 =$	6 in.^{3}	
Sum of areas	16 in.^{2}		
Algebraic sum of moments of areas		28 in.^{3} =	$16\bar{x}$

$$\bar{x} = \frac{28}{16} = 1.75 \text{ in.}$$

Now, the location of the center of gravity can be fixed in the xz plane by finding the other coordinate, the $\bar{z}$. This distance, which is similar in character to the $\bar{x}$, is measured parallel to the z axis. It may be obtained by taking moments about the x axis, since the coordinate to the vectors and to the resultant will be z distances.

	Moments of Vectors		Moment of Resultant
Areas 1 to 10	$10 \times 2.5 =$	25.0	$16 \times \bar{z}$
Areas 11 to 14	$4 \times 1.0 =$	4.0	
Areas 15 to 16	$2 \times 4.5 =$	9.0	
		38.0 in.^{3} =	$16\bar{z}$

$$\bar{z} = \frac{38}{16} = 2.38 \text{ in.}$$

A-3 Advantageous Selection of Axes

It will be noted that the moments of the vectors and the resultant were all positive in the previous problem. Whenever possible, it is suggested that the axes be so selected that this simplification be obtained.

Should some of the moments of the component areas be positive and others negative, owing to the necessity of choosing internal axes, the location of the center of gravity will be determined by using the same principles as previously given and, of course, will be at the same place in the area.

To illustrate this point, let us choose a different set of axes, shown in Fig. A-2. Taking moments about the z axis, we obtain

Moments of Vectors		Moment of Resultant
$10 \times -1.0 =$	-10 in.^3	$16 \times \bar{x}$
$4 \times 1.0 =$	4 in.^3	
$2 \times 1.0 =$	2 in.^3	
16 in.^2	$-4\text{ in.}^3 =$	$16\bar{x}$

whence

$$\bar{x} = \frac{-4}{16} = -0.25\text{ in.}$$

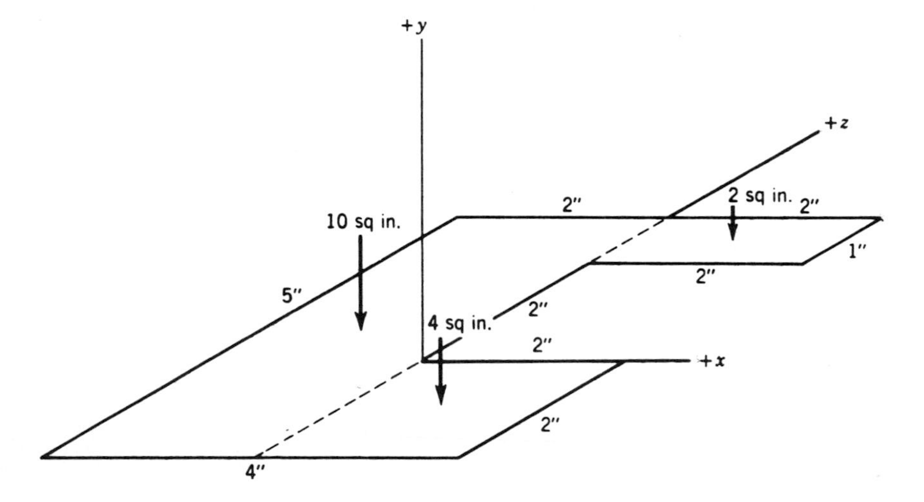

Figure A-2

The moment about the x axis will be

Moments of Vectors		Moment of Resultant
$10 \times 0.5 =$	5.0	$16 \times \bar{z}$
$2 \times 2.5 =$	5.0	
$4 \times -1.0 =$	-4.0	
	$6.0 \quad =$	$16\bar{z}$

$$\bar{z} = \frac{6}{16} = 0.38\text{ in.}$$

It will be seen that these new $\bar{x}$ and $\bar{z}$ values locate the center of gravity at exactly the same location as found in our previous calculation.

In each of the above equations the $\bar{x}$ and $\bar{z}$ distances were considered positive until by computation they were proven otherwise. A change in the location of the axes will make no difference in the ultimate answer, as long as the moments on one side of an axis are given a sign opposite to those on the other side.

A-4 Further Development in Center of Gravity; Areas with Holes

An altogether different approach to the previous problem on center of gravity can be made if we consider its area to consist of a rectangle 4 in. wide and 5 in. long with a 2-in. square cut out of its side. If, then, in addition to the area vectors shown in Fig. A-3, we were to add two opposing area

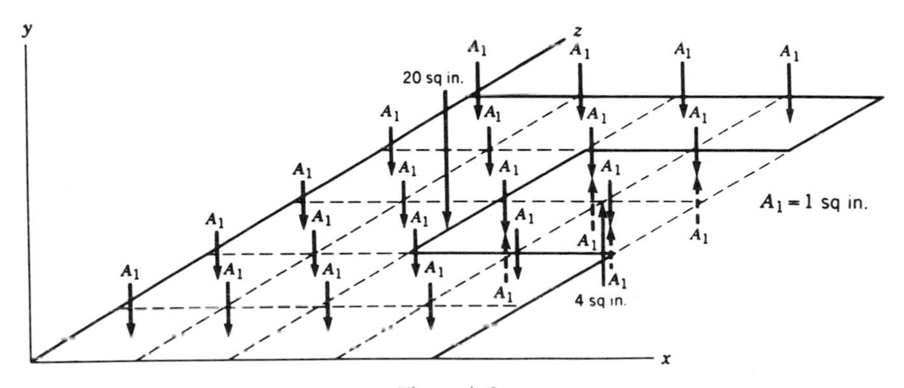

Figure A-3

vectors at the centers of each of the four squares included in the cutout section, we would have a downward-acting-area vector system representing the complete rectangle 4 in. wide and 5 in. long, and an upward-acting-area vector system representing the area of the cutout square. Since the opposing vectors are equal and annul one another, the moment of the downward vectors representing the cutout section and included expediently in the moment of the total rectangle is balanced by the moment of similar vectors acting in the upward direction. Stated briefly, then, the moment of the net area about any axis would be equal to the moment of the large rectangle less the moment of the square. A simplified and composite vector diagram consisting of the 20- and 4-in.^{2} vectors is shown in Fig. A-3, superimposed on the more elementary vector diagram.

It has been previously stated that the resultant of a parallel force system is equal to the algebraic sum of the component forces. Vectorially speaking, the resultant area in our problem would therefore be 16 in.^{2}, which is the algebraic sum of the 20- and 4-in.^{2} vectors or, in other words, the net area of the figure. Since the moment of the resultant equals the moment of its components, we may write, using the simplified diagram, the following moments about the z axis.

Moments of Vectors		Moment of Resultant
$20 \times 2 = \quad 40 \text{ in.}^3$		$16 \times \bar{x} \text{ in.}^3$
$-4 \times 3 = -12 \text{ in.}^3$		
28 in.^3	$=$	$16\bar{x}$

$$\bar{x} = \frac{28}{16} = 1.75 \text{ in.}$$

TABLE 11

IMPORTANT CENTERS OF GRAVITY

	Area	Center of Gravity
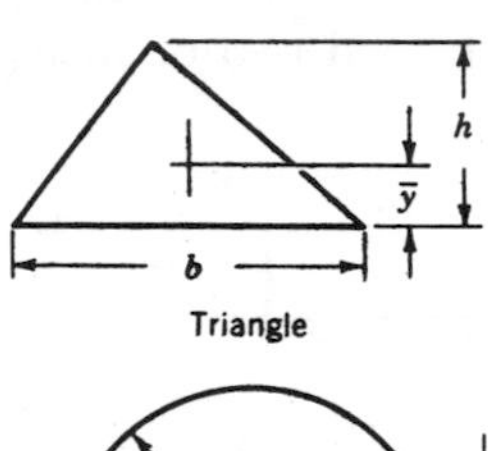 Triangle	$A = \frac{1}{2} bh$	$\bar{y} = \frac{h}{3}$
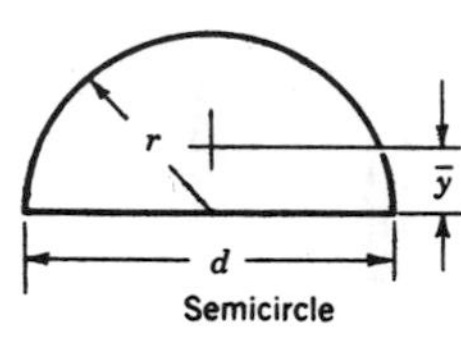 Semicircle	$A = \frac{\pi r^2}{2}$	$\bar{y} = \frac{4r}{3\pi}$
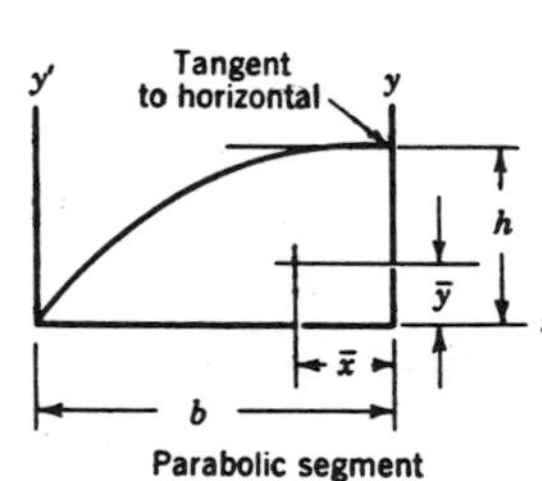 Parabolic segment	$A = \frac{2}{3} bh$	$\bar{x} = \frac{3}{8} b, \quad \bar{y} = \frac{2}{5} h$
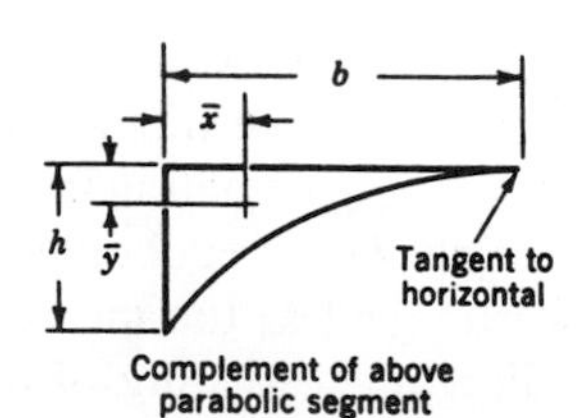 Complement of above parabolic segment	$A = \frac{1}{3} bh$	$\bar{x} = \frac{1}{4} b, \quad \bar{y} = \frac{3}{10} h$

The moments of the vectors taken about the x axis are

Moments of Vectors		Moment of Resultant
$20 \times 2.5 = \quad 50 \text{ in.}^3$		$16 \times \bar{z}$
$-4 \times 3 \quad = -12 \text{ in.}^3$		
38 in.^3	$=$	$16\bar{z}$

$$\bar{z} = \frac{38}{16} = 2.38 \text{ in.}$$

A-5 Summary of Center of Gravity of Composite Areas

To summarize the procedure used in the previous problems on the center of gravity of areas, we may deduce the following general rule:

$$\frac{\text{center of gravity coordinate}}{\text{(measured from moment axis)}} = \frac{\text{algebraic sum of moments of component areas about axis}}{\text{algebraic sum of component areas}}$$

$$\bar{x} = \frac{\sum Ax}{\sum A}, \qquad \bar{z} = \frac{\sum Az}{\sum A} \tag{A.1}$$

PROBLEMS

A-1. Find the centers of gravity of the areas shown in Fig. A-4 about the axes indicated. *Answers:* (a) $\bar{y} = 6.15$ in., $\bar{x} = 2.60$ in.; (c) $\bar{y} = 4.92$ in., $\bar{x} = 6.72$ in.; (e) $\bar{y} = 3.91$ in.; (g) $\bar{y} = 2.68$ in.

A-2. Determine the location of the center of gravity of area (a) in Problem A-1 about axes 2 in. to the left and below the y and x axes, respectively.

A-3. Determine the location of the center of gravity of area (f) in Problem A-1 about a horizontal axis located at the top of the figure.

MOMENT OF INERTIA OF AREAS

A-6 General Principles

When in the study of strength of materials a formula is derived based on the bending or torsional resistance of a cross-sectional area, there is invariably included in that formula a term called the *moment of inertia* of the area. It has the general form of

$$\int z^2 \, dA \tag{A.2a}$$

The magnitude of this moment of inertia term is a measure of the resistance offered by the material included within this cross-sectional area by *virtue of its shape and location.* If we should consider two prismatic beams made of the same material, but of different cross sections, that beam whose cross-sectional area had the greater moment of inertia would have the greater resistance to bending. To have a greater moment of inertia does not necessarily imply, however, a greater cross-sectional area. Less area more advantageously placed may have a greater moment of inertia.

In our discussion of the method of finding the center of gravity of areas, each area was subdivided into small elementary areas which were multiplied by their respective perpendicular distances from the reference axis. The procedure is somewhat similar for determinations of moment of inertia.

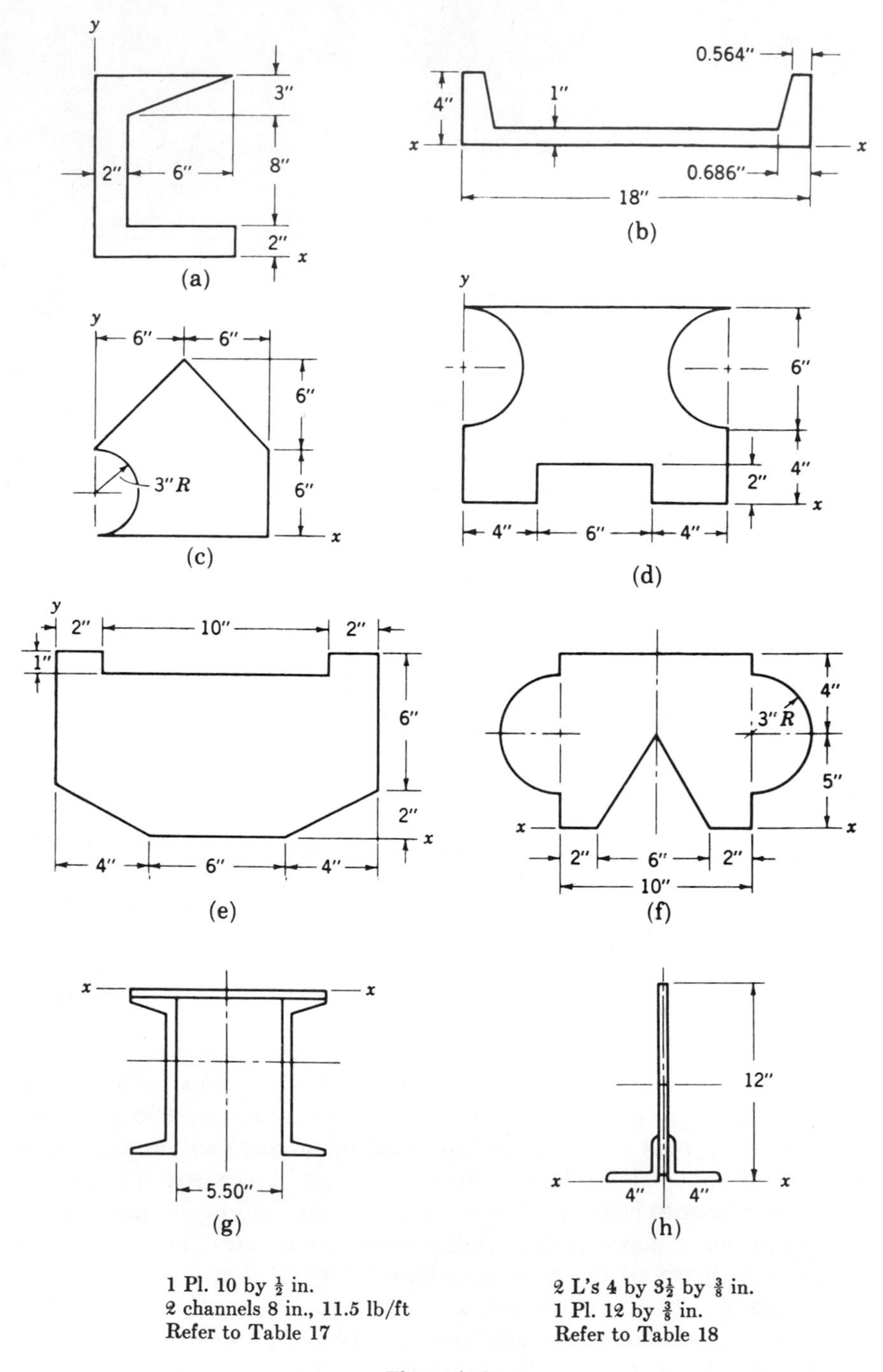

1 Pl. 10 by ½ in.
2 channels 8 in., 11.5 lb/ft
Refer to Table 17

2 L's 4 by 3½ by ⅜ in.
1 Pl. 12 by ⅜ in.
Refer to Table 18

Figure A-4

The moment of inertia about the same axis would require, however, that each elementary area be multiplied by the *square* of the respective perpendicular distances from the reference axis.

Let us illustrate by using Fig. A-5. Suppose we wanted to find the moment of inertia of this irregular area about the x axis. We would first

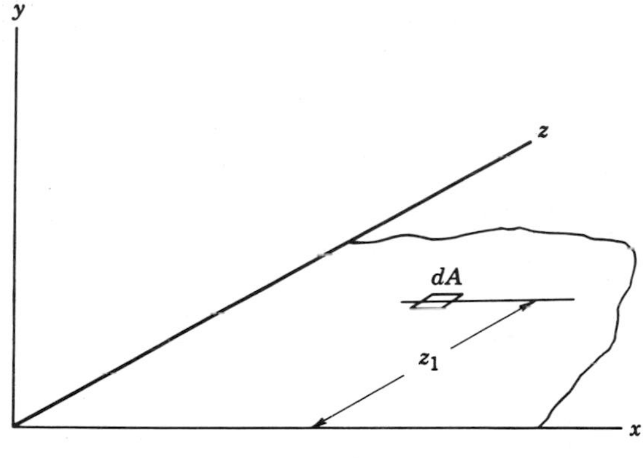

Figure A-5

consider the area to consist of many infinitesimal dA areas. Considering the dA shown, its moment of inertia about the x axis would be $z_1^2\ dA$. But this product is only a minute portion of the whole moment of inertia. Each dA comprising the area, when multiplied by the square of its corresponding moment arm z and added together, will give the moment of inertia of the entire area about the x axis.

The difficulty of this integration process is largely dependent on the equation of the outline of the area and its limits. When the integration becomes too difficult or when a more simple, approximate solution is permissible, the moment of inertia may be expressed in finite terms by

$$I_x = \sum z^2\ \Delta A \qquad \text{(A.2b)}$$

The solution becomes more accurate as the size of ΔA is decreased.

Let us illustrate the approximation involved by the calculation of the moment of inertia of a rectangle about its center-of-gravity axis by the approximate method and by the exact formula derived from the calculus.

Approximate method. Let us consider the elementary areas of Fig. A-6 to be 1 in. wide and 6 in. long. The use of the 6-in. length will in no way detract from the accuracy of the solution, since increasing the length of the rectangle merely adds more area similarly placed with respect to the center-of-gravity axis. It is the necessity of using a 1-in. width rather than one of infinitesimal width which lessens the accuracy.

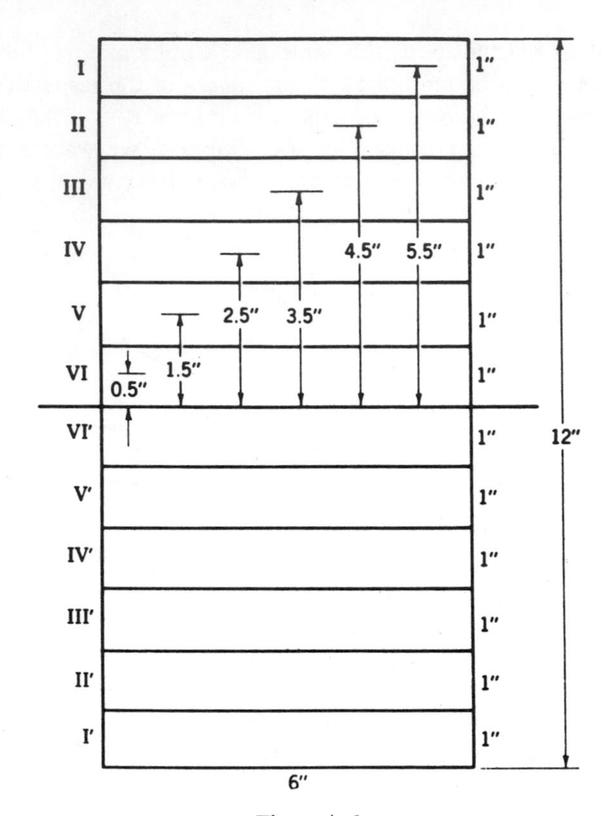

Figure A-6

	ΔA (*in.*^{2})	y (*in.*)	y^2	ΔAy^2
I	6	5.5	30.25	181.5 in.^{4}
II	6	4.5	20.25	121.5
III	6	3.5	12.25	73.5
IV	6	2.5	6.25	37.5
V	6	1.5	2.25	13.5
VI	6	0.5	0.25	1.5

$\sum \Delta Ay^2$ for top of rectangle $= 429.0$

$\sum \Delta Ay^2$ for bottom of rectangle $= 429.0$

$\sum Ay^2 = 858.0$ in.^{4}

Exact method. Moment of inertia of rectangle about its center-of-gravity axis

$$= \tfrac{1}{12} bd^3$$

$$= \tfrac{1}{12} \times 6 \times 12^3 = 864 \text{ in.}^4$$

where b = width of rectangle and d = depth of rectangle.

Although the discrepancy here is less than 1 per cent of the exact moment of inertia, a still greater refinement can be made by using narrower strips.

TABLE 12

MOMENTS OF INERTIA AND RADII OF GYRATION

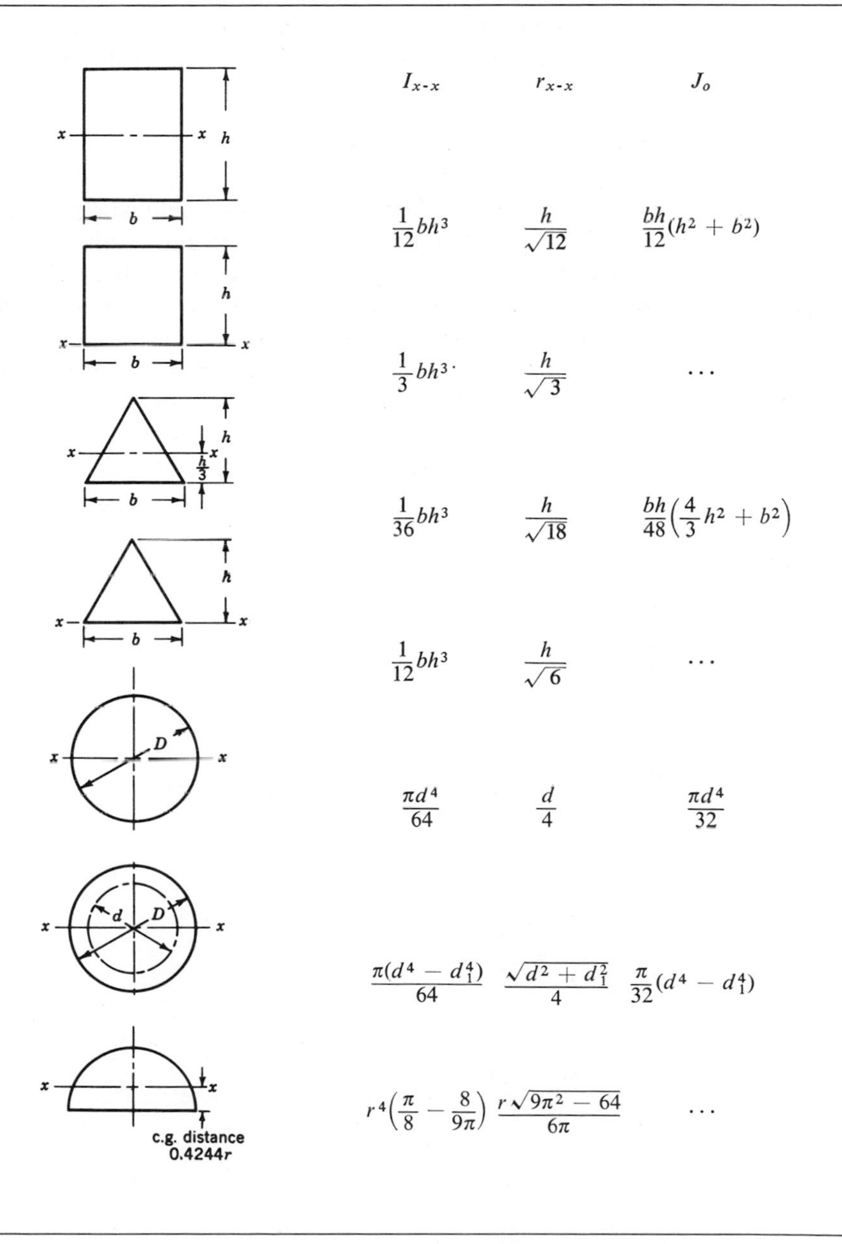

$I_{x\text{-}x}$	$r_{x\text{-}x}$	J_o
$\frac{1}{12}bh^3$	$\frac{h}{\sqrt{12}}$	$\frac{bh}{12}(h^2 + b^2)$
$\frac{1}{3}bh^3$	$\frac{h}{\sqrt{3}}$	. . .
$\frac{1}{36}bh^3$	$\frac{h}{\sqrt{18}}$	$\frac{bh}{48}\left(\frac{4}{3}h^2 + b^2\right)$
$\frac{1}{12}bh^3$	$\frac{h}{\sqrt{6}}$	. . .
$\frac{\pi d^4}{64}$	$\frac{d}{4}$	$\frac{\pi d^4}{32}$
$\frac{\pi(d^4 - d_1^4)}{64}$	$\frac{\sqrt{d^2 + d_1^2}}{4}$	$\frac{\pi}{32}(d^4 - d_1^4)$
$r^4\left(\frac{\pi}{8} - \frac{8}{9\pi}\right)$	$\frac{r\sqrt{9\pi^2 - 64}}{6\pi}$	. . .

Although the approximate method is used to great advantage in determining the moment of inertia of irregular areas, whenever the area is geometrically familiar to the extent that its center of gravity and moment of inertia are or can be ascertained from their component areas, the more exact formulas already derived from the calculus are always used. Those moment-of-inertia formulas of particular value are given in Table 12.

A-7 Moment of Inertia of a Composite Area

Whenever an area can be considered as equal to a combination of other areas whose centers of gravity and moments of inertia are known, the moment of inertia of the composite area about any desired axis can be obtained by an algebraic summation of its component moments of inertia taken about the same axis. To obtain the moments of inertia of the dependent component areas, it frequently becomes necessary to use the *parallel-axis theorem.* By this theorem, one may ascertain the moment of inertia of an area about any axis parallel to a *center-of-gravity* axis, if the moment of inertia of the area about the latter axis is known. For instance, consider the case of the rectangle in Fig. A-7. Knowing the moment of inertia about its center-of-gravity axis y-y, we can obtain the moment of inertia of the rectangle about any axis parallel to the y-y axis—for instance, the y'-y' axis.

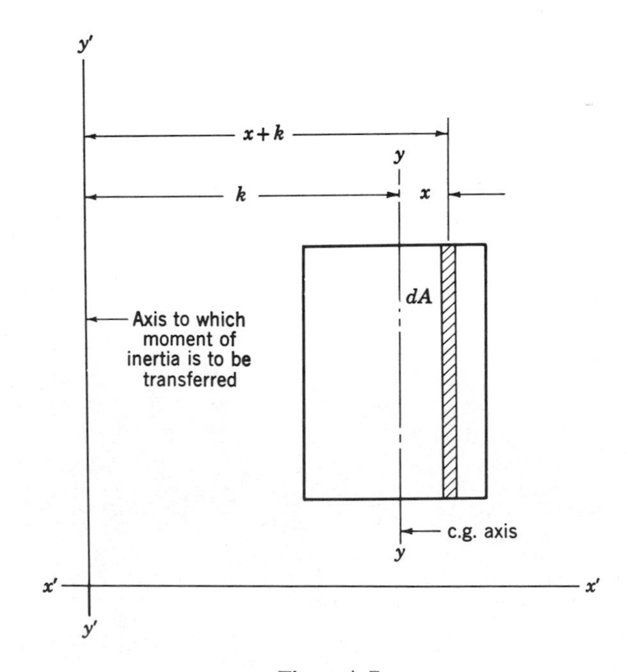

Figure A-7

Proof—parallel-axis theorem. Let the y'-y' axis be k distance from the parallel centroidal axis y-y. Consider the area to be divided into dA strips, any one of which has every portion of its area equidistant from the y'-y' axis. Select that dA area located a distance x from the y-y axis. The expression for the moment of inertia of dA about the y'-y' axis will, as previously discussed, be equal to

$$dI_{y'-y'} = (k + x)^2 \, dA$$

If we were now to require the moments of inertia of the total area about the same axis, knowing that each dA had a moment of inertia similar to that given above, x being a general distance, we would merely sum up the incremental moments of inertia.

$$\int dI_{y'-y'} = \int (k + x)^2 \, dA$$

Squaring the binomial,

$$\int dI_{y'-y'} = \int (k^2 + 2kx + x^2) \, dA$$

or

$$I_{y'-y'} = \int k^2 \, dA + \int 2kx \, dA + \int x^2 \, dA \qquad \text{(A.3a)}$$

Now, k being a constant value—that is, the same for all dA's—we may write

$$\int k^2 \, dA = k^2 \int dA = Ak^2$$

and

$$\int 2kx \, dA = 2k \int x \, dA$$

But since the y-y axis is the center-of-gravity axis and the moment of the areas balance about this axis—that is, the resultant of the area acts through the y-y axis—$\int x \, dA$ must be equal to zero.

We also note that $\int x^2 \, dA$ is equal to the moment of inertia of the rectangle about the y-y axis written in general terms.

Therefore, rewriting Eq. (A.3a), we have

$$I_{y'-y'} = I_{y-y} + Ak^2 \qquad \text{(A.3b)}$$

which is known as the *parallel-axis theorem.*

In using the formula care should be taken to substitute correctly. To avoid misunderstanding, the following facts should be clearly understood.

1. The y-y axis is a centroidal axis, that is, passing through the center of gravity, about which the moment of inertia is known.

2. I_{y-y} is the moment of inertia of the area about the centroidal axis y-y.

3. Axis y'-y' is parallel to the centroidal axis y-y, and $I_{y'-y'}$ is the moment of inertia about the axis y'-y'.

4. Distance k is always the distance between the parallel axes, that is, the centroidal axis y-y and the new axis y'-y', about which the moment of inertia is to be determined.

On the basis of these principles, there may be written an even more general moment-of-inertia formula whose new terms can now be understood:

$$I_p = I_{c.g.} + Ak^2 \tag{A.4}$$

Illustrative Problem 1. Using the area in Fig. A-8, find the moment of inertia about axis x'-x'.

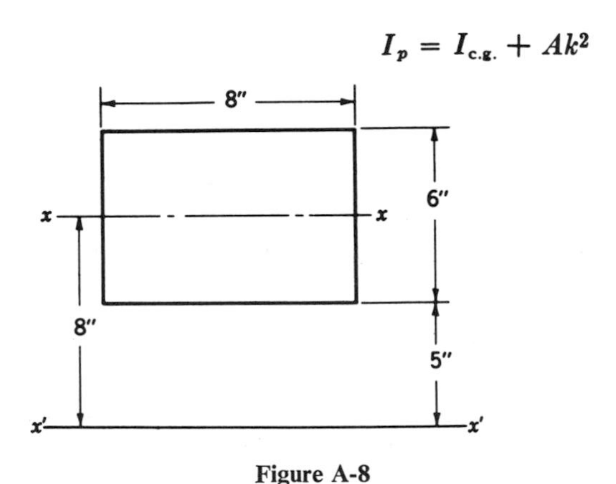

Figure A-8

SOLUTION. There should immediately be located the center of gravity of the figure and an axis x-x drawn through it parallel to axis x'-x'. Then determine the distance k equal to 8 in. Now, by substituting in

$$I_p = I_{c.g.} + Ak^2$$

we obtain

$$\begin{aligned} I_{x'-x'} &= \tfrac{1}{12} \times 8 \times 6^3 + 48 \times 8^2 \\ &= 144 + 3072 = 3216 \text{ in.}^4 \end{aligned}$$

(See Table 12 for moments of inertia of common areas.)

Illustrative Problem 2. Let it be desired to find the moment of inertia of the area previously used in the discussion of center of gravity (Fig. A-1) about the centroidal axes x-x and y-y of the net area.

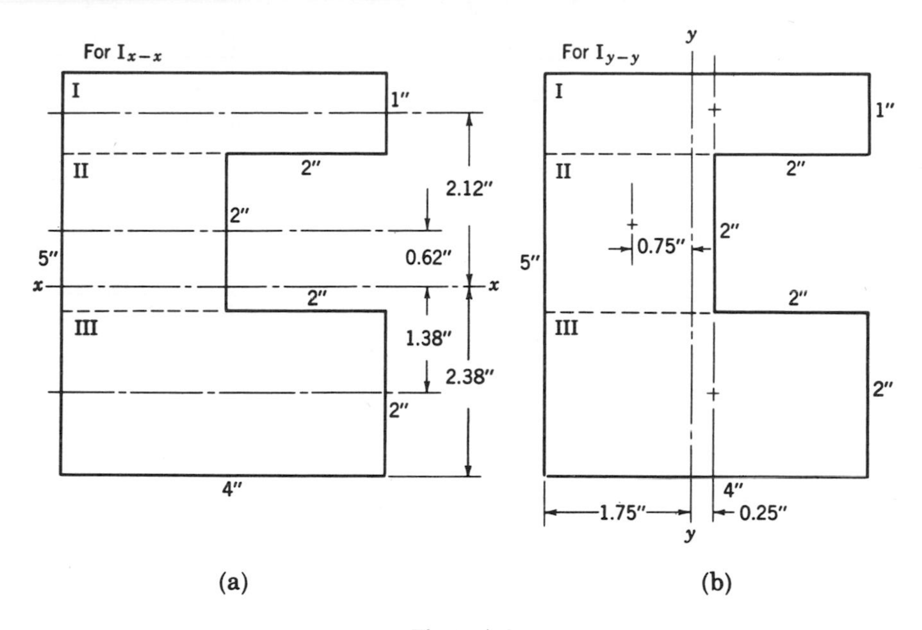

Figure A-9

SOLUTIONS.

Method I. This area is easily divided into two rectangles and a square (Fig. A-9a and b). Locating the centroidal axes of each incremental area and the net area as found in the previous problem, we may now proceed as follows:

Areas	$I_{c.g.}$	A	k	k^2	Ak^2	$I_{c.g.} + Ak^2$
			I_{x-x}			
I	$\frac{1}{12} \times 4 \times 1^3 = 0.33$	4	2.12	4.50	18.00	18.33
II	$\frac{1}{12} \times 2 \times 2^3 = 1.33$	4	0.62	0.39	1.56	2.89
III	$\frac{1}{12} \times 4 \times 2^3 = 2.67$	8	1.38	1.90	15.20	17.87
					$I_{x-x} =$	39.09 in.[4]
			I_{y-y}			
I	$\frac{1}{12} \times 1 \times 4^3 = 5.33$	4	0.25	0.063	0.252	5.58
II	$\frac{1}{12} \times 2 \times 2^3 = 1.33$	4	0.75	0.563	2.252	3.58
III	$\frac{1}{12} \times 2 \times 4^3 = 10.67$	8	0.25	0.063	0.505	11.18
					$I_{y-y} =$	20.34 in.[4]

Method II. This problem may also be solved by considering the area shown to consist of a rectangle 4 in. wide by 5 in. long with a 2-in. square cut from its side (Fig. A-10a and b). By subtracting the moment of inertia of the square—which was expediently added to the irregular area—from the large 4- by 5-in. rectangle, we shall obtain the moment of inertia of the net area as shown.

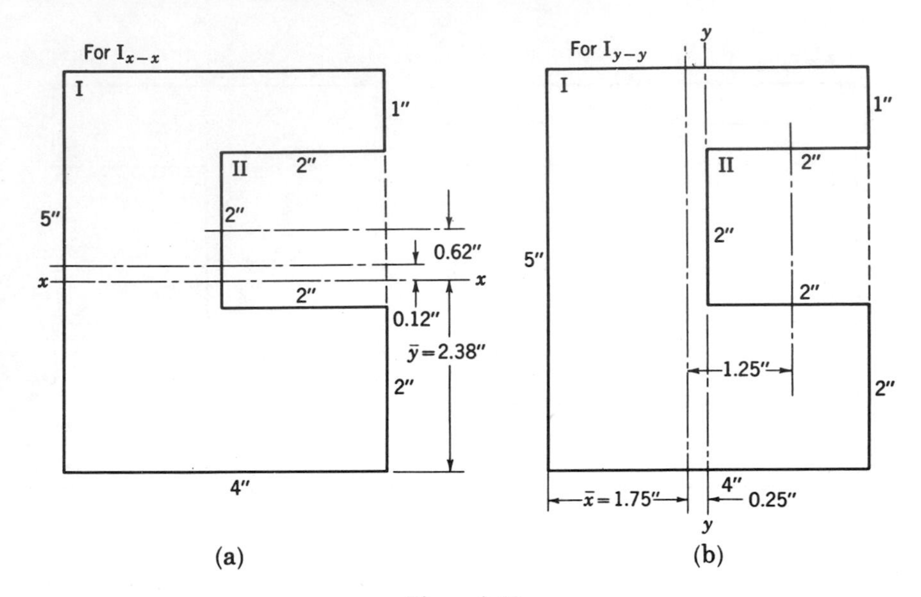

Figure A-10

Areas	$I_{c.g.}$	A	k	k^2	Ak^2	$I_{c.g.} + Ak^2$
			I_{x-x}			
I	$\frac{1}{12} \times 4 \times 5^3 = 41.67$	20	0.12	0.0144	0.288	41.96
II	$\frac{1}{12} \times 2 \times 2^3 = 1.33$	4	0.62	0.384	1.536	−2.87
					$I_{x-x} =$	39.09 in.⁴
			I_{y-y}			
I	$\frac{1}{12} \times 5 \times 4^3 = 26.67$	20	0.25	0.0625	1.25	27.92
II	$\frac{1}{12} \times 2 \times 2^3 = 1.33$	4	1.25	1.56	6.25	−7.58
					$I_{y-y} =$	20.34 in.⁴

A-8 Radius of Gyration

It has previously been stated that the moment of inertia of an area about a certain axis could be expressed in a general way as being equal to

$$I_z = \int x^2 \, dA$$

where x is the distance from the z axis to any incremental area dA. Inasmuch as x is a length and dA is a length squared, the value of I_z is expressed in terms of a length to the fourth power, such as in.^{4} or ft^4.

If, instead of using a summation of incremental areas, we could use the total area in this equation to obtain the same moment of inertia, we would have to multiply the area by the square of a specific distance known as the *radius of gyration*. In other words, $I_z = r_x^2 A$, whence

$$r_x = \sqrt{\frac{I_z}{A}} \tag{A.5}$$

The value of the radius of gyration should be thought of as a convenient term, closely related to the moment of inertia and often appearing in engineering formulas in which the moment of inertia would otherwise appear.

The units of radii of gyration used in strength of materials are usually expressed in inches.

Illustrative Problem 3. Determine the radius of gyration about the centroidal axes of the figure used in Illustrative Problem 2.

SOLUTION

$$r_y = \sqrt{\frac{I_x}{A}} = \sqrt{\frac{39.09}{16}} = \sqrt{2.44} = 1.56 \text{ in.}$$

$$r_x = \sqrt{\frac{I_y}{A}} = \sqrt{\frac{20.34}{16}} - \sqrt{1.27} = 1.13 \text{ in.}$$

PROBLEMS

A-4a, b, c, d, e, f, g, h. Determine the centroidal moments of inertia I_x and I_y of the similarly lettered areas found in Problem A-1. *Answers:* (a) $I_x = 1004$ in.^{4}, $I_y = 216$ in.^{4}; (c) $I_x = 812$ in.^{4}, $I_y = 684$ in.^{4}; (e) $I_x = 1943$ in.^{4}, $I_y = 770$ in.^{4}; (g) $I_x = 114$ in.^{4}, $I_y = 119$ in.^{4}.

A-5. Determine the centroidal moments of inertia I_x and I_y and the corresponding radii of gyration of the angle section shown in Fig. A-11.

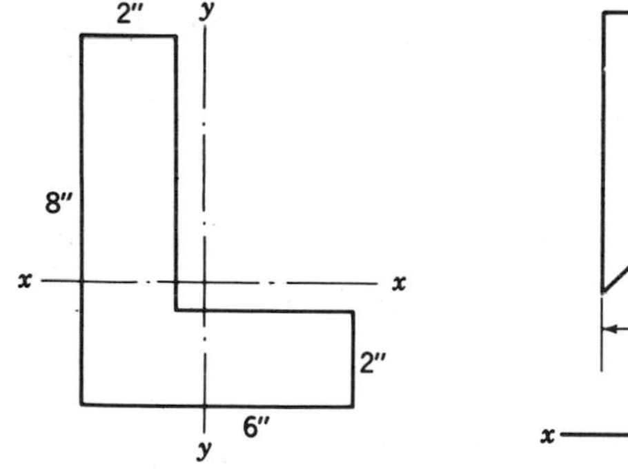

Figure A-11

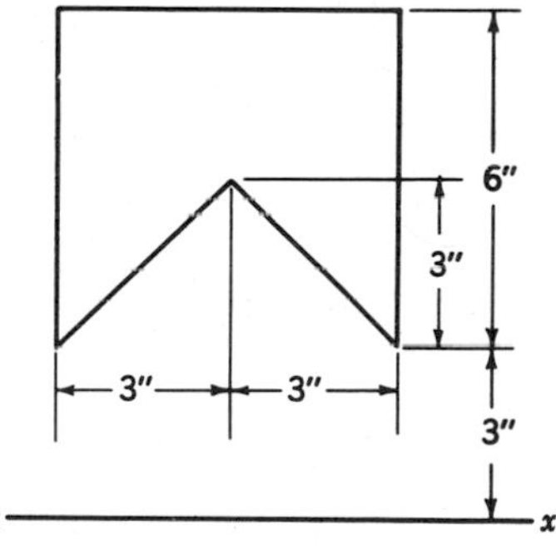

Figure A-12

A-6. (a) Find the moment of inertia of the area shown in Fig. A-12 about the *x-x* axis. (b) Determine the radius of gyration about the same axis. *Answers:* $I_{x-x} = 1256$ in.^{4}, $r_{x-x} = 6.82$ in.

A-7. The base of a triangle is 6 in. and its altitude is 8 in. Find the moment of inertia and radius of gyration of the area of the triangle with respect to the base. *Answers:* $I = 256$ in.^{4}, $r = 3.27$ in.

A-8. Determine the moment of inertia of the area of a circle, with respect to an axis tangent to the circle, in terms of r, the radius of the circle.

POLAR MOMENT OF INERTIA

A-9 General Principles

The polar moment of inertia of an area is distinguished from the moment of inertia previously described by the position of the axis about which the area is assumed to rotate. Whereas previously the axis was located *in* the plane of the area, the axis for the determination of a polar moment of inertia is always located *perpendicular* to the plane. This change, however, does not create a radical difference in the concept of the two types of moments of inertia, since the polar type is also a measure of the resistance of a material within a given area due solely to its shape and location. Its general formula is also quite similar:

$$J = \int \rho^2 \, dA \tag{A.6}$$

An investigation of this general formula reveals that as dA is located further from the polar axis, the value of J, a measure of resistance, increases. (See Fig. A-13.)

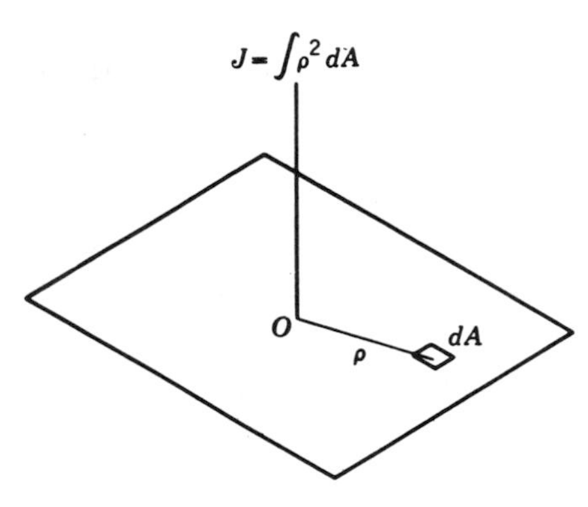

Figure A-13

Typical of the problems requiring a determination of polar moment of inertia is that of the shaft. As a shaft increases in diameter, its cross-sectional area, polar moment of inertia J, and its capacity for resisting torque also increase. The polar moments of inertia of several common cross-sectional areas are given in Table 12.

A-10 A Polar Moment-of-Inertia Relationship

A very important relationship between the two types of moments of inertia previously described is discussed below. Familiarity with this relationship will help simplify the solution of many design problems involving irregular cross sections, and will develop a broader concept of the nature of bending and twisting resistance.

Since $J = \int \rho^2 \, dA$ and, from Fig. A-14, $\rho^2 = x^2 + z^2$, we may write

$$J = \int (x^2 + z^2) \, dA \tag{A.7}$$

or

$$\begin{aligned} J &= \int x^2 \, dA + \int z^2 \, dA \\ J_y &= I_x + I_z \end{aligned} \tag{A.8}$$

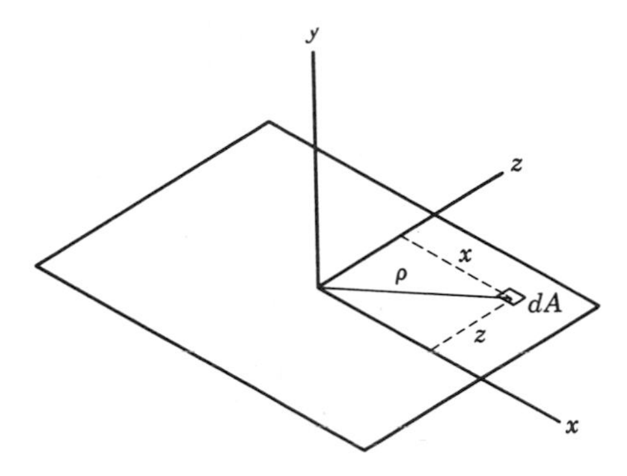

Figure A-14

The two latter values will be recognized as the regular moments of inertia about the x and z axes. Thus, if we know the moments of inertia of an area about two mutually perpendicular intersecting axes located in the plane of the area, the polar moment of inertia about an axis perpendicular to the plane at the point of intersection of the two axes will be equal to the sum of the two.

Could J ever be smaller than either I_{x-x} or I_{y-y}?

PROBLEMS

A-9. Find the polar moment of inertia of the angle shown in Fig. A-11 about its center of gravity. *Answer:* 212 in.^{4}.

A-10. Find the polar moment of inertia of the channel section shown in Fig. A-4b about the intersection of the centroidal axes. *Answer:* 786 in.^{4}.

PRODUCT OF INERTIA

A-11 General Principles

The product of inertia is defined by the integral

$$I_{xy} = \int xy\, dA \qquad \text{(A.9)}$$

in which x and y are the distances of the differential area dA from two perpendicular reference axes (Fig. A-15). Inasmuch as these reference axes may pass through an area, it is obvious that the values of x and y may be positive or negative, providing similar sign possibilities for I_{xy}. Should an area be symmetrical

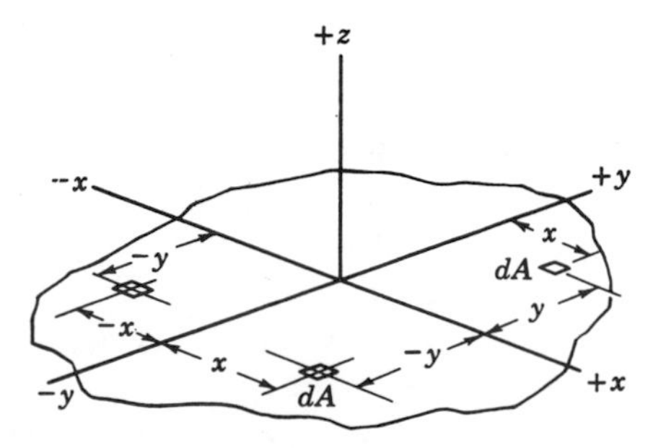

Figure A-15

about one of these axes, say, the x, there will be a positive y value for every similar negative y value, canceling the products $xy\,dA$ and making the product of inertia I_{xy} equal to zero. Thus, the product of inertia of a rectangle with respect to two perpendicular axes, one of which is a centroidal axis, is zero.

A-12 Transfer Theorem—Product of Inertia

If the product of inertia of an area is known with respect to two perpendicular axes x and y passing through its centroid, the product of inertia with respect to any other parallel pair of axes x' and y' may be obtained easily as follows. Consider the area shown in Fig. A-16. Its product of inertia with respect to the x' and y' axes is expressed mathematically by

$$J_{x'y'} = \int x'y'\,dA \qquad \text{(A.10)}$$

But because $x' = x + a$ and $y' = y + b$, the foregoing expression may be written as

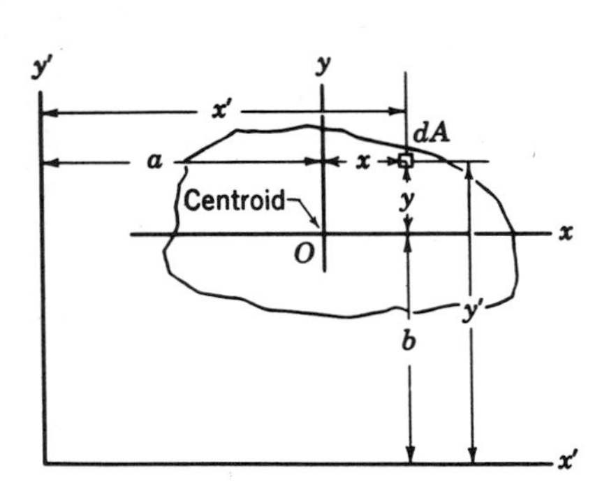

Figure A-16

$$I_{x'y'} = \int (x + a)(y + b)\,dA$$

or

$$I_{x'y'} = \int xy\,dA + \int bx\,dA + \int ay\,dA + \int ab\,dA$$

The middle two integrals, however, equal zero, since the x and y axes are centroidal axes. The equation above may therefore be expressed as

$$I_{x'y'} = \int xy\,dA + \int ab\,dA$$

or

$$I_{x'y'} = I_{\bar{x}\bar{y}} + Aab \qquad \text{(A.11)}$$

in which $I_{\bar{x}\bar{y}}$ is the product of inertia with respect to the centroidal axes.

Illustrative Problem 4. Determine the product of inertia of the area shown in Fig. A-17 about axes x' and y'.

SOLUTION. It is obvious that the two subdivided areas I and II are symmetrical about pairs of rectangular axes. The products of inertia of these areas about their own centroidal axes are therefore zero. Because the product of inertia of the total area is equal to the products of inertia of the several parts, we may write

$$I_{x'y'} = (I_{\bar{x}\bar{y}} + Aab)_{\text{I}} + (I_{\bar{y}\bar{x}} + Aab)_{\text{II}}$$

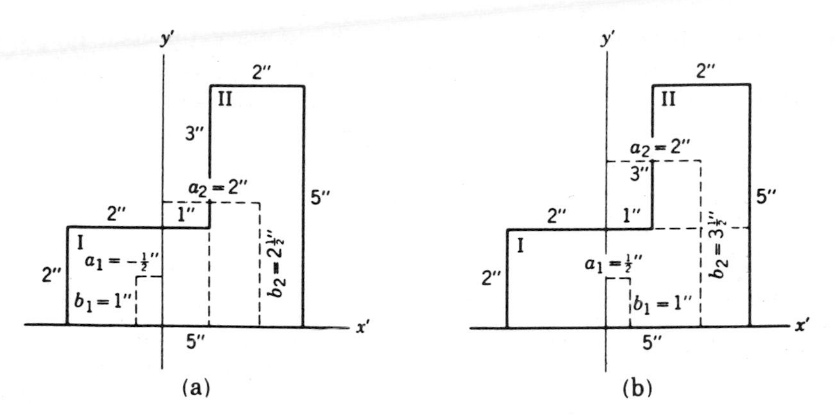

Figure A-17

Simplified, this expression becomes

$$
\begin{aligned}
I_{x'y'} &= (Aab)_{\text{I}} + (Aab)_{\text{II}} \\
&= (6)(-\tfrac{1}{2})(1) + (10)(2)(2.5) \\
&= -3 + 50 = 47 \text{ in.}^4
\end{aligned}
$$

Had the area been subdivided in the manner shown in Fig. A-17b, the answer would have been the same.

$$
\begin{aligned}
I_{x'y'} &= (Aab)_{\text{I}} + (Aab)_{\text{II}} \\
&= (10)(\tfrac{1}{2})(1) + (6)(2)(3\tfrac{1}{2}) \\
&= +5 + 42 = 47 \text{ in.}^4
\end{aligned}
$$

PROBLEMS

A-11. Determine the product of inertia of the area shown in Fig. A-9a and b using the x and y axes shown. *Answer:* $+2.50$ in.^{4}.

A-12. Determine the product of inertia of the angle section shown in Fig. A-11 about its centroidal axes.

A-13. Determine the product of inertia of a right-angle triangle about the axes coincident with its two legs. *Answer:* $b^2h^2/24$.

MOMENTS OF INERTIA ABOUT INCLINED AXES

A-13 General Principles

For unsymmetrical areas, such as shown in Fig. A-1b. moments of inertia may be arbitrarily determined with respect to axes x and y passing through their centroid. There is no guarantee, however, that either of these moments of inertia will be a maximum. On the contrary, the maximum moment of inertia for such an unsymmetrical area will invariably be found to occur about an axis inclined to the x and y axes. Because of the need for knowing the magnitude of the maximum and minimum moments of inertia and the axes on which they act in problems of unsymmetrical bending, the relationship of moments of inertia about inclined axes will now be presented.

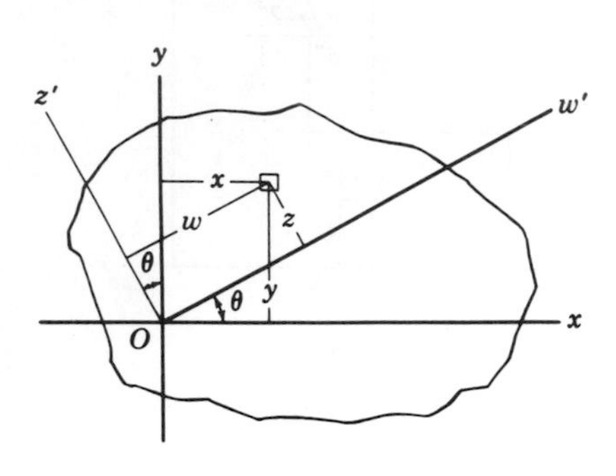

Figure A-18

In order to express the moment of inertia about the w and z axes in terms of the moments of inertia about the x and y axes, the following distance relationships taken from Fig. A-18 will be required.

$$z = y\cos\theta - x\sin\theta, \qquad w = y\sin\theta + x\cos\theta$$

Then

$$I_w = \int z^2\,dA = \int (y^2\cos^2\theta - 2xy\sin\theta\cos\theta + x^2\sin^2\theta)\,dA \tag{A.12a}$$

$$I_z = \int w^2\,dA = \int (y^2\sin^2\theta + 2xy\sin\theta\cos\theta + x^2\cos^2\theta)\,dA \tag{A.12b}$$

But

$$I_x = \int y^2\,dA, \quad I_y = \int x^2\,dA, \quad \text{and} \quad I_{xy} = \int xy\,dA$$

Therefore, Eq. (A.12a) becomes

$$I_w = I_x\cos^2\theta + I_y\sin^2\theta - I_{xy}\sin 2\theta$$

or expressing the $\cos^2\theta$ and $\sin^2\theta$ in double-angle functions,

$$I_w = \frac{I_x + I_y}{2} + \frac{I_x - I_y}{2}\cos 2\theta - I_{xy}\sin 2\theta \tag{A.13a}$$

Similarly, Eq. (A.12b) becomes

$$I_z = \frac{I_x + I_y}{2} - \frac{I_x - I_y}{2}\cos 2\theta + I_{xy}\sin 2\theta \tag{A.13b}$$

Thus, if the values of I_x, I_y, and I_{xy} are known for any pair of perpendicular axes and the angle θ to any inclined pair, the value of I_z may be directly computed.

Differentiating Eqs. (A.13a) and (A.13b) with respect to θ and setting them equal to zero reveals that the maximum or minimum moments of inertia occur on planes whose

$$\tan 2\theta_{\substack{\max\\ \min}} = \frac{2I_{xy}}{I_y - I_x} \tag{A.14}$$

If the two θ angles which obey this function are now substituted in Eqs. (A.13a) and (A.13b), the maximum and minimum moments of inertia are obtained as follows:

$$I_W = \frac{I_x + I_y}{2} + \sqrt{\left(\frac{I_x - I_y}{2}\right)^2 + I_{xy}^2} \qquad \text{(A.15)}$$

$$I_Z = \frac{I_x + I_y}{2} - \sqrt{\left(\frac{I_x - I_y}{2}\right)^2 + I_{xy}^2} \qquad \text{(A.16)}$$

The similarity of Eqs. (A.15) and (A.16) to Eq. (12.26) reveals the applicability of Mohr's circle for combined stresses to finding moments of inertia on inclined planes, as well as their maximum and minimum values.

Another relationship. If Eqs. (A.15) and (A.16) are added, another significant relationship is obtained:

$$I_z + I_w = I_x + I_y \qquad \text{(A.17)}$$

Stated verbally, this equation declares that the sum of the moments of inertia with respect to any pair of rectangular axes having a common origin is a constant quantity.

Illustrative Problem 5. If the values I_x, I_y, and I_{xy} of a certain unsymmetrical section equal, respectively, 200, 1000, and -300 in.^{4}, determine the maximum and minimum moments of inertia and the planes on which they act.

SOLUTION. The direction of the planes requested are defined by Eq. (A.14). Thus,

$$\tan 2\theta = \frac{2(-300)}{1000 - 200} = \frac{-600}{800} = -0.75$$

$$2\theta = 36°52' \quad \text{or} \quad 216°52' \qquad \text{measured clockwise}$$

$$\theta = 18°26' \quad \text{or} \quad 108°26'$$

The maximum and minimum moments of inertia will be obtained by Eqs. (A.15) and (A.16).

$$I_W = \frac{I_x + I_y}{2} + \sqrt{\left(\frac{I_x - I_y}{2}\right)^2 + I_{xy}^2}$$

$$= \frac{200 + 1000}{2} + \sqrt{\left(\frac{-800}{2}\right)^2 + (-300)^2}$$

$$= 600 + 500 = 1100 \text{ in.}^4$$

$$I_Z = 600 - 500 = 100 \text{ in.}^4$$

The solution to this problem may also have been obtained by drawing a Mohr's circle, as shown in Fig. A-19.

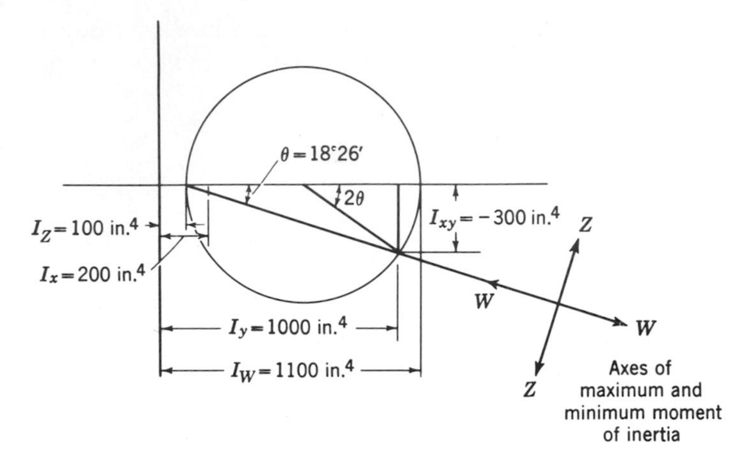

Figure A-19

PROBLEMS

A-14. If an area has I_x, I_y, and I_{xy} equal, respectively, to 300, 600, and 200 in.⁴, determine its maximum and minimum moments of inertia. On what planes will these moments of inertia act with respect to the x and y axes used for I_x and I_y? *Answer:* $I_{max} = 700$ in.⁴, $I_{min} = 200$ in.⁴, $\theta = 26°34'$.

A-15. Solve the previous problem if I_x, I_y, and I_{xy} were, respectively, 400, 800, and −200 in.⁴.

A-16. If the maximum and minimum moments of inertia I_W and I_Z of a certain area are, respectively, 300 and 500 in.⁴, determine (a) the moments of inertia on planes rotated 30° clockwise with respect to the w and z axes, and (b) the maximum product of inertia.

VECTOR ANALYSIS

Vectors can be used with considerable effectiveness in the study of mechanics. A few of the fundamentals, applicable primarily to statics, will now be reviewed.

A-14 Vector Designation

A vector is a quantity having both magnitude and direction and is usually indicated by a capital letter having an arrow over it, e.g., $\vec{A}$. Without the arrow, the letter indicates merely its magnitude.

If the vector $\vec{A}$ in Fig. A-20 is obliquely directed with respect to a rectangular axis system x, y, z, it is made equal to the sum of its rectangular components, each of which is the product of its corresponding i, j, k unit vector

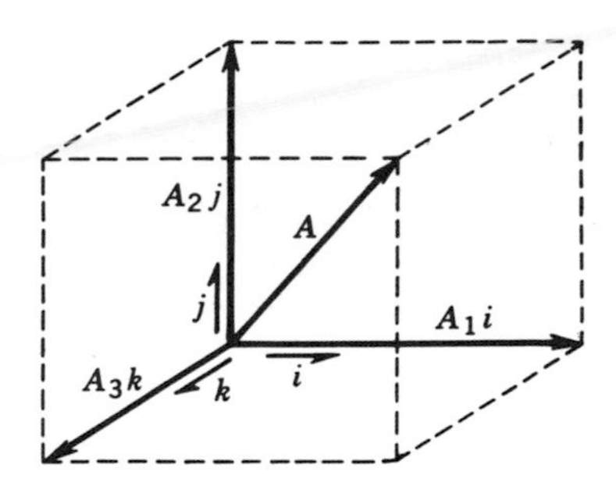

Figure A-20

and multiplication factor A_1, A_2, and A_3. Thus, the vector $\vec{A} = A_1 i + A_2 j + A_3 k$, and has a magnitude equal to $A = \sqrt{A_1^2 + A_2^2 + A_3^2}$.

A-15 Resultant—Concurrent Force System

Find the resultant and magnitude of the four forces of a concurrent force system: $\vec{A} = 2i + 3j - 5k$, $\vec{B} = +3i + 4j - 2k$, $\vec{C} = -4i - j + 3k$, and $\vec{D} = i - 2j + 4k$.

SOLUTION. The resultant is obtained by adding the several similar components:

$$\vec{R} = (2 + 3 - 4 + 1)i + (3 + 4 - 1 - 2)j + (-5 - 2 + 3 + 4)k$$
$$= 2i + 4j$$

The magnitude is therefore $R = \sqrt{2^2 + 4^2} = 2\sqrt{5}$.

A-16 Vector Multiplication

Two types of products are obtained from the multiplication of vectors, the dot product and the cross product.

Dot product. The dot product of the two vectors $\vec{A}$ and $\vec{B}$ (Fig. A-21) having an angle θ between them is equal to the product of their magnitudes and the cosine of the angle between them.

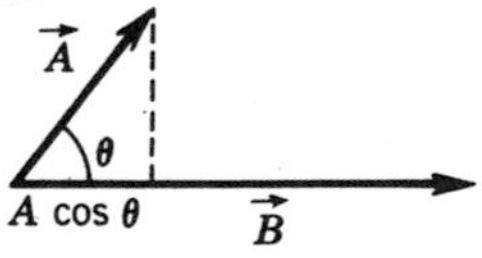

Figure A-21

$$\vec{A} \cdot \vec{B} = AB \cos \theta$$

Note that this type of product would be obtained in multiplying a force by a distance to obtain work. It is always a scalar value.

It follows that $i \cdot i = j \cdot j = k \cdot k = 1$ and $i \cdot j = j \cdot k = k \cdot i = 0$. The algebraic multiplication of $\vec{A} = A_1 i + A_2 j + A_3 k$ and $\vec{B} = B_1 i + B_2 j + B_3 k$

simplified by introducing the known products of $i \cdot i$, $j \cdot j$ and $k \cdot k$, reveals that

$$\vec{A} \cdot \vec{B} = A_1 B_1 + A_2 B_2 + A_3 B_3$$

Illustrative Problem 6. Obtain the dot product of $\vec{A} = 3i + 5j - 4k$ and $\vec{B} = -3i + 5j + 4k$ (Fig. A-22).

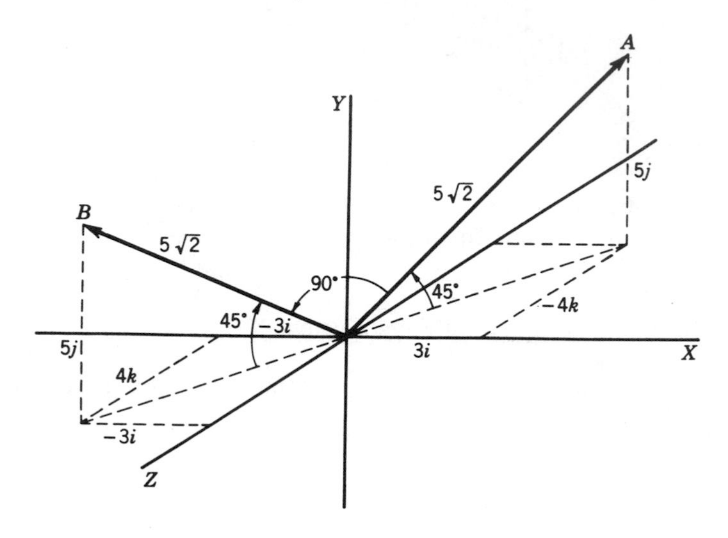

Figure A-22

SOLUTION

$$\vec{A} \cdot \vec{B} = -9 + 25 - 16 = 0$$

Note the diagram reveals $\vec{A}$ to be perpendicular to $\vec{B}$—a corroboration of $\vec{A} \cdot \vec{B} = 0$.

Illustrative Problem 7. Find the angle between $\vec{A} = 3i + 5j - 4k$ and $\vec{B} = -3i + 5j + 4k$.

SOLUTION

$$\vec{A} \cdot \vec{B} = AB \cos\theta, \qquad A = \sqrt{3^2 + 5^2 + (-4)^2} = 5\sqrt{2}$$

$$B = \sqrt{(-3)^2 + 5^2 + 4^2} = 5\sqrt{2}$$

$$AB = 50$$

Thus,

$$\vec{A} \cdot \vec{B} = A_1 B_1 + A_2 B_2 + A_3 B_3 = (3)(-3) + (5)(5) + (-4)(+4) = 0$$

and

$$\cos\theta = \frac{0}{50} = 0, \qquad \theta = 90°$$

Obtaining Components. Vector analysis is extremely useful in obtaining components of forces.

To illustrate, the component of $\vec{A}$ in the direction of vector $\vec{B}$ in Fig. A-21 is equal to $A \cos \theta$. If now the unit vector $\vec{b}$ is defined as

$$\vec{b} = \frac{\vec{B}}{B}$$

the dot product of $\vec{A}$ and $\vec{b}$ is

$$\vec{A} \cdot \vec{b} = |A||b| \cos \theta$$

But since b is unity, $\vec{A} \cdot \vec{b} = |A| \cos \theta =$ component of $\vec{A}$ in direction $\vec{B}$. Thus, the component of a vector in any direction is equal to the dot product of that vector and the unit vector in the desired direction.

Illustrative Problem 8. Find the projection of the vector $\vec{A} = i - 2j + k$ on the vector $\vec{B} = 4i - 4j + 7k$.

SOLUTION. The unit vector in the B direction is

$$\frac{\vec{B}}{B} = \vec{b} = \frac{4i - 4j + 7k}{\sqrt{(4)^2 + (-4)^2 + (7)^2}} = \frac{4}{9}i - \frac{4}{9}j + \frac{7}{9}k$$

Thus, the projection of $\vec{A}$ on the vector $\vec{B}$ is

$$\vec{A} \cdot \vec{b} = (i - 2j + k)\left(\frac{4}{9}i - \frac{4}{9}j + \frac{7}{9}k\right)$$

$$= (1)\left(\frac{4}{9}\right) + (-2)\left(-\frac{4}{9}\right) + (1)\left(\frac{7}{9}\right) = 2\frac{1}{9}$$

Illustrative Problem 9. Find the component of the 424.2-lb force in Fig. A-23 on the z axis.

SOLUTION. The scale for the 424.2-lb force is

$$\frac{424.2}{\sqrt{(-4)^2 + (5)^2 + (-3)^2}} = \frac{424}{5\sqrt{2}}$$
$$= 60 \text{ lb/ft}$$

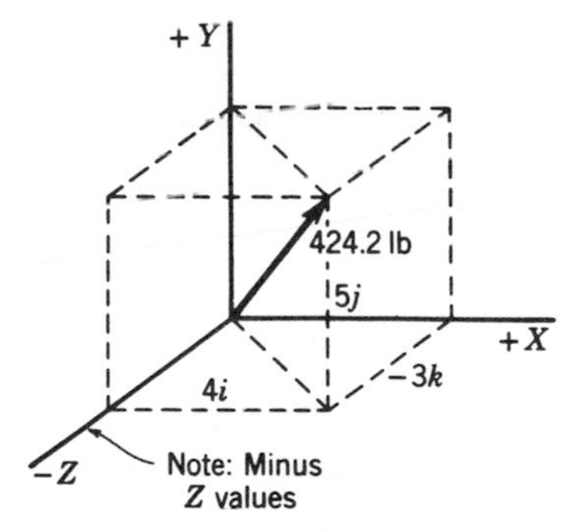

Figure A-23

and

$$\vec{F} = 60(4i + 5j - 3k)$$
$$= 240i + 300j - 180k$$

The unit vector along the z axis is k. The dot product of $\vec{F}$ and $\vec{k}$ equals

$$\vec{F} \cdot \vec{k} = (240i + 300j - 180k) \cdot (k)$$
$$= (-180)(1) = -180 \text{ lb}$$

Work—Illustrative Problem 10. What is the work performed by a force $F = 2i - 3j - k$ moving through a vector distance $d = 3i - 4j + k$.

SOLUTION. Work, being the product of the magnitude of force in the direction moved and the distance moved, equals

$$(F \cos \theta)(d) = \vec{F} \cdot \vec{d} = (2i - 3j - k) \cdot (3i - 4j + k) = 6 + 12 - 1 = 17$$

Cross product. The cross product of two intersecting vectors $\vec{A}$ and $\vec{B}$ (Fig. A-24), identified as $\vec{A} \times \vec{B}$, is a vector whose magnitude is obtained by multiplying their magnitudes by the sine of the angle between them.

$$(\vec{C}) = (\vec{A}) \times (\vec{B}) = AB \sin \theta$$

It is an operation similar to that used in obtaining an area—or a moment of a force.

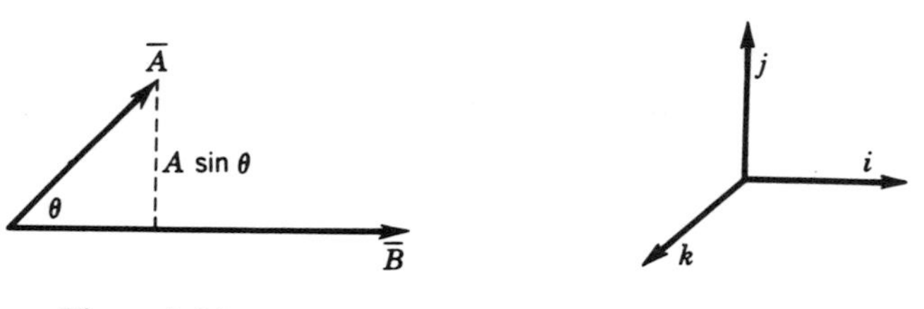

Figure A-24

Figure A-25

The vector $\vec{C} = \vec{A} + \vec{B} = AB \sin \theta \, n$, where n is a unit vector perpendicular to the plane containing vectors $\vec{A}$ and $\vec{B}$ and pointing in the direction of an advancing screw as it is rotated from A to B. Thus, in Fig. A-25, $i \times j = k$. In contrast, $j \times i = -k$. It follows that $j \times k = i$ and $k \times i = j$. Also $i \times i = j \times j = k \times k = 0$.

If $\vec{A}$ is parallel to $\vec{B}$, then $\sin \theta = 0$ and $\vec{A} \times \vec{B} = 0$. The cross product of $\vec{A} = A_1 i + A_2 j = A_3 k$ and $\vec{B} = B_1 i + B_2 j + B_3 k$ is indicated as follows:

$$\begin{aligned} \vec{A} \times \vec{B} &= (A_1 i + A_2 j + A_3 k) \times (B_1 i + B_2 j + B_3 k) \\ &= A_1 B_2 i \times i + A_1 B_2 i \times j + A_1 B_3 i \times k + A_2 B_1 j \times i + A_2 B_2 j \times j \\ &\quad + A_2 B_3 j \times k + A_3 B_1 k \times i + A_3 B_2 k \times j + A_3 B_3 k \times k \end{aligned}$$

Using previously indicated products of unit vectors,

$$\begin{aligned} \vec{A} \times \vec{B} &= (A_2 B_3 - A_3 B_2) i + (A_3 B_1 - A_1 B_3) j + (A_1 B_2 - A_2 B_1) k \\ &= \begin{vmatrix} i & j & k \\ A_1 & A_2 & A_3 \\ B_1 & B_2 & B_3 \end{vmatrix} \end{aligned}$$

Applications of Cross Products
Moment of force about a point:

$$M = Fr \sin \theta = \vec{F} \times \vec{r} \qquad \text{See Fig. A-26.}$$

Area of parallelogram:

$$\text{area} = |A| \sin \theta \, |B| = \vec{A} \times \vec{B} \qquad \text{See Fig. A-27.}$$

Area of triangle:

$$\text{area} = \frac{|B| \sin \theta \, |B|}{2} = \frac{\vec{A} \times \vec{B}}{2} \qquad \text{See Fig. A-28.}$$

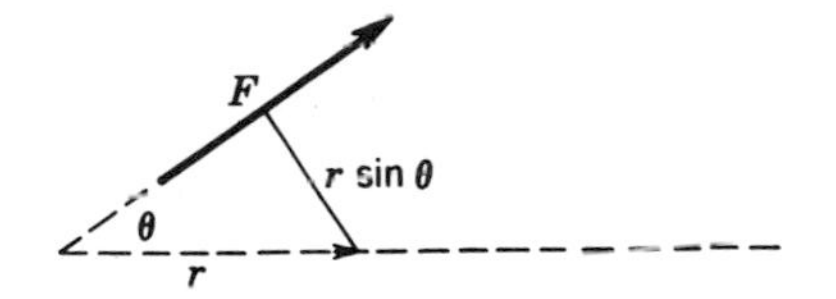

Figure A-26

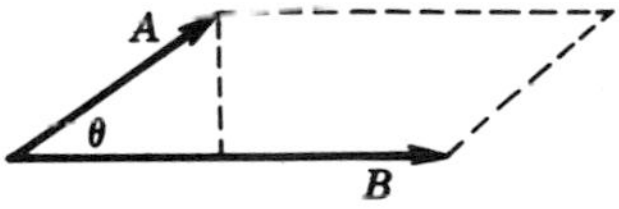

Figure A-27

Figure A-28

PROBLEMS—VECTOR ANALYSIS

A-17. Find the angle between $\vec{A} = 2i + 2j - k$ and $\vec{B} = 6i - 3j + 2k$.

A-18. For what value of a will $\vec{A} = 4i + aj + 2k$ be perpendicular to $\vec{B} = 8i - 4j - 4k$? *Answer:* 6.

A-19. Prove that vectors $\vec{A} = 4i - 2j + 3k$, $\vec{B} = 2i - 3j + k$, and $\vec{C} = 2i + j + 2k$ form a triangle.

A-20. Find the angles which the vector $\vec{A} = 3i - 6j + k$ makes with the x, y, and z axes. *Answers:* 63.8°, −28.0°, 81.5°.

A-21. Find the projection of the vector $\vec{A} = 2i - 2j + 3k$ on the vector $\vec{B} = 5i - 3j + 7k$. *Answer:* 4.06.

A-22. If $\vec{A} = 3i - 2j + k$ and $\vec{B} = 2i + 3j - 2k$, find (a) $\vec{A} \times \vec{B}$; (b) $\vec{B} \times \vec{A}$; (c) $(\vec{A} + \vec{B}) \times (\vec{A} - \vec{B})$.

A-23. Find the area of the triangle having vertices at $P(2, 4, 2)$, $Q(2, -2, -1)$, $R(-2, 4, 3)$. *Answer:* 13.75.

A-24. In a tetrahedron with faces F_1, F_2, F_3, and F_4 show that the vectors representing these areas and erected perpendicular to them have a sum equal to zero.

A-25. Find the magnitude and direction of the resultant of four non-coplanar forces concurrent at the origin and terminating at points $A(2, 4, 4)$, $B(-2, 3, 5)$, $C(-3, -4, 5)$, and $D(3, 4, -5)$, respectively.

Appendix Tables

TABLE 13

I

S SHAPES
PROPERTIES FOR DESIGNING

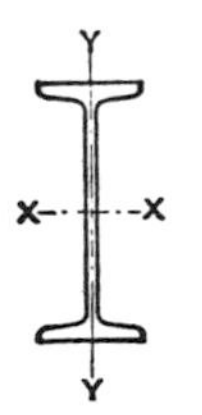

Designation	Area A	Depth d	Flange Width b_f	Flange Thickness t_f	Web Thickness t_w	Elastic Properties Axis X-X I	Axis X-X S	Axis X-X r	Axis Y-Y I	Axis Y-Y S	Axis Y-Y r
	In.^{2}	In.	In.	In.	In.	In.^{4}	In.^{3}	In.	In.^{4}	In.^{3}	In.
S24×120	35.3	24.00	8.048	1.102	0.798	3030	252	9.26	84.2	20.9	1.54
×105.9	31.1	24.00	7.875	1.102	0.625	2830	236	9.53	78.2	19.8	1.58
S24×100	29.4	24.00	7.247	0.871	0.747	2390	199	9.01	47.8	13.2	1.27
×90	26.5	24.00	7.124	0.871	0.624	2250	187	9.22	44.9	12.6	1.30
×79.9	23.5	24.00	7.001	0.871	0.501	2110	175	9.47	42.3	12.1	1.34
S20×95	27.9	20.00	7.200	0.916	0.800	1610	161	7.60	49.7	13.8	1.33
×85	25.0	20.00	7.053	0.916	0.653	1520	152	7.79	46.2	13.1	1.36
S20×75	22.1	20.00	6.391	0.789	0.641	1280	128	7.60	29.6	9.28	1.16
×65.4	19.2	20.00	6.250	0.789	0.500	1180	118	7.84	27.4	8.77	1.19
S18×70	20.6	18.00	6.251	0.691	0.711	926	103	6.71	24.1	7.72	1.08
×54.7	16.1	18.00	6.001	0.691	0.461	804	89.4	7.07	20.8	6.94	1.14
S15×50	14.7	15.00	5.640	0.622	0.550	486	64.8	5.75	15.7	5.57	1.03
×42.9	12.6	15.00	5.501	0.622	0.411	447	59.6	5.95	14.4	5.23	1.07
S12×50	14.7	12.00	5.477	0.659	0.687	305	50.8	4.55	15.7	5.74	1.03
×40.8	12.0	12.00	5.252	0.659	0.472	272	45.4	4.77	13.6	5.16	1.06
S12×35	10.3	12.00	5.078	0.544	0.428	229	38.2	4.72	9.87	3.89	0.98
×31.8	9.35	12.00	5.000	0.544	0.350	218	36.4	4.83	9.36	3.74	1.00
S10×35	10.3	10.00	4.944	0.491	0.594	147	29.4	3.78	8.36	3.38	0.90
×25.4	7.46	10.00	4.661	0.491	0.311	124	24.7	4.07	6.79	2.91	0.95
S 8×23	6.77	8.00	4.171	0.425	0.441	64.9	16.2	3.10	4.31	2.07	0.79
×18.4	5.41	8.00	4.001	0.425	0.271	57.6	14.4	3.26	3.73	1.86	0.83
S 7×20	5.88	7.00	3.860	0.392	0.450	42.4	12.1	2.69	3.17	1.64	0.73
×15.3	4.50	7.00	3.662	0.392	0.252	36.7	10.5	2.86	2.64	1.44	0.76
S 6×17.25	5.07	6.00	3.565	0.359	0.465	26.3	8.77	2.28	2.31	1.30	0.67
×12.5	3.67	6.00	3.332	0.359	0.232	22.1	7.37	2.45	1.82	1.09	0.70
S 5×14.75	4.34	5.00	3.284	0.326	0.494	15.2	6.09	1.87	1.67	1.01	0.62
×10	2.94	5.00	3.004	0.326	0.214	12.3	4.92	2.05	1.22	0.809	0.64
S 4×9.5	2.79	4.00	2.796	0.293	0.326	6.79	3.39	1.56	0.903	0.646	0.56
×7.7	2.26	4.00	2.663	0.293	0.193	6.08	3.04	1.64	0.764	0.574	0.58
S 3×7.5	2.21	3.00	2.509	0.260	0.349	2.93	1.95	1.15	0.586	0.468	0.51
×5.7	1.67	3.00	2.330	0.260	0.170	2.52	1.68	1.23	0.455	0.390	0.52

TABLE 14

W SHAPES
PROPERTIES FOR DESIGNING
(SELECTED LIST)

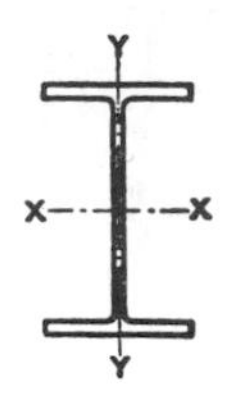

Designation	Area A	Depth d	Flange		Web Thickness t_w	Elastic Properties					
						Axis X-X			Axis Y-Y		
			Width b_f	Thickness t_f		I	S	r	I	S	r
	In.^{2}	In.	In.	In.	In.	In.^{4}	In.^{3}	In.	In.^{4}	In.^{3}	In.
W14×219	64.4	15.87	15.825	1.623	1.005	2800	353	6.59	1070	136	4.08
×211	62.1	15.75	15.800	1.563	0.980	2670	339	6.56	1030	130	4.07
×202	59.4	15.63	15.750	1.503	0.930	2540	325	6.54	980	124	4.06
×193	56.7	15.50	15.710	1.438	0.890	2400	310	6.51	930	118	4.05
×184	54.1	15.38	15.660	1.378	0.840	2270	296	6.49	883	113	4.04
×176	51.7	15.25	15.640	1.313	0.820	2150	282	6.45	838	107	4.02
×167	49.1	15.12	15.600	1.248	0.780	2020	267	6.42	790	101	4.01
×158	46.5	15.00	15.550	1.188	0.730	1900	253	6.40	745	95.8	4.00
×150	44.1	14.88	15.515	1.128	0.695	1790	240	6.37	703	90.6	3.99
×142	41.8	14.75	15.500	1.063	0.680	1670	227	6.32	660	85.2	3.97
W14×136	40.0	14.75	14.740	1.063	0.660	1590	216	6.31	568	77.0	3.77
×127	37.3	14.62	14.690	0.998	0.610	1480	202	6.29	528	71.8	3.76
×119	35.0	14.50	14.650	0.938	0.570	1370	189	6.26	492	67.1	3.75
×111	32.7	14.37	14.620	0.873	0.540	1270	176	6.23	455	62.2	3.73
×103	30.3	14.25	14.575	0.813	0.495	1170	164	6.21	420	57.6	3.72
×95	27.9	14.12	14.545	0.748	0.465	1060	151	6.17	384	52.8	3.71
×87	25.6	14.00	14.500	0.688	0.420	967	138	6.15	350	48.2	3.70
W14×84	24.7	14.18	12.023	0.778	0.451	928	131	6.13	225	37.5	3.02
×78	22.9	14.06	12.000	0.718	0.428	851	121	6.09	207	34.5	3.00
W14×74	21.8	14.19	10.072	0.783	0.450	797	112	6.05	133	26.5	2.48
×68	20.0	14.06	10.040	0.718	0.418	724	103	6.02	121	24.1	2.46
×61	17.9	13.91	10.000	0.643	0.378	641	92.2	5.98	107	21.5	2.45
W14×53	15.6	13.94	8.062	0.658	0.370	542	77.8	5.90	57.5	14.3	1.92
×48	14.1	13.81	8.031	0.593	0.339	485	70.2	5.86	51.3	12.8	1.91
×43	12.6	13.68	8.000	0.528	0.308	429	62.7	5.82	45.1	11.3	1.89
W14×38	11.2	14.12	6.776	0.513	0.313	386	54.7	5.88	26.6	7.86	1.54
×34	10.0	14.00	6.750	0.453	0.287	340	48.6	5.83	23.3	6.89	1.52
×30	8.83	13.86	6.733	0.383	0.270	290	41.9	5.74	19.5	5.80	1.49

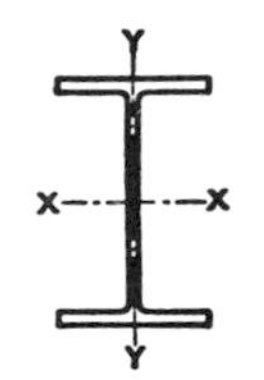

W SHAPES
PROPERTIES FOR DESIGNING
(SELECTED LIST)

Designation	Area A	Depth d	Flange		Web Thickness t_w	Elastic Properties					
						Axis X-X			Axis Y-Y		
			Width b_f	Thickness t_f		I	S	r	I	S	r
	In.^{2}	In.	In.	In.	In.	In.^{4}	In.^{3}	In.	In.^{4}	In.^{3}	In.
W12×106	31.2	12.88	12.230	0.986	0.620	931	145	5.46	301	49.2	3.11
×99	29.1	12.75	12.192	0.921	0.582	859	135	5.43	278	45.7	3.09
×92	27.1	12.62	12.155	0.856	0.545	789	125	5.40	256	42.2	3.08
×85	25.0	12.50	12.105	0.796	0.495	723	116	5.38	235	38.9	3.07
×79	23.2	12.38	12.080	0.736	0.470	663	107	5.34	216	35.8	3.05
×72	21.2	12.25	12.040	0.671	0.430	597	97.5	5.31	195	32.4	3.04
×65	19.1	12.12	12.000	0.606	0.390	533	88.0	5.28	175	29.1	3.02
W12×58	17.1	12.19	10.014	0.641	0.359	476	78.1	5.28	107	21.4	2.51
×53	15.6	12.06	10.000	0.576	0.345	426	70.7	5.23	96.1	19.2	2.48
W12×50	14.7	12.19	8.077	0.641	0.371	395	64.7	5.18	56.4	14.0	1.96
×45	13.2	12.06	8.042	0.576	0.336	351	58.2	5.15	50.0	12.4	1.94
×40	11.8	11.94	8.000	0.516	0.294	310	51.9	5.13	44.1	11.0	1.94
W12×36	10.6	12.24	6.565	0.540	0.305	281	46.0	5.15	25.5	7.77	1.55
×31	9.13	12.09	6.525	0.465	0.265	239	39.5	5.12	21.6	6.61	1.54
×27	7.95	11.96	6.497	0.400	0.237	204	34.2	5.07	18.3	5.63	1.52
W10×66	19.4	10.38	10.117	0.748	0.457	382	73.7	4.44	129	25.5	2.58
×60	17.7	10.25	10.075	0.683	0.415	344	67.1	4.41	116	23.1	2.57
×54	15.9	10.12	10.028	0.618	0.368	306	60.4	4.39	104	20.7	2.56
×49	14.4	10.00	10.000	0.558	0.340	273	54.6	4.35	93.0	18.6	2.54
W10×45	13.2	10.12	8.022	0.618	0.350	249	49.1	4.33	53.2	13.3	2.00
×39	11.5	9.94	7.990	0.528	0.318	210	42.2	4.27	44.9	11.2	1.98
×33	9.71	9.75	7.964	0.433	0.292	171	35.0	4.20	36.5	9.16	1.94
W10×29	8.54	10.22	5.799	0.500	0.289	158	30.8	4.30	16.3	5.61	1.38
×25	7.36	10.08	5.762	0.430	0.252	133	26.5	4.26	13.7	4.76	1.37
×21	6.20	9.90	5.750	0.340	0.240	107	21.5	4.15	10.8	3.75	1.32
W 8×48	14.1	8.50	8.117	0.683	0.405	184	43.2	3.61	60.9	15.0	2.08
×40	11.8	8.25	8.077	0.558	0.365	146	35.5	3.53	49.0	12.1	2.04
×35	10.3	8.12	8.027	0.493	0.315	126	31.1	3.50	42.5	10.6	2.03
×31	9.12	8.00	8.000	0.433	0.288	110	27.4	3.47	37.0	9.24	2.01
W 8×28	8.23	8.06	6.540	0.463	0.285	97.8	24.3	3.45	21.6	6.61	1.62
×24	7.06	7.93	6.500	0.398	0.245	82.5	20.8	3.42	18.2	5.61	1.61
8×20	5.89	8.14	5.268	0.378	0.248	69.4	17.0	3.43	9.22	3.50	1.25
×17	5.01	8.00	5.250	0.308	0.230	56.6	14.1	3.36	7.44	2.83	1.22

TABLE 15

CHANNELS
AMERICAN STANDARD
PROPERTIES FOR DESIGNING

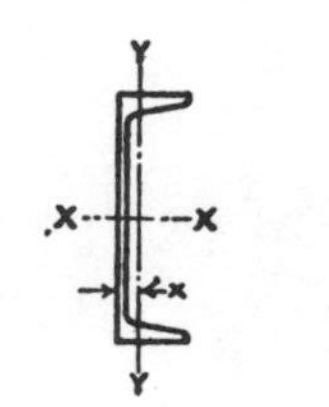

Designation	Area A	Depth d	Flange		Web Thickness t_w	$\frac{d}{A_f}$	Axis X-X			Axis Y-Y			$\bar{x}$	Shear Center Location E_0	Torsional Constant J	Warping Constant C_w
			Width b_f	Average Thickness t_f			I	S	r	I	S	r				
	In.^{2}	In.	In.	In.	In.		In.^{4}	In.^{3}	In.	In.^{4}	In.^{3}	In.	In.	In.	In.^{4}	In.^{6}
C15×50	14.7	15.00	3.716	0.650	0.716	6.21	404	53.8	5.24	11.0	3.78	0.867	0.799	0.941	2.66	492
×40	11.8	15.00	3.520	0.650	0.520	6.56	349	46.5	5.44	9.23	3.36	0.886	0.778	1.03	1.46	410
×33.9	9.96	15.00	3.400	0.650	0.400	6.79	315	42.0	5.62	8.13	3.11	0.904	0.787	1.10	1.01	358
C12×30	8.82	12.00	3.170	0.501	0.510	7.55	162	27.0	4.29	5.14	2.06	0.763	0.674	0.873	0.865	151
×25	7.35	12.00	3.047	0.501	0.387	7.85	144	24.1	4.43	4.47	1.88	0.780	0.674	0.940	0.541	131
×20.7	6.09	12.00	2.942	0.501	0.282	8.13	129	21.5	4.61	3.88	1.73	0.799	0.698	1.01	0.371	112
C10×30	8.82	10.00	3.033	0.436	0.673	7.55	103	20.7	3.42	3.94	1.65	0.669	0.649	0.705	1.22	79.5
×25	7.35	10.00	2.886	0.436	0.526	7.94	91.2	18.2	3.52	3.36	1.48	0.676	0.617	0.757	0.690	68.4
×20	5.88	10.00	2.739	0.436	0.379	8.36	78.9	15.8	3.66	2.81	1.32	0.691	0.606	0.826	0.370	57.0
×15.3	4.49	10.00	2.600	0.436	0.240	8.81	67.4	13.5	3.87	2.28	1.16	0.713	0.634	0.916	0.211	45.5
C 9×20	5.88	9.00	2.648	0.413	0.448	8.22	60.9	13.5	3.22	2.42	1.17	0.642	0.583	0.739	0.429	39.5
×15	4.41	9.00	2.485	0.413	0.285	8.76	51.0	11.3	3.40	1.93	1.01	0.661	0.586	0.824	0.209	31.0
×13.4	3.94	9.00	2.433	0.413	0.233	8.95	47.9	10.6	3.48	1.76	0.962	0.668	0.601	0.859	0.169	28.2

CHANNELS
AMERICAN STANDARD
PROPERTIES FOR DESIGNING

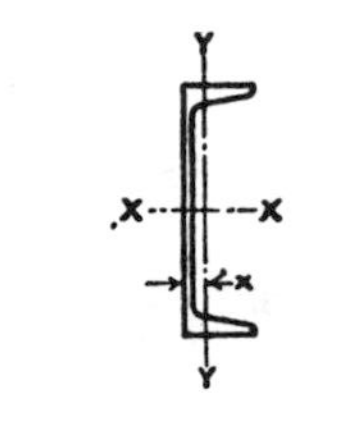

Designation	Area A	Depth d	Flange		Web Thickness t_w	$\frac{d}{A_f}$	Axis X-X			Axis Y-Y			$\bar{x}$	Shear Center Location E_0	Torsional Constant J	Warping Constant C_w
			Width b_f	Average Thickness t_f			I	S	r	I	S	r				
	In.^{2}	In.	In.	In.	In.		In.^{4}	In.^{3}	In.	In.^{4}	In.^{3}	In.	In.	In.	In.^{4}	In.^{6}
C 8×18.75	5.51	8.00	2.527	0.390	0.487	8.12	44.0	11.0	2.82	1.98	1.01	0.599	0.565	0.674	0.436	25.1
×13.75	4.04	8.00	2.343	0.390	0.303	8.75	36.1	9.03	2.99	1.53	0.853	0.615	0.553	0.756	0.187	19.3
×11.5	3.38	8.00	2.260	0.390	0.220	9.08	32.6	8.14	3.11	1.32	0.781	0.625	0.571	0.807	0.131	16.5
C 7×14.75	4.33	7.00	2.299	0.366	0.419	8.31	27.2	7.78	2.51	1.38	0.779	0.564	0.532	0.651	0.268	13.1
×12.25	3.60	7.00	2.194	0.366	0.314	8.71	24.2	6.93	2.60	1.17	0.702	0.571	0.525	0.695	0.161	11.2
×9.8	2.87	7.00	2.090	0.366	0.210	9.14	21.3	6.08	2.72	0.968	0.625	0.581	0.541	0.752	0.100	9.16
C 6×13	3.83	6.00	2.157	0.343	0.437	8.10	17.4	5.80	2.13	1.05	0.642	0.525	0.514	0.599	0.241	7.21
×10.5	3.09	6.00	2.034	0.343	0.314	8.59	15.2	5.06	2.22	0.865	0.564	0.529	0.500	0.643	0.131	5.94
×8.2	2.40	6.00	1.920	0.343	0.200	9.10	13.1	4.38	2.34	0.692	0.492	0.537	0.512	0.699	0.075	4.73
C 5×9	2.64	5.00	1.885	0.320	0.325	8.29	8.90	3.56	1.83	0.632	0.449	0.489	0.478	0.590	0.109	2.93
×6.7	1.97	5.00	1.750	0.320	0.190	8.93	7.49	3.00	1.95	0.478	0.378	0.493	0.484	0.647	0.055	2.22
C 4×7.25	2.13	4.00	1.721	0.296	0.321	7.84	4.59	2.29	1.47	0.432	0.343	0.450	0.459	0.546	0.082	1.24
×5.4	1.59	4.00	1.584	0.296	0.184	8.52	3.85	1.93	1.56	0.319	0.283	0.449	0.458	0.594	0.040	0.923
C 3×6	1.76	3.00	1.596	0.273	0.356	6.87	2.07	1.38	1.08	0.305	0.268	0.416	0.455	0.500	0.073	0.463
×5	1.47	3.00	1.498	0.273	0.258	7.32	1.85	1.24	1.12	0.247	0.233	0.410	0.438	0.521	0.043	0.380
×4.1	1.21	3.00	1.410	0.273	0.170	7.78	1.66	1.10	1.17	0.197	0.202	0.404	0.437	0.546	0.027	0.307

TABLE 16

ANGLES
UNEQUAL LEGS
PROPERTIES FOR DESIGNING

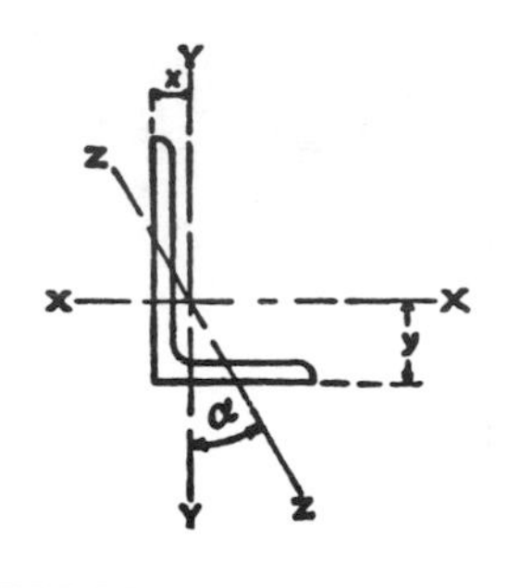

Size and Thickness	k	Weight per Foot	Area	AXIS X-X				AXIS Y-Y				AXIS Z-Z	
				I	S	r	y	I	S	r	x	r	Tan α
In.	In.	Lb	In.^{2}	In.^{4}	In.^{3}	In.	In.	In.^{4}	In.^{3}	In.	In.	In.	
L9×4×1	1-1/2	40.8	12.0	97.0	17.6	2.84	3.50	12.0	4.00	1.00	1.00	.834	.203
7/8	1-3/8	36.1	10.6	86.8	15.7	2.86	3.45	10.8	3.56	1.01	.953	.836	.208
3/4	1-1/4	31.3	9.19	76.1	13.6	2.88	3.41	9.63	3.11	1.02	.906	.841	.212
5/8	1-1/8	26.3	7.73	64.9	11.5	2.90	3.36	8.32	2.65	1.04	.858	.847	.216
9/16	1-1/16	23.8	7.00	59.1	10.4	2.91	3.33	7.63	2.41	1.04	.834	.850	.218
1/2	1	21.3	6.25	53.2	9.34	2.92	3.31	6.92	2.17	1.05	.810	.854	.220
L8×6×1	1-1/2	44.2	13.0	80.8	15.1	2.49	2.65	38.8	8.92	1.73	1.65	1.28	.543
7/8	1-3/8	39.1	11.5	72.3	13.4	2.51	2.61	34.9	7.94	1.74	1.61	1.28	.547
3/4	1-1/4	33.8	9.94	63.4	11.7	2.53	2.56	30.7	6.92	1.76	1.56	1.29	.551
5/8	1-1/8	28.5	8.36	54.1	9.87	2.54	2.52	26.3	5.88	1.77	1.52	1.29	.554
9/16	1-1/16	25.7	7.56	49.3	8.95	2.55	2.50	24.0	5.34	1.78	1.50	1.30	.556
1/2	1	23.0	6.75	44.3	8.02	2.56	2.47	21.7	4.79	1.79	1.47	1.30	.558
7/16	15/16	20.2	5.93	39.2	7.07	2.57	2.45	19.3	4.23	1.80	1.45	1.31	.560
L8×4×1	1-1/2	37.4	11.0	69.6	14.1	2.52	3.05	11.6	3.94	1.03	1.05	.846	.247
7/8	1-3/8	33.1	9.73	62.5	12.5	2.53	3.00	10.5	3.51	1.04	.999	.848	.253
3/4	1-1/4	28.7	8.44	54.9	10.9	2.55	2.95	9.36	3.07	1.05	.953	.852	.258
5/8	1-1/8	24.2	7.11	46.9	9.21	2.57	2.91	8.10	2.62	1.07	.906	.857	.262
9/16	1-1/16	21.9	6.43	42.8	8.35	2.58	2.88	7.43	2.38	1.07	.882	.861	.265
1/2	1	19.6	5.75	38.5	7.49	2.59	2.86	6.74	2.15	1.08	.859	.865	.267
7/16	15/16	17.2	5.06	34.1	6.60	2.60	2.83	6.02	1.90	1.09	.835	.869	.269
L7×4×7/8	1-3/8	30.2	8.86	42.9	9.65	2.20	2.55	10.2	3.46	1.07	1.05	.856	.318
3/4	1-1/4	26.2	7.69	37.8	8.42	2.22	2.51	9.05	3.03	1.09	1.01	.860	.324
5/8	1-1/8	22.1	6.48	32.4	7.14	2.24	2.46	7.84	2.58	1.10	.963	.865	.329
9/16	1-1/16	20.0	5.87	29.6	6.48	2.24	2.44	7.19	2.35	1.11	.940	.868	.332
1/2	1	17.9	5.25	26.7	5.81	2.25	2.42	6.53	2.12	1.11	.917	.872	.335
7/16	15/16	15.8	4.62	23.7	5.13	2.26	2.39	5.83	1.88	1.12	.893	.876	.337
3/8	7/8	13.6	3.98	20.6	4.44	2.27	2.37	5.10	1.63	1.13	.870	.880	.340

TABLE 16 (Continued)

ANGLES
UNEQUAL LEGS
PROPERTIES FOR DESIGNING

Size and Thickness	k	Weight per Foot	Area	AXIS X-X				AXIS Y-Y				AXIS Z-Z	
				I	S	r	y	I	S	r	x	r	Tan α
In.	In.	Lb	In.^{2}	In.^{4}	In.^{3}	In.	In.	In.^{4}	In.^{3}	In.	In.	In.	
L6 × 4 × 7/8	1-3/8	27.2	7.98	27.7	7.15	1.86	2.12	9.75	3.39	1.11	1.12	.857	.42
3/4	1-1/4	23.6	6.94	24.5	6.25	1.88	2.08	8.68	2.97	1.12	1.08	.860	.42
5/8	1-1/8	20.0	5.86	21.1	5.31	1.90	2.03	7.52	2.54	1.13	1.03	.864	.43
9/16	1-1/16	18.1	5.31	19.3	4.83	1.90	2.01	6.91	2.31	1.14	1.01	.866	.43
1/2	1	16.2	4.75	17.4	4.33	1.91	1.99	6.27	2.08	1.15	.987	.870	.44
7/16	15/16	14.3	4.18	15.5	3.83	1.92	1.96	5.60	1.85	1.16	.964	.873	.44
3/8	7/8	12.3	3.61	13.5	3.32	1.93	1.94	4.90	1.60	1.17	.941	.877	.44
5/16	13/16	10.3	3.03	11.4	2.79	1.94	1.92	4.18	1.35	1.17	.918	.882	.44
1/4	3/4	8.3	2.44	9.27	2.26	1.95	1.89	3.41	1.10	1.18	.894	.887	.45
L6 × 3-1/2 × 1/2	1	15.3	4.50	16.6	4.24	1.92	2.08	4.25	1.59	.972	.833	.759	.34
3/8	7/8	11.7	3.42	12.9	3.24	1.94	2.04	3.34	1.23	.988	.787	.767	.35
5/16	13/16	9.8	2.87	10.9	2.73	1.95	2.01	2.85	1.04	.996	.763	.772	.35
1/4	3/4	7.9	2.31	8.86	2.21	1.96	1.99	2.34	0.847	1.01	.740	.777	.35
L5 × 3-1/2 × 3/4	1-1/4	19.8	5.81	13.9	4.28	1.55	1.75	5.55	2.22	.977	.996	.748	.46
5/8	1-1/8	16.8	4.92	12.0	3.65	1.56	1.70	4.83	1.90	.991	.951	.751	.47
1/2	1	13.6	4.00	9.99	2.99	1.58	1.66	4.05	1.56	1.01	.906	.755	.47
7/16	15/16	12.0	3.53	8.90	2.64	1.59	1.63	3.63	1.39	1.01	.883	.758	.48
3/8	7/8	10.4	3.05	7.78	2.29	1.60	1.61	3.18	1.21	1.02	.861	.762	.48
5/16	13/16	8.7	2.56	6.60	1.94	1.61	1.59	2.72	1.02	1.03	.838	.766	.48
1/4	3/4	7.0	2.06	5.39	1.57	1.62	1.56	2.23	.830	1.04	.814	.770	.49
L5 × 3 × 1/2	1	12.8	3.75	9.45	2.91	1.59	1.75	2.58	1.15	.829	.750	.648	.35
7/16	15/16	11.3	3.31	8.43	2.58	1.60	1.73	2.32	1.02	.837	.727	.651	.36
3/8	7/8	9.8	2.86	7.37	2.24	1.61	1.70	2.04	.888	.845	.704	.654	.36
5/16	13/16	8.2	2.40	6.26	1.89	1.61	1.68	1.75	.753	.853	.681	.658	.36
1/4	3/4	6.6	1.94	5.11	1.53	1.62	1.66	1.44	.614	.861	.657	.663	.37

TABLE 16 (Continued)

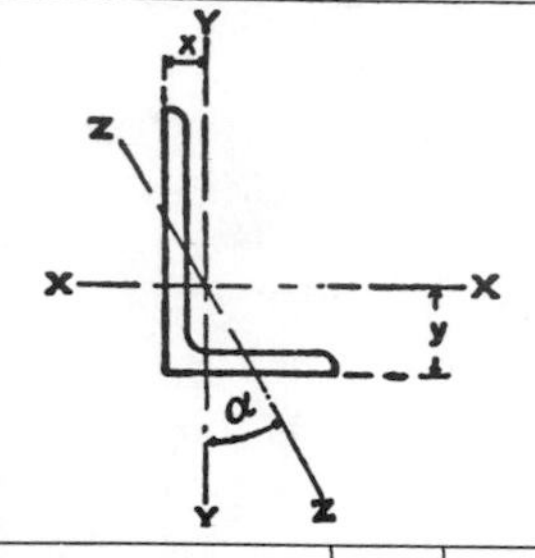

ANGLES
UNEQUAL LEGS
PROPERTIES FOR DESIGNING

Size and Thickness	k	Weight per Foot	Area	AXIS X-X				AXIS Y-Y				AXIS Z-Z	
				I	S	r	y	I	S	r	x	r	Tan α
In.	In.	Lb	In.^{2}	In.^{4}	In.^{3}	In.	In.	In.^{4}	In.^{3}	In.	In.	In.	
L4×3-1/2×5/8	1-1/16	14.7	4.30	6.37	2.35	1.22	1.29	4.52	1.84	1.03	1.04	.719	.745
1/2	15/16	11.9	3.50	5.32	1.94	1.23	1.25	3.79	1.52	1.04	1.00	.722	.750
7/16	7/8	10.6	3.09	4.76	1.72	1.24	1.23	3.40	1.35	1.05	.978	.724	.753
3/8	13/16	9.1	2.67	4.18	1.49	1.25	1.21	2.95	1.17	1.06	.955	.727	.755
5/16	3/4	7.7	2.25	3.56	1.26	1.26	1.18	2.55	.994	1.07	.932	.730	.757
1/4	11/16	6.2	1.81	2.91	1.03	1.27	1.16	2.09	.808	1.07	.909	.734	.759
L4×3×5/8	1-1/16	13.6	3.98	6.03	2.30	1.23	1.37	2.87	1.35	.849	.871	.637	.534
1/2	15/16	11.1	3.25	5.05	1.89	1.25	1.33	2.42	1.12	.864	.827	.639	.543
7/16	7/8	9.8	2.87	4.52	1.68	1.25	1.30	2.18	.992	.871	.804	.641	.547
3/8	13/16	8.5	2.48	3.96	1.46	1.26	1.28	1.92	.866	.879	.782	.644	.551
5/16	3/4	7.2	2.09	3.38	1.23	1.27	1.26	1.65	.734	.887	.759	.647	.554
1/4	11/16	5.8	1.69	2.77	1.00	1.28	1.24	1.36	.599	.896	.736	.651	.558
L3-1/2×3×1/2	15/16	10.2	3.00	3.45	1.45	1.07	1.13	2.33	1.10	.881	.875	.621	.714
7/16	7/8	9.1	2.65	3.10	1.29	1.08	1.10	2.09	.975	.889	.853	.622	.718
3/8	13/16	7.9	2.30	2.72	1.13	1.09	1.08	1.85	.851	.897	.830	.625	.721
5/16	3/4	6.6	1.93	2.33	.954	1.10	1.06	1.58	.722	.905	.808	.627	.724
1/4	11/16	5.4	1.56	1.91	.776	1.11	1.04	1.30	.589	.914	.785	.631	.727
L3-1/2×2-1/2×1/2	15/16	9.4	2.75	3.24	1.41	1.09	1.20	1.36	.760	.704	.705	.534	.486
7/16	7/8	8.3	2.43	2.91	1.26	1.09	1.18	1.23	.677	.711	.682	.535	.491
3/8	13/16	7.2	2.11	2.56	1.09	1.10	1.16	1.09	.592	.719	.660	.537	.496
5/16	3/4	6.1	1.78	2.19	.927	1.11	1.14	.939	.504	.727	.637	.540	.501
1/4	11/16	4.9	1.44	1.80	.755	1.12	1.11	.777	.412	.735	.614	.544	.506
L3×2-1/2×1/2	7/8	8.5	2.50	2.08	1.04	.913	1.00	1.30	.744	.722	.750	.520	.667
7/16	13/16	7.6	2.21	1.88	.928	.920	.978	1.18	.664	.729	.728	.521	.672
3/8	3/4	6.6	1.92	1.66	.810	.928	.956	1.04	.581	.736	.706	.522	.676
5/16	11/16	5.6	1.62	1.42	.688	.937	.933	.898	.494	.744	.683	.525	.680
1/4	5/8	4.5	1.31	1.17	.561	.945	.911	.743	.404	.753	.661	.528	.684
3/16	9/16	3.39	.996	.907	.430	.954	.888	.577	.310	.761	.638	.533	.688

TABLE 16 (Continued)

ANGLES
UNEQUAL LEGS
PROPERTIES FOR DESIGNING

Size and Thickness	k	Weight per Foot	Area	AXIS X-X				AXIS Y-Y				AXIS Z-Z	
				I	S	r	y	I	S	r	x	r	Tan α
In.	In.	Lb	In.²	In.⁴	In.³	In.	In.	In.⁴	In.³	In.	In.	In.	
L3 × 2 × 1/2	13/16	7.7	2.25	1.92	1.00	.924	1.08	.672	.474	.546	.583	.428	.414
7/16	3/4	6.8	2.00	1.73	.894	.932	1.06	.609	.424	.553	.561	.429	.421
3/8	11/16	5.9	1.73	1.53	.781	.940	1.04	.543	.371	.559	.539	.430	.428
5/16	5/8	5.0	1.46	1.32	.664	.948	1.02	.470	.317	.567	.516	.432	.435
1/4	9/16	4.1	1.19	1.09	.542	.957	.993	.392	.260	.574	.493	.435	.440
3/16	1/2	3.07	.902	.842	.415	.966	.970	.307	.200	.583	.470	.439	.446
L2-1/2 × 2 × 3/8	11/16	5.3	1.55	.912	.547	.768	.831	.514	.363	.577	.581	.420	.614
5/16	5/8	4.5	1.31	.788	.466	.776	.809	.446	.310	.584	.559	.422	.620
1/4	9/16	3.62	1.06	.654	.381	.784	.787	.372	.254	.592	.537	.424	.626
3/16	1/2	2.75	.809	.509	.293	.793	.764	.291	.196	.600	.514	.427	.631
L2-1/2 × 1-1/2 × 5/16	5/8	3.92	1.15	.711	.444	.785	.898	.191	.174	.408	.398	.322	.349
1/4	9/16	3.19	.938	.591	.364	.794	.875	.161	.143	.415	.375	.324	.357
3/16	1/2	2.44	.715	.461	.279	.803	.852	.127	.111	.422	.352	.327	.364
L2 × 1-1/2 × 1/4	1/2	2.77	.813	.316	.236	.623	.663	.151	.139	.432	.413	.320	.543
3/16	7/16	2.12	.621	.248	.182	.632	.641	.120	.108	.440	.391	.322	.551
1/8	3/8	1.44	.422	.173	.125	.641	.618	.085	.075	.448	.368	.326	.558
L2 × 1-1/4 × 1/4	1/2	2.55	.750	.296	.229	.628	.708	.089	.097	.344	.333	.269	.378
3/16	7/16	1.96	.574	.232	.177	.636	.686	.071	.075	.351	.311	.271	.387
1/8	3/8	1.33	.391	.163	.122	.645	.663	.050	.052	.359	.287	.274	.396
L1-3/4 × 1-1/4 × 1/4	7/16	2.34	.688	.202	.176	.543	.602	.085	.095	.352	.352	.267	.486
3/16	3/8	1.80	.527	.160	.137	.551	.580	.068	.074	.359	.330	.269	.496
1/8	5/16	1.23	.359	.113	.094	.560	.557	.049	.051	.368	.307	.272	.506

TABLE 17

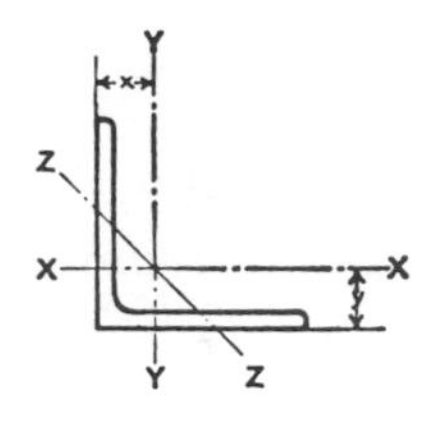

ANGLES
EQUAL LEGS
PROPERTIES FOR DESIGNING

Size and Thickness	k	Weight per Foot	Area	AXIS X-X AND AXIS Y-Y				AXIS Z-Z
				I	S	r	x or y	r
In.	In.	Lb	In.^{2}	In.^{4}	In.^{3}	In.	In.	In.
L8×8×1-1/8	1-3/4	56.9	16.7	98.0	17.5	2.42	2.41	1.56
1	1-5/8	51.0	15.0	89.0	15.8	2.44	2.37	1.56
7/8	1-1/2	45.0	13.2	79.6	14.0	2.45	2.32	1.57
3/4	1-3/8	38.9	11.4	69.7	12.2	2.47	2.28	1.58
5/8	1-1/4	32.7	9.61	59.4	10.3	2.49	2.23	1.58
9/16	1-3/16	29.6	8.68	54.1	9.34	2.50	2.21	1.59
1/2	1-1/8	26.4	7.75	48.6	8.36	2.50	2.19	1.59
L6×6×1	1-1/2	37.4	11.0	35.5	8.57	1.80	1.86	1.17
7/8	1-3/8	33.1	9.73	31.9	7.63	1.81	1.82	1.17
3/4	1-1/4	28.7	8.44	28.2	6.66	1.83	1.78	1.17
5/8	1-1/8	24.2	7.11	24.2	5.66	1.84	1.73	1.18
9/16	1-1/16	21.9	6.43	22.1	5.14	1.85	1.71	1.18
1/2	1	19.6	5.75	19.9	4.61	1.86	1.68	1.18
7/16	15/16	17.2	5.06	17.7	4.08	1.87	1.66	1.19
2/8	7/8	14.9	4.36	15.4	3.53	1.88	1.64	1.19
5/16	13/16	12.4	3.65	13.0	2.97	1.89	1.62	1.20
L5×5×7/8	1-3/8	27.2	7.98	17.8	5.17	1.49	1.57	.973
3/4	1-1/4	23.6	6.94	15.7	4.53	1.51	1.52	.975
5/8	1-1/8	20.0	5.86	13.6	3.86	1.52	1.48	.978
1/2	1	16.2	4.75	11.3	3.16	1.54	1.43	.983
7/16	15/16	14.3	4.18	10.0	2.79	1.55	1.41	.986
3/8	7/8	12.3	3.61	8.74	2.42	1.56	1.39	.990
5/16	13/16	10.3	3.03	7.42	2.04	1.57	1.37	.994
L4×4×3/4	1-1/8	18.5	5.44	7.67	2.81	1.19	1.27	.778
5/8	1	15.7	4.61	6.66	2.40	1.20	1.23	.779
1/2	7/8	12.8	3.75	5.56	1.97	1.22	1.18	.782
7/16	13/16	11.3	3.31	4.97	1.75	1.23	1.16	.785
3/8	3/4	9.8	2.86	4.36	1.52	1.23	1.14	.788
5/16	11/16	8.2	2.40	3.71	1.29	1.24	1.12	.791
1/4	5/8	6.6	1.94	3.04	1.05	1.25	1.09	.795

TABLE 17 (Continued)

ANGLES
EQUAL LEGS
PROPERTIES FOR DESIGNING

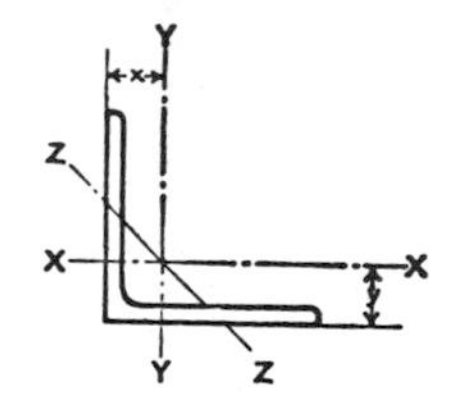

Size and Thickness	k	Weight per Foot	Area	AXIS X-X AND AXIS Y-Y				AXIS Z-Z
				I	S	r	x or y	r
In.	In.	Lb	In.²	In.⁴	In.³	In.	In.	In.
L3-1/2×3-1/2×1/2	7/8	11.1	3.25	3.64	1.49	1.06	1.06	.683
7/16	13/16	9.8	2.87	3.26	1.32	1.07	1.04	.684
3/8	3/4	8.5	2.48	2.87	1.15	1.07	1.01	.687
5/16	11/16	7.2	2.09	2.45	.976	1.08	.990	.690
1/4	5/8	5.8	1.69	2.01	.794	1.09	.968	.694
L3×3×1/2	13/16	9.4	2.75	2.22	1.07	.898	.932	.584
7/16	3/4	8.3	2.43	1.99	.954	.905	.910	.585
3/8	11/16	7.2	2.11	1.76	.833	.913	.888	.587
5/16	5/8	6.1	1.78	1.51	.707	.922	.869	.589
1/4	9/16	4.9	1.44	1.24	.577	.930	.842	.592
3/16	1/2	3.71	1.09	.962	.441	.939	.820	.596
L2-1/2×2-1/2×1/2	13/16	7.7	2.25	1.23	.724	.739	.806	.487
3/8	11/16	5.9	1.73	.984	.566	.753	.762	.487
5/16	5/8	5.0	1.46	.849	.482	.761	.740	.489
1/4	9/16	4.1	1.19	.703	.394	.769	.717	.491
3/16	1/2	3.07	0.92	.547	.303	.778	.694	.495
L2×2×3/8	7/8	4.7	1.36	.479	.351	.594	.636	.389
5/16	13/16	3.92	1.15	.416	.300	.601	.614	.390
1/4	3/4	3.19	.938	.348	.247	.609	.592	.391
3/16	11/16	2.44	.715	.272	.190	.617	.569	.394
1/8	5/8	1.65	.484	.190	.131	.626	.546	.398
L1-3/4×1-3/4×1/4	11/16	2.77	.813	.227	.186	.529	.529	.341
3/16	5/8	2.12	.621	.179	.144	.537	.506	.343
1/8	9/16	1.44	.422	.126	.099	.546	.484	.347
L1-1/2×1-1/2×1/4	5/8	2.34	.688	.139	.134	.449	.466	.292
3/16	9/16	1.80	.527	.110	.104	.457	.444	.293
5/32	9/16	1.52	.444	.094	.088	.461	.433	.295
1/8	1/2	1.23	.359	.078	.072	.465	.421	.296
L1-1/4×1-1/4×1/4	5/8	1.92	.563	.077	.091	.369	.403	.243
3/16	9/16	1.48	.434	.061	.071	.377	.381	.244
1/8	1/2	1.01	.297	.044	.049	.385	.359	.246
L1×1×1/4	5/8	1.49	.438	.037	.056	.290	.339	.196
3/16	9/16	1.16	.340	.030	.044	.297	.318	.195
1/8	1/2	.80	.234	.022	.031	.304	.296	.196

TABLE 18

TIMBER
AMERICAN STANDARD SIZES
PROPERTIES FOR DESIGNING
NATIONAL LUMBER MANUFACTURES ASSOCIATION

Nominal Size	American Standard Dressed Size	Area of Section	Weight per Foot	Moment of Inertia	Section Modulus	Nominal Size	American Standard Dressed Size	Area of Section	Weight per Foot	Moment of Inertia	Section Modulus
In.	In.	In.^{2}	Lb	In.^{4}	In.^{3}	In.	In.	In.^{2}	Lb	In.^{4}	In.^{3}
2 × 4	1-5/8×3-5/8	5.89	1.64	6.45	3.56	10×10	9-1/2×9-1/2	90.3	25.0	679	143
6	5-5/8	9.14	2.54	24.1	8.57	12	11-1/2	109	30.3	1204	209
8	7-1/2	12.2	3.39	57.1	15.3	14	13-1/2	128	35.6	1948	289
10	9-1/2	15.4	4.29	116	24.4	16	15-1/2	147	40.9	2948	380
12	11-1/2	18.7	5.19	206	35.8	18	17-1/2	166	46.1	4243	485
14	13-1/2	21.9	6.09	333	49.4	20	19-1/2	185	51.4	5870	602
16	15-1/2	25.2	6.99	504	65.1	22	21-1/2	204	56.7	7868	732
18	17-1/2	28.4	7.90	726	82.9	24	23-1/2	223	62.0	10274	874
3 × 4	2-5/8×3-5/8	9.52	2.64	10.4	5.75	12×12	11-1/2×11-1/2	132	36.7	1458	253
6	5-5/8	14.8	4.10	38.9	13.8	14	13-1/2	155	43.1	2358	349
8	7-1/2	19.7	5.47	92.3	24.6	16	15-1/2	178	49.5	3569	460
10	9-1/2	24.9	6.93	188	39.5	18	17-1/2	201	55.9	5136	587
12	11-1/2	30.2	8.39	333	57.9	20	19-1/2	224	62.3	7106	729
14	13-1/2	35.4	9.84	538	79.7	22	21-1/2	247	68.7	9524	886
16	15-1/2	40.7	11.3	815	105	24	23-1/2	270	75.0	12437	1058
18	17-1/2	45.9	12.8	1172	134	14×14	13-1/2×13-1/2	182	50.6	2768	410
4 × 4	3-5/8×3-5/8	13.1	3.65	14.4	7.94	16	15-1/2	209	58.1	4189	541
6	5-5/8	20.4	5.66	53.8	19.1	18	17-1/2	236	65.6	6029	689
8	7-1/2	27.2	7.55	127	34.0	20	19-1/2	263	73.1	8342	856
10	9-1/2	34.4	9.57	259	54.5	22	21-1/2	290	80.6	11181	1040
12	11-1/2	41.7	11.6	459	79.9	24	23-1/2	317	88.1	14600	1243
14	13-1/2	48.9	13.6	743	110	16×16	15-1/2×15-1/2	240	66.7	4810	621
16	15-1/2	56.2	15.6	1125	145	18	17-1/2	271	75.3	6923	791
18	17-1/2	63.4	17.6	1619	185	20	19-1/2	302	83.9	9578	982
6 × 6	5-1/2×5-1/2	30.3	8.40	76.3	27.7	22	21-1/2	333	92.5	12837	1194
8	7-1/2	41.3	11.4	193	51.6	24	23-1/2	364	101	16763	1427
10	9-1/2	52.3	14.5	393	82.7	18×18	17-1/2×17-1/2	306	85.0	7816	893
12	11-1/2	63.3	17.5	697	121	20	19-1/2	341	94.8	10813	1109
14	13-1/2	74.3	20.6	1128	167	22	21-1/2	376	105	14493	1348
16	15-1/2	85.3	23.6	1707	220	24	23-1/2	411	114	18926	1611
18	17-1/2	96.3	26.7	2456	281	26	25-1/2	446	124	24181	1897
20	19-1/2	107.3	29.8	3398	349	20×20	19-1/2×19-1/2	380	106	12049	1236
8 × 8	7-1/2×7-1/2	56.3	15.6	264	70.3	22	21-1/2	419	116	16150	1502
10	9-1/2	71.3	19.8	536	113	24	23-1/2	458	127	21089	1795
12	11-1/2	86.3	23.9	951	165	26	25-1/2	497	138	26945	2113
14	13-1/2	101.3	28.0	1538	228	28	27-1/2	536	149	33795	2458
16	15-1/2	116.3	32.0	2327	300	24×24	23-1/2×23-1/2	552	153	25415	2163
18	17-1/2	131.3	36.4	3350	383	26	25-1/2	599	166	32472	2547
20	19-1/2	146.3	40.6	4634	475	28	27-1/2	646	180	40727	2962
22	21-1/2	161.3	44.8	6211	578	30	29-1/2	693	193	50275	3408

All properties and weights given are for dressed size only.
The weights given above are based on assumed average weight of 40 pounds per cubic foot.

Bibliography

Texts and articles which explain in greater detail the more advanced material of this test.

Procurement of Experimental Data:

Davis, H. E., Troxell, G. E., and Wiskocil, C. T., *Testing and Inspection of Engineering Materials*. New York: McGraw-Hill Book Company, 1941.

Marin, J., *Mechanical Properties of Materials and Design*. New York: McGraw-Hill Book Company, 1942.

Tatnall, F., *Tatnall on Testing*. Metals Park, Ohio: American Society for Metals, 1966.

Stress in Riveted and Welded Joints:

Baron, F., and Larson, E. W., "Comparative Behavior of Bolted and Riveted Joints." *Proc, ASCE*, **80**, Separate No. 470, August 1954.

Churchill, H. D., and Austin, J. B., *Weld Design*. Englewood Cliffs, N.J.: Prentice-Hall, Inc., 1949.

Davis, R. E., Woodruff, G. R., and Davis, H. E., "Tension Tests of Large Riveted Joints." *Trans. ASCE*, **105**, 1193, 1243 (1940).

General—Advanced Mechanics of Materials:

Durelli, A. J., Phillips, E. A., and Tsao, C. H., *Introduction to the Theoretical and Experimental Analysis of Stress and Strain.* New York: McGraw-Hill Book Company, 1958.

Murray, G., *Advanced Mechanics of Materials.* New York: McGraw-Hill Book Company, 1946.

Plastic Theory:

Beedle, L. S., "Plastic Strength of Steel Frames." *Proc. ASCE,* **81**, Paper No. 764, August 1955.

Neal, B. G., *The Plastic Methods of Structural Analysis.* New York: John Wiley & Sons, Inc., 1963.

Van den Broek, J. A., "Theory of Limit Design." *Trans. ASCE,* **105,** 638 (1940).

Membrane Analogy:

Marin, J., "Evaluating Torsional Stresses by Membrane Analogy." *Machine Design,* **15,** 118–123 (1943).

Mindlin, R. D., and Salvadori, M. G., "Analogies" in *Handbook of Experimental Stress Analysis* (M. Hetenyi, ed.), Chap. 16, p. 700. New York: John Wiley & Sons, Inc., 1950.

Electric Strain Gages:

Baldwin-Lima-Hamilton Co., Bulletins on Strain Gages.

Perry, C. C., and Lissner, H. R., *The Strain Gage Primer*, 2nd ed. New York: McGraw-Hill Book Company, 1962.

Photoelasticity:

Frocht, M., *Photoelasticity,* Vols. I and II. New York: John Wiley & Sons, Inc., 1941.

General—Experimental Stress Analysis:

Dove, R. C., and Adams, P. H., *Experimental Stress Analysis and Motion Measurement.* Columbus, Ohio: Charles E. Merrill Publishing Company, 1964.

Hetenyi, M., *Handbook of Experimental Stress Analysis.* New York: John Wiley & Sons, Inc., 1950.

Lee, G. H., *An Introduction to Experimental Stress Analysis*. New York: John Wiley & Sons, Inc., 1950.

Prestressed Concrete:

Chi, M., and Biberstein, F. A., *Theory of Prestressed Concrete*. Englewood Cliffs, N.J.: Prentice-Hall, Inc., 1963.

Lin, T. Y., *Design of Prestressed Concrete Structures*. New York: John Wiley & Sons, Inc., 1956.

Materials of Engineering:

Keyser, C. A., *Materials of Engineering*. Englewood Cliffs, N.J.: Prentice-Hall, Inc., 1956.

Mills, A. P., Hayward, H. W., and Rader, L. F., *Materials of Construction*. New York: John Wiley & Sons, Inc., 1955.

Fatigue:

Osgood, C. C., *Fatigue Design*. New York: John Wiley & Sons, Inc., 1970.

Stress Corrosion:

Logan, H. L., *The Stress Corrosion of Metals*. New York: John Wiley & Sons, Inc., 1966.

Index

D

E

F

G

H

I

J

K

L

M

N

O

P

R

S

T

U

V

W

Y